INTRODUCTION TO OPTICS

INTRODUCTION TO OPTICS

Anchal Srivastava
Reader in Physics
Lucknow University

R.K. Shukla
Reader in Physics
Lucknow University

T.P. Pandya
Member, Executive Council
Lucknow University

PUBLISHING FOR ONE WORLD

NEW AGE INTERNATIONAL (P) LIMITED, PUBLISHERS
New Delhi • Bangalore • Chennai • Cochin • Guwahati • Hyderabad
Jalandhar • Kolkata • Lucknow • Mumbai • Ranchi
Visit us at www.newagepublishers.com

Published by New Age International (P) Ltd., Publishers
First Edition: 2011
Reprint: 2011

Branches:

- No. 37/10, 8th Cross (Near Hanuman Temple), Azad Nagar, Chamrajpet, **Bangalore** - 560 018. Tel.: (080) 26756823, Telefax: 26756820, E-mail: bangalore@newagepublishers.com
- 26, Damodaran Street, T. Nagar, **Chennai** - 600 017. Tel.: (044) 24353401, Telefax: 24351463 E-mail: chennai@newagepublishers.com
- CC-39/1016, Carrier Station Road, Ernakulam South, **Cochin** - 682 016. Tel.: (0484) 2377004, Telefax: 4051303 E-mail: cochin@newagepublishers.com
- Hemsen Complex, Mohd. Shah Road, Paltan Bazar, Near Starline Hotel, **Guwahati** - 781 008.Tel.: (0361) 2513881 Telefax: 2543669, E-mail: guwahati@newagepublishers.com
- No. 105, 1st Floor, Madhiray Kaveri Tower, 3-2-19, Azam Jahi Road, Nimboliadda, **Hyderabad** - 500 027. Tel.: (040) 24652456, Telefax: 24652457, E-mail:hyderabad@newagepublishers.com
- RDB Chambers (Formerly Lotus Cinema)106A, 1st Floor, S.N. Banerjee Road, **Kolkata** - 700 014. Tel.: (033) 22273773, Telefax: 22275247, E-mail:kolkata@newagepublishers.com
- 16-A, Jopling Road, **Lucknow** - 226 001. Tel.: (0522) 2209578, 4045297, Telefax: 2204098 E-mail: lucknow@newagepublishers.com
- 142C, Victor House, Ground Floor, N.M. Joshi Marg, Lower Parel, **Mumbai** - 400 013.Tel.: (022) 24927869 Telefax: 24915415, E-mail: mumbai@newagepublishers.com
- 22, Golden House, Daryaganj, **New Delhi** - 110 002. Tel.: (011) 23262370, 23262368, Telefax: 43551305 E-mail: sales@newagepublishers.com

ISBN : 978-81-224-3081-3

₹ 350.00

C-11-08-5809

Printed in India at Ajit Printers, Delhi.
Typeset at Akriti Graphics, Delhi

PUBLISHING FOR ONE WORLD
NEW AGE INTERNATIONAL (P) LIMITED, PUBLISHERS
4835/24, Ansari Road, Daryaganj, New Delhi-110002
Visit us at **www.newagepublishers.com**

PREFACE

Optics has acquired newer dimensions with the advent of lasers which occupies a major position among the outstanding achievements of science and engineering in the present era. It is not feasible to cover the vast field of optics starting from classical phenomenon to modern discoveries in a single book. Authors have tried to include all the topics which are dealt with in undergraduate courses in Indian Universities. The latter part of this book dwells upon some aspects of lasers, holography, and optical fibers at introductory level. Fundamentals of optics have been presented in a simple and gradual manner keeping continuity among different topics dealt in successive chapters. This book would be equally useful for a wide variety of readers including those appearing for competitive examinations.

Despite best efforts, errors would have occurred in this first edition. All suggestions for improvement of book's content, in every aspect, would therefore, be appreciated and acknowledged in subsequent editions. We wish to thank Mr. L.N. Mishra and team members of New Age International (P) Limited for their skilled and enthusiastic help throughout the project.

Authors

CONTENTS

A BRIEF REVIEW OF THE THEORY OF LIGHT

1.1 HISTORICAL THEORIES ABOUT LIGHT

Hindu Theories

In ancient India, the Hindu schools of Samkhya and Vaisheshika, from around the 6th–5th century BC, developed theories on light. According to the Samkhya school, light is one of the five fundamental "subtle" elements (tanmatra) out of which emerge the gross elements. The atomicity of these elements is not specifically mentioned and it appears that they were actually taken to be continuous.

On the other hand, the Vaisheshika school gives an atomic theory of the physical world on the non-atomic ground of ether, space and time. The basic atoms are those of earth (*prthvi*), water (*apas*), fire (*tejas*), and air (*vayu*), that should not be confused with the ordinary meaning of these terms. These atoms are taken to form binary molecules that combine further to form larger molecules. Motion is defined in terms of the movement of the physical atoms and it appears that it is taken to be non-instantaneous. Light rays are taken to be a stream of high velocity of tejas (fire) atoms. The particles of light can exhibit different characteristics depending on the speed and the arrangements of the tejas atoms. Around the first century BC, the Vishnu Purana refers to sunlight as the "the seven rays of the sun".

Later in 499, Aryabhata, who proposed a heliocentric solar system of gravitation in his Aryabhatiya, wrote that the planets and the Moon do not have their own light but reflect the light of the Sun.

The Indian Buddhists, such as Dignāga in the 5th century and Dharmakirti in the 7th century, developed a type of atomism that is a philosophy about reality being composed of atomic entities that are momentary flashes of light or energy. They viewed light as being an atomic entity equivalent to energy, similar to the modern concept of photons, though they also viewed all matter as being composed of these light/energy particles.

Greek and Hellenistic Theories

In the fifth century BC, Empedocles postulated that everything was composed of four elements;

fire, air, earth and water. He believed that Aphrodite made the human eye out of the four elements and that she lit the fire in the eye which shone out from the eye making sight possible. If this were true, then one could see during the night just as well as during the day, so Empedocles postulated an interaction between rays from the eyes and rays from a source such as the sun.

In about 300 BC, Euclid wrote Optica, in which he studied the properties of light. Euclid postulated that light travelled in straight lines and he described the laws of reflection and studied them mathematically. He questioned that sight is the result of a beam from the eye, for he asks how one sees the stars immediately, if one closes one's eyes, then opens them at night. Of course if the beam from the eye travels infinitely fast this is not a problem.

In 55 BC, Lucretius, a Roman who carried on the ideas of earlier Greek atomists, wrote:

"The light & heat of the sun; these are composed of minute atoms which, when they are shoved off, lose no time in shooting right across the interspace of air in the direction imparted by the shove." – On the nature of the Universe

Despite being similar to later particle theories, Lucretius's views were not generally accepted and light was still theorised as emanating from the eye.

Ptolemy (c. 2nd century) wrote about the refraction of light in his book Optics, and developed a theory of vision whereby objects are seen by rays of light emanating from the eyes.

Optical Theory

The Muslim scientist Ibn al-Haytham (965–1040), known as Alhacen or Alhazen in the West, developed a broad theory of vision based on geometry and anatomy in his 1021 Book of Optics. Al-Haytham postulated that every point on an illuminated surface radiates light rays in all directions, but that only one ray from each point can be seen: the ray that strikes the eye perpendicularly. The other rays strike at different angles and are not seen. He described the pinhole camera and invented the camera obscura, which produces an inverted image, using it as an example to support his argument. This contradicted Ptolemy's theory of vision that objects are seen by rays of light emanating from the eyes. Alhacen held light rays to be streams of minute energy particles that travelled at a finite speed. He improved Ptolemy's theory of the refraction of light, and went on to discover the laws of refraction.

He also carried out the first experiments on the dispersion of light into its constituent colours. His major work Kitab al-Manazir (Book of Optics) was translated into Latin in the Middle Ages, as well his book dealing with the colours of sunset. He dealt at length with the theory of various physical phenomena like shadows, eclipses, the rainbow. He also attempted to explain binocular vision, and gave an explanation of the apparent increase in size of the sun and the moon when near the horizon, known as the moon illusion. Because of his extensive experimental research on optics, Ibn al-Haytham is considered the "father of modern optics".

Ibn al-Haytham also correctly argued that we see objects because the sun's rays of light, which he believed to be streams of tiny energy particles travelling in straight lines, are reflected from objects into our eyes. He understood that light must travel at a large but finite velocity, and that refraction is caused by the velocity being different in different substances. He also studied spherical and parabolic mirrors, and understood how refraction by a lens will allow images to be focused and magnification to take place. He understood mathematically why a spherical mirror produces aberration.

Avicenna (980–1037) agreed that the speed of light is finite, as he "observed that if the perception of light is due to the emission of some sort of particles by a luminous source, the speed of light must be finite. Abū Rayhān al - Birūni (973–1048) also agreed that light has a finite speed, and he was the first to discover that the speed of light is much faster than the speed of sound. In the late 13th and early 14th centuries, Qutb al-Din al-Shirazi (1236–1311) and his student Kamāl al - Din al-Fārisi (1260–1320) continued the work of Ibn al-Haytham, and they were the first to give the correct explanations for the rainbow phenomenon.

The 'Plenum'

René Descartes (1596–1650) held that light was a disturbance of the plenum, the continuous substance of which the universe was composed. In 1637 he published a theory of the refraction of light that assumed, incorrectly, that light travelled faster in a denser medium than in a less dense medium. Descartes arrived at this conclusion by analogy with the behaviour of sound waves. Although Descartes was incorrect about the relative speeds, he was correct in assuming that light behaved like a wave and in concluding that refraction could be explained by the speed of light in different media. As a result, Descartes' theory is often regarded as the forerunner of the wave theory of light.

Particle Theory

Ibn al-Haytham (Alhazen, 965–1040) proposed a particle theory of light in his Book of Optics (1021). He held light rays to be streams of minute energy particles that travel in straight lines at a finite speed. He states in his optics that "the smallest parts of light," as he calls them, retain only properties that can be treated by geometry and verified by experiment; they lack all sensible qualities except energy. Avicenna (980–1037) also proposed that "the perception of light is due to the emission of some sort of particles by a luminous source".

Pierre Gassendi (1592–1655), an atomist, proposed a particle theory of light which was published posthumously in the 1660s. Isaac Newton studied Gassendi's work at an early age, and preferred his view to Descartes' theory of the plenum. He stated in his Hypothesis of Light of 1675 that light was composed of corpuscles (particles of matter) which were emitted in all directions from a source. One of Newton's arguments against the wave nature of light was that waves were known to bend around obstacles, while light travelled only in straight lines. He did, however, explain the phenomenon of the diffraction of light (which had been observed by Francesco Grimaldi) by allowing that a light particle could create a localised wave in the aether.

Newton's theory could be used to predict the reflection of light, but could only explain refraction by incorrectly assuming that light accelerated upon entering a denser medium because the gravitational pull was greater. Newton published the final version of his theory in his *Opticks* (optic) of 1704. His reputation helped the particle theory of light to hold sway during the 18th century. The particle theory of light led Laplace to argue that a body could be so massive that light could not escape from it. In other words it would become what is now called a black hole. Laplace withdrew his suggestion when the wave theory of light was firmly established. A translation of his essay appears in The large scale structure of space-time, by Stephen Hawking and George F. R. Ellis.

Wave Theory

In the 1660s, Robert Hooke published a wave theory of light. Christiaan Huygens worked out his own wave theory of light in 1678, and published it in his Treatise on light in 1690. He proposed that light was emitted in all directions as a series of waves in a medium called the Luminiferous ether. As waves are not affected by gravity, it was assumed that they slowed down upon entering a denser medium.

Thomas Young's sketch of the two-slit experiment showing the diffraction of light. Young's experiments supported the theory that light consists of waves.

The wave theory predicted that light waves could interfere with each other like sound waves (as noted around 1800 by Thomas Young), and that light could be polarized, if it were a transverse wave. Young showed by means of a diffraction experiment that light behaved as waves. He also proposed that different colours were caused by different wavelengths of light, and explained colour vision in terms of three-coloured receptors in the eye.

Another supporter of the wave theory was Leonhard Euler. He argued in Nova theoria lucis et colorum (1746) that diffraction could more easily be explained by a wave theory.

Later, Augustin-Jean Fresnel independently worked out his own wave theory of light, and presented it to the Académie des Sciences in 1817. Simeon Denis Poisson added to Fresnel's mathematical work to produce a convincing argument in favour of the wave theory, helping to overturn Newton's corpuscular theory. By the year 1821, Fresnel was able to show via mathematical methods that polarization could be explained only by the wave theory of light and only if light was entirely transverse, with no longitudinal vibration whatsoever.

The weakness of the wave theory was that light waves, like sound waves, would need a medium for transmission. A hypothetical substance called the luminiferous aether was proposed, but its existence was cast into strong doubt in the late nineteenth century by the Michelson-Morley experiment.

Newton's corpuscular theory implied that light would travel faster in a denser medium, while the wave theory of Huygens and others implied the opposite. At that time, the speed of light could not be measured accurately enough to decide which theory was correct. The first to make a sufficiently accurate measurement was Léon Foucault, in 1850. His result supported the wave theory, and the classical particle theory was finally abandoned.

Electromagnetic Theory

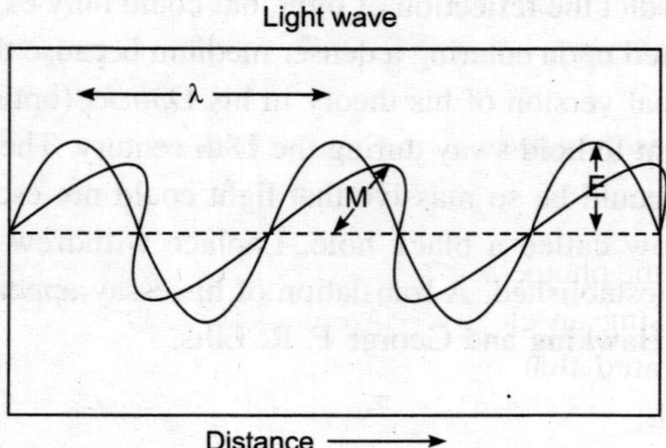

A linearly-polarised light wave frozen in time and showing the two oscillating components of light; an electric field and a magnetic field perpendicular to each other and to the direction of motion (a transverse wave).

In 1845, Michael Faraday discovered that the plane of polarisation of linearly polarised light is rotated when the light rays travel along the magnetic field direction in the presence of a transparent dielectric, an effect now known as Faraday rotation. This was the first evidence that light was related to electromagnetism. In 1846 he speculated that light might be some form of disturbance propagating along magnetic field lines. Faraday proposed in 1847 that light was a high-frequency electromagnetic vibration, which could propagate even in the absence of a medium such as the ether.

Faraday's work inspired James Clerk Maxwell to study electromagnetic radiation and light. Maxwell discovered that self-propagating electromagnetic waves would travel through space at a constant speed, which happened to be equal to the previously measured speed of light. From this, Maxwell concluded that light was a form of electromagnetic radiation: he first stated this result in 1862 in On Physical Lines of Force. In 1873, he published A Treatise on Electricity and Magnetism, which contained a full mathematical description of the behaviour of electric and magnetic fields, still known as Maxwell's equations. Soon after, Heinrich Hertz confirmed Maxwell's theory experimentally by generating and detecting radio waves in the laboratory, and demonstrating that these waves behaved exactly like visible light, exhibiting properties such as reflection, refraction, diffraction, and interference. Maxwell's theory and Hertz's experiments led directly to the development of modern radio, radar, television, electromagnetic imaging, and wireless communications.

The Special Theory of Relativity

The wave theory was widely successful in explaining nearly all optical and electromagnetic phenomena, and was a great triumph of nineteenth century physics. By the late nineteenth century, however, a handful of experimental anomalies remained that could not be explained by or were in direct conflict with the wave theory. One of these anomalies involved a controversy over the speed of light. The constant speed of light predicted by Maxwell's equations and confirmed by the Michelson-Morley experiment contradicted the mechanical laws of motion that had been unchallenged since the time of Galileo, which stated that all speeds were relative to the speed of the observer. In 1905, Albert Einstein resolved this paradox by revising the Galilean model of space and time to account for the constancy of the speed of light. Einstein formulated his ideas in his special theory of relativity, which advanced humankind's understanding of space and time. Einstein also demonstrated a previously unknown fundamental equivalence between energy and mass with his famous equation

$$E = mc^2$$

where E is energy, m is rest mass, and c is the speed of light in a vacuum.

Particle Theory Revisited

Another experimental anomaly was the photoelectric effect, by which light striking a metal surface ejected electrons from the surface, causing an electric current to flow across an applied voltage. Experimental measurements demonstrated that the energy of individual ejected electrons was proportional to the frequency, rather than the intensity, of the light. Furthermore, below a certain minimum frequency, which depended on the particular metal, no current would flow regardless of the intensity. These observations appeared to contradict the wave theory, and for years physicists tried in vain to find an explanation. In 1905, Einstein solved this puzzle as well, this time by resurrecting the particle theory of light to explain the observed effect. Because of the preponderance of evidence in favour of the wave theory, however, Einstein's ideas were met initially by great skepticism among established physicists. But eventually Einstein's explanation of the photoelectric effect would triumph, and it ultimately formed the basis for wave–particle duality and much of quantum mechanics.

Quantum Theory

A third anomaly that arose in the late 19th century involved a contradiction between the wave theory of light and measurements of the electromagnetic spectrum emitted by thermal radiators, or so-called black bodies. Physicists struggled with this problem, which later became known as the ultraviolet catastrophe, unsuccessfully for many years. In 1900, Max Planck developed a new theory of black-body radiation that explained the observed spectrum correctly. Planck's theory was based on the idea that black bodies emit light (and other electromagnetic radiation) only as discrete bundles or packets of energy. These packets were called quanta, and the particle of light was given the name photon, to correspond with other particles being described around this time, such as the electron and proton. A photon has an energy, E, proportional to its frequency, f, by

$$E = hf = \frac{hc}{\lambda}$$

where h is Planck's constant, λ is the wavelength and c is the speed of light. Likewise, the momentum p of a photon is also proportional to its frequency and inversely proportional to its wavelength:

$$p = \frac{E}{c} = \frac{hf}{c} = \frac{h}{\lambda}$$

As it originally stood, this theory did not explain the simultaneous wave- and particle-like natures of light, though Planck would later work on theories that did. In 1918, Planck received the Nobel Prize in Physics for his part in the founding of quantum theory.

Wave-Particle Duality

The modern theory that explains the nature of light includes the notion of wave-particle duality, described by Albert Einstein in the early 1900s, based on his study of the photoelectric effect and Planck's results. Einstein asserted that the energy of a photon is proportional to its frequency. More generally, the theory states that everything has both a particle nature and a wave nature, and various experiments can be done to bring out one or the other. The particle nature is more easily discerned if an object has a large mass, and it was not until a bold proposition by Louis de Broglie in 1924 that the scientific community realised that electrons also exhibited wave–particle duality. The wave nature of electrons was experimentally demonstrated by Davission and Germer in 1927. Einstein received the Nobel Prize in 1921 for his work with the wave–particle duality on photons (especially explaining the photoelectric effect thereby), and de Broglie followed in 1929 for his extension to other particles.

Quantum Electrodynamics

The quantum mechanical theory of light and electromagnetic radiation continued to evolve through the 1920s and 1930s, and culminated with the development during the 1940s of the theory of quantum electrodynamics, or QED. This so-called quantum field theory is among the most comprehensive and experimentally successful theories ever formulated to explain a set of natural phenomena. QED was developed primarily by physicists Richard Feynman, Freeman Dyson, Julian Schwinger, and Shin-Ichiro Tomonaga. Feynman, Schwinger and Tomonaga shared the 1965 Nobel Prize in Physics for their contributions.

1.2 SPEED OF LIGHT

The speed of light in a vacuum is presently defined to be exactly 299,792,458 m/s (about 186,282 miles per second). This definition of the speed of light means that the metre is now defined in terms of the speed of light. Light always travels at a constant speed, even between particles of a substance through which it is shining. Photons excite the adjoining particles that in turn transfer the energy to the neighbour. This may appear to slow the beam down through its trajectory in real time. The time lost between entry and exit accounts to the displacement of energy through the substance between each particle that is excited.

Different physicists have attempted to measure the speed of light throughout history. Galileo attempted to measure the speed of light in the seventeenth century. An early experiment to measure the speed of light was conducted by Ole Rømer, a Danish physicist, in 1676. Using a telescope, Ole observed the motions of Jupiter and one of its moons, Io. Noting discrepancies in the apparent period of Io's orbit, Rømer calculated that light takes about 22 minutes to traverse the diameter of Earth's orbit. Unfortunately, its size was not known at that time. If Ole had known the diameter of the Earth's orbit, he would have calculated a speed of 227,000,000 m/s.

Another, more accurate, measurement of the speed of light was performed in Europe by Hippolyte Fizeau in 1849. Fizeau directed a beam of light at a mirror several kilometers away. A rotating cog wheel was placed in the path of the light beam as it traveled from the source, to the mirror and then returned to its origin. Fizeau found that at a certain rate of rotation, the beam would pass through one gap in the wheel on the way out and the next gap on the way back. Knowing the distance to the mirror, the number of teeth on the wheel, and the rate of rotation, Fizeau was able to calculate the speed of light as 313,000,000 m/s.

Léon Foucault used an experiment which used rotating mirrors to obtain a value of 298,000,000 m/s in 1862. Albert A. Michelson conducted experiments on the speed of light from 1877 until his death in 1931. He refined Foucault's methods in 1926 using improved rotating mirrors to measure the time it took light to make a round trip from Mt. Wilson to Mt. San Antonio in California. The precise measurements yielded a speed of 299,796,000 m/s.

1.3 CLASSIFICATION

All the known properties of light are described in terms of the experiments by which they were discovered and the many and varied demonstrations by which they are frequently illustrated. Numerous though these properties are, their demonstrations can be grouped together and classified under one of three heads: geometrical optics, wave optics, and quantum optics, each of which may be subdivided as follows:

Geometrical Optics

- Rectilinear propagation
- Finite speed
- Reflection
- Refraction
- Dispersion

Wave Optics

- Interference
- Diffraction
- Electromagnetic character
- Polarisation
- Double refraction

Quantum Optics

- Atomic orbits
- Probability densities
- Energy levels
- Quanta
- Lasers

1.4 UNITS AND MEASURES

Light is measured with two main alternative sets of units: radiometry consists of measurements of light power at all wavelengths, while photometry measures light with wavelength weighted with respect to a standardised model of human brightness perception. Photometry is useful, for example, to quantify illumination intended for human use. The SI units for both systems are summarised in the following tables.

SI radiometry units

Quantity	Symbol	SI unit	Abbr.	Remark
Radiant energy	Q	joule	J	energy
Radiant flux	ϕ	watt	W	radiant energy per unit time, also called *radiant power*
Radiant intensity	I	watt per steradian	$W \cdot sr^{-1}$	power per unit solid angle
Radiance	L	watt per steradian per square metre	$W \cdot sr^{-1} \cdot m^{-2}$	power per unit solid angle per unit *projected* source area. called *intensity* in some other fields of study.
Irradiance	E, I	watt per square metre	$W \cdot m^{-2}$	power incident on a surface. sometimes confusingly called "intensity".
Radiant exitance / Radiant emittance	M	watt per square metre	$W \cdot m^{-2}$	power emitted from a surface.
Radiosity	J or J_λ	watt per square metre	$W \cdot m^{-2}$	emitted plus reflected power leaving a surface
Spectral radiance	L_λ or L_v	watt per steradian per metre^3 *or* watt per steradian per square metre per hertz	$W \cdot sr^{-1} \cdot m^{-3}$ *or* $W \cdot sr^{-1} \cdot m^{-2} \cdot Hz^{-1}$	commonly measured in $W \cdot sr^{-1} \cdot m^{-2} \cdot nm^{-1}$
Spectral irradiance	E_λ or E_v	watt per metre^3 *or* watt per square metre per hertz	$W \cdot m^{-3}$ *or* $W \cdot m^{-2} \cdot Hz^{-1}$	commonly measured in $W\, m^{-2} \cdot nm^{-1}$

The SI photometry units

Quantity	Symbol	SI unit	Abbr.	Notes
Luminous energy	Q_v	lumen second	lm · s	units are sometimes called talbots
Luminous flux	F	lumen (= cd·sr)	lm	also called *luminous power*
Luminous intensity	I_v	candela (= lm/sr)	cd	an SI base unit
Luminance	L_v	candela per square metre	cd/m^2	units are sometimes called "nits"
Illuminance	E_v	lux (= lm/m^2)	lx	Used for light incident on a surface
Luminous emittance	M_v	lux (= lm/m^2)	lx	Used for light emitted from a surface
Luminous efficacy		lumen per watt	lm/W	ratio of luminous flux to radiant flux
SI • Photometry				

The photometry units are different from most systems of physical units in that they take into account how the human eye responds to light. The cone cells in the human eye are of three types which respond differently across the visible spectrum, and the cumulative response peaks at a wavelength of around 555 nm. Therefore, two sources of light which produce the same intensity (W/m^2) of visible light do not necessarily appear equally bright. The photometry units are designed to take this into account, and therefore are a better representation of how "bright" a light appears to be than raw intensity. They relate to raw power by a quantity called luminous efficacy, and are used for purposes like determining how to best achieve sufficient illumination for various tasks in indoor and outdoor settings. The illumination measured by a photocell sensor does not necessarily correspond to what is perceived by the human eye, and without filters which may be costly, photocells and CCDs tend to respond to some infrared, ultraviolet or both.

1.5 LIGHT PRESSURE

Light pushes on objects in its path, just as the wind would do. This pressure is most easily explainable in particle theory: photons hit and transfer their momentum. Light pressure can cause asteroids to spin faster, acting on their irregular shapes as on the vanes of a windmill. The possibility to make solar sails that would accelerate spaceships in space is also under investigation.

Although the motion of the Crookes radiometer was originally attributed to light pressure, this interpretation is incorrect; the characteristic Crookes rotation is the result of a partial vacuum. This should not be confused with the Nichols radiometer, in which the motion is directly caused by light pressure.

1.6 OPTICS

The study of light and the phenomena associated with its generation, transmission, and detection. In a broader sense, optics includes all the phenomena associated with visible, infrared and ultraviolet radiation. Geometrical optics assumes that light travels in straight lines and is concerned with the laws controlling the reflection and refraction of rays of light. Physical optics deals with phenomena that depend on the wave nature of light, *e.g.*, diffraction, interference, and polarisation.

REFLECTION AND REFRACTION

The term *light* in common phraseology is used to denote that aspect of radiant energy by which objects are made visible, due to stimulation of the retina of the eye. Nowadays it has become customary to include in this term certain other kinds of radiations called *ultraviolet* and *infra-red* which, although incapable of exciting the sense of sight, nevertheless show other effects similar to those of visible light. For example, the radiation in the near ultraviolet region affects the photographic plate even more markedly than that in the visible. The ultraviolet region lies beyond the violet end of the visible spectrum while the infra-red region lies beyond the red one.

The study of nature and properties of light forms a subject itself in Physics called *Optics*, which can be conveniently divided into three distinct branches: (*a*) *Geometrical Optics*, in which many facts concerning light can be investigated from the standpoint of the ray theory, that is, on the supposition that light travels in straight lines or rays; while no assumptions are made in it regarding the nature of light. (*b*) *Physical Optics*, in which certain phenomena like propagation, reflection, refraction, interference, diffraction and polarisation exhibited by light are studied from the standpoint of the wave theory; one studies in it the nature of light in detail, (*c*) *Quantum Optics*; in which one studies the interaction of light with the atomic entities of matter.

Geometrical Optics mainly deals with the image formation by mirrors, lenses and prisms. The whole subject can be developed on the basis of the following four fundamental properties of light and with the aid of geometrical or trigonometrical calculations.

(*i*) In a homogeneous medium, light travels in straight lines.

(*ii*) Rays of light are independent and any two of them can intersect each other without in any way affecting their future paths, which are the same as if each ray existed separately.

(*iii*) A beam of light is reflected from any optical surface in accordance with the following laws:

(*a*) The reflected ray lies in the plane of incidence, which is the plane containing the incident ray and the normal at the point of incidence.

(*b*) The angles of incidence and reflection are equal, the reflected ray and the incident ray being on the opposite sides of the normal.

(*iv*) The refraction of light from one medium to another is governed according to the following laws:

(*a*) The refracted ray lies in the plane of incidence.

(*b*) The sine of angle of incidence bears a constant ratio to the sine of angle of refraction. The value of this ratio depends upon the media involved and on the wavelength of light. This law is called *Snell's Law* and in symbols it is expressed in the form

$$\frac{\sin i}{\sin r} = \frac{V_1}{V_2} = \frac{\mu_2}{\mu_1} = {}_1\mu_2$$

where V_1 is the velocity of light in medium in which incident ray travels while V_2 is the velocity of light in the medium into which the light is refracted; μ_1 and μ_2, the indices of refraction of the first and the second medium with respect to vacuum, are defined as

$$\mu_1 = \frac{\text{Velocity of light in vacuo}}{\text{Velocity of light in first medium}}$$

$$\mu_2 = \frac{\text{Velocity of light in vacuo}}{\text{Velocity of light in second medium}}$$

With the help of these fundamental laws it is possible to trace the path of light rays through any optical system and the formation of images studied. Accordingly, this branch of optics has important applications in the design, manufacture, selection and use of excellent optical instruments.

The ray theory, however, fails to explain interference and diffraction phenomena. Moreover, a closer examination of the fact reveals *that the rectilinear propagation of light is only approximate*. These phenomena are explained on the basis of wave nature of light. But owing to the small wavelength of the visible light, the approximate limits being 4×10^{-5} cm and 7.5×10^{-5} cm rectilinear propagation of light is observed under normal circumstances and hence assumptions made in Geometrical Optics are nearly true. Geometrical Optics can, therefore, be regarded as an asymptotic form of Wave Optics when the wavelength of light is taken to be negligibly small and under this limitation many results of Wave Optics pass over into those of Ray Optics.

The laws of Geometrical Optics, which have all been experimentally verified, can be derived from one basic principle called Fermat's Principle of Least Time.

2.1 FERMAT'S PRINCIPLE

It states that "when light travels from one point to another it always follows the path along which the time taken is the least". In other words the path of light between two given points is determined by the condition that the variation in optical path is zero, *i.e.*

$$\delta\Sigma(\mu \cdot l) = 0$$

for any infinitesimal deviation from the actual path.

Stated in this form the law holds even if in going from one point to another light suffers one or more reflections or refractions. Moreover it is only the optical path and not the direction of propagation of light which is involved. The principle that the path of light is reversible as long as the media and the order in which light passes through these remains unchanged is, therefore, also included in this statement.

If we consider a ray of light travelling from one point to another in a homogeneous medium Fermat's law implies that the length of its path between the two points should be the least. This path is clearly a straight line. It is, therefore, obvious that in an isotropic medium light always travels in straight lines and the law of rectilinear propagation results as a direct consequence of Fermat's Principle.

Let us now apply the principle to deduce the laws of reflection and refraction.

Laws of Reflection

Let PQ (Fig. 2.1) be a plane mirror. Suppose a ray of light starting from A reaches D after reflection from the mirror, along the path ABD.

Draw AO normal to the mirror and produce it to A' such that $AO = OA'$. It follows, therefore, that whatever be the position of B we shall always have $AB = A'B$ and hence the length of the path ABD is always equal to $A'BD$, *i.e.*, $ABD = A'BD$.

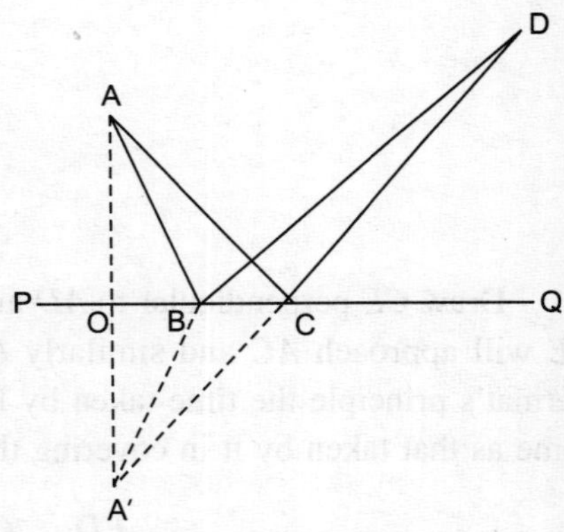

Fig. 2.1. Law of reflection

Since light is travelling in the same medium the time taken by light in going from A to D via the mirror will be the least if the length of the path itself is a minimum.

If light goes along ABD, the length of the path (as seen above) is $A'BD$. It is clear from the triangle $A'BD$ that

$$A'B + BD > A'D$$

or in other words the minimum length from A' to D is $A'CD$. Hence according to Fermat's Principle light must follow the path ACD [since $AC = A'C$].

Now the triangles, ACO and $A'CO$ are congruent because

(*i*) $AO = A'O$, (*ii*) $\angle AOC = \angle A'OC$, (*iii*) OC is common to both.

It follows, therefore, that

$$\angle ACO = \angle OCA' \qquad ...(2.1\ a)$$

Since the straight lines PCQ and $A'CD$ intersect at C

$$\angle PCA' = \angle DCQ \qquad ...(2.1\ b)$$

So that Eqns. (2.1 *a*) and (2.1 *b*) on combination give

$$\angle ACO = \angle DCQ$$

which is the law of reflection.

Laws of Refraction

Suppose a ray of light starting from a point A in the first medium arrives at a point B in the second after suffering a refraction at the plane surface of separation PQ (Fig. 2.2). If D is a point on PQ very near C it follows from Fermat's principle that the time taken by the ray of light along the path ADB will, in the limit, be equal to that taken along ACB.

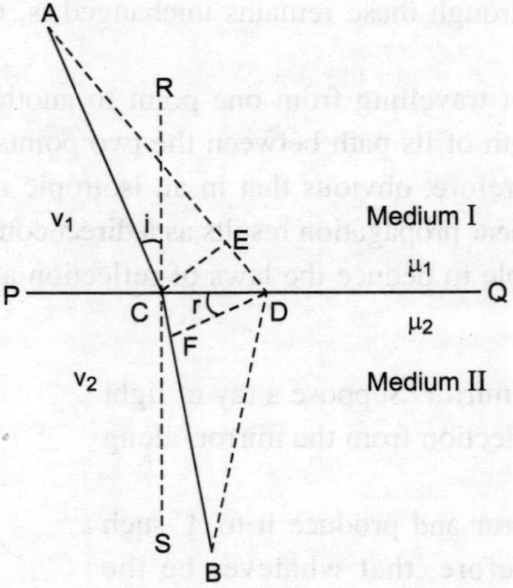

Fig. 2.2. Law of refraction

Draw CE perpendicular to AD and DF perpendicular to CB. Then as D approaches C, the length AE will approach AC and similarly BD becomes more and more equal to BF. Hence according to Fermat's principle the time taken by light in covering the distance DE with a velocity V_1 will be the same as that taken by it in covering the distance CF with a velocity V_2. Therefore

$$\frac{ED}{V_1} = \frac{CF}{V_2} \quad ...(2.1\ c)$$

or
$$\frac{ED}{CF} = \frac{V_1}{V_2} = \frac{\mu_2}{\mu_1}$$

where μ_1 and μ_2 are the refractive indices of media I and II respectively. This equation can be written as

$$\frac{ED/CD}{CF/CD} = \frac{\mu_2}{\mu_1}$$

or
$$\frac{\sin ECD}{\sin CDF} = \frac{\mu_2}{\mu_1} \quad ...(2.1\ d)$$

Since $\angle ECD = \frac{\pi}{2} - \angle RCE = i$ the angle of incidence and

$\angle CDF = \frac{\pi}{2} - \angle DCF = r$ the angle of refraction.

Equation (2.1*d*) gives

$$\frac{\sin i}{\sin r} = \frac{\mu_2}{\mu_1}$$

If the first medium be air $\mu_1 = 1$ and the above Eqn. reduces to

$$\frac{\sin i}{\sin r} = \mu$$

which is Snell's law. Here μ refers to the refractive index of the second medium with respect to air.

It has been experimentally observed that when light passes from a rarer to a denser medium, *e.g.*, from air to water, the refracted ray bends towards the normal or the angle of refraction '*r*' is less than the angle of incidence *i* (Fig. 2.3) and hence μ is greater than unity. It has also been proved that the refractive index μ is the ratio of velocities of light in the two media. Thus

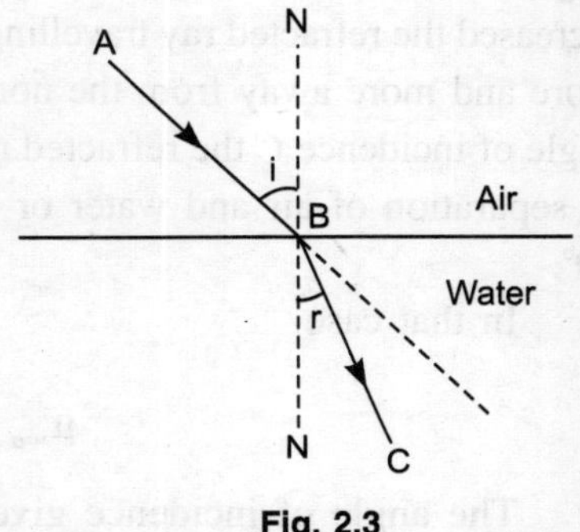

Fig. 2.3

$$\frac{\sin i}{\sin r} = \mu_{aw} = \frac{\text{Velocity of light in air}}{\text{Velocity of light in water}} = \frac{V_{\text{air}}}{V_{\text{water}}}$$

If instead of air the first medium be vacuum

$$\mu_w = \frac{V}{V_{\text{water}}}$$

Where 'V' is the velocity of light in vacuum and μ_w is the absolute refractive index of water. Also

$$\mu_{aw} = \frac{V_{\text{air}}}{V_{\text{water}}} = \frac{V}{V_{\text{water}}} \div \frac{V}{V_{\text{air}}} = \frac{\mu_w}{\mu_a}$$

In practice μ_a is taken as unity and hence μ_{aw} does not appreciably differ from μ_w.
It can be readily seen that

$$\mu_{aw} = \frac{1}{\mu_{wa}}$$

or $$\mu_{aw} \times \mu_{wa} = 1$$

In general for more than two media

$$\mu_{12} \times \mu_{23} \times \mu_{34} \times \mu_{41} = 1$$

Breakdown of Snell's Law – Total Internal Reflection

It has been seen above that sin $i = \mu$ sin r where μ is the refractive index of the denser medium with respect to air and is greater than unity. If light be now considered to be travelling from water to air (Fig. 2.4), the angle of incidence i will be less than the angle of refraction r and consequently μ_{wa} will be less than unity. As the angle of incidence i is increased the refracted ray travelling in the rarer medium bends more and more away from the normal NN, till at a particular angle of incidence C the refracted ray travels along the surface of separation of air and water or the angle of refraction r = 90º.

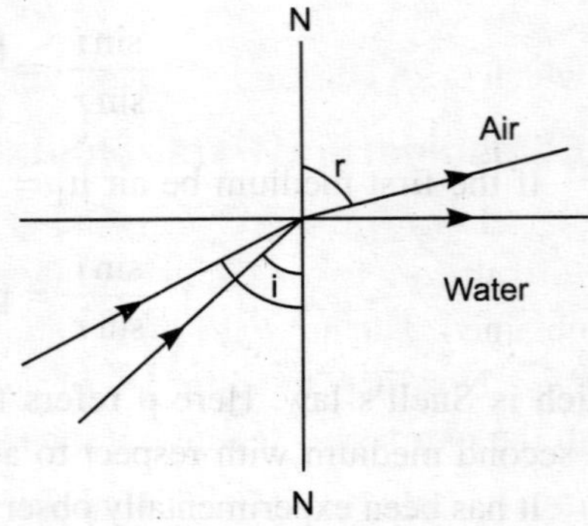

Fig. 2.4. Critical angle

In that case

$$\mu_{wa} = \frac{\sin C}{\sin \pi/2} = \sin C$$

The angle of incidence given by μ_{wa} = sin C is known as the critical angle. If this angle is increased further 'sin r' must be greater than unity which is impossible. Consequently there can be no refracted ray and Snell's law breaks down. In this case light is reflected back in the (denser) medium and the phenomenon is known as total internal reflection.

2.2 APLANATIC SURFACES

In the foregoing discussion the reflecting and refracting surfaces were considered plane. But if the surfaces be curved it is no longer necessary that the length of the path followed by light may always be a minimum; it may sometimes be maximum as well. In such cases Fermat's principle of least time is not applicable. A consideration of aplanatic surfaces throws further light on such situations.

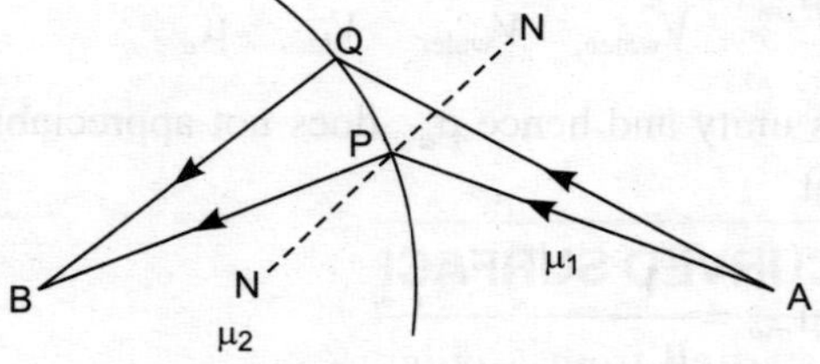

Fig. 2.5. Definition of aplanatic surface

An aplanatic surface is such that from any point on it, the sum of the optical paths to any two points is a constant quantity. If P be any point such that

$$\mu_1 AP + \mu_2 BP = \text{constant} \qquad \text{...(2.2 } a\text{)}$$

then the locus of the point P will be an aplanatic surface.

The curve represented by Eqn. (2.2 a) in one plane is called a Cartesian oval and is shown in Fig. 2.5. If the figure be rotated about the axis AB the corresponding aplanatic surface is obtained.

In the case of reflection, light is confined to a single medium and Eqn. (2.2 *a*) reduces to

$$AP + BP = \text{constant} \qquad ...(2.2\ b)$$

which is an ellipsoid of revolution with points *A* and *B* as foci (Fig. 2.6).

Ellipsoidal and Paraboloidal Mirrors

Let *PQR* (Fig. 2.6) be an arc of the ellipse whose major axis is *MN* and whose foci are *A* and *B*. If the arc is rotated about the major axis a trace of the surface called the ellipsoid of revolution is obtained. The arc *PQR* may, therefore, be considered as the principal section of an ellipsoidal mirror.

Now it is the property of an ellipse that the sum of the distances from any point on the ellipse to the two foci is a constant quantity

i.e., $$AP + BP = \text{constant}$$

Hence if a point source of light be placed at *A* all the rays starting from *A* will after reflection from *PQR* pass through the point *B*. *PQR* is then an aplanatic surface and the two conjugate points *A* and *B* are called aplanatic foci.

As the distance between the foci of the ellipse increases it approximates more and more to a parabola. If, therefore, it is desired to converge all the rays coming from infinity to one single point, the mirror to be used must have a principal section which is a parabola. The mirror itself is called a paraboloidal mirror (Fig. 2.7).

It may be mentioned here that a spherical mirror cannot be used for bringing a parallel beam of light to a point focus.

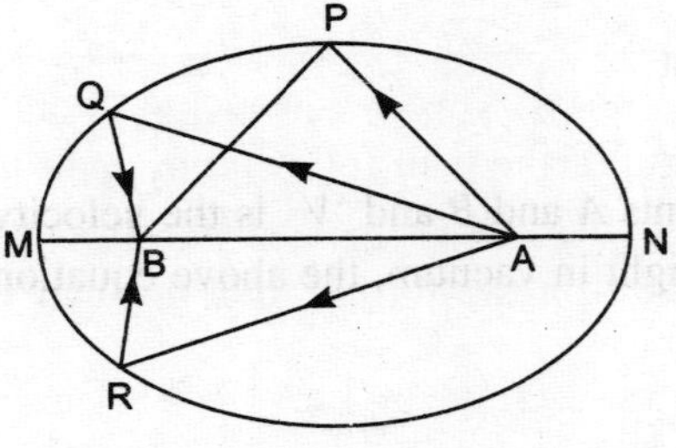

Fig. 2.6. Aplanatic surface

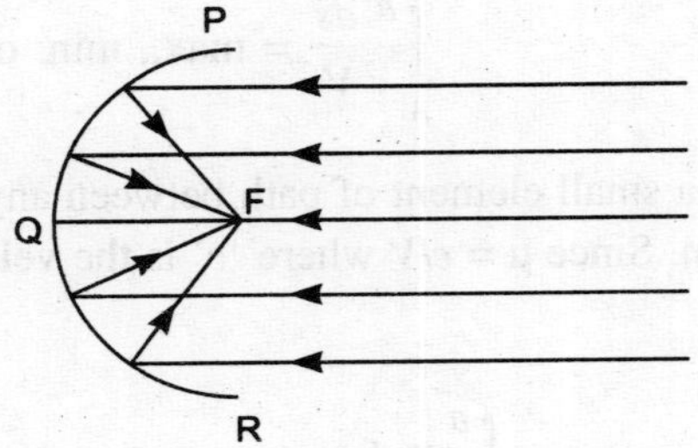

Fig. 2.7. Paraboloidal mirror

2.3 REFLECTION AT CURVED SURFACES – LAW OF EXTREME PATHS

Suppose the rays of light starting from a point object at *A* arrive at *B* after reflection from the mirror *MM* (Fig 2.8). If an aplanatic surface, be drawn with points *A* and *B* as its foci its equation will be

$$AP + BP = \text{constant.}$$

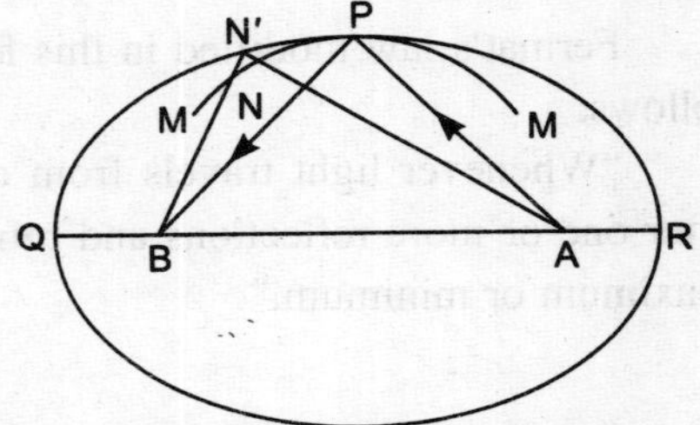

Fig. 2.8. Law of extreme paths

Such a surface will be tangential to the mirror at the point *P*. Take any point *N* on *MM* which is very near *P*. Join *AN* and produce it to meet the surface *PQR* at *N'*. Since *N'* lies on *PQR* it must satisfy the condition

$$AN' + N'B = AP + BP \qquad \text{...(2.3 a)}$$

According to Fermat's principle

$$AN + NB < AP + BP$$

Now $$AN' + N'B = AN + NN' + N'B \qquad \text{...(2.3 b)}$$

Also, in the triangle $N'NB$

$$NN' + N'B > NB \qquad \text{...(2.3 c)}$$

Hence it follows that

$$AP + BP = AN + NN' + N'B > AN + NB$$

It is thus seen that the time taken by light along the path APB instead of being a minimum is maximum.

It may be pointed out that if the curvature of MM be less than that of the aplanatic surface PQR, Fermat's principle of least time will hold good.

Similarly it can be shown that if a convex refracting surface is more convex than the corresponding aplanatic surface, the path followed by light in going from a less refracting medium to a more refracting medium is a maximum. On the other hand, if it is less convex or concave, the path is a minimum.

But whether the total optical path is a maximum or a minimum its first derivative will always vanish and hence the law can be written in a modified form *i.e.*,

$$\int_A^B \frac{ds}{V} = \text{max., min. or constant}$$

where 'ds' is a small element of path between any two points A and B and 'V' is the velocity of light in the medium. Since $\mu = c/V$ where 'c' is the velocity of light in vacuum, the above equation may be written as

$$\int_A^B \mu\, ds = \text{max., min. or constant.}$$

or $$\delta \int_A^B \mu\, ds = 0 \qquad \text{...(2.3 d)}$$

Fermat's law modified in this form is known as the *law of extreme paths* and may be stated as follows:

"Whenever light travels from one point to another (it may reach the second point directly or after one or more reflections and refractions) it always follows the path which is extreme *i.e.*, either maximum or minimum."

2.4 REFRACTION AT CURVED SURFACES

Let RR (Fig. 2.9) be a curved refracting surface of radius r and centre of curvature C. Suppose a ray of light OP starting from O placed at a distance 'u' from V reaches I after suffering refraction at the surface.

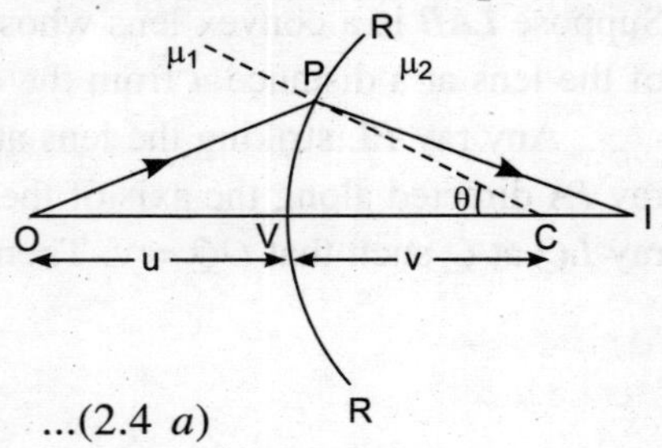

Let $VI = v$.

Fig. 2.9

In the triangle OPC

$$OP^2 = r^2 + (u+r)^2 - 2r(u+r)\cos\theta \quad ...(2.4\ a)$$

Similarly in the triangle PCI

$$PI^2 = r^2 + (v-r)^2 + 2r(v-r)\cos\theta \quad ...(2.4\ b)$$

According to the law of extreme path, the first derivative of the total optical path should vanish and hence.

$$\delta(\mu_1 OP + \mu_2 PI) = 0 \quad ...(2.4\ c)$$

Equation (2.4 c) when combined with (2.4 a) and (2.4 b) yields

$$\delta[\mu_1\{r^2 + (u+r)^2 - 2r(u+r)\cos\theta\}^{\frac{1}{2}} + \mu_2\{r^2 + (v-r)^2 + 2r(v-r)\cos\theta\}^{\frac{1}{2}}] = 0 \quad ...(2.4\ d)$$

Differentiating partially with respect to θ (because θ varies if the path is changed) it is seen that

$$\frac{\mu_1 r(u+r)\sin\theta}{[r^2 + (u+r)^2 - 2r(u+r)\cos\theta]^{\frac{1}{2}}} + \frac{\mu_2[-r(v-r)\sin\theta]}{[r^2 + (v-r)^2 + 2r(v-r)\cos\theta]^{\frac{1}{2}}} = 0 \quad ...(2.4\ e)$$

If the angle θ be small cos θ will be nearly unity and Eqn. (2.4 e) gives

$$\frac{\mu_1(u+r)}{u} = -\frac{\mu_2(v-r)}{v}$$

or

$$\frac{\mu_1}{u} + \frac{\mu_2}{v} = \frac{\mu_2 - \mu_1}{r} \quad ...(2.4\ f)$$

In this case, however, 'v' and 'r' are positive (see Convention of signs, Chapter 3) and, therefore,

$$\frac{\mu_2}{v} - \frac{\mu_1}{u} = \frac{\mu_2 - \mu_1}{r} \quad ...(2.4\ g)$$

If the medium to the left of the refracting surface be air, the above equation simplifies to

$$\frac{\mu}{v} - \frac{1}{u} = \frac{\mu - 1}{r}$$

2.5 REFRACTION AT A LENS

Suppose *LAB* is a convex lens whose refractive index is μ. Let *P* be a point object placed on the axis of the lens at a distance *u* from the optical centre *i.e.*, $OP = u$. (Fig. 2.10)

Any ray *PL* striking the lens at *L* will after refraction converge at the point *Q*. Similarly another ray *PA* directed along the axis of the lens will go undeviated after refraction and cut the first refracted ray *LQ* at *Q* such that $OQ = v$. Then *Q* is the image of *P*.

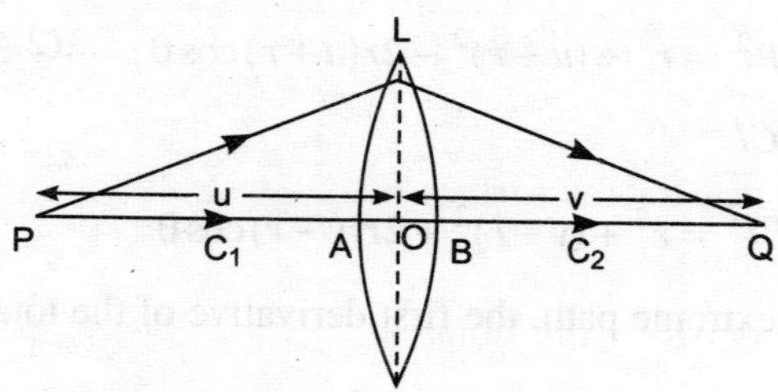

Fig. 2.10

Since the rays starting from *P* intersect at *Q*, the optical paths along the two directions according to the law of extreme paths, must be the same and hence

$$PL + LQ = PA + \mu AB + BQ = \text{max., min or constant.} \qquad ...(2.5\ a)$$

[Since the ray *PA* travels a distance AB in glass its equivalent optical path in air will be μ.AB.]

Also, in the triangle *PLO*

$$PL^2 = PO^2 + OL^2$$

and if *OL* be small as compared to *OP*

$$PL^2 = PO^2\left(1 + \frac{OL^2}{OP^2}\right)$$

or

$$PL = PO + \frac{OL^2}{2OP} \quad \text{approximately} \qquad ...(2.5\ b)$$

Similarly the triangle *OLQ* gives

$$QL = QO + \frac{OL^2}{2OQ} \quad \text{approximately} \qquad ...(2.5\ c)$$

If these values of *PL* and *QL* are introduced in Eqn. (2.5 *a*) the latter reduces to

$$PO + QO + \frac{OL^2}{2}\left(\frac{1}{OP} + \frac{1}{OQ}\right) = PA + \mu AB + BQ$$

or

$$PO + QO + \frac{1}{2}OL^2\left(\frac{1}{OP} + \frac{1}{OQ}\right) = PO - OB + QO - OA + \mu(AO + OB)$$

or $$\frac{1}{2}OL^2\left(\frac{1}{OP}+\frac{1}{OQ}\right)=(\mu-1)(AO+BO)$$

$\therefore$ $$\frac{1}{v}+\frac{1}{u}=2(\mu-1)\frac{AO+BO}{OL^2} \qquad ...(2.5\ d)$$

For the surface whose radius of curvature is r_1 and centre is C_1 it can be seen that

$$OL^2 = OB\ (2r_1 - OB)$$

If OB be small as compared to r_1 *i.e.*, if the lens be thin,

$$OL^2 = OB \cdot 2r_1 \qquad ...(2.5\ e)$$

Similarly it can be shown that for the surface of radius of curvature r_2 and centre C_2

$$OL^2 = OA \cdot 2r_2 \qquad ...(2.5\ f)$$

On introducing the values of OL^2 from Eqns. (2.5 *e*) and (2.5 *f*) in Eqn. (2.5 *d*) it becomes

$$\frac{1}{v}+\frac{1}{u}=(\mu-1)\left(\frac{1}{r_1}+\frac{1}{r_2}\right) \qquad ...(2.5\ g)$$

where the proper signs are to be inserted.

PROBLEMS

1. State Fermat's principle and use it to establish the laws of reflection and refraction of light.
2. State and explain Fermat's law of extreme paths and analyse a case where the actual path of light may be maximum. Use Fermat's law to deduce Snell's law of refraction.
3. Show that if the refracting angle of a glass prism is larger than twice the critical angle of refraction, no light can be passed through it by refraction.
4. Discuss the variation of deviation with angle of incidence for a prism of lower refractive index than the surrounding medium.
5. Calculate the angle of minimum deviation for a 60° prism (μ = 1.5) and the angle of incidence i at which it occurs. Hence find the deviation for rays at incident angles i + 15° and i + 10° and then calculate the angular separation between the pair of rays on emergence. Also calculate the angular separation of rays on emergence which struck the prism at i – 15° and i – 10°. Hence explain Schuster's method of focusing the spectrometer.
6. Write notes on :
   (a) Fermat's Principle.
   (b) Dispersive power of a medium.
   (c) Irrationality of prismatic dispersion.

REFRACTION AT SPHERICAL SURFACE

The study of refraction of light through a single spherical surface separating two media of different indices of refraction, is of great fundamental importance in Geometrical Optics. Therefore we shall begin the study of this branch of optics by developing the theory of image formation by a single spherical refracting surface. It will be assumed at the outset, of course, for simplicity in discussion, that a luminous object point is monochromatic, otherwise owing to light rays of different colours being deviated by different amounts in passing from one medium into another, a very confusing image would result. Furthermore, the detailed discussion will be limited to *paraxial rays*, *i.e.* rays inclined at small angles to the axes, for the reason that the results of the paraxial theory hold good also in the optics of "*corrected*" systems.

3.1 CONVENTION OF SIGNS

The derivations of various formulae in this branch of optics are based upon the measurement of angles and distances from a suitably chosen axis and origin in every optical system. It is therefore essential to adopt a convention of signs for distances and angles in order to ensure consistency in the derivations and use of various formulae. We shall adopt the following set of conventions which agree with usual convention of *Cartesian system of coordinates* used in *Coordinate Geometry*.

Convention of Distances

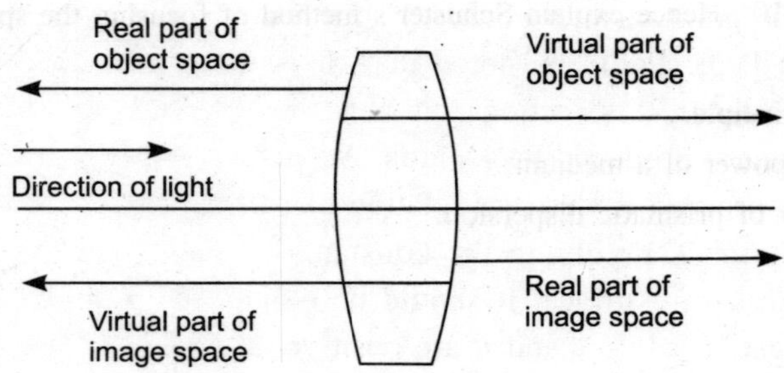

Fig. 3.1. Object space and Image space exist through all the space

(*i*) Draw all figures with light advancing from left to right.

(*ii*) Consider all distances measured to the right of suitably chosen *origin positive* and all distances to the *left negative*. Thus all longitudinal distances measured from some origin will be considered positive when they are along the initial direction of propagation of light, which is assumed to be always from left to right. In other words, according to this convention, the object distance *u* of a real object point is negative whereas it is positive for a virtual object point. Similarly, the image distance *v* is positive for a real image point and negative for a virtual image point.

(*iii*) Consider a distance transverse to the axis *positive* when measured *upwards* from it and *negative* when measured downwards.

Convention for Angles

(*i*) Consider the slope angle that a ray makes with the axis positive when an anticlockwise turn through this acute angle will bring a ruler from the principle axis direction into coincidence with that of the ray.

(*ii*) Consider angles of incidence and refraction positive when a ruler from a normal direction at the point of incidence on the surface must be rotated in the anticlockwise direction through this angle to bring it into coincidence with the ray.

It is advisable that this convention of signs should be memorised at this very stage because this would make it easier to grasp and to derive consistent relations in this branch of optics.

3.2 REFRACTION AT A SPHERICAL SURFACE

Let us suppose that in Fig. 3.2., the common boundary of two transparent homogeneous media of refractive indices μ_1 and μ_2, (where $\mu_1 < \mu_2$) is spherical in form, being convex towards the medium of index μ_1, in which is situated a luminous object point *P*. A line joining *P* and the centre *C* of the spherical surface is called *the axis of the system and its intersect O* with the refracting surface is called the pole of the surface. We shall consider this point *O* as the origin of system. A ray *PS* from the axial object point *P*, inclined at an angle θ_1 with the axis, is incident on the surface at an angle *i* with the normal *SC*. Since $\mu_1 < \mu_2$, the incident ray *PS* on refraction at *S* bends towards the normal in accordance with the fundamental laws of refraction and pursues the path inclined at an angle *r* with the normal. The slope angle of the refracted ray *SP'* with the axis is θ_2. Another ray *PO*, which is coincident with the axis and therefore incident normally on the surface, enters undeviated in the second medium. The point of intersection *P'* of the ray *SP'* and the axial ray is, therefore, the image of the object point *P*. It should be noted that for some object positions with respect to refracting surface, discussed later in this section and illustrated in Fig. 3.3, refracted ray intersects axial ray only when the former is produced backwards. Accordingly in such cases, a virtual image is formed.

We shall now proceed to derive a relation between *u* and *v*, the distances of the object point *P* and its image *P'* from *O*, involving the constants of the system viz. μ_1, μ_2 and *R*, the radius of curvature of the refracting surface. It should be pointed out that according to convention of signs, in Fig. 3.2, *u* is negative while *v* and *R* are positive, the pole *O* being the origin. The slope angle θ_1

of incident ray *PS* with the axis is positive whereas the slope angle θ_2 of the refracted ray *SP'* is negative.

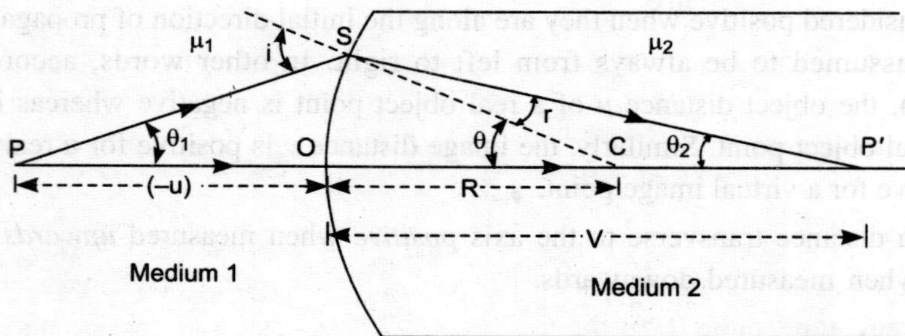

Fig. 3.2. Refraction at convex surface forming a real image

Now in triangle *PSC*, the law of sines gives

$$\frac{\sin\theta_1}{R} = \frac{\sin(\pi - i)}{R + (-u)} = \frac{\sin i}{R - u}$$

Solving for *i*,

$$\sin i = \frac{R-u}{R}\sin\theta_1 \qquad \text{...(3.2 a)}$$

The angle of refraction *r* conjugate to angle of incidence *i* may be found from Snell's law.

$$\frac{\sin i}{\sin r} = \frac{\mu_2}{\mu_1}$$

Solving for *r*,

$$\sin r = \frac{\mu_1}{\mu_2}\sin i \qquad \text{...(3.2 b)}$$

Slope angle θ_2 of the refracted ray should now be computed. In triangle *PSP'* the sum of the angles is π radians, that is

$$\theta_1 + (\pi - i) + r + (-\theta_2) = \pi$$

Since θ_2 by convention is negative. On solving for θ_2

$$\theta_2 = r + \theta_1 - i \qquad \text{...(3.2 c)}$$

The image distance *v* from the pole *O* should now be computed. In triangle *P'SC*, the law of sines gives

$$\frac{\sin r}{v - r} = \frac{\sin(-\theta_2)}{R}$$

$$\sin\theta_2 = \frac{R}{R - v}\sin r \qquad \text{...(3.2 d)}$$

Solving for v,

$$v = R - R\frac{\sin r}{\sin\theta_2}$$

and with the help of Eqns. (3.2 *a*), (3.2 *b*) and (3.2 *c*) the above equation can be easily given as

$$v = R - \frac{\mu_1}{\mu_2}\frac{(R-u)\sin\theta_1}{\sin(r+\theta_1-i)}$$

Equations (3.2 *a*), (3.2 *b*), (3.2 *c*) and (3.2 *d*) are quite sufficient to determine the image distance v and the slope angle θ_2 conjugate to quantities u and θ_1 respectively. In general, the image distance v will be different for different slope angle θ_1 of the incident ray because the ratio $\sin\theta_1/\sin(r+\theta_1-i)$ is not constant for a spherical surface. Therefore, the image of the object point will not be a point image. This departure from the ideal point image of the axial object point is called *Spherical Aberration* and will be discussed in detail in the chapter 4 on *Aberrations of Optical Images.*

First Order Theory

The sine of any angle can be expanded by *Maclaurin's Theorem* into power series in the angle, for example,

$$\sin\theta = \theta - \frac{\theta^3}{3!} + \frac{\theta^5}{5!} - \frac{\theta^7}{7!} + \ldots\ldots..$$

For incident rays restricted to the region close to the axis, called the paraxial region, the slope angle θ_1 becomes very small and as a consequence the slope angle θ_2 of the conjugate refracted ray, the angle i and r also become small to the same order of approximation. Rays which satisfy these restrictions are called paraxial rays and for the refraction of these rays a very simple relationship between v, u, μ_1, μ_2 and R can be obtained by replacing, to a close approximation, the *sine* of the angle in Eqns. (3.2 *a*), (3.2 *b*), (3.2 *d*) with the angle (in radians) itself. To assume $\sin\theta_1 = \theta_1$ is to neglect in the series all terms of higher order in θ_1 beyond the first order term and therefore the theory based on this approximation is called the first order theory.

Equations (3.2 *a*), (3.2 *b*), (3.2 *d*) under paraxial conditions become

$$i = \frac{R-u}{R}\theta_1 \qquad \text{...(3.2 } e\text{)}$$

$$r = \frac{\mu_1}{\mu_2}i \qquad \text{...(3.2 } f\text{)}$$

$$\theta_2 = \frac{R}{R-v}r \qquad \text{...(3.2 } g\text{)}$$

while Eqn. (3.2 *c*)

$$\theta_2 = r + \theta_1 - i$$

remains as such. On substituting the values of θ_2, r and θ_1, this equation becomes after cancelling i and rearranging

$$\mu_1 v(R-u) = \mu_2 u(R-v)$$

Dividing by $v\,u\,R$ and rearranging, we get

$$\frac{\mu_2}{v} - \frac{\mu_1}{u} = \frac{\mu_2 - \mu_1}{R} \quad \text{...(3.2 } h)$$

or

$$\frac{f'}{v} + \frac{f}{u} = 1 \quad \text{...(3.2 } i)$$

where $f = -\mu_1 R/(\mu_2 - \mu_1)$ and $f' = \mu_2 R/(\mu_2 - \mu_1)$

From a theoretical point of view, since the slope angles θ_1 and θ_2 do not appear in Eqn. (3.2 *h*) it follows that all rays diverging from an axial point *P*, making small angles with the axis, after refraction at a spherical surface cross the axis at a common point *P'*, which is, therefore, the image corresponding to object point *P*. This equation, therefore, applies only to rays in paraxial region but it does apply to rays making large angles with the axis in optical systems which have been corrected for spherical aberration.

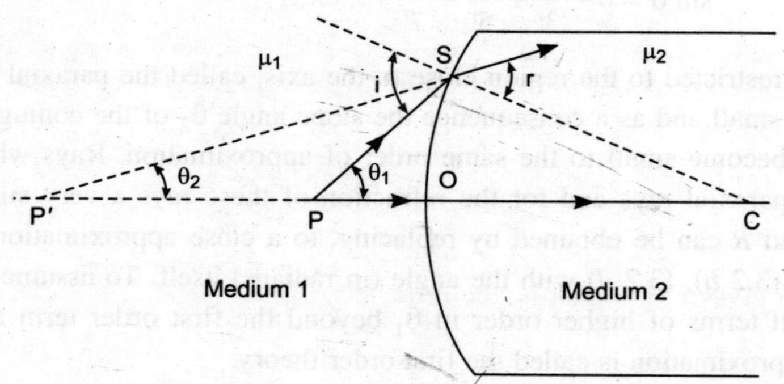

Fig. 3.3. Refraction at convex surface forming a virtual image

Although we have derived Eqn. (3.2 *h*) for the case of a convex refracting surface (*R positive*) when the object distance is such that a real image is formed, it holds equally well unchanged for object distance, illustrated in Fig. 3.3 which gives a virtual image and also for the formation of a virtual image by a concave refracting surface (*R negative*), provided consistent sign convention for *u*, *v* and *R* is employed, that is, correct sign is attached to the numerical values on substitution for the symbols. Further, Eqn. (3.2 *h*) is of fundamental importance in *Geometrical Optics* because it can be applied successively to each of the number of co-axial refracting surfaces, treating the image formed by the first surface as an object for the second and so on, in order to locate the final image of a given object.

In most of the practical cases of image formation object point *P* is in air or vacuum, that is, $\mu_1 = 1$ and index of refraction of the second medium with respect to first is $\mu = \mu_2/\mu_1$. Therefore,

Eqn. (3.2 *h*) applied to this case becomes

$$\frac{\mu}{v} - \frac{1}{u} = \frac{\mu - 1}{R} \qquad \text{...(3.2 j)}$$

The conditions governing the object distance *u* for the formation of a real or virtual image due to refraction at a convex refracting surface may be very easily derived. Let *u'* be some object distance from the pole of the refracting surface. Putting $u = -u'$ in (3.2 *h*) and keeping the sign of *R* positive, we get

$$\frac{\mu_2}{v} = \frac{u'(\mu_2 - \mu_1) - \mu_1 R}{u'R} \qquad \text{...(3.2 k)}$$

Therefore, *v* is positive when

$$u' > \frac{\mu_1 R}{\mu_2 - \mu_1}$$

which is consequently the condition for the object distance for the formation of a real image. This condition corresponds to the case illustrated in Fig. 3.2. The image distance *v* is negative and indicates a virtual image when

$$u' < \frac{\mu_1 R}{\mu_2 - \mu_1}$$

This condition corresponds to the case illustrated in Fig. 3.3. However, when the object distance is

$$u = -u' = -\frac{\mu_1 R}{\mu_2 - \mu_1}$$

Eqn. (3.2 *k*) gives $v = \infty$, that is, the refracted rays are parallel to the axis and image is formed at infinity.

3.3 PRINCIPAL FOCAL POINTS AND PRINCIPAL FOCAL LENGTHS

The axial object point which is imaged at infinity by a refracting surface is called the *principal focal point* of the *object space* or *the first principal focal point F* of the refracting surface. The distance of *F* from the pole of the surface is called *the first principal focal length* of the surface and is denoted by *f*. Therefore, *f* is the value of *u* conjugate to $v = \infty$ and hence its substitution in Eqn. (3.2 *h*) yields

$$f = -\frac{\mu_1 R}{\mu_2 - \mu_1} \qquad \text{...(3.3 a)}$$

The axial image point conjugate to infinitely distant axial object point is called the *principal focal point* of the *image space* or simply the *second principal focal point F'* of the system. Its distance from the pole of the surface is called the *second principal focal length* of the refracting surface and is denoted by f'. Accordingly, f' is the value of v conjugate to $u = -\infty$ and its substitution in Eqn. (3.2 *h*) gives

$$f' = \frac{\mu_2 R}{\mu_2 - \mu_1} \qquad \text{...(3.3 } b\text{)}$$

whence $-\dfrac{f'}{f} = \dfrac{\mu_2}{\mu_1}$ or $-\dfrac{f'}{f} = \mu$

Since μ, the index of refraction of the second medium with respect to the first, is always positive, it follows that *the two focal lengths of the refracting surface are always of opposite signs.*

This discussion obviously explains the physical significance of the symbols f and f' introduced in Eqn. (3.2 *i*).

3.4 LINEAR LATERAL MAGNIFICATION

Let PQ be a small object of height y_1 placed transversely to the axis in front of the convex refracting surface. Ray PS from the foot of the object suffers refraction at S and follows the path SP' and intersects the axial ray PO, which suffers no deviation, at P'. Thus P' is the image of P. Ray QS' is directed towards the centre of curvature C and therefore it is incident at S' normally to the surface; as a consequence it passes undeviated. Now a line drawn at P', perpendicular to the principal axis, intersects the ray QS' at the point Q', which is consequently the image of Q. Thus $P'Q'$ is the image of PQ.

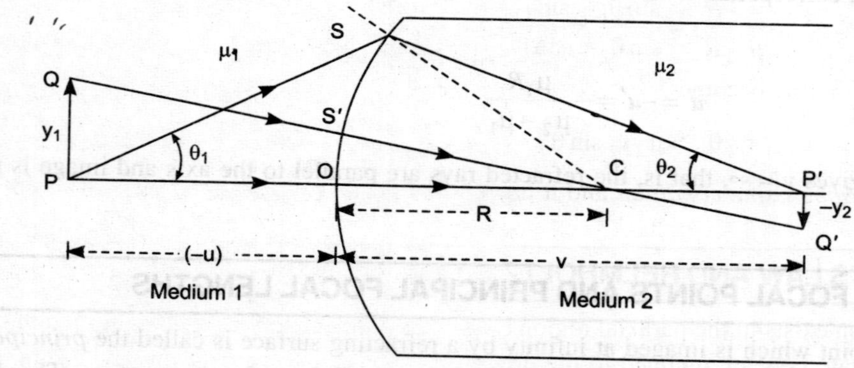

Fig. 3.4. Illustrates linear lateral magnification

The linear lateral magnification is defined by

$$m = \frac{P'Q'}{PQ} = \frac{y_2}{y_1} \qquad \text{...(3.4 } a\text{)}$$

where y_2 is the image height $P'Q'$. From similar triangles $CP'Q'$, and CPQ, we write

$$\frac{(-y_2)}{y_1} = \frac{CP'}{PC} = \frac{v-R}{R+(-u)} = \frac{v-R}{R-u}$$

$$\frac{y_2}{y_1} = \frac{v - R}{u - R} \quad ...(3.4\ b)$$

Now, Eqn. (3.2 *h*) can be rewritten as

$$\mu_2\left(\frac{1}{v} - \frac{1}{R}\right) = \mu_1\left(\frac{1}{u} - \frac{1}{R}\right)$$

or

$$\frac{v - R}{u - R} = \frac{\mu_1 v}{\mu_2 u}$$

whence, Eqn. (3.4 *b*) becomes

$$m = \frac{y_2}{y_1} = \frac{\mu_1 v}{\mu_2 u} \quad ...(3.4\ c)$$

This relation determines the height y_2 of an image conjugate to height y_1 of object, formed by refraction at a single spherical surface, whether convex or concave, provided a consistent sign convention is used. This relation shows that if v and u are of opposite signs, y_1 and y_2 are also of opposite signs, that is, an inverted image of the object is formed, and magnification m is negative. On the other hand, if v and u have same signs, the image is erect with respect to object and magnification m is positive.

Incidentally we may derive *Abbe's Sine Condition* even at this stage very easily. From Eqns. (3.2 *a*), (3.2 *b*) and (3.2 *d*), we have the relation

$$\frac{v - R}{u - R} = \frac{\sin\theta_1}{\sin\theta_2} \times \frac{\sin r}{\sin i} = \frac{\mu_1}{\mu_2} \times \frac{\sin\theta_1}{\sin\theta_2}$$

Whence Eqn. (3.4 *b*) becomes

$$\mu_1 y_1 \sin\theta_1 = \mu_2 y_2 \sin\theta_2 \quad ...(3.4\ d)$$

which is known as *Abbe's sine condition* and it holds good for all rays whether paraxial or not.

3.5 LAGRANGE'S LAW AND HELMHOLTZ'S LAW

An important relation exists in the paraxial region between the object and image heights and the angles at which the conjugate rays are inclined to the principal axis. Let the incident ray *PS*, in Fig. 3.4, be inclined at an angle θ_1 with the axis and the conjugate refracted ray *SP′* be inclined at θ_2 with the axis. According to sign convention θ_2 is negative. Then, referring to Fig. 3.4 and using Eqn. (3.4 *c*), we have

$$\frac{y_1}{(-y_2)} = \frac{\mu_2}{\mu_1}\frac{(-u)}{v}$$

In the paraxial region, the point *S* in Fig. 3.4 is close to the pole *O*. We, therefore, write

$$\frac{\tan\theta_1}{\tan(-\theta_2)} = \frac{-OS/u}{OS/v}$$

On multiplying, $$\frac{y_1}{y_2}\frac{\tan\theta_1}{\tan\theta_2} = \frac{\mu_2}{\mu_1}$$

or $$\mu_1 y_1 \tan\theta_1 = \mu_2 y_2 \tan\theta_2 \quad ...(3.5\ a)$$

This equation was first given by Lagrange and is frequently called *Lagrange's Law* after its discoverer. For paraxial region, since θ_1 and θ_2 are small, Eqn. (3.5 *a*) reduces to a more familiar form,

$$\mu_1 y_1 \theta_1 = \mu_2 y_2 \theta_2 \quad ...(3.5\ b)$$

Now consider $(n - 1)$ co-axial refracting surfaces separating n media of indices μ_1, μ_2, μ_3 ....μ_n in succession. Then the image formed by the first surface becomes the object for the second surface and so on. Therefore applying Eqn. (3.5 *a*) successively to each of $(n - 1)$ refracting surfaces, we get

$$\mu_1 y_1 \tan\theta_1 = \mu_2 y_2 \tan\theta_2$$

$$\mu_2 y_2 \tan\theta_2 = \mu_3 y_3 \tan\theta_3$$

$$\cdots\cdots\cdots\cdots$$

$$\mu_{n-1} y_{n-1} \tan\theta_{n-1} = \mu_n y_n \tan\theta_n$$

Hence on addition,

$$\mu_1 y_1 \tan\theta_1 = \mu_n y_n \tan\theta_n$$

This equation, therefore, holds good for a system of co-axial refracting surfaces. In this form it was first investigated by *Helmholtz*. Accordingly, it is now called *Helmholtz's law of magnification.*

3.6 REFLECTION AS A SPECIAL CASE OF REFRACTION

The only mathematical difference between the phenomena of refraction and reflection of light is that *Snell's Law*

$$\frac{\sin i}{\sin r} = \frac{\mu_2}{\mu_1}$$

replaces the law of reflection $i = r$ (here r is the angle of reflection). If we employ consistent sign convention for angles, stated in *Sec.* 3.1, in both the phenomena, the law of reflection should be written as

$$i = -r \quad ...(3.6\ a)$$

where r is the angle of reflection [See Fig. 3.5]. It is therefore, evident that the law of reflection, represented by Eqn. (3.6 *a*) can be derived from Snell's law by using the artifice of replacing μ_2 by $-\mu_1$, that is, $\mu_2 = -\mu_1$ and interpreting the angle r as the angle of reflection. The steps of this procedure are

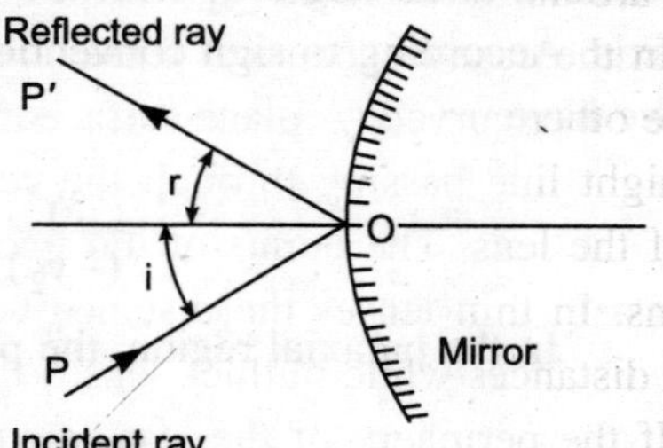

Fig. 3.5. Reflection at a mirror

$$\mu_1 \sin i = \mu_2 \sin r = (-\mu_1)\sin r$$

$$\sin i = -\sin r$$

$$i = -r$$

As a consequence, all formulae for reflection of paraxial rays at spherical mirror can be obtained from those for refraction of paraxial rays at a spherical refracting surface simply by putting $\mu_2 = -\mu_1$ in the formulae. Thus Eqn. (3.2 *h*) becomes

$$\frac{(-\mu_1)}{v} - \frac{\mu_1}{u} = \frac{(-\mu_1) - \mu_1}{R}$$

or $$\frac{1}{u} + \frac{1}{v} = \frac{2}{R} \quad \text{.....(3.6 } b\text{)}$$

which is therefore the relation between the conjugate object and image distances for a spherical mirror. Similarly, (3.4 *c*) becomes

$$m = \frac{y_2}{y_1} = \frac{\mu_1 v}{(-\mu_1)u} = -\frac{v}{u} \quad \text{...(3.6 } c\text{)}$$

where y_2 is the image height conjugate to height y_1 of the object.

This expression for the magnification shows that if v and u are of the same sign, as in the case of a real image formed by reflection at a *concave mirror*, y_1 and y_2 are of opposite signs, that is, an inverted image of the object is formed. On the other hand, if v and u are opposite signs, as in the case of reflection at a *convex mirror*, y_1 and y_2 are of the same sign, that is, the *image* is erect with respect to the object.

Equations (3.6 *b*) and (3.6 *c*) are sufficient to locate the position and determine the size of the image of an object due to reflection at a spherical reflecting mirror. The convention of signs defined in *Sec*. 3.1 may be used in the image formation by mirrors.

3.7 LENSES

A lens is a portion of a refracting medium bounded by two curved surfaces which themselves may either be spherical or cylindrical. If both the bounding surfaces are spherical the lens is said to be spherical. Similarly if both the surfaces are cylindrical, the lens goes by the corresponding name. If one surface is spherical and the other cylindrical the lens is said to be spherocylindrical. The spherical lenses are more commonly used and hence they shall be discussed in detail in this chapter.

In the case of spherical lenses both the bounding surfaces may be curved or one may be plane and the other curved (A plane surface may be taken as a curved surface of infinite radius of curvature). A straight line passing through the centres of curvature of the two surfaces is called the principal axis of the lens. The points of intersection of the two surfaces with this axis are called the poles of the lens. In thin lenses the distance between the poles is negligible as compared to the object and image distances while in thick lenses this is not so.

If the periphery of the lens is circular in contour its diameter is known as the aperture of the lens.

Some of the more common types of lenses, and their names are given in Fig. 3.6.

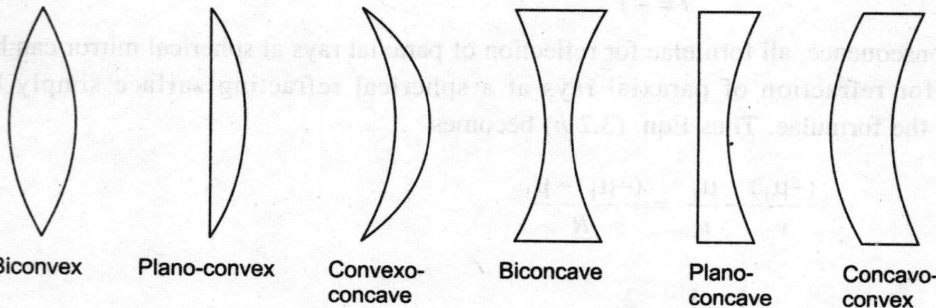

Fig. 3.6. Spherical lenses

3.8 REFRACTION THROUGH A THICK LENS

Suppose an object O is placed on the axis of a thick lens at a distance 'u' from the first surface (Fig. 3.7). Rays of light starting from O are refracted through the first surface and form an image, say, at a distance v' from it. If μ be the refractive index of the lens and r_1 the radius of curvature of the first surface, then

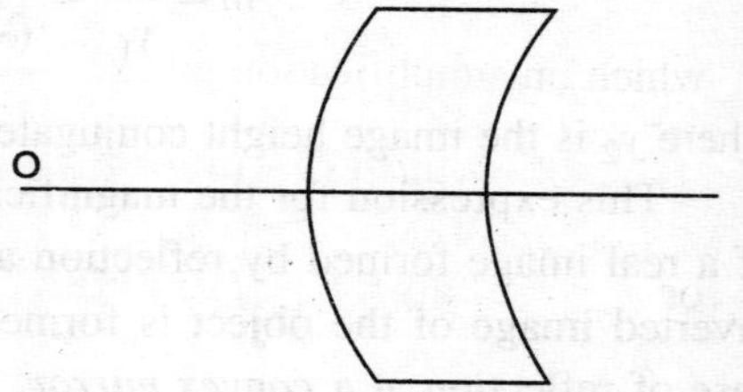

Fig. 3.7. Thick lens

$$\frac{\mu}{v'} - \frac{1}{u} = \frac{\mu - 1}{r_1} \quad \text{...(3.8 a)}$$

If in this expression $v' = \infty, u = \dfrac{r_1}{(\mu - 1)}$

If this value of u is denoted by f_1, then f_1 is called the first focal distance of the first surface of the lens and is given by

$$1/f_1 = -(\mu - 1)/r_1 \quad \text{...(3.8 b)}$$

Eqns. (3.8 a) and (3.8 b) on combination give

$$\mu/v' - 1/u = -1/f_1 \quad \text{...(3.8 c)}$$

This image at a distance v' from the first surface serves as an object for the second refracting surface of the lens. The distance of this object as measured from the second surface will be $(v' + t)$ where t is the thickness of the lens. If the final image be formed in air at a distance v from the second surface of radius of curvature r_2 then

$$\frac{1/\mu}{v} - \frac{1}{v' + t} = \frac{(1/\mu) - 1}{r_2}$$

(since light travels from glass to air in this case)

This equation may be written as

$$\frac{1}{v}-\frac{\mu}{v'+t}=\frac{1-\mu}{r_2} \qquad \text{...(3.8 } d\text{)}$$

As before if f_2 is the value of v for $v' = \infty$, f_2 is the second focal distance of the second surface and is given by

$$1/f_2 = (1-\mu)/r_2 \qquad \text{...(3.8 } e\text{)}$$

It can be seen from Eqns. (3.8 *d*) and (3.8 *e*) that

$$\frac{1}{v}-\frac{\mu}{v'+t}=\frac{1}{f_2} \qquad \text{...(3.8 } f\text{)}$$

Eliminating v' from Eqns. (3.8 *c*) and (3.8 *f*) we get

$$\frac{\mu u f_1}{f_1-u}=-t+\frac{\mu v f_2}{f_2-v} \qquad \text{...(3.8 } g\text{)}$$

which on simplification gives

$$\mu u f_1(f_2-v)+t(f_1-u)(f_2-v)-\mu v f_2(f_1-u)=0$$

or $$u(\mu f_1 f_2 - f_2 t)-v(\mu f_1 f_2+f_1 t)+uv\{u(f_2-f_1)+t\}+f_1 f_2 t=0$$

Dividing throughout by $\{\mu(f_2 - f_1) + t\}$ it can be seen that

$$uv+\frac{\mu f_1 f_2-f_2 t}{\mu(f_2-f_1)+t}u-\frac{\mu f_1 f_2+f_1 t}{\mu(f_2-f_1)+t}v+\frac{f_1 f_2 t}{\mu(f_2-f_1)+t}=0 \qquad \text{...(3.8 } h\text{)}$$

This Eqn. is of the form

$$\frac{1}{v-\beta}-\frac{1}{u-\alpha}=1/F \qquad \text{...(3.8 } i\text{)}$$

for on multiplying both the sides of Eqn. (3.8 *i*) by $F(u-\alpha)(v-\beta)$ and rearranging we get

$$uv-(F+\beta)u+(F-\alpha)v-F(\beta-\alpha)+\alpha\beta=0 \qquad \text{...(3.8 } j\text{)}$$

For Eqns. (3.8 *h*) and (3.8 *j*) to be identical the following identities must hold

$$-(F+\beta)=\frac{\mu f_1 f_2-f_2 t}{\mu(f_2-f_1)+t} \qquad \text{...(3.8 } k\text{)}$$

$$-(F-\alpha)=\frac{\mu f_1 f_2+f_1 t}{\mu(f_2-f_1)+t} \qquad \text{...(3.8 } l\text{)}$$

and $$\alpha\beta-F\beta+F\alpha=\frac{f_1 f_2 t}{\mu(f_2-f_1)+t} \qquad \text{...(3.8 } m\text{)}$$

The values of α, β and F can be obtained by multiplying Eqns. (3.8 *k*) and (3.8 *l*) and adding the product to Eqn. (3.8 *m*). Thus

$$F^2 - \alpha\beta - F\alpha + F\beta = \frac{(\mu f_1 f_2 + f_1 t)(\mu f_1 f_2 - f_2 t)}{[\mu(f_2 - f_1) + t]^2} \quad \text{...(3.8 } n\text{)}$$

and

$$F^2 = \frac{(\mu f_1 f_2)^2 + \mu f_1 f_2 t(f_1 - f_2) - f_1 f_2 t^2}{[\mu(f_2 - f_1) + t]^2} + \frac{f_1 f_2 t}{\mu(f_2 - f_1) + t}$$

$$= \frac{(\mu f_1 f_2)^2 + \mu f_1 f_2 t(f_1 - f_2) - f_1 f_2 t^2 + f_1 f_2 t[\mu(f_2 - f_1) + t]}{[\mu(f_2 - f_1) + t]^2}$$

$$= \left[\frac{\mu f_1 f_2}{\mu(f_2 - f_1) + t}\right]^2$$

$$\therefore \quad F = \pm \frac{\mu f_1 f_2}{\mu(f_2 - f_1) + t} \quad \text{...(3.8 } o\text{)}$$

It has now to be decided as to which of the two signs is admissible. It is clear that this equation will hold for all values of t, including $t = 0$ which is the case of a thin lens. In such a lens if a luminous object is placed at a distance f_1 from the first surface the rays after refraction from the surface will be rendered parallel and hence after refraction at the second surface of the lens they will come to a focus at a distance f_2 from the latter. Moreover in the case of a thin lens both α and β are zero. Thus on putting $\alpha = \beta = 0$ and $v = f_2$ and $u = f_1$ in Eqn. (3.8 *i*), it reduces to

$$1/f_2 - 1/f_1 = 1/F$$

$$F = -f_1 f_2 \neq (f_2 - f_1) \quad \text{...(3.8 } p\text{)}$$

It can at once be seen that the values of F obtained by putting $t = 0$ in Eqn. (3.8 *o*) will agree with that given by Eqn. (3.8 *p*) only when the negative sign is taken into account. As such the positive sign becomes inadmissible and Eqn. (3.8 *o*) becomes

$$F = -\frac{\mu f_1 f_2}{\mu(f_2 - f_1) + t} \quad \text{...(3.8 } q\text{)}$$

The values of α and β can now be easily obtained from Eqns. (3.8 *k*), (3.8 *l*) and (3.8 *q*). Thus

$$\alpha = \frac{f_1 t}{\mu(f_2 - f_1) + t} \quad \text{...(3.8 } r\text{)}$$

$$\beta = \frac{f_2 t}{\mu(f_2 - f_1) + t} \quad \text{...(3.8 } s\text{)}$$

such that $\alpha f_2 = \beta f_1$ ...(3.8 *t*)

a relation which is of great importance.

If V is measured from a point at a distance β in the positive direction from the second surface and U from a point at a distance α in the positive direction from the first surface and

$$u - \alpha = U \text{ and } v - \beta = V$$

Equation (3.8 *i*) may then be written as

$$\frac{1}{V} - \frac{1}{U} = \frac{1}{F} \quad \text{...(3.8 } u\text{)}$$

a relation which is similar to that used in the case of a thin lens.

3.9 PRINCIPAL POINTS AND PRINCIPAL PLANES

In the case of thin lenses the distance are always measured from the centre of the lens. The distance from the centre of the lens to either focus gives the focal length. But if a thick lens is used the distance from its centre to one focus is different from the distance to the other focus and neither distance equals the focal length of the lens. Also if an attempt is made to determine the focal length by say U, V method, there is no fixed point from which these distances can be measured.

It has, however been shown above that there are two points or planes from which the distances may be measured in all cases and the simple formula for lens may be applied as usual. These points are called *principal points*. Thus the point at a distance α from the first surface is called the *first principal point* and similarly the point at a distance β from the second surface is called the *second principal point*. Planes drawn perpendicular to the axis and passing through these points are called *Principal Planes*.

In Eqn. (3.8 *u*) it is seen that if $V = \infty$, $U = -F$ so that $-F$ is the first focal distance of the lens. Its magnitude is given by Eqn. (3.8 *q*) and is measured from the first principal point. Similarly if $U = \infty$, $V = F$ then F is the second focal distance of the lens and is measured from the second principal point. A point at a distance F from the first principal point, as defined in *Sec.* 3.3, is called the first focal point of the lens. Similarly the point at a distance F from the second principal point is termed as the second focal point of the lens.

If the lens is placed in a medium of uniform refractive index, the first and second focal distances are found to be equal in magnitude but opposite in sign.

3.10 PROPERTIES OF PRINCIPAL PLANES AND PRINCIPAL POINTS

(*a*) *Principal points are conjugate foci.* Equation (3.8 *i*) may be written as

$$v - \beta = \frac{F(u-\alpha)}{F+(u-\alpha)}$$

If $u = \alpha$, whatever be the value of F, $v = \beta$, so that if an object is placed at a distance α from the first surface or at the first principal point, its image will be formed at a distance β from the second surface, *i.e.*, at the second principal point and *vice versa*. If the first principal point is outside the lens on the positive side, a real object can be placed, if it is on the negative side a virtual object must be placed.

In general the two principal planes of a thick lens are conjugate planes.

(*b*) *Principal planes are planes of unit magnification* Let O (Fig. 3.8) be an object placed on the axis of a thick lens at a distance u from the first surface, and AB, any ray passing through the first principal focus A of the first surface. This ray which touches the upper extremity of the object will after refraction from the first surface travels parallel to the axis inside the lens. After refraction from the second surface this ray will pass through D, the second principal focus of the second surface. This ray, it is clear, will pass through the extremity of the image whose position is given by Eqn. (3.8 *i*). Let I be this image. Drop perpendiculars BE and CF on the axis. When the rays AB and CD are nearly parallel to the axis the points E and F coincide with the poles of the curved surfaces. Let the object distance $EO = -u$ and the image distance $v = +FI$. Also let $EA = -f_1$ and $FD = +f_2$. If $BE = CF = h$, then $h/y_1 = f_1/(f_1 - u)$ from the similar triangles ABE and $AB'O$ and $h/y_2 = -f_2/+(v - f_2)$ from similar triangles CFD and DID' so that the magnification m is given by

$$m = \frac{y_2}{y_1} = \frac{f_1(f_2 - v)}{f_2(f_1 - u)} \qquad \text{...(3.10 } a)$$

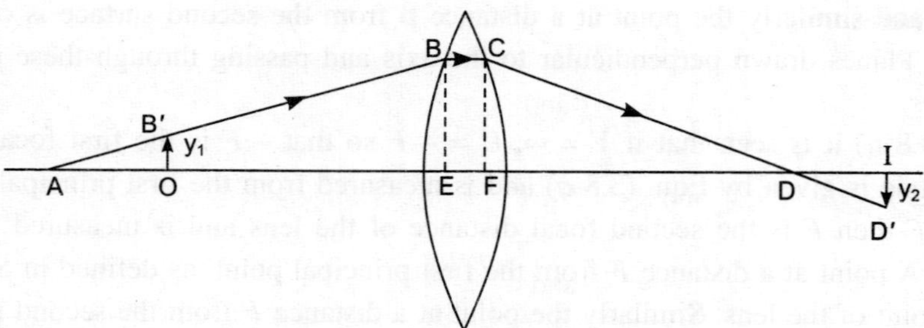

Fig. 3.8. Planes of unit magnification

This is the expression for magnification in any case. If the value of m is positive the image I will be erect, if negative the image I will be inverted.

If now the object O is supposed to lie in the first principal plane $u = \alpha$, the image will lie in the second principal plane for which $v = \beta$. With these values the magnification becomes

$$m = \frac{f_1(f_2 - \beta)}{f_2(f_1 - \alpha)} = \frac{f_1 f_2 - f_1\beta}{f_1 f_2 - f_2\alpha} = +1 \qquad \text{...(3.10 } b)$$

since $f_1\beta = f_2\alpha$ from Eqn. (3.8 *t*).

Thus the dimensions of the image seen in the second principal plane are equal to those of the object in the first principal plane and also the lateral magnification is unity and positive.

3.11 LATERAL MAGNIFICATION REFERRED TO THE FOCAL POINTS

Let Q_1R_1 and Q_2R_2 be the principal planes of a thick lens and P_1, P_2 be the two principal points. Let an object A_1B_1 of size y_1 be placed at a distance x_1 from the first focal point F_1. If A_2B_2 be a plane conjugate to A_1B_1, then the former will be the image of the latter.

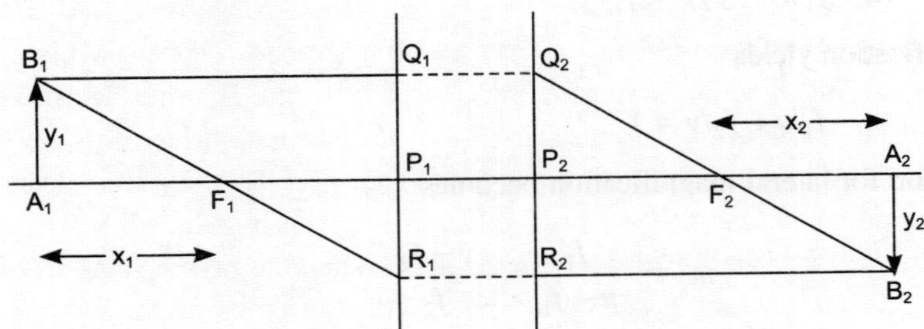

Fig. 3.9. Lateral magnification

Suppose the first and second focal distances P_1F_1 and P_2F_2 are respectively f_1 and f_2. Any ray B_1R_1 passing through F_1 and striking Q_1R_1 at R_1 will emerge out on the same direction of the axis at the point R_2 such that $P_1R_1 = P_2R_2$ (lateral magnification is unity and positive for principal planes) and will be rendered parallel to the axis. Similarly a ray B_1Q_1 parallel to the axis and striking Q_1R_1 at Q_1 will emerge out on the same side of the axis at Q_2 such that $P_1Q_1 = P_2Q_2$. It is evident from the figure that the triangles $A_1B_1F_1$ and $P_1F_1R_1$ are similar and so are the triangles $P_2Q_2F_2$ and $F_2A_2B_2$. Therefore

$$\frac{y_1}{-P_1R_1} = \frac{x_1}{f_1} \text{ and } \frac{-y_2}{P_2Q_2} = \frac{x_2}{f_2} \qquad \text{...(3.11 } a\text{)}$$

Since $y_1 = P_1Q_1 = P_2Q_2$ and $y_2 = P_1R_1 = P_2R_2$; Eqn. (3.11 a) gives

$$\frac{P_1Q_1}{-P_1R_1} = \frac{x_1}{f_1} \text{ and } \frac{-P_2R_2}{P_2Q_2} = \frac{x_2}{f_2}$$

or

$$\frac{P_1Q_1}{P_1R_1} \times \frac{P_2R_2}{P_2Q_2} = \frac{x_1x_2}{f_1f_2}$$

$$\therefore \qquad x_1x_2 = f_1f_2 \qquad \text{...(3.11 } b\text{)}$$

This is known as Newton's formula.

3.12 LATERAL MAGNIFICATION REFERRED TO PRINCIPAL POINTS

In the preceding section the distances of the image and object were measured from the focal points. Sometimes it is convenient to measure the distances from the principal points P_1 and P_2 (Fig. 3.9).

Let $P_1A_1 = u$ and $P_2A_2 = v$. Then according to our convention of signs

$$x_1 = -(u - f_1) \text{ and } x_2 = +(v - f_2)$$

Substituting these values of x_1 and x_2 in Eqn. (3.11 *b*) and remembering that f_1 is negative it is seen that

$$(u - f_1)(v - f_2) = f_1 f_2$$

which on simplification yields

$$f_1/u + f_2/v = 1 \quad \text{...(3.12 } a\text{)}$$

and the expression for lateral magnification becomes

$$m = -\frac{f_1}{u - f_1} = -\frac{v - f_2}{f_2} \quad \text{...(3.12 } b\text{)}$$

A more useful expression for the lateral magnification can be obtained by applying Lagrange's law (*Sec.* 3.5) to the principal planes. If θ_1 and θ_2 be the slope angles in the object and image spaces respectively then

$$\mu_1/\mu_2 = \theta_2/\theta_1 \quad \text{...(3.12 } c\text{)}$$

since lateral magnification is unity in principal planes. If θ_1 and θ_2 are small (Fig. 3.10)

$$-\theta_1 = y_1/-u \text{ and } -\theta_2 = -y_2/+v$$

so that

$$m = \frac{y_2}{y_1} = +\frac{v}{u} \cdot \frac{\theta_2}{\theta_1} = +\frac{\mu_1}{\mu_2} \cdot \frac{v}{u} \quad \text{...(3.12 } d\text{)}$$

because of Eqn. (3.12 *c*)

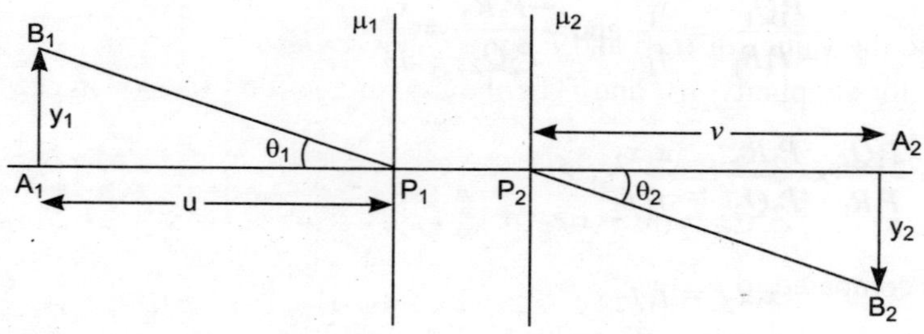

Fig. 3.10

It will be found on comparison, that Eqns. (3.4 *c*) and (3.12 *d*) are of the same form. It may be recalled that the distances *u* and *v* [in Eqn. (3.4 *c*)] where measured from the vertex while here these distances have been measured from the principal points. It is, therefore, inferred that the principal points of a single refracting surface coincide with its vertex.

3.13 DETERMINATION OF THE PRINCIPAL AND FOCAL POINTS

Equations (3.8 *q*) to (3.8 *s*) are used to determine the positions of the principal and focal points. According to Eqn. (3.8 *q*)

$$F = \frac{\mu f_1 f_2}{\mu(f_2 - f_1) + t} \quad \text{where } f_1 = -\frac{r_1}{\mu - 1}, f_2 = -\frac{r_2}{\mu - 1}$$

If $\mu = 1.5$ then

$$F = \frac{1.5 f_1 f_2}{1.5(f_2 - f_1) + t} \quad \text{and } f_1 = -\frac{r_1}{0.5}, f_2 = -\frac{r_2}{0.5}$$

$$\therefore \quad F = \frac{1.5 r_1 r_2}{0.25[1.5 \times (1/0.5)(r_1 - r_2) + t]}$$

$$= \frac{6 r_1 r_2}{t + 3(r_1 - r_2)} \quad \text{...(3.13 } a\text{)}$$

Similarly

$$\alpha = \frac{f_1 t}{\mu(f_2 - f_1) + t} = \frac{-(r_1/0.5)t}{1.5 \times (1/0.5)(r_1 - r_2) + t}$$

and if t is small,

$$\alpha = \frac{-2 r_1 t}{3(r_1 - r_2)} \quad \text{...(3.13 } b\text{)}$$

and

$$\beta = \frac{-2 r_2 t}{3(r_1 - r_2)} \quad \text{...(3.13 } c\text{)}$$

Let us calculate the values of α, β and F for a biconvex lens.

Let $r_1 = r_2 = r$ for simplicity. If f and f' be the first and second focal distances

$$f = -f' = \frac{6r^2}{6r - t} = r$$

if t is small as compared to r

$\alpha = -\,t/3$ and

$\beta = t/3$ if t is small.

Thus if the thickness of the lens be known the principal and focal points of a lens can be calculated. Further if the lens is placed in air the nodal points coincide with the principal points. Some typical cases have been shown in Fig. 3.11.

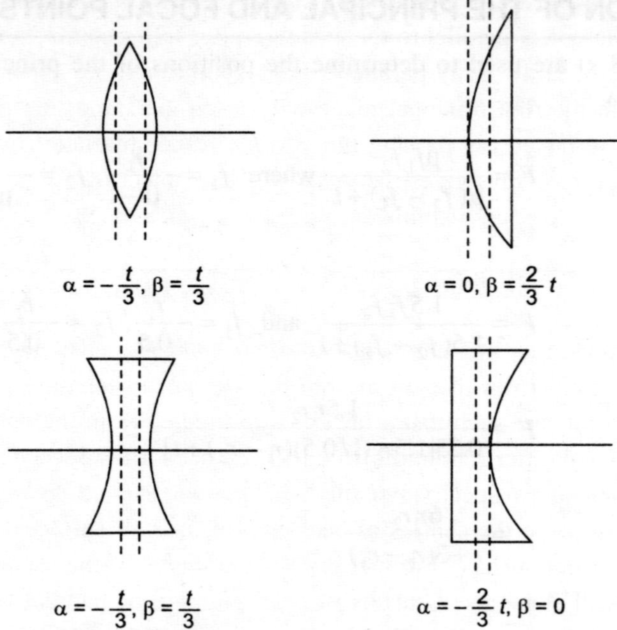

Fig. 3.11. Positions of principal points in thick lenses

3.14 GRAPHICAL CONSTRUCTION OF IMAGES

Let LL' be a thick lens whose principal points are P_1 and P_2 (Fig. 3.12). These points can be calculated with the help of the formula given above. Through these points draw planes perpendicular to the axis. These will be the principal planes of the lens. Suppose an object O is placed at a certain distance from the lens.

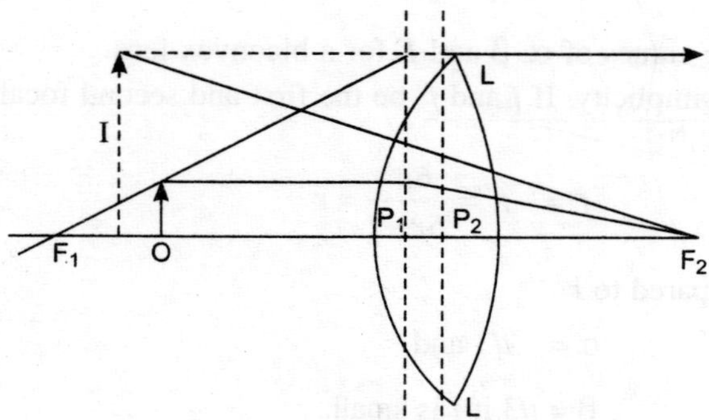

Fig. 3.12. Construction of images

To construct the image, take a ray passing through the upper extremity of the object and the first focal point F_1 of the lens. Then this ray will be rendered parallel to the axis after refraction by both surfaces of the lens. Also this ray will emerge out of the second principal plane on the same side of the axis and as far from it as the point in which the incident ray cuts the first principal plane. Further another ray from the object parallel to the axis will after refraction from the lens pass through F_2 the second principal focus. Draw a ray parallel to the axis through the upper extremity of O. The intersection of this ray with the first principal plane will be on the same side of the axis and at the same height as that of the refracted ray passing through F_2. Hence the latter ray can be drawn. The point of intersection of both these rays gives the image I of the object O.

The image can be constructed in a similar manner for a concave lens.

3.15 NODAL POINTS

Until now it has been assumed that the lens is placed in air or in one medium but if the medium on one side of the lens is different from that on the other it becomes necessary to consider two more points on the axis. These points are such that a ray of light directed from the positive side of the lens towards one of these points will after refraction from the lens appear to proceed in a parallel direction from the second. These points are respectively called the *first* and *second nodal points*.

Let A and B (Fig. 3.13 *a*) be the centres of curvature of the two surfaces of a thick lens. From A draw a radius AQ of the front surface. From B draw a radius BR of the second surface parallel to AQ. Then it is clear that small elements of surface at Q and Ra will be parallel to each other (because the normals to these surfaces are parallel). Hence a ray of light PQ directed towards N_1 which is incident at Q will after refraction through the lens proceed along RS in a direction parallel to that of PQ. Produce SR back to cut the axis at N_2. Then N_1 and N_2 are the two nodal points.

In the above construction the media on both the sides of the lens were the same. However, if the two media are different two nodal points exist in this case as well.

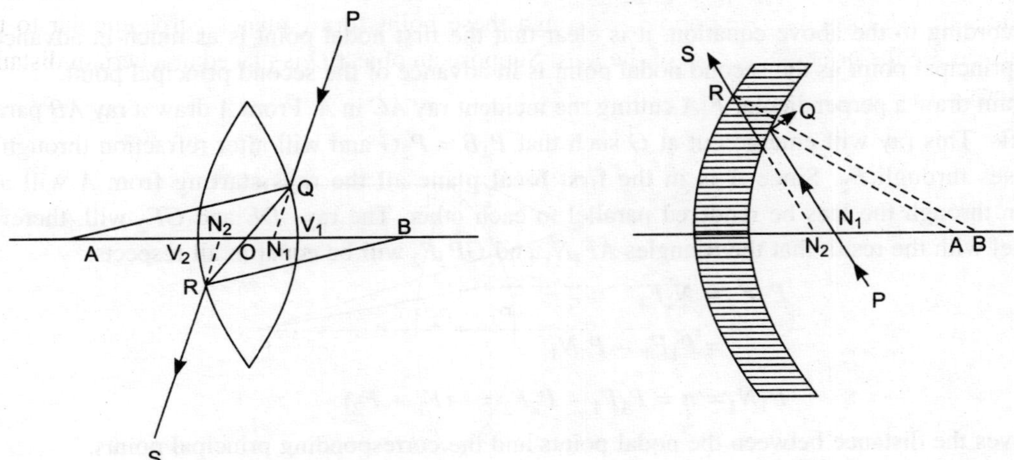

Fig. 3.13 (*a*). Nodal points **Fig. 3.13 (*b*).** Nodal points

Let P_1B and P_2G be the principal planes of a lens and F_1 and F_2 be the first and the second principal focus respectively. Let a ray of light AC directed towards the first nodal point N_1 cut the first principal plane at C (Fig. 3.14). It will then emerge out at the second principal plane on the same side and at the same distance from the axis along DE such that $P_1C = P_2D$. Also the ray DE will be parallel to the incident ray AC and appear to proceed from the second nodal point N_2 (obtained by producing DE backwards).

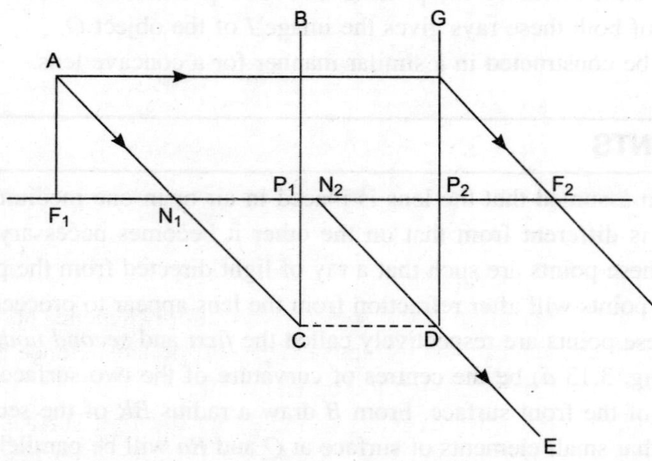

Fig. 3.14. Nodal points

The two parallelograms N_1N_2DC and P_1P_2DC stand on the common base CD and between the two parallel lines P_1P_2 and CD and hence they are equal in area.

$$\therefore \quad \text{area } P_1P_2DC = \text{area } N_1N_2DC$$

$$\therefore \quad N_1P_1 = N_2P_2$$

According to the above equation, it is clear that the first nodal point is as much in advance of the first principal point as the second nodal point is in advance of the second principal point.

Again draw a perpendicular F_1A cutting the incident ray AC in A. From A draw a ray AB parallel to the axis. This ray will emerge out at G such that $P_1B = P_2G$ and will after refraction through the lens passes through F_2. Since A is in the first focal plane all the rays starting from A will after refraction through the lens be rendered parallel to each other. The rays DE and GF_2 will, therefore, be parallel with the result that the triangles AF_1N_1 and GP_2F_2 will be equal in all respects.

$$P_2F_2 = N_1F_1$$

$$= P_1F_1 - P_1N_1$$

or

$$P_1N_1 = n = P_1F_1 - P_2F_2 = -(F_1 + F_2)$$

which gives the distance between the nodal points and the corresponding principal points.

If the medium on both the sides of the lens is air – $F_1 = + F_2$ and in that case $n = 0$ or in other words the nodal points coincide with the principal points when the lens is placed in air. In the case of a thin lens these points are evidently coincident with the centre of the lens.

3.16 OPTICAL CENTRE

In Fig. (3.13 *a*) the ray *PQ* after refraction by the surface follows the path *QR* and cuts the axis at *O*. This point *O* is called the optical centre of the lens.

Since QN_1 is parallel to RN_2 and *AQ* is parallel to *BR* the triangles AQN_1 and BRN_2 are similar and hence

$$OQ/OR = AQ/RB$$

If only the paraxial rays are considered

$$OQ = OV_1 \text{ and } OR = OV_2 \text{ approximately}$$

and if the radii of curvature *AQ* and *RB* be r_1 and r_2 then

$$OV_1/OV_2 = r_1/r_2$$

which gives the position of the optical centre.

It will be noticed that the position of the optical centre simply depends upon the radii of curvature of the surfaces and is independent of the refractive index of the lens. Hence all the rays irrespective of their wavelengths or colour pass through the optical centre.

Since the ray *PQ* directed towards N_1 is refracted by the first surface along *QR* cutting the axis at *O* it is clear that *O* is the image of N_1. Similarly it can be seen that N_2 is the image of *O* formed by refraction at the second surface.

3.17 CARDINAL POINTS

Thus it has been seen that there are certain points for a thick lens which simplify the consideration of its problem to a great extent. There are three pairs of these points: (*i*) principal points or Gauss points, (*ii*) focal points and (*iii*) nodal points. These sets of points are called *cardinal points* of a lens system.

A comparatively less important set of points is also occasionally encountered in literature. These points simplify the problems in rare cases. They are as follows:

(*i*) *Anti-principal points*: They are conjugate points for which lateral magnification is unity and negative. They lie as far from the focal points as the corresponding principal points but on the negative side.

(*ii*) *Anti-nodal points*: They are points for which the angular magnification is unity and negative. They lie as far from the focal points as the corresponding nodal points but on the negative side.

(*iii*) *Bravais points*: They are self-conjugate points. These points do not exist at all in some cases. In some cases one or even two pairs may exist.

3.18 THE THIN LENS

The focal length of a thick lens placed in air is given by Eqn. (3.8 *q*)

$$F = -\frac{\mu f_1 f_2}{\mu(f_2 - f_1) + t}$$

where $$1/f_1 = -\frac{\mu-1}{r_1} \quad \text{and} \quad 1/f_2 = \frac{1-\mu}{r_2}$$

With these values of f_1 and f_2, F becomes

$$f = F = -\frac{r_1 r_2}{(\mu-1)\left[r_1 - r_2 + \frac{\mu-1}{\mu}t\right]} \qquad \text{... (3.18 } a)$$

In a thin lens the thickness t is negligible. Hence Eqn. (3.18 *a*) gives

$$1/f = (\mu-1)(1/r_1 - 1/r_2) \qquad \text{...(3.18 } b)$$

Also when t is small, it can be seen from Eqns. (3.8 *r*) and (3.8 *s*) that

$$\alpha = \beta = 0$$

i.e., in the case of a thin lens placed in air, the principal points as well as the nodal points coincide with the vertices of the surface whose mutual separation can be neglected. As a consequence a thin lens may be represented by a straight line perpendicular to the axis.

3.19 FOCAL LENGTHS OF THIN LENSES IN SPECIAL CASES

(*i*) *Biconvex lens.* The radius of curvature r_1 of the surface facing the incident ray is positive, that of the other surface r_2 is negative and hence

$$1/f = +(\mu-1)(1/r_1 + 1/r_2)$$

The focal length of a biconvex lens is, therefore, positive.

(*ii*) *Plano-convex lens.* If the plane surface faces the incident light $r_1 = \infty$ and

$$\frac{1}{f} = +\frac{\mu-1}{r_2}$$

If the convex surface faces the incident light r_1 is positive. Even then the focal length is positive.

(*iii*) *Concavo-convex lens.* Here both the radii of curvature are on the same side so that if $r_1 > r_2$ the focal length will be negative and if $r_2 > r_1$ it will be positive.

(*iv*) *Biconvex lens.* In this case the radius of curvature r_1 of the surface facing the incident ray is negative while r_2 is positive so that

$$1/f = -(\mu-1)(1/r_1 + 1/r_2)$$

(*v*) *Plano-concave lens.* If the plane surface faces the incident ray $r_1 = \infty$, while r_2 is positive. Consequently

$$\frac{1}{f} = -\frac{\mu-1}{r_2}$$

It can be shown that if the concave surface faces the incident light the focal length will remain negative.

In general if the thickness of a lens increases towards its periphery the focal length is negative, if it increases towards its central region, the focal length is positive.

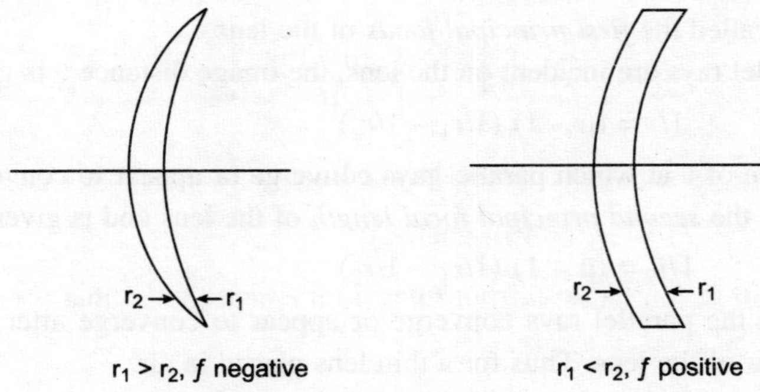

Fig. 3.15

Power of a Lens

The reciprocal of the focal length of a lens expressed in metres is called its power and is expressed in dioptres. A dioptre is defined as the power of a convex lens whose focal length is one metre, *i.e.*,

$$+1D = +1/100 \text{ cm}$$

Thus the power of a convergent lens is positive while that of a divergent lens is negative.

Principal Foci

Putting $t = 0$ in Eqn. (3.8 *g*) one gets

$$\frac{\mu u f_1}{f_1 - u} = \frac{\mu v f_2}{f_2 - v}$$

where $\qquad f_1 = \frac{1}{\mu - 1}$ and $f_2 = -\frac{r_2}{\mu - 1}$

With these values the above equation becomes

$$1/v - 1/u = (\mu - 1)\ (1/r_1 - 1/r_2) = 1/f \qquad \text{...(3.19 } a)$$

This is the general equation for the conjugate distances in the case of a thin lens placed in air. The distances u and v are measured from the optical centre of the lens which coincides with the two principal points for all practical purposes.

If in Eqn. (3.19 *a*) $v = \infty$, then

$$1/u \;= -(\mu - 1)(1/r_1 - 1/r_2) = 1/f \qquad \text{...(3.19 } b)$$

Expressed in words it means that if rays starting from an object are rendered parallel after refraction through the lens, the particular object distance given by Eqn. (3.19 *b*) is termed as the first *Principal focal distance* of the lens and is given by

$$1/f_1 = -(\mu - 1)(1/r_1 - 1/r_2)$$

The point on the axis at a distance f_1 or in other words, the point at which the object is placed in this particular case is called the *first principal focus* of the lens.

If $u = \infty$ *i.e.*, parallel rays are incident on the lens, the image distance v is given by

$$1/v = (\mu - 1)(1/r_1 - 1/r_2) \quad ...(3.19\ c)$$

This particular value of v at which parallel rays converge or appear to converge after refraction through the lens is called the *second principal focal length* of the lens and is given by

$$1/f_2 = (\mu - 1)(1/r_1 - 1/r_2)$$

The point at which the parallel rays converge or appear to converge after refraction is called the *second principal focus* of the lens. Thus for a thin lens placed in air.

$$f_1 = -f_2$$

3.20 THE CARDINAL POINTS OF THIN LENSES

A method for locating the cardinal points of a combination of lenses may also be given. Before considering the combination of two lenses, we shall show that the deviation of any ray due to refraction through a thin lens is independent of the angle of incidence and depends upon the height h of the point where total deviation occurs above the principal axis and the value of the second principal focal length. In Fig. 3.16, I is the real image of the axial point object O. A ray OA on refraction through the lens undergoes a deviation, which is evidently given by

$$\delta = \alpha + (-\beta)$$

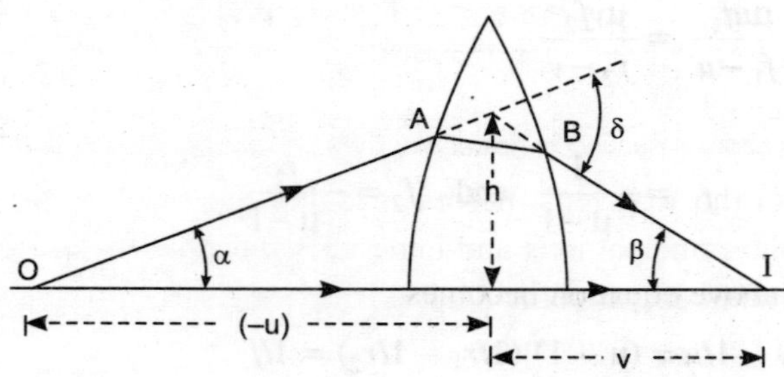

Fig. 3.16

In the paraxial region, since inclinations of the incident and refracted rays to the principal axis are small, we can write

$$\alpha = h/(-u) \quad \text{and} \quad (-\beta) = h/v$$

Hence $$\delta = h\left[\frac{1}{v} - \frac{1}{u}\right] = \frac{h}{f'} \qquad ...(3.20\ a)$$

We, therefore, conclude that *all rays incident at the same point of the lens, from whatever directions, undergo the same deviation on refraction through it.* This statement is true for a divergent lens also.

Consider now two thin convergent lenses 1 and 2, as shown in Fig. 3.17 separated by a distance d. Let f_1 and f_1' be the first and the second principal focal lengths of the first lens, and let f_2 and f_2' denote the same quantities for the second lens. A ray of monochromatic light, parallel to the principal axis, is shown incident at the height h_1 above the axis of the first lens. Due to refraction through the first lens it is deviated towards the second principal focal point F_1' of this lens. The deviation δ_1 thus produced is given by

$$\delta_1 = h_1/f_1'$$

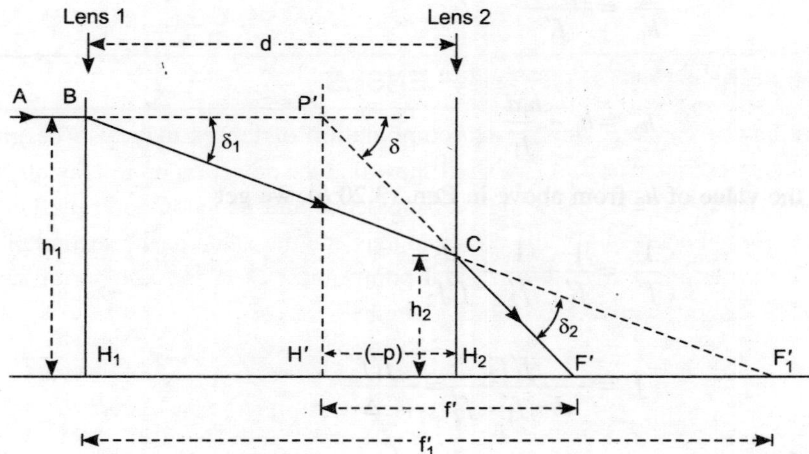

Fig. 3.17. Location of second principal plane P'H' of combination of two lenses

The emergent ray from the first lens, before reaching the focal point F_1', meets the second lens at the height h_2 above the principal axis and undergoes deviation δ_2 in refraction through it and finally crosses the principal axis at the point F'. Evidently, the deviation δ_2, for reasons similar to those used in deriving Eqn. (3.20 *a*), is given by

$$\delta_2 = h_2/f_2'$$

The total deviation δ of the incident ray due to refraction through the combination of lenses, is equal to the sum of δ_1 and δ_2, because deviations are in the same direction.

The ray *AB* incident on the optical system is parallel to the principal axis; hence *F'*, the point where the emergent ray *CF'* intersects the principal axis, is termed as the second principal focus of the optical system. The conjugate rays *AB* and *CF'* when produced intersect at the point *P'*. As a consequence, it would appear that the combination of lenses functions as if the deviation of the incident ray occurs at P'. We may, therefore, replace the optical system by a single lens of second principal focal length equal to $H'F'$, placed in position $P'H'$, where $P'H'$ is a plane through P' transverse to the principal axis. Under this condition rays parallel to the principal axis, after refraction through this hypothetical lens, would also come to focus at F'. Accordingly, this lens is termed as the equivalent lens for parallel rays incident on lens 1. However, for parallel rays incident or emergent from the second lens we shall have another equivalent lens in another position. $H'F'$ is called the equivalent focal length of the combination and $P'H'$ is called the second principal plane of the optical system.

Now $\delta = \delta_1 + \delta_2$

Substituting the values of δ, δ_1 and δ_2, we get

$$\frac{h_1}{f'} = \frac{h_1}{f_1'} + \frac{h_2}{f_2'} \qquad \text{...(3.20 } b)$$

From the similar triangles CH_2F_1' and BH_1F_1', we conclude

$$\frac{h_2}{h_1} = \frac{f_1' - d}{f_1'} \qquad \text{...(3.20 } c)$$

or

$$h_2 = h_1 - \frac{h_1 d}{f_1'}$$

Substituting the value of h_2 from above in Eqn. (3.20 b), we get

$$\frac{1}{f'} = \frac{1}{f_1'} + \frac{1}{f_2'} - \frac{d}{f_1'f_2'} \qquad \text{...(3.20 } d)$$

or

$$f' = -\frac{f_1'f_2'}{d - f_1' - f_2'} = -\frac{f_1'f_2'}{\Delta} \qquad \text{...(3.20 } e)$$

where we have written,

$$\Delta = d - f_1' - f_2' \qquad \text{...(3.20 } f)$$

$$\Delta = d + f_1 + f_2 \qquad \text{...(3.20 } g)$$

To derive the first principal focal length f of the combination of lenses we proceed as shown in Fig. 3.18. A ray of light from an axial point F after suffering deviations δ_1 and δ_2 due to refraction through the first and the second lens respectively is rendered parallel to the principal axis. As a consequence, F is the first principal focal point of the optical system. The total deviation δ is the sum of δ_1 and δ_2 and the point where the deviation appears to have occurred is P, the intersection of the emergent ray parallel to the principal axis and its conjugate incident ray on the optical system. PH is called the first principal plane and HF is called the first principal focal length of the optical system. By convention of signs it is negative, since F is to the left of H, the first principal point of the optical system.

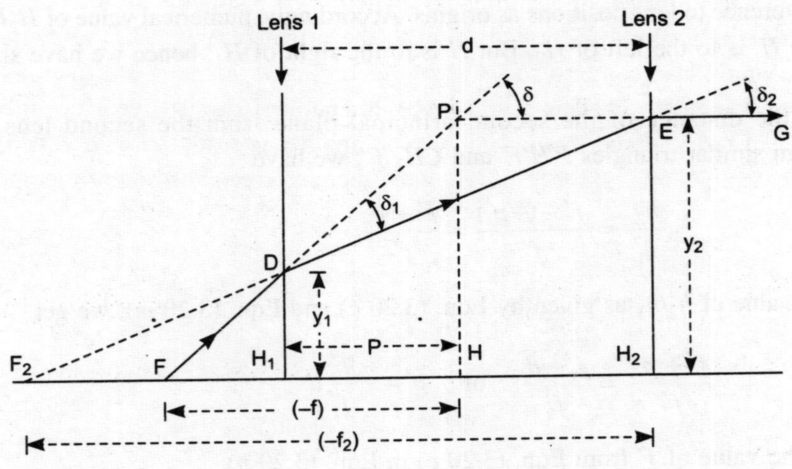

Fig. 3.18. Location of first principal plane PH

Now $$\delta = \delta_1 + \delta_2 \quad ...(3.20\ h)$$

But $$\delta = \frac{y_2}{(-f)},\ \delta_1 = \frac{y_1}{(-f_1)} \text{ and } \delta_2 = \frac{y_2}{(-f_2)}$$

On substituting the values of δ, δ_1 and δ_2 in Eqn. (3.20 *h*), we get

$$\frac{y_2}{f} = \frac{y_1}{f_1} + \frac{y_2}{f_2} \quad ...(3.20\ i)$$

From the similar triangles DH_1F_2 and EH_2F_2, we have

$$\frac{y_1}{y_2} = \frac{(-f_2)-d}{(-f_2)} = \frac{f_2+d}{f_2} \quad ...(3.20\ j)$$

or $$y_1 = y_2 + y_2\,(d/f_2)$$

Substituting the value of y_1 in Eqn. (3.20 *i*), we get

$$\frac{1}{f} = \frac{1}{f_1} + \frac{1}{f_2} + \frac{d}{f_1 f_2} \quad ...(3.20\ k)$$

or $$f = \frac{f_1 f_2}{f_1 + f_2 + d} = \frac{f_1 f_2}{\Delta} \quad ...(3.20\ l)$$

Situation of Principal Points

Let *p* denote the distance of the first principal plane *PH* from the first lens and *p′* that of the second principal plane *P′H′* from the second lens. Owing to the reason that the principal planes are located from the given position of the lens, we should assign signs to p and p' according to our sign convention, with reference to lens positions as origins. Accordingly, numerical value of H_2H' is assigned negative sign since H' is to the left of H_2. But H is to the right of H_1, hence we have shown H_1H as $(+p)$.

To evaluate the distance of the second principal plane from the second lens we refer to Fig. 3.17. Now from similar triangles $P'H'F'$ and CH_2F', we have

$$\frac{h_2}{h_1} = \frac{f' - (-p')}{f'} = \frac{f' + p'}{f'} \qquad ...(3.20\ m)$$

Equating the value of h_2/h_1 as given by Eqn. (3.20 c) and Eqn. (3.20 m), we get

$$\frac{f' + p'}{f'} = \frac{f_1' - d}{f_1'} \quad \text{or} \quad p' = -\frac{f'}{f_1'}d \qquad ...(3.20\ n)$$

Substituting the value of f' from Eqn. (3.20 e) in Eqn. (3.20 n),

$$p' = f_2'd/\Delta \qquad ...(3.20\ o)$$

To evaluate p we refer to Fig. 3.18 in which from similar triangles PHF and DH_1F, we have

$$\frac{y_1}{y_2} = \frac{(-f) - p}{(-f)} = \frac{f + p}{f} \qquad ...(3.20\ p)$$

Equating the value of y_1/y_2 as given by Eqns. (3.20 j) and (3.20 p), we get

$$\frac{f + p}{f} = \frac{f_2 + d}{f_2}d$$

or

$$p = +\frac{f}{f_2}d = +\frac{f'}{f_2'}d \qquad ...(3.20\ q)$$

Substituting the value of f from Eqn. (3.20 l) in Eqn. (3.20 q), we get

$$p = f_1 d/\Delta \qquad ...(3.20\ r)$$

SOLVED EXAMPLES

Example 1. *Two convex lenses each of 20 cm focal length are placed 5 cm apart. A tower of height 100 meters and 200 meters distant is viewed through them. Find the position and the size of the image.*

Suppose light from the tower is incident on the given optical system from the left. The lenses are situated in air.

Here, $f_1' = +20$ cm, $f_2' = +20$ cm and $d = 5$ cm

Hence, $$\frac{1}{f'} = \frac{1}{f_1'} + \frac{1}{f_2'} - \frac{d}{f_1'f_2'} = \frac{2}{20} - \frac{5}{20\times 20} = \frac{7}{80}$$

or $$f' = 80/7 \text{ cm}$$

Similarly $$f = -80/7 \text{ cm}$$

and $$p' = -\frac{df'}{f_1'} = -\frac{5\times 80}{20\times 7} = -\frac{20}{7}\text{ cm}$$

and $$p = +\frac{df'}{f_2'} = +\frac{20}{7}\text{cm}$$

Thus, the first principal point H is 20/7 cm. to the right of the first lens and the second principal point H' is 20/7 cm to the left of the second lens. The tower is 20000 cm in front of the first lens or 20000 + (20/7) = 140020/7 cm in front of the first principal plane. Hence, for the tower $u = -140020/7$ when referred to principal points.

Now, $$\frac{f'}{v} + \frac{f}{u} = 1$$

or $$\frac{80}{7\times v} + \frac{80\times 7}{7\times 140020} = 1$$

or $$v = +11.4 \text{ cm}$$

The positive sign indicates that the image of the tower is formed at 11.4 cm to the right of the second principal point H' or at 11.4 – (20/7) = **8.54 cm** to the right of the second lens.

Also, since the lenses are situated in air, hence we have

$$\frac{y'}{y} = \frac{v}{u} = -\frac{11.4\times 7}{140020}$$

or $$y' = -\frac{11.4\times 7\times 10000}{140020} = -5.699 \text{ cm}$$

The image of the tower is of height **5.699 cm** and the negative sign indicates that the image is inverted with respect to the object.

Example 2. *Obtain an expression for the focal length of the combination of two thin co-axial lenses of focal lengths* $\mathbf{f_1}$ *and* $\mathbf{f_2}$ *separated by a distance 'a', the refractive index of the medium enclosed between the lenses being 'n'.*

It will be recalled, as proved in *Sec.* 3.3 that the ratio of the second principal focal length to the first principal focal length of a thin lens is numerically equal to the ratio of the refractive index of the real image space to the index of the real object space, *i.e.*,

$$f_1' = nf_1$$

where n is the refractive index of the real image space with respect to the index of the real object space. Since, we assume $n > 1$, hence $f_1' > f_1$. A reverse of this holds good if the refractive index of the real object space with respect to that of the real image space be *n, i.e.*, in this case the second principal focal length will be smaller than the first principal focal length.

In our problem of two lenses separated by a medium of index 'n', we assume light to be incident on the combination from the left in air. Then, for the first lens real object space is air medium while for the second lens real image space is air medium. *We further assume that the given focal lengths* $\mathbf{f}_1$ *and* $\mathbf{f}_2$ *are the numerical values of the smaller focal lengths of the two lenses.* Accordingly, the second principal focal length, f_1', of the first lens is given by

$$f_1' = n\mathbf{f}_1$$

and the second principal focal length of the second lens is given by

$$f_2' = \mathbf{f}_2$$

and its first principal focal length is given by

$$f_2 = nf_2' = n\mathbf{f}_2$$

Now, the optical separation, Δ, between the two lenses is given by

$$\Delta = a - f_1' + f_2 = a - n\mathbf{f}_1 - n\mathbf{f}_2$$

Therefore, the second principal focal length, f' of the combination, is

$$f' = -\frac{f_1'f_2'}{\Delta} = -\frac{n\mathbf{f}_1\mathbf{f}_2}{a - n\mathbf{f}_1 - n\mathbf{f}_2} = \frac{\mathbf{f}_1\mathbf{f}_2}{\mathbf{f}_1 + \mathbf{f}_2 - (a/n)}$$

and the first principal focal length of the combination is given by

$$f = \frac{f_1 f_2}{\Delta} = \frac{(-\mathbf{f}_1)(-n\mathbf{f}_2)}{a - n\mathbf{f}_1 - n\mathbf{f}_2} = -\frac{\mathbf{f}_1\mathbf{f}_2}{\mathbf{f}_1 + \mathbf{f}_2 - (a/n)}$$

PROBLEMS

1. Define cardinal points of a system of co-axial lenses. Describe how you would determine the principal planes of a combination of two thin lenses separated by a distance.
2. What are the properties of cardinal points of a co-axial lens system? Plot the cardinal points of a Huygens' eyepiece. How can they be determined experimentally?
3. Show that for a co-axial lens system $xx' = ff'$ where x and x' are the respective distances of the object and image from the first and second focal points and f and f' are two focal lengths. What form does the expression take when the media on the two sides of the system are the same.
4. Find the expression for the equivalent focal length of two thin lenses separated co-axially by a finite distance.
5. How would you determine experimentally the positions of the focal and principal points of a co-axial lens system? Explain and illustrate how to trace the image of an object formed by a lens system, given the positions of cardinal points.
6. Give graphical construction of the formation of an image by a system of lenses. How are the nodal points of a system of lenses determined experimentally.
7. Two thin lenses of focal lengths 12 cm and 4 cm are kept separated by a distance of 8 cm. Plot the positions of the cardinal points for the combination. Derive the formula used.

8. A telephoto lens consists of a convergent lens of focal length 12 cm facing the object and a divergent lens of focal length 5 cm placed 8 cm behind the former. Find the position where the plate be placed to photograph a distant object.
9. What are different kinds of magnifications associated with a pair of conjugate positions of a co-axial optical system? Establish a relation between them.
10. Two thin convergent lenses each of 20 cm focal length are set co-axially 5 cm. apart. An image of an upright pole 200 meters distant and 10 meters high is formed by the combination. Find the position of the unit and focal planes and the image. Also find the size of the image.
11. Prove that for a combination of two thin lenses of focal lengths f_1 and f_2 separated by a distance d, the focal length of the combination is given by

$$\frac{1}{f} = \frac{1}{f_1} + \frac{1}{f_2} - \frac{d}{f_1 f_2}$$

12. Two convex lenses of focal lengths 20 cm and 5 cm are 10 cm apart. Calculate the power of such a combination.
13. Derive an expression for the focal length of a system of two thin lenses separated by a distance and calculate the position of principal points.

    A thin converging lens and a thin diverging lens are placed co-axially in air at a distance of 5 cm. If the focal length of each be 10 cm, find for the combination (a) the focal length (b) the power and (c) the position of the principal points.
14. Derive an expression for the equivalent focal length of two thin lenses of focal lengths f_1 and f_2 arranged co-axially at a distance apart.
15. Two thin convex lenses of focal lengths 12 cm and 6 cm are placed co-axially 8 cm apart. Determine the position of cardinal points of the system.
16. An object is placed at a distance 60 cm from a thin convex lens of focal length 20 cm. There is a second thin convex lens of 30 cm focal length 10 cm away from the first. Calculate the distance of the final image from the second lens.
17. Two plano-concave lenses of refractive index 1.50 have radii of curvature of 20 cm and 30 cm respectively. They are placed in contact with the curved surfaces facing each other and the space between them is filled with a liquid of refractive index 1.33. What is the focal length of the combination?
18. A concavo-convex lens has an index of refraction 1.5 and the radii of curvature of its surface are 10 cm and 20 cm The concave surface is upward and filled with an oil of refractive index 1.60. Calculate the focal length of oil-glass combination.
19. What do you mean by conjugate planes? How many conjugate planes are there for an optical system?
20. Define principal planes. Where do these principal points lie in case of a glass-sphere.
21. Define nodal planes. Where do the nodal planes lie in case of a glass hemisphere.
22. A refracting surface separates two media, one called incident medium having r.i. n_1, the other having r.i. n_2. If f_1 and f_2 are image side focal length and object side focal length respectively, then prove, $\left(\frac{f_1}{n_1}\right)+\left(\frac{f_2}{n_2}\right)=0.$

ABERRATIONS OF OPTICAL IMAGES

Chapter

The theory of image formation by mirrors and lenses, developed in the previous chapters, was based on two fundamental assumptions: (*i*) the object and the image points were considered as if they were situated on or very close to the principal axis and (*ii*) the rays of light diverging from the object point were assumed to be confined to a narrow cone of a small angular opening so as to allow us to replace sine of the slope angle, θ, of the ray by the angle θ. These paraxial conditions lead to a point image of a point object in the *First order theory*, so called because of the inclusion of the first order of the slope angle only.

In reality objects have finite dimensions and, moreover, considerations of the brightness of the image demand the use of reflecting and refracting systems having large apertures. In actual practice, therefore, the rays from object points are confined not simply to a region close to the principal axis, termed the paraxial region, but they form a cone of such a wide angular opening that in the expansion of sine of the slope angle θ into a series,

$$\sin\theta = \theta - \frac{\theta^3}{3!} + \frac{\theta^5}{5!} - \frac{\theta^7}{7!} + \cdots\cdots$$

it becomes essential to take into account, in our discussions of the image formation, at least the first two terms of this series. This theory, called the *Third order theory* due to the inclusion of the third order term in θ, was developed by *Ludwig Von Seidal* in 1855 and gives various departures of the actual image from the one predicted by the *First order theory*. These departures, the so called *Seidal aberrations*, may be classified as (i) *Spherical aberration*, (ii) *Coma*, (iii) *Astigmatism*, (iv) *Curvature* and (v) *Distortion*. These aberrations are present in the images formed by ordinary lenses, even when light of only one wavelength is employed. They are, therefore, also known as *monochromatic aberrations* (*monos* is a Greek word for *single* and *Chroma* for *Colour*). Each of the five monochromatic aberrations depends upon the wavelength of light but, leaving aside spherical aberration, the variation is always negligibly small in the others.

Furthermore, the focal length of the lens depends upon its index of refraction, which varies with wavelength of light. Therefore, even in the first order theory, if the light diverging from the object is not monochromatic, a lens will form a number of coloured images of different sizes and at different positions. Thus, we shall encounter two new aberrations known as the *lateral chromatic aberration* and the *longitudinal chromatic aberration*. These aberrations are, however, absent in the images formed by mirrors.

A rigorous mathematical study of these aberrations is beyond the scope of this book, but wherever possible simple mathematical deductions will be given while describing them in detail.

4.1 SPHERICAL ABERRATION OF A LENS

When the ratio of the aperture to the focal length of the lens is relatively large, a cone of rays from any axial object point P, after refraction, is not focussed at a single point in a manner demanded by the first order theory, but produces an image as illustrated in Fig. 4.1. *This effect is known as spherical aberration and arises due to the fact that the different annular zones of the lens have different focal lengths — the greater the radius of the zone, the smaller is the focal length along the axis.* Consequently in Fig. 4.1 the marginal rays *i.e.*, those which are refracted at the boundary of the lens, converge to an image point P_m which is considerably closer to the lens than the point P_p, the focus of the paraxial rays. In general, the rays which are refracted at any other circular zone cross the axis at one image point between P_m and P_p. However, a different axial image point corresponds to each annular zone. There is evidently no plane in which a sharp image of P is formed. The cross-section of the refracted beam is everywhere circular and it will be seen from the figure that there is one plane, AB, at which the cross-section has least diameter. This smallest cross-section is known as the *circle of least confusion* and the best image of P is obtained if a screen is placed perpendicular to the lens axis at this point.

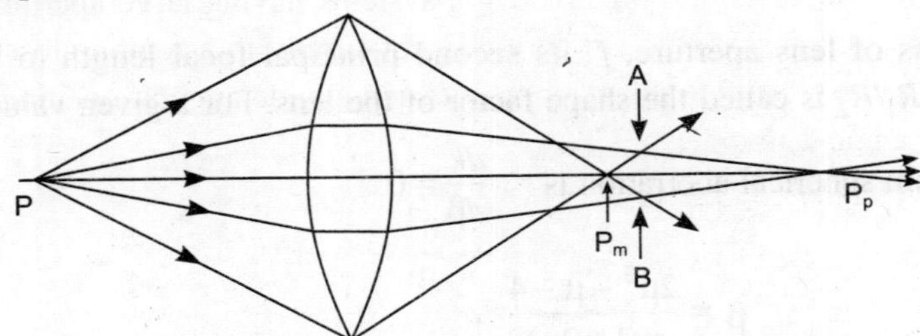

Fig. 4.1. Spherical aberration of a convergent lens for an axial point object

The spherical aberration of a biconcave lens for parallel incident beam is illustrated in Fig. 4.3 while that for virtual object in Fig. 4.2. The separation between the marginal focus P_m and the paraxial focus P_p is taken as a measure of *longitudinal spherical aberration* of the lens, while the radius of the circle of least confusion is taken as a measure of *lateral* or *transverse spherical aberration*. The spherical aberration varies approximately as the square of the height of the incident ray above the axis and it also depends upon the distance of the object point. The longitudinal spherical aberration is considered positive or negative according as the marginal rays focus P_m lies on the left or on the right of the paraxial rays focus P_p. A convergent lens produces positive aberration while a divergent lens produces a negative aberration.

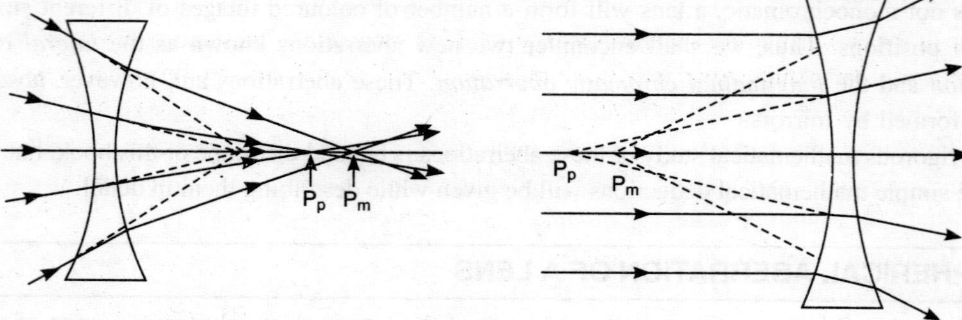

Fig. 4.2. Spherical aberration of a concave lens

Fig. 4.3

Reduction or Elimination of Spherical Aberration

The following devices are usually employed for reducing spherical aberration in images:

(*a*) The spherical aberration of a given lens depends on the radius of lens aperture. Therefore it may be reduced by *stopping down* the lens aperture with the help of a coaxial aperture stop, thus using only the central portion of the lens. Of course, this would reduce the amount of light transmitted by the lens.

(*b*) A single lens of given focal length can also be designed so that the spherical aberration is minimum. The longitudinal spherical aberration of a thin lens for parallel incident rays is given by

$$= +k\,\frac{h^2}{f'}\left\{\frac{\beta^2\mu^3 + \beta(\mu + 2\mu^2 - 2\mu^3) + \mu^3 - 2\mu^2 + 2}{2\mu(\mu-1)^2(1-\beta)^2}\right\} \quad \text{...(4.1 } a)$$

where h is the radius of lens aperture, f' its second principal focal length to be substituted with proper sign and $\beta = R_1/R_2$ is called the shape factor of the lens. For a given value of h, f' and μ, the condition for minimum spherical aberration is $\frac{dk}{d\beta} = 0$

which leads to

$$\beta = \frac{2\mu^2 - \mu - 4}{\mu(1+2\mu)} \quad \text{...(4.1 } b)$$

Since β has only one turning value, it obviously gives a minimum value of spherical aberration k, because when $\beta = 1$, k becomes infinite. If $\mu = 1.5$, Eqn. (4.1 *b*) gives $\beta = R_1/R_2 = -1/6$. Thus the form of a lens giving minimum spherical aberration is biconvex or biconcave, the radius of curvature R_1 of the surface facing the object is about one sixth of that of the other face. In general, the more curved face of the lens should face towards the incident or emergent rays whichever are more nearly parallel to the lens axis. A lens having $R_1/R_2 = -1/6$ is termed as the *crossed lens* and this process of changing the shape of the

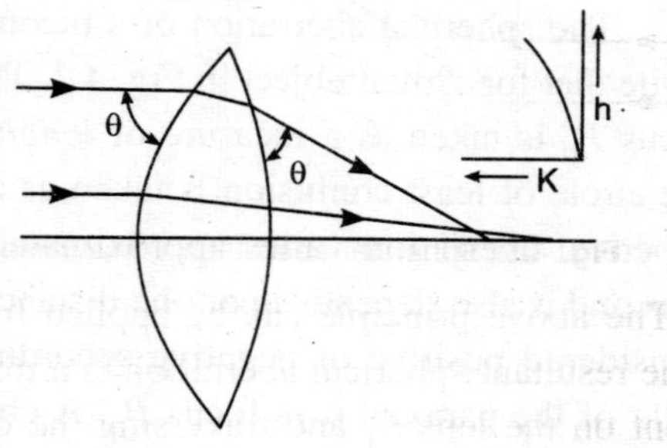

Fig. 4.4. Crossed lens

lens without change of focal length is known as *bending the lens* for minimum spherical aberration. It should be emphasized that the spherical aberration cannot be completely removed by any bending of lens with spherical surfaces.

Spherical aberration can also be reduced in a single lens of given focal length by constructing it from material of high refractive index. For example, suppose $f' = 100$ cm, $h = 10$ cm and $\mu = 1.5$. For minimum spherical aberration $\beta = -1/6$ and from Eqn. (4.1 *a*), $k = 1.07$ cm. Similarly, with $\mu = 2$, $\beta = +1/5$ and k is diminished to 0.44 cm.

(*c*) In some optical instruments plano-convex lenses are employed because they are cheaper to make and when their curved side faces the parallel incident (or emergent) light, the spherical aberration is very nearly the same as that of a crossed lens. However, if the plane side is turned towards the object the spherical aberration is very large. The reason is not far to seek. The spherical aberration being the result of greater deviation of the marginal rays as compared to that of paraxial rays, it is obvious that if the deviation of marginal rays be made minimum, the marginal focal point P_m will be at its farthest to the right and thus the least value of spherical aberration will result. Just as in the case of a prism the deviation is minimum when the incident and emergent rays make equal angles with its faces, similarly in the case of a lens the deviation of marginal rays will be minimum when they enter the first lens-surface and leave the second surface at more or less equal angles. Thus, as a general rule, spherical aberration will be minimised when a lens is so designed or used that the total deviation of a given ray is divided equally between the two refractions. In a plano-convex lens with plane side towards the distant object, as shown in Fig. 4.5 the deviation of a given ray is simply produced at the curved surface and therefore the condition of minimum spherical aberration is violated, thus accounting for the presence of large spherical aberration. But when the convex side faces the distant object, as shown in Fig. 4.5, the deviation is divided between the two surfaces, thus accounting for the presence of minimum spherical aberration.

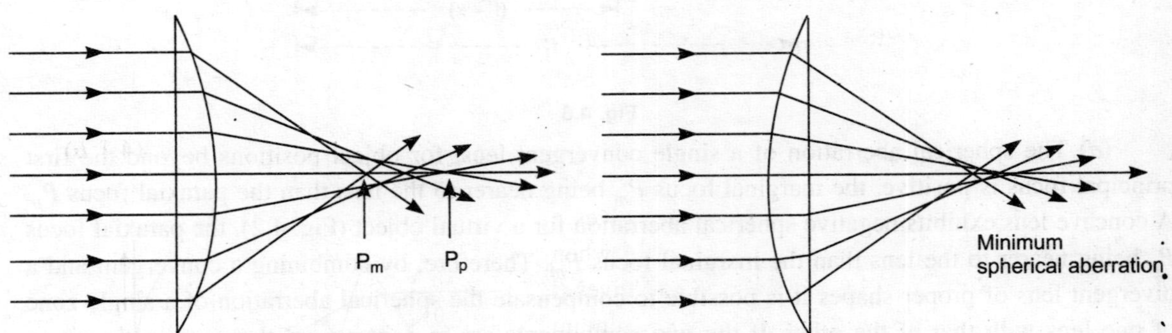

Fig. 4.5. Spherical aberration is minimised if the deviation is equally shared between the surfaces

The above principle can be applied for finding the distance apart of two convergent lenses so that the resultant spherical aberration is a minimum. Consider a ray *PQ*, parallel to the principal axis, incident on the lens L_1 and traversing the optical system as shown in Fig. 4.6. It will be seen from Fig. 4.6 that the deviation produced in the ray by L_1 is δ_1 while that produced by L_2 is δ_2. Hence the spherical aberration will be minimum when

$$\delta_1 = \delta_2$$

or
$$\frac{h_1}{f_1'} = \frac{h_2}{f_2'} \quad \text{...(4.1 } c\text{)}$$

also from similar triangles QL_1F_1' and $R_2L_2F_1'$, we have

$$\frac{h_2}{f_1' - x} = \frac{h_1}{f_1'} \quad \text{...(4.1 } d\text{)}$$

Equating the values of h_1/h_2 as given by Eqns. (4.1 *c*) and (4.1 *d*), we get

$$x = f_1' - f_2' \quad \text{...(4.1 } e\text{)}$$

Thus the spherical aberration of a combination of two convergent lenses is minimum when their distance apart is equal to the difference of their principal focal lengths. It is essential in this case that the incident ray should suffer refraction first through the lens of larger focal length and then through the one of smaller focal length.

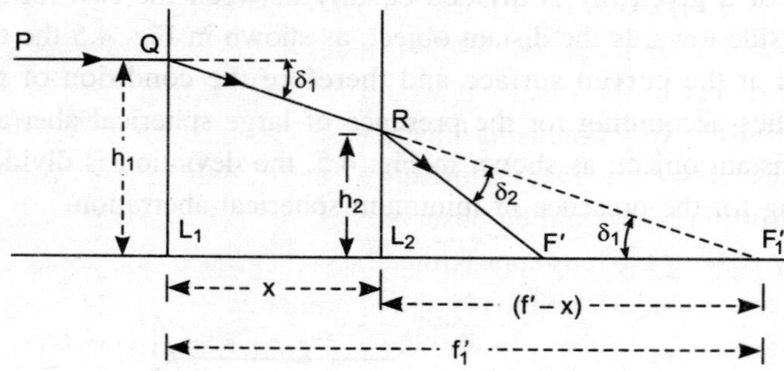

Fig. 4.6

(*d*) The spherical aberration of a single convergent lens, for object positions beyond the first principal focus is positive, the marginal focus P_m being nearer to the lens than the paraxial focus P_p. A concave lens exhibits negative spherical aberration for a virtual object (Fig. 4.2), the paraxial focus P_p being nearer to the lens than the marginal focus P_m. Therefore, by combining a convergent and a divergent lens of proper shapes it is possible to compensate the spherical aberration of a single zone of one lens with that of the other. If the two components are in contact and the combination is to act as a convergent lens, the principal focal length of the convergent component should be smaller than that of the divergent one. Since the spherical aberration increases with decrease in focal length, it is obvious that the spherical aberration of the convergent lens will be greater than that of the divergent one. Consequently the convergent lens should be so shaped as to have minimum spherical aberration while the shape of the divergent one should be such as to have increased aberration by just an amount to make the resultant aberration zero. Thus the system is said to be *corrected* if the aberration is zero for one zone and negligible for others. Figure 4.7 gives a graph and the spherical aberration of a cemented doublet corrected for marginal zone. For other zones the focal length is slightly less than the paraxial focal length. Combinations of lenses of this type are usually made, in such a way as to eliminate both the spherical aberration and the chromatic aberration.

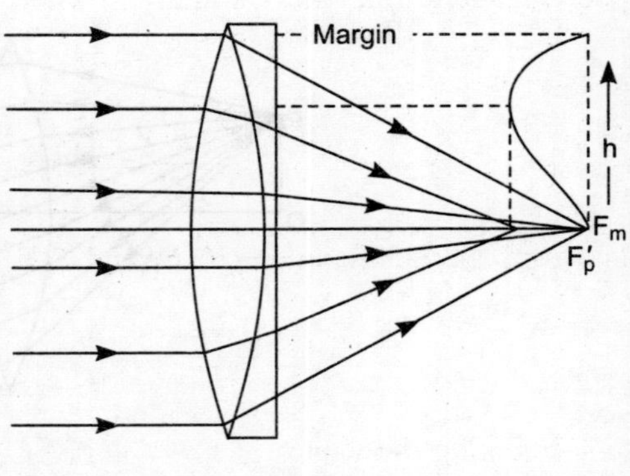

Fig. 4.7

(*e*) Although *spherical aberration* cannot be eliminated, in general, from a single lens, it is absent for one pair of conjugate points, called the aplanatic points of the lens. For this pair of points the lens is also free from another aberration called *Coma*. We shall refer to this point in the next section.

4.2 COMA

The aberration known as *Coma* (Greek *kome*, meaning *hair*) affects rays that come from object points even if located very close to the axis of the lens. *It arises from the fact that for these non-axial point objects, there is either a decrease [Fig. 4.8 (b)] or increase [Fig. 4.8 (a)] of lateral magnification with the height of the narrow circular zone of the lens through which the rays are refracted.* In order to discuss the comatic defect in the image, it would be assumed that the lens system is free from other aberrations, otherwise a very confusing image would result.

Figure 4.8 (*a*) illustrates, in a lens, the effect of coma in the image of an object point situated close to the axis but at infinity. It would be observed that the ray below the optical centre strikes the equivalent prism in any zone of the lens approximately in a symmetrical condition of refraction (equal angles of incidence and emergence) than the ray above it; therefore the former ray has nearly minimum deviation while the latter has comparatively greater deviation. As a consequence, the rays from the upper and the lower portions of the same zone, after refraction, intersect below the principal ray and thus the lateral magnification for outer zone is greater than that for the central zone and as explained in the sequence, the 'coma' is said to be positive.

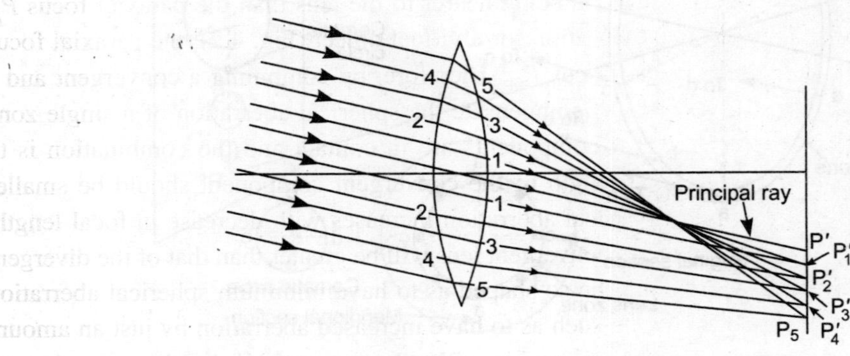

Fig. 4.8 (*a*). Illustrating the formation of "Positive Coma" in the meridional plane

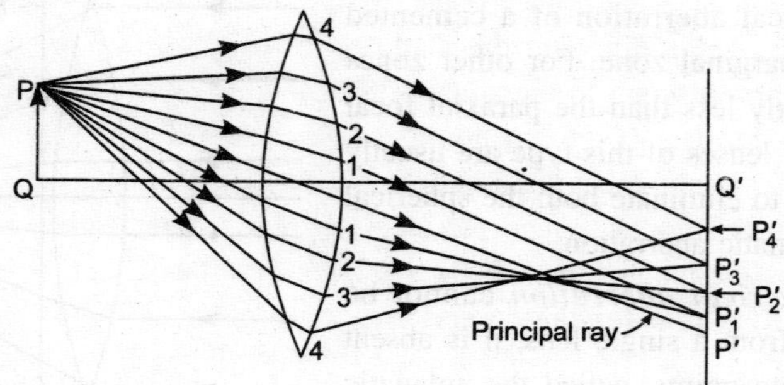

Fig. 4.8 (*b*). Illustrating the formation of "Negative Coma" in the meridional plane

For the non-axial object point *P* at finite distance from the lens, it would be observed, as shown in Fig. 4.8 (*b*), that the ray above the optical centre strikes the equivalent prism in any zone of the lens approximately in the symmetrical condition of refraction than the ray below it. As a consequence, the rays from the upper and lower portions of the same zone, after refraction, intersect above the principal ray and thus the lateral magnification for outer zones is less than that for the central zone. The coma, under this condition as explained in sequence, is said to be negative. In reality, a cone of rays diverges from the point *P*. Therefore, we should also consider the refraction of rays not lying in the meridional plane. This is illustrated by Fig. 4.9 which shows the refraction of rays through different points of a single zone. By tracing the rays it is found that the meridional rays *Pa*, after refraction through the points *a*, *i* of the lens zone, come in focus at a point *a* of the image plane while the rays *Pc*, in a plane perpendicular to the meridional plane, after refraction through the points *c*, *c* of the lens zone come in focus at the point *c* of the image plane and so on. The annular cone of rays from *P*, which is refracted through the lens zone, thus comes to a focus in a circle *a*, *b*, *c*, *d*, called by *H*.

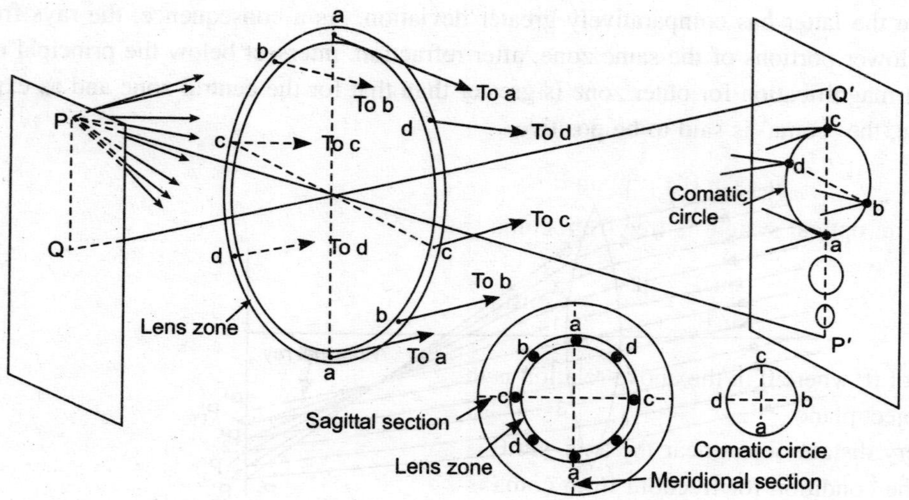

Fig. 4.9. Each zone of the lens forms a ring shaped image called comatic circle

Denis Taylor a *comatic circle*. In the image plane, there will be such a comatic circle corresponding to each zone of the lens. The radius and the distance of the centre of the comatic circle from the paraxial image point is proportional to the square of the radius of lens zone. The image of the non-axial point P, therefore, consists of an expanding series of overlapping *comatic circles*. The totality of these circles produces an image having a balloon type of flare of light like a comet, with bright nucleus at P' and the tail becoming fainter and fainter as it grows, hence the name *Coma.* If the lateral magnification for outer zones is smaller than that for the central zone, as shown is Fig. 4.8 (*b*), the tail of the comatic image extends inward towards the principal axis and coma is said to be negative. But if the lateral magnification for outer zones is greater than that for the central zone, as illustrated in Fig. 4.8 (*a*), the tail of the comatic image extends outwards from the principal axis and the coma is to be *positive*.

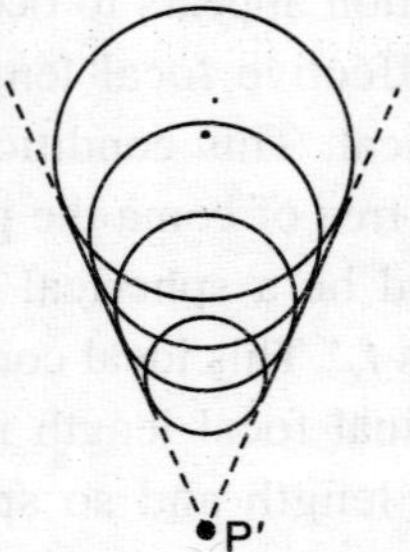

Fig. 4.10. Comatic flare

Reduction or Elimination of Coma

We see that coma is the result of the variation of the lateral magnification for rays passing through different zones of a lens. Therefore to eliminate the coma the lens should be so designed that the lateral magnification y'/y is the same for all zones. From Abbe's Sine Condition (Eqn. 3.4 *d*)

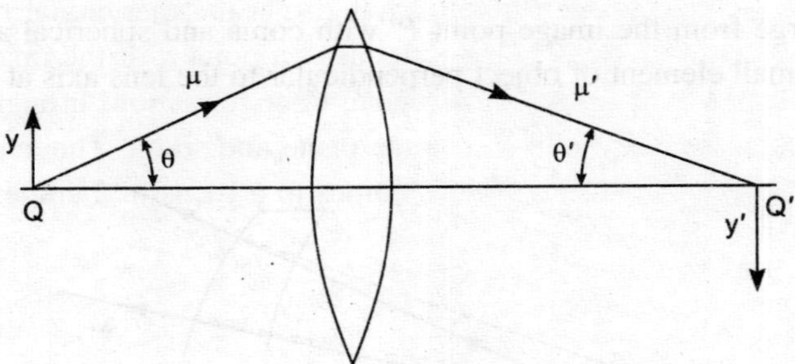

Fig. 4.11. Abbe's sine condition

$$\mu\, y \sin\theta = \mu' y' \sin\theta'$$

it follows that an optical system is free from coma when the ratio

$$\frac{\sin\theta}{\sin\theta'} = \frac{\mu' y'}{\mu y} = \text{Constant}$$

for all values of θ, where θ in the above relation is the angle subtended by the zone radius at the axial point in the object plane.

For a very distant object near the axis, sin θ is proportional to h the height of the incident ray. In this case, the condition for freedom from coma is then

$$\frac{h}{\sin\theta'} = \text{Constant}$$

From Fig. 4.12 it is evident that the ratio $h/\sin\theta'$ is the effective focal length f_x' of the zone, which is the distance measured along the ray from the focal point F' to the point A where the ray deviation appears to occur. To eliminate the coma, the effective focal lengths of the zones must be identical. This condition demands that for a lens to be free of coma the principal surface of the lens should be a spherical one with centre at F' and radius f_x'. This ideal condition also ensures that the marginal focal length is the same as the paraxial focal length and so spherical aberration is also absent.

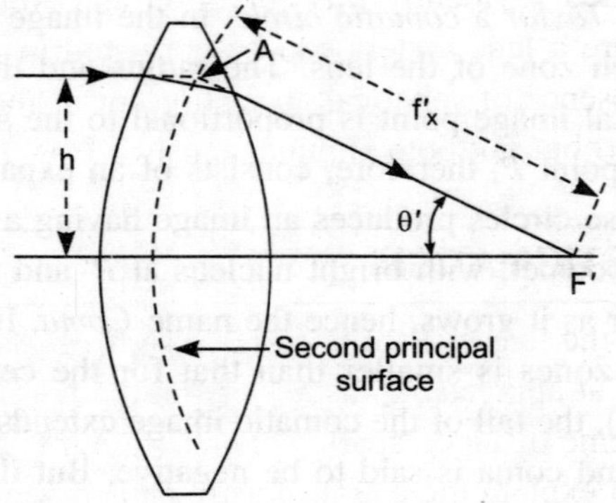

Fig. 4.12

The coma, like the spherical aberration, may be minimized by the proper choice of radii of curvature of the lens surface and this is investigated by the ray tracing method. But a lens designed for minimum spherical aberration, however, cannot be free from coma because the necessary shape of the lens for minimum spherical aberration is not the same as required for zero coma. However, the so called *aplanatic lens* is simultaneously free from the coma and the spherical aberration for one particular pair of object and image points called the *aplanatic points* of the lens. All rays diverging from an object placed at the centre of curvature of the first surface, after refraction at the second surface appear to diverge from the image point P' with coma and spherical aberration missing. This also holds good for a small element of object perpendicular to the lens axis at the aplanatic point.

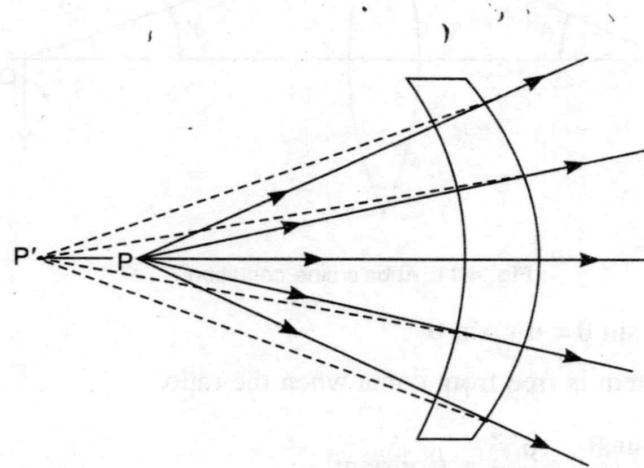

Fig. 4.13. Aplanatic lens

Sir John Hershel showed that a lens is free from spherical aberration for two neighbouring axial points if the condition

$$\frac{\mu\sin(\theta/2)}{\mu'\sin(\theta'/2)} = \text{Constant}$$

is satisfied. Evidently this condition and Abbe's sine condition cannot be satisfied simultaneously. Therefore a lens system can be corrected to form a true image of either a small element along the axis (absence of spherical aberration) or a small element perpendicular to the principal axis (absence of coma) but not both simultaneously.

4.3 ASTIGMATISM

Even if the lens system is corrected for *spherical aberration* and *coma*, the image of an object point situated at appreciable distance from the lens axis is not a point but a pair of mutually perpendicular lines, some distance apart. This aberration is therefore known as *astigmatism*.

To explain the cause of astigmatism we shall first express the focal length of a thin convergent lens

$$\frac{1}{f'} = +(\mu - 1)\left(\frac{1}{R_1} + \frac{1}{R_2}\right)$$

in terms of the diameter d of the lens aperture and the axial thickness t of the lens. By elementary geometrical considerations applied to Fig. 4.14, we have

$$OE^2 = OA \cdot OC$$

$$\frac{1}{4}d^2 = t_1(2R_1 - t_1) = 2R_1/t_1$$

approximately

Hence $$t_1 = d^2/8R_1$$

Similarly, $$t_2 = d^2/8R_2$$

where $$t = OA + OB = t_1 + t_2$$

whence $$\frac{1}{R_1} + \frac{1}{R_2} = \frac{8t}{d^2}$$

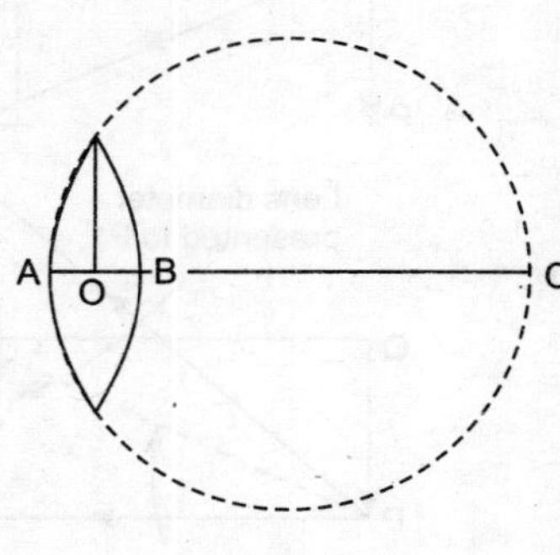

Fig. 4.14

Hence $$f' = +\frac{d^2}{8(\mu - 1)t}$$

Thus the focal length of a thin convergent lens is proportional to d^2/t. Now, in Fig. 4.15, Q' is the image formed by a convergent lens of an axial object point Q. Owing to symmetry of the lens with respect to Q, every section of the lens through its axis presents the same diameter to the axial point Q and so the focal length is the same in all sections through the axis, and hence the refracted beam converges to one point. However, for the object point P, far away from the axis, the lens is not perfectly symmetrical. In the *meridional section*, the lens diameter presented to P is less than d. Moreover, the principal ray now passes inclined to the principal axis of lens and therefore traverses a thickness greater than the axial thickness t of the lens. As a consequence of both the foregoing reasons, the focal length of lens in the meridional section is decreased for an off axial object point. Consequently the image of P in this section is at P_1' nearer to the lens than the image point which would be obtained by assuming the focal length in this section to be the same as along the axis. In the *sagittal section* of the lens, that is, the section by a plane through the principal ray and perpendicular to the meridional section, the lens diameter presented to non-axial object point P is the same as for the axial point, but the principal ray traverses greater thickness of the lens. The focal length in the sagittal section is, therefore, smaller than the axial focal length but the decrease is not so much as in the meridional section; because in the former section the lens diameter presented to P remains d. The image of P, due to refraction of rays in this section is formed at P_2', which is slightly away from the lens than P_1' but closer to the lens than the image point which would be given by the axial focal length. These considerations show that *for an off axial object point the focal length of any narrow coaxial circular lens zone is not the same at every point of it, being least in the meridional section and maximum in the sagittal but in each case less than the axial focal length.*

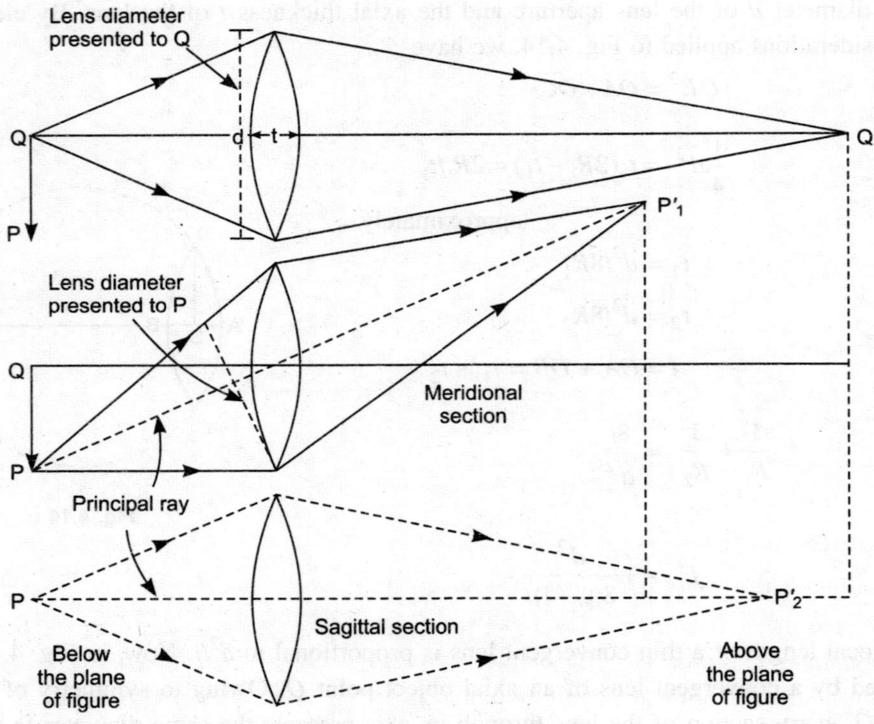

Fig. 4.15. Illustrates the origin of astigmatism

The effect of astigmatism in the image of an off axial object point P is illustrated in Fig. 4.16, where it is assumed that other aberrations are absent. The meridional section and the sagittal section of the cone of rays diverging from P are shown by vertical and horizontal shading respectively, before and after refraction through the lens. The rays in the meridional plane *GHQ*, owing to shorter focal length of the lens in this section, after refraction come to focus at P_1', while those in the sagittal plane *PAB*, come to focus at another point P_2', farther away from the lens than P_1'. Similar considerations also show that all rays diverging from *P*, after refraction through the lens, do not pass through a single point but they pass through a horizontal line *ST*, which is perpendicular to the principal ray $PPP_1'P_2'$ and later on pass through a vertical line *UW*. Thus, a small refracted beam is, in general, not stigmatic but astigmatic.

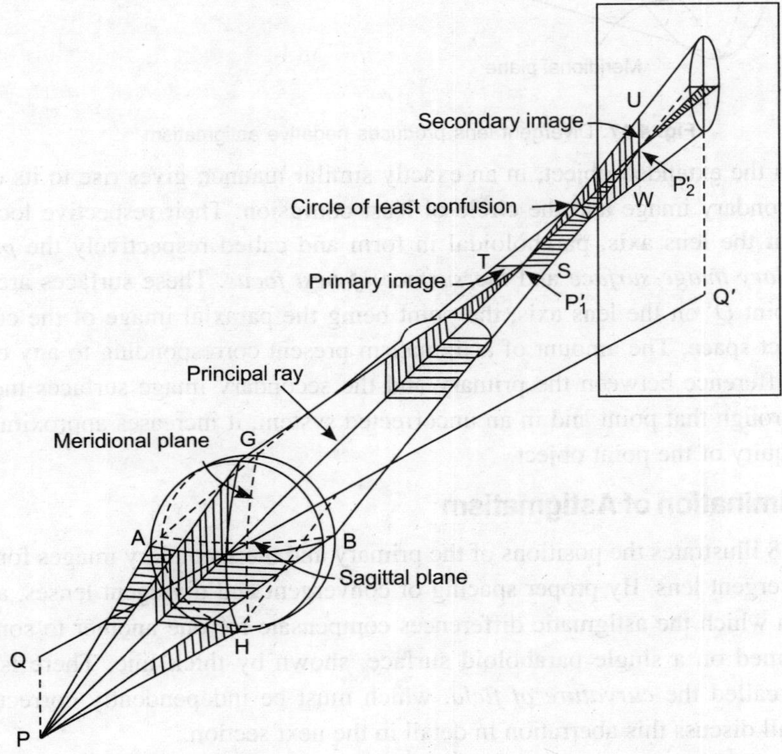

Fig. 4.16. Diagram shows two focal lines which constitute the image of an off axial object point P

The *meridional line focus* $SP_1'T$, which is perpendicular to the meridional plane, is called the *primary image*, and the *sagittal line focus* $UP_2'W$, which *is vertical in the meridional plane*, is *called the secondary image*. The cross-section of the refracted beam is, in general, elliptical, the ellipse degenerating into straight lines at positions of the primary and secondary foci, and into a circle, somewhere between these positions, where the major and minor axis of the ellipse are equal. This circle is known as the *circle of least confusion* and gives the best focus of the astigmatic pencil. For a simple convergent lens, the meridional focus P_1' is nearer the lens than sagittal focus P_2' as shown in Fig. 4.16 and the astigmatism is said to be *positive*. But for simple divergent lens as shown in Fig. 4.17 (which illustrates only the refracted rays when produced backwards) the sagittal focus P_2' is nearer the lens than the meridional focus P_1' and astigmatism is said to be *negative*.

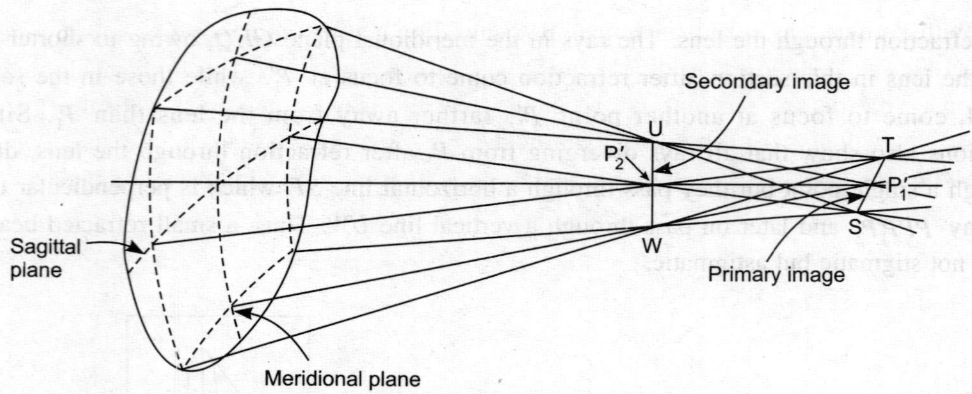

Fig. 4.17. Divergent lens produces negative astigmatism

Each point on the extended object, in an exactly similar manner, gives rise to its corresponding primary image, secondary image and the circle of least confusion. Their respective loci are surfaces of revolution, about the lens axis, paraboloidal in form and called respectively the *primary image surface, the secondary image surface* and *the surface of best focus*. These surfaces are tangential to one another at a point Q' on the lens axis, the point being the paraxial image of the conjugate axial point Q in the object space. The amount of astigmatism present corresponding to any object point is measured by the difference between the primary and the secondary image surfaces measured along the principal ray through that point and in an uncorrected system, it increases approximately as $\tan^2\theta$ where θ is the obliquity of the point object.

Reduction or Elimination of Astigmatism

(*a*) Figure 4.18 illustrates the positions of the primary and the secondary images for a convergent as well as for a divergent lens. By proper spacing of convergent and divergent lenses, a combination may be designed in which the astigmatic differences compensate for one another to some extent and the images are formed on a single paraboloid surface, shown by thick line. There is thus present another aberration called the *curvature of field*, which must be independently corrected for in the lens design. We shall discuss this aberration in detail in the next section.

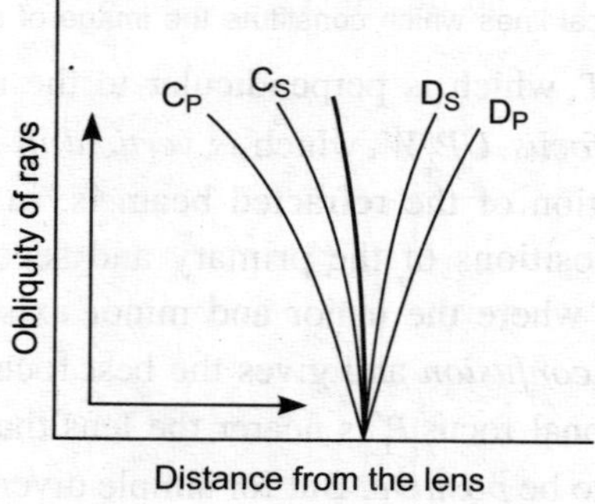

Fig. 4.18

(*b*) With a single lens, astigmatism may be reduced by cutting out the oblique rays by means of coaxial apertures called stops, placed judiciously. The essential condition for complete elimination of astigmatism is that the primary and the secondary image surfaces must have the same curvature. The principal ray will now pass eccentrically through the lens and the astigmatism is markedly reduced. Moreover with lens shape, as shown in Fig. 4.19, astigmatism may be eliminated completely.

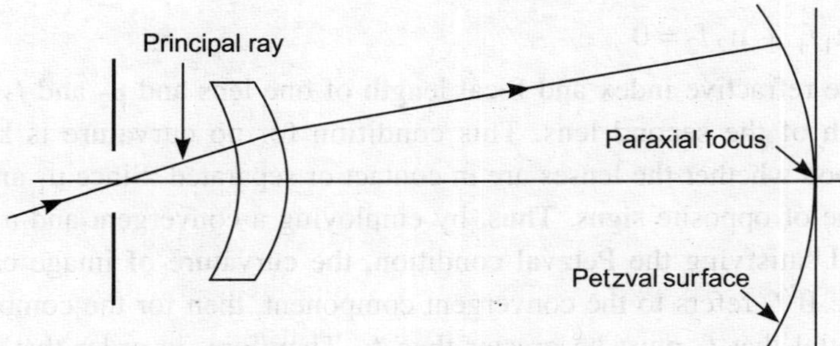

Fig. 4.19. Correction for astigmatism

4.4 CURVATURE OF FIELD

Even if the lens system is free from spherical aberration, coma, and astigmatism, that is, the zonal focal length of the system does not vary from its axial focal length for off axial object points, the image of an object plane is, in general, a curved surface. This is due to the reason that since off-axial object points are far away from the lens system than the axial point, therefore, in the case of convergent lens forming a real image, the image of an off-axial object point is formed closer to the lens than the image of axial object point, thus accounting for the curvature of image shown in Fig. 4.20. In a convergent system the radius of curvature of the image surface is negative and hence the curvature is designated as negative, while the converse of this holds true in a divergent system, the concave side being towards the object in the former system while away from the object in the latter system. As a consequence of this by their suitable combination, as explained later on, the image can be flattened. The curved image surface due to single lens refraction, is called Petzval surface (after J. Petzval who investigated the subject) and the aberration is known as *Curvature of Field*.

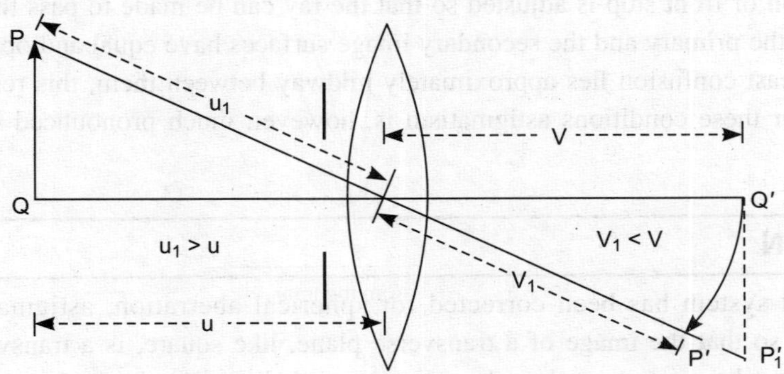

Fig. 4.20. Curvature of field

Reduction or Elimination of Curvature of Field

(*a*) For a system of thin lenses, the curvature in the final image is given by

$$1/R = \Sigma(1/\mu f) \quad ...(4.4\ a)$$

Hence, for no curvature the condition is that $\Sigma\ (1/\mu f) = 0$.

In the case of two lenses this condition reduces to

$$\mu_1 f_1 + \mu_2 f_2 = 0 \quad ...(4.4\ b)$$

where μ_1 and f_1 are the refractive index and focal length of one lens and μ_2 and f_2 are the refractive index and focal length of the second lens. This condition for no curvature is known as Petzval condition and holds good whether the lenses are in contact or separated. Since μ_1 and μ_2 are positive, hence f_1 and f_2 must be of opposite signs. Thus, by employing a convergent and a divergent lens of different materials and satisfying the Petzval condition, the curvature of image can be eliminated, resulting in a flat field. If f_1 refers to the convergent component, then for the combination to be also convergent, it is essential that f_2 must be greater than f_1. Therefore, in order that Eqn. (4.4 *b*) may be satisfied it is essential that μ_2 must be less than μ_1. This condition cannot be satisfied in the ordinary achromatic doublet consisting of crown convergent lens in contant with a divergent lens of flint glass because the flint glass has a higher refractive index and a higher dispersion, the latter condition being essential, as explained later on in *Sec.* 4.9, for the elimination of chromatic aberration. However, certain kinds of glasses were developed by Abbe at Jena, so that in a given pair one having a lower index had a higher dispersive power. Therefore the achromatic doublets constructed of these glasses can be made free from curvature simply by choosing the focal lengths of the components so as to satisfy Petzval condition. It should be emphasised that under this condition, astigmatism is also simultaneously minimised. High quality photographic objectives are designed on the principles discussed above and are known as an astigmat.

(*b*) Petzval condition is also satisfied by forming a combination of convergent and divergent lenses, of equal focal lengths, made of the same glass and separated by a distance less than their individual focal lengths. Under such conditions, their combined focal length f is negative, that is, the combination is convergent free from astigmatism and curvature.

(*c*) With a single lens, it is possible to minimise either curvature of field or astigmatism, by proper location of stops on the lens axis. To eliminate curvarture, the lenses are usually of meniscus type and the position of front stop is adjusted so that the ray can be made to pass through the lens in such a manner that the primary and the secondary image surfaces have equal and opposite curvatures. As the surface of least confusion lies approximately midway between them, this results in flattening of the image. Under these conditions astigmatism is, however, much pronounced in the outer parts of the field.

4.5 DISTORTION

Even if the optical system has been corrected for spherical aberration, astigmatism, coma, and curvature of image so that the image of a transverse plane, like square, is a transverse plane in the image space there can be present another aberration termed *distortion*, in the image. This aberration refers to the departure from strict geometrical relationship to each other of points in the image plane as their corresponding object points have in the transverse object plane. *This departure arises due to variation of magnification with the lateral distance of an object point from the lens axis.* If the magnification increases with increase in the lateral distance then the image of a square like object situated transverse to the principal axis, owing to the disproportionately high magnification of the corners as compared with other points, is of the form resembling a pin-cushion. This distortion, shown in Fig. 4.21, is termed as *Pin-Cushion Distortion*. On the other hand, if the magnification decreases with increasing axial distance, the images of the diagonal is shortened relatively more than the images of the side of square, that is, the opposite effect is produced, and therefore the resulting image of the square is of the form resembling a barrel. This distortion shown in Fig. 4.21 is, therefore, termed *Barrel Distortion*. The ideal image in each case is shown by a dotted square. Distortion is measured by the distance between the actual image point P'' from the ideal image point P' of an off-axial object point P, the latter image being obtained by the paraxial condition. It would be observed from Fig. 4.21 that the Pin-cushion distortion and the Barrel distortion are of opposite signs.

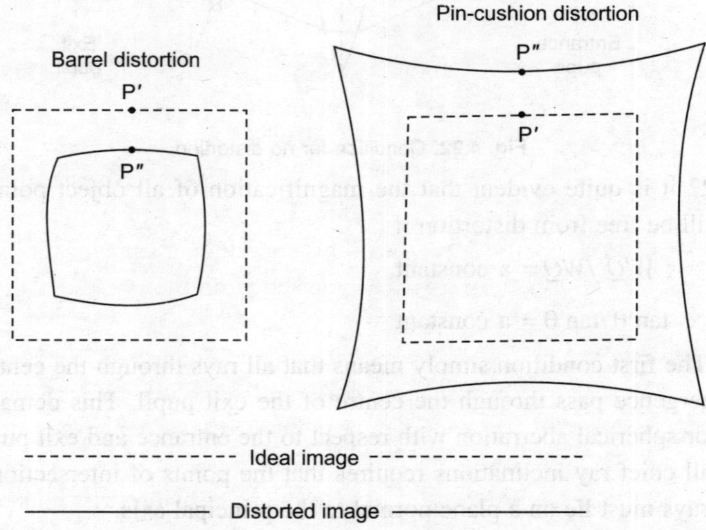

Fig. 4.21. Illustrates two possible distortions in the images

Before explaining the manner in which the Barrel distortion or the Pin-cushion distortion arises, we shall derive the conditions which any optical system must satisfy for freedom from distortion. A converging system, with an aperture stop in front of it, is shown in Fig. 4.22 forming an image at Q' of a plane normal to the axis at Q, where it is assumed that the lens system is corrected for the first four aberrations with respect to the points Q and Q' so that the image at Q' is also a plane normal to the axis. The aperture stop serves as an entrance pupil W in the object space and in Fig. 4.22, it is shown to the left of the first focal point so that the exit pupil W', which is the image of the entrance pupil W for paraxial rays, is real. Of course, this has been done for convenience to simplify the diagram by bringing the exit pupil on the other side of the lens, otherwise, it is not essential in the discussion. Consider a ray from an object point *R*, passing through the centre of the entrance pupil *W* and after refraction, through the centre of the exit pupil *W′*. This ray is known as the chief ray. It cuts the image plane at *R′*, which is consequently the paraxial image of *R*. The magnification produced by the optical system is given by

$$m = \frac{Q'R'}{QR} = \frac{W'Q'}{WQ}\frac{\tan\theta'}{\tan\theta}$$

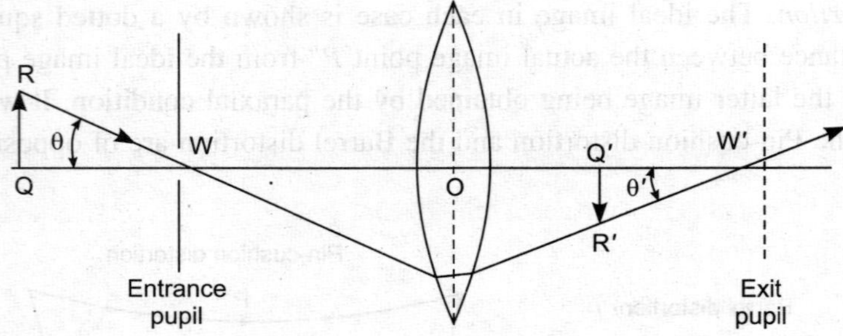

Fig. 4.22. Condition for no distortion

From Fig. 4.22 it is quite evident that the magnification of all object points will be constant, that is, the image will be free from distortion if

$$W'Q'/WQ = \text{a constant}$$

and $$\tan\theta'/\tan\theta = \text{a constant}$$

for all values of θ. The first condition simply means that all rays through the centre of entrance pupil *W* must also on emergence pass through the centre of the exit pupil. This demands that the system must be corrected for spherical aberration with respect to the entrance and exit pupils. This constancy of tan θ′/tan θ for all chief ray inclinations requires that the points of intersection of the entrant and the emergent chief rays must lie on a plane normal to the principal axis.

Now, to explain the manner in which barrel distortion arises we shall consider two object points *R* and *P*, near and remote respectively from the axis. Since the point *R* lies in the paraxial region, the incident chief ray *RW*, after refraction will pass through the centre *W′* of the exit pupil. But the principal ray *PW*, being more steeply inclined to the axis, after refraction will not cross the axis at *W′*, unless the system is corrected for spherical aberration with respect to *W* and *W′*, but will cross the axis at W_1 which is closer to the lens than *W′*, the paraxial image of *W*. If the system were corrected for spherical aberration with respect to *W* and *W′*, that is, if the magnification were constant the chief ray *PW* on emergence would have pursued the path shown by dotted line *P′W′* in Fig. 4.23 and would have crossed the image plane at *P′*. Hence, in such a case, we would have the relation

$$\frac{Q'P'}{QP} = \frac{Q'R'}{QR}$$

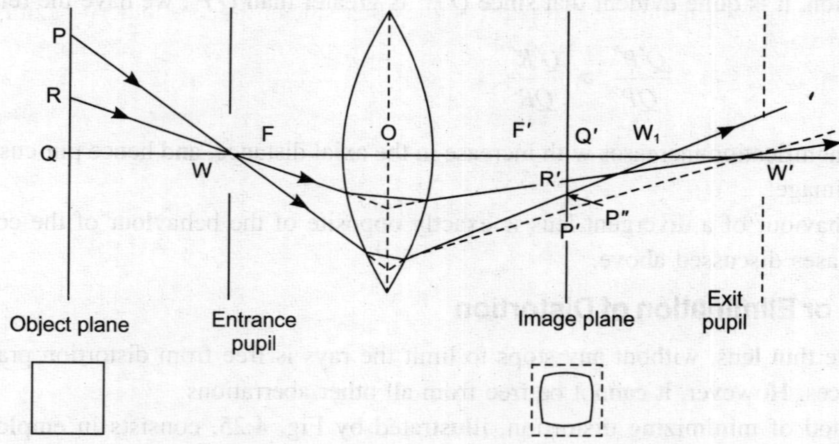

Fig. 4.23. Illustrates the origin of barrel shaped distortion

Now, it is quite evident that, in virtue of the system being not correct for spherical aberration with respect to W and W' the relation

$$Q'P'' < Q'P'$$

holds good. As a consequence, we have another relation

$$\frac{Q'P''}{QP} < \frac{Q'R'}{QR}$$

that is, the magnification decreases with increase in the axial distance of the object point, and hence Barrel distortion arises in the image.

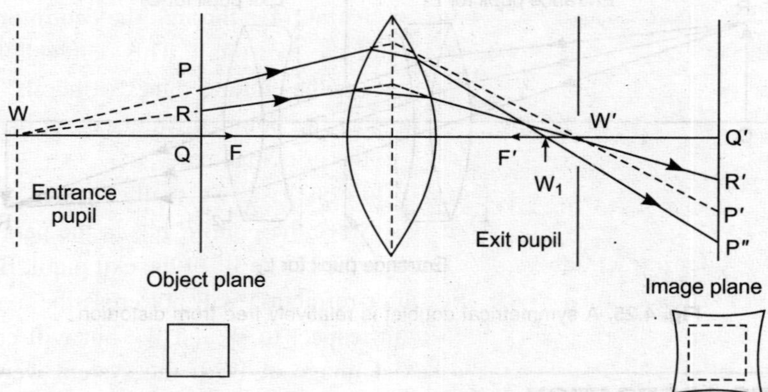

Fig. 4.24. Illustrates the origin of pin-cushion distortion

The manner in which pin cushion distortion arises is shown in Fig. 4.24, where the aperture stop is in the rear of the system so that it becomes the exit pupil W', the entrance pupil W being its paraxial image in the object space. By reasoning along the lines similar to that given in the case of barrel distortion, it is quite evident that since $Q'P''$ is greater than $Q'P'$, we have the relation

$$\frac{Q'P''}{QP} > \frac{Q'R'}{QR}$$

that is, the magnification increases with increase in the axial distance, and hence pin-cushion distortion arises in the image.

The behaviour of a divergent lens is exactly opposite of the behaviour of the convergent lens, in both the cases discussed above.

Reduction or Elimination of Distortion

A single thin lens, without any stops to limit the rays is free from distortion practically for all object distances. However, it cannot be free from all other aberrations.

A method of minimizing distortion, illustrated by Fig. 4.25, consists in employing two lens systems well corrected for other aberrations and placed symmetrically on opposite sides of a central aperture. This system is quite distortionless at unit magnification, otherwise, it must of course be corrected for spherical aberration with respect to the centre of the aperture stop. In effect, the pin-cushion distortion introduced by the first lens system is compensated by the barrel distortion produced by the second. Many camera and projection lenses are constructed in this way.

It should be emphasized that the combination of lenses sketched in Fig. 4.25 cannot be simultaneously corrected for spherical aberration with respect to the object and image positions as well as with respect to the entrance and exit pupils. Therefore the lens system, if free from distortion, will suffer with spherical aberration and astigmatism.

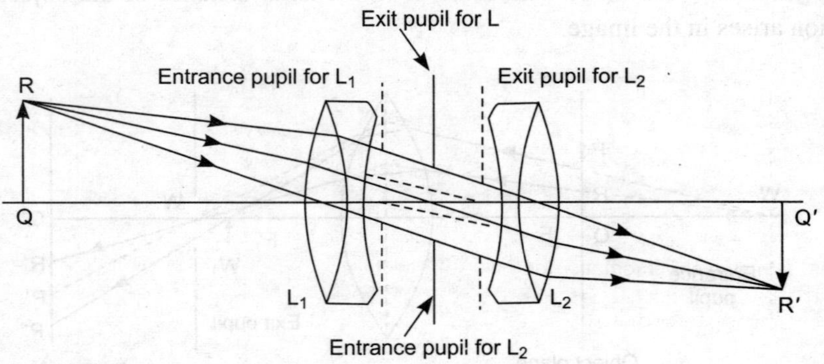

Fig. 4.25. A symmetrical doublet is relatively free from distortion

4.6 CHROMATIC ABERRATION

Even if the lens system is somehow corrected simultaneously for spherical aberration, coma, curvature, astigmatism and distortion (which is, of course, practically impossible), there will be present another aberration called *chromatic aberration* in the images formed by lenses, when, instead of monochromatic light, the light diverging from object is heterogeneous one. Light waves of different colours have different velocities in a refracting medium. Every refracting medium, therefore, has a different refractive index for each colour or wavelength of the spectrum, the index being least for the red and maximum for the violet colour. Now, the focal length of a thin lens is related to the refractive index of the optical material, of which the lens is composed by the relation.

$$1/f' = (\mu - 1)\left(\frac{1}{R_1} - \frac{1}{R_2}\right)$$

As a consequence of this relationship it follows that since $\mu_v > \mu_R$, hence $f'_v < f'_R$. *Thus the focal length decreases as we pass from red to the violet end of the spectrum.* A single lens, therefore, forms not merely one image of an object point but a series of coloured images at varying distances from the lens, one for each of the colours constituting the incident beam. Such a series of coloured images of an axial point object *P* are shown in Fig. 4.26. Furthermore, as a consequence of dependence of magnification on the focal length and, therefore, on the colour of light, the size of image of extended object varies with the wavelength. This is illustrated in Fig. 4.27 which shows for simplicity of the diagram only the red and the violet images of an extended white object *PQ*. Since $f'_v < f'_R$ the violet rays are focussed nearest the lens and therefore the violet image is smallest, while the red rays are focussed farthest away and therefore the red image is the largest. The images of other colours, although not shown in the diagram, are formed at the intermediate positions and are of intermediate sizes. Evidently there is no one plane in which all images are simultaneously in good focus. Therefore on a screen, we shall get a blurred image due to superposition of numerous coloured images out of focus.

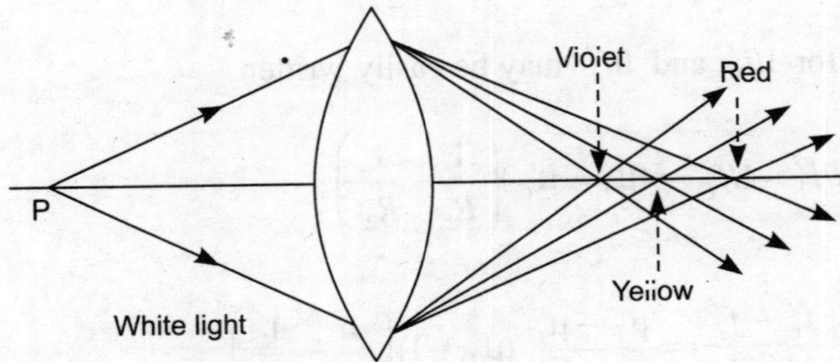

Fig. 4.26. Longitudinal chromatic aberration (for simplicity of Fig. other refracted colours are not shown)

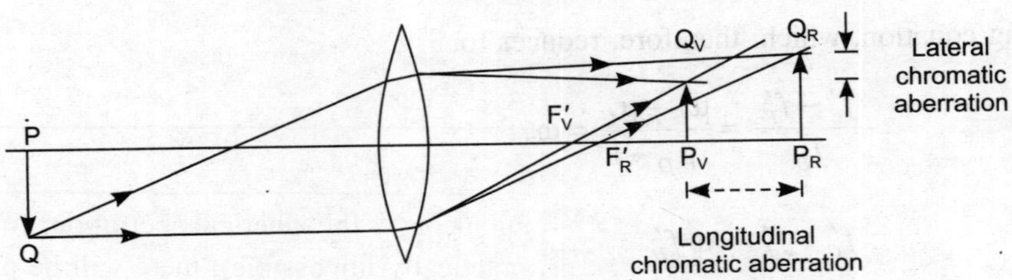

Fig. 4.27. Two types of chromatic aberrations of an image

The variation of image distance along the axis of lens due to change in wavelength, dv/dλ*, is called the axial or longitudinal chromatic aberration, while the variation in the image size with wavelength, dy/d*λ*, is called the lateral chromatic aberration*. The former is measured by the linear distance along the axis between the extreme images and is said to be positive when the violet image is situated to the left of the red image; while the latter is measured by the difference in the lateral sizes of extreme images and is said to be positive when the red image is more magnified than the violet. Divergent lens, as shown in Fig. 4.28, produces negative longitudinal chromatic aberration but positive lateral chromatic aberration.

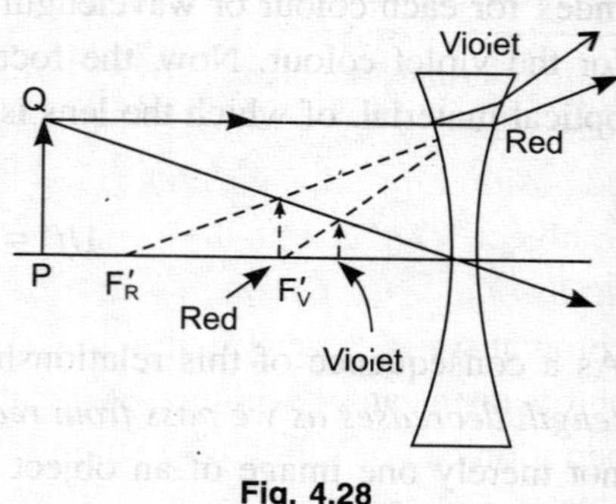

Fig. 4.28

4.7 AXIAL CHROMATIC ABERRATION FOR OBJECT AT INFINITY

The refractive indices of glass are usually given for *Fraunhofer lines* viz. *C—Red* (λ_C = 6563 *A.U.*), *D—Yellow* (λ_D = 5893 *A.U.*), *F—Blue* (λ_F = 4862 *A.U.*)

Let f'_C, f'_D and f'_F be the focal length of the lens for *C*, *D* and *F* lines. Then we write for a thin lens

$$1/f'_D = (\mu_D - 1)\left(\frac{1}{R_1} - \frac{1}{R_2}\right)$$

and similar expressions for $1/f'_C$ and $1/f'_F$ may be easily written,

Hence $$1/f'_F - 1/f'_C = (\mu_F - \mu_C)\left(\frac{1}{R_1} - \frac{1}{R_2}\right)$$

or $$\frac{f'_C - f'_F}{f'_F f'_C} = \frac{\mu_F - \mu_C}{\mu_D - 1}(\mu_D - 1)\left(\frac{1}{R_1} - \frac{1}{R_2}\right)$$

since f'_D is intermediate in value between f'_C and f'_F we may write as an approximation

$$f'^2_D = f'_F f'_C$$

in the preceding equation, which, therefore, reduces to

$$\frac{f'_C - f'_F}{f'_D} = \frac{\mu_F - \mu_C}{\mu_D - 1} = \omega$$

or $$f'_C - f'_F = \omega f'_D$$

The axial separation, $(f'_C - f'_F)$, between the foci for red and violet rays is called the longitudinal chromatic aberration of the lens for parallel rays and it is equal to the product of geometric mean of extreme focal lengths and the dispersive power, ω, of the material of which the lens is composed.

4.8 ACHROMATISM

An optical instrument free from chromatic aberrations is called an achromatic instrument. This simply means that the lens system of the achromatic instrument should form images of same size and at one position for all wavelengths not only for any one specified object position but for all positions of the object. *This complete achromatism demands that not only focal points must be made the same but also the principal planes must be made to coincide for different wavelengths.* However, this ideal achromatism is extremely difficult to accomplish in the case of any actual lens system. Only partial achromatism can be accomplished, namely, either longitudinal chromatic aberration or lateral chromatic aberration can be eliminated but not both simultaneously. Moreover this, too, is possible only for two or three wavelengths at one time. Therefore in making the lens system of the optical instrument partially achromatic, it is essential to keep in view the aperture and field of view of the optical system in deciding the kind of chromatic aberration to the eliminated because the kind of achromatism which is advantageous in one type of the instrument may be entirely disadvantageous in the other. For example, a telescope objective, which has a large aperture but a small field of view, should be corrected for longitudinal chromatic aberration. In this case correction would bring rays of all wavelengths in focus in the same plane, that is the achromatism is with respect to the position of its focal point, but as the principal points of the lens are different for different wavelengths, the focal length would not be equal with the result that magnification would not be constant. But owing to the close proximity of images to the principal axis, the variation in their sizes is not annoying and consequently comparatively unimportant. On the other hand, if the optical system has a small aperture and a large field of view, it should be corrected for lateral chromatic aberration. Here the correction would make the focal length the same for all wavelengths and hence magnification would be constant but the images would lie in different planes. It should be emphasized that no lens, composed of two kinds of glass only, can be achromatic for light of all colours. The achromatism in eyepieces is such that the coloured virtual images subtend the same angle at the eye, no matter whether their actual sizes and positions are different or not, and thus eye fails to detect any colour effect in the image.

4.9 CONDITION OF ACHROMATISM FOR TWO LENSES IN CONTACT

The index of refraction of glass is usually specified for the following Fraunhofer lines.

	Colour	Wavelength
C	Red	6563 A.
D	Yellow	5893 A.
F	Blue	4862 A.
G	Violet	4308 A.

For accurate work, yellow *D* line is disadvantageous because it is a doublet line. Therefore, for accurate work yellow *d* line λ_d = 5875.6 A., of the helium spectrum is employed as *intermediate wavelength* between the *C* and *F* lines.

An *achromatic doublet* consisting of two thin lenses in contact, one made of crown glass and the other of flint glass, may be designed so as to make the resultant focal length of the combination the same for any two wavelengths of the spectrum. We shall now derive the condition to be satisfied by two thin lenses in contact so as to be achromatic for two wavelengths, namely, the *C* and the *F* lines, the combination being itself considered as a thin lens.

Let f_1 and f_2 be the mean focal lengths of two thin lenses placed coaxially in contact with each other. If *F* be the focal length of the combination; then

$$1/F = 1/f_1 + 1/f_2$$

which on partial differentiation leads to

$$\frac{\partial F}{F^2} = \frac{\partial f_1}{f_1^2} + \frac{\partial f_2}{f_2^2}$$

Since $\partial f / f = \omega$, the dispersive power of the lens, the above equation can be written as

$$\partial F/F^2 = \omega_1/f_1 + \omega_2/f_2$$

where ω_2 and ω_1 are the dispersive powers of the two lenses. For no chromatic aberration $\partial F = 0$, hence

$$\omega_1/f_1 + \omega_2/f_2 = 0 \qquad \text{...(4.9 } a\text{)}$$

As the dispersive power ω_1 and ω_2 are always positive [$\omega = (\mu_F - \mu_C) / (\mu - 1)$, $\mu_F > \mu_C$ and $\mu > 1$], in order that Eqn. (4.9 *a*) be satisfied the focal lengths f_1, f_2 must have opposite signs. In practice a convergent lens of crown glass and a divergent lens of flint glass are used. The convergent lens is more strongly convergent for the blue rays than for the red rays while divergent lens is more strongly divergent for the blue rays than for the red rays. Hence if a doublet is made of these lenses which satisfy Eqn. (4.9 *a*) the coloured image formed by a single lens will fold on itself and the effects of chromatic aberration will be absent.

If $\omega_1 = \omega_2$, Eqn. (4.9 *a*) simplifies to $1/f_1 + 1/f_2 = 0$ which means that the focal length of the combination is infinite. Thus it will be obvious that the removal of chromatic aberration is impracticable if the lenses are made of same material. To eliminate chromatic aberration the two lenses must be so chosen that

(*i*) One is convergent and the other divergent;

(*ii*) The lenses should be made of different materials;

(*iii*) The choice of the dispersive power and focal length is governed by Eqn. (4.9 *a*).

It may be mentioned that if a combination is designed such that

$$\omega_1/f_1 + \omega_2/f_2 = 0$$

for two colours, the images formed by lights of these colours coincide in size and position. Such a combination is said to be achromatic for these colours. The amount of chromatic aberration remaining after a combination has been made achromatic for two colours is called *secondary spectrum*. This defect, however, has been considerably reduced by the introduction of Jena glass which is nowadays generally used for this purpose.

Achromatic Doublet

Since the removal of chromatic aberration only restricts the choice of focal lengths and not the radii of curvature of the lenses, the latter may be chosen to make spherical aberration minimum. If loss of light is to be avoided two surfaces of the lens should be cemented together by Canada balsam and thus the radii of curvature of these surfaces should be the same. Only two surfaces are thus left free to be manipulated.

In practice a convergent lens of crown glass and a divergent lens of flint glass are conected together by Canada balsam (Fig. 4.29). To minimise spherical aberration the free surface of the crown glass lens is more curved than the free surface of the flint glass lens both being convex outwards. The free surface of the crown glass lens is made to face the beam. Such a combination is called an achromatic doublet.

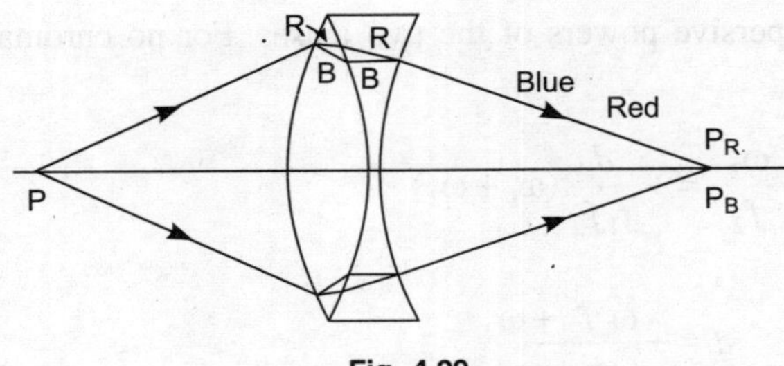

Fig. 4.29

If instead of two, three lenses are combined together the combination will be achromatic for three colours.

The condition in that case is

$$\frac{\omega_1}{f_1} + \frac{\omega_2}{f_2} + \frac{\omega_3}{f_3} = 0$$

This device is made use of in achieving visual or optical achromatism. Such achromatism is necessary for instruments which are to be used as adjuncts to the eye. In these cases, therefore chromatic aberration should be removed with respect to the brightest part of the spectrum. This part is the yellowish green region lying between the blue and red rays. The images formed by the red and the blue rays are made to coincide with that formed by the yellowish-green region. Thus, chromatic aberration should be removed for three colours. This can be done by taking three lenses satisfying the above equation.

Even when the combination has been made achromatic for three colours, there is still some chromatic aberration present due to the remaining colours. The remaining aberration is termed as *tertiary spectrum*. As mentioned above, glasses have been developed which reduce the tertiary and remaining dispersion without increasing the number of lenses. This higher kind of achromatism is called "apochromatism".

4.10 CONDITION OF ACHROMATISM OF SEPARATED DOUBLET

If two lenses of mean focal lengths f_1 and f_2 are separated by a certain distance 'd', the focal length of the equivalent lens is given by

$$\frac{1}{F} = \frac{1}{f_1} + \frac{1}{f_2} - \frac{d}{f_1 f_2}$$

The condition for achromatism can then be obtained by a partial differentiation of the above equation which gives

$$\frac{\partial F}{F^2} = \frac{\partial f_1}{f_1^2} + \frac{\partial f_2}{f_2^2} - \frac{d[f_1 \partial f_2 + f_2 \partial f_1]}{f_1^2 f_2^2}$$

$$= \frac{\omega_1}{f_1} + \frac{\omega_2}{f_2} - \frac{d}{f_1 f_2}(\omega_1 + \omega_2) \qquad \text{...(4.10 } a)$$

where ω_1 and ω_2 are the dispersive powers of the two lenses. For no chromatic aberration $\partial F = 0$. Hence

$$\frac{\omega_1}{f_1} + \frac{\omega_2}{f_2} = + \frac{d}{f_1 f_2}(\omega_1 + \omega_2)$$

$$\therefore \qquad d = + \frac{\omega_1 f_2 + \omega_2}{\omega_1 + \omega_2} \qquad \text{...(4.10 } b)$$

Thus in order that the combination be achromatic, the distance between the two lenses must be determined by Eqn. (4.10 *b*). If the lenses are made of the same material $\omega_1 = \omega_2 = \omega$ and the condition simplifies to

$$d = \frac{1}{2}(f_1 + f_2) \qquad \text{...(4.10 } c)$$

Hence if the lenses are made of the same material, chromatic aberration can be eliminated by keeping them separated by half the sum of their focal lengths. The combination will then be achromatic for all the colours near those for which the mean focal lengths f_1 and f_2 have been calculated.

It may be noted that both the errors of chromatic aberration cannot be removed in the case of two coaxial lenses not in contact with each other.

Such a system of lenses whose distance apart is given by Eqn. (4.10 *c*) is generally used in the construction of eyepieces or oculars. It has been mentioned before that an eyepiece should be achromatic in the sense that different coloured images should subtend equal angles at the eye. This is very easily achieved in the case of separated doublet since the lateral chromatic aberration is highly corrected through constancy in focal length. However, the longitudinal chromatic aberration is relatively large because the equivalent planes are not in the same position in different colours.

4.11 THE BALANCING OF ABERRATIONS

We have seen in the preceding pages that a single lens, in general, is afflicted with seven major aberrations—five monochromatic aberrations and two chromatic aberrations. It is quite evident from the discussion of these aberrations that it is not possible to eliminate or even to minimise simultaneously all of the seven aberrations for a single lens. However, it is possible to design a compound lens from a number of lenses in such a way that the aberrations of one part of the field are balanced against those of the other. The greater the number of component lenses in the compound lens, the greater the degree of accuracy which may be secured. Even then no compound lens can ever contain sufficient number of elements permitting the simultaneous elimination of all the seven aberrations even for a single position of the object. In practice, therefore, the attention of a lens designer is always directed towards minimising those aberrations in the lens system which are likely to be most detrimental to the work for which the lens system is designed. For example, a telescope objective, which is required to cover only a small angular field, should always be so designed as to minimise spherical aberration. On the other hand, a photographic objective, which is always required to cover large angular field, should be so designed as to minimise astigmatism, coma, and distortion while other aberrations should be partly corrected and so on.

SOLVED EXAMPLES

Example 1. *The focal lengths of a thin convex lens are 100 cm and 96.8 cm for red and blue rays respectively. Calculate the dispersive power of the material of the lens.*

The dispersive power of the material of the lens in terms of its principal focal lengths is given by the expression.

$$\omega = (f'_C - f'_F)/f'_D$$

In the given problem, f'_C = 100 cm, the focal length for red rays and f'_F = 96.8 cm, the focal length for blue rays. Since f'_D is intermediate in value between f'_C and f'_F, we write

$$f'_D = \sqrt{f'_C \times f'_F} = \sqrt{100 \times 96.8} = 98.38 \text{ cm.}$$

Hence $\omega = (100 - 96.8)/98.38 = \mathbf{0.0325.}$

Example 2. *An achromatic objective of focal length 50 cm is to be made of different kinds of glass shown below. Find the focal length of each lens, stating whether it is convergent or divergent.*

	Glass A	*Glass B*
μ_{Red}	1.51	1.64
μ_{Blue}	1.52	1.66

The mean ray in glass *A* will have a refractive index

$$\mu_M = (1.51 + 1.52)/2 = 1.515$$

The dispersive power ω_1 of glass *A* is given by

$$\omega_1 = \frac{\mu_B - \mu_R}{\mu_M - 1} = \frac{1.52 - 1.51}{1.515 - 1} = 0.0194$$

Similarly, $$\omega_2 = \frac{1.66 - 1.64}{1.65 - 1} = 0.0307$$

Now, for the combination of two lenses in contact to be achromatic for red and blue light.

$$f'_1 / f'_2 = \omega_1/\omega_2$$

where symbols have their usual meaning.

Also the focal length f' of the combination is given by

$$\frac{1}{f'} = \frac{1}{f_1'} + \frac{1}{f_2'}$$

or

$$\frac{f_1'}{f'} - 1 = \frac{f_1'}{f_2'} = -\frac{\omega_1}{\omega_2}$$

or

$$f_1' = \frac{\omega_2 - \omega_1}{\omega_2} f' = \frac{(0.0307 - 0.0194)}{0.0307} \times 50 = 18.4 \text{ cm}$$

$$f_2' = \frac{\omega_1 - \omega_2}{\omega_1} f' = \frac{(0.0194 - 0.0307)}{0.0194} \times 50 = -29.12 \text{ cm}$$

Thus the lens of focal length **18.4 cm** is **convergent** while that of focal length **29.12 cm** is a **divergent** one.

Example 3. *A convex lens of crown glass is perfectly cemented to a plano-concave lens of flint glass to form an achromatic combination of power +5D. Calculate the radii of curvature of the convex lens from the following data:*

	Refractive index	*Dispersive power*
Crown glass	*1.50*	*0.01*
Flint glass	*1.60*	*0.02*

The power of achromatic lens is +5*D*. Hence it is a convergent lens of focal length 20 cm. formed by cementing two lenses together.

Let f_1' be the focal length of the crown glass and f_2' the focal length of flint glass lens. Then, as explained in Example 2, if the combination is achromatic, we have

$$f_1' = \frac{\omega_2 - \omega_1}{\omega_2} f' = \frac{0.02 - 0.01}{0.02} \times 20 = 10 \text{ cm}$$

and

$$f_2' = \frac{\omega_1 - \omega_2}{\omega_1} f' = \frac{0.01 - 0.02}{0.01} \times 20 = -20 \text{ cm}$$

Let the concave surface of the plano-concave divergent lens be of radius of curvature R'_1; we write

$$-\frac{1}{20} = \frac{1.60 - 1}{R_1'}$$

or

$$R_1' = -20 \times 0.6 = -12 \text{ cm}.$$

Let R_1 and R_2 be the radii of curvature of the surfaces of the crown glass convergent lens. Since the lenses are perfectly cemented together, we have

$$R_2 = R_1' = -12 \text{ cm}$$

Now $$1/f_1' = (\mu_1 - 1)\left(\frac{1}{R_1} - \frac{1}{R_2}\right)$$

Substituting the values with proper sign, we get

$$1/10 = (1.5 - 1)\left(\frac{1}{R_1} + \frac{1}{12}\right)$$

or $$R_1 = 8.57 \text{ cm}$$

The radii of curvature of surfaces of the convex lens are **8.57 cm** and **12 cm**.

PROBLEMS

1. Describe and explain with the help of suitable diagrams (i) longitudinal spherical aberration, (ii) lateral spherical aberration.

   How can the spherical aberration be minimised in the case of an ordinary lens.
2. Describe astigmatism, coma, curvature and distortion. How they may be reduced to a minimum?
3. What is spherical aberration? How is it minimised when two thin lenses of the same medium are placed at a distance from each other? Show that a Huygens' eyepiece satisfies the condition of minimum spherical aberration.
4. What is spherical aberration? How is it minimised in a microscope objective and an eyepiece?
5. Explain chromatic aberration and derive an expression for the axial chromatic error for a thin lens.
6. A combination of two thin lenses, one of crown and the other of flint glass, has the same focal length for a given pair of wavelengths. Derive expressions for the powers of the crown and flint lenses in terms of the combination and dispersive powers of glasses.
7. What do you understand by the term achromatism? Derive and discuss the conditions of achromatism for two thin lenses of focal lengths f_1 and f_2 (a) when they are made of different materials but placed in contact and (b) when they are made of same material but separated by a distance.
8. A thin crown glass lens is in contact with a thin flint glass lens, the radius of curvature of the common surface being 25 cm. If the combination forms an achromatic combination of 40 cm focal length. Find the radii of curvature of the second face of the two lenses.

   μ for crown glass $= 1.50$

   μ for flint glass $= 1.60$

   Dispersive power of crown glass $= .021$

   Dispersive power of flint glass $= .045$
9. Show that a combination of two lenses separated by a distance cannot be achromatic in the same sense as a telescope objective is achromatic.
10. Describe and explain chromatic aberration. Deduce a condition for achromatism of two lenses separated by a distance.

    Explain what is meant by an achromatic system.

and $$2e = (n+250)\times 4000\times 10^{-8}$$

Solving the two equations we have $n = 400$ and hence

$$e = \mathbf{0.013\ cm.}$$

Example 2. *White light is incident on a soap film at an angle of incidence $\sin^{-1} 4/5$ and the reflected light on examination by a spectroscope shows dark bands. Two consecutive dark bands correspond to wavelength 6.1×10^{-5} and 6.0×10^{-5} cm. If the refractive index of soap-film be 4/3, calculate its thickness.*

Suppose the thickness of soap film is e. Neglecting the small variation in the refractive index of the soap film for the spectral range under consideration, we write the conditions for consecutive dark bands corresponding to λ_1 and λ_2,

$$2\mu e\cos\theta = n\lambda_1 = (n+1)\lambda_2$$

Solving for n we have, $$n = \lambda_2/(\lambda_1 - \lambda_2)$$

Hence $$2\mu e\cos\theta = \lambda_1\lambda_2/(\lambda_1 - \lambda_2)$$

or $$2e\sqrt{(\mu^2 - \sin^2 i)} = \lambda_1\lambda_2/(\lambda_1 - \lambda_2)$$

In the given problem $\mu = 4/3$, $\sin i = 4/5$, $\lambda_1 = 6.1 \times 10^{-5}$ cm and $\lambda_2 = 6.0 \times 10^{-5}$cm. Substituting these values in the above equation and solving for e, we get

$$e = \mathbf{0.00172\ cm.}$$

Example 3. *Two rectangular pieces of plane glass are laid one upon the other and a thin wire is placed between them at one edge so that very thin wedge of air is formed. The plates are illuminated with sodium light (λ = 5893 A.U.) at normal incidence. Bright and dark bands are formed, there being ten of each per cm length of the wedge measured normal to the edges in contact. Find the angle of wedge.*

The distance between any two consecutive dark fringes is expressed by Eqn. (12.3 *i*), which is

$$\overline{X} = \frac{\lambda}{2\tan\alpha\sqrt{(\mu^2 - \sin^2 i)}}$$

where symbols have their usual meaning.

In the given problem $\overline{X}$ = 1/10 cm, $\mu = 1$, $i = 0$, $\lambda = 5893 \times 10^{-8}$ cm.

Hence $$\tan\alpha = \frac{5893\times 10^{-8}}{2\times 1/10} = 2.95\times 10^{-4}$$

$$\alpha = \mathbf{2.95\times 10^{-4}\ radian}$$

was found at a distance of ¼ λ from the mirror. Hence the mirror was a nodal plane. This simply means that the light waves underwent a phase change of π due to reflection so that they continuously annual each other at the mirror surface.

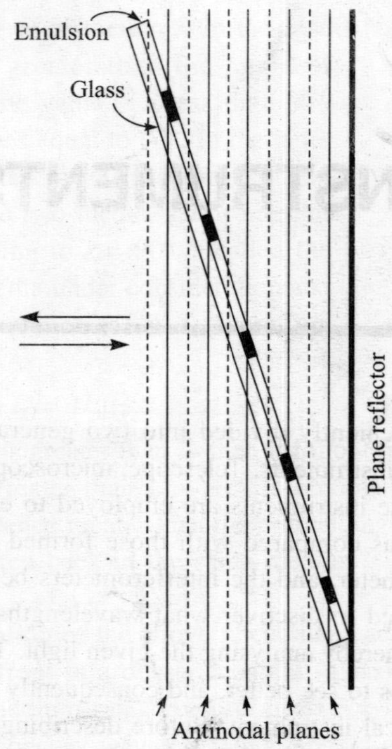

Fig. 12.23. Wiener's experiment

The *electromagnetic theory* indicates that the electric vectors of the direct and reflected waves interfere destructively, thereby forming a node at the reflecting mirror while magnetic vectors produce antinode at the mirror, *Wiener,* therefore, concluded that the photographic action is brought about by the electric vector and not by the magnetic vector present in the electromagnetic light waves. *In general, the electric vector is effective in every optical phenomenon. Wiener's* experiment constituted another experimental confirmation of *Maxwell's electromagnetic theory of light.*

12.8 THE LIPPMAN PROCESS OF COLOUR PHOTOGRAPHY

Soon after the publication of *Wiener's* experiment *Lippman*, in 1891, invented a process of photography in natural colours, which is based on the phenomenon of *stationary light waves.* A specially prepared photographic plate, having an extremely fine grain emulsion, finer than that of the commercial plates, is placed in a camera. The glass side is towards the objective while the emulsion side is backed with a layer of mercury to form a reflecting surface. This of course requires a special form of the plate holder.

OPTICAL INSTRUMENTS

Chapter

Optical instruments may be conveniently divided into two general classes: (i) the image forming instruments, and (ii) the analysing instruments. Telescope, microscope, prismatic binoculars etc. come under the first class, because these instruments are employed to examine objects by forming their images, much magnified in size as compared with those formed by the unaided eye. The prism-spectrometer, the grating spectrometer, and the interferometers belong to class two, because these instruments are primarily employed to discover what wavelengths are present in a given beam of light, incident on the instrument, thereby analysing the given light. The function of most of the image forming instruments is to enable us to see better, and consequently the eye should be regarded as an essential part of every visual optical instrument. Before describing the optical system of any visual instrument, therefore, it is best to describe the essential parts constituting the eye and discuss its image forming properties, limiting our discussion only to the physical principles involved in the process of image formation.

5.1 THE EYE

The essential parts of the human eye, considered as an optical instrument, are illustrated by the sectional diagram 5.1 of the eye. The eye is roughly spherical in shape, the front part being slightly more curved and is covered with a tough transparent membrane *C*, called the *Cornea*. Behind the cornea is a weak salt solution known *as the aqueous humour A*, backed by *a crystalline lens L* of varying optical density. The lens is of fibrous jelly, harder at the centre and then gradually becoming softer towards the surface, the refractive index varying between 1.38 and 1.41. It is held in position by *the suspensory ligaments I* through which it is connected to *the ciliary muscle M*, which is spread all round the lens. The remainder of the eye ball is filled with a thin transparent jelly consisting largely

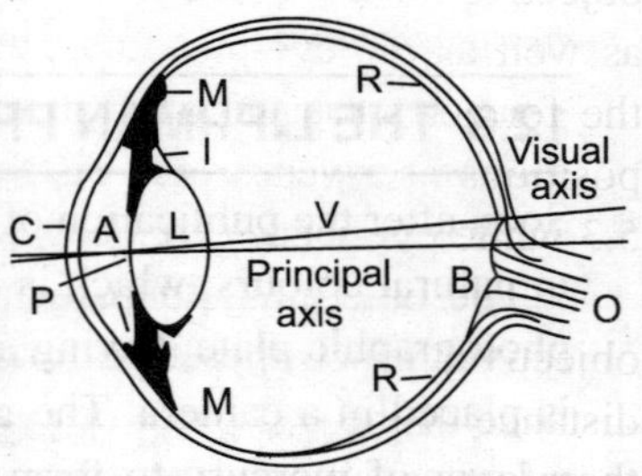

Fig. 5.1. The eye

of viscous liquid, called *the vitreous humour V*. The refractive index of both the vitreous humour and the aqueous humour is 1.336. The cornea and the crystalline lens essentially constitute a compound convergent lens system which forms on sensitive nerve fibres and cells, called *the retina R*, inverted images of objects, situated near or far away in front of the eye. The nerve fibres constituting the retina are in fact the branches of the optic nerve *O* and terminate in minute structure called *rods and cones*. In and about them circulates a bluish liquid known as *Visual purple*. The optic nerve is connected to the mind and thus the image formed on the retina is transmitted via the optic nerve to the mind. The point at which optic nerve enters the eye is called *the blind spot B*, because it is not possible to see an object if its image is formed at this point.

Immediately in front of the crystalline lens is *the Iris I*, having the central hole called *the pupil P* and this serves the purpose of a diaphragm in admitting light into the eye. The size of the pupil automatically increases if the intensity of the object is low, and decreases if the intensity is high. The crystalline lens is also flexible, its shape and hence its focal length can be varied by a change in the tension of the suspensory ligaments by the ciliary muscles. As a consequence, focal length of the compound lens changes and the eye can be made to focus on objects at different distances. This process is known as accommodation and the extremes of the range of vision over which the eye can bring objects in focus are called the far point and the near point of the eye. The object is said to be at the far point when it can be seen by the completely relaxed eye. The second focal point of the normal eye under relaxed ciliary muscles is at the retina. The object is said to be at the near point when the lens acquires maximum curvature by the appropriate contraction of the muscles in accommodation. The focal length of the lens decreases to focus the near object. For the normal eye, the far point is at infinity while the near point is at 25 cm and is called the least distance of distinct vision. However, the far point and the near point depend upon the age. The range of accommodation gradually diminishes with advancing age on account of the lens becoming less flexible. This loss of accommodation is known as presbyopia. There are three other common defects of vision, namely, (1) myopia or shortsightedness, (2) hyperopia or farsightedness and (3) astigmatism. These defects are corrected by the use of appropriate spectacle lenses. It is needless to describe these defects in detail at this place for it can be easily presumed that the students already possess a working knowledge of them.

5.2 THE COMPOUND MICROSCOPE

The compound microscope was invented by *Galileo* in 1610. The optical parts of a compound microscope, illustrated by Fig. 5.2, are (*i*) an objective and (*ii*) an eye-piece. The focal length of the objective is very small while that of the eye-piece is somewhat larger. Figure 5.2 shows the objective as well as the eye-piece as single lens but in reality they are highly corrected compound lenses; in the former spherical aberration, coma and chromatic aberrations are specially corrected for object positions at a very close distance. The actual microscope objectives are described in detail in *Sec*. 5.3 while eye-pieces are dealt with in *Sec*. 5.8.

The object *PQ* to be examined is placed just outside the first principal focal point F_o of the objective, which therefore forms a real magnified image *P'Q'* either within the first principal focal distance of the eye-piece or exactly at the first principal focal point F_e of the eye-piece. Eye-piece is, therefore, positioned as a simple magnifier with respect to the image *P'Q'*. As a consequence, the eye-piece forms a large virtual image either somewhere between the infinity and the least distance of

Example 10. *Calculate in nearest whole number the fringes that must be counted for green cadmium light for shortest etalon used with Michelson interferometer. The length of shortest etalon is 0.390 mm and the wavelength of green cadmium light is 5085.8 A.*

Suppose N fringes cross the field of view in the evaluation of shortest etalon in terms of cadmium green light. The path difference introduced between the two interfering beams of light is $2d = 2 \times 0.039$ cm. Hence we have the relation [See *Sec.* 12.6.4]

$$2 \times 0.039 = N \times 5085.8 \times 10^{-8}$$

$$N = \mathbf{1533} \text{ fringes.}$$

Example 11. *In an experiment with Michelson interferometer, the small readings for maximum indistinctness were found to be 0.6939 mm and 0.9884 mm. If the mean wavelength for the two components of D lines be 5893 A.U., deduce the difference between the two wavelengths.*

The difference between the two wavelengths of D lines is given by Eqn. (12.6 *q*) which is

$$\lambda_1 - \lambda_2 = \lambda_{av}^2 / 2(d_1 - d_2)$$

In the given problem $\lambda_{av} = 5893 \times 10^{-8}$ cm, $d_1 = 0.09884$ cm and $d_2 = 0.06939$ cm

Hence $\lambda_1 - \lambda_2 = 6 \times 10^{-8}$ cm = **6 A.U.**

Example 12. *Two wavelengths $\lambda_1 = 5890$ A.U. and $\lambda_2 = 5896$ A.U. are emitted by a source of light. Interference fringes are observed with a certain arrangement when the paths of the interfering beams and exactly equal. How much will the path difference have to be increased so that a bright fringe for λ_2 coincides with a dark fringe of λ_1?*

When the paths of the interfering beams are exactly equal, the fringe systems due to λ_1 and λ_2 are coincident *i.e.* the bright or dark fringes due to λ_1 fall over the bright or dark fringes corresponding to λ_2. If Δ be the path difference introduced when a bright fringe for λ_2 coincides with a dark fringe for λ_1, then we must have the relation

$$\Delta = n\lambda_2 = \left(n + \frac{1}{2}\right)\lambda_1$$

or

$$\Delta = \frac{\lambda_1 \lambda_2}{2(\lambda_2 - \lambda_1)} = \frac{5890 \times 10^{-8} \times 5896 \times 10^{-8}}{2 \times 6 \times 10^{-8}}$$

$$\Delta = \mathbf{0.0289\ cm.}$$

Example 13. *Circular fringes are observed in a Michelson interferometer illuminated with light of wavelength 5896 A. If the path difference between the mirrors M_1 and M_2 is 0.3 cm the central fringe is bright. Calculate the angular diameter of the 7th bright fringe.*

Suppose the central bright fringe ($\theta = 0$) is of nth order. Hence we have the relation

$$2e = n\lambda$$

We have seen in *Sec.* 12.6.2 that the orders of fringes decrease as we consider fringes outwards from the central fringe. The second bright fringe is accordingly of $(n - 1)$th order, ..., the seventh

Example 4. *White light is incident on two parallel glass plates separated by an air film of 0.001 cm thickness and the reflected light is examined by the spectroscope. Find the number of dark bands seen in the spectrum between the wavelengths 4 × 10^{-5} cm and 7 × 10^{-5} cm when light is incident at an angle of 30° to the normal to surfaces.*

According to theory of colours of thin films exhibited due to interference of reflected light waves, the dark bands are given by the condition

$$2\mu e\cos\theta = n\lambda$$

In the given problem, $\mu = 1$, $e = 0.001$ cm, $\theta = 30°$ and the range of visible spectrum is from 4×10^{-5} to 7×10^{-5} cm. Substituting the values for μ, e, θ in the above equation, we have

$$n_1 \times 4\times10^{-5} = 2\times0.001\times\left(\frac{1}{2}\sqrt{3}\right) \qquad \therefore n_1 = 43$$

Similarly, $$n_2 \times 7\times10^{-5} = 2\times0.001\times\left(\frac{1}{2}\sqrt{3}\right) \qquad \therefore n_2 = 24$$

Hence the number dark bands seen within the given spectral range = $(n_1 - n_2)$ = **19.**

Example 5. *If the angle of wedge is 0.25 deg. of an arc and the wavelength of sodium D lines are 5890 × 10^{-8} and 5896 × 10^{-8} cm find the distance from the apex of the wedge at which the maximum due to each wavelength first coincide.*

Suppose e represents the thickness of the wedge at a point where the maximum due to $\lambda_1 = 5896 \times 10^{-8}$ cm and $\lambda_2 = 5890 \times 10^{-8}$ cm coincide when viewed by the reflected light. The condition for this is expressed by the relation

$$2e = \left(n+\frac{1}{2}\right)\lambda_1 = \left(n+\frac{3}{2}\right)\lambda_2$$

Hence $$n = \frac{(3\lambda_2 - \lambda_1)}{2(\lambda_1 - \lambda_2)}$$

Substituting the value of n in the first of the above two equations, we have

$$2e = \lambda_1\lambda_2/(\lambda_1 - \lambda_2)$$

Suppose the distance from the apex of the wedge at which the maximum due to λ_1 and λ_2 coincide is x cm; hence

$$e = x\tan\alpha = x\alpha$$

where α is the angle of the wedge, expressed in radians [180° = π radians].

$$\therefore 2x\frac{\pi\times0.25}{180} = \frac{5896\times10^{-8}\times5890\times10^{-8}}{6\times10^{-8}}$$

$$x = \mathbf{6.63\ cm.}$$

the instrument is simply the image of its objective formed by the eye-piece. To have a full field of view, all the rays emerging from the microscope must enter the pupil of the eye. Therefore, it is evident from Fig. 5.2 that the eye must be placed at the position of the exit pupil because all the rays entering the objective and emerging from the eye-piece also pass through the exit pupil of the instrument. Moreover, the diameter of the exit pupil should be equal to that of the pupil of the eye. Under these conditions the magnifying power of the microscope is called *normal magnifying power*. In order to make the full use of the available resolving power of the microscope objective the overall magnification must be at least equal to the normal magnifying power.

5.3 MICROSCOPE OBJECTIVE

We have seen that the objects, examined by the microscope, are to be very close to the objective, and considerations of the brightness of the image and the limit of resolution of the microscope objective [see chapter on Resolving power] demand that wider cone of rays should enter the objective. Consequently if a single lens is employed as an objective, the *spherical aberration* and the *coma* will be much pronounced in the image formed by it. Moreover, as the objects are usually illuminated by the white light, a single lens objective will also exhibit the *chromatic aberrations*. The microscope objectives must be, therefore, carefully designed to eliminate, as far as possible, the *chromatic aberrations, the spherical aberration and the coma.*

In a low power microscope, the objective is a doublet achromatised for two wavelengths and corrected for spherical aberration and coma. A high power microscope objective consists of a number of achromatic doublets and the deviation of any ray is distributed into small amounts at each surface; thus the spherical aberration is eliminated. One such objective, called *apochromat*, sketched in Fig. (5.3 *b*), is achromatised for three wavelengths and is corrected for spherical aberration at two wavelengths. The spherical aberration and the chromatic aberrations introduced by the front hemispherical lens must be corrected by the latter lenses in the objective.

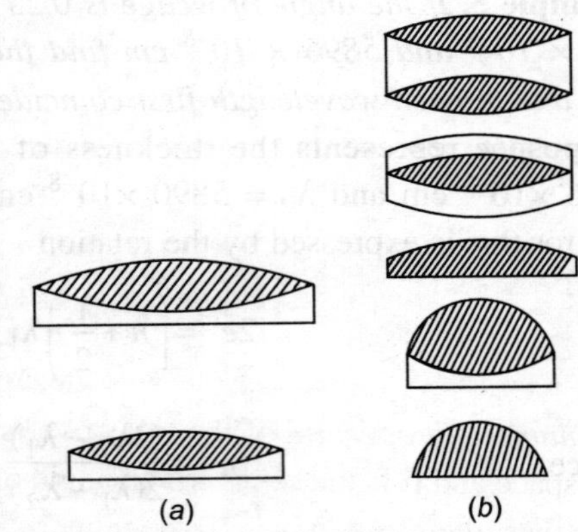

Fig. 5.3. Microscope objectives (*a*) Low power (*b*) High power-apochromatic

High quality microscope objectives are of *oil immersion type*. One such objective, designed by *Amici* about the middle of the last century, is illustrated by Fig. 5.4. The space between the cover glass and the front lens is filled with a drop of cedarwood oil which has a refractive index, $\mu = 1.51414$ and dispersive power very nearly the same as for the glass. The front lens and the cover glass have the same refractive index as the oil. As a consequence, the object is virtually imbedded within the first lens and therefore the first refraction occurs only at the spherical surface of the front lens. The movement of the objective in focussing brings the object at the nearest *aplanatic point* L_1 of the spherical surface of the front lens. The rays of light after refraction by this hemispherical surface are rendered less divergent, and whatever may be the inclination of the incident rays to the axis, after refraction appear to come from the conjugate *second aplanatic point* L_2. The second lens in the objective is a meniscus lens, the centre of curvature of its first surface being at L_2. Therefore, the rays enter the meniscus lens without any deviation. The second surface of this lens is of such a curvature that L_2 becomes the *first aplanatic point* for it. The rays after refraction through the second surface, therefore, appear to come from the conjugate *aplanatic point* L_3. The rays are rendered still less divergent and finally convergent by the other achromatic lenses which also serve to eliminate the chromatic aberrations introduced by the first two lenses.

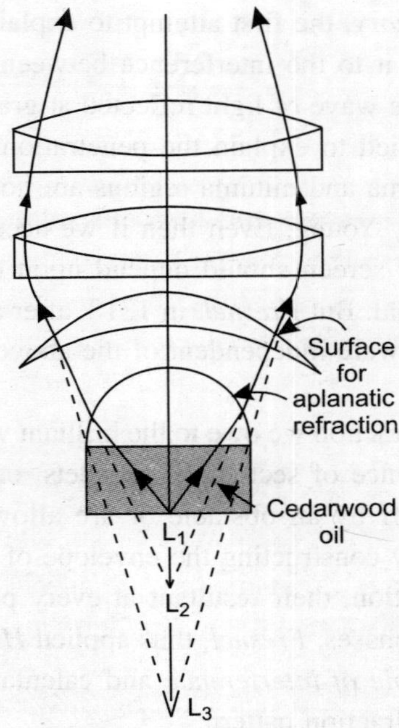

Fig. 5.4. High-power oil immersion objective

The *numerical aperture* of the objective is defined as μ sin α, where μ is the refractive index of the object space and α is the semivertical angle of the cone of rays entering the objective. Oil immersion type objectives, therefore, have a large numerical aperture as compared with that of objectives having air in the object space. As a consequence, the least distance between two close points resolvable by the microscope objective, defined by

$$s = \frac{1.22\lambda}{2\,N.A.}$$

decreases by employing an oil immersion objective. For the sake of convenience we may postpone the derivation of the above expression and its discussion to the chapter on Resolving Power.

The term *diffraction* is applied to all phenomena like this which are produced whenever there is any limitation on the width of the beam of light. Since in most diffraction patterns some light penetrates within the region of geometrical shadow, diffraction is sometimes defined as "*the bending of light around an obstacle*". This phenomenon was first discovered in 1665 by an *Italian* Scientist named *Grimaldi* and was latter on studied by Newton. In the days, when *corpuscular theory* was prevalent attempts were made by *Newton* to interpret diffraction effects as due to attractive or repulsive forces exerted by the edges of obstacles on flying corpuscles, so as to deflect them from their rectilinear path. In terms of the *wave theory*, the first attempt to explain the diffraction phenomenon was made by *Dr. Young*, who attributed it to the interference between the direct light waves which pass near the edge of the obstacle and the wave of light reflected at grazing incidence from the edge. This explanation could scarcely be applied to explain the penetration of light within the region of geometrical shadow. Moreover the maxima and minima regions are not equally spaced as demanded by the mode of interference imagined by Young. Even then if we accept Young's explanation to be correct, the intensity distribution on the screen should depend upon (*i*) the sharpness of the edge, (*ii*) its degree of polish and (*iii*) its material. But *Fresnel*, in 1818, after a series of experiments proved that the details of the diffraction pattern were independent of the above mentioned factors so long as the material of the obstacle is opaque.

The correct interpretation of the diffraction we owe to the brilliant work of *Fresnel*, who attributed this phenomenon to the mutual interference of secondary wavelets, originating from various points of wave-front which are not blocked off by an obstacle or are allowed to pass by a slit. That is instead of finding the new wave-front by constructing the envelope of these secondary wavelets, we must find, by the principle of superposition, their resultant at every point of the screen, taking due account of their relative amplitudes and phases. *Fresnel*, thus applied *Huygen's principle of secondary wavelets* in conjunction with the *principle of interference* and calculated the position of fringes, in general agreement with the observed diffraction pattern.

It should be emphasised, however, that the process responsible for the production of *diffraction* phenomenon, is going on continuously during the propagation of every wave-front. But the diffraction effects are observed only when a portion of the wave-front is cut off by some obstacle. Every optical instrument, is fact, makes use of only a limited portion of the wave-front. A telescope or a microscope, for example, utilises only a limited portion of a wave-front transmitted by the objective lens. Therefore some diffraction is always present in the image. Hence a clear comprehension of the nature of diffraction is of fundamental importance for a complete understanding of practically all the optical phenomena.

13.1 FRESNEL'S EXPLANATION OF THE RECTILINEAR PROPAGATION OF LIGHT

One of the greatest difficulties encountered by the supporters of the *wave theory* of light in its early days was the explanation of the observed fact that light appears to be propagated in straight lines. The correct interpretation in terms of the *wave-theory* was given by *Fresnel* in 1815 by applying *principles of interference* in conjunction with the *Huygen's principle of secondary wavelets* but at the same time he showed that the rectilinear propagation of light is only approximate. To explain this let us first discuss Fresnel's method of finding the effect of a plane wave-front at a point ahead of it.

bright fringe is of $(n-6)$th order. Hence, if θ be the angular radius of 7th bright fringe, we have the relation

$$2e\cos\theta = (n-6)\lambda$$

Hence $\quad 2e(1-\cos\theta) = 6\lambda$

$$2\times 0.3\left[1-\left(1-\frac{1}{2}\theta^2\right)\right] = 6\times 5896\times 10^{-8}$$

$$\theta^2 = 2\times 5896\times 10^{-7}$$

$$\theta = 0.344 \text{ radian} = 19.7°$$

Hence the angular diameter of 7th bright fringe = **39.4°**.

PROBLEMS

1. Describe and explain the phenomenon of interference in thin films. Give the necessary theory. Why do colours appear in a thin film in white light?
2. Discuss the formation of colours in thin films and show that with monochromatic light, the interference patterns of reflected and transmitted light are complementary.
3. (a) Explain why different colours are exhibited by thin films in white light. When seen by reflected light, why an excessively thin film appears to be perfectly black? Explain this perfect blackness with the help of Stoke's treatments.
   (b) With suitable diagrams explain why broad source of light is needed to observe the above phenomenon.
   (c) Why a thick film seen by reflected light shows no colours but appears white?
4. Explain the formation of Newton's rings. Show how you would use them to find (a) refractive index of a liquid and (b) radius of curvature of a lens.
5. (a) Describe and explain the formation of Newton's rings in reflected light. Prove that in reflected light (i) diameters of dark rings are proportional to the square roots of natural numbers and (ii) diameters of bright rings are proportional to the square roots of odd numbers.
   (b) Account for the perfect blackness of the central spot in Newton's rings.
   (c) What will happen if (i) white light is used and (ii) the lens is lifted slowly off the flat?
6. Describe Newton's rings method for measuring the wavelength of light and give the necessary theory. What will happen if a little water is introduced between the lens and the plate?
7. Explain with necessary theory how you can determine the refractive index of a liquid by means of Newton's rings. Derive the formula used. Why are the rings circular?
8. Describe the construction of a Michelson's interferometer and explain its working.
9. Outline the theory of Michelson's interferometer and discuss the nature of the interference pattern produced. Under what conditions will it give circular fringes? Show how it may be used to measure the wavelength of light.
10. (a) How can Michelson's interferometer be used for measuring the wavelength of light?
    (b) How will you use Michelson's inteferometer to determine the difference between two wavelengths very close together?

the object itself. A distant object (not shown) subtends an angle θ at the objective and would subtend essentially the same angle at the unaided eye. Also, since the pupil of the eye is at the exit pupil, the angle subtended at the eye by the final magnified image is θ'. Accordingly the angular magnifying power is

$$\gamma = \frac{\tan\theta'}{\tan\theta} \quad \text{...(5.6 } a\text{)}$$

Let the height of the real image $P'Q'$ at F_o' be denoted by y', which according to our sign convention is negative. From the right triangle $O_o\ F_o'Q'$, since θ and y' are negative and $O_oF_o' = f_o'$ is positive, we get

$$\tan\theta' = y'/f_o'$$

From the right triangle $O_e\ F_eQ'$, since y' and f_e are negative and θ' is positive, we get

$$\tan\theta' = y'/f_e = -y'/f_e'$$

Since $f_e = -f_e'$ the medium being air on both sides of the eye-piece. Substitution in Eqn. (5.6 *a*) yields

$$\gamma = \frac{\tan\theta'}{\tan\theta} = \frac{-y'/f_e'}{y'/f_e'} = -\frac{f_o'}{f_e'} \quad \text{...(5.6 } b\text{)}$$

where it will be recalled that f_o' and f_e' are the second principal focal lengths of the objective and the eye-piece respectively. From Eqn. (5.6 *b*) it follows that a *short focal length eye-piece* and a *long focal length objective* are essential to impart a large angular magnifying power to the instrument. Since f_o' and f_e' are positive, hence γ is negative. Hence, the final image is inverted with respect to the object under examination.

We may also express the angular magnifying power γ in terms of the diameter of the entrance pupil (objective) and that of the exit pupil of the instrument. Applying the lens equation,

$$1/v - 1/u = 1/f'$$

to the image formation of the objective (the entrance pupil) by the eye-piece, we have

$$u/v = 1 + u/f_e' \quad \text{...(5.6 } c\text{)}$$

Also, if D and d represent the diameters of the objective (entrance pupil) and its image (exit pupil), we have from similar triangles $O'_1O_eO'_2$ and $O_1O_eO_2$,

$$u/v = D/d \quad \text{...(5.6 } d\text{)}$$

From Eqns. (5.6 *c*) and (5.6 *d*), it follows that

$$D/d = 1 + u/f_e'$$

According to sign convention u is negative and its magnitude is

$$u = f_o' - f_e = f_o' + f_e'$$

Substituting the value of *u* with negative sign in the preceding equation, we have

$$D/d = 1 - \frac{f_o' + f_e'}{f_e'} = -f_o'/f_e' \qquad ...(5.6\ e)$$

As a consequence of Eqns. (5.6 *b*) and (5.6 *e*), we get

$$\gamma = D/d \qquad ...(5.6\ f)$$

as an alternative expression for the angular magnifying power of a telescope. Equation (5.6 *f*) offers a convenient method of determining the angular magnifying power simply by measuring the diameters of the entrance pupil (objective) and of the exit-pupil, the image of the objective formed by the eye-piece. When the diameter, *d*, of the exit pupil becomes equal to the diameter, d_E, of the eye pupil, the magnification is called the *normal magnification* and is denoted by γ_N.

$$\gamma_N = D/d_E \qquad ...(5.6\ g)$$

In order to make full use of the available resolving power of the objective the overall magnification must be the normal angular magnification. This statement will be established in the chapter on *Resolving Power*.

5.7 OCULARS OR EYE-PIECES

An *Ocular* or *eye-piece* is employed to magnify an image formed by a lens or lenses preceding it in any optical instrument. For example, a microscope or telescope objective forms a real image of an object under examination, and an ocular functions as a simple magnifier in enlarging this image. This final magnified image, as seen by the eye placed behind the ocular, must be free from the various aberrations. It is impossible for a single lens magnifier to satisfy this requirement. Moreover, there is another disadvantage in employing a single lens as an eye-piece, arising from an entirely different cause. A real object diverges rays in all directions, whereas when an image is used as an object, rays from any point of it are simply confined to a narrow cone. As a consequence, each point of the image is seen through a single lens, only by a limited cone of rays. The full image can be, therefore, seen only when all such cones simultaneously enter the pupil of the eye in any given position for it. It is quite evident from Fig. 5.7 that the whole image can be visible only when the eye is at a point *A*, which of course is at a considerable distance from the lens. However, to shield the eye from the light not coming through the telescope it is desirable to place it immediately behind the magnifier. In this position of the eye the field of view becomes limited simply due to the reason that rays from every point of the real image cannot enter the pupil of the eye.

To overcome these defects, it is usual to employ in every optical instrument a compound system of lenses, instead of a single lens, as an eye-piece. The most popular eye-pieces consists of two convergent lenses, constructed from the same glass and separated by a distance so chosen that the *spherical aberration* and the *chromatic aberrations* are minimised. For spherical aberration to be minimum, the separation between the component lenses should be

$$d = f_1' - f_2'$$

If the original series is such that each term is less than mean of two adjacent terms, the bracketed terms are all positive. Now the following inequalities must hold,

$$R_1 - \frac{1}{2}\ (R_2 + R_{n-1}) + R_n > R > \frac{1}{2}(R_1 + R_n)$$

But $R_1 = R_2$ and $R_n = R_{n-1}$, approximately. Making these substitutions in the inequalities given above, we get the equality.

$$\frac{1}{2}R_1 + \frac{1}{2}R_n = R = \frac{1}{2}R_1 + \frac{1}{2}R_n$$

Similarly, when n is even, we get

$$\frac{1}{2}R_1 - \frac{1}{2}R_n = R = \frac{1}{2}R_1 - \frac{1}{2}R_n$$

If n becomes very large, the amplitude due to nth zone, on account of obliquity factor $(1 + \cos \theta_n)$ becomes negligibly small. The resultant amplitude at P due to the whole wave, whether n is odd or even, is therefore merely ½ R_1 *i.e.*, one half of that which could be produced by the first Fresnel-zone alone.

Rectilinear Propagation of Light

Let us now apply the above theory to explain the rectilinear propagation of light. Let us suppose a plane wave-front of monochromatic light to be incident normally on a square aperture *EFGH* in a screen. Let A be the pole of the wave-front with respect to the point P at which it is desired to find the resultant intensity due to entire wave-front. The pole A is within the aperture, sufficiently far away from its edges. Suppose the wave-front is now divided into zones around A. By the time these intersect the edges their number will be fairly large and all the effective zones will be exposed. The resultant amplitude at P, according to above theory, will be ½ R_1, that is, it will be just the same as if the screen with aperture removed. In the same way if the point P_3 is so far inside the geometrical shadow of the aperture that the pole A_3 corresponding to it is outside the aperture, well far off from its edges, the effect at P_3 reduces to that of one half of the Fresnel first zone around A_3 which is obviously zero. The point P_3, will be just as dark as if light travelled strictly in straight lines. It is only when the pole is in the region enclosed by dotted squares, the above theory does not apply. This is simply due to the reason that for points like P_1 and P_2, only some of the effective zones are exposed and the rest blocked. Therefore, Fresnel's theory becomes inadequate in deciding the intensity at these points. The wavelength of light being extremely small, the sides of the two dotted squares lie extremely close to the edges of the aperture.

We have thus arrived at the law of rectilinear propagation for all points whose poles lie within the dotted square $E_1F_1G_1H_1$ and for all points whose poles lie outside $E_2F_2G_2H_2$. The law fails in the case of those points whose poles lie between the two dotted squares—in fact when light passes close to the edges of obstacles, there is some deviation from its rectilinear path. Thus the propagation of light is approximately linear.

Fresnel's Half-period Zones

In Fig. 13.2, let *ABCD* represent a plane wave-front of monochromatic light of wavelength λ, travelling from left to right. According to *Huygen's principle, every point in this plane may be regarded as the origin of the secondary wavelets and at any given instant, every one of these secondary wavelets pass through the point P*. The resultant effect at *P* due to the whole wave-front will simply be equal to the resultant of all these secondary wavelets. To find this resultant, we divide the entire wave-front into concentric zones by the following construction. We drop a perpendicular from *P* on *ABCD*. The point *O, the foot of perpendicular from P, is called the pole of the wave with respect to P*. Let $PO = b$ and λ be the wavelength of light waves. With *P* as centre and radius $b + \lambda/2$ we construct a sphere which intersects the wave-front in a circle M_1. Then

$$PM_1 = b + \lambda/2 \quad \text{and} \quad PM_1 - PO = \lambda/2$$

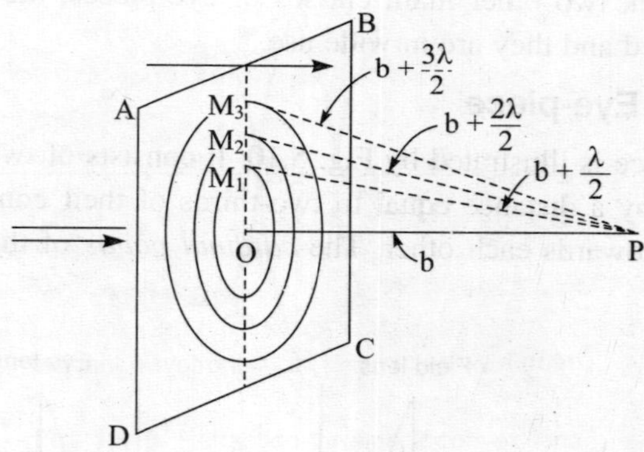

Fig. 13.2. Half-period zones on a plane wave-front

This simply means that the secondary wavelets originating from O and from the points on the circumference of the circle M_1, on reaching P simultaneously, will differ in their phase by

$$\frac{2\pi}{\lambda}(PM_1 - PO) = \pi \text{ radians}$$

But π radians is equivalent to phase difference of T/2. Consequently, the area enclosed by the circle M_1 is called Fresnel's first half-period zone. Similarly, we construct other spheres of radii

$$b + 2\lambda/2,\ b + 3\lambda/2,\ b + 4\lambda/2 \text{ etc.}$$

which intersect the wave-front in circles M_2, M_3, M_4 etc. This construction divides the entire wave-front into a number of *Fresnel's half-period zones*. These zones are so situated that the distance of the point *P* increases by $\lambda/2$ as we pass from the inner to the outer boundary of each zone. The first half-period zone, as explained above, is a circle, while the second half-period zone is an annular space between the circles M_1 and M_2 and so on.

The area A_n of nth zone, is from Fig. 13.3,

$$A_n = \pi\left(r_n^2 - r_{n-1}^2\right) = \pi\{(PM_n^2 - b^2) - (PM_{n-1}^2 - b^2)\}$$

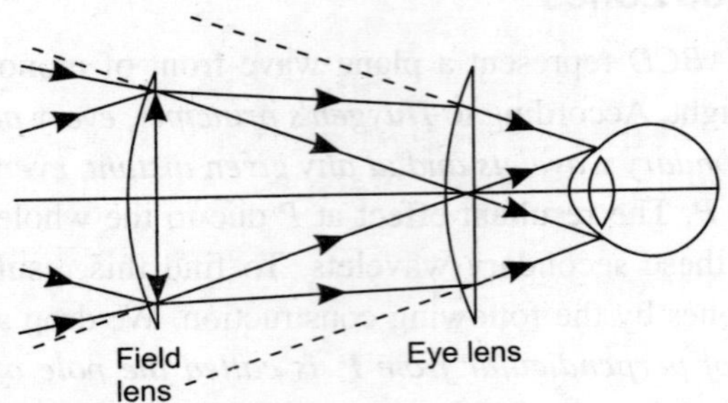

Fig. 5.9. The Kellner eye-piece

A serious disadvantage in Kellner's eye-piece lies in the circumstance that any dust particles or scratches present on the field-lens would be also magnified by the eye-lens along with the real image. To remedy this drawback two other main classes of eye-pieces, the Huygens' and the Ramsden's types, have been designed and they are in wide use.

5.7.2 The Ramsden Eye-piece

Ramsden's eye-piece is illustrated by Fig. 5.10. It consists of two plano-convex lenses of equal focal length, separated by a distance equal to two-thirds of their common focal length *f*, with their curved surfaces turned towards each other. The *cardinal points* of this eye-piece may be located as follows:

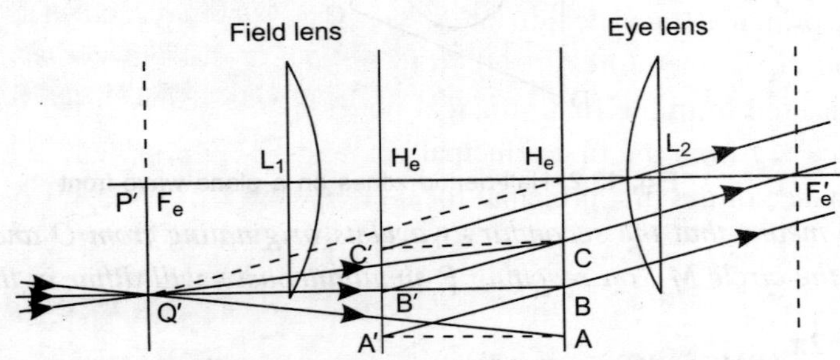

Fig. 5.10. Cardinal points of Ramsden's eye-piece

In *Sec.* 3.20, it is shown that the *optical separation* Δ between the two lenses constituting a compound lens, having same medium on either side of the component lenses is expressed by Eqn. (3.20 *f*), which is

$$\Delta = d - f'_1 - f'_2$$

where f_1' and f_2' are the second principal focal lengths of the lenses and d is the actual separation between them. In Ramsden's eye-piece,

$$d = \frac{2}{3}f; \quad f_1' = f \quad \text{and} \quad f_2' = f$$

since the second principal focal lengths f_1' and f_2', according to our sign convention, are positive. f is the numerical value of focal length.

Hence, the *optical separation* Δ of lenses in Ramsden's eye-piece is given by

$$\Delta = \frac{2}{3}f - f - f = -\frac{4}{3}f$$

The *second principal focal length* f_e' of the eye-piece, according to Eqn. (3.20 *e*) is

$$f_e' = \frac{f_1' f_2'}{\Delta} = -\frac{f \times f}{-\frac{4}{3}f} = \frac{3}{4}f$$

Since the medium is the same (air) on either sides of the component lenses, the *first principal focal length* f_e, *of the eye-piece is*

$$f_e' = -f_e' = -\frac{3}{4}f$$

The *distance of the first principal plane from the field lens* L_1 according to Eqn. (3.20 *r*), is

$$L_1 H_e = \frac{f_1 d}{\Delta} = \frac{(-f)\left(\frac{2}{3}f\right)}{\left(\frac{4}{3}-f\right)} = \frac{f}{2}$$

where f_1, the first principal focal length of L_1, according to our sign convention is negative. The first principal point H_e is, therefore, at ½ *f* from the field lens in the rear of it. The first principal focal length is measured from the first principal point. Therefore, the first principal focal point F_e is situated it a distance ¾ *f* from the first principal point and in front of it.

Also, the distance of the first principal focal point from the field lens is

$$-L_1 F_e = F_e H_e - L_1 H_e = \frac{3}{4}f - \frac{1}{2}f = \frac{1}{4}f$$

or

$$L_1 F_e = -\frac{1}{4}f$$

i.e., F_e is numerically at ¼ *f* in front of the field lens.

The *distance between the eye-lens and the second principal plane* of the eye-piece, according to Eqn. (3.20 *o*), is

$$L_2 H_e' = \frac{f_2' d}{\Delta} = \frac{(f)\left(\frac{2}{3}f\right)}{\left(-\frac{4}{3}f\right)} = -\frac{f}{2}$$

Thus, the second principal point H_e' is at ½ f in front of the eye-lens. The second principal focal length of the eye-piece is measured from the second principal plane. Therefore, the second principal focal point F_e' is situated at a distance ¾ f from the second principal point H_e' and in the rear of it. It can be easily shown that the distance of the second principal focal point from the eye-lens is ¼ f, in the rear of it.

The nodal points of the eye-piece obviously coincide with its principal points, since the medium is the same on both sides of the eye-piece. The focal points and its principal points are shown in Fig. 5.10.

We may now explain the image formation by the Ramsden eye-piece on the basis of the properties of cardinal points. In order that the final magnified image may be formed at infinity it follows from the property of the principal focal points that the objective of any instrument must form a real image in the first principal focal plane of the eye-piece. The real image $P'Q'$ is, therefore, at a distance ¼ f in front of the field lens. Three rays $Q'A$, $Q'B$ and $Q'C$ through the tip Q' of the real image are shown, in Fig. 5.10, intersecting the first principal plane at points A, B, C respectively. The path of the ray conjugate to $Q'A$ must be drawn through the point A' in the second principal plane at the same distance from the principal axis as A, because of the unit positive lateral magnification characterising the principal planes of the optical system; similar consideration also applies to rays conjugate to $Q'B$ and $Q'C$. Rays conjugate to $Q'A$, $Q'B$ and $Q'C$ must form a parallel beam since the real image $P'Q'$ which serves as an object for the eye-piece is situated in its first principal focal plane. Moreover, the direction of this parallel beam must be parallel to the line joining Q' to H_e, the first principal point, because if we imagine the ray $Q'H_e$ to exist its conjugate ray would emerge parallel to it from H_e' by virtue of H_e and H_e' being also the nodal points of the eye-piece. This explains the ray construction in Fig. 5.10.

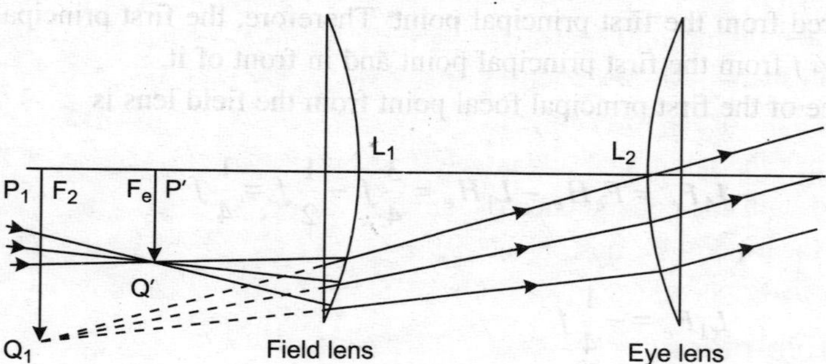

Fig. 5.11. The Ramsden eye-piece

In reality the final image is obtained by two steps. The real image $P'Q'$ formed in the first principal focal plane of the eye-piece by the objective serves as a real object for the field lens and as obtained above, $L_1 F_e = u = -$ ¼ f. In the lens equation

$$\frac{1}{v} - \frac{1}{u} = \frac{1}{f_1'}$$

if we substitute $u = -f/4$ and $f_e' = f$ it is easy to deduce $v = -f/3$. In other words, the field lens forms a virtual image P_1Q_1 of $P'Q'$ and the former is situated at $f/3$ in front of the field lens or at $\frac{2f}{3}+\frac{f}{3}=f$ in front of the eye-lens. The virtual image P_1Q_1 which lies in the first principal focal plane of the eye-lens serves as an object for it. Hence P_1Q_1 is imaged at infinity by the eye-lens.

The first principal focal point F_e, of the Ramsden eye-piece, as explained above, lies in front of the field lens. As a consequence, it is more convenient to introduce a micrometer scale or cross-wires, movable by a fine calibrated screw, in the plane in which the objective forms a real image $P'Q'$ of an object under examination. Accordingly, the cross-wires or micrometer scale will be magnified by the eye-piece to the same extent as the real image $P'Q'$. Hence it becomes practicable to take micrometer measurements of the image with sufficient accuracy with the help of this eye-piece. This is the main reason which lead to the designing of Ramsden's eye-piece.

Various aberrations in the final image are minimised to a great extent as follows: Since the final image is seen by refraction at four surfaces, the *spherical aberration* and the *coma* are reduced by providing total deviation, as far as possible, in four equal increments at the four refracting surfaces. *Spherical aberration* is further reduced by employing plano-convex lenses as the field lens and the eye-lens. The circle of least confusion due to *spherical aberration* and *astigmatism* is also very small owing to the narrow pencil of rays transmitted through the eye-piece. Finally, in order to correct for lateral chromatic aberration the eye-lens and the field lens must be separated numerically by the distance

$$d=\frac{1}{2}(f_1'+f_2')=f'=f$$

This condition of achromatism would, however, bring the field lens in the plane of real image formed by the objective. Thus the combination would be simply Kellner's eye-piece, associated with the undesirable feature of bringing into focus dust particles and scratches, if any, on the field lens. As a consequence, in Ramsden's eye-piece lenses are separated by $2/3f$ instead of f, thus sacrificing some lateral achromatism in order to eliminate the undesirable features of Kellner's eye-piece. Moreover, this sacrifice has the additional advantage of making the first principal focal plane of the eye-piece *real*, associated with the convenience of introducing cross-wires in this focal plane.

The first principal focal point F_e lies on the negative side of field lens, hence this eye-piece is often termed as *Negative Eye-piece*.

5.7.3 The Huygens Eye-piece

The kind of eye-piece which enjoys the widest use is the one designed by *Huygens* in order to improve the corrections for the spherical aberration as well as for the chromatic aberrations present in the Ramsden eye-piece. This eye-piece is constructed of two plano-convex lenses, made of the same material, the focal length of the field-lens being three times that of the eye-lens. The two lenses are co-axially mounted with their curved faces turned towards the incident light and separated by a distance of $2f$, where f is the focal length of the eye-lens. The combination of lenses in this eye-piece satisfies the condition of achromatism, viz.,

$$d=\frac{1}{2}(f_1'+f_2')=\frac{1}{2}(3f+f)=2f$$

as well as it satisfies that condition of minimum spherical aberration, viz.

$$d = f_1' - f_2' = 3f - f = 2f$$

The *cardinal points* of the Huygens eye-piece may be located as follows:

The *optical separation* between the field lens and the eye-lens can be obtained by Eqn. (3.20 *f*), which is

$$\Delta = d - f_1' - f_2'$$

by substituting in it $d = 2f$, $f_1' = 3f$ and $f_2' = f$; according to our sign convention the second principal focal length of the convergent lens is positive. Accordingly, the *optical separation* Δ of lenses in Huygen's eye-piece is

$$\Delta = 2f - 3f - f = -2f$$

The *second principal focal length* of the eye-piece, according to Eqn. (3.20 *e*), is

$$f_e' = -\frac{f_1' f_2'}{\Delta} = \frac{(3f)(f)}{(-2f)} = \frac{3}{2}f$$

Since the medium is the same (air) on either side of the component lenses, *the first principal focal length* f_e of the eye-piece is

$$f_e = -f_e' = -\frac{3}{2}f$$

The *distance of the first principal plane from the field lens* L_1, according to Eqn. (3.20 *r*), is

$$L_1H_e = \frac{f_1 d}{\Delta} = \frac{(-3f)(2f)}{(-2f)} = 3f$$

Thus the first principal point H_e is at $3f$ from the field-lens and in the rear of it. The separation between the field-lens and the eye-lens is $2f$, hence H_e lies at a distance f from the eye-lens in the rear of it. The first principal focal length is measured from the first principal point. Therefore, the first principal focal point F_e is situated at a distance $3/2f$ from H_e and in front of it or at a distance $1/2f$ in front of the eye-lens. Thus the first principal focal point F_e lies in between the two lenses.

The *distance between the eye-lens and the second principal plane* of the eye-piece, according to Eqn. (3.20 *o*) , is

$$L_2H_e' = \frac{f_2' d}{\Delta} = \frac{(f)(2f)}{(-2f)} = -f$$

Thus, the second principal point H_e' is at f in front of the eye-lens. Therefore, the second principal focal point F_e' lies at a distance

$$L_2F_e' = H_e'F_e' - H_e'L_2 = \frac{3}{2}f - f = \frac{1}{2}f$$

from the eye-lens, that is, F_e' is at $1/2f$ in the rear of the eye-lens.

The nodal points of the eye-piece obviously coincide with the corresponding principal points, since the medium is the same on both sides of the eye-piece. The principal points and the focal points of this eye-piece are shown in Fig. 5.12.

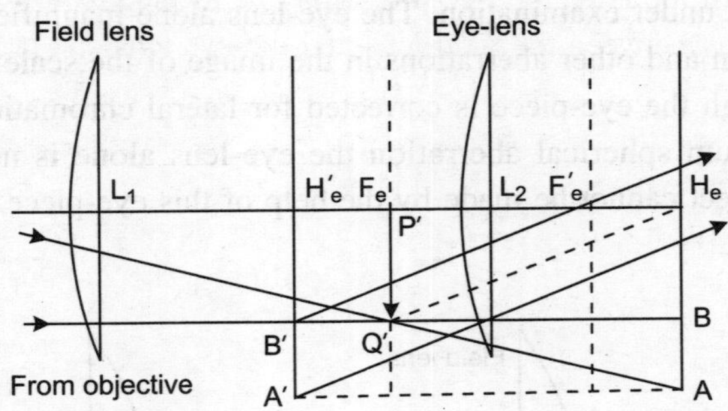

Fig. 5.12. Cardinal points of Huygens eye-piece

We may now explain the image formation by this eye-piece, utilising the properties of the cardinal points of the optical system. The first principal focal point F_e as explained above lies in between the two lenses at $f/2$ in front of the eye-lens. In order that the final magnified image may be formed at infinity by the eye-piece, it follows from the property of the principal focal planes that the objective of an instrument must form an image $P'Q'$ in the first principal focal plane of the eye-piece. Two rays $Q'B$ and $Q'A$ are shown through the tip Q' of the image $P'Q'$; $Q'B$ is parallel to the principal axis of the lens system and intersects the first principal plane at B while the second ray intersects this plane at A. By virtue of the positive unit lateral magnification, the property characterising the principal planes, the ray conjugate to $Q'A$ is drawn through a point A' in the second principal plane so that $H_eA = H_e'A'$, and a similar consideration applies to the ray conjugate to $Q'B$. Rays emerging from the eye-piece must form a parallel beam, since $P'Q'$ being in the first principal focal plane it must be imaged at infinity by the eye-piece. The direction of this beam must be parallel to the line joining Q' to H_e, the first principal point of the eye-piece, because if we imagine the ray $Q'H_e$ to exist, its conjugate ray would emerge parallel to it through H_e' by virtue of H_e and H_e' being also the nodal points of the eye-piece. This explains the ray construction shown in Fig. 5.12.

In reality the final image is obtained by two steps. The image $P'Q'$ which the objective would form, if the field lens were absent, in the first principal focal plane of the eye-piece, serves as a virtual object for the field lens and according to our sign convention, $L_1P' = u = 3f/2$, since F_e is in the rear of the field-lens. In the lens equation

$$\frac{1}{v} - \frac{1}{u} = \frac{1}{f_1'}$$

if we substitute $u = 3f/2$, $f_1' = +3f$, it is easy to deduce $v = f$. In other words, the field lens forms, at f in the rear of it, a real image P_1Q_1 of $P'Q'$. Since the separation between the field-lens and eye-lens is $2f$, therefore, the real image P_1Q_1 is situated in the first principal focal plane of the eye-lens.

Consequently P_1Q_1 is imaged at infinity by the eye-lens. A field stop should be placed in the plane of the image P_1Q_1 to bring into effect a well defined final image.

In order to carry out the measurements of the image, a micrometer scale or cross-wires should be employed between the eye-lens and the field-lens and it must be mounted in the plane of the real image P_1Q_1 of the object under examination. The eye-lens alone magnifies the scale, so there would be present some distortion and other aberrations in the image of the scale or cross wires. This is due to the reason that although the eye-piece is corrected for lateral chromatic aberration and it satisfies the condition for minimum spherical aberration the eye-lens alone is not corrected. The accurate measurements of any object cannot be made by the help of this eye-piece.

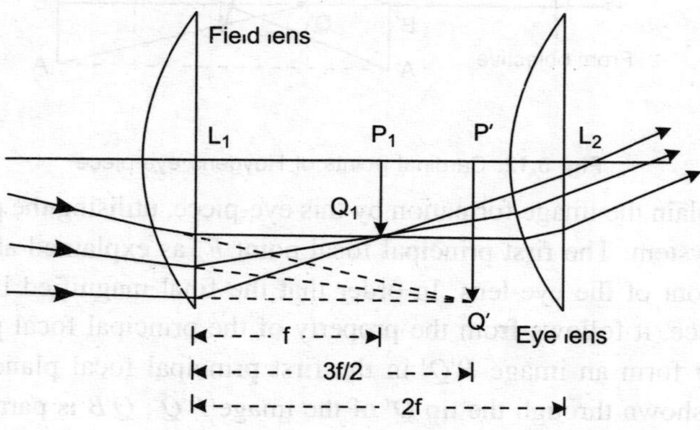

Fig. 5.13. The Huygens eye-piece

The Huygens eye-piece exhibits some spherical aberration, astigmatism and curvature of field but is free from lateral chromatic aberration, although longitudinal chromatic aberration is not completely absent. The type of achromatism accomplished in this eye-piece is described in *Sec.* 4.10. The image is of pincushion distortion type.

The first principal focal point F_e lies on the *positive side* of field lens, hence this eye-piece is often termed as *Positive Eye-piece.*

5.7.4 Comparison of Ramsden's Eye-piece and Huygens' Eye-piece

The chief advantage of Ramsden's eye-piece over Huygens' type lies in the fact that in the former the first principal focal plane lies in front of the field lens while in the latter the principal focal plane lies between the two lenses. As a consequence of this, Ramsden's eye-piece can be used to carry out accurate measurements of an image with the help of a micrometer scale or movable cross-wires. Moreover, it can also be employed as a simple magnifier to examine real objects, while Huygens' eye-piece can only examine images. On the other hand, the chief advantage of Huygens' eye-piece is the elimination of the lateral chromatic aberration but other aberrations are more satisfactorily reduced in Ramsden's eye-piece. Of the two types, Huygens, eye-piece is chosen where the object to be examined is illuminated by white light and the residual lateral chromatic aberration, if Ramsden's eye-piece were used, would be objectionable, which is, however, specially the case while examining biological slides. Thus Huygens' eye-piece is invariably employed in microscopes used in biological work. In telescopes fitted in spectrometers and in other kinds of spectroscopes Ramsden's eye-piece is employed. Since only one wavelength region is viewed at a time in such instruments hence the residual lateral chromatic aberration becomes unimportant. Huygens' eye-piece also gives slightly higher field than given by Ramsden's type.

In the case of Ramsden's eye-piece the final magnified image is almost flat but so far as Huygens' eye-piece is concerned it is convex towards the eye.

5.7.5 The Gauss Eye-piece

In a telescope used in spectrometer it is sometimes convenient to employ the Gauss eye-piece. This eye-piece is illustrated by Fig. 5.14. It is merely a Ramsden's eye-piece except that between the two lenses a thin transparent glass plate is mounted at 45° to the axis of the tube and there is an opening in the side of the tube carrying the lenses through which light can be admitted within the eye-piece. This eye-piece is used to set the optical axis of the spectrometer telescope at right angles to its axis of rotation in the spectrometer.

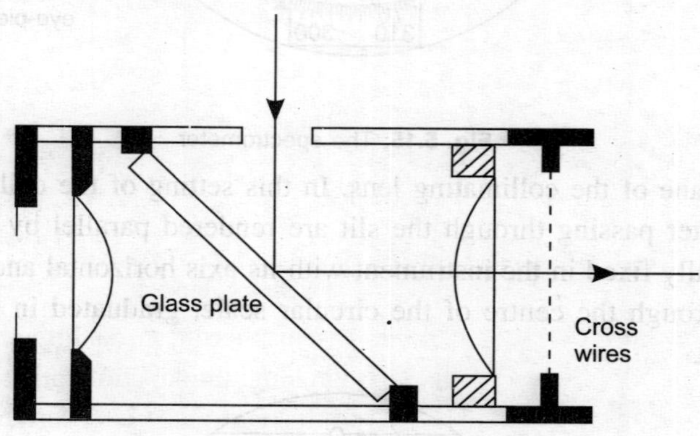

Fig. 5.14. The Gauss eye-piece

5.8 THE SPECTROMETER

The spectrometer is one of the most important optical instruments. It is usually employed for the study of spectrum produced by the transmission of light through dispersion producing devices like a prism or a grating. Quantitatively, it is always employed for the measurement of an angle, for example, the angle of prism, the angle of minimum deviation and the angle of diffraction of light due to transmission of light through a plane diffraction grating.

The essential parts of the spectrometer, illustrated by Fig. 5.15, are

(*i*) A *Collimating device* or collimator which serves the purpose of rendering parallel the rays of light to be examined by the spectrometer.

(*ii*) A *turn table* on which a dispersion producing device, like a prism or a grating, can be conveniently mounted.

(*iii*) A *telescope* for examining the spectrum.

The collimator simply employs a well constructed achromatic lens at one end of a tube, and sliding within it at its other end is another tube carrying a vertical slit of adjustable width. The relative position of the two tubes can be adjusted by a rack and pinion arrangement and in this way the slit can be adjusted in the

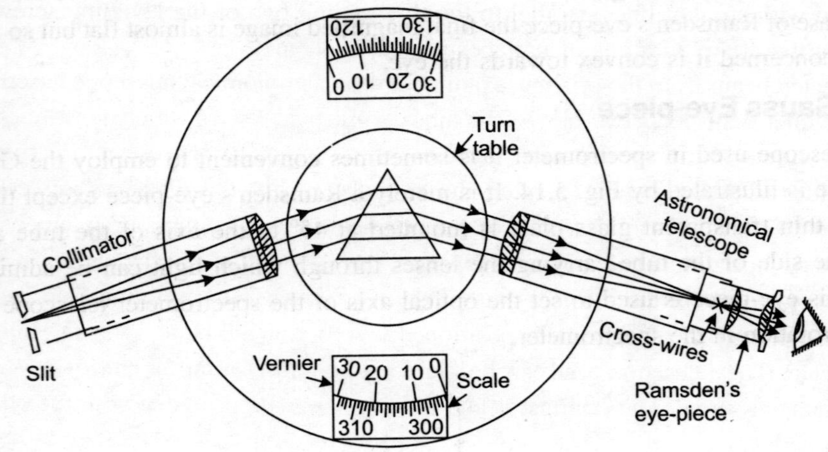

Fig. 5.15. The spectrometer

first principal focal plane of the collimating lens. In this setting of the collimator, the rays of light, under examination, after passing through the slit are rendered parallel by the collimator lens. The collimator tube is usually fixed in the instrument with its axis horizontal and intersecting the vertical axis, which passes through the centre of the circular scale, graduated in half-degree, fixed at the base of the instrument.

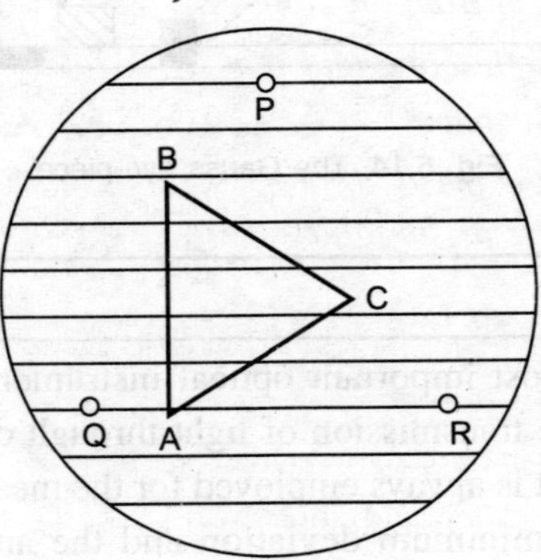

Fig. 5.16. Turn table

The parallel beam of light emerging from the collimator falls on the prism resting on a table, which can be rotated, independently of the collimator and the telescope, about a vertical axis which passes through its centre as well as through the centre of the circular scale. The rotation of the table can be read by the help of the diametrically positioned verniers sliding over the circular scale. The turn table can also be adjusted to any desired height and clamped. The slow rotation can be now given with a tangent screw provided at the base of the instrument. The table is provided with three screws P, Q, and R, which form the three corners of an equilateral triangle. By the help of these screws the plane of the table can be made as nearly horizontal as possible. The surface of the table is usually ruled with straight lines parallel to the line joining two of the levelling screws. These lines assist in setting the prism with one of its faces normal to them.

The refracted beam from the prism is examined by an astronomical telescope described in *Sec.* 5.6 and provided with a Ramsden's eye-piece. The tube carrying the lenses which constitute the Ramsden's eye-piece can slide within the tube carrying the cross-wires. The two tubes are fixed to the third tube which slides within the tube carrying the objective. Thus the distance between the plane of cross-wires and the objective can be varied by the help of a rack and pinion arrangement. The telescope tube with its axis horizontal and pointing towards the axis of the spectrometer, is fixed in a arm which can be rotated about the vertical axis of the instrument. This rotation can also be read with the help of another pair of diametrically positioned verniers which slide, with the telescope arm, on the circular scale. The telescope arm can be clamped and slow rotation can be accomplished with the help of a tangent screw. The collimator and the telescope are provided with the levelling, as well as the locking screws.

Adjustment of the Spectrometer

Before using the spectrometer for any experiment the following adjustments must be made in the instrument.

(*a*) *The axis of the telescope and that of the collimator must intersect the principal vertical axis of rotation of the telescope.*

(*b*) *The telescope must be focussed for parallel rays and the collimator must be adjusted for rendering the rays of light from the illuminated slit parallel.*

(*c*) *The optical axis of the telescope and that of the collimator must be perpendicular to the principal axis of the instrument.*

(*d*) *The refracting edge of the prism, if one is employed, must be made parallel to the vertical axis of the rotation of the prism table and the telescope.*

PROBLEMS

1. What is an eye-piece? What are its advantages over a single lens?
   Describe Huygens' eye-piece and show why it is free from both spherical and chromatic aberrations.
2. An ocular consists of two identical positive thin lenses, each of focal length 2 in. separated by a distance of one inch. Determine the positions of the focal points and principal points of the combination.
3. The lenses in a Huygens' eye-piece have focal lengths of 2 and 4 cm. Find the distance between them and locate the cardinal points.

4. Describe the construction of higher-power microscope and discuss the relation between magnifying and resolving powers.
5. The focal length of the more convergent lens of a Huygens' eye-piece is 0.5 cm. Calculate the focal length of the eye-piece and locate on a diagram the positions of its focal points.
6. Prove that when two lenses are separated by the algebraic sum of their focal lengths, the linear magnification is independent of the position of the object, and is equal to the reciprocal of the angular magnification for the case of an object at infinity.
7. A telescope with an objective of focal length F and an eye-piece of focal length f is focussed on an object at a finite distance d from the objective. Calculate its magnifying power and show that it varies upto the limiting value.
8. A distant star is seen with relaxed accommodation through an astronomical telescope fitted with Huygens' eye-piece. Draw a neat diagram of the path of the rays from star to the eye. Use the diagram to explain that the eye-piece is *negative*.
   [N.B.—According to our sign convention it is *positive*.]
9. Mention the characteristic features of the Ramsden's eye-piece. Draw a neat diagram of the path of rays through telescope fitted with this eye-piece when the telescope is focussed with relaxed accommodation on a distant point source of light on its axis.
10. Write notes on:
    (a) Oil immersion objectives
    (b) Eye-pieces
    (c) Huygens' eye-piece and its achromatism
    (d) Ultra-microscope
    (e) Ramsden's eye-piece
    (f) Numerical aperture
11. Explain the function of an eye-piece in an optical instrument. Explain the construction and theory of Ramsden and Huygens' eye-pieces. Discuss their respective merits.
12. Describe fully the construction and working of Huygens' eye-piece.
    Show that this eye-piece is achromatic in the same sense as an ordinary magnifier. Why is Huygens' eye-piece also called negative eye-piece?
    [N.B.—According to our sign convention, it is positive]

NATURE OF LIGHT

6

Chapter

In the preceding chapters we have studied the image forming property of lenses and mirrors, the aberrations of optical images and the action of many optical instruments. We have based our arguments on the theory that light is propagated accurately in straight lines, without, however, concerning ourselves with the actual nature of light. After this long study, the question would naturally arise to an inquisitive reader as to what is light. The answer to this question and experimental evidences in support of it constitute what is called *Physical Optics*.

Speculations about the true nature of light go by centuries. The *Pythagoreans* believed light to consist of minute particles shot out at high speed from luminous objects. *Plato* and his co-thinkers regarded light as some kind of emanations from the eye by means of which objects are scanned and made visible when struck by it. *Aristotle* through of light as something non-material occurring in the space intervening the eye and the object examined.

Light is really a form of energy and so it is regarded as the transfer of energy from the luminous source to the eye, either directly or through the object seen. Energy can be transferred from one point to the other either by means of wave disturbance travelling through the intervening medium or by the motion of material particles between the two points. Accordingly, two different theories as to the nature of light were brought forward in the seventeenth centurey, *i.e., the wave* (*or undulatory*) *theory* and *the emission* (*or corpuscular*) *theory*. *Huygens*', *Hooke* and *Descartes* championed the wave theory. The corpuscular theory was upheld by Huygens' great contemporary *Newton* although *Young*, who in 1801 developed further the wave theory, gave credit of his ideas to Newton's writings which contained some basic notions pointing towards the wave picture of light.

6.1 THE CORPUSCULAR THEORY OF LIGHT

In 1675 Newton sent to the Royal Society a paper in which he presented the fundamental postulates of his corpuscular (emission) theory of the Nature of Light. According to this theory, *light consists of a stream of minute invisible particles, called corpuscles, moving at a great speed. These corpuscles are emitted in straight lines from a luminous source and their mechanical impact on retina stimulates the sensation of vision. Different colours were ascribed to different sized corpuscles*. Newton also postulated "an ethereal medium much of the same constitution with air but far rarer, subtler and more strongly elastic." Ether "pervades the pores of all material bodies but with a greater degree of rarity in these pores than in the free ethereal spaces." When corpuscles "impinge on any refracting or reflecting superficies (a surface layer), must as necessarily excite vibrations in the ether, as stones do in water when thrown into it." Nevertheless, Newton rejected the theory that vibrations of the ether should be regarded in themselves to constitute light, since it appeared to him to conflict with the fact of rectilinear propagation of light. He arrived at these conclusions from considerations which are best expressed in his own words.

"If it (light) consisted in pression (pressure in ether) or motion, propagated either in an instant or in time, it would bend into the shadow......... But light is never known to follow crooked passages nor to bend into the shadows. For the fixed stars by the interposition of any of the planets cease to be seen......... The rays of light are very small bodies emitted every way from shining substances. For such bodies will pass through uniform mediums in right lines without bending into the shadows, which is the nature of rays of light........ Pellucid (transparent) substances act upon the rays of light at a distance in refracting, reflecting, and inflecting them, and the rays mutually agitate the parts of those substances at a distance for heating them; and this action and reaction at a distance very much resemble an attractive force between bodies. If refraction be performed by attraction of the rays, the sines of incidence must be to the sines of refraction in a given proportion........"

By effecting the prismatic decomposition of white light into its constituent colours, Newton recognised that *colour* is an important characteristic of light and it must be associated with definite quality of the corpuscles or the ether vibrations. He writes "......... the most refrangible rays excite the shortest vibrations for making a sensation of deep violet, the least refrangible the largest for making a sensation of deep red, and the several intermediate sorts of rays vibrations of intermediate bigness to make sensation of the several intermediate colours." This is the first statement of the principle that monochromatic light is essentially periodic in its nature and that different periods correspond to different colours. It appears that Newton had also in his mind some form of wave theory regarding the nature of light but he was not committed to it, as he never computed the wavelength of ether vibrations.

In the latter part of his life Newton enunciated the Law of Gravitation which involves no reference to ether hypothesis and he also developed the mechanics of motion of heavenly bodies and concluded that the ether hypothesis was totally superfluous, for he writes "Against filling the Heaven with fluid medium, unless they be exceedingly rare, a great objection arises from the regular and very lasting motion of the Planets and Comets. For thence it is manifest, that Heavens are void of all sensible resistance, and by consequence of all sensible matter." Thus, in the latter part of his life, he was definitely inclined to a purely corpuscular conception as to the nature of light.

6.2 WAVE THEORY OF LIGHT

In 1690, a Dutch Physicist *Christian Huygens* published a *Treatise on Light* in which he presented a theory, which was first developed in a definite form in 1678, bringing out clearly the necessity of wave conception, instead of corpuscular conception, regarding the nature of light. According to his ideas a beam of light consists of a large number of *longitudinal pulses* which are propagated as condensations and rarefactions, like sound waves, from the point of origin to the point of observation of light. He totally rejected the theory that light consists of the motion of small corpuscles, emitted from the luminous sources.

Huygens arrived at these conclusions from the following considerations which are best expressed in his own words: "When we consider........ the very great speed with which light is propagated in all directions, and the fact that when rays come from different directions, even those directly opposite, they cross without disturbing each other, it must be evident that we do not see luminous objects by means of matter translated from the object to us, as a shot or an arrow travels through the air. For certainly this would be in contradiction to the two properties of light which we have just mentioned and specially to the latter. Light is then propagated in some other manner, like that in which sound travels through the air .... Now there can be no doubt that light also comes from the luminous body to us by means of some motion impressed upon the matter which lies in the intervening space: for we have already seen that this cannot occur through the translation of matter from one point to other. If, in addition, light requires time for its passage – a point we shall presently consider – it will then follow that this motion is impressed upon the matter gradually, and hence is propagated, as that of sound, by surfaces and spherical waves."

For the propagation of every wave we require some medium as a *vehicle*. As, however, the propagation of light occurs also in vacuum, Huygens postulated as vehicle for these waves an omnipresent, all pervading, universal and continuous medium called *luminiferous ether*, which according to him was composed structurally of tiny elastic particles. According to Huygens, every luminous body is a source of disturbance in ether particles whose mutual impacts transmitted from one particle to next in continuous succession are propagated with tremendous velocity to the point of observation. Thus he thought of light as consisting of longitudinal waves in ether. We must emphasise here that the conception of a transverse wave motion, which is the true nature of light waves had not then occurred to Huygens. Different colours of light are attributed to light waves of different wavelengths.

6.3 CORPUSCULAR THEORY VERSUS WAVE THEORY

Obviously, the test of the adequacy of any theory consists in its ability to account for the known experimental facts with a minimum of postulates. In this light we must admit that the corpuscular theory triumphs above all prejudices as it can explain many experimental facts like rectilinear propagation, reflection and refraction at plane boundary separating two media, of course, with a minimum of hypothesis.

Twelve years after the presentation of his wave theory, *Huygens* was also successful in explaining reflection and refraction and he arrived at the conclusion theoretically that the speed of light should be greater in an optically rarer medium, whereas the theory of *Newton* gave the opposite result. Obviously, a measurement of the speed of light in water and air would decide the nature of light. Huygens also gave a suitable explanation of the phenomenon of double refraction, discovered in 1669 by *Erasmus Bertholinus* in uniaxial crystals like calcite, by attributing to ether present in the surface layer of uniaxial crystals the property of sending out two waves propagated with different speeds, within the crystal, under the influence of incident light waves. The corpuscular theory was found to be totally inadequate to explain double refraction. But Huygens' discovery, in 1690, of the phenomenon of *Polarisation of Light* namely that the two beams of light emerging from calcite crystal have sides, *i.e.* they are not perfectly symmetrical about the direction of propagation, presented an unsurmountable difficulty in respect of the then prevalent nature of light waves. Of course, it is inconceivable for a longitudinal wave, which is perfectly symmetrical about the direction of propagation (due to vibration of the medium along the direction of the propagation) to acquire one sidedness or a symmetry on transmission through a calcite crystal. Besides this Huygens could not give any satisfactory explanation of the rectilinear propagation of light as another phenomenon called *Interference of Light* had not yet been discovered. Polarisation was first interpreted by Newton on the supposition that the corpuscles were not perfectly spherical. Accordingly, they would present different sides depending on their orientation to the direction of propagation. As a consequence, the new wave theory could not receive immediate acceptance. In fact, Newton argued that let us compare the properties of light with those of sound, which is definitely a wave motion and has the property of bending round corners and other obstacles in its path. This sideway spreading of the waves or their bending round the corners is called diffraction. But up to 1821 no significant diffraction effects were observed with light, although it should be pointed out that the phenomenon of diffraction of light was first observed by *Grimaldi* as early as 1665. But the true significance of his observations was not realised at that time, presumably due to our inability to repeat his observations. It should be emphasised that diffraction effects arise only when the dimensions of the obstacle become comparable with the wavelength of the wave but in Huygens time smallness of the wavelength of light was not known. *Newton*, therefore, argued: "I know sound is waves, how then can light be waves if it is so different from sound in this matter of diffraction." Later, Newton also succeeded in observing diffraction fringes outside the geometrical shadow of the obstacle in the path of light. He argued that it arises as due to deflection of light corpuscles from their straight path in passing the region of variable ether density close to the edges of opaque body. Had he not fatefully overlooked the diffraction fringes within the geometrical shadow, first noticed by Grimaldi, they would have set him on the right path. Furthermore, Newton considered the rectilinear propagation as an essential property of light and since he was successful in explaining it by the emission theory, he criticised the wave theory on this ground also. His views were, unfortunately, interpreted by his followers with a much narrower significance than he intended to attach to them and they rejected the wave theory in favour of the corpuscular theory.

Nevertheless, by the end of 18th century, the theories of light propounded by Newton and Huygens had acquired following in the scientific world, although the controversy was still there whether a corpuscular or a wave picture best fitted in the experimental facts. Then came the discovery of the phenomenon of *Interference of Light* in 1801 by *Thomas Young* (1773–1829). He championed the wave theory to explain the observed uneven distribution of light on a screen placed in front of another screen with two extremely close pinholes, when the sunlight was allowed to fall on these holes after traversing through another pinhole. The corpuscular theory was found to be totally inadequate to explain the observed uneven distribution of light because on this conception, the intersection of two rays at a point could only produce an increased brightness equal to twice the intensity of each one separately but never a partial or complete darkening. In fact, the observed intensity in bright regions was found to be four times the intensity caused by each beam acting separately. According to the wave theory, since the intensity at any point is proportional to the square of the amplitude of the light wave, the addition of two light waves of equal amplitude, originating from the same source as found in Young's experiment could produce under suitable conditions (arriving in equal phase) a wave of twice the amplitude and hence the intensity would be four times that due to component waves alone. In Young's own words, "When two undulations from different origins coincide........ their joint effect is a combination of motion belonging to each." Young's experiment was thus a crucial one at that time because it added a further evidence to the growing belief in the wave nature of light. Besides, he also explained colours of thin films, a phenomenon discovered by Newton himself by attributing it to the interference between light waves reflected from the upper and lower surfaces of thin films.

Owing to Newton's great authority, even Young's experiment and his explanation of the interference of light could not lead to the triumph of the wave theory. It was, however, left to the brilliant contribution of young French engineer *Augustin Fresnel* (1788–1827), to bring about the general conversion of the scientific world to the wave nature of light. He employed Huygen's wave theory together with the principles of interference of light waves and succeeded not only in giving a satisfactory quantitative explanation of uneven distribution of light in all diffraction phenomenon then known to scientist but above all he also explained rectilinear propagation (approximate) of light thereby refuting one of the chief arguments against the wave nature of light.

Although Fresnel's explanation of interference and diffraction effects paved the way for the success of the wave theory still there were scientists who were hesitating to accept this theory, chiefly on account of its inability to account for the phenomenon of polarisation of light waves on the prevalent nature (viz. longitudinal) of light waves. Credit again goes to Fresnel, who, in 1816, after studying the interference of polarised light, put forward the highly revolutionary hypothesis that light must be regarded as a transverse wave motion in ether analogous to those in elastic solid, in contradiction to Huygens' conception of longitudinal nature of light waves In the case of transverse waves vibrations of the medium are executed at right angles to the direction of propagation but if the vibrations be confined to one place then this kind of wave would definitely exhibit one sidedness. Thus the phenomenon of polarisation of light could be easily explained on Fresnel's hypothesis. No doubt, this revolutionary hypothesis proved to be a great triumph for the wave theory over the corpuscular theory.

The death below to the corpuscular theory and the universal acceptance of wave theory were however, achieved simultaneously when *Leon Foucault*, in 1850, showed experimentally that the speed of light is less in water than in the outside atmosphere.

The wave theory of Huygens, although radically modified by Fresnel, finally triumphed over the rival emission theory of Newton.

6.4 ELECTROMAGNETIC THEORY OF LIGHT

Huygens and Fresnel regarded light as simply the propagation of mechanical energy. A completely different conception, regarding the nature of light waves developed in 1873 chiefly due to the theoretical researches of a Scotch Scientist, *James Clerk Maxwell*, Professor of Physics in Cambridge University. He put forward the hypothesis that light must be thought essentially as on electromagnetic wave propagation so that a beam of light propagates not the mechanical but electromagnetic energy.

Maxwell explained by theoretical reasoning that the oscillatory electric circuit must produce, in the medium occupying the space, the periodic variations of electric and magnetic intensities having all the characteristics of transverse waves. The vectors representing are electric and magnetic field strengths are mutually perpendicular and also at right angles to the direction of propagation of waves.

Furthermore, he deduced that the velocity of propagation of electromagnetic waves through this medium (electric ether) is numerically equal to the number of electrostatic units of electric charge which are equivalent to one electromagnetic unit of charge. Light travels as a wave motion in luminiferous ether. To the inquisitive mind of Maxwell, the question naturally arose whether light waves and electromagnetic waves are identical? The problem and method of solution were stated by Maxwell himself in his *Treatise on Electricity and Magnetism, as follows*:

"In several parts of this treatise an attempt has been made to explain electromagnetic phenomenon by means of mechanical action transmitted from one body to another by means of medium occupying the space between them. The undulatory theory of light also assumes the existence of a medium. We have now to show that the properties of the electromagnetic medium are identical with those of the luminiferous ether…………

But the properties of bodies are capable of quantitative measurement. We therefore obtain the numerical value of some property propagated through it which can be calculated from electromagnetic experiments, and also directly in the case of light. If it should be found that the velocity of propagation of electromagnetic disturbance is the same as the velocity of light ……….. we shall have strong reasons for believing that light is an electromagnetic phenomenon."

Maxwell's predictions were remarkably confirmed in 1888 by Helmholtz's brilliant pupil *Heinrich Hertz*. He experimentally detected electromagnetic waves in space, generated by the rapidly accelerated charges in an oscillatory spark discharge. At the same time, he also demonstrated that these waves could be reflected, refracted focussed by a lens, polarised and so on. In short, they exhibited properties similar to those of light. In addition to this, it was found *experimentally that electromagnetic waves are propagated precisely with the same velocity as those of light in vacuum. The evidence went far to show that the light consisted of electromagnetic waves. Fresnel's ideas did not meet any opposition, only they were interpreted not in terms of mechanics but in terms of electrodynamics.*

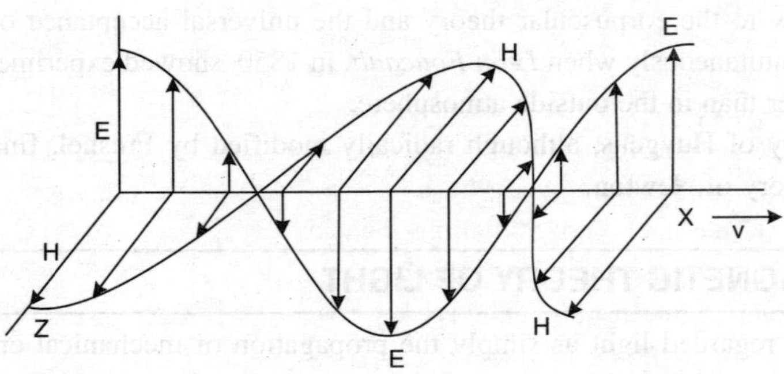

Fig. 6.1. Fields E and H in a plane transverse electromagnetic wave

By electromagnetic theory, *light is, therefore, regarded as the propagation of a vibration of an electric vector* **E** *coupled with the propagation of a vibration of a magnetic vector* **H**. *Both vectors in synchronism to be always in equal phase, mutually at right angles and at right angles to the direction of the velocity vector* **V** of the wave. Wiener's experiment on stationary light waves has shown that the electric vector plays the role of the *light vector*. The experimental confirmation of electromagnetic nature of light waves has since then accumulated *e.g.*, the magnetic splitting of spectral lines (*Zeeman Effect*), and the rotation of the plane of polarisation under the influence of magnetic field (*Faraday Effect*). The electromagnetic theory completely obliterated the corpuscular conception of light and the former reigned supreme in the domain of optics, of course, only upto the end of the 19th century, when the corpuscular conception was revived to explain the new discoveries *e.g.*, *Photoelectric effect Compton's scattering of X-rays, Emission of light.*

6.5 QUANTUM THEORY OF LIGHT

By strange irony of fate, *Hertz*, who provided an experimental confirmation of *Maxwell's electromagnetic theory of light*, discovered another phenomenon called *Photoelectric effect*, which was destined to provide the greatest set-back to the established supremacy of the wave theory and which, ultimately again brought into light the corpuscular picture of radiant energy but in slightly modified form. This phenomenon consists in instantaneous ejection of loosely bound electrons – valence electron – from alkali atoms (Na, K, Li, etc.) under the stimulus of light, *X*-rays and γ-rays. An expulsion of an electron requires an absorption of energy by the atom, equal to the *binding energy* of the electron. According to wave conception, it would take several days for the atom to absorb the requisite amount of energy from a beam of light of moderate intensity to bring about its expulsion. Actually, the expulsion of electrons starts the moment light strikes the metal surface. *Einstein* in 1905 suggested that in order to give an adequate explanation of these experimental observations, it is essential to postulate that energy in a beam of light is not uniformly distributed over the whole wave-front, but is made up of localized centres of energy travelling through free space and each of them is capable of being absorbed by the atoms as a whole. The localised concentrations of energy, which he called *photons*, are propagated like particles with the speed of light. The prestige of the wave picture was retained in as much as the energy of *photon* was assumed to be proportional to the frequency, ν, of the associated light wave, *i.e.*,

$$E = h\nu$$

where h is universal constant known as Planck's constant and its value is 6.624×10^{-27} ergs-sec. In reality, Einstein took his cue from the epoch making hypothesis put forward on 11th Dec. 1900 by *Max Planck*, Professor of Mathematical Physics at Berlin, in connection with the explanation of the energy distribution in the black body radiation. Planck's hypothesis, which was destined to represent one of the outstanding achievements in Physics and to win for him *Nobel Prize in 1918*, may be stated as follows: Every atom, usually called the oscillator, absorbs or gives off energy in intermittent, and discontinuous amounts equal to integral multiple of a certain energy unit, $h\nu$, which he called *quanta* (or energy), where ν is the oscillation frequency of the oscillator. According to wave theory the process of absorption or radiation of energy is a continuous one.

Planck's quantum hypothesis, which constitutes a clear break with the doctrines of classical physics, along with Einstein's ideas of photon character of light constitutes what is now called *Quantum Theory of Light*. Another striking confirmation of the photon character of light is provided by the interpretation of emission and absorption line spectra in 1913 by Neils Bohr and Compton's experimental observation and interpretation in 1921, of the scattering of *X*-rays.

6.6 CONCLUSION

Our present view is that light has a dual character. In experiments on propagation, reflection, refraction, interference, diffraction and polarisation, light exhibits those properties which are definitely attributable to waves while in experiments on interaction with matter (Photoelectric Effect), in the process of emission, absorption and scattering it exhibited those properties which are definitely of particle or photon character. The far reaching consequences of this dual character of radiant energy were, however, not fully recognised until *de-Broglie* announced his new view on the subject which is best expressed in his own words as follows: "A consideration of these problems led me, in 1923, to the conviction that in the theory of matter as in the theory of radiation it was essential to consider corpuscles and waves simultaneously, if it were desired to reach a single theory, permitting of the simultaneous interpretation of the properties of light and those of matter. It then became clear at once that, in order to predict the motion of a corpuscle it is necessary to construct a new mechanics, a theory closely related to that dealing with wave phenomenon, and one in which motion of corpuscle is inferred from the motion in space of a wave. In this way, there will be, for example, light corpuscles, *photons* but their motion will be connected with that of *Fresnel's waves*, and will provide an explanation of the phenomena of interference and diffraction, Meanwhile, it will no longer be possible to consider the material corpuscles or electrons as discrete isolated entities, it will on the contrary, have to be assumed in each case that they are accompanied by a wave which is bound up with their own motion." *Thus, in this unified theory, known as Quantum Mechanics, photon is not to be through as a mere concentration of energy but is to be associated with a periodic phenomenon extended over the space. With the present day knowledge, we, therefore, conclude that the photon and wave ideas are in reality complementary rather than rival conceptions.*

PROBLEMS

1. Discuss "Photon and wave ideas are in reality complementary rather than rival conceptions."
2. Write a short essay on the Nature of light.

SIMPLE HARMONIC MOTION

It is apparent from the introductory remarks about the nature of light (chapter 6) that the phenomena of interference, diffraction and polarization of light can be satisfactorily explained only by postulating that light is a form of wave motion. The physicists concerned with the earlier development of such a concept indeed thought that light was a form of mechanical energy transported through space and matter very much like the flow of acoustic energy. Although this concept turned out to be erroneous in as much as the light waves were subsequently shown to be associated with the transport of electromagnetic energy, it still went a long way in our understanding of the above phenomena and many more. It so happens that the transport of biotic mechanical and electromagnetic energies obeys the same type of differential equations, but when it comes to finding the details about their solutions, subtle but far reaching differences are noticed between the two types of wave propagation. So long as we do not bother about these details the common differential equation suffices to explain in a satisfactory manner the common properties of each wave motion. For a subsequent understanding of transport of electromagnetic energy, it is, therefore, worthwhile to study wave motion in general terms and see how far the parallelism mentioned above takes us along.

7.1 MOTION ABOUT A CENTRAL POINT

We shall start with one of the simplest and most useful concepts of Physics namely the linear to and fro motion of a particle about a central point. So long as the maximum displacement of the vibrating particle, as measured from its mean position is small, its motion is termed as simple harmonic. The motion of a simple pendulum, the vibrations of a flat spiral spring and a stretched string are a few examples of simple harmonic motion encountered in everyday life.

Consider a particle P moving along the circumference of a circle with a constant velocity v (Fig. 7.1). Suppose we get interested in studying its dynamics the moment it crosses the position x ($t = 0$). At this instant its displacement from that 'centre O' measured along the horizontal direction is equal to the radius 'a' of the circle, while that measured along the vertical direction is zero. When it arrives at P, after a lapse of a few seconds, its respective displacements along the X and Y axes, measured from the central point O are a cos ωt and a sin ωt where ω is the angular frequency of rotation and is defined as the angle traced by the radius vector in unit time. On arrival at 'Y' the horizontal displacement vanishes but the vertical displacement attains its maximum value namely a. In the second quadrant of the circle, the vertical displacement gradually decreases, vanishing completely when the particle arrives at X' but the horizontal displacement simultaneously increases becoming a maximum at X'. This maximum is nevertheless in the negative X-direction. Thus, we find that as the particle completes one half-cycle from X to X', to the displacement measured along the Y-axis, starts from zero, rises to a maximum and falls back to zero. In the next half-cycle (from X' to X) the vertical displacement again increases (though in the negative direction), becoming maximum when the particle reaches Y' and starts decreasing till it becomes zero when the particle arrives at X. At each instant the displacement in the Y-direction as measured from the central point O is given by

$$Y = a \sin \omega t \qquad \text{...(7.1 } a\text{)}$$

where 't' is measured from the instant we become interested in following this motion. A plot of Y against 't' gives the sine curve expressed in Eqn. (7.1) and has been shown to the right in Fig. 7.1.

Similarly the displacement measured along the X-axis is given by

$$x = a \cos \omega t \qquad \text{...(7.1 } b\text{)}$$

and has been shown by the dotted curve to the right of Fig. 7.1 and can be obtained by displacing the continuous curve to the left by an interval $T/4$ where 'T' is the time required by the particle to complete one full cycle along the circumference.

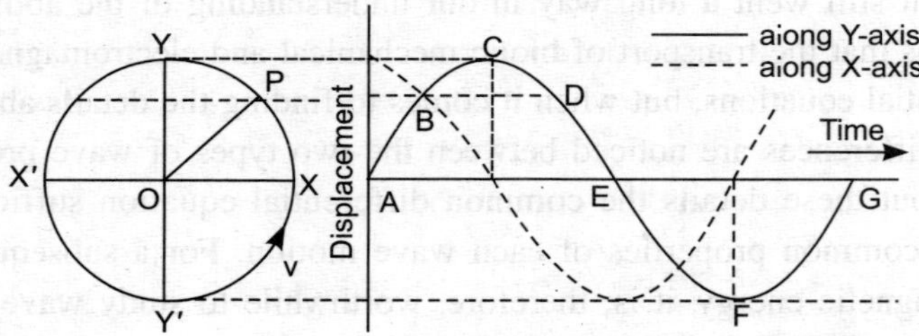

Fig. 7.1

Thus, we find that as P moves along the circumference of the circle its coordinates along the X and Y directions execute a vibratory motion along the two axes about the central point O. Moreover it can be seen from a double differentiation of Eqns. (7.1 a) and (7.1 b) with respect to time that the accelerations along the two axes are directly proportional to the instantaneous displacements of the coordinates along those directions. Thus from Eqn. (7.1 a)

$$\frac{dy}{dt} = y' = a\,\omega \cos \omega t \quad \text{and} \quad \frac{d^2 y}{dt^2} = y'' = -a\omega^2 \sin \omega t$$

or

$$y'' = -\omega^2 y \qquad \text{...(7.1 } c\text{)}$$

Similarly

$$x'' = -\omega^2 x \qquad \text{...(7.1 } d\text{)}$$

The negative sign in Eqns. (7.1 *c*) and (7.1 *d*) indicates that the instantaneous accelerations are directed along directions opposite to that of increasing displacements from the central point O.

These then are the necessary and sufficient conditions for a motion to be called simple harmonic. The motion of a particle along the circumference of a circle can thus be resolved into two mutually perpendicular simple harmonic motions of the same amplitude obeying certain conditions. So long as these conditions are satisfied the converse of this statement is also true.

7.2 VARIABLES OF A S.H.M.

The most general equation of a S.H.M. is given by

$$y = a \sin (\omega t - \delta) \quad \text{...(7.2 } a)$$

where a is the amplitude of vibration, ω the angular frequency and $(\omega t - \delta)$ the phase of the vibration. It may be mentioned that δ is a constant depending upon the particular motion under study and may be evaluated with the help of certain known conditions satisfied by it.

The maximum value of displacement y [obtained by putting $\sin (\omega t - \delta) = 1$] is called amplitude. The quantity

$$\nu = \omega/2\pi = 1/T$$

is called the frequency and is defined as the number of vibrations executed by the particle in one second. The angular frequency ω is the number of vibrations completed by the particle in 2π seconds. If t in Eqn. (7.2 *a*) is replaced by $t + T$, we find that

$$y = a \sin [\omega(t + T) - \delta]$$

$$= a \sin [\omega(t + 2\pi/\omega) - \delta]$$

$$= a \sin [2\pi + (\omega t - \delta)]$$

$$= a \sin [\omega t - \delta]$$

The nature of the function thus remains unaltered and consequently T is called the periodic time of the S.H.M. The phase of the vibration which is the argument of the sine (or cosine) function in Eqn. (7.2 *a*) can be expressed in terms of

(1) the time (fraction of T) elapsed since the vibrating particle passed through its mean position

or

(2) the angle through which the particle has turned after passing through its mean position.

Thus, when the vibrating particle is at its extreme position in the positive direction it has turned through an angle $\pi/2$ radians and the time expressed as function of T is $T/4$ seconds, both being reckoned with reference to its mean position. If there are two particles vibrating along the same straight line with the same frequency the following possibilities may arise:

(1) They pass through their mean positions simultaneously moving in the same direction. They are then said to be in the same phase. Their phase difference may be 0, 2π, 4π etc. since the function expressed in Eqn. (7.2 *a*) repeats itself after every 2π radians.

(2) They pass through their mean positions simultaneously moving in opposite directions. They are then said to be in opposite phase and their phase difference may be π, 3π etc.

(3) They pass through their mean positions at different times, either in the same or opposite directions. Their phase difference will then be determined according to the conditions of the problem and may have a value lying between 0 and 2π radians.

7.3 PARTICLE VELOCITY, ACCELERATION AND ENERGY IN A S.H.M.

We have seen that the instantaneous displacement of a particle executing S.H.M. is given by

$$y = a \sin (\omega t - \delta) \quad \text{...(7.3 } a\text{)}$$

A differentiation of this equation with respect to time gives the particle velocity at any instant. Thus

$$\text{Velocity} = y' = a\, \omega \cos (\omega t - \delta) \quad \text{...(7.3 } b\text{)}$$

The particle velocity therefore undergoes a sinusoidal variation with respect to time. The frequency of this variation however remains unchanged but the amplitude *i.e.*, the maximum value of velocity is ω times the amplitude of displacement. Moreover, there is a phase difference of $\pi/2$ between the *two functions represented by Eqns. (7.3 a) and (7.3 b). Thus when one becomes a maximum, the other is a minimum and vice-versa.*

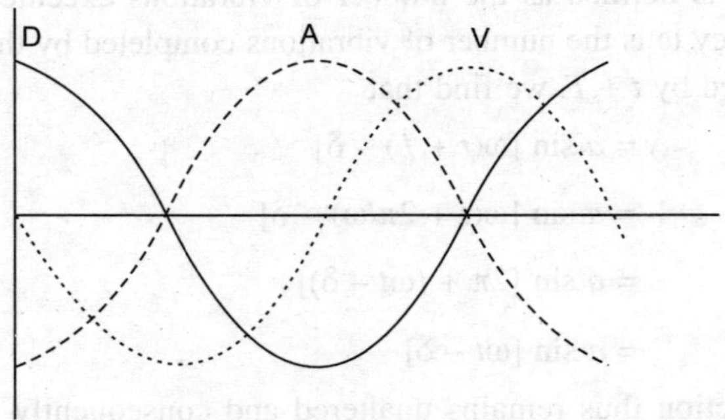

D--displacement, A--acceleration, V--velocity

Fig. 7.2. Particle velocity, displacement and acceleration in a S.H.M.

A second differentiation of Eqn. (7.3 *b*) with respect to time gives particle acceleration. Thus

$$y'' = -a\, \omega^2 \sin (\omega t - \delta)$$
$$= -\omega^2 y \quad \text{...(7.3 } c\text{)}$$

The acceleration of the particle also undergoes periodic variations of the same frequency as its displacement but its amplitude is ω^2 times the amplitude of displacement. Moreover the negative sign in Eqn. (7.3 *c*) indicates that the two variations are in opposite phase.

The graphs of these variations are shown in Fig. 7.2.

Thus we find from Eqn. (7.3 *b*) that as the particle passes through its mean position (δ/ω, $T + \delta/\omega$, $2T + \delta/\omega$ etc.) its velocity has a maximum value $a\omega$. Its kinetic energy $\frac{1}{2}mv^2$ will therefore also have a maximum value. Its acceleration while passing through the mean position is zero [Eqn. (7.3 *c*)] and will go on increasing as it moves farther and farther away but is directed towards the central point (hence deceleration). On reaching the extreme position, the particle loses all its kinetic energy, the deceleration attains its maximum value and the particle starts its journey back towards its mean position experiencing all the time an ever decreasing acceleration till it arrives at its mean position once again where its kinetic energy becomes maximum. It overshoots this position and starts experiencing an ever increasing deceleration till its other extreme position and so on.

At any instant the energy of the vibrating particle is made up of

(1) its kinetic energy

(2) its potential energy.

If m is the mass of the vibrating particle its kinetic energy at any instant (assume $\delta = 0$ for simplicity)

$$E_{\text{kin}} = \frac{1}{2} mv^2 = \frac{1}{2} my'^2$$

$$= \frac{1}{2} ma^2 \omega^2 \cos^2 \omega t \qquad \text{...(7.3 } d)$$

It will be always experiencing a force my'' trying to bring it back to its mean position. If the displacement of the particle at any instant is y it will have a certain energy by virtue of its position in the field of this restoring force. The local potential energy can thus be found out in terms of the work done in displacing the particle through a further distance dy against this restoring force. Hence,

$$dW = \text{Force} \times \text{distance}$$

$$= m\omega^2 y \, dy \qquad \text{[see Eqn. (7.3 } c)]$$

Therefore, the total work done in displacing the body through a distance y from its mean position

$$W = E_{\text{pot}} = \int_o^y m\omega^2 y \, dy$$

$$= \frac{1}{2} m\omega^2 y^2$$

$$= \frac{1}{2} ma^2 \omega^2 \sin^2 \omega t \qquad \text{...(7.3 } e)$$

Hence the total energy of the particle

$$E = E_{\text{kin}} + E_{\text{pot}}$$

$$= \frac{1}{2} ma^2 \omega^2 (\sin^2 \omega t + \cos^2 \omega t)$$

$$= \frac{1}{2} ma^2 \omega^2 \qquad \text{...(7.3 } f)$$

Thus, the total energy of the particle executing a S.H.M. remains independent of its coordinates and depends only on its angular frequency and amplitude of vibration. It also follows from Eqns. (7.3 *d*) and (7.3 *e*) that when the kinetic energy is maximum, the potential energy is zero and vice-versa. But the individual energies, kinetic and potential are seen to be functions of time and hence the position of the vibrating particle. Their average values however are $\frac{1}{4}ma^2\omega^2$ each (since the average values of $\sin^2\theta$ and $\cos^2\theta$ are each equal to half).

7.4 DIFFERENTIAL EQUATION OF S.H.M. AND ITS SOLUTION

Equation (7.3 *c*) gives the differential equation of S.H.M. and can in fact be obtained from the definition of such a motion. We know that the restoring force acting on the vibrating particle is directly proportional to the instantaneous displacement from its mean position and is always directed towards this central point. Hence according to Newton's second law for *linear motion*

$$my'' = -\mu y$$

where μ is the constant of proportionality and gives the restoring force for unit displacement. On putting $\mu/m = \omega^2$ this equation can be reduced to Eqn. (7.4 *a*) *i.e.*,

$$y'' = -\omega^2 y$$

or $$y'' + \omega^2 y = 0 \qquad ...(7.4\ a)$$

The easiest method of solving such equations is by substitution. We suppose that the solution is of the form

$$y = Ae^{\alpha t}$$

Hence $$y' = A\,\alpha\, e^{\alpha t} \text{ and } y'' = A\,\alpha^2\, e^{\alpha t} \qquad ...(7.4\ b)$$

Substituting the value of y'' in Eqn. (7.4 *a*), we get

$$A\alpha^2 e^{\alpha t} + \omega^2 A e^{\alpha t} = 0$$

Since $y = Ae^{\alpha t}$ is not equal to zero, we get

$$\alpha^2 + \omega^2 = 0$$

so that $$\alpha = \pm\, i\omega \qquad \text{where} \qquad i = \sqrt{-1}$$

The possible solution of Eqn. (7.4 *a*) are therefore

$$y = A_1 e^{i\omega t} \text{ or } y = A_2 e^{-i\omega t}$$

It is the property of linear differential equations that the algebraic sum of all admissible solutions is also one of its solutions. Hence the most general solution of Eqn. (7.4 *a*) is of the form

$$y = A_1 e^{i\omega t} + A_2 e^{-i\omega t} \qquad ...(7.4\ c)$$

It can be shown that

$$e^{\pm i\omega t} = \cos \omega t \pm i \sin \omega t$$

So that Eqn. (7.4 *c*) can be written as

$$y = (A_1 + A_2)\cos\omega t + i(A_1 - A_2)\sin\omega t$$

$$= A\cos\omega t + B\sin\omega t \qquad \text{...(7.4 } d)$$

where $A = A_1 + A_2$ and $B = i\,(A_1 - A_2)$. Equation (7.4 *d*) can be written in another form by putting

$$A = R\sin\theta \quad \text{and} \quad B = R\cos\theta$$

where $$R = \sqrt{A^2 + B^2} \quad \text{and} \quad \tan\theta = A/B$$

With these values Eqn. (7.4 *d*) becomes

$$y = R\sin(\omega t + \theta) \qquad \text{...(7.4 } e)$$

Equations (7.4 *d*) and (7.4 *e*) are the most general solutions of Eqn. (7.4 *c*). The pairs of constants *A* and *B* or *R* and θ can be evaluated from the conditions of the problem under study. For instance, if for a motion

$$y = y_o \quad \text{and} \quad y' = 0 \quad \text{at} \quad t = 0 \quad i.e., \text{ initially}$$

then it can be shown that $B = 0$, $\theta = \pi/2$ and $R = A = y_o$. This can be easily recognised as the case of a simple pendulum.

7.5 COMPOSITION OF S.H. MOTIONS

Very often problems are encountered in Physics where the same particle is under the simultaneous action of two or more simple harmonic motions acting either along the same straight line or at right angles to each other. There are two general methods for solving such problems based on (*i*) a graphical treatment and (*ii*) analytical treatment of the dynamics of the particle. The graphical treatment is based on the rectilinear projection of uniform circular motion while the analytical treatment is based on finding the vector sum of the individual motions either with the help of trigonometric functions or writing them as complex quantities. The last of these methods is usually the easiest to handle and is at the same time most informative. It will nevertheless be worthwhile to understand the general outline of the other two.

(*a*) *Composition of S.H. Motions (Graphical Method).* Let the vector x_1 (Fig. 7.3 *a*) represent the amplitude of a S.H.M. By convention of such representation the vector is assumed to

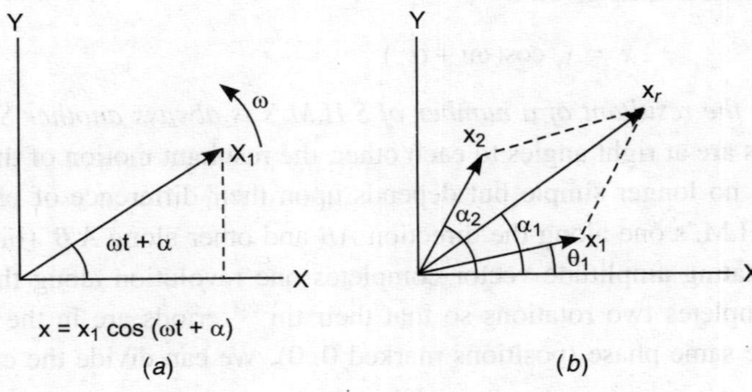

Fig. 7.3

rotate counter clock-wise with an angular frequency ω. Its projection along the X-axis will give the displacement equation of the S.H. Motion.

$$x = x_1 \cos(\omega t + \alpha)$$

where α is the starting angle at $t = 0$. If there are several collinear S.H. Motions of different amplitude, frequency and phase acting simultaneously on a particle, each of these can be represented by an appropriate rotating vector like x_1. The projections of all these vectors will give the individual displacements as above. If all these vectors are rotating at the same angular frequency, the angles between them will remain unchanged at all times and consequently their composition becomes a simple matter. Suppose the amplitude vectors of two S.H. Motions at say $t = 0$ are represented by x_1 and x_2 (Fig. 7.3 *b*).

Resolving them along two mutually perpendicular directions *X* and *Y*, we get

$$\left.\begin{aligned} X_x &= x_1 \cos\alpha_1 + x_2 \cos\alpha_2 \\ \text{and} \quad X_y &= x_1 \sin\alpha_1 + x_2 \sin\alpha_2 \end{aligned}\right\} \qquad \text{...(7.5 a)}$$

Hence if the amplitude vector of the resultant be represented by x_r, we have

$$X_r = \sqrt{X_x^2 + X_y^2}$$

$$= \sqrt{x_1^2 + x_2^2 + 2x_1x_2\cos(\alpha_2 - \alpha_1)} \qquad \text{...(7.5 b)}$$

The phase of the resultant at $t = 0$ will therefore be

$$\alpha_r = \tan^{-1}(X_y/X_x) \qquad \text{...(7.5 c)}$$

In general, the magnitude of the resultant vector can be obtained from the parallelogram law, as above and is given by

$$X_r = \sqrt{(\Sigma X_x)^2 + (\Sigma X_y)^2} \qquad \text{...(7.5 d)}$$

and

$$\tan\alpha_r = \Sigma X_y/\Sigma X_x \qquad \text{...(7.5 e)}$$

where ΣX_x and ΣX_y are respectively the sums of the *x* and *y* components of the amplitude vectors at $t = 0$. The resultant motion is then given by

$$x = x_r \cos(\omega t + \alpha_r) \qquad \text{...(7.5 f)}$$

Thus we find that the resultant of a number of S.H.M.'s is always another S.H.M.

If the two motions are at right angles to each other, the resultant motion of the particle executing them simultaneously is no longer simple but depends upon their difference of phase and frequency. Let us consider two S.H.M.'s one along the direction *AB* and other along *A'B'* (Fig. 7.4) and suppose that by the time the rotating amplitude vector completes one revolution along the lower circle, that on the upper circle completes two rotations so that their tim eriods are in the ratio 2 : 1. Let the two motions start in the same phase (positions marked 0, 0). We can divide the circumference of the two circles into parts such that the two vectors reach the points 0, 1, 2, 3, ..., 8 at the same time. The points of intersection of the perpendiculars drawn from corresponding points (1, 1), (2, 2), ... (8, 8) will then give the locus of the vibrating particle if it were under the simultaneous action of the two S.H.M.'s. This is seen to be a figure of eight in this particular case. If the initial conditions of the two vibrations are changed, the resulting curve may have a different shape. Such curves are called Lissajou's figures.

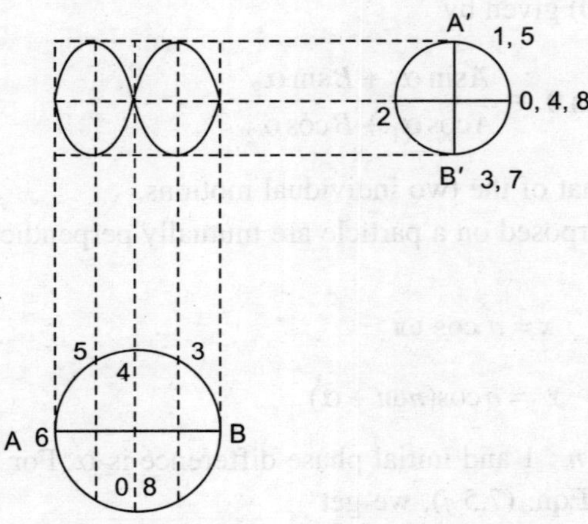

Fig. 7.4. Composition of SHM's at right angles

(*b*) *Analytical Method.* Let a particle execute two collinear S.H.M.'s of the same frequency represented by

$$x_1 = A\cos(\omega t + \alpha_1)$$

and
$$x_2 = B\cos(\omega t + \alpha_2)$$

The resultant displacement of the particle will be given by the vector sum of the individual motions, so that

$$x = x_1 + x_2$$

Therefore

$$X = A \cos (\omega t + \alpha_1) + B \cos (\omega t + \alpha_2)$$

$$= A[\cos \omega t \cos \alpha_1 - \sin \omega t \sin \alpha_1]$$

$$+ B[\cos \omega t \cos \alpha_2 - \sin \omega t \sin \alpha_2]$$

$$= (A \cos \alpha_1 + B \cos \alpha_2) \cos \omega t$$

$$- (A \sin \alpha_1 + B \sin \alpha_2) \sin \omega t$$

If we put

$$R \cos \theta = A \cos \alpha_1 + B \cos \alpha_2$$

and
$$R \sin \theta = A \sin \alpha_1 + B \sin \alpha_2$$

the resultant motion can be written as

$$X = R\,(\cos \omega t \cos \theta - \sin \omega t \sin \theta)$$
$$= R \cos (\omega t + \theta) \qquad ...(7.5\ g)$$

This is the equation of a S.H.M. of amplitude

$$\tan \theta = \sqrt{A^2 + B^2 + 2AB\cos(\alpha_1 - \alpha_2)} \qquad ...(7.5\ h)$$

and initial phase angle (at $t = 0$) given by

$$\tan \theta = \frac{A \sin \alpha_1 + B \sin \alpha_2}{A \cos \alpha_1 + B \cos \alpha_2} \qquad ...(7.5\ i)$$

whose period is the same as that of the two individual motions.

If the two S.H.M.'s superposed on a particle are mutually perpendicular they can be represented by equations of the form

$$x = a \cos \omega t \qquad ...(7.5\ j)$$

$$y = b \cos(n\omega t + \alpha) \qquad ...(7.5\ k)$$

whose periods are in the ratio $n : 1$ and initial phase difference is α. For finding the resultant motion we proceed as follows: From Eqn. (7.5 *j*), we get

$$x/a = \cos \omega t$$

and hence $$\sqrt{1 - x^2/a^2} = \sin \omega t$$

These values of the sine and cosine functions can be substituted in the expansion of $\cos (n\omega t + \alpha)$ on the R.H.S. of Eqn. (7.5 *k*) to get the equation of the resultant motion. For instance, if $n = 2$, we have from Eqn. (7.5 *k*)

$$y/b = \cos 2\,\omega t \cos \alpha - \sin 2\omega t \sin \alpha$$
$$= (2\cos^2 \omega t - 1) \cos \alpha - 2 \sin \omega t \cos \omega t \sin \alpha$$

or $$y/b = (2x^2/a^2 - 1) \cos \alpha - 2x/a\sqrt{(1 - x^2/a^2)}\,\sin \alpha$$

or $$[y/b - (2x^2/a^2 - 1)\cos \alpha]^2 = 4x^2/a^2 (1 - x^2/a^2) \sin^2 \alpha$$

which on simplification gives

$$y^2/b^2 + 4x^4/a^4 - 4x^2/a^2(1 + y/b \cos \alpha) + 2y/b \cos \alpha + \cos^2\alpha = 0 \qquad ...(7.5\ l)$$

This is the general equation of the resultant motion of a particle under the simultaneous action of two mutually perpendicular S.H.M.'s whose periods are in the ratio 2:1. Depending upon the value of the phase difference α between the two motions, the curve will be a figure of eight, parabola etc.

We can obtain the locus of the particle vibrating under the simultaneous action of two mutually perpendicular S.H.M.'s of any other frequency ratio in a similar manner. Two specific cases are treated in detail in the following pages.

(c) *Use of Complex Quantities.* Calculations with harmonic motions are simplified by the use of complex quantities instead of the usual trigonometric functions. Thus if a S.H.M. is represented by the displacement equation

$$y = b \cos(\omega t + \delta) \quad \text{...(7.5 } m\text{)}$$

we can also represent it in the form

$$y = R\{be^{-i(\omega t - \delta)}\} \quad \text{...(7.5 } n\text{)}$$

where $i = \sqrt{-1}$ and R denotes the real part of the exponential function. (We know that $\cos x - i \sin x = e^{-ix}$). If the operations on 'y' are linear we may drop the symbol R in Eqn. (7.5 n) and work with the exponential function only provided we extract out the real part from the final expression which will give the result under study. But if the operations on 'y' are non-linear such as squaring etc., we should be careful to take the real part first and then operate with these alone. It may be mentioned that such representations of harmonic motions cause considerable simplicity in calculations and have other advantages which shall be discussed in a different context.

(i) *Composition of Collinear S.H.M.'s:* Let two S.H.M.'s be represented by

$$x_1 = a \cos(\omega t + \alpha_1) \quad \text{...(7.5 } o\text{)}$$

$$x_2 = b \cos(\omega t + \alpha_2) \quad \text{...(7.5 } p\text{)}$$

These motions have the same period but different amplitudes and phase; $\alpha_1 \sim \alpha_2$ is called the phase difference between the two motions.

As outlined above we can write

$$x_1 = R\{ae^{-i(\omega t + \alpha_1)}\}$$

$$x_2 = R\{be^{-i(\omega t + \alpha_2)}\}$$

It may be mentioned that the amplitudes 'a' and 'b' here are not complex quantities. Since the resultant motion is the vector sum of individual motions we can write

$$x = x_1 + x_2 = e^{-i\omega t}(ae^{-i\alpha_1} + be^{-i\alpha_2}) \quad \text{...(7.5 } q\text{)}$$

Since the operation performed here has been linear we can afford to drop the symbol R on the understanding that only the real part of the final expression will give the desired result. Eqn. (7.5 q) can now be written as

$$x = e^{-i\omega t}\{(a\cos\alpha_1 + b\cos\alpha_2) - i(a\sin\alpha_1 + b\sin\alpha_2)\}$$

$$= e^{-i\omega t}\{A(\cos\phi - i\sin\phi)\} \quad \text{...(7.5 } r\text{)}$$

where

$$A = \sqrt{(a\cos\alpha_1 + b\cos\alpha_2)^2 + (a\sin\alpha_1 + b\sin\alpha_2)^2}$$

$$= \sqrt{a^2 + b^2 + 2ab\cos(\alpha_1 - \alpha_2)} \quad \text{...(7.5 } s\text{)}$$

and $$\tan\phi = \frac{a\sin\alpha_1 + b\sin\alpha_2}{a\cos\alpha_1 + b\cos\alpha_2} \qquad ...(7.5\ t)$$

Hence the resultant motion can be derived from

$$x = Ae^{-i\omega t}e^{-i\phi} = Ae^{-i(\omega t+\phi)}$$

Taking the real part of this expression, we get

$$x = A\cos(\omega t + \phi) \qquad ...(7.5\ u)$$

This is in accordance with Eqn. (7.5 *g*) and the remarks following it.

(ii) Composition of Rectangular S.H.M.'s: Let the two motions be represented by

$$x = a\cos\omega t \qquad ...(7.5\ v_1)$$
$$y = b\cos(n\omega t + \alpha) \qquad ...(7.5\ v_2)$$

Equation (7.5 v_1) represents a vibratory motion of amplitude 'a' along the X-axis while Eqn. (7.5 v_2) another similar motion of amplitude 'b' along the Y-axis. The periods of the two motions are in the ratio $n : 1$. In terms of complex quantities we can write these motions as

$$\frac{x}{a} = R(e^{-i\omega t})$$

$$\frac{y}{b} = R(e^{-i(n\omega t+\alpha)})$$

$$= R[e^{-in\omega t}e^{-i\alpha}]$$

$$= R[(\cos\alpha - i\sin\alpha)e^{-in\omega t}] \qquad ...(7.5\ w)$$

According to De Moivre's theorem, if 'n' is real, one value of $(\cos\phi \pm i\sin\phi)^n$ is $\cos n\phi \pm i\sin n\phi$. Hence the real part of Eqn. (7.5 *w*) will be

$$\cos\alpha\cos n\omega t - \sin\alpha\sin n\omega t$$

So that

$$y/b = \cos\alpha\cos n\omega t - \sin\alpha\sin n\omega t$$

or $$(y/b - \cos\alpha\cos n\omega t) = -\sin\alpha\sin n\omega t \qquad ...(7.5\ x)$$

We had to extract the real part because the next operation is going to be non-linear. Squaring both the sides of Eqn. (7.5 *x*) and simplifying, we get

$$y^2/b^2 - 2y/b\cos\alpha\cos n\omega t + \cos^2 n\omega t = \sin^2\alpha \qquad ...(7.5\ y)$$

where the value of $\cos n\omega t$ has to be found with the help of Eqn. (7.5 *y*). This is the general equation of the locus of the particle under the simultaneous action of two rectangular S.H.M.'s.

Special Cases: *(i) Frequency ratio 1 : 1*

If the time periods of the two S.H.M.'s are equal $n = 1$ and Eqn. (7.5 *y*) can be written as

$$y^2/b^2 - 2y/b\cos\alpha\cos\omega t + \cos^2\omega t = \sin^2\alpha$$

Since $\cos \omega t = x/a$, $\cos^2 \omega t = x^2/a^2$ and we get

$$y^2/b^2 - \frac{2xy}{ab}\cos\alpha + \frac{x^2}{a^2} = \sin^2\alpha \qquad ...(7.5\ z)$$

as the equation of the locus in this case. This is the general equation of a conic whose shape will depend upon the value of the phase difference α between the two motions.

(*a*) If $\alpha = 0$ the equation of the Lissajou's figure is seen to be

$$(y/b - x/a)^2 = 0$$

which is a pair of coincident straight lines inclined at an angle $\tan^{-1}(b/a)$ to the *X*-axis.

(*b*) If $\alpha = \pi/4$, $\sin\alpha = \cos\alpha = 1/\sqrt{2}$, the equation of the curve becomes

$$y^2/b^2 - \sqrt{2}xy/ab + x^2/a^2 = \frac{1}{2}$$

This is the equation of an ellipse inscribed in a rectangle whose length parallel to the *X*-axis is 2*a* and breadth 2*b*. The ellipse touches this rectangle at points $(\pm a, \pm b/\sqrt{2})$ and $(\pm a/\sqrt{2} \pm b)$.

(*c*) If $\alpha = \pi/2$, Eqn. (7.5 *z*) reduces to

$$y^2/b^2 + x^2/a^2 = 1$$

which is the familiar equation of an ellipse whose centre coincides with the origin and whose axes are parallel to the coordinate axes *X* and *Y*. If the two S.H.M.'s have the same amplitude the locus of the particle becomes a circle [$a = b$, hence $x^2 + y^2 = a^2$].

(*d*) If $\alpha = 3\pi/4$, the Lissajou's figure is represented by the equation

$$y^2/b^2 + \sqrt{2}xy/ab + x^2/a^2 = \frac{1}{2}$$

which is again an ellipse with its axes rotated by $\pi/2$ with respect to that in case (*b*).

(*e*) If $\alpha = \pi$, we get again get a pair of coincident straight lines given by

$$(y/b + x/a)^2 = 0$$

But this time they are inclined at an angle $\tan^{-1}(-b/a)$ to the axis of *X*.

All these cases have been summarised in Fig. (7.5)

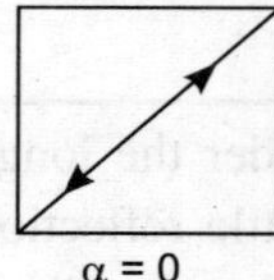

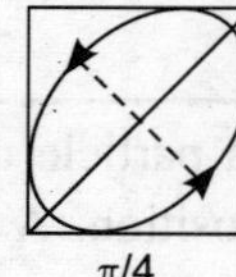

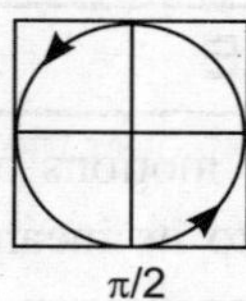

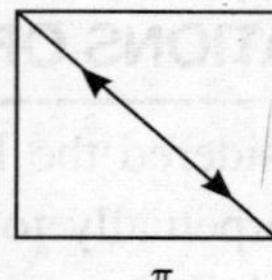

Fig. 7.5. Lissajou's figures (Period ratio 1:1)

(*ii*) *Frequency Ratio 1:2*

The general equation of the Lissajou's figure when the time periods are in the ratio 2 : 1 can be obtained by putting $n = 2$ in Eqn. (7.5 *y*). Thus, we get

$$y^2/b^2 - 2y/b\cos\alpha\cos 2\omega t + \cos^2 2\omega t = \sin^2\alpha \qquad ...(7.5\ z_1)$$

where $\cos \omega t = x/a$ [see Eqn. 7.5 v_1]

Now $\cos 2\omega t = 2\cos^2 \omega t - 1$

$= (2x^2/a^2 - 1)$

and $\cos^2 2\omega t = 4x^4/a^4 - 4x^2/a^2 + 1$

Substituting these values in Eqn. (7.5 z_1), we get

$$y^2/b^2 - 2y/b(2x^2/a^2 - 1)\cos\alpha + (2x^2/a^2 - 1)^2 = \sin^2\alpha$$

which on re-arranging gives

$$y^2/b^2 + 4x^4/a^4(1 + y/b\cos\alpha) + 2y/b\cos\alpha + \cos^2\alpha = 0 \qquad \text{...(7.5 } z_2\text{)}$$

An inspection of this equation shows that the general conic is symmetrical about the X-axis only; the actual shape of the curve will of course depend upon the phase difference 'α' between the two motions. The resulting Lissajou's figures for different values of α are shown in Fig. 7.6.

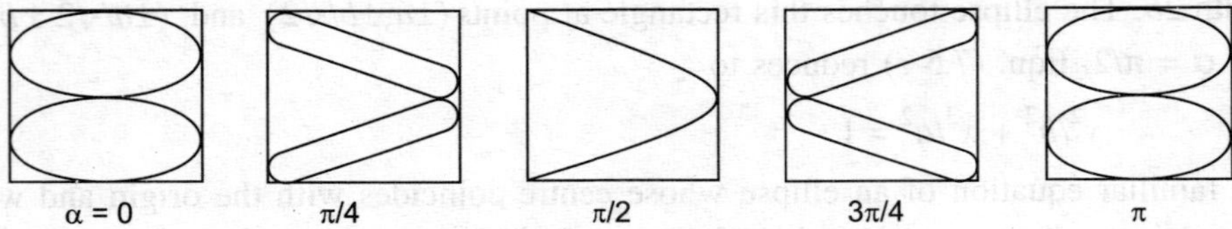

Fig. 7.6. Lissajou's figures (Period ratio (2:1)

It may be mentioned that these figures can be demonstrated on the screen of a cathode ray oscillograph by applying audio-frequency voltage signals on the X and Y plates. Stable figures can be obtained only if the two frequencies bear a whole number ratio. As this ratio goes on increasing, the figures get more and more crowded but always remain closed curves. These figures provide a very sensitive test for determining any unknown frequency by combining it with a known frequency in a perpendicular direction. It may be mentioned that composition of S.H.M.'s is an important problem in Physics and is encountered very frequently in Wave Optics and Acoustics; the general approach remains the same as given above.

7.6 DAMPED VIBRATIONS OF A PARTICLE

Uptil now we have considered the linear vibratory motions of a particle under the long action of a restoring force trying perpetually to bring it back to its mean position. A little reflection will show that this situation is an ideal case of the general phenomenon. In actual practice the vibratory motion is also opposed by frictional or dissipative forces generated in the fluid in which the motion is taking place. Such dissipative forces are usually proportional to some power of the velocity of the particle; for small values of velocity the power can be taken as unity without loss of any significant accuracy. With this restriction in view the dissipative forces can be written as

$$F = -ry'$$

where the constant of proportionality r is the dissipative force per unit velocity of the particle and the negative sign signifies a restraining influence on the vibration of the particle.

According to Newton's second law, the equation of motion in this case will be

$$my'' = -ry' - \mu y$$

or $$y'' + \left(\frac{r}{m}\right)y' + \left(\frac{\mu}{m}\right)y = 0 \qquad ...(7.6\ a)$$

The motion of the particle defined by Eqn. (7.6 *a*) is called damped harmonic motion. Such equations arise many times in different branches of Physics, the most important being that describing the transient behaviour of an LCR circuit. It may be added that the solution to Eqn. (7.6 *a*) in the present context is of a similar nature as that encountered for an LCR circuit, but in order to obtain any meaningful results from its application to the circuit theory the similarity conditions or analogy between the two cases must be closely studied. Moreover there are very many equations in electrodynamics which have a close parallel in mechanics and a solution of the former leads to solution of the latter provided due attention is paid to the analogy between the two fields. In fact, this has led to the development of analogue computers.

Coming back to Eqn. (7.6 *a*) we find that it can be reduced to a simpler form by proper substitutions. Let

$$y = xe^{-bt} \qquad ...(7.6\ b)$$

where 'b' is an arbitrary constant, then

$$y' = x'e^{-bt} - bxe^{-bt}$$

and $$y'' = x''e^{-bt} - bx'e^{-bt} - bx'e^{-bt} + b^2xe^{-bt}$$

Substituting these values in Eqn. (7.6 *a*), we get

$$x'' + (r/m - 2b)x' + (\mu/m + b^2 - br/m)x = 0 \qquad ...(7.6\ c)$$

Since b is an arbitrary constant we can take it equal to $r/2m$ so that the coefficient of y' may vanish. Hence Eqn. (7.6 *c*) becomes

$$x'' + (\omega^2 - b^2)x = 0 \qquad ...(7.6\ d)$$

an equation of the type we have solved before. Knowing the solution to Eqn. (7.6 *d*) we can write down the value of y with the help of Eqn. (7.6 *b*). There are however three important cases which are discussed below.

Case I: Suppose $\omega^2 < b^2$ *i.e.*, $\mu/m < r^2/4m^2$

In the case of large frictional losses, the difference $(\omega^2 - b^2)$ is negative say $-K^2$. Then Eqn. (7.6 *d*) can be written as

$$(D^2 - K^2)\,x = 0$$

where $$D = d/dt \text{ and } D^2 = d^2/dt^2$$

or $$(D - K)(D + K)\,x = 0$$

The solution of this differential equation is of the form (see any book on integral calculus)

$$x = A_1\, e^{+Kt} + A_2 e^{-Kt}$$

Hence $$y = A_1 e^{(-b+\sqrt{b^2-\omega^2})\,t} + A_2 e^{(-b-\sqrt{b^2-\omega^2})\,t} \qquad ...(7.6\ e)$$

where A_1 and A_2 are the constants of integration to be determined from the conditions of each problem.

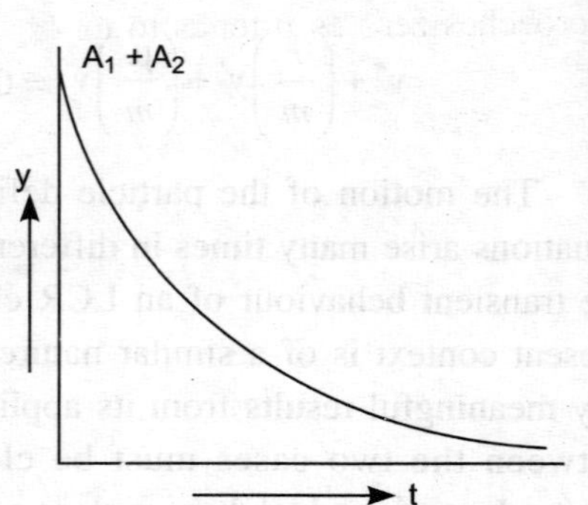

Fig. 7.7 Damped motion of a particle

It will be noticed that $b > \sqrt{b^2 - \omega^2}$ and hence the powers of the exponential will be negative in both the terms of Eqn. (7.6 *e*). Consequently the displacement of the particle will fall exponentially to zero from its initial value of $(A_1 + A_2)$ at the start of its damped motion. How soon the motion dies out will depend upon the magnitude of difference between 'b' and 'ω'.

Case II: Suppose $\omega^2 = b^2$

or $$\mu/m = r^2/4m^2$$

This is the condition of a critically damped motion. In this case Eqn. (7.6 *d*) reduces to $x'' = 0$ which is the equation of the straight line $x = A_1 t + A_2$ where A_1 and A_2 are again the constants of integration, both assumed positive as before.

Hence $$y = e^{-bt}\,(A_1 t + A_2)$$

A plot of this equation has been shown in Fig. 7.8. But for an initial increase (provided $A_1 > A_2$), the amplitude of motion goes on decreasing as time progresses.

This case is of significance in the design of scientific equipments.

Case III: Suppose $\omega^2 > b^2$

or $$\mu/m > r^2/4m^2$$

In this case, where the dissipative forces are small Eqn. (7.6 *d*) reduces to the form of a S.H.M.

$$x'' = -\,(\omega^2 - b^2)\,x$$

The solution of such an equation has been seen to be of the form

$$x = A\cos(\sqrt{\omega^2 - b^2}\,t + \theta)$$

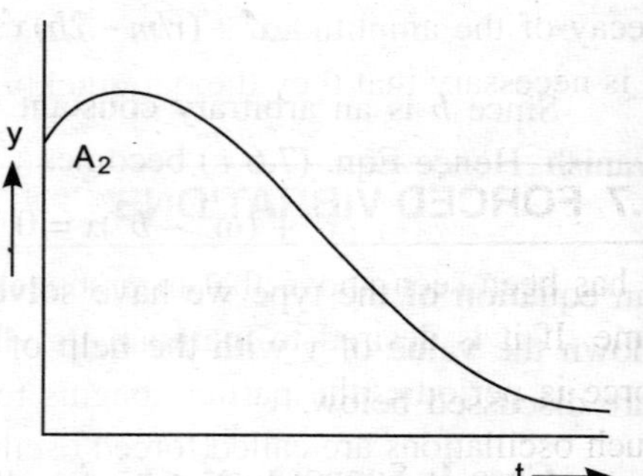

Fig. 7.8. Critically damped motion

where A and θ are the two constants of integration. Therefore the displacement of the damped vibration will be given by

$$y = Ae^{-bt}\cos(\sqrt{\omega^2 - b^2}\,t + \theta) \qquad ...(7.6\ f)$$

From Eqn. (7.6 *f*), we find that the effective amplitude Ae^{-bt} of a damped motion dies out exponentially with time. The rate at which this happens is of course governed by 'b'. If the dissipative forces increase, the exponential decay becomes much more rapid. In addition to this the dissipative forces also affect the time period of the vibrations of the particle. Whereas in the case of free vibrations the periodic time is $2\pi/\omega$, in the case of damped vibrations this is increased to $2\pi/\sqrt{\omega^2 - b^2}$. The frequency of the oscillations is therefore lowered. As 'b' increases, this oscillation frequency approaches zero as b tends to ω.

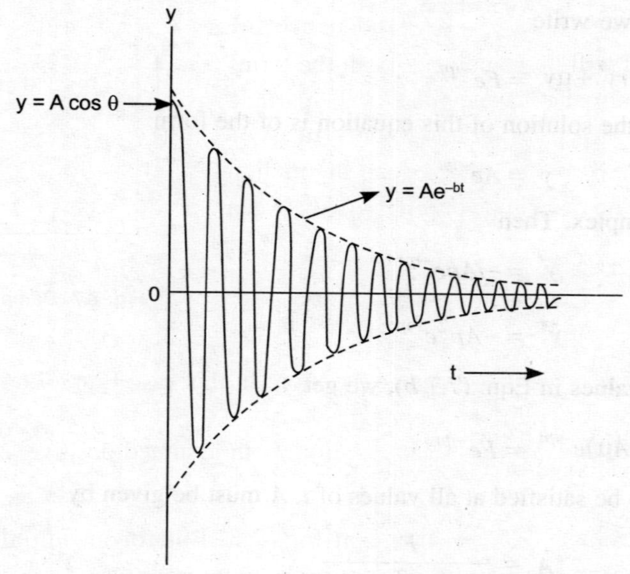

Fig. 7.9. Damped oscillations of a particle

The damped oscillations are shown in Fig. 7.9. The dotted curve represents the exponential decay of the amplitude λ. These damped vibrations are of importance in musical instruments where it is necessary that they die out quickly. This is accomplished by arranging for a large value of b.

7.7 FORCED VIBRATIONS

It has been seen above that in practical cases, the natural vibrations of particle die out with lapse of time. If it is desired to maintain its vibrations, extra force has to be supplied from outside. If this force is periodic, the particle begins to vibrate with a frequency equal to that of the applied force. Such oscillations are called forced oscillations. Let the externally applied force be

$$F \cos pt \;= R(Fe^{-ipt})$$

The equation of motion of the particle then becomes

$$my'' + ry' + \mu y \;= F \cos pt \qquad \text{...(7.7 a)}$$

There are several methods of solving this equation, but we shall try to solve it with the help of complex quantities. The solution to such equations consists of two parts:

(1) *The complementary function*: This is obtained by putting the R.H.S. equal to zero and solving.

This solution is of course exactly the same as that of Eqn. (7.6 *a*) and has two arbitrary constants. This part of the solution is called the transient term.

(2) *The particular integral*: This is obtained by solving Eqn. (7.7 *a*) as it is and does not contain any arbitrary constants. This is called the steady state solution as we shall presently see.

To obtain the steady state solution, we shall replace the R.H.S. of Eqn. (7.7 *a*) by the complex quantity Fe^{-ipt} and solve the equation with the understanding that the real part of the solution will be of relevance to us. Thus we write

$$my'' + ry' + \mu y = Fe^{-ipt} \qquad \text{...(7.7 } b\text{)}$$

Let us assume that the solution of this equation is of the form

$$y = Ae^{-ipt}$$

where A is in general complex. Then

$$y' = -iApe^{-ipt}$$

and

$$y'' = -Ap^2e^{-ipt}$$

Substituting these values in Eqn. (7.7 *b*), we get

$$(Ap^2m - iApr + A\mu)e^{-ipt} = Fe^{-ipt}$$

For this equation to be satisfied at all values of t, A must be given by

$$A = \frac{F}{(\mu - mp^2) - ipr}$$

The complex displacement will then be

$$y = \frac{F}{(\mu - mp^2) - ipr}e^{-ipt}$$

$$= \frac{iF}{p[r + i(mp - \mu/p)]}e^{-ipt} \qquad \text{...(7.7 } c\text{)}$$

Suppose we put

$$\left.\begin{aligned} Z_M &= r + i(mp - \mu/p) = r + iS = Z_o e^{i\phi} \\ \text{where} \quad Z_o &= \sqrt{r^2 + S^2} \\ \text{and} \quad \tan\phi &= S/r \end{aligned}\right\} \qquad \text{...(7.7 } d\text{)}$$

then Eqn. (7.7 *c*) becomes

$$y = iF / pZ_o e^{-i(pt-\phi)}$$

The solution to Eqn. (7.7 *a*) is the real part of the above solution.

Thus, we find that

$$y = F/pZ_o \sin(pt - \phi)$$

$$= \frac{F}{p\sqrt{[r^2 + (mp - \mu/p)^2]}} \sin(pt - \phi) \qquad \text{...(7.7 e)}$$

is the steady state solution of the problem under study.

An examination of Eqn. (7.7 *e*) reveals that the steady state displacement of the particle executing forced oscillations is not in step with the driving force; there is a phase difference of ϕ between the two. The amplitude of these forced vibrations is less than that of the driving force but depends upon F. The frequency of forced vibrations is equal to that of the externally applied force which is different from either that of its free vibrations or that of its damped vibrations. These are given below:

$\nu = \omega/2\pi$ Free vibrations

$\nu = \sqrt{\frac{\omega^2 - b^2}{2\pi}}$ Damped vibrations

$\nu = p/2\pi$ Forced vibrations

The complete solution of Eqn. (7.7 *a*) is therefore

$$y = Ae^{-bt} \cos(\sqrt{\omega^2 - b^2}\, t + \theta) + F/pZ_o \sin(pt - \phi) \qquad \text{...(7.7 f)}$$

where $$Z_o = \sqrt{[r^2 + (mp - \mu/p)^2}$$

The constants of integration A and θ in the transient term of Eqn. (7.7 *f*) are determined from the initial conditions of the problem.

When a particle is subjected to forced vibrations its displacement in the beginning is made up of the transient as well as the steady state components. After a certain time t such that $bt >> 1$, the transient term rapidly fades away leaving only the steady state components in action. Before proceeding with a discussion of forced vibrations it is interesting to note the striking analogy of Eqn. (7.7 *a*) with that for an *LCR* circuit under an impressed alternating voltage. If a voltage $E_o \cos pt$ is impressed on a circuit containing an inductance L, capacity C and resistance R, the current i satisfies the following equation.

$$L\frac{di}{dt} + Ri + 1/c\int idt = E_o \cos pt$$

and the charge $q = \int idt$. The equation then reduces to

$$Lq'' + Rq' + 1/cq = E_o \cos pt \qquad \text{...(7.7 g)}$$

Comparison of Eqns. (7.7 *a*) and (7.7 *g*) leads to the solution

$$q = \frac{E_o}{p[R^2 + (Lp - 1/cp)^2]} \sin(pt - \phi)$$

Hence $$i = \frac{E_o}{[R^2 + (Lp - 1/cp)^2]} \cos(pt - \phi)$$

where $$\tan\phi = \frac{Lp - 1/cp}{R}$$

By analogy with the electrical impedance $\sqrt{[R^2 + (Lp - 1/cp)^2]}$ the quantity under the square root sign in Eqn. (7.7 *e*) is called the mechanical impedance. The analogy can be carried still further the mechanical resistance r is analogous to electrical resistance R, the mass m to the electrical inductance L and the restoring force per unit displacement μ (called the stiffness factor) to the reciprocal of capacitance. Hence we can call the quantity $(mp - \mu/p)$ as mechanical reactance.

7.8 RESONANCE

We have seen that the steady state displacement of a particle undergoing forced vibrations is given by

$$y = \frac{F}{p\sqrt{[r^2 + (mp - \mu/p)^2]}} \sin(pt - \phi)$$

Hence the particle velocity under this condition is

$$y' = \frac{F}{\sqrt{[r^2 + (mp - \mu/p)^2]}} \cos(pt - \phi)$$

The power dissipated in working against dissipative forces will be

$$\text{Force} \times \text{velocity} = ry'^2 = \frac{rF^2}{[r^2 + (mp - \mu/p)^2]} \cos^2(pt - \phi)$$

The average value of this power dissipation over one complete cycle is (since the average value of $\cos^2\theta = \frac{1}{2}$)

$$P_{av} = \frac{rF^2}{2[r^2 + (mp - \mu/p)^2]} \qquad \text{...(7.8 a)}$$

We can say that the amplitude and phase of the particle so adjust themselves that the average power being supplied by the driving force is just equal to that being dissipated in overcoming the frictional forces. The maximum power transfer will therefore take place when the denominator of Eqn. (7.8 *a*) has a minimum value or when

$$mp = \mu/p \quad \text{or} \quad p = \sqrt{\mu/m} = \omega \qquad \text{...(7.8 b)}$$

Hence we find that when the natural frequency of the vibrating particle is equal to that of the externally applied force, the former will pick up maximum power from the latter. Under this condition the mechanical reactance vanishes and the equation for displacement becomes

$$y = (F/\omega r)\sin \omega t \quad ...(7.8\ c)$$

Since $\tan \phi = \dfrac{mp - \mu/p}{r} = 0$

or $\phi = 0$

This phenomenon is known as resonance. This resonance displacement amplitude is therefore $F/\omega r$ and the resonance velocity amplitude is simply F/r. But the resonance displacement amplitude should not be mixed up with maximum displacement amplitude of the forced vibrations. This can be obtained by finding the condition under which the denominator of Eqn. (7.7 *e*) has a minimum value. For this

$$\frac{d}{dp}\left\{p\sqrt{\left[r^2 + \left(mp - \frac{\mu}{p}\right)^2\right]}\right\} = 0$$

which gives

$$p = \sqrt{\omega^2 - 2b^2} \quad \text{where} \quad b = r/2m$$

It may be verified that for this value of p the second differential of the quantity inside curved brackets above leads to a positive value.

The maximum displacement amplitude of the forced vibrations will therefore be

$$y_{max} = F/r\sqrt{p^2 + b^2} = F/r\sqrt{\omega^2 - b^2} \quad ...(7.8\ d)$$

Sharpness of Resonance. It has been seen above that the average power supplied to the vibrating particle by the external force is given by

$$P_{av} = \frac{rF^2}{2[r^2 + (mp - \mu/p)^2]}$$

At resonance, maximum power transfer takes place *i.e.*, at $\omega = p = \sqrt{\mu/m}$, the average power taken up by the particle is $F^2/2r$. Thus it is seen that the power transfer is primarily determined by the value of the mechanical resistance r. This has been shown in Fig. 7.10 which gives the resonance curves for two values of r. It is seen that for low values of r the resonance curve has a sharp peak. Increasing r by a factor of three considerably broadens the resonance peak. In the former case resonance is said to be sharp while in the latter case it is said to be flat.

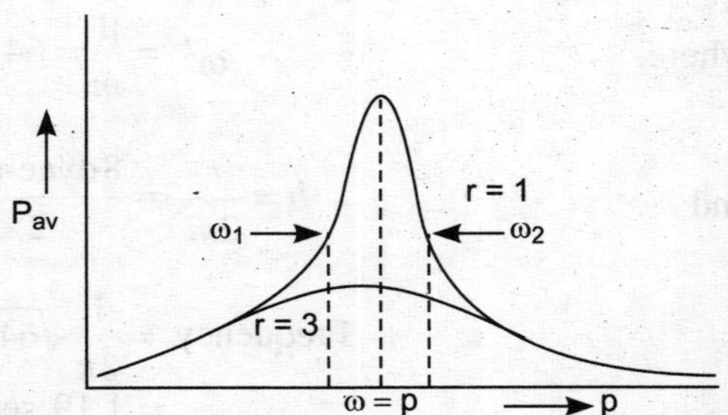

Fig. 7.10. Resonance curves

A quantitative measure of sharpness of resonance is the quality factor Q defined by

$$Q = \frac{\omega}{p_2 - p_1} \qquad \text{...(7.8 } e\text{)}$$

where p_1 and p_2 are two angular frequencies, (one less and the other greater than resonance frequency) at which the power transferred to the particle falls to half its value at resonance. The difference $p_2 - p_1$ can thus be found out from the following equation

$$\frac{F^2}{4r} = \frac{rF^2}{2[r^2 + (mp - \mu/p)^2]}$$

which gives p

$$p_2 - p_1 = r/m \qquad \text{...(7.8 } f\text{)}$$

Combining Eqns. (7.8 e) and (7.8 f) the quality factor is seen to be

$$Q = \frac{m\omega}{r} \qquad \text{...(7.8 } g\text{)}$$

The sharpness of resonance is therefore inversely proportional to the mechanical resistance r. When the damping is high as for example in a resonance column, conditions near to resonance are obtained over an appreciable length but when it is small, as in a sonometer wire a sharp resonance is obtained.

SOLVED EXAMPLES

Example 1. *A particle has a mass of 2 gms. It is free to vibrate under the action of an elastic force of 128 dyne-cm^{-1} and a damping force of 8 dyne-cm^{-1} sec. A periodically varying outside force of maximum value 256 dynes is applied to the particle. Find the frequency for displacement resonance, frequency for velocity resonance and the approximate amplitude at resonance.*

For displacement resonance,

$$\text{Frequency} = \frac{p}{2\pi} = \frac{1}{2\pi}\sqrt{\omega^2 - 2b^2}$$

where,

$$\omega^2 = \frac{\mu}{m} = 64\,\text{sec}^{-2} = \left[\frac{128\ \text{dyne-cm}^{-1}}{2\ \text{gm}}\right]$$

and

$$b = \frac{r}{2m} = \frac{8\,\text{dyne-cm}^{-1}\cdot\text{sec}}{2 \times 2\,\text{gm}} = 2\,\text{sec}^{-2}$$

$\therefore$

$$\text{Frequency} = \frac{1}{2\pi}\sqrt{64 - 8}$$
$$= 1.19\ \text{sec}^{-1}$$

For velocity resonance,

$$\text{Frequency} = \frac{1}{2\pi}\sqrt{\frac{\mu}{m}} = 1.27\,\text{sec}^{-1}$$

$$\text{Amplitude at resonance} = \frac{F}{\omega r}$$

$$= \frac{256}{8\times8} = 4.0 \text{ cms.}$$

Example 2. *A particle of mass 3 gm. is subject to an elastic force of 48 dyne-cm^{-1} and a damping force of 12 dyne-cm^{-1} sec. If the motion is oscillatory, find its period.*

$$\omega = \sqrt{\frac{\mu}{m}} = \sqrt{\frac{48 \text{ dyne-cm}^{-1}}{3 \text{ gm}}}$$

$$= 4 \text{ sec}^{-1}$$

and
$$b = \frac{r}{2m} = \frac{12}{2\times3} \text{sec}^{-1}$$

$$= 2 \text{ sec}^{-1}$$

Since $\omega > b$, the motion is oscillatory.

$$\text{Its periodic time} = \frac{2\pi}{\sqrt{\omega^2 - b^2}} = \frac{2\pi}{\sqrt{16-4}} = \frac{2\pi}{\sqrt{12}}$$

$$= 1.81 \text{ secs approx.}$$

PROBLEMS

1. Two collinear harmonic motions of the same frequency have amplitudes of 2 cm and 3cm respectively and corresponding phase angles of +10° and +30°. Find (*a*) *the amplitude and* (*b*) *the phase angle of the resultant vibration.*
2. Obtain an expression for the resultant motion of a particle under the simultaneous action of two rectangular motions of the same period.
3. What is sharpness of resonance? Obtain an expression for the sharpness of resonance for a mechanical system and discuss its variation with mechanical resistance.
4. Distinguish between forced vibrations and resonance. Discuss the oscillatory motion of a particle subjected to an external periodic force.
5. A certain mass is acted upon by an elastic force.
   (i) *Find its period of motion.*
   (ii) *How will this period be modified if the vibrating particle also experiences a dissipative force?*

WAVE MOTION

When a particle executes SHM motion in an elastic fluid, the energy and phase changes are not confined to its immediate vicinity but are progressively communicated to distant parts of the medium. Such disturbances propagating in the medium are called waves. We are familiar with many *types of waves* as for example the water waves travelling outwards when a pebble is dropped into a pond, the sound waves spreading through concert or cinema halls or the electromagnetic waves comprising the entire solar spectrum travelling through space. Then there are waves of larger amplitudes like sonic bangs or shock waves, detonations and earthquake waves. All the waves of small amplitude have certain properties in common irrespective of their nature: they travel with finite velocities, are reflected, and refacted, show interference and diffraction effects, exhibit the phenomenon of beats and Doppler shift, only the scale of these properties is different for different type of waves. It is not surprising, therefore, that they can be represented by the same general type of equation subject to conditions that differ in each field. The importance of wave motion therefore cannot be overemphasised by pointing out other common properties like dispersion and the existence of phase and group velocities, as for example in water and light waves or the significance of longitudinal and transverse waves in determining the specific heats of solids.

8.1 DIFFERENTIAL EQUATION OF WAVE MOTION

Consider the propagation of *pressure waves* in a homogeneous isotropic fluid medium which is perfectly elastic (*i.e.*, there are no dissipative forces such as those due to viscosity). Suppose the mean density ρ and mean pressure 'p' have uniform values in the medium. In the presence of a wave propagating from left to right the instantaneous values of pressure and density at any point will, in general, be different from their mean values; let them be p' and ρ' respectively. These variables, instantaneous pressure and density, will have reasonably constant values over an elementary volume of the

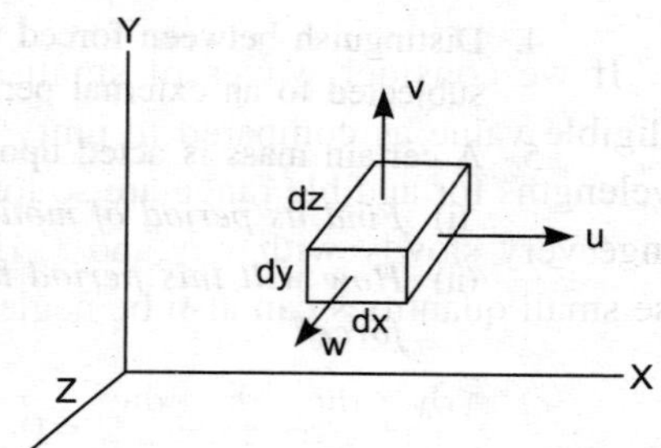

Fig. 8.1

fluid. We can call such volume elements as particles. On account of the changes in pressure and density, these particles will have instantaneous velocities. Let u, v and w be the components of the velocity of any such particle in the x, y and z directions respectively (Fig. 8.1).

Consider a volume element of the medium $dx\,dy\,dz$. The rate of change of the mass inside $dx\,dy\,dz$ will be given by the difference between rates of influx and efflux of fluid from this element. For the X-direction this flow is given by

$$\left[\rho' u - \left\{\rho' u + \frac{\partial}{\partial x}(\rho' u)dx\right\}\right]dy\,dz = -\frac{\partial}{\partial x}(\rho' u)dx\,dy\,dz$$

Similar expressions can be written for flow along the other two directions so that the net influx in the volume element will be

$$-\left[\frac{\partial}{\partial x}(\rho' u) + \frac{\partial}{\partial y}(\rho' v) + \frac{\partial}{\partial z}(\rho' w)\right]dx\,dy\,\partial z$$

According to the law of conservation of mass, this expression must be equal to the rate of change of mass of fluid inside the volume $dx\,dy\,dz$.

Hence $$\frac{\partial \rho'}{\partial t} + \frac{\partial}{\partial x}(\rho' u) + \frac{\partial}{\partial y}(\rho' v) + \frac{\partial}{\partial z}(\rho' w) = 0 \qquad \text{...(8.1 } a)$$

This is usually known as the equation of continuity.

Since ρ' is changing we like to express Eqn. (8.1 *a*) in terms of the mean density ρ. Suppose we write

$$\frac{\rho'}{\rho} = 1 + s \qquad \text{...(8.1 } b)$$

where s is called the condensation, then

$$\frac{\partial \rho'}{\partial t} = \rho \frac{\partial s}{\partial t} \qquad \text{...(8.1 } c)$$

Substituting Eqn. (8.1 *c*) into Eqn. (8.1 *a*) we therefore get

$$\rho\frac{\partial s}{\partial t} + \rho(1+s)\left(\frac{\partial u}{\partial x} + \frac{\partial v}{\partial y} + \frac{\partial w}{\partial z}\right) + \rho\left(u\frac{\partial s}{\partial x} + v\frac{\partial s}{\partial y} + w\frac{\partial s}{\partial z}\right) = 0 \qquad \text{...(8.1 } d)$$

If we consider waves of small amplitude, *e.g.*, acoustic waves, the condensation 's' has a negligible value as compared to unity (for speech it is of the order of 10^{-7}). Moreover the acoustic wavelengths for audible range are so long that all the quantities u, v and w along with their derivatives change very slowly with x, y and z. Hence all the terms in Eqn. (8.1 *d*) involving the products of these small quantities can also be neglected so that we have

$$\frac{\partial s}{\partial t} + \frac{\partial u}{\partial x} + \frac{\partial v}{\partial y} + \frac{\partial w}{\partial z} = 0 \qquad \text{...(8.1 } e)$$

in place of Eqn. (8.1 *d*) on taking these factors into account.

We shall now write u, v and w in terms of the derivatives of a scalar quantity ϕ called the velocity potential. Suppose $\phi = \phi(x, y, z, t)$

$$u = \frac{\partial \phi}{\partial x}, \quad v = \frac{\partial \phi}{\partial y}, \quad w = \frac{\partial \phi}{\partial z}$$

then the gradients of ϕ along the three axes are all vector quantities although ϕ itself is scalar. The introduction of velocity potential at this stage will considerably simplify matters as we proceed. From Eqn. (8.1 *e*), we therefore get

$$\frac{\partial s}{\partial t} + \nabla^2 \phi = 0 \qquad \text{...(8.1 } f\text{)}$$

where the symbol ∇^2 is called the Laplacian operator (∇ is pronounced as del and ∇^2 as del square) and has been written in place of $\frac{\partial^2}{\partial x^2} + \frac{\partial^2}{\partial y^2} + \frac{\partial^2}{\partial z^2}$.

There is an unbalanced force acting along each of the three directions x, y and z. From the definition of force (pressure x area), the net force acting in the positive X-direction is, given by

$$\left[p' - \left(p' + \frac{\partial p'}{\partial x} dx \right) \right] dy\, dz = -\frac{\partial p'}{\partial x} dx\, dy\, dz$$

This must be equal to the rate of change of momentum.

$$\frac{\partial}{\partial t} (\rho' u\, dx\, dy\, dz)$$

so that the equation of motion becomes

$$\frac{\partial p'}{\partial x} + \frac{\partial (\rho' u)}{\partial t} = 0 \qquad \text{...(8.1 } g\text{)}$$

under the conditions mentioned in connection with the continuity equation

$$\frac{\partial}{\partial t} (\rho' u) = \rho \frac{\partial u}{\partial t} \qquad \text{...(8.1 } h\text{)}$$

Hence we have from Eqns. (8.1 *g*) and (8.1 *h*)

$$\left.\begin{aligned} &\frac{\partial p'}{\partial x} + \rho \frac{\partial u}{\partial t} = 0 \\ \text{Similarly} \quad &\frac{\partial p'}{\partial y} + \rho \frac{\partial v}{\partial t} = 0 \\ \text{and} \quad &\frac{\partial p'}{\partial z} + \rho \frac{\partial w}{\partial t} = 0 \end{aligned}\right\} \qquad \text{...(8.1 } i\text{)}$$

We can then write

$$\frac{\partial p'}{\partial x}dx+\frac{\partial p'}{\partial y}dy+\frac{\partial p'}{\partial z}dz+\rho\frac{\partial}{\partial t}(udx+vdy+wdz)=0 \quad ...(8.1\ j)$$

or $$dp'+\rho\frac{\partial}{\partial t}(d\phi)=0$$

where $$dp'=\frac{\partial p'}{\partial x}dx+\frac{\partial p'}{\partial y}dy+\frac{\partial p'}{\partial z}dz$$

and $$d\phi=\frac{\partial \phi}{\partial x}dx+\frac{\partial \phi}{\partial y}dy+\frac{\partial \phi}{\partial z}dz$$

An integration of Eqn. (8.1 *j*) gives

$$p'+\rho\frac{\partial \phi}{\partial t}=C$$

where the constant of integration C can be evaluated from the conditions of the problem. Thus, when no waves are propagating through the medium $\frac{\partial \phi'}{\partial t}=0$ and $p'=p$ so that, we get

$$-\rho\frac{\partial \phi}{\partial t}=p'-p \quad ...(8.1\ k)$$

where the RHS of Eqn. (8.1 *k*) is called the excess pressure (p_o).

On account of the excess pressure acting upon the volume element, the medium will be under volume strain given by

$$\frac{dV}{V}=-\frac{d\rho'}{\rho}=-\frac{\rho'-\rho}{\rho}=-s$$

Hence, the RHS of Eqn. (8.1 *k*) can be expressed in terms of the negative condensation and bulk modulus of elasticity K of the medium.

$$p'-p=dp=-\frac{dV}{KV}=Ks$$

so that $$\frac{\partial \phi}{\partial t}=-\frac{Ks}{\rho}=-C^2s \quad ...(8.1\ l)$$

where $$C^2=K/\rho \quad ...(8.1\ m)$$

Combining Eqns. (8.1 *l*) and (8.1 *f*), we get

$$\frac{\partial^2 \phi}{\partial t^2}=C^2\nabla^2\phi \quad ...(8.1\ n)$$

and the excess pressure

$$p_o = \rho C^2 s \qquad ...(8.1\ o)$$

Equation (8.1 *n*) is the three dimensional acoustic wave equation applicable to fluids. Its solutions give the propagation of the velocity potential ϕ, the velocity of propagation C being given by

$$C = \sqrt{\frac{\text{Bulk modulus}}{\text{Density}}}.$$

8.2 PLANE WAVE EQUATION

If the deformations in the medium are a function of one cartesian coordinate, the wave is said to be plane. In such waves conditions are uniform over a plane specified by the above space coordinate. Thus the differential equation of a plane wave of infinite width propagating in the X-direction can be written as

$$\frac{\partial^2 \phi}{\partial t^2} = C^2 \frac{\partial^2 \phi}{\partial x^2} \qquad ...(8.2\ a)$$

The general solution of this wave equation may be written in the form

$$\phi = f_1(ct - x) + f_2(ct + x) \qquad ...(8.2\ b)$$

It can be seen by substitution that irrespective of the nature of functions f_1 and f_2 Eqn. (8.2 *b*) is always a solution of Eqn. (8.2 *a*). Thus a single scalar quantity ϕ is sufficient to represent the propagation of a ***longitudinal disturbance*** in any fluid medium. If this disturbance is harmonic, the solution is written in the form

$$\phi = Ae^{-i(\omega t - kx)} + Be^{-i(\omega t + kx)} \qquad ...(8.2\ c)$$

where A is the complex amplitude of a plane wave travelling in the positive X-direction and B that of a wave travelling in the negative X-direction both propagating with the velocity $C = \omega/k$.

In the case of acoustic waves where A and B are real we take the real part of Eqn. (8.2 *c*). We thus have

$$\left.\begin{aligned} \phi &= A\cos(\omega t - kx) + B\cos(\omega t + kx) \\ p_o = -\rho\frac{\partial \phi}{\partial t} &= \rho\omega\, A\sin(\omega t - kx) + \rho\omega\, B\sin(\omega t + kx) \\ u = \partial\phi/\partial x &= KA\sin(\omega t - kx) - KB\sin(\omega t + kx) \end{aligned}\right\} \qquad ...(8.2\ d)$$

The particle displacement ξ defined by the equation $d\xi/dt = u$ is thus seen to be

$$\xi = \int u dt = \int KA \sin(\omega t - kx)dt - \int KB \sin(\omega t + kx)dt$$

or

$$\xi = -A/C \cos(\omega t - kx) + B/C \cos(\omega t + kx).$$

8.3 DEFINITIONS

A wave travelling in the positive X-direction has been seen to be represented by the equation

$$\phi = A \cos(\omega t - kx)$$

The time taken by one complete wave to pass any point is called its periodic time T. Thus

$$\omega T = 2\pi \quad \text{or} \quad T = \frac{2\pi}{\omega}$$

The frequency v is the number of waves passing a stationary observer in one second. So that

$$v = 1/T = \omega/2\pi$$

The wave profile repeats itself at regular intervals λ given by

$$K\lambda = 2\pi$$

or

$$\lambda = \frac{2\pi}{k} = \frac{2\pi C}{\omega}$$

which on combination with the expression for frequency gives

$$C = v\lambda$$

All these relations can be used in writing down alternative forms of the wave equation. These are summarised below

$$\left.\begin{aligned} \phi &= A\cos(\omega t - kx) \\ &= A\cos 2\pi(t/T - x/\lambda) \\ &= A\cos(2\pi/\lambda)(ct - x) \\ &= A\cos 2\pi(vt - x/\lambda) \end{aligned}\right\} \qquad \text{...(8.3 } a\text{)}$$

An inspection of Eqn. (8.3 *a*) reveals that at any time t different particles of the medium (different x's) are in different states of disturbance. As time passes every single particle of the medium goes through the same cycle of change of state. Such waves are called progressive waves. A few features of plane progressive waves have been shown in Fig. 8.2. The analytical expressions for these variations have already been derived. Figure 8.2 indicates that the particle velocity, the excess pressure and the condensation are all in phase. But all these quantities are $\pi/2$ out of phase with the displacement of the particle. Accordingly particle velocity, condensation and excess pressure shall attain their maximum values only when the local cross-section of the medium is passing through its central position. The dilatation is of course π out of phase with condensation.

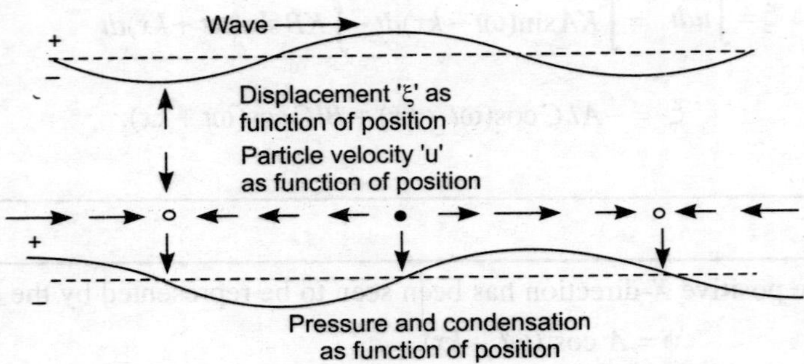

Fig. 8.2. Features of plane progressive waves

8.4 ENERGY DENSITY OF PLANE WAVES

When a progressive wave travels through a fluid, successive particles of the medium possess local velocities at any instant. In addition the medium is in a state of compression or expansion due to the passage of the acoustic wave. The energy involved in wave propagation is therefore of two forms—Kinetic and Potential. The kinetic energy of the moving particles is

$$\Delta E_{\text{kin}} = \frac{1}{2}\rho u^2 \Delta V$$

where ΔV is the volume element which can be treated effectively as a particle. The potential energy of this volume element is

$$\Delta E_{\text{pot}} = -\int p_o dV$$

where the negative sign indicates an increase in potential energy due to compression and decrease due to expansion. In this expression the excess pressure p_o has been seen to be

$$p_o = \rho C^2 s$$

and the condensation s is defined as

$$-\frac{\text{Change in volume}}{\text{Original volume}} = s \quad \text{or} \quad -\left(\int dV\right)/\Delta V = s$$

Hence, the expression for potential energy becomes

$$\Delta E_{\text{pot}} = +\int \rho C^2 s \, ds(\Delta V) = \frac{1}{2}\rho C^2 s^2 (\Delta V)$$

where dV has been written as

$$dV = ds\ (\Delta V)$$

The total acoustic energy of the volume element is therefore

$$\Delta E = \Delta E_{\text{kin}} + \Delta E_{\text{pot}}$$

$$= \frac{1}{2}\rho \Delta V (u^2 + C^2 s^2)$$

and the energy density per unit volume is

$$e = \Delta E / \Delta V = \frac{1}{2}\rho(u^2 + C^2 s^2) \text{ ergs./c.c.} \qquad ...(8.4\ a)$$

It can be shown that

$u = Cs$ for a positively travelling wave

and $u = -Cs$ for a negatively travelling wave.

The instantaneous energy density corresponding to the presence of both the waves is accordingly

$$e = e_+ + e_- = \rho(u_+^2 + u_-^2) \qquad ...(8.4\ b)$$

where the subscripts + and – refer to the two waves travelling in opposite directions.

According to Eqn. (8.2 *e*), the particle velocity of either wave is a function of both space and time. The energy density of either wave is therefore not constant throughout the medium. We can nevertheless calculate either the time average of energy density at any point or the space average of energy density at any time due to motion of either wave. Thus for example the time average density at a point is given by

$$\mathbf{e} = 1/T \int_o^T e\, dt$$

where the integration on the RHS is to be performed over one complete cycle. Substituting the value of *e*, we get

$$\mathbf{e} = 1/T \int_o^T [kA \sin(\omega t - kx)]^2\, dt$$

$$= 1/2\, \rho\, k^2 A^2 \qquad ...(8.4\ c)$$

Similarly it can be shown that the time average of energy density due to a negatively travelling wave is given by

$$\mathbf{e} = \frac{1}{2}\rho k^2 B^2 \qquad ...(8.4\ d)$$

If we obtain the space average densities at any time due to individual waves we find that these averages are the same as the respective time averages.

There are alternative expressions for space and time averages of energy densities at a point in terms of the velocity amplitude kA, pressure amplitude $P = KA\rho C$ and displacement amplitude $\varepsilon = kA/\omega$ which readily follow from Eqns. (8.4 *c*) and (8.4 *d*).

8.5 ACOUSTIC INTENSITY

The acoustic intensity of a pressure wave is defined as the average amount of energy flowing per second through a unit area held normal to the direction of propagation. It is evident that all the energy contained in a length C of the medium will pass through a unit area in one second and consequently sound intensity can be written as

$$I = Ce = \frac{1}{2}\rho C k^2 A^2 \quad \text{or} \quad \frac{1}{2}\rho C k^2 B^2 \qquad ...(8.5\ a)$$

as the case may be. This intensity can be measured in ergs per second per square cm. or Watts per sq. cm. But it is the usual to measure it with reference to some acoustic intensity standard. Since the range of sound intensities encountered in practice is rather large, it is more convenient to use a logarithmic scale. An added reason for this is the fact that the response of human ears to sound intensity follows an approximately logarithmic scale.

Thus the intensity level of a sound of intensity I is defined as

$$n = 10 \log (I/I_o)$$

where I_o is the reference intensity and n is measured decibels (*db*). Intensity levels are also sometimes expressed in terms of pressure amplitude of the sound wave

$$n = 20 \log (P/P_o)$$

where P_o is some reference pressure. This definition is usually followed in recording and reproduction of sound.

The intensity of a 1000 cps pure note just audible to human ears is sometimes taken as reference intensity I_o for plane waves propagating in air. This works out to be 10^{-16} watt/cm^2. But the commonly used standard for underwater measurements is a root mean square pressure of 1 dyne/cm^2 ($P = \rho CkA$). This has also been adopted for airborne sound. A reference pressure of 1 dyne/cm^2 corresponds to a reference intensity of 6.75×10^{-6} watts/sec. in fresh water and 2.41×10^{-2} watts/sec. in air.

8.6 SPECIFIC ACOUSTIC IMPEDANCE

The ratio of acoustic pressure in a medium to the associated particle velocity is called the specific acoustic impedance Z of the medium. Thus

$$Z = \frac{-\rho\partial\phi/\partial t}{\partial\phi/\partial x}$$

These derivatives can be calculated for the positively or negatively travelling plane waves with the help of Eqn. (8.2 *c*). For plane waves travelling in the positive X-direction

$$Z_+ = \rho C$$

and for those travelling in opposite direction

$$Z_- = -\rho C$$

while the specific acoustic impedance is seen to be a real quantity. In the case of plane waves, it is, in general a complex quantity of the form

$$Z = r + is$$

where r is the specific acoustic resistance and 's' the specific acoustic reactance of the medium for the wave motion under study. Thus for plane waves specific acoustic reactance is seen to be zero.

The importance of this quantity will become apparent as we proceed further.

8.7 SPHERICAL ACOUSTIC WAVES

The wave Eqn. (8.1 *n*) can be expressed in coordinate systems other than cartesian. Each of these systems is suitable under different situations. We have already seen how rectangular coordinate system can be used for describing plane progressive waves. If there is a long cylindrical source of sound, it is usually more advantageous to express Eqn. (8.1 *n*) in cylindrical coordinates. On the other hand if the source approximates a point we should choose spherical coordinates. All that needs to be done is the expressing of $\nabla^2\phi$ in either cylindrical coordinates (r, θ, Z) or spherical coordinates (r, θ, ψ). In the spherical system for example (Fig. 8.3)

$$x = r\sin\theta\cos\psi$$

$$y = r\sin\theta\sin\psi$$

$$z = r\cos\theta$$

and
$$\nabla^2\phi = \frac{\partial^2\phi}{\partial t^2} + \frac{2}{r}\frac{\partial\phi}{\partial t} + \frac{1}{r^2\sin\theta}\frac{\partial}{\partial\theta}\left(\sin\partial\frac{\partial\phi}{\partial\theta}\right) + \frac{1}{r^2\sin^2\theta}\frac{\partial^2\phi}{\partial\psi^2}$$

If ϕ is a function of r and t only, the system is said to have spherical symmetry. We shall then have

$$\nabla^2\phi = \frac{\partial^2\phi}{\partial r^2} + \frac{2}{r}\frac{\partial\phi}{\partial r} = \frac{1}{r}\frac{\partial^2}{\partial r^2}(r\phi)$$

and Eqn. (8.7 *a*) reduces to

$$\frac{1}{r}\frac{\partial^2(r\phi)}{\partial t^2} = \frac{C^2}{r}\frac{\partial^2}{\partial r^2}(r\phi) \qquad \text{...(8.7 a)}$$

where the factor 'r' on the LHS of this equation is not a function of time so that

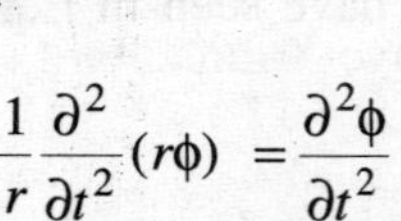

$$\frac{1}{r}\frac{\partial^2}{\partial t^2}(r\phi) = \frac{\partial^2\phi}{\partial t^2}$$

Fig. 8.3

The differential equation of spherical waves is therefore

$$\frac{\partial^2}{\partial t^2}(r\phi) = C^2\frac{\partial^2}{\partial r^2}(r\phi) \qquad \text{...(8.7 b)}$$

The solution of Eqn. (8.7 *b*) can be written at once if we consider '$r\phi$' as a single variable so that

$$r\phi = f_1(ct - r) + f_2(ct + r)$$

or
$$\phi = (1/r)f_1(ct - r) + (1/r)f_2(ct + r) \qquad \text{...(8.7 c)}$$

where the first term of the RHS represents a wave diverging from a central point and the other another wave converging on this point. A harmonic solution of the wave equation can be written as

$$\phi = A/r\, e^{-i(\omega t-kr)} + B/r\, e^{-i(\omega t+kr)} \qquad ...(8.7\ d)$$

As before we can derive expressions for energy density, specific acoustic impedance etc. for spherical waves. It may be mentioned that spherical waves tend to behave like plane waves at considerable distances from the source.

8.8 EQUATION OF TELEGRAPHY

In the derivation and discussions of the wave equations and (8.7 *c*) we had assumed that there were no dissipative forces present in the medium. In the presence of such dissipative forces we have to take into account the effect of damping by introducing a term $r\,(\partial\phi/\partial t)$ in the wave equation. Thus we write

$$\nabla^2\phi = \frac{1}{C^2}\left(\frac{\partial^2\phi}{\partial t^2} + r\frac{\partial\phi}{\partial t}\right) \qquad ...(8.8\ a)$$

An equation similar in form to Eqn. (8.8 *a*) is of great importance in electromagnetic theory (there the meanings of C and ϕ are of course different). This equation can be solved by proper substitutions and gives an amplitude factor decaying exponentially with time. Thus the solution for one dimensional equation is of the form

$$\phi = e^{-rt/2} f(x-ct).$$

8.9 USE OF COMPLEX QUANTITIES IN WAVE THEORY

We shall be dealing mostly with harmonic waves and therefore it is appropriate at this stage to understand a notation used for representing such waves. We have seen in Eqn. (8.2 *e*) that a plane progressive wave may be represented by

$$\phi = a\cos(\omega t - kx)$$

where the velocity potential ϕ has an amplitude 'a'. If we also introduce a phase δ we can write

$$\phi = a\cos[(\omega t - kx) + \delta] \qquad ...(8.9\ a)$$

The trigonometric function on the RHS is seen to be the real part of the complex quantity

$$\phi = ae^{\pm i[(\omega t - kx)+\delta]} \qquad ...(8.9\ b)$$

It is more convenient to use the negative sign in the above expression. We can also write Eqn. (8.9 *b*) as

$$\phi = ae^{-i\delta}e^{-i(\omega t-kx)}$$

or

$$\phi = Ae^{-i(\omega t-kx)} \qquad ...(8.9\ c)$$

where the apparent amplitude A is a complex quantity defined by

$$A = ae^{-i\delta} \quad \text{...(8.9 d)}$$

So that the true amplitude is $|A| = a$ and the negative argument ‡ of A *i.e.*, δ gives the phase. The intensity which is proportional to a^2 can be obtained by finding the product of A and its complex conjugate A^*. Thus

$$AA^* = ae^{-i\delta}ae^{+i\delta} = a^2$$

The scalar quantity ϕ expressed as in Eqn. (8.9 *b*) is itself a solution of the one dimensional wave equation (This we can easily verify by direct substitution). Hence in any equation in which ϕ appears in the first degree, we can use Eqn. (8.9 *c*) as it is without reference to its real or imaginary parts and later on extract the part of the final result of interest to us.

The great advantage of using Eqn. (8.9 *c*) is that if we want to compose a large number of coherent disturbances (disturbances in the same phase or having a constant phase difference) we have simply to add up the complex amplitude because the other term $e^{-i(\omega t - kx)}$ will be common to all of them. But for composing incoherent disturbances we shall have to add individual intensities.

8.10 REFLECTION AND REFRACTION OF WAVES

Normal Incidence. When a plane progressive wave propagating in a fluid medium strikes its boundary with a second medium, a reflected wave is produced in the first medium and a transmitted or refracted wave in the second. The ratio of the intensities of the reflected and transmitted waves is determined by the properties (say acoustic impedance in the case of pressure waves) of the two media and the angle of incidence at the boundary. Suppose BB' represents the boundary between two media *I* and *II* at which an acoustic plane wave is incident normally. This wave may be represented *in terms of the acoustic pressure*** P_i (rather than velocity potential) by an equation of the form

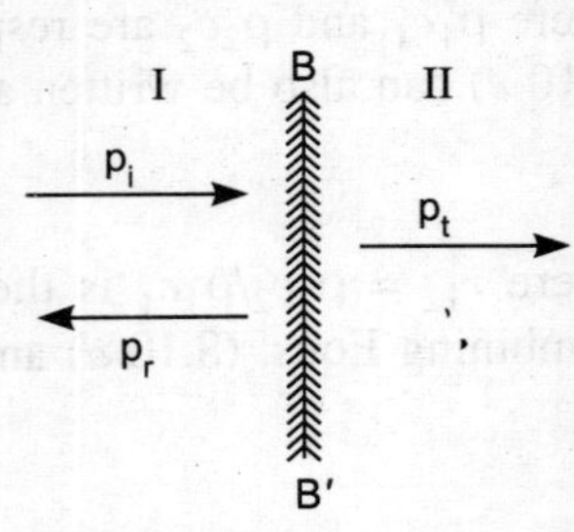

Fig. 8.4. Reflection at normal incidence

$$p_i = A_i e^{-i(\omega t - k_1 x)} \quad \text{...(8.10 a)}$$

where k_1 is the reduced wave number and A_i the pressure amplitude in the first medium. The reflected and transmitted waves will then be given by

$$p_r = A_r e^{-i(\omega t + k_1 x)} \quad \text{...(8.10 b)}$$

and

$$p_t = A_t e^{-i(\omega t - k_2 x)} \quad \text{...(8.10 c)}$$

‡ The modulus $|z|$ of the complex quantity $Z = A + iB$ is defined as $\sqrt{A^2 + B^2}$ and the argument θ is given by $\tan\theta = B/A$.

** This discussion could as well have been in terms of particle displacement (instead of acoustic pressure) but in that case the phase changes on reflection would have been just the opposite of those described below.

where k_2 is the reduced wave number in the second medium and the subscripts r and t stand for the reflected and transmitted waves. It may be noted that the frequency of the transmitted wave is the same as that of incident wave; it is only the wavelength $\lambda = 2\pi/k$ which undergoes a change with change in medium.

We are at liberty to choose our coordinate system. Let us suppose that the boundary coincides with the coordinate $x = 0$. The conditions at the boundary can then be written as

$$p_i = A_i e^{-i\omega t}; \quad p_r = A_r e^{-i\omega t}; \quad p_t = A_t e^{-i\omega t}$$

Since there cannot exist any discontinuity of pressure in a fluid medium, the pressure in the first medium must be equal to that in the second medium at the surface of separation between the two. We thus have

$$p_i + p_r = p_t \quad \text{or} \quad A_i + A_r = A_t \qquad \text{...(8.10 } d\text{)}$$

Also if the two media are to remain in contact with each other, the resultant particle velocity in the first medium should be equal to that in the second. Thus

$$u_i + u_r = u_t$$

or

$$\frac{1}{\rho_1 c_1}(p_i - p_r) = \frac{1}{\rho_2 c_2} p_t \quad \text{(see Section 8.6)} \qquad \text{...(8.10 } e\text{)}$$

where $\rho_1 c_1$ and $\rho_2 c_2$ are respectively the specific acoustic impedances in media *I* and *II*. Equation (8.10 *e*) can also be written as

$$r_{12} = (A_i - A_r) = A_t \qquad \text{...(8.10 } f\text{)}$$

where $r_{12} = \rho_2 c_2/\rho_1 c_1$ is the relative impedance of the second medium with respect to the first. Combining Eqns. (8.10 *d*) and (8.10 *f*), we get

$$\frac{A_r}{A_i} = \frac{r_{12} - 1}{r_{12} + 1} \quad \text{and} \quad \frac{A_t}{A_i} = \frac{2r_{12}}{r_{12} + 1} \qquad \text{...(8.10 } g\text{)}$$

If $r_{12} > 1$, the reflected wave at the boundary is in the same phase as the incident wave but if $r_{12} < 1$ the two waves will be in opposite phase. Expressed differently it means that if the characteristic impedance of the second medium is greater than that of the first, reflection will take place without any change of phase at the boundary. This condition is encountered when a wave is reflected at the boundary while going from a rarer to a denser medium (*e.g.*, air and water). On the contrary, if reflection takes place when a wave travelling from a denser to a rarer medium encounters the boundary, a 180° change of phase is introduced.

There are a few special cases of the phenomenon. If $r_{12} \to 0$, the amplitudes of the incident and reflected waves tend to be the same and the two waves will be 180° out of phase. If $r_{12} \to \infty$, as for example in the case of reflection at a very rigid boundary, the two waves will again have the same amplitude and will be in the same phase as well. In both these cases standing waves will be produced but the state of disturbance at the boundary will not be the same. If $r_{12} = 1$ *i.e.*, the two media have the same acoustic impedance there is no reflection at all. Under this condition all the incident energy will be totally transmitted into the second medium. This important result is often made use of as for example, in developing special rubbers for protecting the surface of transducers against the corrosive action of water in the equipment for underwater sounding.

The transmitted wave, however remains always in the same phase as the incident wave and its amplitude may range from $2A$ to zero depending upon the value of r_{12} (see Eqn. 8.10 *g*).

8.11 OBLIQUE INCIDENCE

Let us now take the general case. Suppose AO is the direction of propagation of plane waves which lies in the XY plane. A plane such as PP', perpendicular to AO will then give the locus of all points in the same phase of vibration (This is called a wavefront). The equation of this plane is

$$d = x\cos\theta + y\sin\theta \qquad ...(8.11\ a)$$

where d is the distance of the plane from the Z-axis measured in the direction of propagation of the waves. The excess pressure in the wavefront can then be written as

$$p_i = A_i e^{-i(\omega t - k_1 d)}$$

or

$$p_i = A_i e^{-i(\omega t - k_1 x\cos\theta - k_1 y\sin\theta)} \qquad ...(8.11\ b)$$

where k_1 is the reduced wave number in the first medium. This plane wavefront on striking the boundary between the two media will be reflected and transmitted.

Let the excess pressure p_r and p_t in the reflected and transmitted waves be represented by

$$p_r = A_r e^{-i(\omega t - k_1 d_r)}$$

and

$$p_t = A_t e^{-i(\omega t - k_2 d_t)}$$

Fig. 8.5. Reflection at oblique incidence

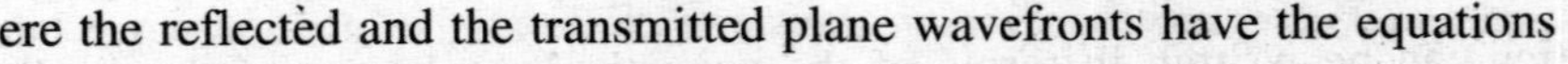

where the reflected and the transmitted plane wavefronts have the equations

$$d_r = l_1 x + m_1 y + n_1 z$$

and

$$d_t = l_2 x + m_2 y + n_2 z$$

respectively. We then have

$$p_r = A_r e^{-i[\omega t - k_1(l_1 x + m_1 y + n_1 z)]} \qquad ...(8.11\ c)$$

and

$$p_t = A_t e^{-i[\omega t - k_1(l_2 x + m_2 y + n_2 z)]} \qquad ...(8.11\ d)$$

for the reflected and transmitted components.

Applying the condition for continuity of pressure at the boundary ($x = 0$), we have

$$A_i e^{-i(\omega t - k_1 y \sin\theta)} + A_r e^{-i[\omega t - k_1(m_1 y + n_1 z)]}$$
$$= A_t e^{-i[\omega t - k_2(m_2 y + n_2 z)]} \quad \text{...(8.11 } e\text{)}$$

This identity can only be satisfied if the indices of all the three terms are identical: *i.e.*

$$\omega t - k_1 y \sin\theta = \omega t - k_1(m_1 y + n_1 z) = \omega t - k_2(m_2 y + n_2 z)$$

or
$$k_1 \sin\theta = k_1 m_1 = k_2 m_2 \quad \text{...(8.11 } f\text{)}$$

and
$$0 = k_1 n_1 = k_2 n_2 \quad \text{...(8.11 } g\text{)}$$

Equation (8.11 *g*) shows that $n_1 = n_2 = 0$ so that the reflected and the transmitted rays are found to be in the plane of incidence *XY*. The first part of Eqn. (8.11 *f*) shows that

$$m_1 = \sin\theta$$

Hence the angles of incidence and reflection are seen to be equal. The second of Eqn. (8.11 *f*) reveals that

$$k_1 \sin\theta = k_2 \sin\phi$$

which is the well known Snell's law of refraction.

Under these conditions, Eqn. (8.11 *e*) reduces to

$$A_i + A_r = A_t \quad \text{...(8.11 } h\text{)}$$

As before we can write the condition for continuity of the *X*-component of velocity at the boundary.

$$U_i \cos\theta + U_r \cos(\pi - \theta) = U_t \cos\phi \quad \text{...(8.11 } i\text{)}$$

where the particle velocities can be written in terms of the acoustic impedances of the media. Eqns. (8.11 *h*) and (8.11 *i*) may then be combined to give the reflected and transmitted fractions of the incident pressure amplitude.

8.12 SUPERPOSITION OF WAVES

As mentioned in *Sec.* 7.4, it is the property of linear differential equations that the algebraic sum of all its solutions is also one of its solutions. We shall prove this interesting result for the harmonic waves. Such waves satisfy the differential equation

$$\partial^2\phi/\partial t^2 = c^2\nabla^2\phi \quad \text{...(8.12 } a\text{)}$$

which is linear as it does not contain any powers higher than unity of the velocity potential ϕ or any of its derivatives viz., $\partial\phi/\partial t$, $\partial^2\phi/\partial t^2$ or $\partial\phi/\partial x$, $\partial^2\phi/\partial x^2$ etc. In addition, it is a homogeneous equation as it does not contain any term independent of ϕ. If ϕ_1, ϕ_2 etc. are the solutions of this equation we can write

$$\left.\begin{aligned}\phi_1 &= A_1 e^{-i(\omega_1 t - k_1 x)}\\ \phi_2 &= A_2 e^{-i(\omega_2 t - k_2 x)}\end{aligned}\right\} \qquad ...(8.12\ b)$$

If the individual harmonic disturbances ϕ_1, ϕ_2 etc. travel through the medium with the same velocity

$$\frac{\omega_1}{k_1} = \frac{\omega_2}{k_2} = ... = \frac{\omega_n}{k_n} = c$$

Equation (8.12 *b*) can then be written as

$$\phi_1 = A_1 e^{-ik_1(ct-x)}$$

$$\phi_2 = A_2 e^{-ik_2(ct-x)}$$

and so on. Therefore, the algebraic sum of all the individual solutions may be written as

$$\phi_1 + \phi_2 + ... = A_1 e^{-ik_1(ct-x)} + A_2 e^{-ik_2(ct-x)} + \cdots$$

If the LHS of this equation is represented by ψ and the RHS by a function $F(ct - x)$ we can write

$$\psi = F(ct - x) \qquad ...(8.12\ c)$$

We now have to show that ψ given by Eqn. (8.12 *c*) is a solution of the classical wave Eqn. (8.12 *a*). Successive partial differentiations of Eqn. (8.12 *c*) lead to the following results.

$$\frac{\partial \psi}{\partial t} = \frac{dF}{d(ct-x)} \cdot \frac{\partial (ct-x)}{\partial t} = c\frac{dF}{d(ct-x)}$$

$$\frac{\partial^2 \psi}{\partial t^2} = c\frac{d^2 F}{d(ct-x)^2} \cdot \frac{\partial (ct-x)}{\partial t} = c^2 \frac{d^2 F}{d(ct-x)^2}$$

Similarly, $$\frac{\partial \psi}{\partial x} = \frac{dF}{d(ct-x)} \cdot \frac{\partial (ct-x)}{\partial x} = -\frac{dF}{d(ct-x)}$$

and $$\frac{\partial^2 \psi}{\partial x^2} = -\frac{d^2 F}{d(ct-x)^2} \cdot \frac{\partial (ct-x)}{\partial x} = \frac{d^2 F}{d(ct-x)^2}$$

We thus find that

$$\frac{\partial^2 \psi}{\partial t^2} = c^2 \frac{\partial^2 \psi}{\partial x^2}$$

so that ψ satisfies the wave Eqn. (8.12 *a*). Hence we arrive at the interesting result that if two or more disturbances propagate through the same medium simultaneously with the same velocity, the algebraic sum of the individual disturbances shall also travel through the medium with the same velocity and hence may be taken as the resultant disturbance. This is the general enunciation of the Principle of

Superposition first propounded by *Young*. In other words, when a non-dispersive medium is disturbed simultaneously by any number of waves *the instantaneous resultant displacement of the medium at every point at every instant is the algebraic sum of the displacements taken independently.*

8.13 SUPERPOSITION OF WAVES OF THE SAME FREQUENCY

(i) Travelling along the same direction. Consider two plane progressive waves of the same frequency and wavelength travelling *approximately* along the same direction in a medium. Let their plane wave fronts be located at any instant at A and B. For reaching P, they have to travel distances x_1 and x_2 and hence the disturbance at P due to the individual waves may be written as

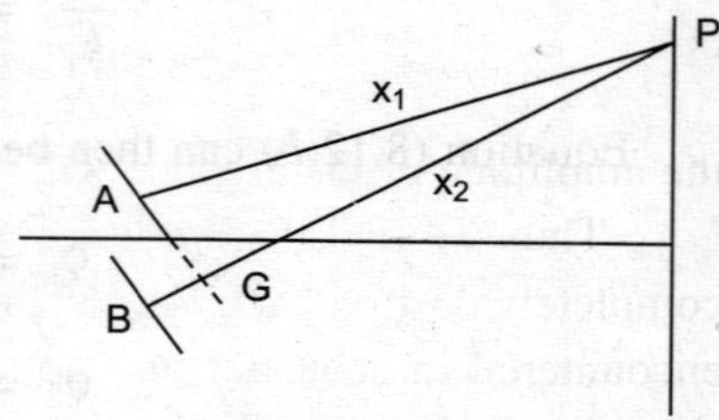

Fig. 8.6. Superposition of waves travelling along the same direction

$$\xi_1 = A\cos(\omega t - kx_1)$$

and

$$\xi_2 = A\cos(\omega t - kx_2)$$

where ξ represents the instantaneous displacement at any point of the medium whose maximum value A has been taken to be the same for the both the waves for the sake of simplicity. According to the principle of superposition, the resultant displacement at P is given by

$$\xi = \xi_1 + \xi_2 = R\{Ae^{-i(\omega t - kx_1)} + Ae^{-i(\omega t - kx_2)}\}$$

$$= R\{Ae^{-i\omega t}(e^{ikx_1} + e^{ikx_2})\} \quad \text{...(8.13 } a)$$

If $x = \frac{1}{2}(x_1 + x_2)$, Eqn. (8.13 *a*) reduces to

$$\xi = R[Ae^{-i(\omega t - kx)}\{e^{+i\frac{1}{2}k(x_2 - x_1)} + e^{-i\frac{1}{2}k(x_2 - x_1)}\}]$$

or

$$\xi = R[2A\cos\frac{1}{2}k(x_2 - x_1)e^{-i(\omega t - kx)}] \quad \text{...(8.13 } b)$$

The resultant displacement at P is therefore obtained by extracting the real part from Eqn. (8.13 *b*) and may be written as

$$\xi = 2A\cos\frac{1}{2}k(x_2 - x_1)\cos(\omega t - kx) \quad \text{...(8.13 } c)$$

Equation (8.13 *c*) represents a plane harmonic wave of the same frequency and wavelength as the individual waves. Its amplitude however varies periodically due to the cosine term. The resultant amplitude at P, given by

$$\text{Amp.} = 2A\cos\frac{1}{2}k(x_2 - x_1) \quad \text{...(8.13 } d)$$

will therefore depend upon the two path lengths which the wave trains cover to reach P.

If $$k/2(x_2 - x_1) = (\pi/\lambda)(x_2 - x_1) = n\pi$$

or $$x_2 - x_1 = n\lambda \quad \text{where} \quad n = 0, 1, 2\ldots$$

the amplitude of disturbance will be a maximum. But if

$$k/2(x_2 - x_1) = (\pi/\lambda)(x_2 - x_1) = (2n+1)\pi/2$$

or $$x_2 - x_1 = (2n+1)\lambda/2 = \left(n + \frac{1}{2}\right)\lambda$$

the amplitude of disturbance will be zero at the point under reference.

Thus we find that *under certain conditions, two disturbances* travelling through a medium, can completely destroy each other. This is a very important property of all wave motions and is frequently encountered in acoustics and optics.

(ii) Stationary Waves. Another important example of the superposition principle is that of two waves travelling in the same medium simultaneously but in opposite directions. This may be the case when a direct wave is reflected at either a rigid surface (closed end) or a free surface (open end). In terms of the particle displacement these waves may be represented as

$$\xi_1 = R\{ae^{-i(\omega t - kx)}\} \qquad \ldots(8.13\ e)$$

and $$\xi_2 = \pm R\{ae^{-i(\omega t + kx)}\} \qquad \ldots(8.13\ f)$$

where the positive sign in Eqn. (8.13 f) is to be taken in the case of reflection at an open end and negative in the case of reflection at a closed end. (It may be recalled that while acoustic pressure in the direct and reflected waves remains in the same phase on reflection at a rigid surface, the particle displacement becomes 180° out of phase (Fig. 7.7). We shall take the case of reflection at a rigid surface. The displacement at any point in front of the surface under the simultaneous action of the two waves will be given by

$$\xi = \xi_1 + \xi_2 = R\{(ae^{-i\omega t})(e^{ikx} - e^{-ikx})\}$$

$$= R\{(2ia \sin kx)e^{-i\omega t}\}$$

or $$\xi = +2a \sin kx \sin \omega t \qquad \ldots(8.13\ g)$$

The particle velocity u and condensation s will then be given by

$$u = \xi' = 2a\omega \sin kx \cos \omega t \qquad \ldots(8.13\ h)$$

and $$s = -d\xi/dx = -2ak \cos kx \sin \omega t \qquad \ldots(8.13\ i)$$

It is thus seen from Eqns. (8.13 g, h, i) that the displacement, particle velocity and condensation at various points of the medium change harmonically with time. Also, at any instant, these physical properties have different values at different points and repeat themselves at regular intervals. Whatever be the value of t, the particle velocity 'u' is always zero for points at which

$$\sin kx = 0 \quad \text{or} \quad \cos kx = 1$$

or $$kx = \pm n\pi \quad \text{where} \quad n = 0, 1, 2, 3, \text{etc.}$$

that is, $$x = 0, \pm\lambda/2, \pm 2\lambda/2, \pm 3\lambda/2 \text{ etc.}$$

If the distance x is to be measured from the rigid surface, the particle velocity u and also the displacement ξ will be zero at the surface itself and at points $-n\lambda/2$ in front of the surface. These values and positions are independent of *time*; such points are called *nodes* and the planes, passing through them in a direction perpendicular to the direction of propagation of the incident wave, are called *nodal planes*.

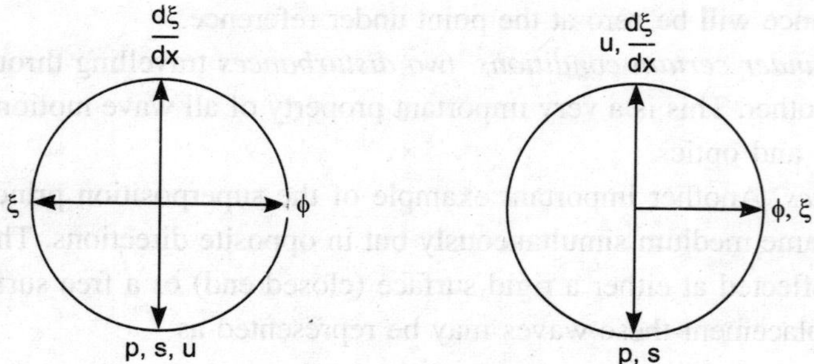

Fig. 8.7. Phase relationship for plane waves travelling in the positive and negative X-directions respectively

On the other hand there will also be points of the medium at which the particle velocity and displacement have maximum values. Their positions are obtained by putting

$$\sin kx = \pm 1 \quad \text{or} \quad \cos kx = 0$$

or $$kx = \pm(2n+1)\pi/2 \quad \text{where} \quad n = 0, 1, 2, \ldots$$

that is $$x = \pm\lambda/4, \pm 3\lambda/4 \text{ etc.}$$

Thus at distances ($-\lambda/4$, $-3\lambda/4$ etc.) in front of the surface the particles of the medium will have maximum velocity as well as displacement. These will give the positions of the *antinodal planes*.

Both nodal and antinodal planes are localised in space since their positions are independent of time. In other words the positions of maximum and zero displacement do not move forward or backward. Such waves are accordingly called *standing* or *stationary waves.*

Consider the nodal planes at any fixed instant. Let this instant be so chosen that $\sin \omega t = 1$. Then the particle velocity at any point of the medium is given by

$$u = 2a\omega \sin kx$$

Since at successive nodal plane $\sin kx = 0$, while x goes on increasing, it follows that the particle velocity changes sign at each of these nodes. Also when $\sin kx = 0$, $\cos kx = 1$ and therefore from Eqn. (8.13 *i*) the condensation has got a maximum value

$$s = -2ak \sin \omega t$$

at these positions.

Thus while the locations $x = 0, -\lambda/2, -2\lambda/2$ etc. are the nodal planes for velocity and displacement, they are also antinodal planes for condensation and excess pressure ($p_o = \rho c^2 s$) and vice versa.

Let us now consider changes with time. From Eqn. (8.13 *g*) we find that if

$$\sin \omega t = 0$$

or $$\omega t = \pm n\pi \text{ where } n = 0, 1, 2, \text{ etc.}$$

that is at $$t = 0, T/2, T, 3T/2 \text{ etc.}$$

all the particles of the medium will have no displacement. The condensation s and the excess pressure p_o will also be zero at these instants. But the particle velocity u will have a maximum value

$$u = 2a\omega \sin kx \qquad (\because \cos \omega t = 1)$$

which will vary from place to place. It therefore follows that twice during each period all the particles of the medium pass through their mean positions with different velocities but this does not produce any compression or rarefaction of the medium.

On the contrary if $\sin \omega t = \pm 1$

or $$\omega t = (2n + 1)\,\pi/2$$

that is at $$t = T/4, 3T/4, 5T/4 \text{ etc.}$$

the displacement of every particle has a maximum value given by

$$\xi = 2a \sin kx$$

This displacement will nevertheless vary with x and will be zero at nodal planes and maximum at antinodal planes. Similarly the *local* condensation and excess pressure will both have their maximum values. But the particle velocity at all the points will be zero ($\because \cos \omega t = 0$).

It may be mentioned that the above discussion is based on the assumption that the amplitude of the reflected wave is the same as that of the incident wave. Also we can lay down a similar treatment if the standing waves are formed due to reflection at an open end. In this case the *displacements* due to the incident and reflected waves will be in the same phase.

Energy in Stationary Waves

We have seen that the particle velocity at any point of the medium in the presence of standing waves is given by

$$u = 2a\omega \sin kx \cos \omega t$$

$$= u_x \cos \omega t \qquad [u_x = 2a\omega \sin kx]$$

where u_x is the amplitude of the particle velocity at that point. Also the local excess pressure p_o is given by

$$p_o = \rho c^2 s$$

$$= -2ak\,\rho\, c^2 \cos kx \sin \omega t$$

or $$p_o = p_x \sin \omega t \qquad [p_x = -2ak\,\rho c^2 \cos kx]$$

where p_x is the amplitude of the excess pressure at any point of the medium.

Hence the work done per unit area (of an elementary volume of the medium called the particle) per unit time by the standing waves is given by

$$\frac{dw}{dt} = p_o u = u_x p_x \sin \omega t \cos \omega t$$

or
$$W = \int_o^T u_x p_x \sin \omega t \cos \omega t \, dt$$

where the integration has to be performed over one time period to obtain the total work done per unit area during this interval of time. This will also be the amount of energy transferred to the next particle. The rate of energy transfer is therefore given by

$$\frac{u_x p_x}{2T} \int_o^T \sin 2\omega t \, dt$$

This is easily seen to be zero. Hence we find that there is no transfer of energy by the standing waves.

Although no transfer of energy takes place in stationary waves, different particles of the medium have their own kinetic and potential energies. Thus at nodes where particle velocity has zero value the energy is wholly potential. This is because the particles at the nodes are under maximum strains or in other words the condensation has a maximum value at these points. On the other hand, at antinodes the particle velocity has a maximum value but condensation (or strain) has zero value. At these points the energy is wholly kinetic. At other points of the medium the energy is partly kinetic and partly potential.

When the particles of the medium pass through their mean positions at times $t = 0, T/2, T$ etc., the energy is wholly kinetic because they have maximum velocity and no strain. But at time $t = T/4$, $3T/4$, etc., when the particles pass through their extreme positions, their velocities have zero value and will be under maximum strain. The energy at these moments will be wholly potential. We thus find that in stationary waves energy is being continuously changed from kinetic to potential and vice versa.

It is of interest to see that the equation of a stationary wave (of Eqn. 8.13 *g*) is a solution of the wave equation.

$$\frac{\partial^2 \phi}{\partial x^2} = \frac{1}{C^2}\frac{\partial^2 \phi}{\partial t^2}$$

For this we suppose that the solution is of the form

$$\phi = X(x)T(t)$$

where X and T are functions of x and t respectively. Differentiating ϕ twice with respect to x and t and substituting the derivatives in the wave equation, we get

$$\frac{1}{X}\frac{d^2X}{dx^2} = \frac{1}{c^2 T}\frac{d^2T}{dt^2}$$

The LHS of this equation is independent of t and the RHS is independent of x. But since the two sides are equal it follows that both of them are independent of x as well as t. We can thus write

$$\frac{1}{X}\frac{d^2X}{dx^2}=\pm p^2 \quad \text{and} \quad \frac{1}{c^2T}\frac{d^2T}{dt^2}=\pm p^2$$

where p^2 is an arbitrary constant usually called the separation constant. The original wave equation has thus been reduced to the above two simultaneous equations by the process called separation of variables. If we are interested in solutions of the harmonic type, we have to solve the equations

$$\frac{d^2X}{dx^2}+p^2X=0 \quad \text{and} \quad \frac{d^2T}{dt^2}+c^2p^2T=0$$

The solution of these equations will be of the form

$$X=e^{\pm ipx} \quad \text{and} \quad T=e^{\pm icpt}$$

where the constants of integration have been omitted. The solution of the wave equation will therefore be of the form

$$\phi=e^{\pm ipx}e^{\pm icpt}$$

A typical solution of the *stationary* type is therefore

$$\phi=a\cos px\cos cpt$$

The phenomenon of standing waves has found many useful applications wherever transfer of energy can be shown to take place through the vehicle of propagating waves. Its applications in the performance of musical instruments and Lippman method of colour photography are only indicative of the wide range to uses to which it has been put.

8.14 SUPERPOSITION OF WAVES OF SLIGHTLY DIFFERENT FREQUENCIES

Suppose that there are two plane progressive waves of the same amplitude but slightly different frequencies and wave numbers propagating in the x-direction. Let them be represented by

$$\phi_1=ae^{-i(\omega t-kx)}$$

$$\phi_2=ae^{-i[(\omega+\delta\omega)t-(k+\delta k)]x}$$

We may recall that

$$\omega=2\pi\nu \quad \text{or} \quad \omega/2\pi=\nu \quad \text{[Frequency]}$$

and

$$k=2\pi/\lambda \quad \text{or} \quad k/2\pi=1/\lambda \quad \text{[Wave number]}$$

give the frequency and wave number of any wave motion with a velocity of propagation $\omega/k=\nu\lambda=C$. The average frequency and average wave number of the two wave motions can be defined as

$$\bar{\omega}=\frac{1}{2}(\omega+\omega+\delta\omega)=\omega+\frac{1}{2}\delta\omega$$

and

$$\bar{k}=\frac{1}{2}(k+k+\delta k)=k+\frac{1}{2}\delta k$$

According to the principle of superposition the resultant wave will be given by

$$\phi=a[e^{-i(\omega t-kx)}+e^{-i\{(\omega+\delta\omega)t-(k+\delta k)x\}}]$$

$$= a\left[e^{-i\{(\bar{\omega}-\frac{1}{2}\delta\omega)t-(\bar{k}-\frac{1}{2}\delta k)x\}} + e^{-i\{(\bar{\omega}+\frac{1}{2}\delta\omega)t-(\bar{k}+\frac{1}{2}\delta k)x\}} \right]$$

$$= ae^{-(\bar{\omega}t-\bar{k}x)}\left[e^{\frac{1}{2}i(t\delta\omega-x\delta k)} + e^{-\frac{1}{2}i(t\delta\omega-x\delta k)} \right]$$

or
$$\phi = 2a\cos\frac{1}{2}(t\delta\omega - x\delta k)e^{-i(\bar{\omega}t-\bar{k}x)} \qquad \text{...(8.14 } a\text{)}$$

The resultant wave motion is thus seen to have a frequency $\bar{\omega}/2\pi$ and wavelength $2\pi/\bar{k}$ propagating along the x-direction with a velocity $\bar{\omega}/\bar{k}$. The amplitude of this wave does not remain constant but varies according to the cosine term in Eqn. (8.14 *a*).

(*i*) *Velocities of Propagation Equal:* Let us suppose that the velocities of propagation of the two individual waves are equal, that is

$$\omega/k = \frac{\omega+\delta\omega}{k+\delta k} = \delta\omega/\delta k$$

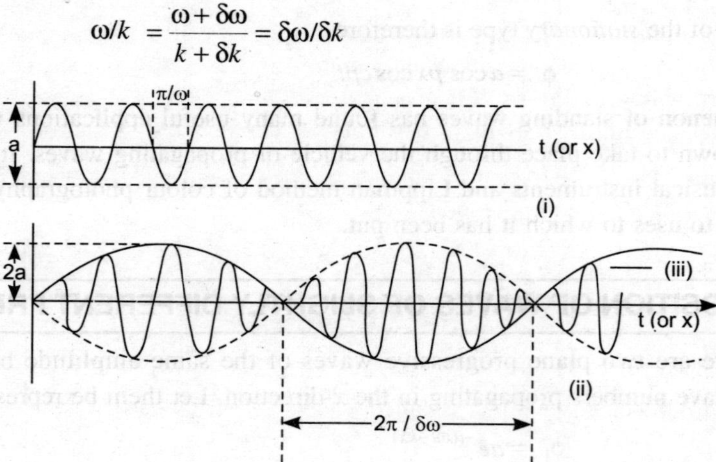

Fig. 8.8. (i) The wave $a\cos(\bar{\omega}t - \bar{k}x)$ (ii) The wave $2a\cos\frac{1}{2}(t\delta\omega - x\delta k)$ (iii) The wave $2a\cos\frac{1}{2}(t\delta\omega - x\delta k)\cos(\bar{\omega}t - \bar{k}x)$

We find that the resultant wave has a frequency which is the mean of the two frequencies but whose amplitude fluctuates with both time and distance. This is shown by the curve (*ii*) [the dotted curve is its reflection in the t or x-axis]. The actual disturbance ϕ lies between these two boundaries. The wave system shown in Fig. 8.8 will steadily move forward without change of shape. The successive maxima of the amplitude function are obtained at intervals

$$\delta t = 4\pi/\delta\omega \text{(with } x \text{ fixed)} \quad \text{or} \quad \delta x = 4\pi/\delta k \text{ (with } t \text{ fixed)}$$

and the maxima of the phase function are obtained at intervals of

$$\delta t = 2\pi/\bar{\omega} \text{ (with } x \text{ fixed)} \quad \text{and} \quad \delta x = 2\pi/\bar{k} \text{ (with } t \text{ fixed).}$$

If we are dealing with sound waves the fluctuating amplitude produces the well known phenomenon of beats. The frequency of intensity fluctuation which is then the beat frequency will be twice that of curve (*ii*) and is thus seen to be just the difference between the individual frequencies. Thus we see from Fig. 8.8 that the amplitude of the resultant wave becomes maximum in $\delta\omega/4\pi$ seconds.

Hence beat frequency $= 2\delta\omega/4\pi$

$$= 2\delta(2\pi v)/4\pi = \delta v$$

where δv is the difference in frequencies of the individual wave trains.

(*ii*) *Velocities of Propagation Unequal.* Let us now take the case where the two individual waves do not move with the same velocity through the medium. In that case

$$\frac{\omega}{k} \neq \frac{\omega+\delta\omega}{k+\delta k} \neq \frac{\delta\omega}{\delta k}$$

and the profile curve (*ii*) will move through the medium with a velocity different from that of the more rapidly oscillating part. The individual waves in Fig. 8.8 will then advance through the profile with gradual change of amplitude. This happens when the velocity of the waves, called the wave velocity depends upon frequency, a phenomenon usually known as dispersion. In a dispersive medium therefore only a single harmonic wave train will propagate without change of shape; any other wave group which may be analysed into two or more harmonic waves will undergo a change in shape.

We find from Eqn. (8.14 *a*) that the planes of constant amplitude and in particular the maxima of amplitude are propagated with a velocity

$$U = \delta\omega/\delta k \qquad \text{...(8.14 } b\text{)}$$

The planes of constant phase are propagated with a velocity

$$V = \overline{\omega}/\overline{k} \qquad \text{...(8.14 } c\text{)}$$

U is called group velocity while *V* is called wave or phase velocity. The two velocities are however inter-related. Thus

$$U = (\delta\omega/\delta k) = (d\omega/dk) \text{ if } \delta\omega \text{ is small}$$

and
$$\frac{d\omega}{dk} = \frac{d}{dk}(kV) = V + k\frac{dV}{dk} \text{ where } k = \frac{2\pi}{\lambda}$$

or
$$U = (d\omega/dk) = V - \lambda(dV/d\lambda) \qquad \text{...(8.14 } d\text{)}$$

where all the quantities refer to the mean frequency $\overline{\omega}$. Also,

$$U = \frac{d\omega}{dk} = \frac{2\pi dv}{2\pi d(1/\lambda)} = -\lambda^2\frac{dv}{d\lambda} \qquad \text{...(8.14 } e\text{)}$$

It may be added that importance of group velocity lies in the fact that energy is propagated in the medium with this velocity and it is always the quantity we measure physically.

8.15 SUPERPOSITION OF MANY HARMONIC WAVES

Let us now consider the superposition of many harmonic waves of angular frequencies lying within a range $\Delta\omega = \omega_2 - \omega_1$. Suppose this range of frequencies is emitted by an oscillator located at $x = 0$. The resultant wave originating at the source is given by

$$\phi(t) = A[\cos\omega_1 t + \cos(\omega_1 + \delta\omega)t + \cos(\omega_1 + 2\delta\omega) + \ldots\ldots + \cos\omega_2 t] \qquad \text{...(8.15 } a)$$

where all the disturbances have been assumed to be of the same amplitude A. Equation (8.15 a) may be expressed in terms of complex quantities as

$$\phi(t) = R[Ae^{-i\omega_1 t} + Ae^{-i(\omega_1+\delta\omega)t} + \ldots\ldots + Ae^{-i\omega_2 t}]$$

where

$$\delta\omega = \frac{\omega_2 - \omega_1}{N-1} = \frac{\Delta\omega}{(N-1)} \qquad \text{...(8.15 } b)$$

and N is the total number of vibrations. Equation (8.15 b) may be reduced to a geometric series of N terms by putting $e^{-i\omega t} = \alpha$. Thus

$$\phi(t) = R[Ae^{-i\omega_1 t}\{1 + \alpha + \alpha^2 + \ldots\ldots\ldots + \alpha^{(N-1)}\}]$$

$$= R\left[Ae^{-i\omega_1 t}\frac{\alpha^{N-1}}{\alpha - 1}\right]$$

$$\phi(t) = R\, Ae^{\left[-i\omega_1 t \frac{\alpha^{N/2}(\alpha^{N/2} - \alpha^{-N/2})}{\alpha^{1/2}(\alpha^{1/2} - \alpha^{-1/2})}\right]} \qquad \text{...(8.15 } c)$$

Remembering the definition of α we find that

$$\alpha^{N/2} - \alpha^{-N/2} = e^{-iN/2\, t\delta\omega} - e^{+iN/2\, t\delta\omega} = -2i\sin\frac{1}{2}Nt\delta\omega$$

and

$$\alpha^{1/2} - \alpha^{-1/2} = e^{-i\frac{1}{2}t\delta\omega} - e^{+i\frac{1}{2}t\delta\omega} = -2i\sin\frac{1}{2}t\delta\omega$$

Also

$$\alpha^{N/2} / \alpha^{1/2} = \alpha^{\frac{1}{2}(N-1)} = e^{-\frac{1}{2}(N-1)t\delta\omega}$$

Substituting these values in Eqn. (8.15 c) we see that

$$\phi(t) = Re\left[Ae^{-i\omega_1 t}e^{-(i/2)(N-1)t\delta\omega}\frac{\sin\frac{1}{2}Nt\delta\omega}{\sin\frac{1}{2}t\delta\omega}\right] \qquad \text{...(8.15 } d)$$

From the definition of $\delta\omega$ we find that

$$\omega_1 + \frac{1}{2}(N-1)\delta\omega = \frac{1}{2}[\omega_1 + \omega_1 + (N-1)\delta\omega]$$

$$= \frac{1}{2}[\omega_1 + \omega_1 + \Delta\omega]$$

$$= \frac{1}{2}[\omega_1 + \omega_2] = \omega_{av}$$

where ω_{av} denotes the average angular frequency. Also

$$\frac{\sin \frac{1}{2} Nt\, \delta\omega}{\sin \frac{1}{2} t\, \delta\omega} = \frac{\sin \frac{1}{2} \frac{t \cdot \Delta\omega \cdot N}{(N-1)}}{\sin \frac{1}{2} \frac{t\Delta\omega}{N-1}}$$

With these values Eqn. (8.15 *d*) becomes

$$\phi(t) = Re\left[Ae^{-i\omega_{av} \cdot t} \frac{\sin \frac{1}{2}(N/N-1)t\Delta\omega}{\sin \frac{1}{2}(1/N-1)t\Delta\omega} \right]$$

Extracting the real part of this expression, we find that the resultant vibration generated by the source is given by

$$\phi(t) = A \frac{\sin \frac{1}{2}[(N/N-1)t\Delta\omega]}{\sin \frac{1}{2}[(1/N-1)t\Delta\omega]} \cos \omega_{av} t \qquad \text{...(8.15 } e\text{)}$$

where the first term on the RHS gives the resultant amplitude at any time *t*. Thus

$$A(t) = A \frac{\sin \frac{1}{2}[(N/N-1)t\Delta\omega]}{\sin \frac{1}{2}[(1/N-1)t\Delta\omega]} \qquad \text{...(8.15 } f\text{)}$$

If *N* is large Eqn. (8.15 *f*) can be simplified further. In such a case, *N* and $(N - 1)$ are nearly the same and the denominator may be replaced by an approximate expression because

$$\sin \frac{1}{2}(1/N-1)t\Delta\omega = (1/2N)t\Delta\omega$$

when the argument of sine is small. With these approximations Eqn. (8.15 *f*) becomes

$$A(t) = NA \frac{\sin \frac{1}{2} t\Delta\omega}{\frac{1}{2} t\Delta\omega} = NA\left(\frac{\sin\theta}{\theta}\right) \qquad \text{...(8.15 } g\text{)}$$

It is thus seen that the amplitude of the resultant disturbance varies with time in accordance with the above expression. At $t = 0$, the resultant amplitude is given by $A(0) = NA$

so that
$$A(t) = A(0)\frac{\sin\frac{1}{2}t\Delta\omega}{\frac{1}{2}t\Delta\omega} \quad ...(8.15\ h)$$

the resultant disturbance defined by

$$\phi(t) = A(t)\cos\omega_{av}t \quad ...(8.15\ i)$$

has been shown in Fig. 8.9. We find that if the source generates a band of frequencies of width $\Delta\omega$ the amplitude of the resultant disturbance called a 'pulse' becomes zero after a time

$$t = \frac{2\pi}{\Delta\omega}.$$

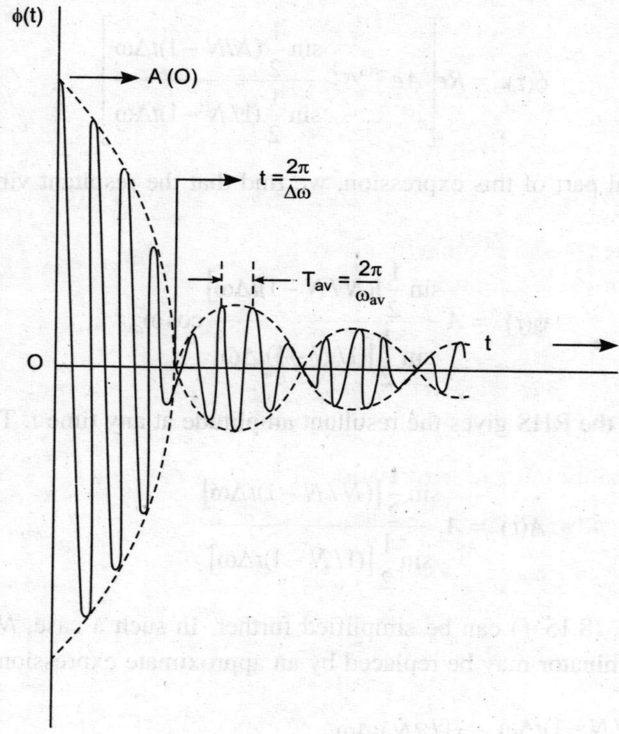

Fig. 8.9. Superposition of many waves

Subsequently the amplitude does not attain any significant values. If the interval for which $\phi(t)$ has a substantial value is denoted by Δt then

$$\Delta t = (2\pi/\Delta\omega) \text{ or } \Delta\omega\Delta t = 2\pi$$

Since $\omega = 2\pi\, v$, where v is the frequency of vibration, it is seen that

$$\Delta v\Delta t = 1 \quad ...(8.15\ j)$$

Equation (8.15 *j*) expresses a very important and general law in Physics of wave motion. In the present context it gives the time interval over which the amplitude of a pulse maintains a substantial value in terms of its bandwidth. More generally, it is interpreted in terms of an intrinsic uncertainty associated with both frequency and time. Applied to Fig. 8.9, it states that although the average frequency ν_{av} predominates the pulse, its exact frequency is not known on account of bandwidth $\Delta\nu$. Expressed differently it means that there is an uncertainty of $\Delta\nu$ in the frequency of the pulse.

8.16 WAVE PACKETS

Let us consider an oscillator switched on at any instant. It takes sometime, howsoever small, in building up to a constant amplitude. Suppose it is allowed to oscillate for sometime and is then switched off so that the oscillations die out. This process is illustrated in Fig. 8.10.

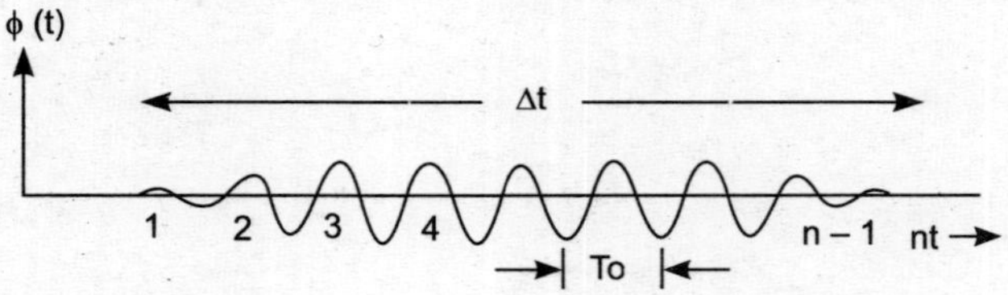

Fig. 8.10

It is seen that the steady angular frequency $\omega = \omega_o$ predominates. This can be determined by counting the number of cycles *n*, during the period Δt for which the oscillator was on.

Thus $$\frac{\omega_o}{2\pi} = \nu_o = \frac{n}{\Delta t} \qquad \text{...(8.16 a)}$$

Nevertheless there is an inherent uncertainty in the determination of *n*. Looking at Fig. 8.10, it is seen that the uncertainty in the determination of *n* amounts to $\pm\frac{1}{2}$ cycles at each end of the pulse. Thus an uncertainty bandwidth $\Delta\nu$ creeps in and this is given by

$$\Delta\nu = \frac{\Delta n}{\Delta t}$$

and because Δn is of the order of one cycle we have

$$\Delta\nu \sim \frac{1}{\Delta t} \qquad \text{...(8.16 b)}$$

which is similar to Eqn. (8.15 *j*).

The travelling waves propagating in the medium in consequence of the pulse shown in Fig. 8.10 are said to form a wave packet. Such wave packets travel with the group velocity *U*. Due to the presence of a band of frequencies $\Delta\omega$ emitted by the source, the wave packet will contain a band of wave numbers, Δk, if the medium is dispersing. This band is centred about k_o, the wave number corresponding to the dominant frequency ω_o and has a width (Eqn. 8.14 *d*).

$$\Delta k = \frac{\Delta \omega}{V_g}$$

A wave packet of length Δx moves across a stationary observer in time Δt such that

$$\Delta x = V_g \Delta t$$

It is thus seen that

$$\Delta k \, \Delta x = \Delta \omega \cdot \Delta t$$

or $$\Delta(k / 2\pi) \, \Delta x = \Delta(\omega / 2\pi) \Delta t$$

or $$\Delta(1/\lambda) \, \Delta x = \Delta \nu \cdot \Delta t \sim 1$$

We once again get the uncertainty relation, this is seen with a wave packet in space.

LIGHT WAVES

Chapter

The study of any wave motion involves an understanding of the phenomena of its emission, propagation and absorption. This becomes a simple matter if the nature of wave motion is known. The two preceding chapters have laid down the basic properties of wave motion. It has also been mentioned in Chapter 6 that light is a form of electromagnetic radiation. We shall now try to put these informations together and see how far the composite picture helps us in understanding the nature and properties of light.

9.1 ELECTROMAGNETIC WAVES

The historical development of the wave theory of light has been outlined in Chapter 6. Prior to the birth of electromagnetic theory, light waves were thought to be associated with transport of mechanical energy through an all pervading medium called luminiferous ether endowed with (properties of) inertia, density and elasticity. Maxwell's theory brought out for the first time the probable nature of light. But even he tried at first to explain his theory in terms of mechanical models. Subsequently it was discovered that in fact no medium is necessary for the propagation of the type of waves postulated by Maxwell. Indeed his electromagnetic waves can travel through empty space or vacuum and the velocity of their propagation is found to be identical with the velocity of light in vacuum.

Maxwell formulated his theory for a homogeneous, isotropic medium on the basis of four physical laws known in his time. These are:

1. Gauss's theorem of total normal induction.
2. The non-existence of isolated magnetic poles.
3. Ampere's rule about the work done in taking a unit pole round a closed circuit carrying current.
4. Lenz's law of electromagnetic induction.

Each of these laws can be written down in the form of mathematical equations. On combination with constitutive relations giving certain physical properties of a dielectric medium, the set of equations leads to the following results:

1. In an electromagnetic field, the strengths of the electric and magnetic fields are given by

$$\left.\begin{aligned}\nabla^2\mathbf{E} &= \frac{\varepsilon\mu}{c^2}\frac{\partial^2\mathbf{E}}{\partial t^2}\\ \nabla^2\mathbf{H} &= \frac{\varepsilon\mu}{c^2}\frac{\partial^2\mathbf{H}}{\partial t^2}\end{aligned}\right\} \quad \ldots(9.1\ a)$$

where E and H are respectively measured in electrostatic and electromagnetic units and 'C' is the ratio between the two sets of units.

Equation (9.1 *a*) is similar to the differential equation of wave motion derived and discussed in Chapter 8. It may therefore be inferred on comparison that an electromagnetic wave propagates in a homogeneous, isotropic dielectric medium with a velocity $c/\sqrt{\varepsilon\mu}$ where ε and μ are respectively the dielectric constant and permeability of the medium. For free space or vacuum $\varepsilon = \mu = 1$; hence the velocity of propagation of electromagnetic waves in free space should just equal the ratio of say the *e. m.* and *e. s.* unit of charge. On determination, this ratio in fact, comes out to be 2.998×10^{10} cms/sec. which is also the velocity of light in free space. It was this fact that led Maxwell to believe that light is a form of electromagnetic radiation. This belief has since been confirmed in many ways and it is now known that X-rays, γ-rays, ultraviolet and infra-red rays and radio-waves are all electromagnetic in character; the entire electromagnetic spectrum is propagated with the same velocity viz., 3×10^{10} cms/sec. through free space; only their wavelengths are different.

For other dielectric media ε is not equal to unity; also μ depends upon the frequency of waves. For the visible region, however, μ may be taken to be unity. Hence the velocity of light in these media is $c/\sqrt{\varepsilon}$ and therefore the refractive index of such media should be $\sqrt{\varepsilon}$. Many substances exhibit this relationship, many others don't. The breakdown of this relation occurs because the electromagnetic theory does not take into account the detailed atomic structure of the dielectric.

2. In an electromagnetic field the rate of flow of energy per unit area is represented by a vector, called Poynting vector, perpendicular to the area under reference. The magnitude and direction of this vector is given by

$$\mathbf{N} = \frac{c}{4\pi}(\mathbf{E}\times\mathbf{H}) \quad \ldots(9.1\ b)$$

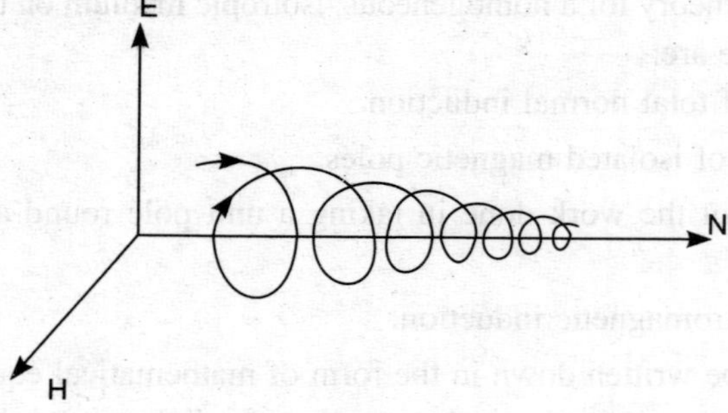

Fig. 9.1. Poynting vector

Equation (9.1 *b*) shows that if there is a flow of electromagnetic energy in say, the X-direction, (Fig. 9.1), it is always accompanied by an electric as well as a magnetic field perpendicular to the direction of energy flow as well as perpendicular to each other. In fact the three mutually perpendicular vectors are so oriented that if a right handed screw is turned from the direction E to H, it will advance in the direction of the Poynting vector. We shall see form the following pages that E and H are quantities varying rapidly in magnitude as well direction but the angles between the three vectors are each equal to ninety degrees at all times.

3. It has been pointed out in Chapter 6, that the electric vector of the electromagnetic radiation is responsible for all observed optic phenomena. Since this vector is always directed perpendicular to the direction of propagation of energy, it is inferred that light waves are transverse in nature. The differential equation of interest in Optics, namely

$$\nabla^2 \mathbf{E} = \frac{\varepsilon\mu}{c^2}\frac{\partial^2 \mathbf{E}}{\partial t^2} \qquad \text{...(9.1 } c\text{)}$$

may also be written in another form by substituting

$$\mathbf{E} = -\frac{1}{c}\frac{\partial \mathbf{A}}{\partial t} \qquad \text{...(9.1 } d\text{)}$$

Thus the quantity analogous to the velocity potential ϕ of Chapter 7 is the electric intensity **E** and that analogous to particle displacement ξ is the vector potential **A** defined by Eqn. (9.1 *d*). We may therefore use the equations developed in that chapter with this understanding keeping in mind the transverse nature of electromagnetic waves. In addition if the electromagnetic wave encounters a boundary between two dielectric media at which it may undergo reflection or refraction, due attention has also to be paid to the conditions at the interface called the boundary conditions. A detailed treatment of these situations is however, beyond the scope of this book.

4. The electromagnetic field stores energy and the density of this energy per unit volume is given by

$$\frac{1}{8\pi}(\varepsilon E^2 + \mu H^2) \qquad \text{...(9.1 } e\text{)}$$

Thus the classical electromagnetic theory of light yields a definite expression for the energy density of such waves.

9.2 IMPLICATIONS OF MAXWELL'S THEORY

If, as predicted by the classical electromagnetic theory (Maxwell's) light is indeed a form of electromagnetic radiation the following consequences would seem to follow from the assumptions implied in the derivation of the theory.

(*i*) The propagation of light through dielectric media should be affected on the application of electric and magnetic fields across them. Such effects (*e.g.* Kerr Effect, Faraday Effect) have been observed for a large number of media.

(*ii*) The electromagnetic waves carry a continuous energy flux of density given by Eqn. (9.1 *b*). The classical theory therefore does not permit the propagation of any packets or lumps of energy through a dielectric medium. On the contrary the modern quantum theory postulates the existence of 'quanta' or packets of energy. The concept of energy quanta has also been borne out by experimental observations (*e.g.*, Raman Effect, Photoelectric Effect and Compton Effect).

This apparent difficulty has been removed by re-interpreting the expression for energy flux (Eqn. 9.1 *b*) in the light of quantum theory. At the same time the results of the two theories may be reconciled if it is recognised that the classical theory leads to results depicting the average effects associated with wave propagation through dielectric media. As soon as we start examining the details of the phenomena *e.g.*, the emission and absorption of electromagnetic radiations, the classical theory is found wanting in many ways.

9.3 PRACTICAL CONSIDERATIONS REGARDING LIGHT WAVES

Let us consider the phenomenon of emission of light as postulated by the classical theory on the understanding that it gives only the overall picture and is found wanting in finer details. Consider an electron bound to a heavy nucleus emitting electromagnetic radiations by way of oscillations induced by forces external to the atom. This is of course a very coarse picture of the probable state of affairs but since at this stage we are not interested in the actual processes of emission of light we may as well omit the details. These oscillations do not build up instantaneously but take a finite interval of time to attain a constant amplitude. This interval may nevertheless be very small as compared to the time for which the electron keeps oscillating and emitting electromagnetic radiations. Since an oscillating electron continuously loses energy by radiation the amplitude of oscillations goes on decreasing tell a stage arises when it becomes hardly detectable; subsequently the oscillations die out completely the amplitude of oscillation. This variation of amplitude with time can be depicted by a curve of the form shown in Fig. 9.2.

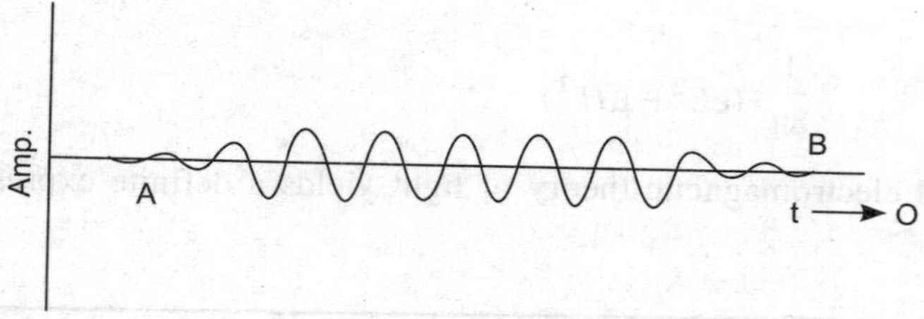

Fig. 9.2

If we wish to determine the frequency of these oscillations we are immediately faced with a difficulty. If we knew the precise instants at which the oscillations had started and died out the determination of frequency would have been an easy matter. We would then have been in a position to locate points *A* and *B* accurately and then to count the number of waves during the time interval *AB* and obtain the desired result. But due to the inherent uncertainty in locating the onset and dying out of the oscillations, their frequency can not be determined accurately. Conversely if the time interval could be somehow determined precisely, the number of waves during this period could not be counted accurately from the graph. No matter what we do we can never do better than counting the number of waves to an in accuracy of ± ½ waves at each end of the pulse. Thus if ΔT be the time for which the electron had emitted radiations the uncertainty in determining the frequency v is given by

$$\Delta v \sim (1/\Delta t) \qquad \text{...(9.3 } a\text{)}$$

In view of this uncertainty we should be naturally curious of knowing the frequency of oscillations. Of course there is no exact value for this frequency. The best we can do is to count the number of waves n without bothering about the uncertainties at two extremes and define a dominant frequency, v_o by the equation

$$v_o = n/\Delta t \qquad \text{...(9.3 } b\text{)}$$

Thus, we find that the radiation emitted out is not strictly monochromatic but contains a band of width Δv centred about the dominant frequency v_o. This band width may be estimated by determining the time interval Δt for which the radiating electron remains in the incited state. An order of magnitude calculation is usually made by expressing Δt in terms of the mean life* τ of the excited electron. Thus if we suppose that the amplitude of oscillations falls to l/eth of its steady value in time τ, it may be reasonable to assume that these oscillations die out for all practical purposes in a time which is a small multiple of τ. In other words the orders of magnitude of Δt and τ may be taken to be the same. Eqn. (9.3 *a*) then yields a measure of the band width. Thus, we find that

$$\Delta v \sim 1/\tau \qquad \text{...(9.3 } c\text{)}$$

The actual state of affairs however does not even correspond to this highly simplified picture. This is so on account of several complicating factors which may be mentioned to indicate the complexity of the phenomenon of emission. Firstly it never happens that atoms take their turn in emitting radiations. In fact there may be several atoms in the excited state simultaneously and hence many electron undergoing oscillations leading to the generation of electromagnetic radiations. Secondly none of these atoms are at rest in gaseous discharge tubes — the most commonly used sources of electromagnetic radiations. They in fact travel with velocities which may be as high as 10^5 cm/sec. On account of this motion of the radiating source the band width may be broadened by at least two orders of magnitude. The Doppler broadening is usually accompanied by another type called 'Collision broadening'. This is brought about when the radiating atom suffers a collision with others in the discharge tube. The combined effect of all these factors and a few more amounts to considerable frequency broadening of the emitted radiations. In real sources therefore the band width is no longer of the order of l/τ but has a much greater value. If Δv_o is the actual band width of a spectral line with a dominant frequency v_o Eqn. (9.3 *c*) has to be replaced by the expression

$$\Delta v_o \sim 1/t_{\text{coh}}$$

where the mean life T has been replaced by a much smaller time interval t_{coh}, called coherence time. The importance of this time interval becomes apparent from the alternative expression

$$2\pi\Delta v_o \sim 2\pi/t_{\text{coh}} \quad \text{or} \quad \Delta\omega_o \cdot t_{\text{coh}} \sim 2\pi$$

* For atoms emitting visible radiations τ is of the order of 10^{-8} sec.

Hence coherence time may be defined as the time interval needed for the extreme frequencies of a band to get out of phase by 2π.

It may be mentioned that in the foregoing discussion regarding band width and coherence time, the emission has been assumed to be spontaneous rather than 'stimulated'.

9.4 COHERENCE TIME AND INTERFERENCE OF LIGHT

We thus find that the so-called line spectrum spontaneously emitted from practical light sources does not consist of a series of strictly monochromatic lines. On the contrary each line has an unavoidable band width which plays a significant role in determining the properties of the emitted radiation. For a discussion of these, let us consider one such line of dominant frequency ν_o and band width $\Delta\nu_o$. Further, let us suppose that $\Delta\nu_o/\nu_o >> 1$.

Such line sources are called quasi-monochromatic. Therefore, during the time interval $(\Delta\nu_o)^{-1}$, *i.e.*, the coherence time, there will be $\nu_o/\Delta\nu_o$ oscillations, a number which is quite large in view of the above inequality. Let us now consider two independent point sources illuminated by light of dominant frequency ν_o. These independent sources may either be two pin holes in a screen surrounding a laboratory source or even two locations within this source at which the atoms are emitting radiations independent of each other. This is usually true if they are separated by a distance d larger than the wavelength under discussion. If a screen is placed at a distance much larger than d the observed intensity distribution will be determined by the relative phases of the wave-trains arriving at various points. For any point P, (Fig. 9.3) the path difference between the waves arriving from A and B can be shown to be $d \sin\theta$ and hence the phase difference will be

$$(2\pi/\lambda)\, d \sin\theta$$

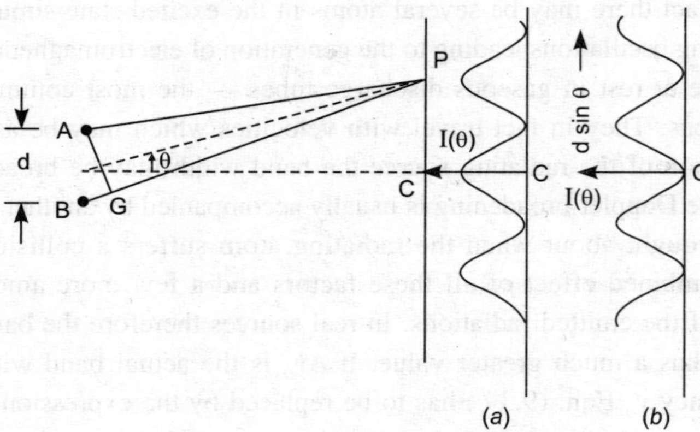

Fig. 9.3

In view of the finite band width of the spectral frequency, this phase difference between the waves arriving at P will change with time. If at any instant the phase difference

$$(2\pi/\lambda)\, d \sin\theta = 2n\pi, \quad n = 0, 1, 2 \ldots$$

then within the time interval t_{coh} this value will be essentially constant. After a greater time the value of phase difference, however, undergoes a change (the geometry of the set up remaining unaltered). Thus if initially the intensity distribution on the screen was as shown at Fig. 9.3 (*a*), after a time larger than t_{coh}, it may change to that shown in Fig. 9.3 (*b*) for which

$$(2\pi/\lambda)\, d \sin\theta = (2n+1)\pi$$

Under this situation the initial maxima intensity give place to minima and vice versa. We therefore, find that if our period of observation is larger than the coherence time of the quasi-monochromatic light, the intensity distribution on the screen will on the average be uniform. But if this period is shorter than t_{coh}, the interference pattern exhibiting maxima and minima will be observed. Thus the probability that the sources. *A* and *B* will be coherent (*i.e.* maintain a constant phase difference at an external point) or incoherent will be entirely determined by the time taken by the receptor for recording the distribution of intensity on the screen. Since for usual laboratory sources t_{coh} is of the order of 10^{-8} or 10^{-9} seconds, the human eye with a much slower response always observes incoherent (*i.e.* uniform) illumination on the screen even if the two independent sources are quasi-monochromatic. Thus there are in fact no coherent or incoherent sources intrinsically; the distinction becomes necessary only due to the limitations of our tools of observations.

9.5 COHERENCE TIME AND POLARISATION

The frequency band width and coherence time also play a significant role in determining the state of polarisation of electromagnetic radiations. Suppose an electron at a given location in the source of light is thrown into vibrations. Confining our attention to the radiation propagating along the direction **X**, we have to consider only the **Y** and **Z** components of the electrons' motion for determining the nature of radiation. These components of motion retain a constant relative phase only for the radiating life time of the excited electron. After some time, this same electron may suffer a second collision, be again excited and start emitting radiations. The amplitudes and phase constants of the **Y** and **Z** motions of the electron, will as before, depend upon the circumstances of the collision. Invariably, as in the case of gaseous discharge tubes, these variables of electron motion are in no way correlated between successive collisions.

In actual practice, however, we should consider not one but many excited electrons for getting an insight into nature of the emitted radiation the **Y** and **Z** components describing the radiation proceeding along. The direction **X** will now be the resultants of the corresponding motions of all electrons in the excited state. In practice, the components e_y and e_z associated with the vibrations of every excited electron may be combined to give the resultants E_y and E_z whose amplitudes and phase constants depend upon those of the contributing atoms; of course, their frequencies will be the same. During a time interval shorter than the coherence time, t_{coh}, the amplitude or phase of either E_y or E_z does not change appreciably inspite of the very large number of oscillations taking place during the period. Technically this state of affairs means that the polarization state of the emitted electromagnetic wave remains unchanged during a time interval shorter than the coherence time.

If we examine the situation after sometime we shall find a new set of atoms in the excited state. The resultants E_y and E_z associated with the excited electron vibrations in the new set of atoms, will on the average, have the same amplitudes as before but the a phase constants will in no way be related to their corresponding values earlier on. Thus we find that although the polarisation state of the emitted radiation (described in terms of the relative phase constants of E_y and E_z) remains unchanged during a time interval shorter than the coherence time, it drifts in an entirely unpredictable manner over periods of observation larger than it. Once again, therefore, coherence time is seen to be a fundamental property of light waves.

HUYGENS' PRINCIPLE AND ITS APPLICATIONS

One of the greatest difficulties encountered in the early development of the wave theory of light lay in its inability to explain the rectilinear propagation of light—a fact which obviously follows on the basis of the corpuscular conception of light. To overcome this difficulty in the progress of the wave theory, *Huygens*, in his *Treatise on Light* enunciated a principle for explaining the wave propagation, which was later used by *Fresnel* in conjunction with the principles of interference to provide a satisfactory explanation of rectilinear propagation of light but at the same time he also showed that this propagation is only approximately rectilinear.

10.1 HUYGENS' PRINCIPLE

One of the most important contributions of *Huygens* to the theory of light and to the wave motion in general lies in his enunciation of a principle, known by his name, which governs the wave propagation in any medium. He declares in his *Treatise on Light* as follows: "*In considering the propagation of waves, we must remember that each particle of the medium through which the wave spreads does not communicate its motion only to that neighbour which lies in straight line drawn from the luminous point but shares it also with all particles which touch it and resist its motion. Each particle is thus to be considered as the centre of a wave*". This statement, known as *Huygens' principle*, affords a geometrical construction for determining, from the given shape and position of the wavefront at any instant, the subsequent position and form of the new wavefront at some later instant. According to Huygens, every point of a wavefront may be regarded as the origin of new disturbance which in its turn emits small *Secondary Wavelets*, propagated in all directions from their respective origins with a speed equal to the speed of propagation of the wave in that medium. The new wavefront at a later stage is obtained by drawing a surface tangential to then existing positions of these elementary secondary wavelets or as it is often called *the new wavefront is simply the envelope of these wavefronts.* He arrived at this principle of wave propagation from the following considerations:

It will be recalled that Huygens' conception of a beam of light was simply a succession of a great number of longitudinal *pulses* propagated at high velocity as condensations and rarefactions. Now, suppose that a luminous point emits a light pulse at a given instant. In a homogeneous isotropic medium, this pulse will spread in all directions with a constant velocity so that after an interval of time t it will have reached a spherical surface $ABC...N$ of radius Vt, where V represents the velocity of propagation of light in the medium under consideration. Thus, at this instant, the form of the wavefront is spherical. A wavefront, it will be recalled, is a surface all points of which are always in the equal phase of vibration. After a lapse of time t' from this moment, the light pulse will have reached the spherical surface $A'B'C'...N'$ of radius $V(t + t')$. He now argued that by virtue of ether on the wavefront $ABC...N$ being disturbed by the light pulse originating from O, one finds disturbance at every point of this wavefront exactly similar to that which originated t seconds earlier at O. Accordingly, every point of this wavefront can be regarded as the origin of ether disturbance, exactly similar to that which originated at O. Thus around every point, after a time t', *a new elementary wavefront* commonly called *secondary spherical wavelet* will have formed with a radius Vt'. The resultant effect of these secondary wavelets must be that the wavefront $ABC...N$ has moved to a position $A'B'C'...N'$ in time t'. Now, it is easy to see that the new wavefront is simply the envelope of the individual secondary wavelets. This explanation clearly establishes the Huygens' principle of wave propagation. We, therefore, conclude that in the space in front of the wavefront $ABC...N$ everything takes place in an exactly similar manner as though the original light source were absent but only a sheet of secondary sources were present in the surface $ABC...N$. Kirchhof has shown that this new interpretation of Huygens' principle helps us to discuss quantitatively diffraction and interference phenomena.

Figure 10.2 illustrates the application of Huygens, principle for the propagation of the wavefront from the luminous point at infinity. In this case, the wavefronts are plane. The lines AA', BB' etc. joining the corresponding points of two wavefronts can be taken as the path traversed by the light or ray of light. In a homogeneous medium rays are normal to the wavefront. If the medium is not homogeneous, the velocity of propagation is different in different directions. Consequently, appropriate velocity must be used at every point of the wavefront to find the new position and shape of the wavefront. The velocity of light, V, has been assumed to be the same at all points and in all directions in Figs. 10.1 and 10.2.

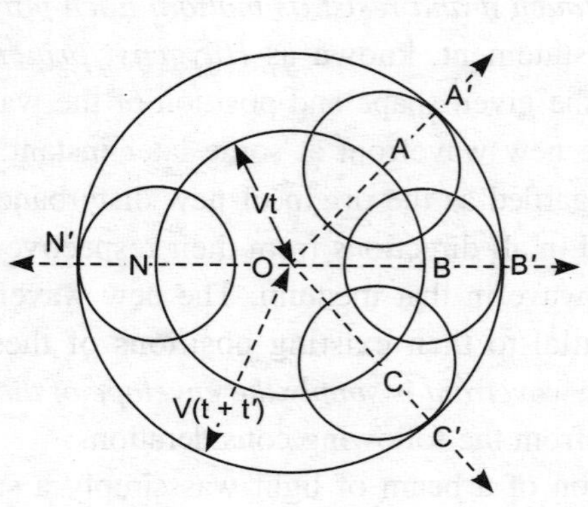

Fig. 10.1. Illustrates Huygens' principle of propagation of wavefronts

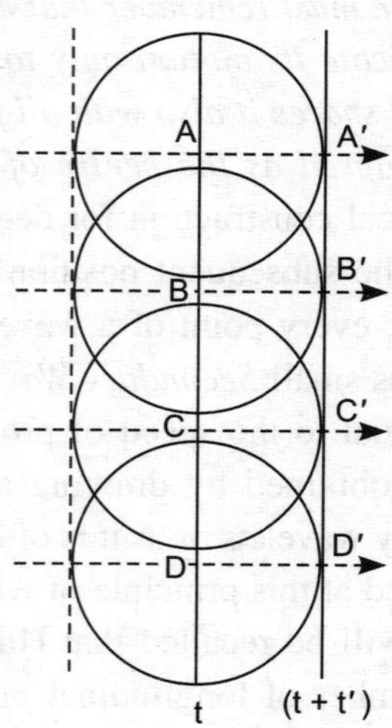

Fig. 10.2

In its simple form, the principle discussed above is, however, not fully satisfactory. The reasons are not far to seek. In the first place, the secondary wavelets, if they spread out in all directions, should also combine to form a *backward wave* moving towards the source O. This, of course, is never observed experimentally. Secondly, the secondary wavelet, which originates from A also reaches the points B', C', D', ... and accordingly light would be present at these points, which had travelled via A, contrary to the rectilinear propagation of light. Lastly, according to this principle, only one point of the secondary wavelet is effective, namely, the point at which they touch their envelope in the forward direction of propagation. There is no direct physical or mathematical explanation in Huygens' original presentation of the theory for this arbitrary decision to ignore all the unwanted part of the secondary wavelets.

We will give a satisfactory explanation of the absence of the *back wave* in the chapter on *Diffraction of Light—Fresnel Class* by following Fresnel's modification of Huygens' idea of secondary wavelets based on the principles of interference and superposition of wave-motions applied to secondary wavelets.

10.2 REFLECTION OF A PLANE WAVE AT A PLANE REFLECTING SURFACE

Consider a plane wavefront ABC advancing in the direction depicted by arrows. Its position is marked in Fig. 10.3 at an instant, say $t = 0$, at which its lower edge just meets a plane reflecting surface MM along a line through A perpendicular to the plane of diagram. The planes of the wavefront and the reflecting surface are also supposed to be at right angles to the plane of figure. At this instant, the point A and all points of the surface on the line through A normal to Fig. 10.3, according to Huygens, will just become origins of secondary wavelets. But as the wavefront ABC advances further, the points of the reflecting surface between A and E, successively struck by corresponding points on the advancing wavefront, become the origins of secondary wavelets which spread out in the upper medium with a speed V. Let us find the shape of the reflected wavefront after a time interval t—the time taken by the end C of the incident wavefront to advance a distance CE, where $CE = Vt$.

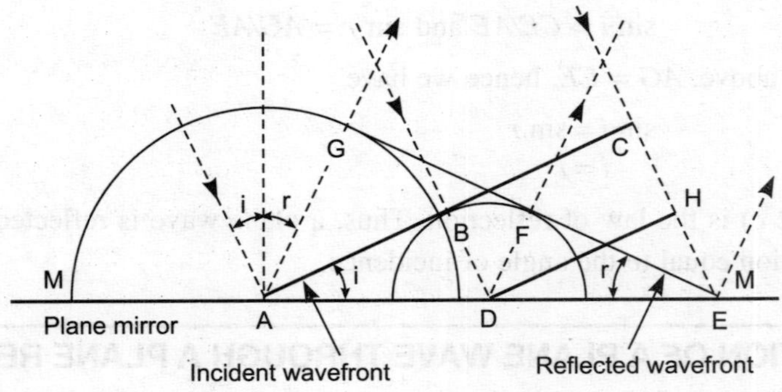

Fig. 10.3. Explanation of laws of reflection by means of Huygens' undulation theory

At this moment, when the point E of the reflecting surface just becomes the origin of the secondary wavelet, the radius acquired by the secondary wavelet which originated t seconds earlier from A would be

$$AG = Vt = CE \qquad \text{...(10.2 } a)$$

and the radius of the secondary wavelet which originated from the point D would be

$$DF = Vt' = HE \qquad \text{...(10.2 } b)$$

where t' is the time interval in which the end H of the incident wavefront from the position DH would just reach E. With points A, D as centres, the hemispheres of radius $AG = CE$ and $DF = HE$ are drawn in the upper medium, which consequently represent the spread of the secondary wavelets which originated from A and D respectively, by the time the end C, of the incident wave has reached E. A plane drawn though E, normally to the plane of diagram, so as to be tangential to the secondary wavelet from A at G, can be easily shown to touch all the secondary wavelets which originated from the points of the surface between A and E. The proof is as follows:

It is easy to see that triangles ACE and DHE are similar. Hence we have

$$AE/DE = CE/HE$$

Also by the help of Eqns. (10.2 a) and (10.2 b), the above relation can be easily written as,

$$AE/DE = AG/DF$$

or
$$\frac{DF}{DE} = \frac{AG}{AE} = \sin r$$

Hence, in triangle DEF the angle DFE must be of 90º. But the radius of the secondary wavelet from D, at the moment when E just becomes the origin of secondary wavelet, is DF. Thus the wavelet from D will touch the line GE at F. The plane GE, normal to the plane of diagram is, therefore, the envelope of all secondary wavelets generated from A to E by the time the incident wave has reached E. The plane GE, therefore, represents the reflected wavefront.

The angle i between the incident wavefront and the reflecting surface is called the angle of incidence; that between the reflected wavefront and the reflecting surface, r, is called the angle of reflection. The relation between them can be easily obtained as follows:

$$\sin i = CE/AE \text{ and } \sin r = AG/AE$$

But as proved above, $AG = CE$, hence we have

$$\sin i = \sin r$$

or
$$i = r \qquad \text{...(10.2 } c)$$

Equation (10.2 c) is the law of reflection. Thus, a plane wave is reflected from a plane surface with angle of reflection equal to the angle of incidence.

10.3 REFRACTION OF A PLANE WAVE THROUGH A PLANE REFRACTING SURFACE

Consider a plane wavefront ABC advancing in the direction depicted by arrows. Its position is marked in Fig. 10.4 at the instant say $t = 0$, at which its lower edge just meets a plane surface MM separating two transparent media (*e.g.*, air and glass), along a line through A perpendicular to the diagram. The speeds of propagation in the two media are V_a and V_g respectively and the refractive indices are $\mu_a = c/V_a$ and $\mu_g = c/V_g$ respectively. The plane of wavefront and that of the refracting surface are also supposed to be perpendicular to the plane of the diagram. At this instant, according to Huygens' principle, two secondary wavelets just originate from the point A of the interface; one, above the interface spreads into the original medium with speed V_a and the second, below the interface, spreads into the second medium with speed V_g, as the incident wavefront advances further. The direction of the reflected wavefront can be obtained by the law of reflection deduced in *Sec.* 10.2. We now proceed to find the direction of propagation of the refracted wavefront, it being assumed that $V_a > V_g$ and $\mu_g > \mu_a$. In what follows we shall simply confine ourselves to wavelets in the lower medium.

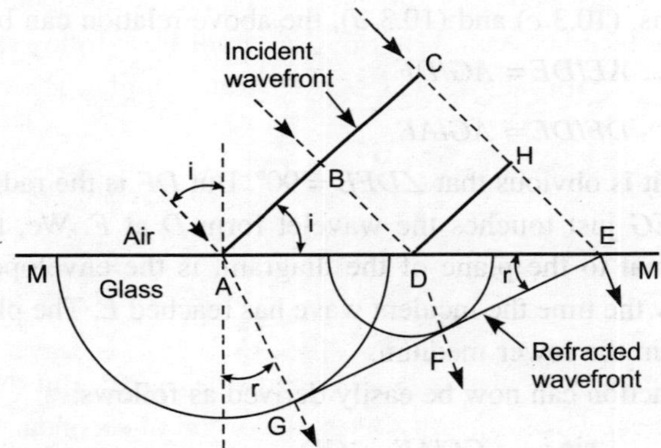

Fig. 10.4. Explanation of Snell's Laws by means of Huygens' wave theory

Now, as the incident wavefront ABC proceeds further, the points $A..D..E$ of the interface are successively disturbed by the advancing wavefront. As a consequence, they successively become the origins of secondary wavelets. Obviously, the point A is the first to act as such and E the last. Let us find the shape of the wavefront after a time interval t, the time taken by the end C of the incident wavefront to advance a distance CE where

$$CE = V_a t \qquad \text{...(10.3 } a\text{)}$$

At the end of this time interval, when the point E of the refracting surface is just struck by the incident wavefront and therefore just becomes the origin of secondary wavelets, the radius acquired by the wavelet, which originated t seconds earlier from A in the lower medium would be

$$AG = V_g t \qquad \text{...(10.3 } b\text{)}$$

By the use of Equation (10.3 *a*), this reduces to

$$AG = (V_g/V_a)\ CE \qquad \text{...(10.3 } c\text{)}$$

Similarly, the radius of the secondary wavelet which originated from D would be

$$DF = (V_g/V_a)\ HE \qquad \text{...(10.3 } d\text{)}$$

Since the secondary wavelets from points intermediate between A and E originate at successively somewhat later times than that which originated at A, hence their radii, at the instant when E just becomes the origin of wavelets, are successively smaller than that of the wavelet which originated at *A*.

With points *A* and *D* as centres hemispheres of radius *AG* given by Eqn. (10.3 *c*), and radius *DF*, given by Eqn. (10.3 *d*), are drawn in the lower medium, which consequently represent the spread of the secondary wavelets which originated from *A* and *D* respectively, by the time the end *C* of the incident wave has reached *E*. A plane drawn through *E*, normally to the plane of diagram, so as to be tangential to the secondary wavelet from *A* at *G* can be easily shown to touch all secondary wavelets, which originated from the points of the refracting surface between *A* and *E*. The proof may be given as follows:

It is easy to see that triangles *ACE* and *DHE* are similar hence, we get

$$AE/DE = CE/HE$$

By the help of Eqns. (10.3 *c*) and (10.3 *d*), the above relation can be easily expressed as,

$$AE/DE = AG/DF$$

or

$$DF/DE = AG/AE$$

From this relation it is obvious that $\angle DFE = 90°$. But *DF* is the radius of the secondary wavelet from *D*; thus the line *EG* just touches the wavelet form *D* at *F*. We, therefore, conclude that the plane through *EG*, normal to the plane of the diagram, is the envelope of all secondary wavelets generated from *A* to *E* by the time the incident wave has reached *E*. The plane *EG*, therefore represents the refracted wavefront in the lower medium.

Snell's law of refraction can now be easily derived as follows:

$$\frac{\sin i}{\sin r} = \frac{CE/AE}{AG/AE} = \frac{CE}{AG}$$

By the help of Eqn. (10.3 *c*), the above equation becomes

$$\frac{\sin i}{\sin r} = \frac{V_a}{V_g} = \frac{c/V_g}{c/V_a} = \frac{\mu_g}{\mu_a} = {}_a\mu_g \qquad ...(10.3\ e)$$

Equation (10.3 *e*) is called as *Snell's law*. With the help of the *wave theory*, we have proved that the refractive index of glass with respect to air must be equal to V_{air}/V_{glass}. The result differs fundamentally from that obtained on *Newton's Corpuscular theory of light* which gives ${}_a\mu_g = V_{glass}/V_{air}$. As a consequence, according to *wave theory*, the speed of light propagation in a denser medium must be less as compared to its value in the rarer medium. The corpuscular theory, however claims just the reverse effect. In 1850 *Foucault* experimentally showed that the speed of light in water is indeed smaller than its speed in air. This resulted in a resounding victory for *Huygens' wave theory.*

10.4 TOTAL REFLECTION

Huygens' principle can be very easily applied to give physical explanation of the phenomenon of total reflection. We may first briefly describe this phenomenon on the basis of the ray theory. Fig. 10.5 shows a number of rays diverging from a monochromatic point source *P* in glass, and striking the interface separating it from air, the refractive index of glass with respect to air being ${}_a\mu_g$. From *Snell's law*, since light is refracted from glass in air, is follows that

$$\sin i/\sin r = {}_g\mu_a = 1/{}_a\mu_g \qquad \text{....(10.4 } a)$$

or

$$\sin r = {}_a\mu_g \sin i$$

Since ${}_a\mu_g > 1$, hence the angle of refraction r, in the case under discussion, is always greater than the angle of incidence i. Therefore, for some value of i less than 90°, the angle of refraction r becomes equal to 90°. In the latter case, as illustrated, the refracted ray 3 just goes grazing the interface and separating the two media. The particular angle of incidence corresponding to r = 90° is called the *critical angle* for the pair of media under consideration and it is usually denoted by i_C. Equation (10.4 *a*), therefore, gives

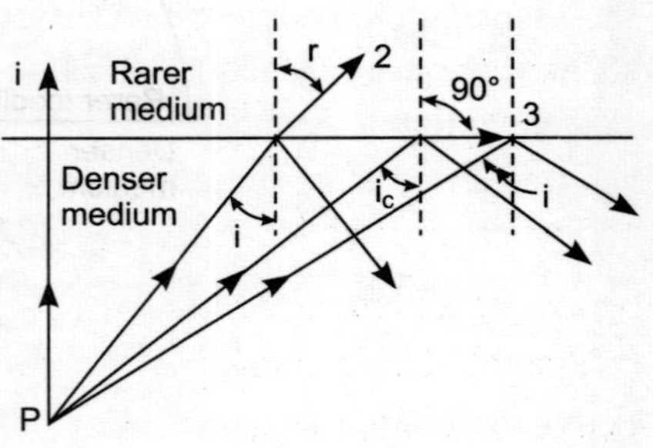

Fig. 10.5

$$\sin i_C = 1/{}_a\mu_g \qquad \text{...(10.4 } b)$$

If the angle of incidence is greater than i_C then it follow from Eqn. (10.4 *a*) that the sine of the angle of refraction becomes greater than 1, which is obviously impossible. This is interpreted as meaning that for angle of incidence greater than the critical angle i_C, the refracted ray does not exist, only the incident ray is reflected back in the same medium. This is known as the *phenomenon of Total Reflection.*

We now give the explanation of the above phenomena according to wave theory. In Fig. 10.6, A_1C_1 represents a plane wavefront in a denser medium (glass), just incident at A_1 on the interface separating it from the rarer medium (air). According to *Huygens' principle*, the points A_1 ... C_2 become the origins of two secondary wavelets which spread out respectively in the upper and the lower medium as the incident wavefront sweeps the interface. As explained in *Sec.* 10.3, the secondary wavelet which originated from A_1 would acquire a radius A_1A_2' in the upper medium, given by the relation

$$A_1A_2' = (V_a/V_g)\ C_1C_2 = {}_a\mu_g\ C_1C_2 \qquad \text{...(10.4 } c)$$

at the instant when C_2 just becomes the origin of secondary wavelets.

But $$C_1C_2 = A_1C_2 \sin i$$

Hence, Eqn. (10.4 *c*) may be written as

$$A_1A_2' = {}_a\mu_g\ A_1C_2 \sin i \qquad \text{...(10.4 } d)$$

Now, if we vary the angle of incidence i, keeping A_1C_2 constant, three possible cases arise:

(*a*) The angle of incidence i is such that we have an inequality,

$$\therefore\ {}_a\mu_g \sin i < 1$$

Hence, Eqn. (10.4 *d*) yields an inequality

$$A_1A_2' < A_1C_2$$

In this case, the radius of the secondary wavelet which originated at A_1 in the upper medium is less than A_1C_2 at the instant when C_2 just becomes the origin of secondary wavelets (Fig. 10.6). As explained in *Sec.* 10.3, A_2A_2', is the refracted wavefront and the ray A_1A_2', normal to A_2C_2', is the refracted ray.

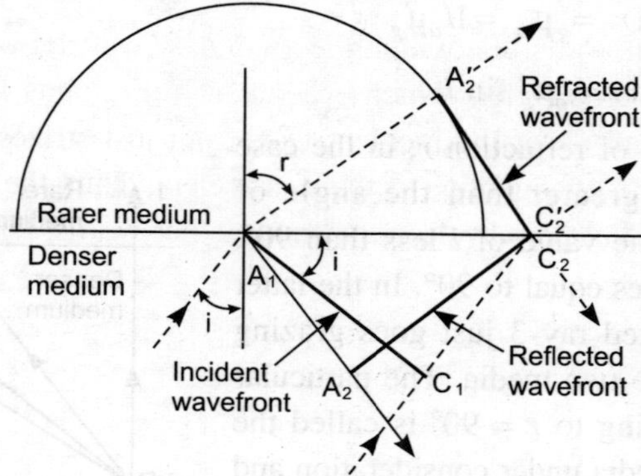

Fig. 10.6. Reflected and Refracted wavefronts

(*b*) If the angle of incidence increases to i_C so that we have

$$_a\mu_g \sin i_C = 1$$

then $$A_1A_2' = A_1C_2$$

In this case, the secondary wavelet which originates from A_1 will pass through C_2 at the instant when the latter just becomes the origin of secondary wavelets, and this obviously holds good for all wavelets generated between A_1 and C_2. In other words, all the secondary wavelets which originate at various points touch each other at C_2 and therefore reinforce each other at C_2. As a consequence, the resultant disturbance is propagated just grazing the interface of two media. Thus, the refracted ray goes just grazing the interface and the angle i_C is the *critical angle* for the two media.

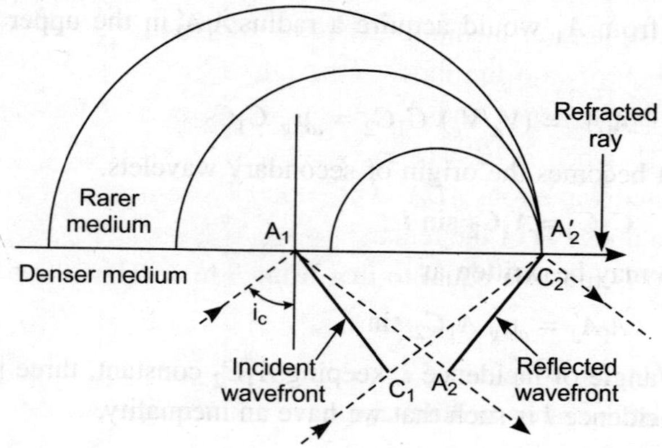

Fig. 10.7. Illustrates refracted ray grazing the refracting surface

(*c*) If the angle of incidence, say i', is greater than i_C so that we have an inequality

$$_a\mu_g \sin i' > 1$$

then, it follows from Eqn. (10.4 *d*), that

$$A_1A_2' > A_1C_2$$

This simply means that the point C_2 will lie within the secondary wavelet which originated from A_1 and this property holds for all other wavelets which originated from points intermediate between A_1 and C_2, at the instant when the end C_1 of the incident wavefront just strikes C_2. Obviously, it is impossible to draw a plane through C_2 tangential to these wavelets. Thus the refracted wavefront does not exist in this case, only the reflected wavefront exists. In other words, light is totally reflected within the same medium.

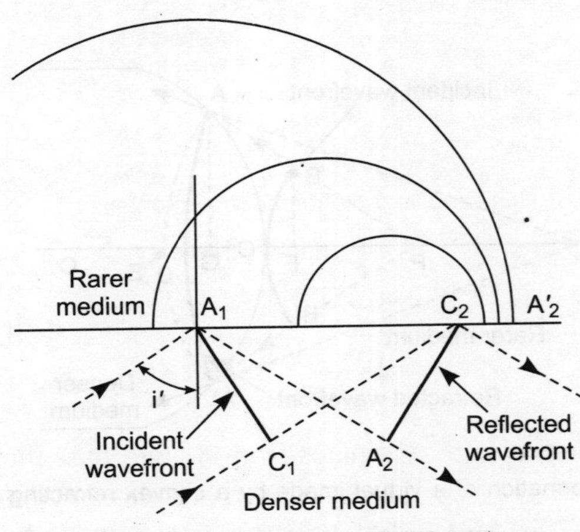

Fig. 10.8. Illustrates total internal reflection

10.5 REFRACTION THROUGH A SPHERICAL SURFACE

Suppose in Fig. 10.9 the circular arc AOA' represents the trace, in the plane of paper, of the spherical refracting surface convex towards the rarer medium. Let BOB' represent the trace of spherical wavefront, diverging from the monochromatic luminous point source P. Now, if P is so close to the refracting surface that at the instant when A and A' become the origins of wavelets, the radius OD of wavelet which originated at O (when BOB' was the position of the incident wavefront) becomes greater than OG, then the refracted wavefront ADA' will be a spherical surface apparently diverging from the axial point Q. Thus, the image Q of the real axial point P is *virtual*.

The optical path from O to D must be equal to that from B to A. Thus the refracted wavefront ADA' satisfies the condition

$$\mu_2 OD = \mu_1 BA$$

where μ_1 is the refractive index of the rarer medium and μ_2 that of the denser medium. In the paraxial region, $BA = FG$ approximately, hence

$$\mu_2 OD = \mu_1 FG$$

or $$\mu_2(OD + GD) = \mu_1(FO + OG)$$

or $$\mu_1 FO - \mu_2 GD = (\mu_2 - \mu_1) OG \qquad \text{...(10.5 } a)$$

Since P and Q are to the left of O while C to the right of O, hence according to our sign convention, we have $OP = -u$, $OQ = -v$ and $OC = R$. Therefore, to a close approximation we write

$$FO = y^2/(-2u); GD = y^2/(-2v); OG = y^2/2R$$

where we have written $AG = y = BF$ approximately. Substituting the values of FO, GD and OG in Eqn. (10.5 *a*)

$$\frac{\mu_2}{v} - \frac{\mu_1}{u} = \frac{\mu_2 - \mu_1}{R} \quad \text{...(10.5 } b\text{)}$$

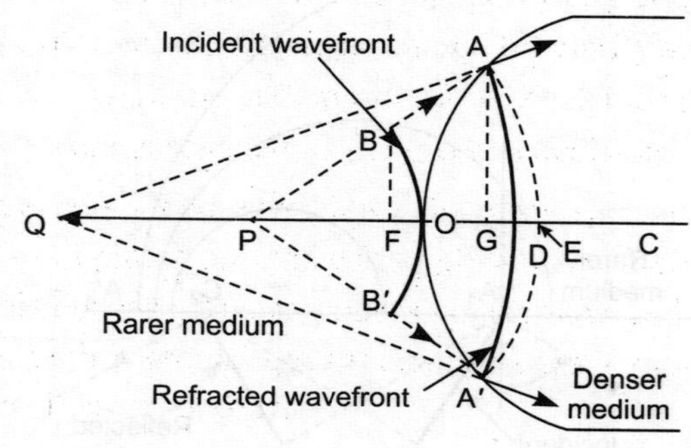

Fig. 10.9. Formation of a virtual image by a convex refracting surface

Dividing Eqn. (10.5 *b*) by μ_1 and writing ${}_1\mu_2$ for μ_2/μ_1, the refractive index of the second medium (denser) with respect to the first (rarer) medium, we get

$$\frac{{}_1\mu_2}{v} - \frac{1}{u} = \frac{{}_1\mu_2 - 1}{R} \quad \text{...(10.5 } c\text{)}$$

10.6 REFRACTION THROUGH A THIN LENS

(*a*) Spherical Waves Incident on a Thin Convex Lens

Let C_1BC_2 and C_1DC_2 in Fig. (10.10) represent two spherical surfaces of a thin convex lens, the radii of curvature of the surfaces being R_1 and R_2 respectively. We suppose the lens material to be denser than the surrounding medium. Let B_1BB_2 represent a spherical wavefront, shown diverging from a luminous axial point P emitting light of one wavelength only, and to have reached a position so as to just touch the pole B of the surface C_1BC_2 of the lens. According to Huygens' principle, at this instant, the pole B just becomes the origin of secondary wavelet and this wavelet traverses an axial thickness BD within the lens medium in the time interval t. On the other hand, by Huygens' principle of propagation of the wavefront, the ends B_1 and B_2 in this interval t practically travel in rarer medium (air) via C_1 and C_2 respectively to respective positions D_1 and D_2,

$\therefore$ Optical path $B_1C_1D_1$ = Optical path BD

i.e., $$B_1C_1 + C_1D_1 = \mu BD$$

where μ is the refractive index of the lens material with respect to the surrounding medium.

In reality in order to trace the formation of the image we should regard every point of the back surface of the lens as new origins of secondary wavelets which are generated when the points are successively struck by the advancing wavefronts from the left. Fig. 10.11 represents a particular case where wavefronts in the lens medium are plane. This has been assumed for simplicity of the figure, nevertheless, the conclusions which will be drawn are quite general and apply to account for the formation of the real image by the lens. If we now confine our attention to the advancing wavefront when in the position C_1CC_2 within the lens (Fig. 10.10), then by the time interval this wavefront reaches D, the secondary wavelets from C_1 and C_2 would acquire the radii, to a close approximation, given by

$$C_1D_1 = C_2D_2 = \mu CD$$

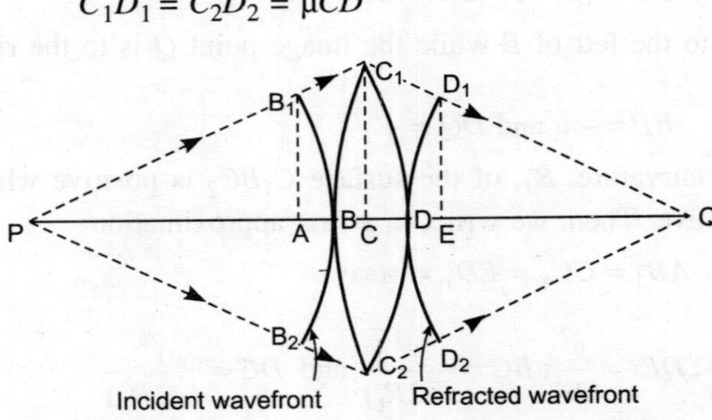

Fig. 10.10. Refraction through a convex lens

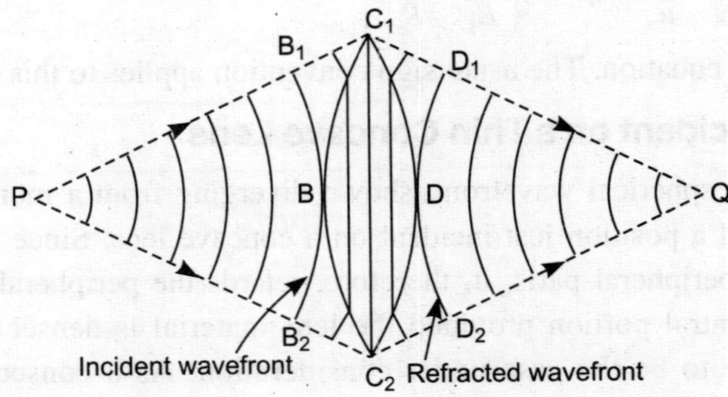

Fig. 10.11. Refraction of waves through a convex lens

Lens is denser than the surrounding medium, that is, $\mu > 1$, therefore it follows from the above relation that C_1D_1 as well as C_2D_2 both are greater than CD. In the paraxial region therefore, the points D_1 and D_2 of the secondary wavelets from C_1 and C_2 lie to the right of D and this applies to every secondary wavelet originated at the back surface except the point D which has just become the origin of the secondary wavelets. The envelope of these secondary wavelets in the paraxial region is a spherical surface, converging towards the axial point Q. Thus a real image of P is formed at Q, due to refraction of monochromatic waves through the lens. In short, the action of the lens is to retard the central portion of the incident wavefront relative to its peripheral parts thereby producing a change in its curvature.

The refracted spherical wavefront D_1DD_2, as explained earlier, satisfies the condition

$$B_1C_1 + C_1D_1 = \mu BD \qquad \text{...(10.6 } a)$$

For a sufficiently small aperture of the lens, we may put

$$B_1C_1 = AC \text{ and } C_1D_1 = CE$$

With this approximation, Eqn. (10.6 *a*) becomes

$$AC + CE = \mu BD$$

or
$$AB + BC + CD + DE = \mu(BC + CD)$$

or
$$AB + DE = (\mu - 1)(BC + CD) \qquad \text{...(10.6 } b)$$

The object point P is to the left of B while the image point Q is to the right of D. Therefore, we write

$$BP = -u \text{ and } DQ = +v$$

Further, the radius of curvature, R_1, of the surface C_1BC_2 is positive while R_2, the radius of curvature of C_1DC_2, is negative. Then, we write, to a first approximation

$$AB_1 = CC_1 = ED_1 = y(\text{say})$$

$$AB = \frac{y^2}{(-2u)};\ DE = \frac{y^2}{2v};\ BC = \frac{y^2}{(+2R_1)} \text{ and } DC = \frac{y^2}{(-2R_2)}$$

Substitution of these values in Eqn. (10.6 *b*) givens

$$\frac{1}{v} - \frac{1}{u} = (\mu - 1)\left(\frac{1}{R_1} - \frac{1}{R_2}\right) \qquad \text{...(10.6 } c)$$

which is called the thin lens equation. The usual sign convention applies to this equation.

(*b*) Spherical Waves Incident on a Thin Concave Lens

Let B_1BB_2 represent a spherical wavefront, shown diverging from a monochromatic luminous point P and to have reached a position just incident on a concave lens. Since the lens is thin in the middle as compared to its peripheral parts, it, therefore, retards the peripheral parts of the incident wavefront relative to its central portion provided the lens material is denser than the surrounding medium, which we assume to be the case under consideration. As a consequence, the emergent spherical wavefront C_1GC_2 is a divergent one with a curvature greater than that of the incident wavefront B_1BB_2.The emergent wavefront appears to diverge from an axial point Q, situated on the same side of the lens as luminous object point P and at a distance from the lens less than that of P. Thus, a virtual image Q of P is formed by a concave lens.

The optical path from B_1 to C_1 must be equal to that between B and G *i.e.*,

$$BC + \mu CD + DG = \mu B_1C_1$$

or
$$(AC - AB) + \mu CD + DE + EF = \mu(AC + CD + DE)$$

or
$$EG - AB = (\mu - 1)(AC + DE) \qquad \text{...(10.6 } d)$$

Let $AB_1 = EC_1 = y$ (say). We, therefore, write to a close approximation,

$$EG = \frac{y^2}{(-2v)}, \quad AB = \frac{y^2}{(-2u)}, \quad AC = \frac{y^2}{(-2R_1)} \quad \text{and} \quad DE = \frac{y^2}{2R_2}$$

Substituting these values in Eqn. (10.6 *d*), we get the familiar relation,

$$\frac{1}{v} - \frac{1}{u} = (\mu - 1)\left(\frac{1}{R_1} - \frac{1}{R_2}\right)$$

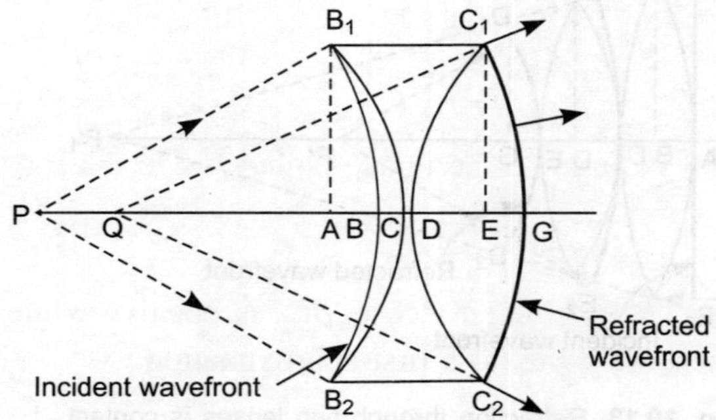

Fig. 10.12. Refraction through a concave lens

10.7 FOCAL LENGTH OF THE COMBINATION OF TWO THIN LENSES IN CONTACT

When two thin lenses of focal lengths f_1' and f_2' are placed in close contact, the focal length f', of this combination, as obtained by the ray theory is given by the relation

$$\frac{1}{f'} = \frac{1}{f_1'} + \frac{1}{f_2'}$$

We shall now derive this formula according to wave theory.

Consider a plane wavefront B_1AB_2, of monochromatic light, just incident on a combination formed of two convergent lenses L_1 and L_2 placed in contact. According to principles described in *Sec.* 10.6, the first lens retards the central part of the incident wavefront relative to its peripheral portions, thereby changing it to a spherical wavefront E_1CE_2, converging towards the second principal focal point F_1' of the first lens. The second lens further retards the central portion of E_1CE_2 with respect to peripheral parts. As a consequence, the emergent wavefront D_1ED_2 from the combination is spherical (in paraxial region), this time, however converging towards the axial point F', which is called as the second principal focal point of the combination of lenses in contact. It is now easy to see that light disturbance from B_1 must traverse the path $(B_1L_1 + L_1L_2 + L_2D_1)$ in air, in the time interval required for the light disturbance from A to reach E, travelling through the lens material.

Therefore, if μ be the refractive index of the first lens and μ′ that of the second lens, both with respect to the surrounding medium, we can easily get

$$B_1L_1 + L_1L_2 + L_2D_1 = \mu AC + \mu' CE \quad \text{...(10.7 } a\text{)}$$

For a sufficiently small aperture of the combination, we may put to a close approximation,

$$L_1L_2 = BD \text{ and } L_2D_1 = DG$$

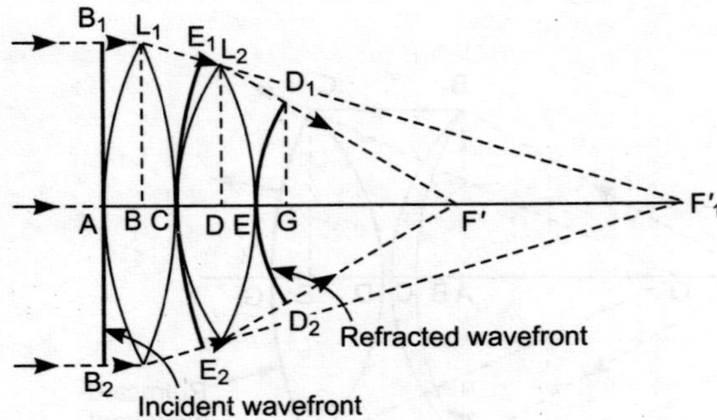

Fig. 10.13. Refraction through two lenses is contact

With this approximation, Eqn. (10.7 *a*) becomes

$$AB + BD + DG = \mu AC + \mu' CE$$

or

$$AB + BC + CD + DE + EG = \mu(AB + BC) + \mu'(CD + DE)$$

or

$$EG = (\mu - 1)(AB + BC) + (\mu' - 1)(CD + DE) \quad \text{...(10.7 } b\text{)}$$

Let the radii of curvature of the surfaces of the first lens be denoted by R_1 and R_2 while that of the second lens by r_1 and r_2. Also, suppose $EF' = f'$. Further, we write to a close approximation,

$$BL_1 = DL_2 = GD_1 = y \text{ (say)}$$

Now, considering the spherical wavefront D_1ED_2, of radius of curvature f' and centred at F', we have

$$(GD_1)^2 = EG\,(2f' \; EG)$$

or

$$EG = y^2/2f' \text{ approximately}$$

Similarly, to a close approximation, we have

$$AB = y^2/2R_1;\; BC = -y^2/2R_2$$

$$CD = y^2/2r_1;\; DE = -y^2/2r_2$$

Substituting these values in Eqn. (10.7 *b*), we get

$$1/f' = (\mu - 1)(1/R_1 - 1/R_2) + (\mu' - 1)(1/r_1 - 1/r_2)$$

or

$$1/f' = 1/f_1' + 1/f_2'$$

PROBLEMS

1. Explain clearly Huygens' principle of wave propagation.

   Deduce the laws of reflection with the help of Huygens' wave theory.

2. (a) State Huygens' principle. Explain on its basis the phenomenon of refraction and obtain the law of refraction. Give the physical significance of refractive index.

   (b) Explain total internal reflection on the basis of wave theory and obtain the value of critical angle.

3. Apply Huygens' theory of light to obtain an expression for refraction of a spherical wave at a spherical surface.

4. Account for the formation of images by refraction through lenses on wave theory and prove the relation

$$\frac{1}{v}-\frac{1}{u}=(\mu-1)\left(\frac{1}{R_1}-\frac{1}{R_2}\right)$$

   where the symbols have their usual meanings.

INTERFERENCE OF LIGHT

(Division of Wavefront)

It is a matter of common experience that two trains of ripples on the surface of water cross each other and proceed onward in their directions undisturbed. In the same way two beams of light cross each other and proceed without being influenced by each other in any way. For example, different observers can view at the same instant different objects through the same narrow aperture with perfect clarity, the beams of light having crossed at the aperture in reaching the observers. However, in the region of crossing where both the beams are acting simultaneously, we expect a modification in their intensity which should be either less or greater than that which would be given by one beam alone. This modification of intensity due to superposition of two or more beams of light is spoken of as *Interference of Light*. This phenomenon demands for its explanation that light must be a wave motion.

11.1 PRINCIPLE OF SUPERPOSITION

The fundamental basis of the explanation of the phenomenon of *Interference of Light* is the *Principle of Superposition* of wavemotions first enunciated by *Thomas Young* in a paper read before the *Royal Society* in 1801. This principle states that when a medium is disturbed simultaneously by any number of waves the instantaneous resultant displacement of the medium at every point at every instant is the algebraic sum of the displacements of the medium due to individual waves, in the absence of the others. In Young's own words "*When two undulations from different origins coincide, either perfectly or very nearly, in direction, their joint effect is a combination of the motions belonging to each.*" Suppose due to a single wave train the displacement of the medium at a certain point at any instant is y_1 in a given direction and that due to another wave-train, in the absence of the first, the displacement is y_2. According to the above principle, the instantaneous resultant displacement R of the medium at that point due to two waves acting together is expressed by

$$R = y_1 + y_2$$

when the two separate displacements y_1 and y_2, as depicted in Fig. 11.1 (*a*), are in the same direction. However, when the two individual displacements y_1 and y_2 are in opposite directions, the instantaneous resultant displacement due to two waves acting together is now given by

$$R = y_1 - y_2$$

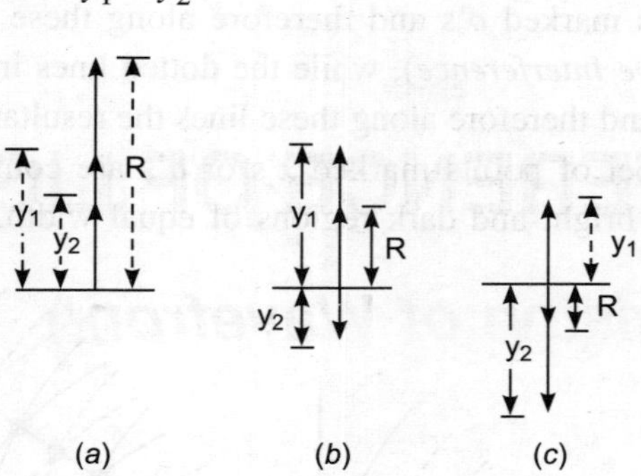

Fig. 11.1. Illustrates principle of superposition of wave motions

In Fig. 11.1 (*b*), y_1 is greater than y_2 but the two are directed in opposite directions. Hence their algebraic sum is $(y_1 - y_2)$ and this is plotted in the direction of y_1. The reverse of this case is depicted in Fig. 11.1 (*c*).

11.2 DISCOVERY OF INTERFERENCE OF LIGHT

Young's Experiment. Historically, the phenomenon of *Interference of light* was first discovered by *Thomas Young* in 1801, when he announced an experiment capable of exhibiting an interference pattern due to the superposition of two beams of light. This experiment was regarded as a crucial one at that time, since it definitely established the wave nature of light. The corpuscular theory was found to be totally inadequate to explain the experimental results.

The apparatus is shown schematically in Fig. 11.2. Young allowed the sun light to pass through a pinhole S and then at some distance through two sufficiently close pinholes S_1 and S_2 in an opaque screen. Finally the light was received on a screen on which he observed an uneven distribution of light intensity. Young found that the illumination on the screen consisted of many alternate bright and dark spots. In accordance with the modern laboratory technique, narrow parallel slits replace pinholes and the slit S is illuminated with monochromatic light of wavelength λ. Light is received on a screen placed at a certain distance to the right and parallel to the plane containing the slits S_1 and S_2. According to Huygens' principle, cylindrical wavelets spread out from slit S and, as the path $SS_1 = SS_2$, the wavelets reach slits S_1 and S_2 at the same instant. A train of Huygens' wavelets, therefore, diverges to the right from both of these slits which have precisely equal phases at the start. Furthermore, their amplitude, wavelength and velocity are also equal. Suppose in two dimensional Fig. 11.2, continuous circular arcs represent the wave crests while dotted circular arcs represent the wave troughs in each wave. At points marked by *o*'s a crest of one wave is superposed on a crest of the other or a trough of one superposed on a trough of the other. In other words, at these points the two waves meet in the same phase. Therefore, according to the principle of superposition, at these points the resultant amplitude is twice that of each component waves. On the other hand at points marked by *x*'s crest of one wave is superposed on the trough of the other and vice-versa *i.e.*, the two waves meet in opposite phase. Hence, according to the principle of superposition, at these points they neutralise each other and the resultant intensity is zero. The solid lines intersecting the screen at P_0 and P_1 connect the points marked o's and therefore along these lines the resultant intensity is always *maximum* (*Constructive Interference*), while the dotted lines intersecting the screen at D_1 are the loci of points marked x's and therefore along these lines the resultant intensity is zero (*Destructive Interference*). Actually the loci of points marked x's or o's are confocal hyperbolae. Thus on the screen a number of alternate bright and dark regions of equal width, called *interference fringes*, is observed parallel to the slits.

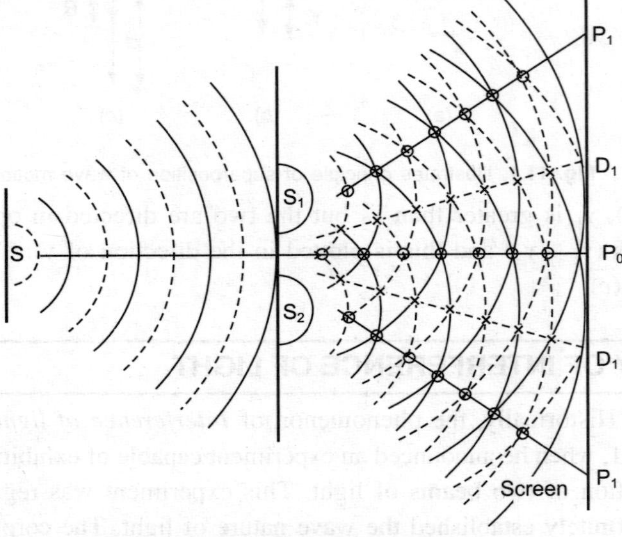

Fig. 11.2. Young's double-slit experiment

That the observed pattern is truly due to interference of two waves of light can be demonstrated by covering one of the double slits. The well defined dark and bright fringes are replaced by a new pattern much coarser due to diffraction of light by uncovered single slit. A comparison of the two patterns indicates that a point on the screen, bright when only one slit is covered changes to dark when both slits are exposed. A corpuscular theory is totally inadequate to account for this fact, which can be easily explained in terms of the wave nature of light.

The interference effects can be conveniently demonstrated in the laboratory with ripples on the surface of water. Two pins are attached to the same prong of an electrically maintained tuning fork and adjusted so that the pins just touch the surface of water placed underneath, while the prongs are capable of vibrating in a vertical plane. The periodic vibratory motion of the prong is also taken up by the attached pins, which are, therefore, always in equal phase. This periodic vibratory up and down motion of the pins, on the surface of water generates two waves or ripple systems of equal amplitude, equal wavelength and equal velocity. The ripples spread out with their centres at the points of action of two pins. In Fig. 11.2, S_1 and S_2 may be now imagined as the points of action of two vibrating pins, continuous and dotted circular arcs may be the traces of *crests* and *troughs* respectively of the ripple systems. It will be observed that along the dotted straight lines water remains permanently calm while along the solid straight lines it permanently vibrates up and down with double the amplitude of the component waves.

11.3 THEORY OF INTERFERENCE

We shall now derive an expression for the resultant intensity an any point P of the screen due to superposition of two waves of light having the same amplitude, frequency and wave number. Suppose that these waves are emitted from two point sources A and B separated by a distance $2d$ (Fig. 11.3) which is of the order of several wavelengths. The distribution of light intensity on a screen placed at a large distance from the sources A and B can then be determined as follows.

If the two point sources are locked in phase or have a constant phase difference this intensity distribution will be independent of time. Further, if the distance D is large, the two waves reaching any field point P will have nearly the same amplitude and the corresponding 'rays' will be effectively parallel and inclined at the same angle θ to the axis ($\theta = 0$) of the system. This amplitude, nevertheless, depends upon the average distance $x = ½(x_1 + x_2)$ which the waves cover to reach the point P. With these reservations, we suppose that the waves reaching P are represented by

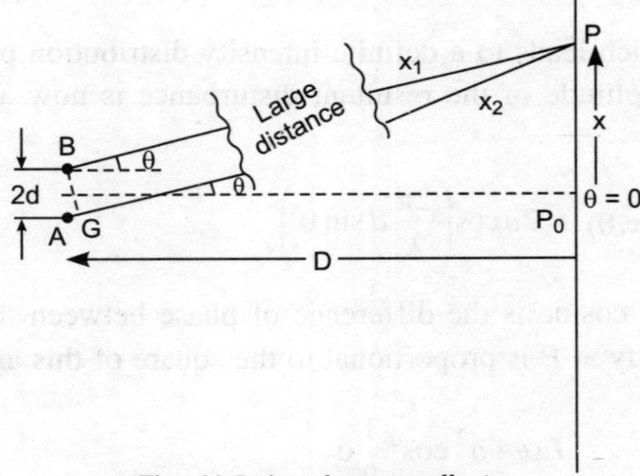

Fig. 11.3. Interference effects

$$\phi_1 = ae^{-i(\omega t - kx_1)}$$

and $$\phi_2 = ae^{-i(\omega t - kx_2)} \qquad ...(11.3\ a)$$

where the amplitude is $a = a\,(x)$. The resultant wave at P is then given by

$$\phi = \phi_1 + \phi_2$$

$$\phi = ae^{-i\omega t}[e^{ikx_1} + e^{ikx_2}] \qquad ...(11.3\ b)$$

Both the distances x_1 and x_2 may be expressed in terms of the average distance x. In this manner Eqn. (11.3 b) may be written as

$$\phi = ae^{-j(\omega t - kx)}\left[e^{ik(x_1 - x)} + e^{ik(x_2 - x)}\right]$$

$$= ae^{-i(\omega t - kx)}\left[e^{\frac{1}{2}ik(x_1 - x_2)} + e^{-\frac{1}{2}ik(x_1 - x_2)}\right] \qquad ...(11.3\ c)$$

The factor $(x_1 - x_2)$ occurring as exponential power in Eqn. (11.3 c) is the path difference BG between the waves reaching P from the sources A and B. This is given by

Path difference $\qquad BG = 2d \sin\theta \qquad ...(11.3\ d)$

Combining Eqns. (11.3 *c*) and (11.3 *d*) we find that the equation of the resultant disturbance at *P* is

$$\phi = ae^{-i(\omega t - kx)}[e^{ikd \sin\theta} + e^{-ikd \sin\theta}]$$

or

$$\phi = 2a \cos(kd \sin\theta)e^{-i(\omega t - kx)}$$

$$= 2a \cos\{(2\pi/\lambda)d \sin\theta\}e^{-i(\omega t - kx)} \qquad ...(11.3\ e)$$

This equation of the resultant disturbance at any field point *P* is in fact true for all wave motions provided we extract the real part from the R.H.S. It will then be equally applicable to acoustic waves or water waves under corresponding situations. Also, the treatment shall be valid even if the point sources *A* and *B* are replaced by line or slit sources of light. Thus

$$\phi = 2a \cos\left(\frac{2\pi}{\lambda} d \sin\theta\right)\cos(\omega t - kx) \qquad ...(11.3\ f)$$

is in fact, the disturbance which leads to a definite intensity distribution pattern on the screen in the case of light waves. The amplitude of the resultant disturbance is now a function of both x and θ and may be written as

$$a(x, \theta) = 2a \cos\left(\frac{2\pi}{\lambda} d \sin\theta\right) \qquad ...(11.3\ g)$$

where twice the argument of cosine is the difference of phase between the waves reaching *P* from *A* and *B*. The resultant intensity at *P* is proportional to the square of this amplitude, hence we write

$$I \propto 4 a^2 \cos^2 \frac{1}{2}\delta$$

where

$$\delta = 2\{(2\pi/\lambda)d \sin\theta\}$$

With k as constant of proportionality, we can also write

$$I = 4ka^2 \cos^2 \frac{1}{2}\delta$$

$$= 2ka^2(1 + \cos\delta) \qquad ...(11.3\ h)$$

which varies with the magnitude of δ.

We shall now study the variation of I with δ.

Constructive Interference

When $\delta = 0, 2\pi, 4\pi$, etc., $\cos\delta = +1$, which is the maximum value of the cosine function. As a consequence, it follows from Eqn. (11.3 *h*) that,

$$I = 4ka^2 \qquad \text{[Maximum]}$$

which is, obviously, greater than what is to be expected from the separate intensities. *The two waves reinforce each other completely and the resultant intensity is thus maximum, when*

$$\delta = 2n\pi \qquad \text{[Maxima]} \qquad ...(11.3\ i)$$

where $n = 0, 1, 2, 3, 4$ etc. Using this value of δ we easily get the condition for maximum intensity in terms of the optical path difference, viz.

$$AP - BP = 2n(\lambda/2) \qquad \text{[Maxima]}$$

The two waves are said to be in the same phase at any point where their phases differ exactly by even multiple of π and their interference is spoken as *Constructive Interference.*

Destructive Interference

When $\delta = \pi, 3\pi, 5\pi, 7\pi$ etc. then $\cos\delta = -1$, which is the least value of the cosine function. As a consequence, Eqn. (11.3 *h*) yields

$$I = 0 \qquad \text{[Minimum]}$$

which is obviously, less than what is to be expected from the separate intensities. *Two waves interfere destructively, when*

$$\delta = (2n + 1)\pi \qquad \text{[Minima]} \qquad \text{...(11.3 } j\text{)}$$

where $n = 0, 1, 2, 3, 4$ etc. In terms of the optical path difference the condition for minimum intensity is

$$AP - BP = (2n + 1)\lambda/2 \qquad \text{[Minima]}$$

where $n = 0, 1, 2, 3$ etc.

Two waves are said to be in opposite phases at any point where their phases differ exactly by odd multiple of π and their interference is spoken of as *Destructive interference.*

We, therefore, conclude that when two waves, having a constant phase difference initially reunite after traversing difference optical paths, they interfere constructively or destructively according as their phases, at the point of observation, differ exactly by even or odd multiple of π. This condition, it should be remarked, is the general one and applies for locating the regions of constructive and destructive interference in all experimental arrangements exhibiting interference effects.

11.3.1 Spacing of Interference Fringes – Fringe Width

In order to plot the intensity distribution in the interference pattern arising from the interference of light waves from two coherent sources S_1 and S_2, it is essential to investigate the spacing of these fringes on a screen placed parallel to the line joining S_1 and S_2. It will be recalled that at any point P on the screen the resultant intensity is maximum or minimum according as its distance from the two coherent sources S_1 and S_2, which are always in equal phase, differs exactly by even or odd multiple of $\lambda/2$, where λ is the wavelength of the light emitted by the sources. In symbols the conditions are expressed as:

$$AP - BP = 2n\frac{\lambda}{2} \qquad \text{[Bright fringes]} \qquad \text{...(11.3 } k\text{)}$$

and

$$AP - BP = (2n+1)\frac{\lambda}{2} \qquad \text{[Dark fringes]} \qquad \text{...(11.3 } l\text{)}$$

where $n = 0, 1, 2, 3, 4$ etc.

Referring to Fig. 11.3, in the plane of diagram, let X be the distance of the point P from the central Point P_o, a point where the perpendicular bisector of AB intersects the screen. We can easily see that

$$(AP)^2 = D^2 + (X + d)^2$$

which can be easily rearranged as

$$AP = D\left[1 + \left(\frac{X+d}{D}\right)^2\right]^{\frac{1}{2}}$$

On expanding the right side of the above relation by *Binomial* theorem and neglecting higher powers of $[(x + d)/D]^2$ than one, which of course is permissible, since, in general, D is some thousand times larger than d and x, we get

$$AP = D\left[1 + \frac{1}{2}\left(\frac{X+d}{D}\right)^2\right]$$

Similarly,

$$(BP)^2 = D^2 + (X - d)^2$$

from which, as above, BP can be easily approximated,

$$BP = D\left[1 + \frac{1}{2}\left(\frac{X-d}{D}\right)^2\right]$$

Whence, $$AP - BP = \frac{1}{2D}[(X+d)^2 - (X-d)^2] = \frac{2Xd}{D}$$

This is the value of the path difference to be substituted in Eqn. (11.3 *k*) or Eqn. (11.3 *l*), according as at the point P a bright or dark fringe is formed. Thus, we have for bright fringes,

$$2Xd/D = n\lambda$$

or $$X = n\lambda\ (D/2d) \qquad \text{[Bright fringes]} \qquad ...(11.3\ m)$$

and for dark fringes, we have the relation

$$2Xd/D = (2n + 1)\lambda/2$$

or $$X = (n + 1)\ \lambda\ (D/2d) \qquad \text{[Dark fringes]} \qquad ...(11.3\ n)$$

The separation, on the screen, between the nth and $(n + 1)$th order bright fringes may now be easily computed, using Eqn. (11.3 *m*) as follows:

$$\bar{X} = X_{n+1} - X_n = (n+1)\lambda\frac{D}{2d} - n\lambda\frac{D}{2d} = \frac{D\lambda}{2d}$$

which is also the separation between any two consecutive dark fringes. Since the separation between two consecutive bright or dark fringes is independent of n, therefore for given values of λ, $2d$ and D the spacing of the fringes is constant. The distance between any two consecutive bright or dark fringes, denoted by $\bar{X}$, is called the *Fringe Width*, which varies directly with slit-screen separation D, inversely with the separation of slits $2d$ and directly with the wavelength λ of light employed. These conclusions are in perfect agreement with the observed interference pattern on the screen.

11.3.2 Shape of the Interference Fringes

We now very easily form an idea regarding the shape of the interference fringes by deriving the equation of the locus of points having a given path difference from two slits S_1 and S_2. Let O, the middle point of S_1S_2, be chosen as the origin of coordinates the axis of x along OX and axis of z perpendicular to the plane of slits. Let P be any point with coordinates (z, x), then from Fig. 11.4, we easily get

$$(S_1P)^2 = z^2 + (x - d)^2 \text{ and } (S_2P)^2 = z^2 + (x + d)^2$$

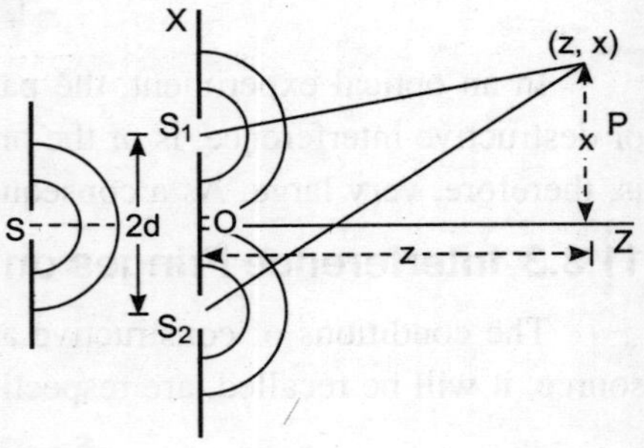

where $2d$ is the separation of slits. The path difference Δ is

Fig. 11.4. Shape of fringes

$$\Delta = S_2P - S_1P = [z^2 + (x+d)^2]^{\frac{1}{2}} - [z^2 + (x-d)^2]^{\frac{1}{2}}$$

On rearranging,

$$\Delta + [z^2 + (x-d)^2]^{\frac{1}{2}} = [z^2 + (x+d)^2]^{\frac{1}{2}}$$

On squaring both sides,

$$\Delta^2 + 2\Delta[z^2 + (x-d)^2]^{\frac{1}{2}} + z^2 + (x-d)^2 = z^2 + (x+d)^2$$

or
$$2\Delta[z^2 + (x-d)^2]^{\frac{1}{2}} = 4xd - \Delta^2$$

On squaring both sides, this equation easily reduces to,

$$4(4d^2 - \Delta^2)x^2 - 4\Delta^2 z^2 = \Delta^2(4d^2 - \Delta^2)$$

or
$$\frac{x^2}{\Delta^2/4} - \frac{z^2}{(4d^2 - \Delta^2)/4} = 1$$

This is the equation of *hyperbola* in a standard form with the foci S_1 and S_2 on the x axis. The loci of points of constant path difference in the XZ plane are thus *hyperbolae*. Furthermore, if instead of slits we have two coherent point sources S_1 and S_2, then the loci of points of constant path difference (hence fringes) are *concentric circles* in any plane parallel to YZ plane. The loci of maxima and minima in space form a system of confocal hyperboloids.

It is left as an exercise for students to establish this statement analytically.

The eccentricities of the hyperbolae are given by,

$$e = \left[\frac{\Delta^2}{4} + \frac{4d^2 - \Delta^2}{4}\right]^{\frac{1}{2}} \div \frac{\Delta}{2} = \frac{2d}{\Delta}$$

In an optical experiment, the path difference Δ, corresponding to the condition of constructive or destructive interference, is of the order of 10^{-8} cm and $2d$ of the order of 10^{-2} cm. The eccentricity is, therefore, very large. As a consequence, hyperbolae are practically straight lines.

11.3.3 Interference Fringes on Screen

The conditions of constructive and destructive interference of two waves derived from a single source, it will be recalled, are respectively,

$$\delta = 2n\pi \qquad \text{[Maxima]} \qquad ...(11.3\ i)$$

and

$$\delta = (2n + 1)\pi \qquad \text{[Minima]} \qquad ...(11.3\ j)$$

Furthermore, it will be recalled that Huygens' wavelets from the source S reach slits S_1 and S_2 at the same instant, when both are equidistant from S. Therefore, the light sources (S_1 and S_2) will have precisely equal phases at all times. The phase difference between two waves on reaching any point P, therefore, depends only on the optical path difference ($S_2P - S_1P$) in this experimental arrangement. At the point P_o on the screen, equidistant from S_1 and S_2, the path difference and hence the phase difference δ between the two waves is zero. A bright fringe is, therefore, formed at P_o. This is the central fringe of the pattern and since $\delta = 0$ can be obtained by putting $n = 0$ in Eqn. (11.3 *i*), this interference is spoken of as *Zero Order*. On either side of P_o, on the screen, the optical path difference gradually increases and we arrive at points D_1, for which the relation $S_2D_1 - S_1D_1 = \frac{1}{2}\lambda$ and hence $\delta = \pi$ hold good. The first dark fringe, on either side of the central bright fringe, is, therefore, formed at D_1. At P_1 the optical path difference increases to, $2\lambda/2$ and hence $\delta = 2\pi$. A bright fringe is, therefore, formed at P_1. Since $\delta = 2\pi$ can be obtained by putting $n = 1$ in Eqn. (11.3 *i*), this bright fringe is spoken of as *First Order Bright Fringe* and the interference is spoken of as *First Order*. In this way it is possible to locate points P_o, P_1, P_2, P_3 etc. on the screen in such a way that their distances from S_1 and S_2 differ exactly by 0, $2\lambda/2$, $4\lambda/2$, $6\lambda/2$ ....in order and the phases of the two waves at these points differ exactly by 0, 2π, 4π, 6π........ in order. Therefore, at P_o, P_1, P_2, P_3 etc. are formed bright fringes of zero order 1st order, 2nd order, 3rd order,..... respectively. In between these points are other points D_1, D_2, D_3, D_4 etc. so that the distance from the two slits differs exactly by $\lambda/2$, $3\lambda/2$, $5\lambda/2$....... in order. Therefore at D_1, D_2, D_3 etc. waves interfere destructively, thereby producing dark fringes at these points.

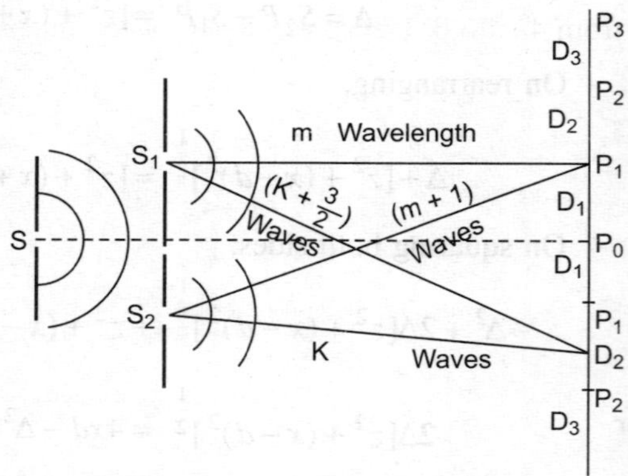

Fig. 11.5

The interference fringes on the screen are, practically straight alternate bright and dark bands. The integer n, it should be emphasised, characterises the order of interference and, therefore, the order of bright fringe. The bright fringes with $n = 0, 1, 2, 3\ldots\ldots$ are called of zero, first, second, …… order.

11.3.4 Intensity Distribution Curve

We are now in a position to plot the intensity distribution for interference fringes from coherent monochromatic beams with the knowledge that the intensity at any point of the screen is expressed by (with k taken as unity)

$$I = 2a^2 (1 + \cos \delta)$$

with further additional knowledge that the spacing of the fringes on the screen is constant for given values of λ, $2d$ and D. Along a horizontal line, the values of δ viz. $0, \pi, 2\pi, 3\pi$ etc. are marked at equidistant points and the axis of I is taken along the normal line through $\delta = 0$. For $\delta = 0, \pm 2\pi, \pm 4\pi, \pm 6\pi$, etc., $I = 4a^2$ and for $\delta = \pm\pi, \pm 3\pi, \pm 5\pi$ etc., $I = 0$. In fact as δ gradually increases from 0 to π, $\cos \delta$ gradually decreases from +1 through 0 to –1 and as a consequence, I gradually diminishes from $4a^2$ to 0. Hence the shape of the intensity distribution curve is as depicted in Fig. 11.6.

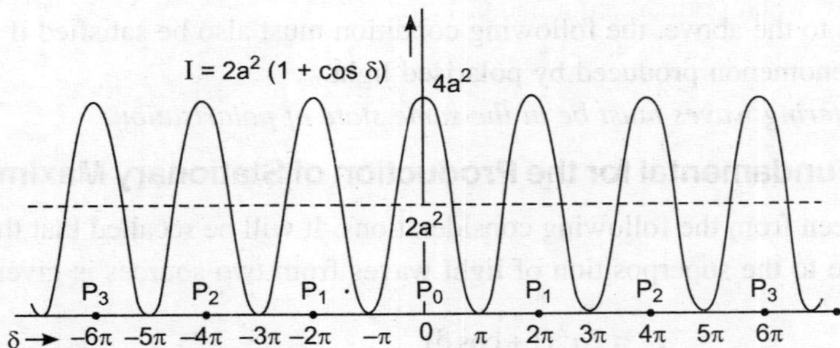

Fig. 11.6. Intensity distribution for the interference fringes from two beams

Before concluding our discussion is should be emphasised that no destruction of light energy occurs in the phenomenon of *Interference of Light*. For interference does not transform light energy into any other form of energy so that the total amount of light energy must remain constant. What has happened is merely a redistribution of energy; the energy which apparently disappear at the dark fringes is actually present in the bright fringes, where the energy is $4a^2$. However, the average value of the energy over any number of fringes is the same as if the interference effects were absent. For example, the average value of the intensity on the screen over the range $\delta = 0$ to $\delta = 2\pi$ is given by

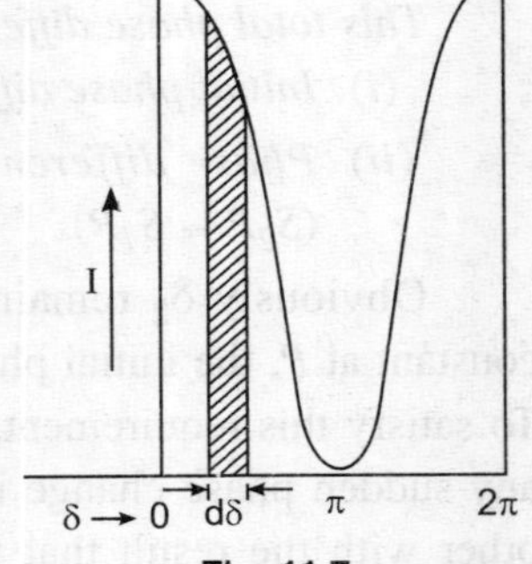

Fig. 11.7

$$I_{\text{Average}} = \frac{\int_o^{2\pi} I d\delta}{\int_o^{2\pi} d\delta} = 2a^2 \frac{\int_o^{2\pi} (1+\cos\delta) d\delta}{\int_o^{2\pi} d\delta} = 2a^2$$

and this justifies the statement made above. There is no violation of the law of conservation of the energy in the interference phenomenon.

11.4 CONDITIONS FOR INTERFERENCE

In a well defined interference pattern, the intensity at regions corresponding to destructive interference must remain zero while at regions corresponding to constructive interference it must remain maximum for all values of time. To accomplish this it is very essential that in every experimental arrangement the following conditions must be fulfilled:

A. *The two beams of light which interfere must originally originate in the same source of light.*

B. *The waves must have the same period and wavelength. Also, their amplitudes must be equal or very nearly equal.*

C. *The original source must emit light of single wavelength or very nearly monochromatic. On the other hand, if the light source is heterogeneous, the optical path difference between the two interfering beams must be very small.*

D. *The two interfering waves must be propagated in almost the same directions or the two interfering wavefronts must intersect at a very small angle.*

E. In addition to the above, the following condition must also be satisfied if we are dealing with the interference phenomenon produced by polarised light.

The two interfering waves must be in the same state of polarisation.

Condition A is Fundamental for the Production of Stationary Maxima and Minima

This can be seen from the following considerations. It will be recalled that the resultant intensity I at any point P due to the superposition of light waves from two sources is given by the expression

$$I = 2a^2(1+\cos\delta)$$

where δ is the phase difference between the two waves at P. If a bright fringe of nth order is formed at P, then $\delta = 2n\pi$. For the sustained fringes the value of δ at P must at all time remain $2n\pi$.

This total phase difference depends on two factors:

(*i*) *Initial phase difference δ_1 between the two sources.*

(*ii*) *Phase difference due to optical path difference $(S_2P - S_1P)$, namely, $\delta_2 = (2\pi/\lambda)(S_2P - S_1P)$.*

Obviously δ_2 remains constant at P throughout the experiment. Hence, in order to keep δ constant at P, the initial phase difference δ_1 between the two vibrating sources must remain constant. To satisfy this requirement, the vibrating sources must be identical in all respects. That is, if there is any sudden phase change in one there must be simultaneously a corresponding phase change in the other with the result that the phase difference between the two interfering waves is not changing with time. Such two sources having point to point phase relationship are said to be coherent. This is only possible when both become the sources of light under the influence of light waves originating from the same source.

Two coherent sources can be accomplished experimentally by either making one source image of the other by reflection [*Lloyd's Single Mirror*] or by dividing the light waves into two parts by reflection [*Fresnel's double mirror*] or by refraction [*Fresnel's Biprism*] or by partial reflection [*Newton's rings, Michelson's Interferometer*]. In these circumstances, any change of phase which the original light wave undergoes is shared by its two parts instantaneously, with the result that the fringes remain stationary.

Sustained interference effects can never be accomplished with two independent sources due to the fact that any two arbitrary beams of light are always *incoherent*. This has been already discussed in *Sec.* 9.4.

Condition B follows from the mathematical analysis given in *Sec.* 11.3. Even with nearly unequal amplitudes, the intensity in the region of destructive interference would be very small so as to render it dark as compared with the maximum intensity in the constructive interference region.

Condition C is essential in order to avoid the complete masking of interference patterns due to presence of different wavelengths in the light emitted by light sources. We shall again refer to this point when discussing *White Light Fringes*.

The necessity of *Condition D* being satisfied can be clearly seen by a comparison of the two systems of plane wavefronts originating from the same source and travelling to the right (Fig. 11.8). The solid lines in the figure represent the regions of maximum positive displacement while dotted lines represent the regions of maximum negative displacement in these waves. As in Fig. 11.2, the regions of constructive and destructive interferences, are marked in the present figure as well. Then angle θ between the two wavefronts is small in Fig. 11.8 (*a*) as compared with that in Fig. 11.8 (*b*). A comparison of these figures clearly illustrates that the larger the angle θ, the smaller becomes the spacing between interference fringes. Although, for larger angle θ, interference effects are sustained yet the fringes may become indistinguishable even under high magnification. Therefore, the condition that the interfering wavefronts must intersect at a small angle with each other, in mathematical language, $2d$, the separation between the two coherent sources must be small so as to produce widely spaced fringes, essentially applies to the observation rather to the production of sustained interference effects.

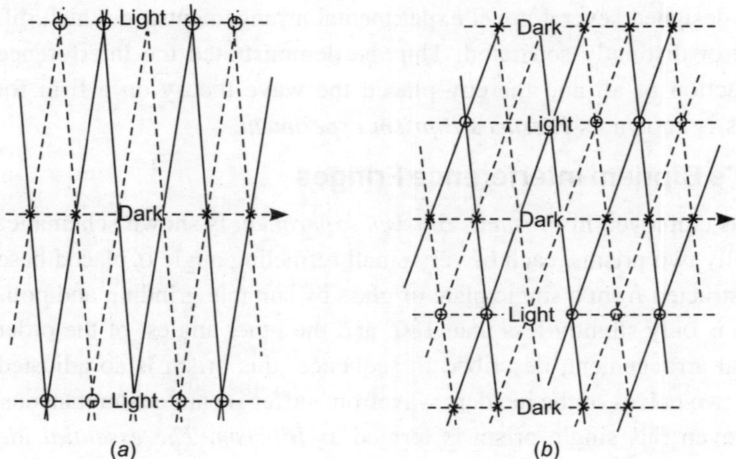

Fig. 11.8. Superposition of two plane waves (*a*) at a small angle, (*b*) at a large angle

11.5 CLASSIFICATION OF INTERFERENCE PHENOMENON

The experimental devices for producing two interfering beams of light from the same source may be conveniently classified under the following two main heads:

A. Division of Wavefront. The devices which divide the incident wavefront into two parts by utilizing the phenomena of reflection, refraction or diffraction in such a way that after traversing different optical paths they eventually reunite at small angle to produce interference bands, come under this class of interference phenomenon. In every experiment employing such devices it is absolutely essential to employ either a point source or a line source *e.g.*, a narrow illuminated slit parallel to the line of division of the incident wavefront. The *Fresnel Biprism and Mirrors, Lloyd's Mirror* described in this chapter, are examples of this class. In all such devices, since limited portions of the wavefront are employed, diffraction effects will also be present along with the interference effects.

B. Division of Amplitude. Devices which divide the amplitude of the incoming wave of light into two or more parts by partial reflection and refraction and thereby give rise to two or more beams which are later made to reunite to produce the interference effects, come under this class. In the case of instruments employing such devices it is not essential to employ a point source or a narrow line source but broad light source may be employed to produce brighter bands. Since a large section of the wavefront is employed diffraction effects are minimized. The interferometer of *Michelson* and *Newton's rings*, described in the next chapter, are examples of this class.

11.6 FRESNEL'S EXPERIMENTAL ARRANGEMENTS

Soon after *Thomas Young* announced his crucial *double slit experiment* on *Interference of Light* and explained the results with reference to the *wave nature*, critics raised the objection that the bright and dark fringes which he had observed were not due to true interference of two beams of lights. On the other hand, they were inclined to attribute them to some complicated modification of light, possibly owing to diffraction at the edges of the double slit (or pinholes). In order to overcome this objection, *Fresnel* designed several novel experimental arrangements, in which diffraction was either largely eliminated or distinctly separated. Thus he demonstrated the interference of two beams of light to the satisfaction of all and thereby placed the wave theory on a firm footing. The first of these arrangements is known as *Fresnel's Biprism experiment.*

11.6.1 Fresnel's Biprism Interference Fringes

The apparatus employed in *Fresnel's Biprism experiment* is shown schematically in Fig. 11.9. A biprism is essentially two prisms, each of very small refracting angle α placed base to base. In reality the biprism is constructed from a single plate of glass by suitable grinding and polishing it; the obtuse angle of the prism is only slightly less than 180° and the other angles, of the order of 30°, are equal. In the experimental arrangement, described in sequence, this prism is so adjusted in relation to the source slit that the two halves of the incident wavefront suffer separate simultaneous refraction through the prism; hence even this single prism is termed as *Biprism. The essential idea is to divide the incident beam into two coherent interfering beams by utilizing the phenomenon of refraction.*

Light from a narrow-slit S, illuminated with monochromatic light of wavelength λ, is allowed to fall symmetrically on a biprism. *The intersection of the two inclined faces forming the obtuse angle must be adjusted accurately parallel to the length of the slit.* In Fig. 11.9, the slit passes through S in a perpendicular direction to the plane of the diagram. Under this condition, this edge B divides the incident wavefront into two parts. Firstly, the one which in passing through the upper half ABD of the biprism is deviated through a small angle towards the lower half of the diagram and appears to diverge from the virtual image S_1. Secondly the one which in passing through the lower half CBD is deviated through a small angle towards the upper half of the diagram and appears to diverge from the virtual image S_2. *The two emergent wavefront, which intersect at a small angle, are derived from the same wavefront and hence the fundamental condition of interference is satisfied.* The virtual images S_1 and S_2, being the image of the slit S, obviously function as coherent sources in this experiment. Since the angles BAC and BCA are small and equal, these images are coplanar with, extremely close to and equidistant from the source slit S, *i.e.* $2d = S_1S_2$ is extremely small. As a consequence, interference fringes are observed on the screen in the overlapping region EF of the two emergent beams of light. Interference pattern can also be seen through a powerful eye-piece in its principal focal plane. The fringes extend into space and are thus non-localised.

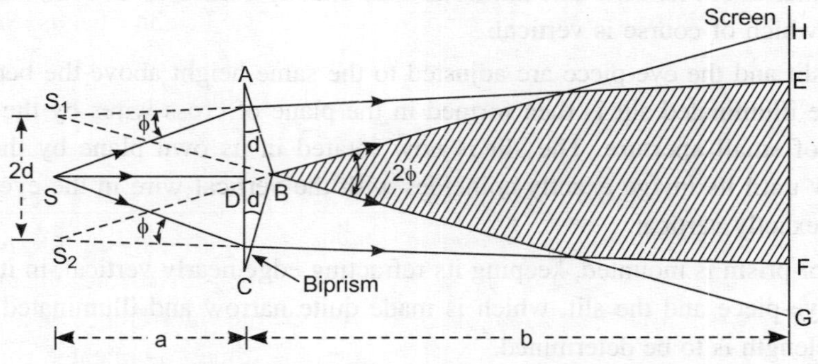

Fig. 11.9. Fresnel biprism

Beyond the overlapping region EF, faint, uneven and widely spaced bands, of different characteristics from the uniformly spaced interference bands of the middle portion, are also observed. These wider bands arise due to diffraction effects at the straight edges present at the apex B and the vertices A and C of the bi-prism. In reality diffraction bands are present over the whole region but equally spaced interference fringes are observed superposed on them in the middle portion.

Determination of the Wavelength of Light

In order to determine the wavelength of monochromatic light with the help of Bi-prism fringes, we employ the formula

$$\lambda = (2d/D)\ \bar{X} \qquad \text{...(11.6 } a\text{)}$$

The value of fringe width $\bar{X}$, the distance $2d$ between the virtual coherent sources S_1 and S_2 and the normal distance D of the plane of observation of the fringes from the slit should be measured after making a few adjustments in the apparatus.

The experiment is performed on a heavy metallic optical bench, about 2 metres in length and supported on four levelling screws at the base. The bench is provided with a scale on one side, graduated in mm. on corrodible material. The bench carries four uprights for supporting the adjustable slit, the bi-prism, a high power micrometer Ramsden's eye-piece and a convergent lens. These uprights are capable of movement along and also perpendicular to the length of the bench and may be adjusted to any desired height. The slit and the bi-prism may be rotated in their own plane with the help of tangent screws provided in uprights. Each upright carries a vernier at the base and thus its position on the bench can be accurately ascertained.

Adjustments. Before carrying out the measurement of the fringe-width it is essential to obtain correct fringes in which the spacing is uniform on the entire field by carrying out the following adjustments in the apparatus:

1. The bed of the optical bench is first levelled with the help of a spirit level and levelling screws.
2. The eye-piece is focussed on the cross-wires by moving the tube containing the lenses in the cross-wires tube until they are distinctly visible. One of the two wires in the cross is then made exactly vertical by observing a plumb line through the eye-piece and rotating the latter about its own axis until one wire exactly coincides with the image of the plumb line, which of course is vertical.
3. The slit and the eye-piece are adjusted to the same height above the bench. A real image of the illuminated slit is then formed in the plane of cross-wires by the help of a convex lens of small aperture. The slit is now rotated in its own plane by the help of tangent screw until its image exactly coincides with the vertical wire in the eye-piece. The slit is then exactly vertical.
4. The bi-prism is mounted, keeping its refracting edge nearly vertical, in its upright between the eye-piece and the slit, which is made quite narrow and illuminated with light whose wavelength is to be determined.
5. The edge formed by the intersection of inclined faces enclosing the obtuse angle in the bi-prism must be now adjusted exactly parallel to the vertical slit. To make this adjustment, two real images of the coherent sources S_1 and S_2 are formed in the focal plane of the eye-piece with the help of a convergent lens. By lateral movement of the prism the images are made equally bright *i.e.*, equally well focussed and of equal height by rotation of the bi-prism in the vertical plane with the help of tangent screw. On removing the lens, this edge can be now made exactly parallel to the slit, by giving finer rotation to the prism until the interference fringes become perfectly distinct and well defined.

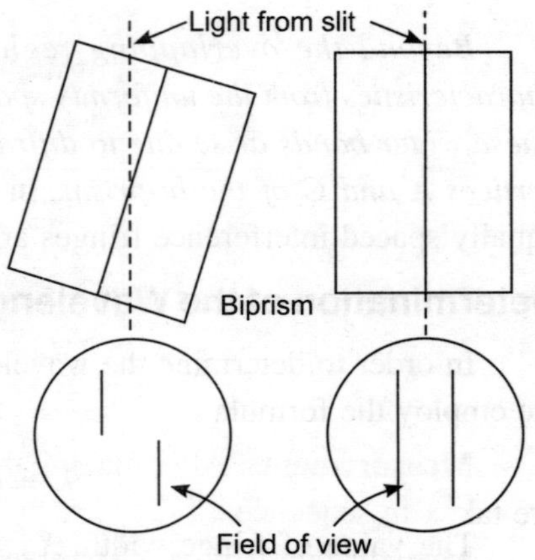

Fig. 11.10

6. **The axis of the experiment should be parallel to the length of the optical bench.** In other words, the line joining the *obtuse edge* of the bi-prism and the slit must be adjusted exactly parallel to the length of the bench.

It will be recalled that the expression for the fringe width $\bar{X}$, viz.,

$$\bar{X} = D\lambda / 2d$$

was derived and shown uniform over the entire field for given D, λ and $2d$ on the supposition that the plane of observation of the fringes is exactly parallel to the plane containing the virtual coherent sources S_1 and S_2. It is absolutely essential that this condition should be satisfied in experimental arrangement based on *Division of Wavefront* to obtain correct fringes. In our experiment, the plane of cross-wires is transverse to the length of the bench. Furthermore, a comparative study of diagrams sketched in Fig. 11.11 leads us to the conclusion that the plane S_1S_2 will be transverse to the bench, as in the lower diagram, only when the line joining the obtuse edge of the prism and the slit is adjusted exactly parallel to the length of the bench. If this adjustment is not perfect, then as shown in the upper diagram, the fringes would shift laterally relative to the cross-wires as the eye-piece is moved backward (or forward) along the bench. To remove this lateral shift, the bi-prism is moved to a small distance transversely to the bench in a direction opposite to the direction of the shift and again the fringes are tested for lateral shift by taking back the eye-piece. When the position of the bi-prism relative to the slit is so adjusted as shown in the lower diagram (Fig. 11.11) the lateral shift of the fringes would vanish.

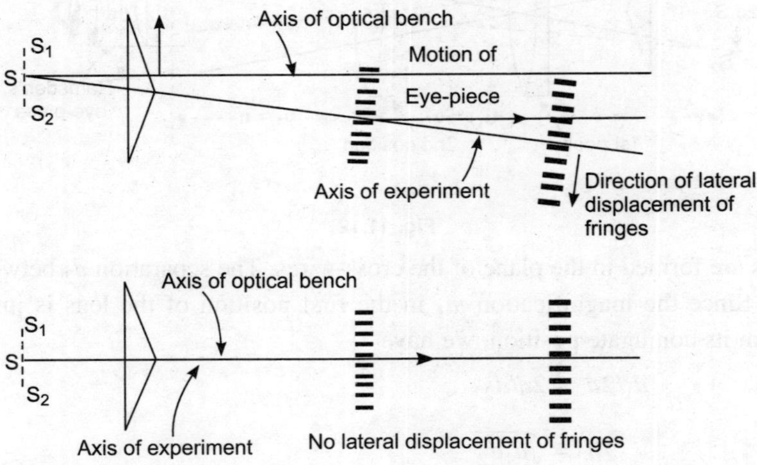

Fig. 11.11

Measurements. After making the above adjustment in the apparatus, the following measurements are taken in sequence.

Measurement of $\bar{X}$. The eye-piece is fixed at a suitable distance from the bi-prism so that the fringes are fairly short and fairly wide in its principal focal plane. The cross wire is set accurately at the centre of the first bright fringe and the reading of the micrometer screw is taken. The cross-wire is then moved in the same direction by the help of the micrometer screw, stopping at the centre of every successive bright fringe and the corresponding readings of the screw are noted down. From these observations, the fringe width $\overline{X}$ may be easily deduced.

Measurement of D. The distance between the eye-piece and the slit can be directly found out by reading the positions of their uprights on the optical bench and taking the difference of the two readings. However this distance is subject to the bench correction in order to obtain correct *D*. This is due to the fact that the slit and the cross-wires are not exactly at the *zero* mark of the vernier of their respective uprights, while *D* is the separation between the slit and the cross-wires. The bench correction should be determined after 2*d* has been measured.

Measurement of 2d. A convex lens is introduced between the bi-prism and the eye-piece and the latter is fixed at a distance from the slit which is greater than four times the focal length of the lens. As in *displacement method of measuring the focal length*, the lens is adjusted to a position, marked 1st position in Fig. 11.14, so that the magnified, distinct and real images of the virtual coherent sources are formed in the plane of cross-wires. The separation d_1 between these images is then measured with the help of cross-wires by giving it a lateral displacement from one image to the other. The lens is then moved to the conjugate position, marked 2nd position in Fig. 11.12, so that distinct

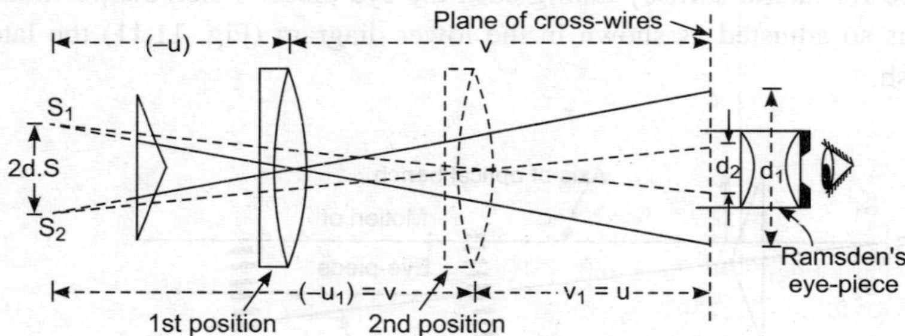

Fig. 11.12

diminished images are formed in the plane of the cross-wires. The separation d_2 between these images is also measured. Since the magnification m_1 in the first position of the lens is just inverse of the magnification m_2 in its conjugate position, we have

$$d_1/2d = 2d/d_2$$

or

$$2d = \sqrt{d_1 d_2}$$

Evaluation of Bench Error. The bi-prism and the lens are removed. A rod of known length is held horizontally with its one end just touching the slit and the other in the plane of the cross-wires. The difference between the length of the rod and the separation between the two uprights is the bench error, which should be applied with proper sign to obtain correct *D*.

The measured values of $\overline{X}$, 2*d* and *D* are substituted in Eqn. (11.6 *a*) to give the wavelength λ of the light employed in the experiment. The unit of wavelength is Angstrom.

$$1 \text{ Angstrom} = 10^{-8} \text{ cm.}$$

11.6.2 Fresnel's Double Mirror

The second device of Fresnel for obtaining the sustained interference effects essentially consists in dividing the incident wavefront into two coherent interfering wavefronts by utilizing the phenomenon of reflection. It is illustrated schematically in Fig. 11.13 and is known as the *Fresnel double-mirror experiment*. Two optically plane mirrors OM_1 and OM_2, highly silvered on their front surfaces or blackened at the back to avoid multiple internal reflections, are mounted vertically inclined at a very small acute angle θ. Light from a narrow slit S, illuminated with monochromatic light of wavelength λ, is allowed to fall on the mirrors, whose line of intersection through O must be adjusted parallel to the length of the vertical slit through S. One half of the incident wavefront is reflected from OM_1 and appears to diverge from the point S_1, which is consequently the virtual image of the source slit S

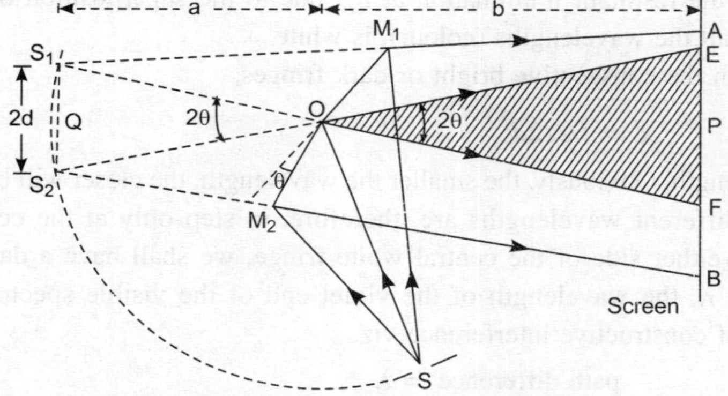

Fig. 11.13. The Fresnel double-mirror experiment

in OM_1. The other half of the incident wavefront is reflected from OM_2, giving rise to an image S_2 of the source slit S. Obviously, S_1 and S_2 are the virtual coherent sources in this experiment, since both are the images of the same source slit S. Thus the fundamental condition of interference is satisfied. Furthermore, since the angle θ between the mirrors is very small, the separation $2d$ between the coherent sources S_1 and S_2, expressed by

$$2d = 2a\theta$$

(where $OS = a$) is also very small. Thus the condition for the observation of widely spaced interference fringes is also satisfied. The light on the screen AB appears to come from two virtual coherent sources S_1 and S_2 and in the region EF where the two beams overlap, interference fringes are observed parallel to the slit. That these bands were actually produced by interference and not by diffraction was demonstrated by Fresnel by covering one of the two mirrors, when all traces of equally spaced fringes disappeared. Therefore, this experiment gives a device evidence in favour of the wave theory of light. However, since the source slit S is employed at a position for almost grazing incidence of the light on the mirrors OM_1 and OM_2, the angle subtended by each mirror at the slit is very small. In other words, the incident wavefront is very narrow and mirrors offer limitation to the wavefront. This results in the production of diffraction fringes, which have approximately the same spacing as the interference fringes. Therefore, the two systems are superposed on each other and cannot be easily distinguished.

11.7 INTERFERENCE WITH WHITE LIGHT

In *Young's double slit*, *Fresnel's Biprism* and *Double mirror experiments* we have employed monochromatic light of single wavelength to obtain large number of bright and dark fringes. Let us now study the changes produced in the fringes when monochromatic light source is replaced by white light in these experiments. White light consists of wavelengths varying between 4000 A. to 7500 A. from the violet to the red end of the white light spectrum.

At any point P_o, situated on the perpendicular bisector of coherent sources S_1 and S_2, the geometrical path difference is zero for the interfering light waves of all wavelengths. At this point, therefore, the condition of constructive interference is satisfied for all the wavelengths present in the white light. Here we get the light of every colour in exactly the same proportion as it exists in white light. As a consequence, the resultant illumination at P_o, due to the superposition of zero order (or central) bright fringes of all the wavelengths (colour), is white.

The spacing between the consecutive bright or dark fringes,

$$\overline{X} = D\lambda/2d$$

is a function of the wavelength. Obviously, the smaller the wavelength, the closer will be corresponding fringes. The fringes of different wavelengths are, therefore, in step only at the central fringe but soon get out of step. On either side of the central white fringe, we shall have a dark fringe tinged with violet colour. Since λ, the wavelength of the violet end of the visible spectrum is the least, therefore, the condition of constructive interference viz.

$$\text{path difference} = \lambda$$

will be first satisfied for the violet colour and then for other colours, in the spectral order, as we move away from the central fringe. As a consequence, the bright fringe nearer to the central fringe shall have strong violet tinge, followed by other bright fringes having a strong tinge of colours in the spectral order. In reality, no bright fringe is of saturated spectral colour. This fact is also illustrated by Fig. 11.17 in which the solid curve represents the intensity distribution for violet light and the dotted curve that for the yellow light, drawn with central fringes superposed on each other. In the region near the central fringe it should be emphasised, the optical path difference between the two interfering waves is small.

After 8 or 10 fringes, the path difference becomes so large that the condition of constructive interference, viz.

$$\text{path difference} = n\lambda$$

may be simultaneously satisfied for the number of wavelengths, for example, we may have the relation,

$$\text{path difference} = 5\lambda_1 = 7\lambda_2 = 9\lambda_3$$

and also at the same point the condition of destructive interference, viz.

$$\text{path difference} = (2n+1)\frac{\lambda}{2}$$

may be simultaneously satisfied for many colours. As a quence, for large optical path difference, the dark fringes of some wavelengths are completely masked by the bright fringes of other wavelengths.

Furthermore, at some points of the screen so many colours are present that the resultant illumination cannot be distinguished from white light with an unaided eye and so the field appears essentially white. It should be noted that the interference is still occurring in this region but is not visible in the ordinary circumstances. However, it can be clearly demonstrated by spectroscopic analysis of this region.

We, therefore, conclude that for observable white light fringes the optical path difference between the two interfering waves of light must be very small and we observe 8 or 10 coloured fringes on either side of the central white fringe.

11.8 DISPLACEMENT OF THE FRINGES

Let us now investigate the effect on the interference fringes of introducing a thin transparent plate in the path of one of the two interfering beams of monochromatic light. Let e be the thickness of the plate and μ its refractive index for the monochromatic light employed. From Fig. 11.14 it is clear that a light wave in travelling from S_1 to P has to traverses a distance e in the plate while the rest $(S_1P - e)$ it traverses in air. The time required for this journey is, therefore, given by

$$t = \frac{S_1P - e}{c} + \frac{e}{V_g} = \frac{1}{c}\left(S_1P - e + \frac{c}{V_g}e\right) = \frac{S_1P + (\mu - 1)e}{c}$$

The physical interpretation of the above relation is that due to introduction of the plate effective path from S_1 to P becomes $[S_1P + (\mu - 1)e]$ in air. Similarly, the effective path from S_1 to P_o, the point equidistant from S_1 and S_2, becomes $[S_1P_o + (\mu - 1)e]$ in air; and since

$$S_1P_o + (\mu - 1)e > S_2P_o$$

the central bright fringe of zero order is not formed at P_o, which of course, is the normal position of the central fringe in the absence of the plate.

To locate the new position of the central fringe we should travel along the screen in such a direction that the left side of the above inequality may decrease while the right side may increase and eventually the two may become equal at some point. Suppose, in the presence of the plate, the central fringe of zero optical path difference is formed at O; we therefore write

Fig. 11.14

$$S_2O = S_1O + (\mu - 1)e$$

or $$S_2O - S_1O = (\mu - 1)e \qquad \text{...(11.8 } a\text{)}$$

If $P_o\, O = x_o$, we can easily obtain the geometrical path difference,

$$S_2O - S_1O = \frac{2x_od}{D}$$

This value of the path difference is to be substituted in Eqn. (11.8 *a*) to get the distance through which the fringe system has been displaced. Thus, we get

$$x_o = \frac{D}{2d}(\mu - 1)e = \frac{\overline{X}}{\lambda}(\mu - 1)e \quad \text{...(11.8 } b)$$

Furthermore, if the central fringe shifts to the *n*th bright fringe of the original system obtained without the plate, then

$$S_2O - S_1O = n\lambda$$

Hence $$(\mu - 1)\, e = n\lambda \quad \text{...(11.8 } c)$$

It can be easily seen that the spacing of the interference fringes remains unaffected due to the introduction of the plate. If at the point *P*, in presence of the plate, bright fringe of the *n*th order is formed, then we must have the relation,

$$S_2P - [S_1P + (\mu - 1)e] = n\lambda$$

or $$S_2P - S_1P = n\lambda + (\mu - 1)e$$

If $P_oP = x_n$, then the geometrical path difference,

$$S_2P - S_1P = 2x_n\, d/D$$

where $$x_n = \frac{D}{2d}[n\lambda + (\mu - 1)e] \quad \text{...(11.8 } d)$$

Similarly, for the $(n + 1)$th bright fringe, we write

$$x_{n+1} = \frac{D}{2d}[(n+1)\lambda + (\mu - 1)e]$$

Hence $$\overline{X} = x_{n+1} - x_n = D\lambda/2d$$

a relation independent of the thickness of the thin plate and exactly similar to that which was derived in the absence of the plate.

On putting $n = 0$ in Eqn. (11.8 *d*) we may again get x_o, the distance through which fringes have been displaced. We conclude that the entire fringe system shifts laterally through a distance $[D(\mu - 1)e/2d]$ towards the side on which the plate is placed while the general shape and the spacing of the fringes remain unaffected provided, of course, the plate is thin.

By the measurement of the fringe shift, it is possible to employ Eqn. (11.8 *b*) to evaluate μ of the plate if we knew its thickness or to evaluate its thickness if μ is known.

11.9 LLOYD'S SINGLE MIRROR

Dr. Lloyd (1837) has devised a very simple arrangement for satisfying the fundamental condition of interference, in which one part of the incident wavefront reaches the screen after suffering reflection while the other part reaches it directly. In addition to simplicity, it has another importance due to the fact that this experiment provides an experimental confirmation of the fact that a sudden phase change of π occurs on reflection in a rarer medium *i.e.*, from the surface backed by a denser medium.

Lloyd's arrangement is shown schematically in Fig. 11.15. Light from a narrow slit S_1, illuminated with monochromatic light, is partly incident at a grazing angle on the surface of an optically plane front-surfaced mirror M_1M_2, while the rest reaches the screen directly. The reflected light appears to diverge from S_2, the virtual image of the source slit S_1, formed by reflection at the mirror. Accordingly, the virtual image S_2 acts as a coherent source with the slit S_1 itself in producing the interference fringes and thus the fundamental condition of interference is satisfied. In the region of the overlapping of the direct and the reflected beams, shown shaded, interference occurs, and equally spaced interference fringes can be observed in the region *EF* of the screen. The mirror employed is one having silvering on its front surface to avoid multiple internal reflections and its surface should be adjusted so as to make it exactly parallel to the vertical source slit S_1.

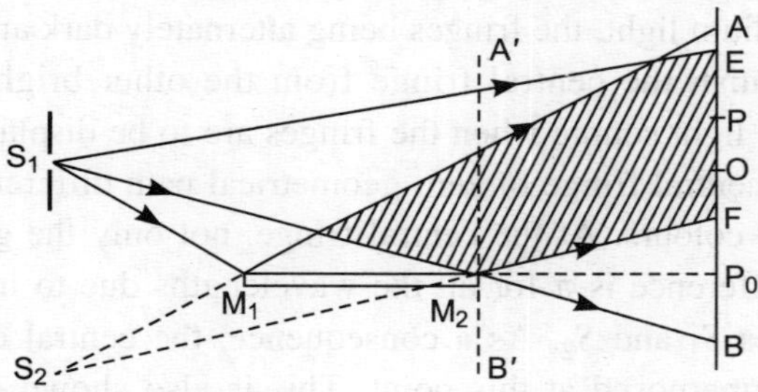

Fig. 11.15. The Lloyd's-mirror experiment

It will be observed that the point P of the screen, equidistant from S_1 and S_2, receives only the direct light from S_1 and not that reaching the screen after reflection. Therefore, the central fringe of zero geometrical path difference is not visible. Not only this, even less than half the interference pattern is visible with the arrangement sketched in Fig. 11.15. However, half the central fringe can be brought into view by displacing the screen to the position $A'B'$ so that it just touches the edge M_2 of the mirror. When this is done, the edge M_2, which is equidistant from S_1 and S_2, passes through the centre of the fringe of zero geometrical path difference. This fringe is dark instead of being a bright one. This can only occur at a point equidistant from S_1 and S_2, provided one of the two interfering beams suffers a sudden phase change of π. Since the direct beam could not suffer a sudden phase change, it must be the reflected beam which should suffer a sudden phase change of π and the only possibility of this, here is on reflection from the surface backed by the denser medium. This may be interpreted as meaning that *the coherent sources S_1 and S_2 differ in phase by π at all times* instead of being in equal phase as in double slit, Fresnel's biprism and double mirror experiments. *As a consequence, the two waves on reaching any point P of the screen interfere constructively or destructively according as the geometrical path difference is odd or even multiple of $\lambda/2$.* In symbols, the conditions can be expressed as:

$$S_2P - S_1P = (2n+1)\frac{\lambda}{2} \qquad \text{[Bright fringe]}$$

$$S_2P - S_1P = 2n\frac{\lambda}{2} \qquad \text{[Dark fringe]}$$

It will be observed that these conditions are just reverse of the conditions of the constructive and destructive interference in the Bi-prism arrangement.

Complete central fringe can also be brought into view by introducing a thin lamina of glass or mica in the path of directly transmitted beam, keeping the screen fixed in the original position. Owing to the retardation of the directly transmitted waves, the complete fringe system shifts in the direction in which plate is introduced, in Fig. 11.15, in the upward direction. If the lamina is of suitable thickness, there can be one point O on the screen in the region of the overlapping of the two beams so that the condition

$$S_2O = S_1O + (\mu - 1)e$$

is fulfilled. The central fringe of zero geometrical path difference will appear at O. But with monochromatic light, say sodium light, the fringes being alternately dark and bright having one colour it is not possible to distinguish the central fringe from the other bright fringes. Therefore, it is customary to employ a white light source when the fringes are to be displaced for locating the central fringe. With white light, the central fringe of zero geometrical path difference will be black while the other fringes will have other colours. At the central fringe, not only the geometrical path difference is zero but also the phase difference is π for all the wavelengths due to initial phase difference of π between the coherent sources S_1 and S_2. As a consequence, the central dark fringes corresponding to all the wavelengths are superposed at this point. This is also shown in Fig. 11.16 in which the intensity distribution curves for the violet and the yellow colours are shown with ordinary and dotted curves respectively, while curves of other colours are not shown for the simplicity of the diagram.

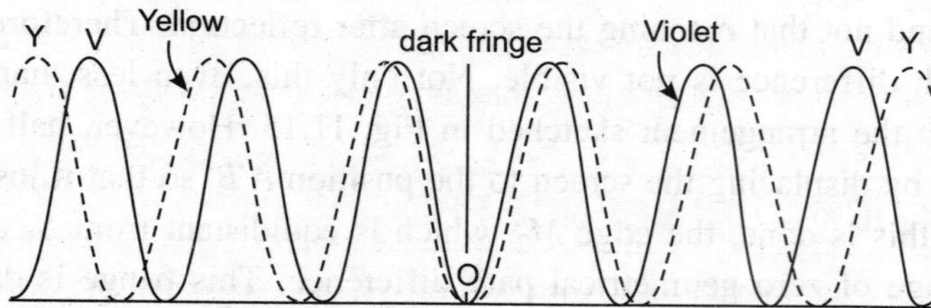

Fig. 11.16. Illustrating the origin of white light fringes with a dark fringe at the centre

It should be remarked that owing to grazing incidence of light on the mirror, the effective aperture on viewing from P is extremely small. The mirror, therefore, appears like a slit, giving rise to diffraction fringes due to narrow slit. But, in this experiment, diffraction effects are of secondary importance.

Determination of the Wavelength

The wavelength of light can also be determined by the measurement of the fringe width $\overline{X}$ of the fringes produced by Lloyd's single mirror arrangement. For the validity of the formula

$$\lambda = (2d/D)\overline{X}$$

it should be again emphasised that the plane containing the coherent sources S_1 and S_2 must be exactly parallel to the plane of the observation of the fringes. Since the line joining the object and its image, formed due to reflection in a plane mirror is always perpendicular to the reflecting surface, hence for the validity of the above equation this reflecting surface must be adjusted exactly perpendicular to the plane of cross-wires of the eye-piece. Furthermore, the reflecting surface must be adjusted exactly parallel to the length of the vertical source slit. We now give the following experimental details to achieve this end.

The experiment is performed on a levelled optical bench. Lloyd's mirror is mounted vertically on an upright, which is provided with screws to effect the rotation of the mirror about a vertical as well as about an horizontal axis. A vertical slit, illuminated with light whose wavelength is to be determined, is mounted on another upright and adjusted close to the line of the mirror so that the light may be reflected as far as possible at grazing incidence. The fringes are observed in the focal plane of the focussed micrometer eye-piece.

It is absolutely essential, as explained above, that M_1M_2 should be normal to AB. To achieve this, a convergent lens is introduced between the mirror and the eye-piece and it is adjusted so that its centre lies opposite the edge M_2 of the mirror and on a level with the centre of the source slit S_1. Finally, the lens is fixed in a position so as to form the real images of S_1 and S_2 in the plane of cross-wires. Both images will be in focus in the plane of cross-wires for one setting of the lens, provided S_1S_2 is parallel to plane (AB) of the cross-wires. In other words, M_1M_2 is normal to AB. We, therefore, rotate M_1M_2 slightly about the vertical axis of rotation until the two images are exactly in focus simultaneously.

The reflecting surface of the mirror may be easily adjusted parallel to the vertical slit. On removing the lens, fringes are observed through the eye-piece. Mirror M_1M_2 is rotated about its horizontal axis to get the position when the fringes become well defined and brightest. The adjustment of M_1M_2 normal to plane of cross-wires may be now tested by the absence of the lateral shift in the fringes as the eye-piece is moved backward (or forward) along the optical bench.

Now the measurement of $\bar{X}$, $2d$ and D may be carried out as in the Bi-prism experiment.

11.10 ACHROMATIC INTERFERENCE FRINGES

An *achromatic fringe is defined as one which shall result due to superposition of the same order fringes of light of all wavelengths emitted by the heterogeneous source.* If it were possible to devise an experimental arrangement in which the geometrical path difference for the two coherent interfering waves from the source to a point of observation is different for different wavelengths, then the geometrical path difference can also be made proportional to the wavelength. As a consequence, the phase difference expressed as

$$\delta = \frac{2\pi}{\lambda} \times \text{Path difference}$$

becomes the same for all wavelengths at the point. In other words, the phase difference is independent of the wavelength over the whole region of the observation. The physical conception of this conclusion is that the first order fringes of all wavelengths are superposed on each other, and obviously the same is true for 2nd order fringes, 3rd order fringes and so on; *i.e.* the fringe width $\bar{X}$ becomes independent of the wavelength. We then have a system of achromatic fringes. But how to devise such an experimental arrangement?

In *Young's experiment* and *Fresnel's biprism* and *double mirror experiments*, there is only one point in the plane of observation viz. the point equidistant from the coherent sources, where the phase difference is the same (zero) for all wavelengths. At this point zero order fringes of all wavelengths coincide giving rise to an achromatic fringe. But the spacing between two consecutive bright or dark fringes is a function of the wavelength,

$$\overline{X} = D\lambda/2d$$

being greatest for the red and least for the violet end of the spectrum. As a consequence, the same order fringes of different wavelengths, excluding zero order, do not coincide. *All the three arrangements, are thus perfectly achromatic only for the central fringe*, provided, we neglect the secondary effects due to finite slit width. This is also true for *Lloyd's interference fringes* but of course independent of the slit width. The fringe of zero geometrical path difference in Lloyd's mirror is not formed under ordinary arrangements. But the central fringe when brought into view by introducing a thin lamina of mica or glass in the path of direct light, is found to be black even with white light source. The other fringes are strongly coloured [Fig. (11.16)]. However, if by some device we render the fringe-width $\overline{X}$ the same for fringes of all wavelengths even with white light, the bright fringes of given order of different wavelengths will be perfectly superposed. The same is also true for dark fringes, thereby giving rise to alternately white and dark fringes, according to our definition the *achromatic fringes*, illustrated in Fig. 11.18.

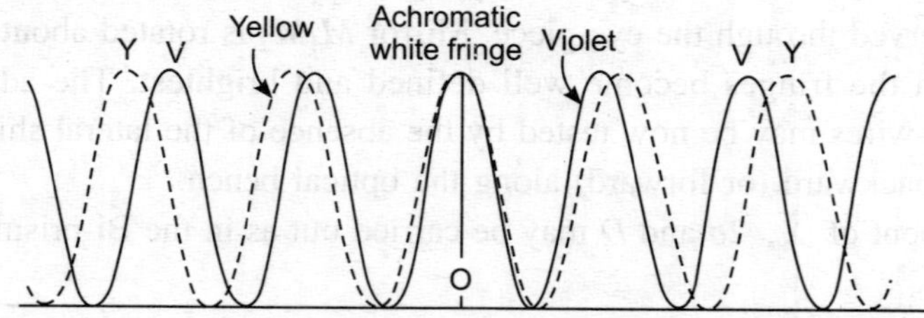

Fig. 11.17. Illustrating the origin of white light fringes with central achromatic white fringe

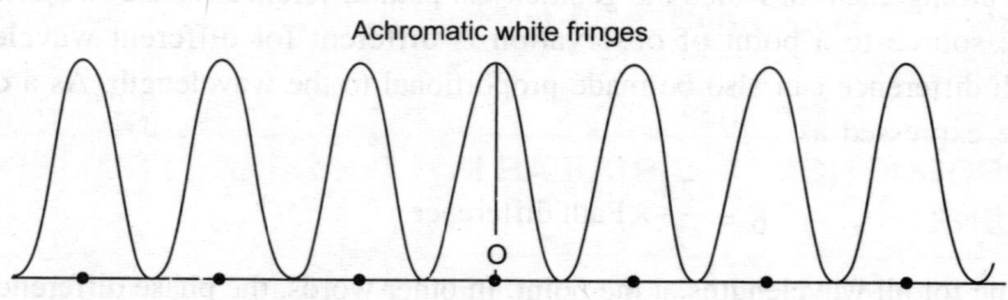

Fig. 11.18. Illustrating the origin of achromatic white fringes

The inspection of the expression for the fringe width $\overline{X} = D\lambda/2d$ gives us a clue towards designing an experimental arrangement for achieving our object. It is obvious that D will be the same for all the wavelengths in any arrangement. Hence for the constancy of $\overline{X}$ it follows from the above expression for the fringe width that by some device the ratio $\lambda/2d$ must be made the same for all wavelengths. Physically, this means that the coherent sources for different wavelengths should be situated at different points in one plane and arranged in such a way that as λ increases, the separation $2d$ between the corresponding coherent sources should also increase in the same proportion. For example, we should satisfy the relation, like

$$\frac{2d_R}{2d_v} = \frac{\lambda_R}{\lambda_v} \quad \text{...(11.10 a)}$$

for every possible pair of different wavelengths emitted by the white light source. Obviously, the fringe width $\overline{X}$ becomes the same for all wavelengths and truly achromatic fringes may be obtained.

11.11 BILLET SPLIT LENS

In 1858 *Billet* obtained two coherent sources by forming two real images S_1 and S_2 of the illuminated slit S with the help of two halves of a lens, obtained by splitting a convergent lens into two parts along a diameter of its circular periphery. The two halves of the lens can be separated or brought closer together by means of a micrometer screw, the motion being along the line perpendicular to the optical axis. The positions of the two halves can be thus very accurately regulated in relation to each other. In other words, the separation $2d$ between the coherent sources S_1 and S_2 can be suitably adjusted as seen with the naked eye. Fringes may be seen on the screen or with the aid of an eye-piece in the overlapping region EF of the two beams of monochromatic light.

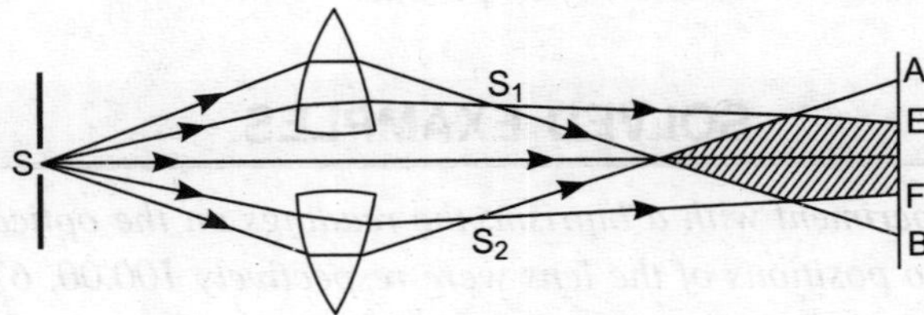

Fig. 11.19. Billet's split lens for producing interference fringes

We may also arrange the split lens with respect to the illuminated slit S so as to form two virtual images S_1 and S_2 which shall function as two coherent sources for the production of interference fringes.

11.12 PRODUCTION OF CIRCULAR INTERFERENCE FRINGES–MESLIN'S SPLIT LENS

It will be recalled, that the loci of constant path difference in space, from two coherent monochromatic point sources S_1 and S_2, are hyperboloids of revolution surfaces generated by revolving hyperbolae about S_1S_2 as an axis. The intersections of these hyperboloids with a screen arranged perpendicularly to S_1S_2 are, therefore, concentric circular fringes with common centre on the prolongation of the line joining the coherent sources S_1 and S_2. The real point sources emit light in all directions; hence we expect full concentric circular fringes on the screen in the above stated position.

But it is impossible to have two real point coherent sources. However, *Meslin* (1893) by an ingenious modification of *Billet* split lens, practically demonstrated alternate bright and dark semi-circular concentric fringes on the screen. His arrangement is sketched in Fig. 11.20. The two halves of the *Billet split lens* are placed a short distance apart, on the optical axis. Two real point images S_1 and S_2 of a pinhole illuminated with monochromatic light, are then formed on the optical axis. The real images function as coherent point sources which are not real point sources, because the images do not radiate light in all directions but only in the shape of a narrow cone. In the region of the overlapping of the two beams shown shaded in Fig. 11.20, interference effects occur. A system of fringes concentric semi-circular arcs in shape, may be then obtained on a screen arranged between S_1 and S_2 normally to the optic axis.

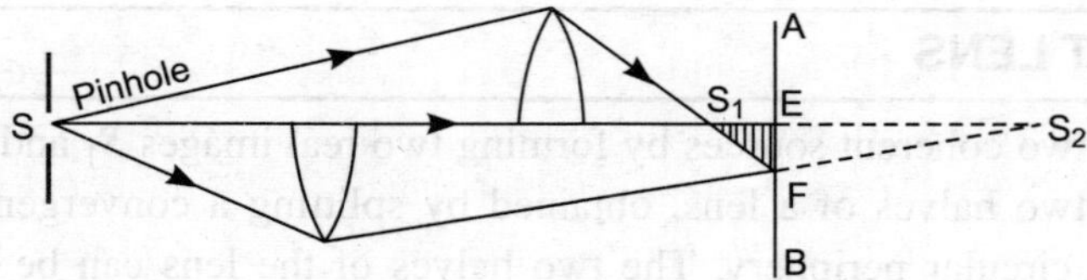

Fig. 11.20. Meslin's split lens for producing circular fringes

In *Meslin's* arrangement, interference is in reality between the waves radiating from a source S_1 with waves converging to a similar source S_2. It should be emphasised that *here circular fringes are obtained by the division of wave-front in contrast to Newton's rings, which arise as a result of interference phenomenon due to the division of amplitude.*

SOLVED EXAMPLES

Example 1. *In an experiment with a biprism, the readings on the optical bench of the position of the eye-piece and the two positions of the lens were respectively 100.00, 67.00 and 34.00 cm. The distances between the two images for the two positions of the lens were respectively 0.3000 and 1.2000 mm and the width of 10 fringes was 9.720 mm. Assuming that there is no index error in any case, calculate (a) the distance between the focal plane of the eye-piece and the plane of interfering sources (b) the wavelength of the light used.*

Referring to Fig. 10.14 and using the symbols marked, the distance between the eye-piece and the second position of the lens is given by

$$v_1 = 100 - 67 = 33 \text{ cm} = u$$

Hence, reading of the slit on the bench is

(1st position of lens) – u = 34 – 33 = 1 cm.

Hence, the distance between the focal plane of the eye-piece and the plane of interfering sources is

$$D = 100 - 1 = \mathbf{99\ cm.}$$

Also, $2d = \sqrt{d_1 \times d_2} = \sqrt{1.2 \times 0.3} = 0.6 \text{ mm} = 0.06 \text{ cm}$

and $\overline{X} = 9.720/10 \text{ mm} = 0.09720 \text{ cm}$

Now $$\lambda = \frac{\overline{X}\,2d}{D} = \frac{0.0972 \times 0.06}{99} = \mathbf{5891 \times 10^{-8}\ cm.}$$

Example 2. *Using sodium light with a Fresnel's biprism the fringes were found to have a width of 0.0196 cm when observed at a distance of 100.0 cm from the slit. When a convex lens was placed between the biprism and the observer to give an image of the source at 100.0 cm from the slit, the distance apart of the images was found to be 0.70 cm; calculate the wavelength. Given distance from slit to lens 30.0 cm.*

In the given problem, $\overline{X}$ = 0.0196 cm and D = 100 cm. To evaluate $2d$, we observe that the distance of the lens from the slit is 30 cm and the images of the coherent sources are formed by lens in a plane at (100 – 30) = 70 cm from the lens. Distance apart of the images of coherent sources is 0.70 cm. Hence applying the Eqn. (3.12 *d*) [For lens in air $\mu_1 = \mu_2 = 1$] for magnification, we have

$$\frac{(-0.70)}{2d} = \frac{70}{(-30)}$$

or $$2d = 0.30 \text{ cm}$$

Hence $$\lambda = \frac{\overline{X}\,2d}{D} = \frac{0.0196 \times 0.30}{100} = \mathbf{5880 \times 10^{-8}\ cm.}$$

Example 3. *When a thin monochromatic source of light was placed at a distance of 50 cm from a Fresnel biprism of μ = 1.5 the distance between two consecutive bands formed on a screen placed at a distance of 100 cm from biprism was found to be 0.012 cm. If the wavelength of light was 5893×10⁻⁸ cm, find the magnitude of obtuse angle of the biprism.*

To solve this problem we employ the equation

$$\overline{X} = \frac{(a+b)\lambda}{2a(\mu-1)\alpha}$$

where, it should be emphasised, α is the refracting angle (acute angle) of the biprism in radian measure.

In the given problem, a = 50 cm; b = 100 cm, μ = 1.5, $\overline{X}$ = 0.012 cm and $\lambda = 5893 \times 10^{-8}$ cm.

Hence $$\alpha = \frac{(50+100) \times 5893 \times 10^{-8}}{2 \times 50 \times (1.5-1) \times 0.012} \text{ radian}$$

But π radian = 180°. Hence α in degree measure is given by

$$\alpha = \frac{150 \times 5893 \times 10^{-8} \times 180}{100 \times 0.5 \times 0.012 \times \pi} = 0°.844$$

Hence, obtuse angle of the biprism = 180 – 2α = **178°31.**

Example 4. *Fringes are produced with monochromatic light of wavelength 5.45×10^{-5} cm. A thin plate of glass of refractive index 1.5 is placed normally in the path of one of the interfering beams and the central bright band of the fringes system is found to move to a position occupied by the third band from the centre. Calculate the thickness of glass plate.*

The central fringe shifts to 3rd bright fringe from the centre of the original system obtained without plate. Hence using Eqn. (11.8 *c*) which is

$$(\mu - 1)e = n\lambda$$

we have $$(1.5 - 1)e = 3 \times 5.45 \times 10^{-5}$$

Hence $$e = \mathbf{3.27 \times 10^{-4}\ cm.}$$

PROBLEMS

1. Consider the superposition of two Simple Harmonic disturbances and show that the resultant intensity is not just the sum of the intensities due to the separate disturbances. On the basis of this result explain why coherent disturbances interfere and incoherent disturbances do not.
2. Discuss the important conditions for the interference of light.
3. Describe, giving experimental details, Fresnel's Biprism method for determining wavelength of light. Derive the formula used.
4. Describe an experimental arrangement for the observation and measurement of Lloyd's mirror fringes. Show, how you would use it to determine the wavelength of monochromatic light. How would you modify the arrangement to obtain achromatic fringes, using a white light source.
5. Calculate the displacement of fringes when a thin transparent lamina is introduced in the path of one of the interfering beams in a biprism. Show how this method is used for finding the thickness of a mica sheet.
6. What are coherent sources? Discuss why two independent sources of light of the same wavelength cannot produce interference fringes. Give diagrams showing clearly how coherent sources are produced in (*i*) Newton's rings arrangement, (*ii*) biprism arrangement.
   Derive the formula for fringe width in biprism experiment.
7. In an experiment with Fresnel's biprism bands 0.196 mm in width are observed at a distance of 100 cm from the slit. A convex lens is then put between the observer and the biprism so as to give an image of the sources at a distance of 100 cm from the slit. The distance apart of the images is found to be 0.70 cm, the lens being 30 cm from the slit. Calculate the wavelength of light used.
8. A thin sheet of glass ($\mu = 1.5$) of 6 microns thickness introduced in the path of one of the interfering beams in a biprism arrangement shifts the central fringe to a position normally occupied by the fifth. Find the wavelength of light use.
9. The central fringe of the interference pattern produced by light of wavelength 6000 A.U. is shifted to the position of the fifth bright fringe by introducing a thin glass plate ($\mu = 1.5$). What is the thickness of the plate?

10. Interference fringes are formed using a biprism having base angles of 4° each and refractive index 1.5. The slit is kept at a distance of 10 cm from the biprism and is illuminated with light of wavelength 5890 A.U. Calculate the fringe width at 100 cm from the biprism.
11. Interference fringes are being formed by a source of light λ = 6000 A.U. It is found that by introducing a transparent plate in one of the paths, the central fringe is shifted to a position occupied by the 6th bright fringe. Find the refractive index of the plate if its thickness is 6 microns. Establish the formula used.
12. Two straight and narrow parallel slits 3 mm apart are illuminated with a monochromatic source ($\lambda = 5.9 \times 10^{-5}$ cm). Fringes are obtained at a distance of 30 cm from the slits. Find the width of fringes.
13. The distance between the slit and the biprism and that between the biprism and the screen are each 50 cm. The angle of biprism is 179° and its refractive index is 1.5. If the distance between the successive bright fringes is 0.0135 cm, calculate the wavelength of light.
14. The inclined faces of a biprism of refractive index 1.5 make angles of 2° with its base. A slit illuminated by a monochromatic light is placed at a distance of 10 cm from the biprism. If the distance between two dark fringes, observed at a distance of 1 metre from the biprism is 0.18 mm, find the wavelength of the light used. Derive the formula used.
15. Fringes are produced by a Fresnel's biprism in the focal plane of reading microscope, which is 100 cm from the slit. A lens inserted between the biprism and microscope gives two images of the slit in two positions. In one case the two images of slit are 4.05 mm and in the other are 2.9 mm apart. If sodium light (λ = 5893 A.) is used, find the distance between the interfering bands.
16. Interference fringes of yellow light (λ = 5800 A.) are formed by Billet split lens. The distance from the source S to the lens L is 25 cm. The focal length of the lens is 15 cm. The lens halves are separated 0.08 mm and the source to screen distance is 200 cm. Find the fringe separation.
17. A Lloyd's mirror of length 5 cm is illuminated with monochromatic light, λ = 5460 A., from a narrow slit 0.1 cm from its plane, and 5 cm, measured in that plane, from its near edge. Find the separation of the fringes at a distance of 120 cm from the slit, and the total width of the pattern produced there.

INTERFERENCE OF LIGHT

(Division of Amplitude)

The phenomenon of *interference of light*, it will be recalled, can be conveniently divided into two general classes. Firstly, the class in which interference effects are obtained by the *division of wavefront* into two parts and in which they unite after they have traversed different optical paths. This class of interference effects, we have studied in the previous chapter. Secondly, we may mention the class in which the incident beam is divided into two portions, by *division of its amplitude*, due to partial reflection and refraction by some optical device and in which the two portions are afterwards recombined to produce interference effects. Such being the case, it should be emphasised, *there is a point to point correspondence between the reflected and transmitted wavefronts. This simply means that any peculiarities present in one are also present in the other wavefront. As a consequence, however complex the incoming wavefront may be, if the corresponding points of reflected and transmitted wavefronts are somehow superposed upon reunion, the clearness of interference bands are not impaired. Consequently, in instruments based on the principle of division of amplitude broad light sources may be employed with profit to produce brighter interference effects*. Furthermore, in some experimental arrangements *e.g. Newton's rings* complementary pattern is also formed elsewhere, the bright regions of which correspond to dark regions of the observed pattern.

12.1 STOKE'S TREATMENT OF REFLECTION AND REFRACTION OF WAVES

Suppose a plane monochromatic light wave of amplitude a travelling in the direction AB is incident from air on the interface GH separating the air and water. Let r be the fraction of the incident amplitude reflected and let t be the fraction transmitted through the interface when the incident wave is in *rarer* of the two media where r and t are each less than one. Accordingly the amplitude of the reflected wave BC will be ar and that of the transmitted wave BD is at, as shown in Fig. 12.1. If somehow or other we reverse the reflected and transmitted waves, then according to the *principle of reversibility*, they should together produce, after striking the interface, a wave in air equal in amplitude to that of the incident wave (illustrated by Fig. 12.1) but travelling in the opposite direction viz. along BA. The wave BC of amplitude ar when reversed gives a reflected wave of amplitude arr in the direction BA in air and a transmitted wave of amplitude art in the direction BE in water. Let r' and t' be the fractions of the incident amplitude reflected and transmitted at the interface when the incident wave lies in the denser of the two media. Accordingly, the wave *BD* of amplitude *at* when reversed gives on striking the interface from water, a reflected wave of amplitude *atr'* in the direction *BE* in water and a transmitted wave of amplitude *att'* in the direction *BA* in air.

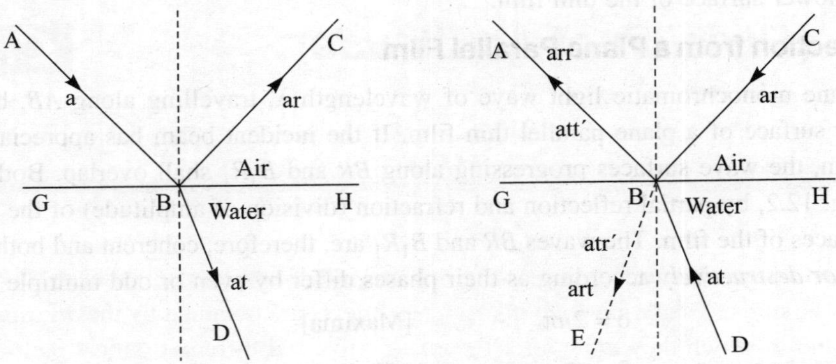

Fig. 12.1. Stoke's treatment of reflection

Now, since the resultant effect of the reversal of *BC* and *BD* must be only a single wave in air of amplitude *a* in the direction *BA*, we have

$$arr + att' = a \qquad \text{...(12.1 } a)$$

and $$art + atr' = 0 \qquad \text{...(12.1 } b)$$

which easily gives, $$tt' = 1 - r^2 \qquad \text{...(12.1 } c)$$

and $$r' = -r \qquad \text{...(12.1 } d)$$

The co-efficients of external and internal reflections are thus numerically equal but of opposite signs. This means a difference of phase of π between the reflections in the rarer and the denser media, since a reversal of sign simply illustrates a displacement in the opposite direction. It was, however, experimentally observed in *Lloyd's* single mirror experiment and in *Wiener's* experiment on stationary light waves that an abrupt phase change of π occurs in the light wave at reflection in the rarer medium *i.e.* from a surface backed by a denser medium. We therefore, conclude that no abrupt phase change occurs in the wave reflected from the surface backed by a rarer medium. Furthermore, since $r'^2 = r^2$, hence the fraction of intensity reflected is the same in the two cases of reflection under consideration. It should however be emphasised that the wave which strikes the interface from the denser medium, must be incident at an angle equal to the angle of refraction of the wave incident on the interface from the rarer medium. Then only the relation

$$r' = -r$$

will hold good.

12.2 COLOURS OF THIN FILMS

Brilliant colours are often exhibited when a beam of white light from an extended source is reflected from the skin of a soap bubble or from a thin layer of oil floating on water or from a thin wedge shaped air film between the two glass plates. The explanation of the origin of this colour phenomenon was given by *Dr. Young,* in 1802, in terms of the interference of light waves reflected from the upper and the lower surface of the thin film.

12.2.1 Reflection from a Plane Parallel Film

Let a plane monochromatic light wave of wavelength λ, travelling along *AB*, be incident on *GH,* the upper surface of a plane parallel thin film. If the incident beam has appreciable width and the film be thin, the wave surfaces progressing along *BR* and B_1R_1 shall overlap. Both originate, as depicted in Fig. 12.2, by partial reflection and refraction (division of amplitude) of the incident wave at the two surfaces of the film. The waves *BR* and B_1R_1 are, therefore, coherent and both will interfere *constructively or destructively* according as their phases differ by even or odd multiple of π.

$$\delta = 2n\pi \qquad \text{[Maxima]} \qquad \text{...(12.2 } a)$$

$$\delta = (2n + 1)\pi \qquad \text{[Minima]} \qquad \text{...(12.2 } b)$$

where $n = 0, 1, 2, 3$ etc. and δ stands for the total phase difference between *BR* and B_1R_1.

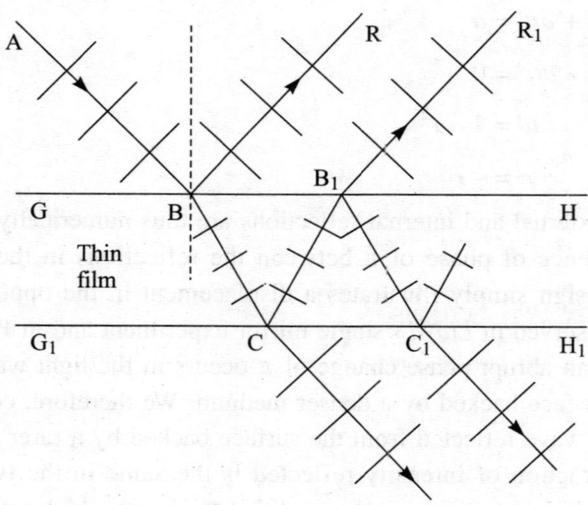

Fig. 12.2. Waves reflected and transmitted by thin film

Calculation of phase difference due to optical path difference

In Fig. 12.3, B_1D and B_1E are perpendiculars drawn from B_1 on *BR* and *BC* respectively. Obviously B_1D is parallel to the wavefronts in the waves *BR* and B_1R_1. Therefore, the optical path difference between the two waves will be simply due to different paths traversed—the path BCB_1 in the film by B_1R_1 and the path BD in air by BR—in reaching the plane B_1D from B. Since the waves pursue parallel paths, hence the path difference beyond B_1D will remain constant.

$$\therefore \quad \text{Path difference,} \quad \Delta = (BC + CB_1)_{\text{film}} - BD_{\text{air}}$$

$$= \mu(BE + EC + CB_1) - BD \qquad \text{...(12.2 } c\text{)}$$

where μ is the refractive index of the film with respect to air.

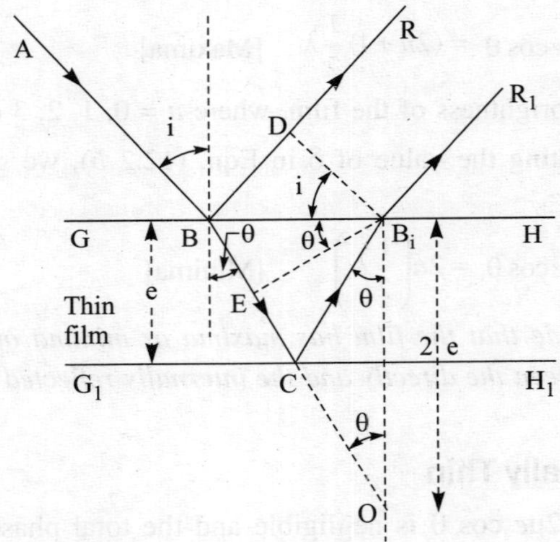

Fig. 12.3

Also, $$CB_1 = CO \qquad \text{...(12.2 } d\text{)}$$

and $$\mu = \frac{\sin i}{\sin \theta} = \frac{BD/BB_1}{BE/BB_1} = \frac{BD}{BE}$$

or $$BD = \mu BE \qquad \text{...(12.2 } e\text{)}$$

The path difference, Δ, by the help the Eqns. (12.2 *d*) and (12.2 *e*) reduces to

$$\Delta = \mu EO = \mu B_1O\cos\theta = 2\mu e\cos\theta$$

The phase difference between BR and B_1R_1 due to this path difference Δ is given by

$$\delta_1 = (2\pi/\lambda)\Delta = (2\pi/\lambda)2\mu e\cos\theta \qquad \text{...(12.2 } f\text{)}$$

where λ is the wavelength of the incident light in air.

Phase change at reflection

The film is optically denser than the surrounding media. Therefore, the wave BR originating by reflection from B a point on the surface backed by a denser medium (film), will differ in phase by π from the incident wave. But the wave B_1R_1 which originates in reflection at C, a point on the surface backed by the rarer medium, will experience no such sudden phase change. The total phase difference

δ between these waves is, therefore, given by

$$\delta = (2\pi/\lambda)2\mu e\cos\theta + \pi \quad \text{...(12.2 } g)$$

Conditions for maxima and minima in terms of the optical path difference

We substitute the value of δ in Eqn. (12.2 *a*) and, we get

$$(2\pi/\lambda)2\mu e\cos\theta + \pi = 2n\pi$$

or

$$2\mu e\cos\theta = (2n-1)\frac{1}{2}\lambda$$

or

$$2\mu e\cos\theta = (2n+1)\frac{1}{2}\lambda \quad \text{[Maxima]} \quad \text{...(12.2 } h)$$

a condition for maxima of brightness of the film, where $n = 0, 1, 2, 3$ etc.

Similarly, on substituting the value of δ in Eqn. (12.2 *b*), we get a condition for minima of brightness of the film.

$$2\mu e\cos\theta = 2n\left(\frac{1}{2}\lambda\right) \quad \text{[Minima]} \quad \text{...(12.2 } i)$$

We, therefore, conclude that the film has maxima or minima of brightness according as the optical path difference between the directly and the internally reflected waves is odd or even multiple of $\lambda/2$ in air.

12.2.2 Film Infinitesimally Thin

When $e << \lambda$, then $2\mu e \cos\theta$ is negligible and the total phase difference between BR and B_1R_1 from Eqn. (12.2 *g*) is simply π. Regardless of wavelength, the waves reflected from the two surfaces of the film are out of step. But the amplitude of BR is greater than that of B_1R_1, since the amplitude drops off sharply on successive reflections. Therefore, waves BR and B_1R_1 cannot completely destroy each other. Poisson, however, pointed out that there will be an infinite series of waves B_2R_2, B_3R_3, B_4R_4, ..., obtained from the same incident wave due to its multiple internal reflections within the film. This series of waves will practically coincide with BR if the film is infinitesimally thin. Further, since every internal reflection is from the surface backed buy the rarer medium, wave B_2R_2, B_3R_3, etc. are formed like B_1R_1 without any sudden phase change. Hence if e is negligible, B_2R_2, B_3R_3 etc. are all in phase with B_1R_1. We are perfectly justified in summing their amplitudes, which are marked in Fig. 12.4 by applying *Stoke's treatment* to successive reflections. We obtain for their resultant amplitude,

$$R = atr't' + atr'^3t' + atr'^5t' + \ldots\ldots$$
$$= att'r'[1 + r'^2 + r'^4 + \ldots\ldots \text{to infinity}]$$

Since $r' < 1$, the geometric progression has a finite sum.

$$R = att't'/(1 - r'^2)$$

But from *Stoke's treatment of reflection*, we write

$$tt' = 1 - r^2 \text{ and } r = -r'$$

Hence, $R = -ar$

Therefore, the overall resultant reflected amplitude is

$$R_{\text{Min}} = -ar + ar = 0 \qquad \text{...(12.2 } j)$$

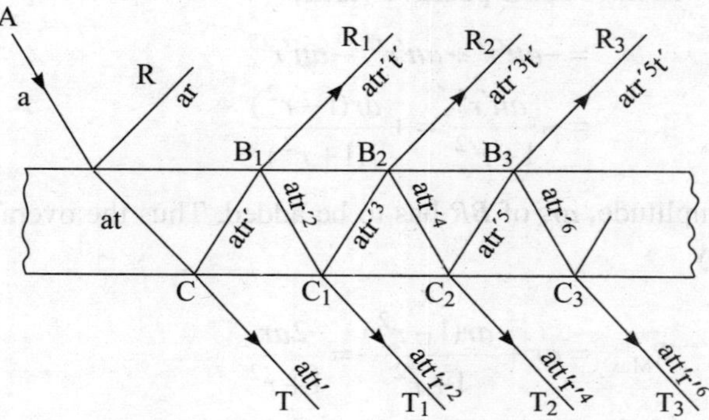

Fig. 12.4. Showing the amplitude factors of successive waves in multiple reflection

Thus, *the film will appear perfectly black by the reflected light when its thickness is small in comparison to the wavelength of light.* It should be remarked that this conclusion is independent of the wavelength. Therefore infinitesimally thin film will also appear perfectly dark even when illuminated with white light and viewed by the reflected light. Now, the question naturally arises where the light will be present, since energy cannot be destroyed and light is a form of energy. We shall see at a later stage that whatever energy is absent in the reflected light, it is present in the light transmitted to the other side of the film.

12.2.3 Plane Parallel Film of Finite Small Thickness

It is clear from Fig. 12.4 that the following relations must be true:

$$BCB_1 = B_1C_1B_2 = B_2C_2B_3 = \ldots$$

Therefore, the optical path difference between each pair of consecutive internally reflected waves, *e.g.*, B_2R_2 and B_3R_3, will be given by the expression

$$\Delta = 2\mu e \cos\theta$$

and phase difference, $\delta' = (2\pi/\lambda)\Delta$

Let us suppose that the film is of such a thickness that the condition $\Delta = n\lambda$ is satisfied. Under this condition, waves BR and B_1R_1, interfere destructively, but waves B_1R_1, B_2R_2, B_3R_3, etc. are all in phase with each other because $\delta = 2n\pi$, which means no phase difference. The amplitudes of the internally reflected waves are to be again added up and their resultant amplitude–$a\,r$ as explained earlier, just cancels the amplitude of BR, thereby producing complete destructive interference, we therefore conclude that under conditions expressed by Eqn. (12.2 *i*) a film of finite small thickness appears perfectly dark when seen by reflected light.

Let us consider a film of such a thickness that the condition

$$\Delta = (2n+1)\lambda/2$$

is satisfied; then wave BR and B_1R_1 interfere constructively but since $\delta' = (2n + 1)\,\pi$, any two consecutive internally reflected waves are in opposite phases. The resultant of the amplitude of the internally reflected waves will have some positive value,

$$= -att'r' + att'r'^3 - att'r'^5$$

$$= -\frac{att'r'}{1+r'^2} = +\frac{ar(1-r^2)}{(1+r^2)}$$

and to this resultant the amplitude, ar, of BR has to be added. Thus the overall resultant amplitude of reflected waves is given by

$$R_{\text{Max}} = ar + \frac{ar(1-r^2)}{1+r^2} = \frac{2ar}{1+r^2} \qquad \text{...(12.2 } k)$$

Thus, the intensity of the film will be maximum under the condition represented by Eqn. (12.2 h). Since r is less than 1, the maximum value of the resultant amplitude of all the reflected waves is less than a, the amplitude of the incident wave. Accordingly, as we shall find at a later stage that the remaining light energy in present in the transmitted side, provided there in no absorption of light in the film itself.

12.2.4 Colours of Thin Films by Transmitted Light

Colour effects analogous to that exhibited by thin films when viewed by reflected light also appear when a thin film is viewed by the light transmitted through it. In Fig. 12.5, transmitted waves CT and C_1T_1 originate from the same incident wave AB and hence they are capable of producing observable interference effects. Both CT and C_1T_1 are obtained without any sudden phase change. The phase difference, δ, between CT and C_1T_1 is, therefore only due to optical path difference, $\Delta = 2\mu e \cos\theta$, in reaching the plane C_1D.

$$\delta = (2\pi/\lambda)2\mu e\cos\theta$$

where λ is the length (in air) of the light incident on the film.

Now, the waves CT and C_1T_1 will interfere constructively when

$$= (2\pi/\lambda)2\mu e\cos\theta = 2n\pi$$

or $$2\mu e\cos\theta = 2n\left(\frac{1}{2}\lambda\right) \quad \text{[Maxima]} \qquad \text{...(12.2 } l)$$

and they will interfere contructively when

$$2\mu e\cos\theta = (2n+1)\frac{1}{2}\lambda \quad \text{[Minima]} \qquad \text{...(12.2 } m)$$

where n = 0, 1, 2, 3 etc., in Eqns. (12.2 l) and (12.2 m). A comparison of Eqns. (12.2 l) and (12.2 m)

with Eqns. (12.2 *h*) and (12.2 *i*) indicates that the conditions for constructive and destructive interference of transmitted waves that when the film is CT and C_1T_1 are just reverse of that of the reflected waves BR and B_1R_1. Hence, it follows of such a thickness that it has maximum illumination when seen by reflected light, it will have minimum illumination when seen by the transmitted light and vice versa. We shall now discuss the statement in detail.

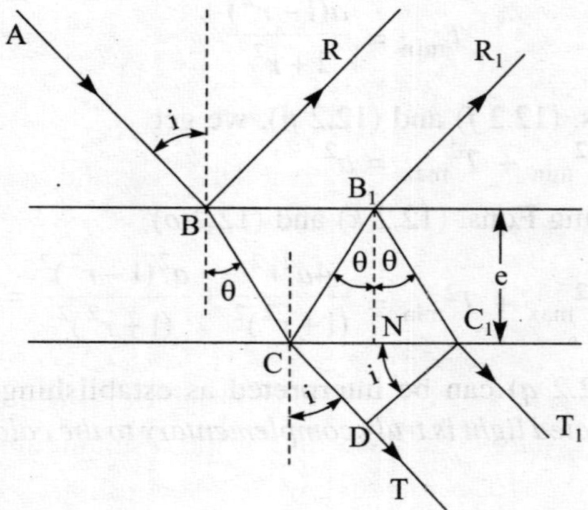

Fig. 12.5

It will be recalled that a film with thickness negligible in comparison with wave-length appears perfectly black when seen by the reflected light. In this case, since e is neglected, the transmitted waves CT, C_1T_1, C_2T_2, etc., which are obtained without any sudden phase jump, are in equal phase.

$$\therefore \quad T_{\text{Max}} = att' + att'r'^2 + att'r'^4 + \ldots\ldots$$

$$= att'[1 + r'^2 + r'^4 + \text{to infinite term}]$$

$$= \frac{att'}{1 - r'^2} = \frac{a(1 - r^2)}{1 - r^2} = a \qquad \ldots(12.2\ n)$$

The transmitted amplitude is exactly equal to the amplitude of the incident wave. So we conclude that an infinitesimal film transmits whole of the incident light to the other side, provided we neglect any absorption of light in the film itself.

A film of finite small thickness, which satisfies the condition represented by Eqn. (12.2 *l*) also transmits whole of the incident light to the other side. The reason is not far to seek. The optical path difference between any pair of consecutive transmitted waves is Δ. The transmitted waves do not suffer any sudden phase change. Therefore, when $\Delta = n\lambda$ is satisfied, waves CT, C_1T_1, C_2T_2 etc. are in phase with each other and thereby reinforce each other completely. The resultant transmitted amplitude is again maximum and equal to a, the amplitude of the incident wave,

$$T_{\max} = a$$

On the other hand, if the film is of such a thickness that the condition $\Delta = (2n + 1)\ \lambda/2$ is satisfied, the pair of waves CT, and C_1T_1, C_2T_2, C_3T_3, C_4T_4 and C_5T_5 etc. are in opposite phase and therefore interfere destructively in pair. The resultant transmitted amplitude, by the principle of superposition, is given by the relation

$$T_{min} = a\,t\,t' - a\,t\,t'\,r'^2 + a\,t\,t'\,r'^4 - \ldots\ldots$$

where the alternate plus and minus signs occur because the transmitted waves are alternately in and out of phase.

$$T_{min} = \frac{a(1-r^2)}{1+r^2} \qquad \ldots(12.2\ o)$$

On squaring and adding Eqns. (12.2 *j*) and (12.2 *n*), we get

$$R^2_{min} + T^2_{max} = a^2 \qquad \ldots(12.2\ p)$$

and similarly, on squaring and adding Eqns. (12.2 *k*) and (12.2 *o*)

$$R^2_{max} + T^2_{min} = \frac{4a^2r^2}{(1+r^2)^2} + \frac{a^2(1-r^2)^2}{(1+r^2)^2} = a^2 \qquad \ldots(12.2\ q)$$

Physically, Eqns. (12.2 *p*) and (12.2 *q*) can be interpreted as establishing the fact that the *colour phenomenon exhibited with the reflected light is truly complementary to the colour phenomenon exhibited with the transmitted light.*

12.3 FILM OF VARYING THICKNESS

Consider a film in the shape of a thin wedge whose sides form a small angle α and illuminated with plane monochromatic light waves. The directly reflected wave *BR* and the internally reflected wave B_1R_1 originate from the same incident wave propagating along *AB* and hence they are capable of producing observable interference effects. The two interfering waves do not reach the eye along parallel paths, but as depicted in Fig. 12.6, they appear to diverge from a point *Q* in the rear of the film. Destructive or constructive interference, therefore, occurs at the point *Q*, which is, however, virtual. [If the two interfering waves *BR* and B_1R_1 fall on a lens, they will cross each other at a real point *Q'*, the focus conjugate to *Q*. As a consequence real reinforcement or destructive interference would occur at *Q'*.]. Thus, *Q* will be the position where the interference fringe appears to be formed. When the light is incident from the thick side of the wedge, the two reflected waves intersect above the wedge. Fringes have their maximum visibility when the eye is focussed on these wave intersection points.

To derive an expression for the optical path difference between *BR* and B_1R_1 we draw with *Q* as centre and QB_1 as radius, an arc B_1D, which will be approximately straight and perpendicular to *BR*. We also draw perpendicular B_1E from B_1 on *BC*. The optical path difference δ, between the waves *BR* and B_1R_1 in reaching the arc B_1D from *B*, the point their origin is expressed by

$$\delta = \mu(BC + CB_1) - BD$$

$$= \mu(BE + EC + CB_1) - BD \qquad \ldots(12.3\ a)$$

It is easy to see that $BD = \mu BE$, hence Eqn. (12.3 *a*) reduces to

$$\delta = \mu\ (EC + CB_1) \qquad \ldots(12.3\ b)$$

In Fig. 12.6, *BN* and *CN* are respectively normals to the upper and lower surfaces of the film, hence we get

$$\angle CNB = \alpha$$

and $$\angle PCB = \angle CBN + \angle CNB = (\theta + \alpha)$$

and $$\angle PCB_1 = \angle PCB = \theta + \alpha$$

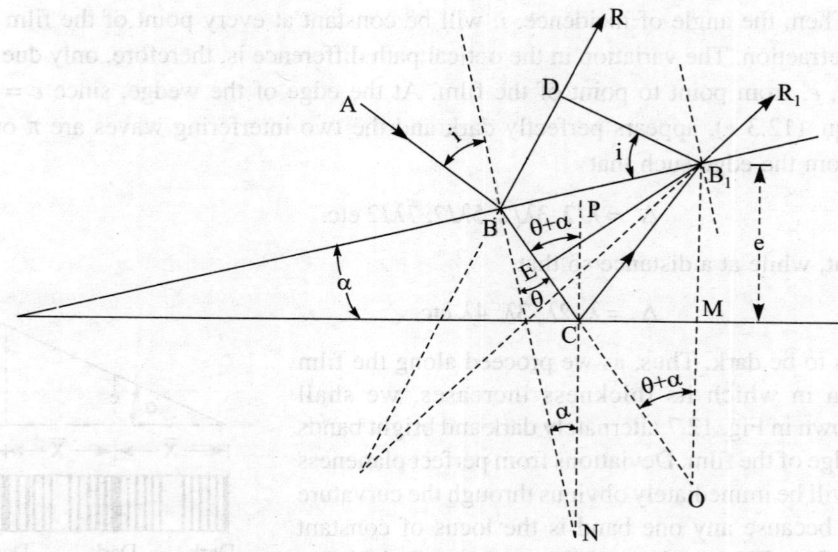

Fig. 12.6

In Fig. 12.6, B_1M is perpendicular from B_1 on the lower surface and *BC* when produced further intersects it at *O*, hence we have the relation

$$\angle B_1OC = \angle PCB = (\theta + \alpha)$$

Also, $$\angle CB_1O = \angle PCB_1 = (\theta + \alpha) = \angle B_1OC$$

Thus B_1OC is an isosceles triangle, with $CB_1 = CO$

The path difference Δ, expressed by Eqn. (12.3 *b*), now reduces to

$$\Delta = \mu(EC + CO) = \mu EO = \mu B_1O\cos(\theta + \alpha)$$

or $$\Delta = 2\mu e\cos(\theta + \alpha) \qquad \text{...(12.3 } c\text{)}$$

where *e* stands for the thickness of the film at the point B_1. The path difference, Δ, thus varies both on account of changing thickness as well as changing angle of incidence, provided the broad light source is at finite distance from the film.

Now, taking into account the abrupt phase change of π in the phase of the wave due to reflection at *B*, the waves *BR* and B_1R_1 will interfere constructively, as in the case of a plane parallel optically denser film, when

$$2\mu e \cos(\theta+\alpha) = (2n+1)\lambda/2 \quad \text{[Maxima]} \qquad ...(12.3\ d)$$

where n = 0, 1, 2, 3, 4 etc. and they will interfere destructively when

$$2\mu e \cos(\theta+\alpha) = 2n(\lambda/2) \quad \text{[Minima]} \qquad ...(12.3\ e)$$

where n = 0, 1, 2, 3 etc.

Consider the wedge shaped film to be illuminated by a parallel beam of monochromatic light of wavelength λ. Then, the angle of incidence, i, will be constant at every point of the film and so also θ, the angle of refraction. The variation in the optical path difference is, therefore, only due to variation in the thickness, e, from point to point of the film. At the edge of the wedge, since e = 0, the film, according to Eqn. (12.3 e), appears perfectly dark and the two interfering waves are π out of phase. At a distance from the edge such that

$$\Delta = \lambda/2, 3\lambda/2, 5\lambda/2, 7\lambda/2 \text{ etc.}$$

the film is bright, while at a distance so that

$$\Delta = \lambda, 2\lambda, 3\lambda, 4\lambda \text{ etc.}$$

the film appears to be dark. Thus, as we proceed along the film in the direction in which its thickness increases, we shall encounter as shown in Fig. 12.7 alternately dark and bright bands parallel to the edge of the film. Deviations from perfect planeness of the surfaces will be immediately obvious through the curvature of these bands, because any one band is the locus of constant thickness of the film. That is why such fringes are called fringes of constant thickness.

Dark Dark Dark Dark

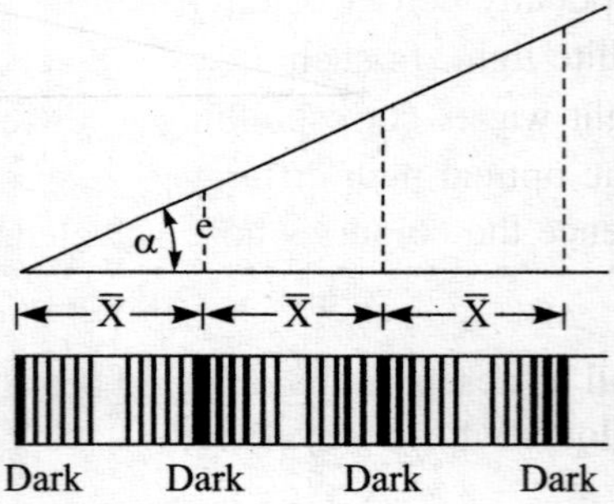

Fig. 12.7

The separation $\bar{X}$ between any two consecutive dark bands can be easily computed. If e is the thickness at the dark band next to that formed at the edge of the film, then we have

$$2\mu e \cos(\theta+\alpha) = \lambda \qquad ...(12.3\ f)$$

and in Fig. 12.7, $\bar{X}$ satisfies the relation

$$e = \bar{X} \tan\alpha \qquad ...(12.3\ g)$$

From Eqn. (12.3 f) and (12.3 g), on eliminating e, we get

$$\bar{X} = \frac{\lambda}{2\mu \tan\alpha \cos(\theta+\alpha)} \qquad ...(12.3\ h)$$

In practice, we deal with wedge shaped films that are so narrow in such a way that α may be neglected on comparison to θ, then

$$\Delta = 2\mu e \cos\theta = 2e\sqrt{\mu^2 - \sin^2 i}$$

and $\bar{X}$ in this case is given by the expression

$$\overline{X} = \frac{\lambda}{2\tan\alpha\sqrt{\mu^2 - \sin^2 i}} \qquad \text{...(12.3 } i\text{)}$$

The distance between any two consecutive dark fringes is thus the same. It can be easily proved in a similar way, the same expression gives the separation between the two consecutive bright bands. The smaller is the wedge angle α and greater the *angle of incidence i*, the greater is the size of $\overline{X}$. Also the separation $\overline{X}$ is the greater, the smaller is μ of the film and the greater is λ. Now, if the wedge shaped film is illuminated successively with light of different colours, it will be seen, other things being equal since $\lambda_R > \lambda_v$, that the spacing of the fringes is greatest with the red light and least with the violet and intermediate for intermediate wavelengths.

12.4 THIN FILM ILLUMINATED WITH A PARALLEL BEAM OF WHITE LIGHT

It is now quite easy to discuss the colour effects exhibited by a plane parallel thin transparent film (optically denser or rarer than the surrounding media) when it is illuminated with a parallel beam of white light. In such a case, neglecting the small variation in the angle of refraction with wavelength, light waves corresponding to different colours pursue approximately the same path within the film. The optical path difference $2\mu e \cos\theta$ will be the same to a close approximation for all the colours. Hence the colours whose wavelengths satisfy the equation

$$2\mu e \cos\theta = n\lambda$$

will be absent in the reflected beam but they will be present in the transmitted beam. On the other hand, colours whose wavelengths satisfy the relation

$$2\mu e \cos\theta = (2n+1)\lambda/2$$

will be present in the reflected beam but practically absent in the transmitted beam. Consequently the resultant colour of the film by reflected light is exactly complementary to its resultant colour by the transmitted light.

Let us now suppose that we have a wedge shaped thin film, whose thickness at the thin edge is very small in comparison to the wavelength of the violet light. At the thin edge, since its thickness is infinitesimally small, a truly achromatic black fringe parallel to the edge will be seen when the film is illuminated with white light and viewed by reflected light. Passing along the film in the direction of increasing thickness, since λ_v is least, we shall first reach a point where the thickness of the film is such that the condition of constructive interference is satisfied for violet colour only and accordingly a violet fringe is seen at that point. Proceeding still further we shall encounter successively blue, green, yellow and red fringes due to the same cause. Beyond this, the condition of constructive interference is satisfied for two or more colours simultaneously and a coloured band due to overlapping of bright fringes of these colours is encountered. Ultimately, we shall arrive at the point where such an overlapping of different colours produces uniform illumination.

The complementary character of reflected and transmitted colour can be easily verified by illuminating a thin vertical soap film with white light and analysing the reflected and the transmitted light from the same portion of film with a prism or grating spectrometer. On examining both the spectra it would be found that exactly at positions where we get dark lines in the spectrum of reflected light we get these missing wavelengths in the transmitted light *i.e.* just those colours which are absent in the reflected light are found to predominate in transmitted light and vice versa.

Before concluding our discussion of the colours of thin films it should be pointed out that in general, reflected waves received by the eye do not pursue exactly parallel paths *i.e.* cos θ in the relation

$$2\mu e \cos\theta = n\lambda$$

is not the same for all of them. For one definite colour to be absent in the reflected light, it is, therefore, absolutely essential that n in the above relation must be a small whole number. This is only possible when the film is extremely thin. Only very thin films upto one wavelength thick exhibit bright colours. The colours begin to fade with thicker films and in such cases the direction of incidence should be kept nearly normal, otherwise, at other angles the directly and internally reflected waves may get separated to such an extent that their wavefronts may not interfere each other and so all colour effects may vanish when viewed with naked eye. It is, however, important to note that in order to observe the interference effects in the latter case we should employ some optical instrument to collect the directly and the internally reflected beams, which is not possible with naked eye owing to the pupil being very small.

Suppose white light from an extended source is incident on a plane parallel thick film, which is viewed by reflected light. For any given value of θ, due to large value of e, values of n can be found to satisfy the condition of constructive interference for every colour in the spectrum. Due to this reason, different coloured fringes of equal inclination will be superposed to produce a general white illumination. Thus a thick film shows no colours but appears white in contrast to excessively thin film which appears black.

Necessity of Broad Source of Light

When a thin film is illuminated with light from a point source and it is observed with a lens of small aperture such as the eye, the rays that can enter the eye are confined to a small range of directions. Thus, the field of view is narrow. But if we employ an extended source, light from every point of the source after reflection from the film enters the eye in the separate direction. If the accommodation is relaxed the waves in each pair are brought together at a different point on the retina. Thus the field of view is wide and the interference effects can be observed over the entire film.

12.5 NEWTON'S RINGS

The phenomenon of Newton's rings is a noteworthy illustration of the interference of light waves reflected from the opposite surfaces of a thin film of variable thickness. A thin air film of progressively increasing thickness in all directions from one point can be very easily formed by placing on a plane glass plate a plano-convex lens of a large radius of curvature so that is convex surface faces the plate. The air film thus possesses a radial symmetry about the point of contact. When it is illuminated normally, preferably with monochromatic light, the observed interference fringes, which are the loci of points of equal optical film thickness, are circular rings concentric with the point of contact. The rings gradually become narrower as their radii increase until the eye or the magnifying instrument can no longer separate them. The circular rings can also be formed by bringing in contact two spherical surfaces of different radii of curvature. The rings are localised in the air film.

These rings are termed Newton's rings, so called because Newton was first to examine them in detail and to measure their radii with some precision, although they were first observed by Robert Hooke in 1635. We owe their explanation to Young in terms of the interference of light waves reflected from the convex lower surface of the lens and the flat upper surface of the glass plate, which respectively constitute the upper and the lower surfaces of thin variable air film. The two waves *BD* and B_1D_1 under consideration originate, as depicted in Fig. 12.8, from the same incident wave *AB* by the division of its amplitude and hence they are capable of producing observable interference effects. Furthermore, we also get a set of waves like *CT* and C_1T_1 transmitted to the other side of the film. They also originate from the same incident wave and as a consequence, their interference also produces circular concentric fringes.

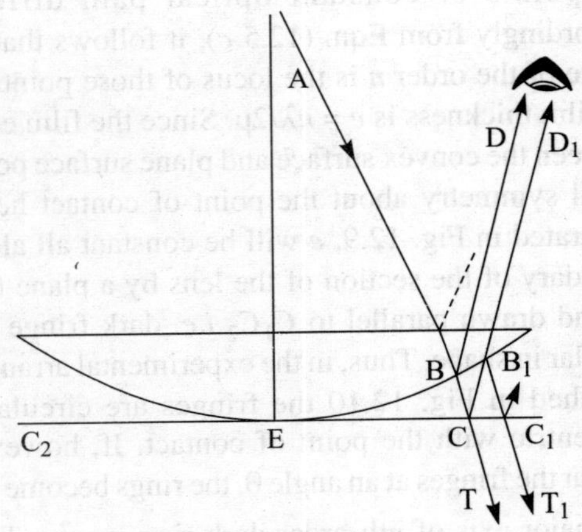

Fig. 12.8. Theory of Newton's rings

To discuss theoretically the formation of Newton's rings, let us suppose that the thickness of the film at the point B_1 is *e* and let the tangent to the convex surface at B_1 be inclined at an angle α with the horizontal. Then the optical path difference, Δ, between the waves *BD* and B_1D_1, as obtained in *Sec*. 12.3, is $2\mu e \cos(\theta + \alpha)$ in air where θ is the angle of refraction at *B* and μ is the refractive index of the film with respect to air. Also assuming the film to be optically rarer than the media above and below it, an abrupt phase change of π occurs in the wave at the instant of its reflection at *C*, a point on the surface backed by a denser medium. No such phase change occurs at the instant of reflection at *B*. As a consequence, waves *BD* and B_1D_1 shall interfere destructively when

$$2\mu e \cos(\theta + \alpha) = (2n)\lambda/2 \qquad \text{[Minima]} \qquad ...(12.5\ a)$$

and they will interfere constructively when

$$2\mu e \cos(\theta + \alpha) = (2n+1)\lambda/2 \qquad \text{[Maxima]} \qquad ...(12.5\ b)$$

where in Eqns. (12.5 *a*) and (12.5 *b*), λ is the wavelength in air and $n = 0, 1, 2, 3$ etc. characterise the order of interference minima or maxima as the case may be.

In practice however, we employ a thin lens of extremely small curvature in order to keep the film enclosed between the lens and plane glass plate extremely thin. As a consequence, the angle α becomes negligibly small as compared to θ. Furthermore, the experimental arrangement is so designed, as sketched in Fig. 12.10 that the light falls on the film almost normally and it is viewed from nearly normal direction by reflected light, so that $\cos\theta = 1$. Accordingly, Eqns. (12.5 *a*) and (12.5 *b*) reduce to their respective forms

$$2\mu e = n\lambda \quad \text{[Minima]} \quad ...(12.5\ c)$$

and

$$2\mu e = (2n+1)\lambda/2 \quad \text{[Maxima]} \quad ...(12.5\ d)$$

An interference fringe of given order n is the locus of the points of constant optical path difference. Accordingly from Eqn. (12.5 *c*), it follows that a dark fringe of the order n is the locus of those points where the film thickness is $e = n\lambda/2\mu$. Since the film enclosed between the convex surface and plane surface possesses radial symmetry about the point of contact hence, as illustrated in Fig. 12.9, e will be constant all along the boundary of the section of the lens by a plane through B_1 and drawn parallel to C_1C_2 *i.e.* dark fringe will be circular in shape. Thus, in the experimental arrangement sketched in Fig. 12.10 the fringes are circular rings concentric with the point of contact. If, however, we look at the fringes at an angle θ, the rings become ellipse, the major axis of nth order dark ring is r_n^2 whilst the minor axis becomes $r_n^2/\cos\theta$.

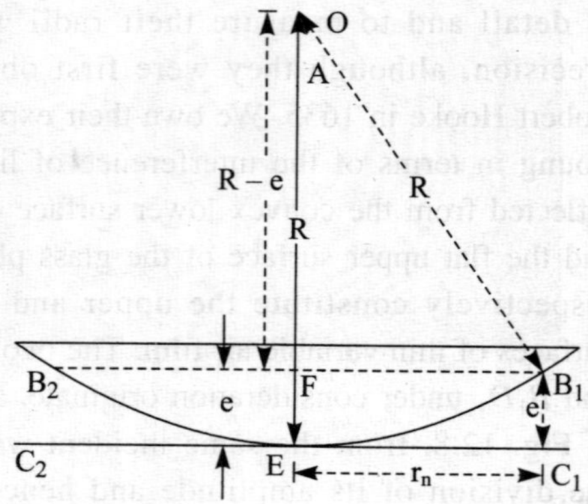

Fig. 12.9. Evaluation of the radii of Newton's rings

The radius of any ring may be computed as follows: Let R be the radius of the curvature of the convex surface which rests on the plane glass surface. From the right triangle OFB_1, we get the relation

$$R^2 = r_n^2 + (R-e)^2$$

or

$$r_n^2 = 2Re - e^2$$

where r_n is the radius of the circular fringe corresponding to the constant film thickness e, which, however, must be extremely small as compared to R so that the condition discussed in *Sec.* 12.4, for observable interference effect may be satisfied. Hence, neglecting e^2, we may state to a sufficient degree of accuracy that

$$e = r_n^2/2R$$

Substituting for e in Eqn. (12.5 *c*), we have r_n^2, the square of the radius of dark ring of nth order given by

$$r_n^2 = n\lambda R/\mu \quad \text{[Dark Ring]} \quad ...(12.5\ e)$$

Similarly, the square of the radius of bright ring of nth order is given by

$$r_n^2 = (2n+1)\lambda R/2\mu \quad \text{[Bright Ring]} \quad ...(12.5\ f)$$

The square of the diameters of dark and bright rings are, therefore, given by the respective expressions

$$D_n^2 = 4n\lambda\ R/\mu \quad \text{[Dark Ring]} \quad ...(12.5\ g)$$

and $$D_n^2 = 2(2n+1)\lambda R/\mu \quad \text{[Bright ring]} \qquad ...(12.5\ h)$$

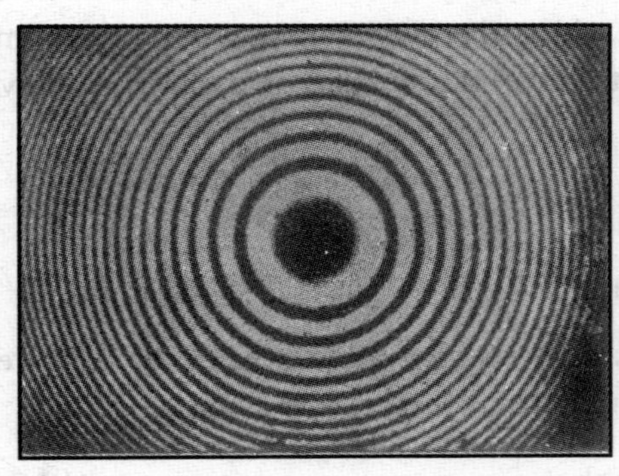

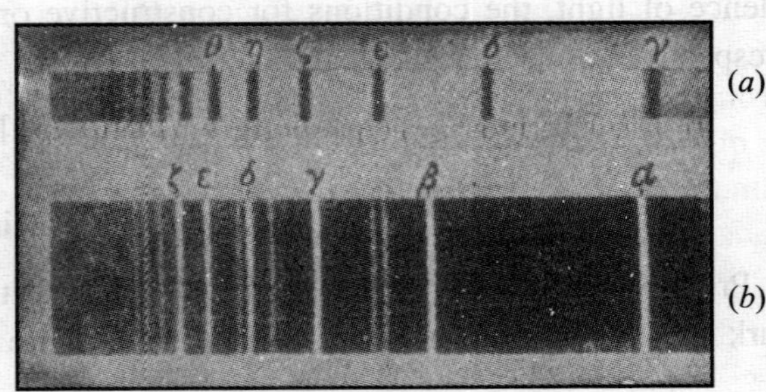

Newton's rings **Balmer lines (*a*) in emission and (*b*) in absorption**

where $n = 0, 1, 2, 3$ ...... for the first, second, third, fourth bright ring etc. respectively. Thus when Newton's rings are seen by reflected light, the diameters of the bright rings are proportional to the square roots of the odd natural numbers 1, 3, 5, 7 etc. and the diameters of the dark rings are proportional to the square roots of the natural numbers 1, 2, 3, 4, 5 etc. The difference of the diameter of dark rings of 4th and 5th order is

$$D_5 - D_4 = 2[\sqrt{5} - \sqrt{4}]\sqrt{\lambda R/\mu} = 0.46\sqrt{\lambda R/\mu}$$

and $$D_{17} - D_{16} = 2[\sqrt{17} - \sqrt{16}]\sqrt{\lambda R/\mu} = 0.26\sqrt{\lambda R/\mu}$$

Thus it is clear that the rings gradually become narrower as their radii increase.

At the point of contact of the lens and the glass plate, since $e = 0$, therefore, the total phase difference, $\left(\frac{2\pi}{\lambda}\cdot 2\mu e + \pi\right)$, between the directly and internally reflected waves reduces to π. As a consequence, in the reflected light, in the interference pattern central spot appears to be dark. Furthermore, it is perfectly dark and the explanation of this is exactly the same as that given in *Sec.* 12.2.2 to explain the perfect blackness of infinitesimally thin film when seen by reflected light. With monochromatic light, the central black spot is surrounded alternately by a large number of bright and dark rings, which are comparatively narrower when blue light is employed as compared to red light. With white light, the central black spot will be surrounded with eight or ten highly coloured rings of several colours or a mixture of colours. Beyond this due to overlapping of rings of several colours, all phenomenon of interference would be obliterated, resulting in uniform illumination.

12.5.1 Newton's Rings by Transmitted Light

We shall now discuss the formation of Newton's rings due to interference of transmitted waves CT and C_1T_1. The wave CT originates from the wave BC by single refraction at C while the wave

C_1T_1 is obtained from BC by two abrupt phase jumps each equal to π due to reflections at C and B followed by a refraction at C_1. The abrupt phase changes when considered together amount to 2π, which physically interpreted means no phase change. Thus, the only phase difference between CT and C_1T_1 is due to different optical paths traversed in reaching the eye. Consequently, for normal incidence of light, the conditions for constructive or destructive interference of transmitted waves are respectively,

$$2\mu e = n\lambda \quad \text{[Maximum]}$$

and

$$2\mu e = (2n+1)\lambda/2 \quad \text{[Minimum]}$$

Proceeding as in the case of reflected system, it is easy to see that the square of the diameters of dark and bright rings in the transmitted system are given by respective expressions.

$$D_n^2 = 4n\lambda\ R/\mu \quad \text{[Bright Ring]} \qquad \text{...(12.5 } i\text{)}$$

and

$$D_n^2 = 2(2n+1)\lambda\ R/\mu \quad \text{[Dark Ring]} \qquad \text{...(12.5 } j\text{)}$$

A comparison of Eqns. (12.5 *i*) and (12.5 *j*) with Eqns. (12.5 *g*) and (12.5 *h*) clearly indicates that the rings observed by reflected light are exactly complementary to those seen by the transmitted light. The central spot which is dark when seen by reflected light is bright when seen by transmitted light. But the dark rings in the transmitted light are not perfectly dark and therefore Newton's rings are not so distinct as observed with reflected light. With white light coloured rings are observed in the transmitted light, but due to the admixture of unmodified white light, colours are not so brilliant as those due to reflected light. Hence, Newton's rings are usually observed by the reflected light.

12.5.2 Determination of Wavelength

In the laboratory, by measuring the diameters of Newton's rings with a travelling microscope it is possible to compute the wavelength of monochromatric light employed to illuminate the film. For an air film, since $\mu = 1$, Eqns. (12.5 *g*) and (12.5 *h*) reduce to respective forms,

$$D_n^2 = 4n\lambda R \quad \text{[Dark Ring]} \qquad \text{...(12.5 } k\text{)}$$

and

$$D_n^2 = 2(2n+1)\lambda R \quad \text{[Bright Ring]} \qquad \text{...(12.5 } l\text{)}$$

Replacing n by $(n + p)$ in Eqns. (12.5 *k*) and (12.5 *l*), we get

$$D_{n+p}^2 = 4(n+p)\lambda R \quad \text{[Dark Ring]} \qquad \text{...(12.5 } m\text{)}$$

and

$$D_{n+p}^2 = 2(2n+2p+1)\lambda R \quad \text{[Bright Ring]} \qquad \text{...(12.5 } n\text{)}$$

Subtracting either Eqn. (12.5 *k*) from Eqn. (12.5 *m*) or Eqn. (12.5 *l*) from Eqn. (12.5 *n*), we get in both the cases,

$$D_{n+p}^2 - D_n^2 = 4p\lambda R \qquad \text{...(12.5 } o\text{)}$$

and $$\lambda = \frac{D_{n+p}^2 - D_n^2}{4pR} \qquad ...(12.5\ p)$$

Equation (12.5 *p*) is invariably employed to compute λ in this experiment. The advantage of this equation over Eqns. (12.5 *k*) and (12.5 *l*) can be seen from the following discussion: Eqns. (12.5 *k*) and (12.5 *l*) have been derived on the supposition that the surfaces of the lens and the plate are perfect and they are in perfect contact *i.e.* at the point of contact thickness of air film is zero and as a consequence, in a reflected system a central dark ring of zero order is formed. In actual practice, however, due to encroachment of some dust particles between lens and the plate or due to some imperfectness of their surfaces, they may not be in perfect contact *i.e.* the thickness of the film may not be zero at the central point. The order, *x*, of the central ring, therefore, is indeterminate and the central ring may even be white. As a consequence, the order of every other dark or bright ring advances by an indeterminate number *x*. For any one of them, the square of the diameter is not given by Eqn. (12.5 *k*) or (12.5 *l*). But in taking the difference of the squares of the diameters of *n*th dark ring and $(n + p)$th dark ring, counting the rings *p*, between them visually, this indeterminacy in their orders does not occur in Eqn. (12.5 *p*).

The experimental arrangement for the production and observation of Newton's rings is sketched in Fig. 12.10. Light from a monochromatic source *S* is rendered nearly parallel by a lens L_1 and then falls on a thin transparent glass plate *G*, fixed at 45° to the vertical, which partially reflects the light in the downward direction. Thus, light falls normally on an air film formed between the convex surface of the lens *L* in contact with the plane glass plate *P*. The directly and internally reflected waves pursue nearly the same path as that of incident light and finally received by the eye or in a travelling microscope. The diameters of bright or dark rings can be now measured by gradually moving the microscope *M* from one side of the ring system to the diametrically opposite side, fixing one of the wires of the cross-wires exactly at the centre of width of the successive rings, care being taken to avoid the back lash in the microscope screw.

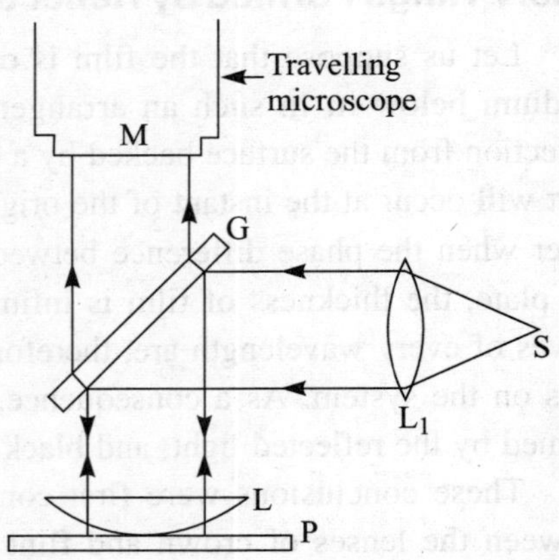

Fig. 12.10. Experimental arrangement for Newton's rings

We may also compute the radius of curvature, *R*, of the lens surface by Eqn. (12.5 *p*), provided the wavelength of light employed has been previously determined by some other method like Michelson's interferometer.

12.5.3 Determination of Refractive Index of a Liquid with Newton's Rings

By the measurement of diameters of Newton's rings with air film and repeating the same by forming a liquid film between the lens and the glass plate it is possible to compute the refractive index of liquid. Let μ be the refractive index of the liquid with respect to air. Then, from Eqn. (12.5 *g*) or Eqn. (12.5 *h*), it easy to derive, as explained earlier with air film, the following relation;

$$(D_{n+p}^2 - D_n^2)_{\text{Liquid}} = \frac{4p\lambda R}{\mu} \quad \text{...(12.5 } q\text{)}$$

and for an air film,

$$(D_{n+p}^2 - D_n^2)_{\text{Air}} = 4p\lambda R \quad \text{...(12.5 } o\text{)}$$

Hence,

$$\mu = \frac{(D_{n+p}^2 - D_n^2)_{\text{Air}}}{(D_{n+p}^2 - D_n^2)_{\text{Liquid}}} \quad \text{...(12.5 } r\text{)}$$

This method is usually employed for determining the refractive index of rarer liquids. After measuring the diameters of bright rings with air film, *n* drop of liquid is carefully introduced on the plate. The liquid is drawn in at the centre by the capillary action. The rings now contract (since $\mu > 1$) in the ratio $\sqrt{(1/\mu)}$. The diameters of bright rings are again measured and finally, μ computed by Eqn. (12.5 *r*).

12.5.4 Rings Formed by Reflected Light with White Centre

Let us suppose that the film is optically denser than the medium above it but rarer than the medium below it. In such an arrangement, each of the waves *BD* and B_1D_1 will be obtained by reflection from the surface backed by a denser medium and as a consequence, an abrupt phase change of π will occur at the instant of the origination of each reflected wave, which obviously balance each other when the phase difference between them is computed. At the point of contact of the lens and the plate, the thickness of film is infinitesimally small, $e = 0$. The directly and internally reflected waves of every wavelength are, therefore, in the same phase at the point of contact when white light falls on the system. As a consequence, white achromatic fringe will be seen at the centre of rings formed by the reflected light, and black achromatic fringe in rings formed by the transmitted light.

These conclusions were first confirmed by Dr. Young by forming a film of oil of sassafras between the lenses of crown and flint glass. This oil has a refractive index intermediate in value between the indices of crown and flint glass.

In concluding our discussion of Newton's rings it should be pointed out that in the optical arrangement sketched in Fig. 12.10 if the plane glass plate is replaced by another plane plate, highly silvered on its front surface whole phenomenon of Newton's rings becomes invisible. This is due to the fact that at the polished reflector reflection is complete. Hence the amplitude of the light wave obtained by reflection at the plane reflector is considerably greater than the amplitude of the light wave obtained by reflection of the incident wave at the lower convex surface of the lens. The two light waves under consideration originate from the same incident wave and hence satisfy the fundamental condition of interference. But they violate the condition, discussed in *Sec.* 12.4 viz., the amplitudes of interfering waves must be equal or nearly equal, which is, however, absolutely essential for observable interference effects. Due to great difference in their amplitudes there is no observable contrast between the regions of constructive and destructive interference of light waves and so rings become invisible.

12.6 MICHELSON INTERFEROMETER

An ingenious arrangement for experimenting on interference of light has been designed by *Michelson*, in which the light from an extended source is divided into two coherent beams by partial reflection (*division of amplitude*) at the lightly silvered surface of a glass plate and then made to reunite at the same plate after traversing widely separated optical paths. The basic principle of this instrument was given in 1881 by *Michelson* in his now famous experiment, called *Michelson Morley experiment*, on the effect of motion of ether on the velocity of light. This instrument is very ingenious in the sense that both equal thickness and equal inclination interference fringes can be produced, the latter with any desired large path difference, say 50 cm when light is truly homogeneous. But this is not possible with any of the experimental arrangements described previously for the production of interference fringes. He called this instrument *interferometer* because it is primarily employed to determine the wavelength of light in terms of some standard of length and vice-versa utilising the phenomenon of *Interference of Light.*

Optical parts of Michelson Interferometer

The essential parts of Michelson interferometer are schematically represented in Fig. 12.11. The main optical part consist of two excellent optically plane mirrors M_1 and M_2, highly silvered on their

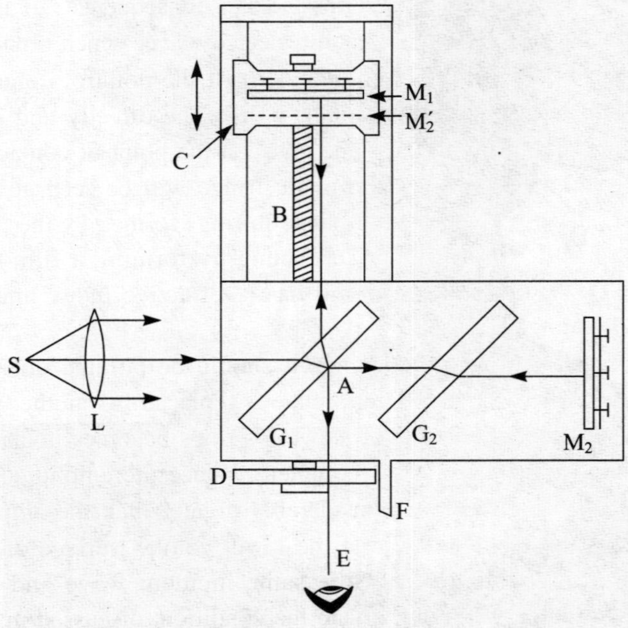

Fig. 12.11. Optical system of Michelson interferometer

front surfaces to avoid multiple internal reflections, and two plane parallel glass plates G_1, G_2, of equal thickness, being cut from a single optically plane-parallel plate to ensure the equality of thickness and the nature of the material. Both the plates are mounted vertically, exactly parallel to each other, on a heavy frame and are usually inclined at 45° to the interferometer arm. The rear side of G_1 is covered with a film of silver, of such a thickness that one half of the amplitude of incident wave is reflected while the rest is transmitted through it. The mirror M_2 is fixed while the other mirror M_1 is mounted on a *carriage* C, which can be moved in the direction of double arrow, along precision machined straight and parallel tracks. During the motion, M_1 remains exactly parallel to its preceding positions. This slow and accurately controlled motion is communicated by means of a calibrated screw B, which is provided with a wheel and a graduated circle D so that the motion of M_1 may be accurately known. Mirrors M_1 and M_2 are capable of slight rotation about the vertical as well as horizontal axis by the help of screws provided at their backs and thus they may be adjusted mutually perpendicular. In Hilger instrument two special screws are provided which alter the angular setting of M_2.

The Michelson Interferometer

Action of the Apparatus

Light from an extended monochromatic source *S*, rendered nearly parallel by the lens *L*, is incident on the instrument, and on entering G_1 it is divided into two parts of equal intensities by partial reflection at its rear side. The reflected wave which proceeds to M_1 and a transmitted wave which proceeds to M_2, are incident normally on the mirrors when instrument is in normal adjustment, and accordingly, after reflection, they retrace their paths. On return, at the lightly silvered surface, a part of the amplitude of the wave from M_2 is reflected while a part of wave from M_1, is transmitted along *AE*. Since, the two waves are derived from the same wave, originally incident at *A* on G_1, hence the fundamental condition of interference is satisfied. Accordingly, one observes the fringes by looking from the position *E* into the mirror M_1 through the plate G_1.

Optical Paths of Two Halves of the Incident Wave

The wave reflected at M_1 crosses the glass plate G_1 twice while the optical path of the other wave, in the absence of G_2, lies wholly in air. Hence due to this cause only, in the former wave an extra optical path $2(\mu - 1)\, t$ (air) is introduced, where *t* is the thickness of the plate and μ its refractive index for the light employed. Although this is immaterial of the production of fringes with monochromatic light, nevertheless this presents the serious difficulty when instrument is used with white light because of the variation of μ with wavelength. Hence, if white light achromatic fringe is to be observed in this instrument, it is essential to compensate for this extra optical path $2\,(\mu - 1)t$, not only for one wavelength but for all wavelengths. This is accomplished by introducing in the beam AM_2 a plate G_2, called *compensating plate*, of the same thickness and material as G_1 and placed exactly parallel to the latter. Light reflected from M_2 passes twice through G_2 and in this way the optical paths AM_1A and AM_2A contain the same thickness of glass.

Phase Changes on Reflection

The phase changes on reflection at M_1 and M_2 are similar, and Edser has shown that when the silver coating on G_1 is thin, the phase changes, due to reflections from it, in air and glass are also similar, each being equal to π. The exact phase changes occurring at plate G_1 will depend on the nature of the semi-reflecting film deposited on the plate. The optical path difference between the two beams is, therefore, simply due to different path traversed in air before reaching the observer. Consequently, the two waves shall interfere constructively or destructively according as the optical path difference Δ between them is even or odd multiple of λ/2.

$$\Delta = 2n(\lambda/2) \quad \text{[Maxima]} \qquad ...(12.6\ a)$$

$$\Delta = (2n+1)(\lambda/2) \quad \text{[Minima]} \qquad ...(12.6\ b)$$

If, however, the plate G_1 be unsilvered the wave AM_1 would undergo internal reflection in G_1 from the surface backed by rarer medium and hence without any abrupt phase jump while the wave AM_2 would undergo reflection at G_1 from the surface backed by denser medium with consequent abrupt phase jump of π. In such a case the condition for constructive or destructive interference of two coherent light waves are respectively,

$$\Delta = (2n+1)(\lambda/2) \quad \text{[Maxima]} \qquad ...(12.6\ c)$$

$$\Delta = 2n(\lambda/2) \quad \text{[Minima]} \qquad \ldots(12.6\ d)$$

The path difference between any two coherent beams can be altered by moving M_1. Thus any point in the field of view appears alternately bright and dark.

Looking in the direction M_1 from E, one observes M_1 and also a virtual image M_2' of M_2 formed in G_1. Therefore, one of the interfering beams comes by reflection from M_1 and the other which is reflected from M_2 functions as if it had been reflected from M_2', the two reflected beams originating from the same incident beam. Michelson inteferometer is, therefore, optically equivalent to air film between M_1 and M_2' but without the phenomenon of multiple reflections within the film. When the interferometer is in normal adjustment M_1 and M_2' are exactly parallel to each other and it is also possible, depending upon the orientation of M_2 with respect to M_1, that a wedge shaped air film be formed between M_1 and M_2'. The interference fringes may be, therefore, of different shapes—*straight, circular, parabolic, elliptical* and *hyperbolic*—depending upon Δ and the angle between M_1 and M_2' but only straight and circular fringes are most commonly employed.

12.6.1 Adjustment of Michelson Interferometer

Michelson interferometer is said to be in normal adjustment when the lightly silvered surface of the dividing glass plate G_1 accurately bisects the angle between the highly silvered front surfaces of mirrors M_1 and M_2, and normals to their surfaces are exactly co-planer. Under this condition the image M_2' of M_2 in G_1 would be exactly parallel to the real mirror M_1 and as explained in sequence, the fringes could be concentric circles. To accomplish this adjustment we proceed as follows: The distances of the silvered surfaces of M_1 and M_2 from the silvered plane of G_1 are adjusted nearly the same within a few millimetres. A metallic screen with a fine pin hole is placed between G_1 and the monochromatic light source. On looking along EG_1, two pairs of images are seen as a rule. This image formation is illustrated in Fig. 12.12 (*a*). One pair, 1 and 2, is formed due to partial reflections at the unsilvered surface of G_1 and reflections at M_1 and M_2 respectively. While the other pair, 3 and 4, is formed due to partial reflections at the lightly silvered surface of G_1 and reflections at M_1 and M_2 respectively. Obviously, the second pair is brighter than the first and is formed by the waves which traverse the glass plates three times (paths in G_1 and G_2 taken together). The waves forming the images 2 and 1 traverse the plates five times and once, respectively. Mirrors M_1 and M_2 are therefore, turned in the proper direction, about vertical as well as horizontal axis, with the help of tilting screws until the brighter images 3 and 4 coincide, [Fig. 12.12 (*b*)]. On removing the screen, fringes should appear. By a further finer tilt of mirrors so as to continuously increase the width of fringes, concentric circular fringes can be brought into view. But the centre of fringes may be outside the field of view. The angular setting of M_2 is now slightly altered with the help of special screws until the centre of circular fringes is in the centre of the field of view. Now M_1 is slightly displaced so as to obtain large number of concentric fringes in the field of view. The perfectness of this adjustment may be tested by observing the fringes while the eye is moved vertically or sideways. If the circular fringes do not expand or contract as the eye is thus moved we conclude that the fringe system is situated at infinity (see discussion under circular fringes); if otherwise, finer tilt should be given to M_2 until the adjustment is perfect.

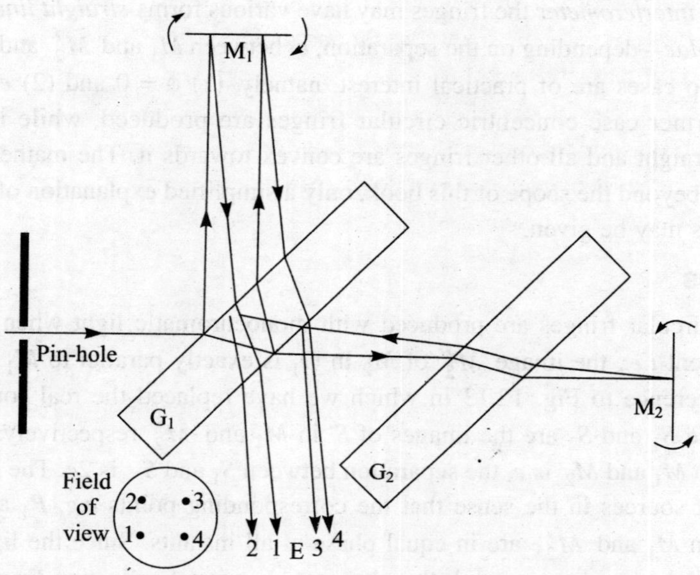

Fig. 12.12 (*a*). Adjustment of Interferometer

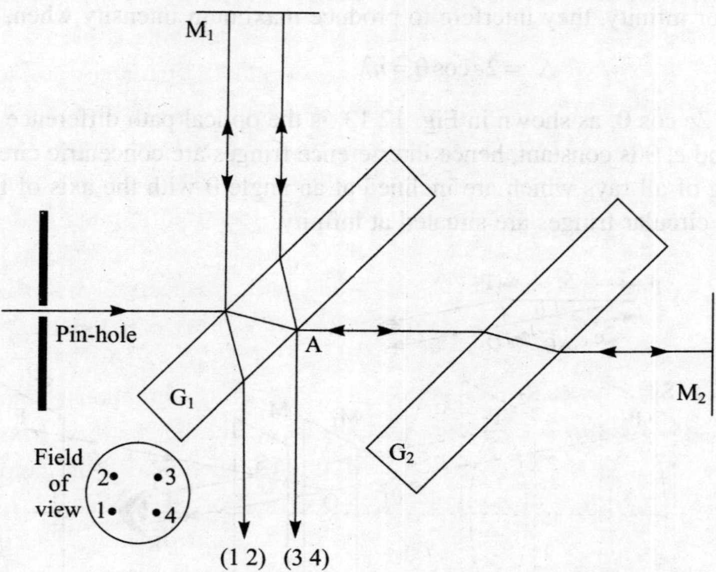

Fig. 12.12 (*b*)

12.6.2 Form of Fringes

In *Michelson interferometer* the fringes may have various forms-*straight lines, circles, parabolas, ellipses or hyperbolae*—depending on the separation, *e*, between M_1 and M_2' and the angle ϕ, between these surfaces. Two cases are of practical interest, namely, (1) $\phi = 0$ and (2) $e = 0$ but ϕ not equal to zero. In the former case concentric circular fringes are produced, while in the latter case the central fringe is straight and all other fringes are convex towards it. The mathematical theory of the shape of fringes is beyond the scope of this book, only a simplified explanation of the origin of circular and straight fringes may be given.

Circular Fringes

Concentric circular fringes are produced with monochromatic light when the interferometer is in normal adjustment *i.e.*, the image M_2' of M_2 in G_1 is exactly parallel to M_1. Their origin may be explained with reference to Fig. 12.13 in which we have replaced the real source by its images S' formed in G_1 while S_1 and S_2 are the images of S' in M_1 and M_2' respectively. Consequently, if the separation between M_1 and M_2' is *e*, the separation between S_1 and S_2 is $2e$. The latter images function as virtual coherent sources in the sense that the corresponding points *e.g.* P_1 and P_2, the images of the point *P* of S' in M_1 and M_2', are in equal phase at all instants. Since the light from an extended source is not parallel some beams reach the observer on normal reflection from M_1 while others on reflection at a small angle. For example, in Fig. 12.13 a light wave from *P* after reflection from M_1 and M_2' at an angle θ appears as if it originated from P_1 and P_2. Hence the two reflected waves *QR* and *ST* are coherent and they pursue parallel paths. Consequently, when brought to focus by a telescope or eye focussed for infinity, they interfere to produce maximum intensity when,

$$\Delta = 2e\cos\theta = n\lambda \qquad \text{...(12.6 } e)$$

where $\Delta = P_1D = 2e\cos\theta$, as shown in Fig. 12.13, is the optical path difference between them. Since for a given *n*, λ and *e*, θ is constant, hence interference fringes are concentric circles. The hollow cone of light consisting of all rays which are inclined at an angle θ with the axis of telescope is seen as a circular ring. The circular fringes are situated at infinity.

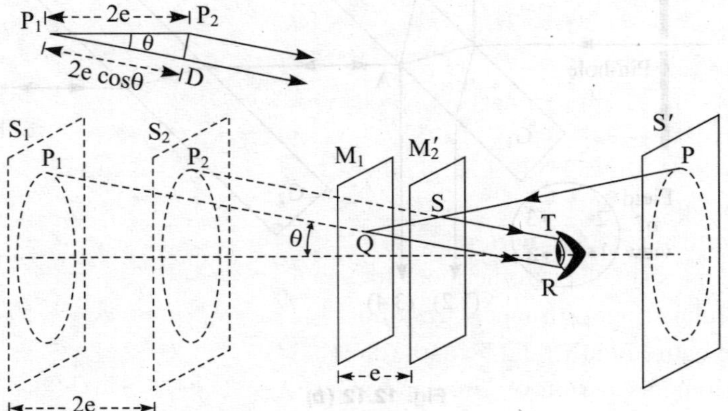

Fig. 12.13. Theory of circular fringes

A given fringe corresponds to a constant value of θ, hence *Lummer* has given them the name *fringes of constant inclination* and owing to this reason they superficially resemble with *Haidinger fringes*. The latter is the name given to concentric circular fringes, any one of them arising due to interference of infinite number of light waves of constant inclination to the axis. Such light waves are obtained by transmission due to multiple partial reflections within plane parallel thick film. But in Michelson interferometer, any ring arises due to interference of two coherent light waves of constant inclination. Hence there is superficial resemblance between the two systems. *Haidinger fringes* are produced in *Fabry Perot interferometer.*

As an interest, we may also derive the analytic equation of a circle from Eqn. (12.6 *e*). In Fig. 12.14, co-ordinates of *S* are (*x*, *y*) when referred to *O*, the foot of perpendicular from eye on M_2' as origin. We, therefore, substitute for the cosine of the angle θ, the expression

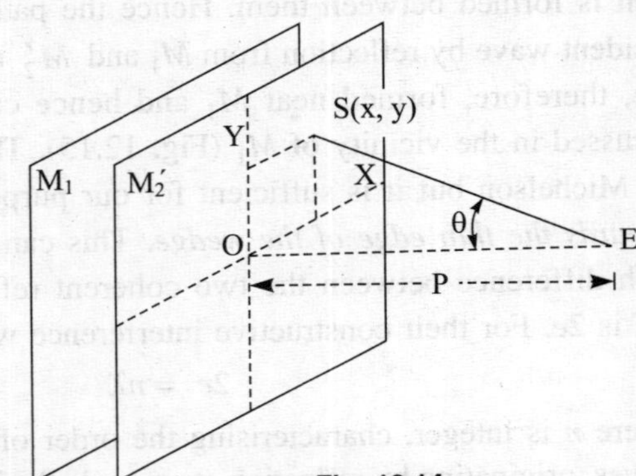

Fig. 12.14

$$\cos\theta = \frac{p}{\sqrt{(p^2 + x^2 + y^2)}}$$

in Eqn. (12.6 *e*) and then on squaring we easily get,

$$x^2 + y^2 = p^2\left[\frac{4e^2}{\Delta^2} - 1\right] \qquad \text{...(12.6 } f)$$

which is the equation of a family of circles with centre at *O*. Substituting $\Delta = n\lambda$ in the above equation, the square of the radius of the bright ring of *n*th order is given by

$$r_n^2 = p^2\left[\frac{4e^2}{n^2\lambda^2} - 1\right] \quad \text{[Bright ring]} \qquad \text{...(12.6 } g)$$

Similarly, for the dark ring of the *n*th order

$$r_n^2 = p^2\left[\frac{4e^2}{\left(n + \frac{1}{2}\right)^2 \lambda^2} - 1\right] \quad \text{[Dark ring]} \qquad \text{...(12.6 } h)$$

From Eqn. (12.6 *g*) it follows that if *e* is increased *i.e.* M_1 is moved away from M_2' the radius of bright ring of the order *n* increases. Thus as *e* increases each ring moves outwards and it is easy to see that the central rings move more rapidly, thereby the angular separation of the fringes in any part of the field becomes smaller. Furthermore for $e = 0$, the path difference is zero for all angles of incidence hence circular fringes disappear, resulting in uniform illumination of the field. However, if G_1 be unsilvered the entire field would be dark (theoretically).

In concluding our discussion of circular fringes it should be emphasised that the order n_o of fringe at the centre is always greater than the order *n* of any of the rings, for $2e > 2e \cos\theta$. Thus as the radii of circular fringes increase the order of interference decreases, while in *Newton's rings* which are superficially similar, the order of interference increases as the radii increase. The former fringes are *loci of constant inclination* whereas the latter are *loci of equal optical thickness.* The circular fringes in *Michelson* interferometer are situated at infinity while *Newton's rings* are localised in the film. Nevertheless, in both the systems, rings get closer as their radii increase.

Localised Fringes

When the mirror M_1 and the virtual mirror M_2' are not exactly parallel, a wedge shaped air film is formed between them. Hence the paths of two reflected waves, originating from the same incident wave by reflection from M_1 and M_2' are no longer parallel but intersect near M_1. The fringes are, therefore, formed near M_1 and hence called localized fringes and to see them eye must be focussed in the vicinity of M_1 (Fig. 12.15). The exact shape of these fringes have been investigated by Michelson but it is sufficient for our purpose to say that *they are curved with their convex side towards the thin edge of the wedge.* This can be easily explained with reference to Fig. 12.16. The path difference between the two coherent reflected waves from O and O', and traversing the path OE is $2e$. For their constructive interference we, therefore, write the condition

$$2e = n\lambda$$

where n is integer, characterising the order of the fringe. The path difference between the coherent waves, originating by reflection at an angle θ of the same wave from A and A', which are vertically over O and O' is $2e \cos \theta$ (of course, assuming ϕ to be negligibly small in comparison to θ). Since $2e \cos \theta < 2e$, hence O and A do not lie on the fringe of the same order. We may, however, located another pair of points B and B' such that the path difference between coherent reflected waves from these points is $2e' \cos \theta'$, which satisfies the relation,

$$2e' \cos\theta' = 2e = n\lambda$$

O and B, therefore, lie on the fringe of the same order. It is easy to see that since $\cos \theta' < 1$, hence $e' < e$. We, therefore, conclude that the fringes are convex towards the thin edge of the wedge.

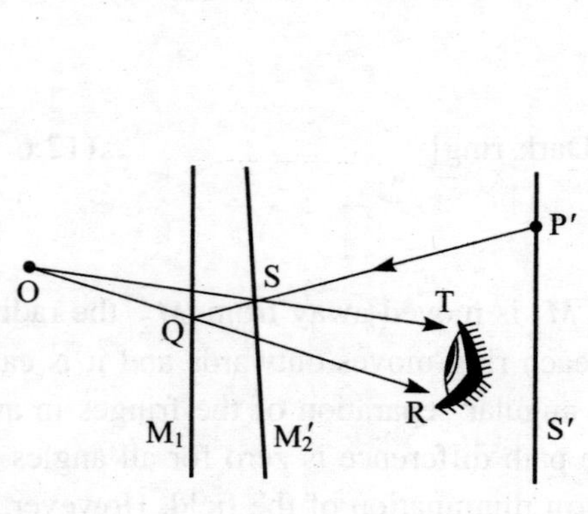

Fig. 12.15

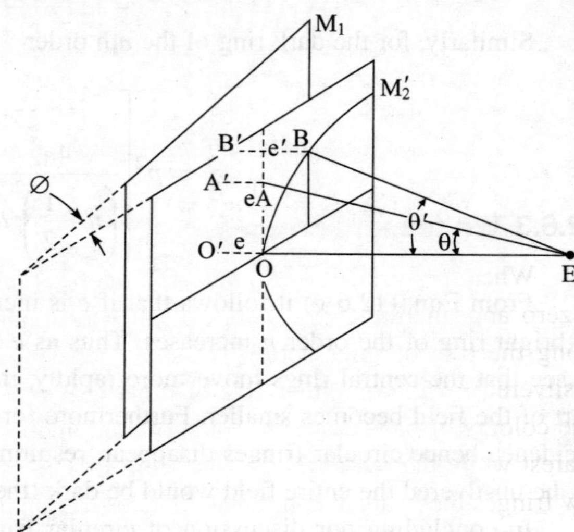

Fig. 12.16

Starting with the real mirror M_1 in the position depicted in Fig. 12.17 (*a*), the thin edge of the wedge is to the left so that observed fringes are convex towards the left. As the separation between M_1 and M_2' is gradually decreased, fringes move across the field away from the thin edge of the wedge and at the same time gradually become straighter. When M_1 intersects M_2', as depicted in Fig. 12.17 (*b*), the path difference is zero all along the line of their intersection and so the fringe at this point is perfectly straight. Furthermore, it is clear from Fig. 12.17 (*b*) that now we have two air wedges formed between M_1 and M_2', with their thin edge coinciding with the line of intersection of M_1 and M_2'. As a consequence, on either side of the central straight fringe, all other fringes are curved with their convex side towards the former fringe *i.e.,* curvature of fringes on one side is opposite to that of the fringes on the other side. For small values of Δ, the curvature of fringes in Fig. 12.17 (*b*) is hardly distinguishable within the very small field of view and so the fringes appear practically straight. When M_1 is moved in the same direction to the position depicted in Fig. 12.17 (*c*), fringes are again curved but with their convex side towards the right. Thus in moving M_1 about the position for Δ = 0, the curvature of fringes changes sign.

The localized fringes, discussed above, become invisible for large path difference of the order of several millimeters.

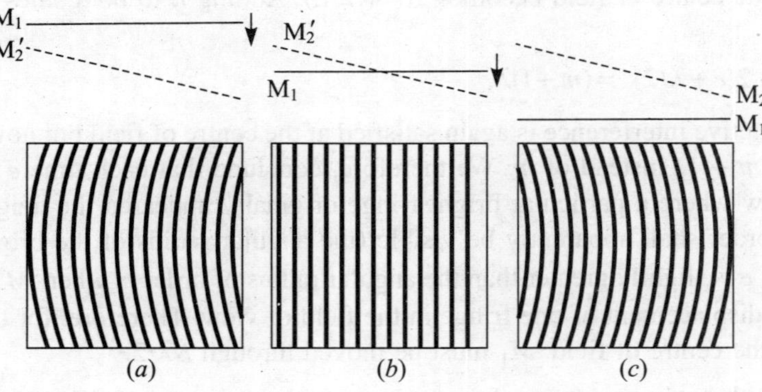

Fig. 12.17. Formation of localised fringes with inclined mirrors M_1 and M_2

12.6.3 White Light Fringes

When the surfaces of M_1 and M_2' intersect the path difference along the line of their intersection is zero and, therefore, the same for all wavelengths. As a consequence, with a white light source, along the line of intersection we shall get central achromatic bright fringe (dark fringe if G_1 is unsilvered), while on either side of it about 3 or 4 coloured practically straight fringes are observed. The colours of bands on either side of achromatic fringe are obviously symmetrical, the violet being nearest while the red farthest away from it. The colours of fringes rapidly become impure until after few fringes they vanish completely due to overlap, yielding essentially white illumination.

To produce white light fringes, the mirrors M_1 and M_2 are tilted slightly as for localized fringes and then we locate with monochromatic light, the position of M_1 which nearly corresponds to zero path difference, so that the localised fringes are practically straight. Now it is only necessary to replace the monochromatic source with a white light source and given a slow motion to the mirror M_1 through this region in the proper direction till white light fringes come into view.

12.6.4 Applications of Michelson Interferometer

(*a*) Measurement of Wavelength

An extraordinary degree of accuracy in the determination of wavelength of spectral lines was made possible with the invention of this interferometer. Concentric circular fringes are invariably employed in this measurement and hence the instrument must be adjusted beforehand in the manner described in *Sec.* 12.6.1.

Suppose the separation between M_2' and the real mirror, M_1 is such that a bright fringe of the order m is formed in the centre of the field of view. We, therefore, write for the central bright fringe, since $\cos\theta = 1$,

$$2e = m\lambda \qquad \text{...(12.6 } i\text{)}$$

By moving M_1 so that e increases by $\lambda/2$, the path difference between the coherent normally reflected waves from the centre of field becomes $2(e + \lambda/2)$. Adding λ to both sides of Eqn. (12.6 *i*) we observe that since,

$$2(e+\lambda/2) = (m+1)\lambda,$$

the condition of constructive interference is again satisfied at the centre of field but now the new bright fringe is of the order $(m + 1)$ instead of m. We therefore, conclude that each time e is increased by $\lambda/2$, in the field of view where a particular bright fringe originally appeared the neighbouring bright fringe of next higher order shall apparently be visible (the angular radius of the bright ring of order $(m - 1)$ when $M_1M_2' = e$ is slightly greater than the angular radius of order m when $M_1M_2' = (e + ½\lambda)$. There will be a lateral displacement of one fringe in the field of view. Therefore, for the displacement of N bright fringes in the centre of field, M_1 must be moved through $N\lambda/2$.

To determine λ, the mirror M_1 is, therefore, moved from one position corresponding to micrometer screw reading say x_1 to another position corresponding to micrometer screw reading x_2 and the number N of the bright monochromatic fringes which cross the centre of the field is counted. Following the reasoning given above it is easy to write the relation,

$$\frac{1}{2}N\lambda = x_2 - x_1$$

from which λ can be easily computed. During the motion of M_1 away from M_2', bright and dark fringes appear to come out in the centre of field while the radii of other rings gradually increases.

This method of wavelength measurement has several advantages over other methods:

1. The distance between the two settings of M_1 is easily and accurately determined. In fact, it is possible to record a motion of mirror M_1 of one ten thousandth (1/10000) of a millimetre.
2. It eliminates the precise measurement of the angle of diffraction in the measurement of wavelength by plane grating.

3. It permits the introduction of any desired path differences which are exact multiples of some unit.

(*b*) Ratio and difference between the Wavelengths of Sodium D Lines

The ratio and the difference between the wavelengths of close doublet lines, such as sodium *D* lines (λ_1 = 5896 A ; λ_2 = 5890 A) can be measured in the laboratory with the help of *Michelson Interferometer.* Concentric circular fringes are usually employed in this measurement. At the centre, the intensity is simultaneously a maximum for λ_1 and λ_2 ($\cos\theta = 1$) when

$$2d_o = n_1\lambda_1 \quad \text{...(12.6 } j\text{)}$$

and

$$2d_o = n_2\lambda_2 \quad \text{...(12.6 } k\text{)}$$

where n_1 and n_2 are the orders of bright rings at the centre due to λ_1 and λ_2 respectively and $2d_o$ is the path difference between the coherent reflected waves focussed at the centre of the field. Under this condition, maxima of λ_1 practically coincide with maxima of λ_2 and so the interference pattern is well defined and has *maximum visibility*. Due to difference in λ_1 and λ_2, as the separation M_1M_2' is increased, the maxima at first fall farther and farther from coinciding, ultimately the maxima of one practically coincides with the minima of other radiation and vice-versa, thereby producing *minimum visibility.* This happens when the extra path difference introduced at the centre due to motion of M_1 contains one half wavelength more of radiation λ_2 than that of the other, for then the phase of the wave of radiation λ_2 advances by π over that of the other. Continuing the motion of M_1 in the same direction again gradually brings the maxima of two patterns close together. Ultimately they again coincide at the centre when the extra optical path difference $2d$ introduced at the centre contains one wavelength more of one radiation than of the other. The visibility of fringes is again maximum and we write.

$$2(d_o + d) = (n_1 + N)\lambda_1 \quad \text{...(12.6 } l\text{)}$$

and

$$2(d_o + d) = (n_2 + N + 1)\lambda_2 \quad \text{...(12.6 } m\text{)}$$

where λ_2 is assumed to be less than λ_1 and d represents the displacement of M_1 between successive visibility maxima or minima and N represents the number of bright fringes of λ_1 emerging at the centre due to displacement d of the mirror M_1.

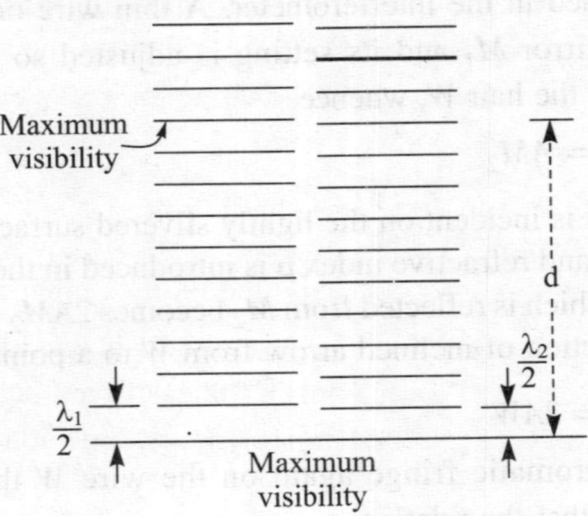

Fig. 12.18

Subtracting Eqn. (12.6 *j*) from Eqn. (12.6 *l*) and Eqn. (12.6 *k*) from Eqn. (12.6 *m*), we get

$$2d = N\lambda_1 \qquad \text{...(12.6 } n\text{)}$$

and

$$2d = (N+1)\lambda_2 \qquad \text{...(12.6 } o\text{)}$$

The ratio of wavelengths is accordingly,

$$\frac{\lambda_1}{\lambda_2} = \frac{N+1}{N} \qquad \text{...(12.6 } p\text{)}$$

It is extremely tedious to count N in this experiment due to the presence of two sets of interference fringes in the field. The better method is to compute N from the measured value of d by the relation

$$N = 2d/\lambda_1$$

if one knows the value of λ_1 by some other experiment.

To compute the difference between the two wavelengths we subtract unity from both sides of Eqn. (12.6 *p*), obtaining

$$\frac{\lambda_1 - \lambda_2}{\lambda_2} = \frac{1}{N} = \frac{\lambda_1}{2d}$$

or

$$\lambda_1 - \lambda_2 = \frac{\lambda_1\lambda_2}{2d} = \frac{\lambda_{av}^2}{2d} \qquad \text{...(12.6 } q\text{)}$$

where λ_{av} is the average wavelength or the wavelength as measured by some other experiment in which two pairs of lines are not resolved.

The visibility minima can be distinguished more accurately than the maxima. The value of d is, therefore, determined by the positions of M_1 for at least ten successive visibility minima and then computing d, the average interval by the usual method of differences.

(c) Determination of Refractive Index

To determine the refractive index of the refracting medium in the form of thin transparent sheet, white light fringes are first obtained in the interferometer. A thin wire or horse-hair, W, is attached vertically at the middle of the mirror M_1 and its setting is adjusted so that the achromatic bright fringe is observed coinciding with the hair W, whence,

$$AW = AM_2$$

where A is the point at which light is incident on the lightly silvered surface of G_1. When the sheet of a refracting material of thickness t and refractive index μ is introduced in the path AM_2 normal to it, the optical path covered by the beam which is reflected from M_2 becomes $2AM_2 + 2(\mu - 1)t$. The achromatic fringe, therefore, shifts in the direction of inclined arrow from W to a point W' so that

$$2AM_2 + 2(\mu - 1)t = 2AW'$$

Obviously, to bring the achromatic fringe again on the wire W the mirror M_1 is displaced backward through a distance d so that the relation

$$2AM_2 + 2(\mu - 1)t = 2AW + 2d$$

is satisfied. This easily simplifies to the form

$$d = (\mu - 1)t \qquad ...(12.6\ r)$$

from which μ can be easily computed.

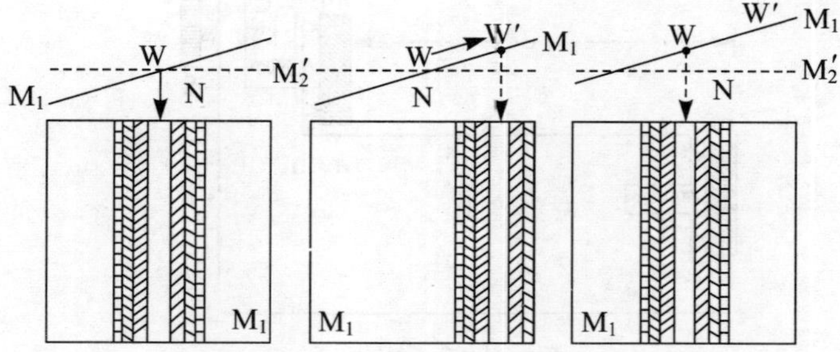

Fig. 12.19. Determination of μ of thin transparent plate

(*d*) Evaluation of Standard Metre in Terms of Light Waves

The standard metre accepted internationally is defined as the distance between two fine lines engraved on a irido-platinum bar, kept at the International Bureau of Weights and Measures at Sevres near Paris, when the bar is at 0°C. Lamont, in 1823, first suggested the desirability of evaluating the length of standard metre once for all in terms of the absolute standard *i.e.* the wavelength of light, so that in the case of accident to this bar, it should be possible to replace it easily. In 1892, *Michelson* and *Benoit* first undertook this task with the help of a modification of the *Michelson interferometer.* Amongst the hundreds of radiations examined by them, red line, λ_R = 6438 A, of cadmium vapour spectrum was selected as the standard light due to its absolute homogeneity in structure.

Michelson foresaw that even with most homogeneous light source the interference fringes in Michelson interferometer would disappear for a displacement of movable mirror from one *fiduciary mark* to another *mark* of the standard metre. Furthermore, in order to eliminate the necessity of counting the large number of fringe displacements due to such large path difference between the interfering beams, he constructed nine intermediate substandards called *etalons,* of the shape sketched in Fig. 12.20, the length of each being represented by the distance between the silvered planes of their respective mirrors m_1 and m_1'. The length of these substandards were 10 cm, 10×2^{-1} cm, 10×2^{-2} cm.........., the smallest one being of 10×2^{-8} cm or 0.0391 cm contains nearly 1200 *cadmium red fringes* in doubled length, which of course, can be easily counted without any fear of error. The determination of the number of light waves in standard metre, therefore, involves three distinct experiments:

(*a*) *The determination of the number of light waves in the smallest etalon.*

(*b*) *Comparison of the intermediate etalons.*

(*c*) *Comparison of 10 cm etalon with the standard metre.*

Every *etalon* consists of two optically plane mirrors m_1 and m_1', highly silvered on their front surfaces, mounted in a *L* shaped bronze holder *H* in such a way that one mirror is at a higher level than the other. The parallelism of their silvered surfaces could be effected with the help of a screw *S*, which acting on the spring *C*, moved the upper supporting pin of m_1' slightly backwards or forwards depending upon the direction of its rotation. *Michelson Interferometer* is slightly modified to a form sketched in Fig. 12.22(a), in which the compensating plate G_2 could be rotated about the vertical axis to introduce any desired small path difference between the intefering beams.

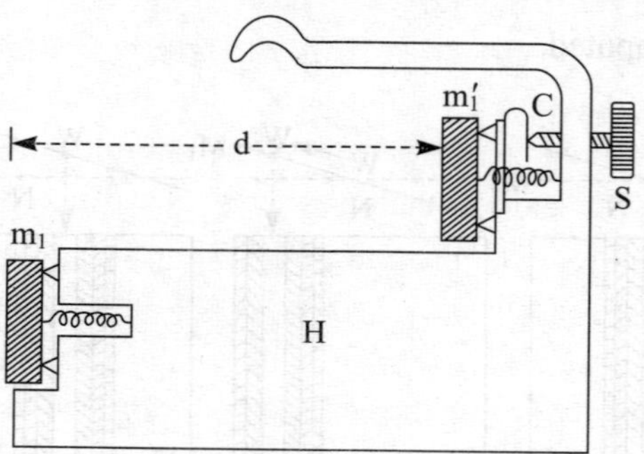

Fig. 12.20. Etalon

Determination of the number of light-waves in the shortest etalon.

The silvered surface of the lower mirror m_1 of the *etalon* is made nearly co-planer with the image of the reference mirror M_1, making, however, a small horizontal angle with it, so that the white light achromatic fringe is formed at the centre of m_1. The fixed mirror M_2 of the interferometer is made exactly parallel to the image of M_1 so that cadmium red light circular fringes are seen on M_2. The mirror M_1 is now steadily moved backward and bright circular fringes are counted as they successively enlarge away from the centre of M_2. We have to stop the motion of M_1 at a point where its image R_2 intersects the upper mirror m_1' of the etalon along the vertical line which indicated the intersection of m_1 and R_1 at the beginning of the motion. Therefore, towards the end of counting, white light source is periodically switched on to see the position of the achromatic fringe. The motion M_1 is finally stopped when the achromatic fringe appears at the same position on m_1' as it had before on m_1. The fractional part of cadmium red fringe is determined by rotating the compensating plate G_2 through a small measurable angle, say θ, in the direction which increases the size of the circular fringes on M_2, the angle of rotation ϕ for one complete fringe displacement having been previously determined. The required fraction to be added to the whole number of fringes is $1 - (\theta/\phi)$.

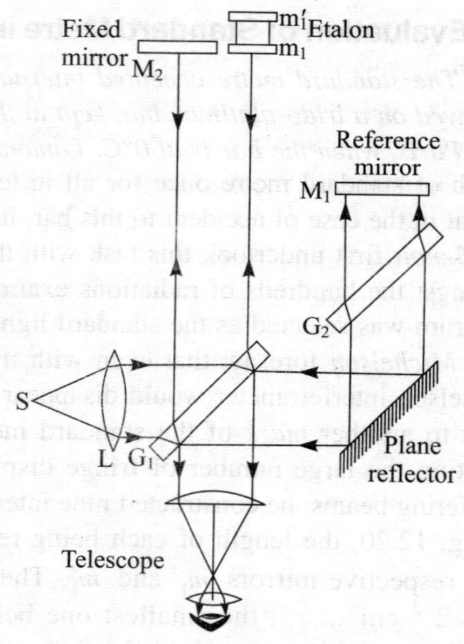

Fig. 12.21(a). Interferometer set up to evaluate the length of shortest etalon

Michelson and *Benoit* found the number of cadmium red waves in the doubled length of shortest etalon to be

1212.61, 1212.38, 1211.83, 1212.56 in 4 separate counts establishing the whole number of waves as 1212.

Comparison of Intermediate Etalons. The *shortest etalon* is now compared with the next largest one. Two *etalons* are mounted side by side with their mirrors mutually parallel and inclined at a small angle with the image R_1 of the *reference mirror* M_1. The position of M_1 is adjusted so that white light achromatic fringe is formed on the lower mirrors m_1 and m_2 of the two etalons as shown in Fig. 12.22(*a*). M_1 is then moved back to a position until its image occupies the plane R_2 so that the achromatic fringe appears on the upper mirror m_1' of the shorter etalon, exactly over its previous position in m_1 [Fig. 12.22 (*b*)]. The shortest etalon is then moved back, keeping the reference mirror stationary, until m_1 occupies the previous position of m_1' which is shown by the reappearance of the achromatic fringe on m_1 [as shown in Fig. 12.22 (*c*)], exactly at the same position as at the beginning of the operation. M_1 is then moved back to a position, its image R_3, so that the achromatic fringe again reappears on the upper mirror m_1' of the shorter etalon, [as shown in Fig. 12.22 (*d*)], exactly in the previous position. In general, the longer etalon is slightly greater than twice the shorter one. As a consequence, with M_1 in a position so that its image is at R_3, the achromatic fringe is formed, [as shown in Fig. 12.22 (*d*)] on m_2' somewhat to the left of the standard position. The difference between twice the length of the shortest etalon and the length of the next largest one being very small, localised red fringes of thin film may be used to evaluate this difference. The cadmium red light is used and fringes counted, as M_1 is moved so that its image shifts from R_3 to R_4, when on testing with white light, achromatic fringe appears on m_2' exactly in the same position as on m_2 at the beginning of the operation. [Compare Fig. 12.22 (*a*) and Fig. 12.22 (*e*)]. The doubled path of twice the shorter etalon should contain 2 (1212.38) = 2424.76 fringes or waves so that the number of next fringe in the measure of m_1' m_2' is 2425 and the final fraction in the second *etalon* is determined as in the previous case, by rotating the compensating plate G_2.

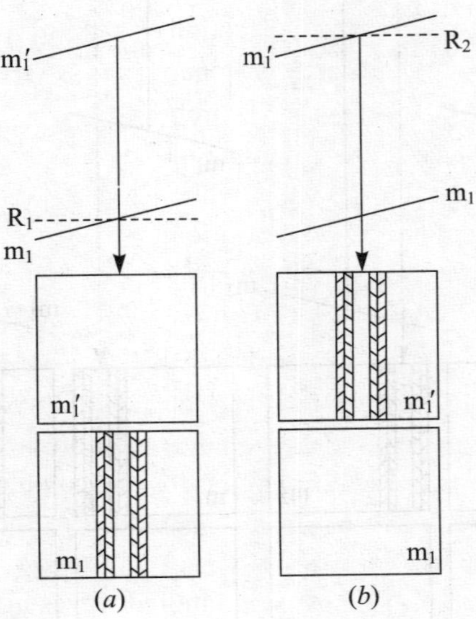

Fig. 12.21 (*b*)

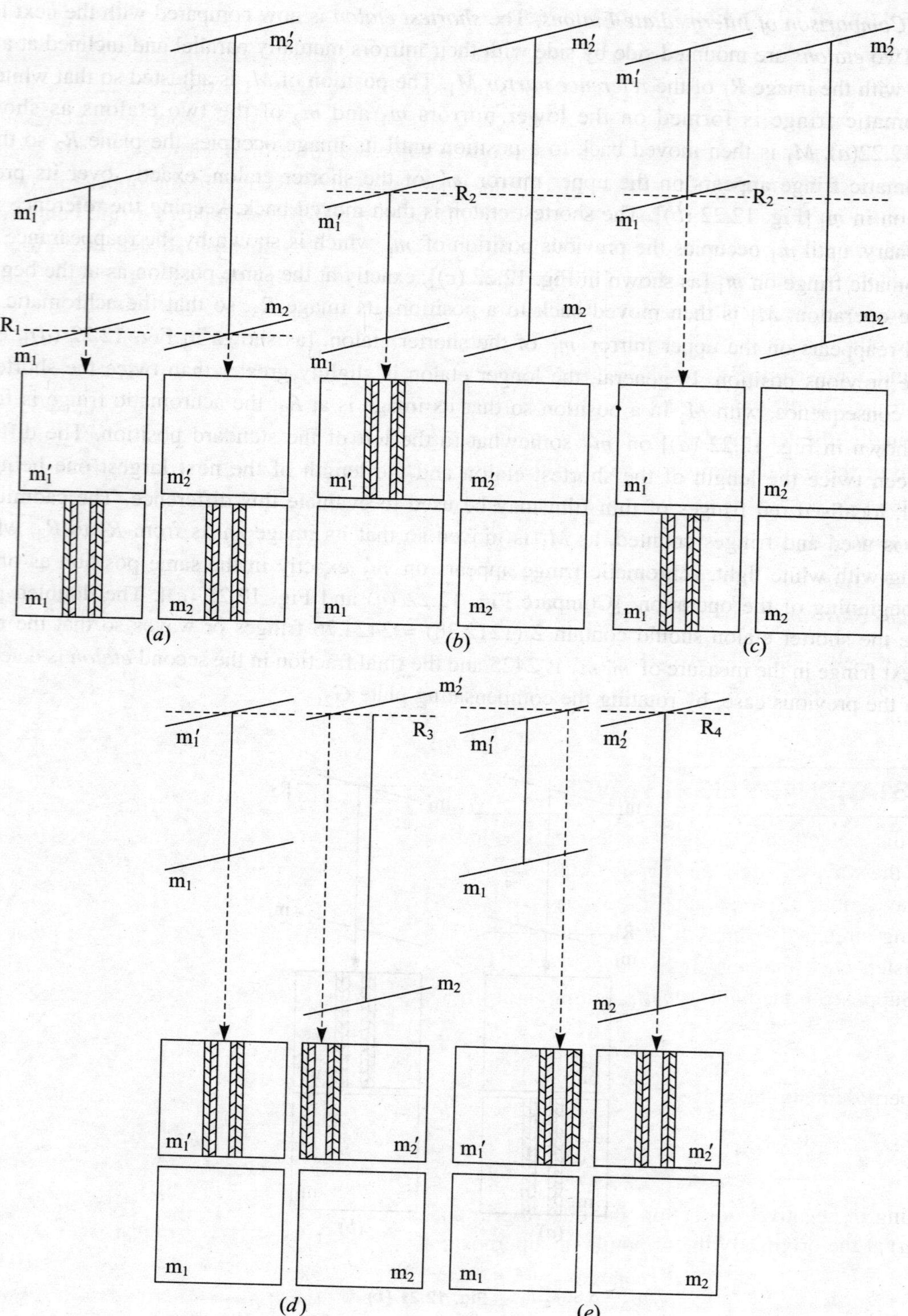

Fig. 12.22. Comparison of one etalon with another of twice its length

Comparison of 10 cm Etalon with the Meter

After determining the number of light waves in 10 cm *etalon,* the final step is its comparison with the working *standard* metre. The *etalon* is provided with a short horizontal arm on which a line is engraved exactly parallel to the silvered surfaces of the mirrors. This line is set in perfect coincidence with the terminal mark of the working metre as observed by the powerful microscope T_1. By adjusting the position of M_1, white light achromatic fringe is obtained at the centre of the upper mirror m' of the etalon. The etalon is then moved back until the achromatic fringe appears at the centre of its lower mirror m. The reference mirror M_1 is then moved back to a position so that the achromatic fringe reappears at the centre of m'. This operation of alternately moving the etalon and M_1 is repeated until after 9 displacements, the indicator line of etalon comes under the microscope T_2 which is fixed over the other terminal mark of the scale. The number of cadmium red fringes required to make the etalon index line coincide with the second fiduciary mark is finally determined. In this way the number of red fringes in the working standard metre is computed.

Finally, the length of this auxiliary standard metre is compared with the *prototype* metre with the help of the travelling microscope and the number of waves of cadmium red light in the metre is computed. Michelson obtained the following results:

Cadmium Red line,	1 metre = 1,553,163.5 λ	or	λ = 6438.4722 A.
Cadmium Green line,	1 metre = 1,966,249.7 λ	or	λ = 5085.8240 A.
Cadmium Blue line,	1 metre = 2,083,372.1 λ	or	λ = 4799.9107 A.

It is estimated that these results are correct to about one part in two millions. *Incidentally, Michelson could also determine the absolute values of the wavelengths of three lines of the cadmium spectrum.*

12.7 STATIONARY LIGHT WAVES

When the wave theory of light was established in the early years of the last century it was natural to regard the waves of light analogous to the waves of sound. Consequently it is but natural that if light is a wave motion, we should come across *stationary light waves* like *stationary waves* formed in vibrating stretched strings. It was not until 1890, however, that *Wiener* demonstrated conclusively the existence of stationary light waves.

Suppose on the light wave travelling in positive x direction and represented by the equation,

$$y_1 = a\sin 2\pi\left(\frac{t}{T}-\frac{x}{\lambda}\right) \quad \text{...(12.7 } a\text{)}$$

we superpose a similar wave

$$y_2 = -a\sin 2\pi\left(\frac{t}{T}+\frac{x}{\lambda}\right) \quad \text{...(12.7 } b\text{)}$$

travelling in negative x direction and differing in phase by π from the wave represented by Equation (12.7 *a*) at the origin. By the principle of superposition, the resultant wave is given by

$$y = a\sin 2\pi\left(\frac{t}{T}-\frac{x}{\lambda}\right) - a\sin 2\pi\left(\frac{t}{T}+\frac{x}{\lambda}\right)$$

which readily simplifies to form

$$y = 2a\sin\left(-\frac{2\pi x}{\lambda}\right)\cos 2\pi\frac{t}{T} \qquad ...(12.7\ c)$$

If this expression is plotted as a function of x for different values of t, it is found that it oscillates from the form $y = 2a \sin (2\pi x/\lambda)$ through a straight line to form $y = -2a \sin (2\pi x/\lambda)$ and back again. The displacement of ether particles is zero for all values of t at the points defined by the solution of the equation

$$\sin(2\pi x/\lambda) = 0$$

i.e. at $$x = n\lambda/2$$

where n = 0, 1, 2, 3, 4, etc. Such points are called *nodes*. At a point midway between any two consecutive nodes, displacement is always greater as compared with its magnitude at any other point between the same two nodes, of course, excluding the instant when displacement would be zero at all points of the medium. Analytically, these points are located by the condition that the amplitude $2a \sin (2\pi x/\lambda)$ attains its maximum value $\pm\ 2a$ only at these points. In other words, these points, called *antinodes,* are given by the solution of the equation

$$\sin(2\pi x/\lambda) = \pm 1$$

i.e. at $$x = (2n+1)\lambda/4$$

where n = 0, 1, 2, 3 etc. Eqn. (12.7 *c*) is said to represent *stationary wave* because the wave does not progress, only the loops keep on opening and shrinking.

We, therefore, conclude that when monochromatic light is reflected from a perfectly reflecting surface, the incident and the reflected waves combine to form a stationary wave in front of the reflector. If a phase change of π occurs on reflection then reflecting surface will be a *nodal plane* and the other nodal planes will be parallel planes at distance equal to $\lambda/2$, λ, $3\lambda/2$, 2λ, ...from the surface. In each of these planes ether will be permanently stationary while at planes midway between any two consecutive nodal planes, ether will alternately suffer maximum displacement.

These conclusions were remarkably confirmed in 1890 by *Wiener*. His experimental arrangement is sketched in Fig. 12.23. A specially prepared photographic emulsion, only about 1/20th of a wavelength thick, is placed inclined at a small angle with the perfectly reflecting surface so that it exactly crosses the *nodes* and loops (*antinodes*) successively, in a stationary wave pattern formed by the interference of normally incident and reflected monochromatic light waves. At *nodal points,* since either is permanently at rest, hence there *silverhalide* is not affected. But at *antinodes* since ether alternately suffers maximum displacement it causes an oscillatory motion of emulsion particles, which is thereby reduced to *silver*. As expected, when the plate is developed and examined under a high power microscope, the film is found to be crossed by a series of equidistant dark bands situated along the lines of the intersection of the plate with the *antinodal planes.* No blackening results along the lines of intersection of the film with the nodal planes.

In Wiener's experiment, along the line of contact of the film with the reflecting surface no blackening resulted, while the first black band which corresponds to a loop in the stationary wave was found at a distance of ¼ λ from the mirror. Hence the mirror was a nodal plane. This simply means that the light waves underwent a phase change of π due to reflection so that they continuously annual each other at the mirror surface.

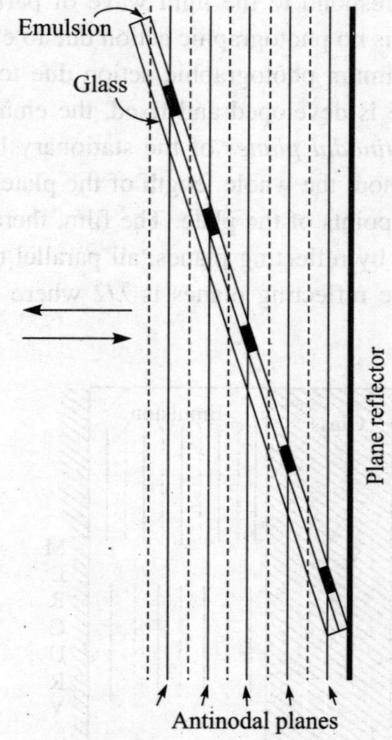

Fig. 12.23. Wiener's experiment

The *electromagnetic theory* indicates that the electric vectors of the direct and reflected waves interfere destructively, thereby forming a node at the reflecting mirror while magnetic vectors produce antinode at the mirror, *Wiener,* therefore, concluded that the photographic action is brought about by the electric vector and not by the magnetic vector present in the electromagnetic light waves. *In general, the electric vector is effective in every optical phenomenon. Wiener's* experiment constituted another experimental confirmation of *Maxwell's electromagnetic theory of light.*

12.8 THE LIPPMAN PROCESS OF COLOUR PHOTOGRAPHY

Soon after the publication of *Wiener's* experiment *Lippman,* in 1891, invented a process of photography in natural colours, which is based on the phenomenon of *stationary light waves.* A specially prepared photographic plate, having an extremely fine grain emulsion, finer than that of the commercial plates, is placed in a camera. The glass side is towards the objective while the emulsion side is backed with a layer of mercury to form a reflecting surface. This of course requires a special form of the plate holder.

Now, suppose that light of various wavelengths, corresponding to various colours present on the object to be photographed, is focussed on the plate as shown in Fig. 12.24. Light, on entering the film from glass, is reflected by mercury layer back through the film to the glass. Due to the interference of direct and reflected waves, stationary wave patterns are set up in the film. The separation of *nodes* and *antionodes* correspond to the light wave of particular wavelength incident on the plate at that point. At *nodes* there is no photographic action due to ether remaining permanently at rest. But at *antinodes* there is a maximum photographic action due to ether suffering alternately maximum displacement. When the plate is developed and fixed, the emulsion contains a number of layers of silver, coinciding with the *antinodal planes* of the stationary light waves. Of course, the antinodal planes are not uniform throughout the whole length of the plate because of the variation in the wavelength of light striking various points of the plate. The film, therefore, contains a number of transparent layers of *gelatine*, separated by reflecting planes, all parallel to the mercury surface. The separation between any two consecutive reflecting planes is $\lambda/2$ where λ is the wavelength of the light incident on the plate at that point.

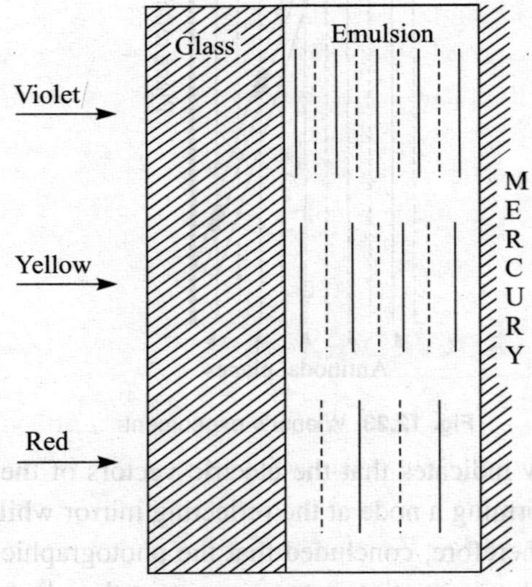

Fig. 12.24. Principle of Lippman colour photography

Now suppose the plate so photographed is illuminated normally by white light and viewed by reflection. The thickness of silver in each reflecting plane is so small that they are practically transparent and therefore they reflect only a small fraction of the incident light. As the white light penetrates the film, it is partially reflected at each successive silver layer and suffers similar phase changes due to reflection. But of all the reflected waves of various wavelengths from any one set of reflecting planes only those reinforce one another's effect, giving rise to intense reflection for this colour, for which half the wavelength equals the distance between the consecutive silver layers. This is due to the fact that the path difference between the successive reflected waves of this colour, being equal to twice the separation between two consecutive nodal planes, is λ, which is obviously the condition of constructive interference. For other colours, the path difference between the successive reflected beams, from reflecting planes of separation $\lambda/2$, is not exactly λ or any other integral multiple of λ.

Therefore, they do not satisfy the condition of constructive interference. As a consequence, their effect will be negligibly small compared with that colour for which the condition of constructive interference is satisfied.

The resultant intensity due to superposition of number of *noncoherent waves* is found simply by adding the intensities of individual waves, while in case of *coherent waves,* all in phase, the resultant intensity is obtained by adding their individual amplitudes and then squaring the resultant amplitude to obtain the resultant intensity. A number of waves are said to be perfectly *coherent* if their phases bear a definite phase relationship to each other, but if the phase relations are quite at random, then the waves are said to be *noncoherent.* The superposition of *n noncoherent waves,* assumed to be of equal amplitude a for the sake of clarity of argument, would produce the resultant intensity equal to na^2, while if the waves were coherent, the resultant intensity would be $(na)^2$. In *Lippman process,* the number, n, of the reflected beams is very large. As a consequence, the intensity of coherent reflection would predominate over that of the noncoherent one.

We, therefore, conclude that when the exposed photographic plate is illuminated with white light and seen by reflection, every set of reflecting planes will reflect back strongly only that colour by which it was originally formed during the exposure.

Lippman colour photographs are extremely brilliant but the process is extremely difficult and no colour prints can be made from the plate, which itself serves as positive when viewed by reflected light. It should be emphasised that angle of incidence of light beam for viewing the plate by reflection must be the same as that during exposure, otherwise true reproduction of colours would not occur.

Lippman's process is not used in modern colour photography which is based on three colour process.

SOLVED EXAMPLES

Example 1. *A parallel beam of white light is allowed to fall axially on the slit of the collimator of a spectrometer and a thin air film of uniform thickness enclosed between two glass plates is placed in front of the slit with the surface of air film perpendicular to the path of light. On viewing through the telescope 250 fringes are observed in the spectral region between wavelengths 4000 A.U. and 6500 A.U. as measured in air. Calculate the thickness of the air film.*

The air film is seen by light transmitted normally through it. Accordingly, for constructive interference of light of wavelength λ, we have the condition

$$2e = n\lambda$$

where e stands for the thickness of the air film.

Suppose that the bright fringe corresponding to 6500 A.U. is of the order n, then the bright fringe corresponding to 4000 A.U. must be of the order $(n + 250)$, since 250 fringes are observed in the spectral region between these two wavelengths. Accordingly, we have two conditions

$$2e = n \times 6500 \times 10^{-8}$$

and $$2e = (n+250)\times 4000\times 10^{-8}$$

Solving the two equations we have n = 400 and hence

$$e = \mathbf{0.013\ cm.}$$

Example 2. *White light is incident on a soap film at an angle of incidence sin^{-1} 4/5 and the reflected light on examination by a spectroscope shows dark bands. Two consecutive dark bands correspond to wavelength 6.1 × 10^{-5} and 6.0 × 10^{-5} cm. If the refractive index of soap-film be 4/3, calculate its thickness.*

Suppose the thickness of soap film is e. Neglecting the small variation in the refractive index of the soap film for the spectral range under consideration, we write the conditions for consecutive dark bands corresponding to λ_1 and λ_2,

$$2\mu e\cos\theta = n\,\lambda_1 = (n+1)\lambda_2$$

Solving for n we have, $$n = \lambda_2/(\lambda_1 - \lambda_2)$$

Hence $$2\mu e\cos\theta = \lambda_1\lambda_2/(\lambda_1 - \lambda_2)$$

or $$2e\sqrt{(\mu^2 - \sin^2 i)} = \lambda_1\lambda_2/(\lambda_1 - \lambda_2)$$

In the given problem $\mu = 4/3$, $\sin i = 4/5$, $\lambda_1 = 6.1\times 10^{-5}$ cm and $\lambda_2 = 6.0\times 10^{-5}$cm. Substituting these values in the above equation and solving for e, we get

$$e = \mathbf{0.00172\ cm.}$$

Example 3. *Two rectangular pieces of plane glass are laid one upon the other and a thin wire is placed between them at one edge so that very thin wedge of air is formed. The plates are illuminated with sodium light (λ = 5893 A.U.) at normal incidence. Bright and dark bands are formed, there being ten of each per cm length of the wedge measured normal to the edges in contact. Find the angle of wedge.*

The distance between any two consecutive dark fringes is expressed by Eqn. (12.3 i), which is

$$\bar{X} = \frac{\lambda}{2\tan\alpha\sqrt{(\mu^2 - \sin^2 i)}}$$

where symbols have their usual meaning.

In the given problem $\bar{X}$ = 1/10 cm, μ = 1, i = 0, $\lambda = 5893\times 10^{-8}$ cm.

Hence $$\tan\alpha = \frac{5893\times 10^{-8}}{2\times 1/10} = 2.95\times 10^{-4}$$

$$\alpha = \mathbf{2.95\times 10^{-4}\ radian}$$

Example 4. *White light is incident on two parallel glass plates separated by an air film of 0.001 cm thickness and the reflected light is examined by the spectroscope. Find the number of dark bands seen in the spectrum between the wavelengths 4 × 10⁻⁵ cm and 7 × 10⁻⁵ cm when light is incident at an angle of 30° to the normal to surfaces.*

According to theory of colours of thin films exhibited due to interference of reflected light waves, the dark bands are given by the condition

$$2\mu e \cos\theta = n\lambda$$

In the given problem, $\mu = 1$, $e = 0.001$ cm, $\theta = 30°$ and the range of visible spectrum is from 4×10^{-5} to 7×10^{-5} cm. Substituting the values for μ, e, θ in the above equation, we have

$$n_1 \times 4 \times 10^{-5} = 2 \times 0.001 \times \left(\frac{1}{2}\sqrt{3}\right) \quad \therefore n_1 = 43$$

Similarly, $$n_2 \times 7 \times 10^{-5} = 2 \times 0.001 \times \left(\frac{1}{2}\sqrt{3}\right) \quad \therefore n_2 = 24$$

Hence the number dark bands seen within the given spectral range = $(n_1 - n_2)$ = **19.**

Example 5. *If the angle of wedge is 0.25 deg. of an arc and the wavelength of sodium D lines are 5890 × 10⁻⁸ and 5896 × 10⁻⁸ cm find the distance from the apex of the wedge at which the maximum due to each wavelength first coincide.*

Suppose e represents the thickness of the wedge at a point where the maximum due to $\lambda_1 = 5896 \times 10^{-8}$ cm and $\lambda_2 = 5890 \times 10^{-8}$ cm coincide when viewed by the reflected light. The condition for this is expressed by the relation

$$2e = \left(n + \frac{1}{2}\right)\lambda_1 = \left(n + \frac{3}{2}\right)\lambda_2$$

Hence $$n = \frac{(3\lambda_2 - \lambda_1)}{2(\lambda_1 - \lambda_2)}$$

Substituting the value of n in the first of the above two equations, we have

$$2e = \lambda_1\lambda_2 / (\lambda_1 - \lambda_2)$$

Suppose the distance from the apex of the wedge at which the maximum due to λ_1 and λ_2 coincide is x cm; hence

$$e = x \tan\alpha = x\alpha$$

where α is the angle of the wedge, expressed in radians [180° = π radians].

$$\therefore 2x \frac{\pi \times 0.25}{180} = \frac{5896 \times 10^{-8} \times 5890 \times 10^{-8}}{6 \times 10^{-8}}$$

$$x = \mathbf{6.63\ cm.}$$

Example 6. *Newton's rings are formed by light reflected normally from a plano-convex lens and a plane glass plate with a liquid between them. The diameter of the nth ring is 2.18 mm and that of (n + 10)th ring is 4.51 mm. Calculate the refractive index of the liquid, given that the radius of curvature of the lens is 90 cm and the wavelength of light used is 5893 ×10⁻⁸ cm.*

In the given problem $\lambda = 5893 \times 10^{-8}$ cm, $R = 90$ cm,

$$p = (n+10) - n = 10 \text{ and } (D_{n+10}^2 - D_n^2)_{\text{Liquid}} = (4.51)^2 - (2.18)^2$$

$$= 12.444 \text{ sq mm} = 0.12444 \text{ sq cm}$$

Now, using Eqn. (12.5 *q*) the refractive index of the liquid is

$$\mu = \frac{4p\lambda R}{(D_{n+p}^2 - D_n^2)_{\text{Liquid}}} = \frac{4 \times 10 \times 5893 \times 10^{-8} \times 90}{0.12444} = \mathbf{1.7.}$$

Example 7. *A Newton's ring arrangement is used with a source emitting two wavelengths $\lambda_1 = 6 \times 10^{-5}$ cm and $\lambda_2 = 4.5 \times 10^{-5}$ cm, and it is found that the nth dark ring due to λ_1 coincides with (n + 1)th dark ring for λ_2. If the radius of curvature of the curved surface is 90 cm find the diameter of nth dark ring for λ_1.*

Let the diameter of *n*th dark ring when light of wavelength λ_1 is employed be *D*. According to the condition given in the problem *viz*., *n*th dark ring due to λ_1 coincides with $(n + 1)$th dark ring for λ_2, *D* is also the diameter of $(n + 1)$th dark ring due to λ_2. Accordingly, we have the relation

$$D^2 = 4n\lambda_1 R = 4(n+1)\lambda_2 R$$

On substituting the values of λ_1 and λ_2 in the above equation, we have

$$4n \times 6 \times 10^{-5} = 4(n+1) \times 4.5 \times 10^{-5}$$

$$\therefore \qquad n = 3$$

Hence, $$D = \sqrt{4n\lambda_1 R} = \sqrt{4 \times 3 \times 6 \times 10^{-5} \times 90} = \mathbf{0.2545\ cm.}$$

Example 8. *The radius of curvature of the convex side of a plano-convex lens is 2 metres. The lens is placed convex side down on the concave surface of a plano-concave lens of radius of curvature 400 cm. The combination of lenses so formed is illuminated with monochromatic light of wavelength λ = 6250 A. Find the diameter of the third bright ring in Newton's rings seen by reflected light and in the pattern seen by the transmitted light.*

Referring to *Sec*. 12.5, it can be easily shown that the thickness of the air film, formed between the plano-convex lens and plano-concave lens, at any point is expressed by

$$e = e_2 - e_1 = (r^2/2R_2) - (r^2/2R_1)$$

where R_2 is the radius of curvature of plano-convex lens and R_1 that of plano-concave lens and *r* is the radius of the section of lens at that point by a plane parallel to plane side.

When the interference pattern is seen nearly is the normal direction by the reflected light, the condition for maxima is given by Eqn. (12.5 *d*), which is

$$2\mu e = (2n+1)\lambda/2$$

Substituting the value of *e* in the above relation, we have after some easy simplication,

$$D = \sqrt{\frac{2(2n+1)\lambda_1 R_1 R_2}{\mu(R_1 - R_2)}};\ n = 0, 1, 2, 3 \text{ etc.}$$

For the third bright ring $n = 2$ and in our problem $\lambda = 6250 \times 10^{-8}$ cm, $R_1 = 400$ cm, $R_2 = 200$ cm, and $\mu = 1$

Hence $$D = \sqrt{\frac{2 \times 5 \times 6250 \times 10^{-8} \times 400 \times 200}{200}} = \mathbf{0.5\ cm.}$$

It is easy to see that the diameters of the bright rings seen by the trasmitted light are given by

$$D = \sqrt{\frac{4n\lambda R_1 R_2}{\mu(R_1 - R_2)}};\ n = 0, 1, 2, 3 \text{ etc.}$$

For the third bright ring $n = 2$, hence

$$D = \mathbf{0.4472\ cm.}$$

Example 9. *Newton's rings by reflection are formed between two biconvex lenses having equal radii of curvature being 100 cm each. Calculate the distance between the 5th and the 15th dark rings using monochromatic light of wavelength 5400 A.U.*

It is easy to see that the thickness of the air film at a point in the given experimental arrangement is given by

$$e = e_2 + e_1 = (r^2/2R_2) + (r^2/2R_1)$$

where symbols have their usual meanings.

Radii of the dark rings seen by the reflected light are given with the help of Eqn. (12.5 *c*) and the above relation. We get

$$r = \sqrt{n\lambda R_1 R_2/\mu(R_1 + R_2)}$$

where $n = 0, 1, 2, 3$, etc. It should be emphasised that $n = 0$ corresponds to the centre of the central black spot and $n = 1, 2, 3$ etc. respectively correspond to the first, second, third,.......... dark rings.

In our problem, $n = 5$ (for 5th dark ring), $n = 15$ (for 15th dark ring), $\lambda = 5400 \times 10^{-8}$ cm, $R_1 = R_2 = 100$ cm, $\mu = 1$.

Hence the distance between the 5th and 15th dark ring is

$$r_{15} - r_5 = \frac{(\sqrt{15} - \sqrt{5}) \times \sqrt{5400 \times 10^{-8} \times 100 \times 100}}{\sqrt{200}} = \mathbf{0.553\ cm.}$$

Example 10. *Calculate in nearest whole number the fringes that must be counted for green cadmium light for shortest etalon used with Michelson interferometer. The length of shortest etalon is 0.390 mm and the wavelength of green cadmium light is 5085.8 A.*

Suppose N fringes cross the field of view in the evaluation of shortest etalon in terms of cadmium green light. The path difference introduced between the two interfering beams of light is $2d = 2 \times 0.039$ cm. Hence we have the relation [See *Sec.* 12.6.4]

$$2 \times 0.039 = N \times 5085.8 \times 10^{-8}$$

$$N = \mathbf{1533} \text{ fringes.}$$

Example 11. *In an experiment with Michelson interferometer, the small readings for maximum indistinctness were found to be 0.6939 mm and 0.9884 mm. If the mean wavelength for the two components of D lines be 5893 A.U., deduce the difference between the two wavelengths.*

The difference between the two wavelengths of D lines is given by Eqn. (12.6 q) which is

$$\lambda_1 - \lambda_2 = \lambda_{av}^2 / 2(d_1 - d_2)$$

In the given problem $\lambda_{av} = 5893 \times 10^{-8}$ cm, $d_1 = 0.09884$ cm and $d_2 = 0.06939$ cm

Hence $$\lambda_1 - \lambda_2 = 6 \times 10^{-8} \text{ cm} = \mathbf{6\ A.U.}$$

Example 12. *Two wavelengths $\lambda_1 = 5890$ A.U. and $\lambda_2 = 5896$ A.U. are emitted by a source of light. Interference fringes are observed with a certain arrangement when the paths of the interfering beams and exactly equal. How much will the path difference have to be increased so that a bright fringe for λ_2 coincides with a dark fringe of λ_1?*

When the paths of the interfering beams are exactly equal, the fringe systems due to λ_1 and λ_2 are coincident *i.e.* the bright or dark fringes due to λ_1 fall over the bright or dark fringes corresponding to λ_2. If Δ be the path difference introduced when a bright fringe for λ_2 coincides with a dark fringe for λ_1, then we must have the relation

$$\Delta = n\lambda_2 = \left(n + \frac{1}{2}\right)\lambda_1$$

or $$\Delta = \frac{\lambda_1\lambda_2}{2(\lambda_2 - \lambda_1)} = \frac{5890\times10^{-8} \times 5896\times10^{-8}}{2\times6\times10^{-8}}$$

$$\Delta = \mathbf{0.0289\ cm.}$$

Example 13. *Circular fringes are observed in a Michelson interferometer illuminated with light of wavelength 5896 A. If the path difference between the mirrors M_1 and M_2 is 0.3 cm the central fringe is bright. Calculate the angular diameter of the 7th bright fringe.*

Suppose the central bright fringe ($\theta = 0$) is of nth order. Hence we have the relation

$$2e = n\lambda$$

We have seen in *Sec.* 12.6.2 that the orders of fringes decrease as we consider fringes outwards from the central fringe. The second bright fringe is accordingly of (n – 1)th order, ..., the seventh bright fringe is of $(n - 6)$th order. Hence, if θ be the angular radius of 7th bright fringe, we have the relation

$$2e\cos\theta = (n-6)\lambda$$

Hence $$2e(1-\cos\theta) = 6\lambda$$

$$2\times 0.3\left[1-\left(1-\frac{1}{2}\right)\theta^2\right] = 6\times 5896\times 10^{-8}$$

$$\theta^2 = 2\times 5896\times 10^{-7}$$

$$\theta = 0.344 \text{ radian} = 19.7°$$

Hence the angular diameter of 7th bright fringe = **39.4°.**

PROBLEMS

1. Describe and explain the phenomenon of interference in thin films. Give the necessary theory. Why do colours appear in a thin film in white light?
2. Discuss the formation of colours in thin films and show that with monochromatic light, the interference patterns of reflected and transmitted light are complementary.
3. (a) Explain why different colours are exhibited by thin films in white light. When seen by reflected light, why an excessively thin film appears to be perfectly black? Explain this perfect blackness with the help of Stoke's treatments.

   (b) With suitable diagrams explain why broad source of light is needed to observe the above phenomenon.

   (c) Why a thick film seen by reflected light shows no colours but appears white?
4. Explain the formation of Newton's rings. Show how you would use them to find (a) refractive index of a liquid and (b) radius of curvature of a lens.
5. (a) Describe and explain the formation of Newton's rings in reflected light. Prove that in reflected light (i) diameters of dark rings are proportional to the square roots of natural numbers and (ii) diameters of bright rings are proportional to the square roots of odd numbers.

   (b) Account for the perfect blackness of the central spot in Newton's rings.

   (c) What will happen if (i) white light is used and (ii) the lens is lifted slowly off the flat?
6. Describe Newton's rings method for measuring the wavelength of light and give the necessary theory. What will happen if a little water is introduced between the lens and the plate?
7. Explain with necessary theory how you can determine the refractive index of a liquid by means of Newton's rings. Derive the formula used. Why are the rings circular?
8. Describe the construction of a Michelson's interferometer and explain its working.
9. Outline the theory of Michelson's interferometer and discuss the nature of the interference pattern produced. Under what conditions will it give circular fringes? Show how it may be used to measure the wavelength of light.
10. (a) How can Michelson's interferometer be used for measuring the wavelength of light?

    (b) How will you use Michelson's inteferometer to determine the difference between two wavelengths very close together?

(c) How will you use Michelson's interferometer to determine the thickness of a thin transparent film or plate?

(d) Mention the various applications of the Michelson's interferometer.

11. Write short notes on:

(a) Colour photography

(b) Coherence

(c) Newton's rings

12. Light of wavelength 5880 A.U. is incident on a thin glass plate ($\mu = 1.5$) such that the angle of refraction is 60°. Calculate the smallest thickness of the plate which will make it dark by reflection.
13. A parallel beam of sodium light strikes a film of oil floating on water. When viewed at an angle of 30° from the normal, the eighth dark band is seen. Determine the thickness of the film.
14. White light is reflected from an oil film of thickness 0.01 mm and $\mu = 1.4$ at an angle of 45° to the vertical. If the reflected light falls on the slit of a spectrometer, calculate the number of dark bands seen between 4000 A.U. and 5000 A.U.
15. A convex lens is placed on a slab of plane glass and illumination by a monochromatic light the diameter of the 10th dark ring is found to be 0.433 cm. Find wavelength of light if the radius of curvature of the lower face of the lens is 70 cm.
16. If the diameter of a bright ring is 0.65 cm and that of the tenth ring beyond is 0.95 cm, calculate the radius of curvature of the face of convex lens in contact with the glass plate.
17. If the diameter of the nth dark ring changes from 0.3 cm to 0.25 cm when air is replaced by a transparent liquid, find the refractive index of the liquid.
18. If in a Newton's ring arrangement, the air is replaced by a liquid of $\mu = 1.3$, in what proportion would the diameters of the ring change?
19. Newton's rings are formed in reflected light of wavelength 5895 A.U. with a liquid between the plane and curved surfaces. If the diameter of 5th bright ring be 3 mm and the radius of the curved surface be 100 cm, calculate the refractive index of the liquid.
20. When the movable mirror of Michelson's interferometer is shifted through 0.0589 mm, a shift of 200 fringes is observed. What is the wavelength of light used?
21. Using a monochromatic light, 100 circular fringes pass through the centre when the movable mirror of the Michelson interferometer is shifted through 0.0295 mm. Calculate the wavelength of light used.
22. Calculate the distance between successive positions of the movable mirror of the Michelson's interferometer giving best fringes in the case of a sodium source having lines 5896 and 5890 A.U.?
23. In an experiment for determining the refractive index of air with Michelson's interferometer, a shift of 150 fringes is observed when all the air was removed from the tube. If the length of the tube is 20 cm, find the refractive index of air. ($\lambda = 4000$ A.U.)
24. A transparent film of glass ($\mu = 1.50$) is introduced normally in the path of one of the beams of a Michelson's interferometer which is illuminated with light of wavelength 4800 A.U. This causes 500 dark fringes of sweep across the field. Determine the thickness of the film.
25. A Michelson interferometer has mirrors which are 4 cm × 4 cm. With sodium light ($\lambda = 5890$ A.U.) 20 vertical fringes are observed. Calculate the angle between M_1 and M_2.

DIFFRACTION OF LIGHT

(*Fresnel Class*)

The *wave theory of light* in its original form, as proposed by *Huygens* was not successful in explaining the observed phenomenon that light appears to travel in straight lines. The phenomenon is easily borne out by corpuscular theory. A minute investigation of the fact, however, reveals that light suffers some deviation from its straight path in passing close to the edges of opaque obstacles and narrow slits. Some of the light does bend into the region of geometrical shadow and its intensity falls off rapidly. This deviation, moreover, is extremely small when the wavelength of the waves of light is small in comparison to the dimensions of the obstacle or aperture. But the deviation becomes much pronounced, though never of the same order as the bending of sound waves round corners, when the dimensions of the aperture or the boundary of obstacle are comparable with the wavelength of light. Even then, only a careful examination of the screen, on which the geometrical shadow of the obstacle is received, reveals this deviation of light from its rectilinear path.

For example, let us suppose, with reference to Fig. 13.1, that waves of light diverging from a narrow slit *S*, which acts as a secondary source under the influence of light from the monochromatic source *O*, pass an obstacle *AB* with a straight edge *A* parallel to the slit. If the shadow of the obstacle is received at *C* on a screen, it is observed that the boundary of the shadow is never sharp. A small portion of light has 'bent' around the edge into the geometrical shadow; outside the shadow parallel to its edge are observed several bright and comparatively dark bands. The distribution of intensity on the screen is shown in Fig. 13.14.

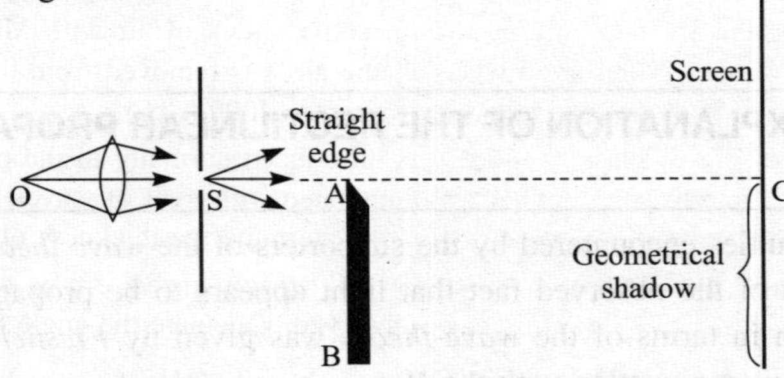

Fig. 13.1. Geometrical shadow of a straight edge

The term *diffraction* is applied to all phenomena like this which are produced whenever there is any limitation on the width of the beam of light. Since in most diffraction patterns some light penetrates within the region of geometrical shadow, diffraction is sometimes defined as "*the bending of light around an obstacle*". This phenomenon was first discovered in 1665 by an *Italian* Scientist named *Grimaldi* and was latter on studied by Newton. In the days, when *corpuscular theory* was prevalent attempts were made by *Newton* to interpret diffraction effects as due to attractive or repulsive forces exerted by the edges of obstacles on flying corpuscles, so as to deflect them from their rectilinear path. In terms of the *wave theory*, the first attempt to explain the diffraction phenomenon was made by *Dr. Young*, who attributed it to the interference between the direct light waves which pass near the edge of the obstacle and the wave of light reflected at grazing incidence from the edge. This explanation could scarcely be applied to explain the penetration of light within the region of geometrical shadow. Moreover the maxima and minima regions are not equally spaced as demanded by the mode of interference imagined by Young. Even then if we accept Young's explanation to be correct, the intensity distribution on the screen should depend upon (*i*) the sharpness of the edge, (*ii*) its degree of polish and (*iii*) its material. But *Fresnel*, in 1818, after a series of experiments proved that the details of the diffraction pattern were independent of the above mentioned factors so long as the material of the obstacle is opaque.

The correct interpretation of the diffraction we owe to the brilliant work of *Fresnel*, who attributed this phenomenon to the mutual interference of secondary wavelets, originating from various points of wave-front which are not blocked off by an obstacle or are allowed to pass by a slit. That is instead of finding the new wave-front by constructing the envelope of these secondary wavelets, we must find, by the principle of superposition, their resultant at every point of the screen, taking due account of their relative amplitudes and phases. *Fresnel*, thus applied *Huygen's principle of secondary wavelets* in conjunction with the *principle of interference* and calculated the position of fringes, in general agreement with the observed diffraction pattern.

It should be emphasised, however, that the process responsible for the production of *diffraction* phenomenon, is going on continuously during the propagation of every wave-front. But the diffraction effects are observed only when a portion of the wave-front is cut off by some obstacle. Every optical instrument, is fact, makes use of only a limited portion of the wave-front. A telescope or a microscope, for example, utilises only a limited portion of a wave-front transmitted by the objective lens. Therefore some diffraction is always present in the image. Hence a clear comprehension of the nature of diffraction is of fundamental importance for a complete understanding of practically all the optical phenomena.

13.1 FRESNEL'S EXPLANATION OF THE RECTILINEAR PROPAGATION OF LIGHT

One of the greatest difficulties encountered by the supporters of the *wave theory* of light in its early days was the explanation of the observed fact that light appears to be propagated in straight lines. The correct interpretation in terms of the *wave-theory* was given by *Fresnel* in 1815 by applying *principles of interference* in conjunction with the *Huygen's principle of secondary wavelets* but at the same time he showed that the rectilinear propagation of light is only approximate. To explain this let us first discuss Fresnel's method of finding the effect of a plane wave-front at a point ahead of it.

Fresnel's Half-period Zones

In Fig. 13.2, let *ABCD* represent a plane wave-front of monochromatic light of wavelength λ, travelling from left to right. According to *Huygen's principle, every point in this plane may be regarded as the origin of the secondary wavelets and at any given instant, every one of these secondary wavelets pass through the point P*. The resultant effect at *P* due to the whole wave-front will simply be equal to the resultant of all these secondary wavelets. To find this resultant, we divide the entire wave-front into concentric zones by the following construction. We drop a perpendicular from *P* on *ABCD*. The point *O, the foot of perpendicular from P, is called the pole of the wave with respect to P*. Let $PO = b$ and λ be the wavelength of light waves. With *P* as centre and radius $b + \lambda/2$ we construct a sphere which intersects the wave-front in a circle M_1. Then

$$PM_1 = b + \lambda/2 \quad \text{and} \quad PM_1 - PO = \lambda/2$$

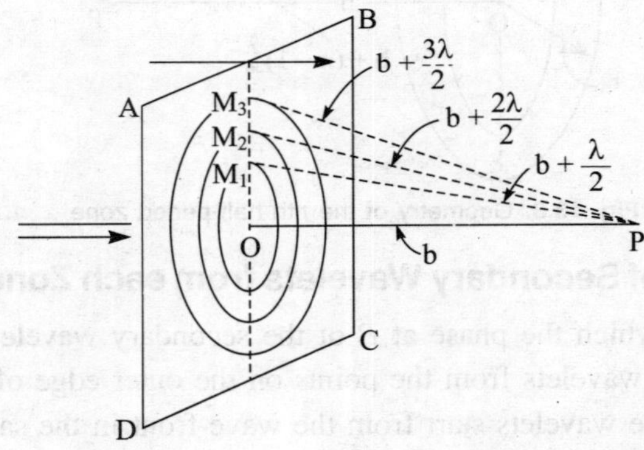

Fig. 13.2. Half-period zones on a plane wave-front

This simply means that the secondary wavelets originating from O and from the points on the circumference of the circle M_1, on reaching P simultaneously, will differ in their phase by

$$\frac{2\pi}{\lambda}(PM_1 - PO) = \pi \text{ radians}$$

But π radians is equivalent to phase difference of T/2. Consequently, the area enclosed by the circle M_1 is called Fresnel's first half-period zone. Similarly, we construct other spheres of radii

$$b + 2\lambda/2,\ b + 3\lambda/2,\ b + 4\lambda/2 \text{ etc.}$$

which intersect the wave-front in circles M_2, M_3, M_4 etc. This construction divides the entire wave-front into a number of *Fresnel's half-period zones*. These zones are so situated that the distance of the point *P* increases by $\lambda/2$ as we pass from the inner to the outer boundary of each zone. The first half-period zone, as explained above, is a circle, while the second half-period zone is an annular space between the circles M_1 and M_2 and so on.

The area A_n of nth zone, is from Fig. 13.3,

$$A_n = \pi\left(r_n^2 - r_{n-1}^2\right) = \pi\left\{(PM_n^2 - b^2) - (PM_{n-1}^2 - b^2)\right\}$$

$$A_n = \pi[(b+n\lambda/2)^2 - \{b+(n-1)\lambda/2\}^2]$$

$$A_n = \pi b\lambda + \pi(2n-1)\lambda^2/4 \quad \text{...(13.1 } a)$$

On neglecting the term in λ^2 as b is large in comparison with λ, we get

$$\boxed{A_n = \pi b\lambda \text{ approximately}} \quad \text{...(13.1 } b)$$

The approximate value of A_n is independent of n, hence all zones are approximately equal in area. Actually the area will increase with the order n of the zone according to Eqn. (13.1 *a*).

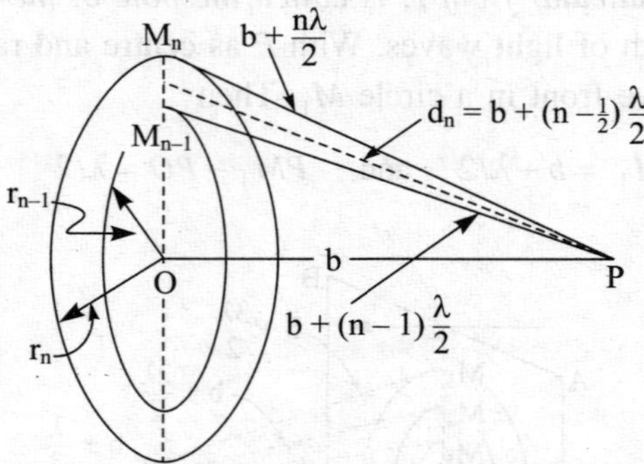

Fig. 13.3. Geometry of the *n*th half-period zone

Phase of the Resultant of Secondary Wavelets from each Zone

Consider an instant at which the phase at P of the secondary wavelet from O is zero. At this instant, the phase at P of the wavelets from the points on the outer edge of the first zone will be π, since $PM_1 - PO = \lambda/2$ and the wavelets start from the wave-front in the same *phase*. Consequently, the phase of the resultant at P of entire wavelets from the first zone, at this instant, will be approximately the mean of the phases of these wavelets and may be taken to be $\pi/2$. At this same instant, the phases at P of the secondary wavelets from the inner and outer boundaries of the second zone are π and 2π respectively, so the mean phase of the resultant due to this zone is $3\pi/2$. Thus the mean phases of resultants of wavelets from the successive outer zones are $\pi/2$, $3\pi/2$, $5\pi/2$, $7\pi/2$, $9\pi/2$, and so on. Let R_1, R_2, R_3, R_4,......, R_n be the resultant amplitudes at P of wavelets reaching P from first, second, third....... and nth zones. Calling the resultant amplitude due to the entire wave-front as R, this may now be expressed, by the principle of superposition, as the sum of the series,

$$R = R_1 - R_2 + R_3 - R_4 + \ldots\ldots(-1)^{n-1} R_n \quad \text{...(13.1 } c)$$

Factors Governing the Magnitude of R_n

The magnitudes of successive terms of the above series depend upon the following three factors:

(*i*) The area of each zone determines the number of secondary wavelets reaching P from that zone and the magnitude of R_n depends on the number of wavelets from the nth zone.

$$\therefore R_n \propto A_n$$

(*ii*) The magnitude of R_n is inversely proportional to the average distance d_n from the point P of the nth zone.

(*iii*) Lastly $R_n \propto (1 + \cos\theta_n)$ where θ_n is the angle which the direction of P from the nth zone makes with OP. It is assumed that the obliquity factor varies slowly and that it may be regarded as constant over a single half period zone.

But
$$\frac{A_n}{d_n} = \frac{\pi\lambda\left[b+(2n-1)\frac{1}{4}\lambda\right]}{\frac{1}{2}\left[b+\frac{1}{2}n\lambda+b+(n-1)\frac{1}{2}\lambda\right]} = \pi\lambda$$

The amplitudes of successive zones, therefore, depend on the obliquity factor $(1 + \cos\theta)$. As θ increases from zero, $\cos\theta$ decreases very slowly at first, but more rapidly for larger value of θ. Therefore, the successive amplitudes R_1, R_2, R_3, R_4...... decrease, at first, slowly but more rapidly for higher value of n.

Summing up of the Series

We have seen above that the successive terms in the series

$$R = R_1 - R_2 + R_3 - R_4 + \ldots\ldots + (-1)^{n-1} R_n$$

are in decreasing order of magnitude. With this knowledge we will try to sum the above series by a method given by *Schuster*. Let n be an odd integer. The terms in the above series can then be grouped in the following two alternate ways:

$$R = \frac{1}{2}R_1 + \left(\frac{1}{2}R_1 - R_2 + \frac{1}{2}R_3\right) + \left(\frac{1}{2}R_3 - R_4 + \frac{1}{2}R_5\right) + \ldots\ldots + \frac{1}{2}R_n$$

or

$$R = R_1 - \frac{1}{2}R_2 - \left(\frac{1}{2}R_2 - R_3 + \frac{1}{2}R_4\right) - \left(\frac{1}{2}R_4 - R_5 + \frac{1}{2}R_6\right) - \ldots\ldots - \frac{1}{2}R_{n-1} + R_n$$

Suppose first that each term of the original series is greater than the arithmetic mean of two adjacent terms, that is,

$$R_2 > \frac{1}{2}(R_1 + R_3),\ R_3 > \frac{1}{2}(R_2 + R_4),\ R_4 > \frac{1}{2}(R_3 + R_5)$$

then the bracketed terms are all negative. Therefore, the above equations may be written as

$$R = \frac{1}{2}(R_1 + R_n) - \alpha$$

and $$R = R_1 - \frac{1}{2}(R_2 + R_{n-1}) + R_n + \beta$$

where α and β represent the sum of all terms within brackets. It is now obvious that the following inequalities must hold good

$$\frac{1}{2}(R_1 + R_n) > R > R_1 - \frac{1}{2}(R_2 + R_{n-1}) + R_n$$

If the original series is such that each term is less than mean of two adjacent terms, the bracketed terms are all positive. Now the following inequalities must hold,

$$R_1 - \frac{1}{2}(R_2 + R_{n-1}) + R_n > R > \frac{1}{2}(R_1 + R_n)$$

But $R_1 = R_2$ and $R_n = R_{n-1}$, approximately. Making these substitutions in the inequalities given above, we get the equality.

$$\frac{1}{2}R_1 + \frac{1}{2}R_n = R = \frac{1}{2}R_1 + \frac{1}{2}R_n$$

Similarly, when n is even, we get

$$\frac{1}{2}R_1 - \frac{1}{2}R_n = R = \frac{1}{2}R_1 - \frac{1}{2}R_n$$

If n becomes very large, the amplitude due to nth zone, on account of obliquity factor $(1 + \cos\theta_n)$ becomes negligibly small. The resultant amplitude at P due to the whole wave, whether n is odd or even, is therefore merely ½ R_1 *i.e.*, one half of that which could be produced by the first Fresnel-zone alone.

Rectilinear Propagation of Light

Let us now apply the above theory to explain the rectilinear propagation of light. Let us suppose a plane wave-front of monochromatic light to be incident normally on a square aperture *EFGH* in a screen. Let A be the pole of the wave-front with respect to the point P at which it is desired to find the resultant intensity due to entire wave-front. The pole A is within the aperture, sufficiently far away from its edges. Suppose the wave-front is now divided into zones around A. By the time these intersect the edges their number will be fairly large and all the effective zones will be exposed. The resultant amplitude at P, according to above theory, will be ½ R_1, that is, it will be just the same as if the screen with aperture removed. In the same way if the point P_3 is so far inside the geometrical shadow of the aperture that the pole A_3 corresponding to it is outside the aperture, well far off from its edges, the effect at P_3 reduces to that of one half of the Fresnel first zone around A_3 which is obviously zero. The point P_3, will be just as dark as if light travelled strictly in straight lines. It is only when the pole is in the region enclosed by dotted squares, the above theory does not apply. This is simply due to the reason that for points like P_1 and P_2, only some of the effective zones are exposed and the rest blocked. Therefore, Fresnel's theory becomes inadequate in deciding the intensity at these points. The wavelength of light being extremely small, the sides of the two dotted squares lie extremely close to the edges of the aperture.

We have thus arrived at the law of rectilinear propagation for all points whose poles lie within the dotted square $E_1F_1G_1H_1$ and for all points whose poles lie outside $E_2F_2G_2H_2$. The law fails in the case of those points whose poles lie between the two dotted squares—in fact when light passes close to the edges of obstacles, there is some deviation from its rectilinear path. Thus the propagation of light is approximately linear.

We can see in another way how rectilinear propagation (approximate) follows from *Fresnel's* theory of half-period zones. Let us first find out the radius of the first zone of the plane wave for a point *P*, 20 cm ahead of it, taking λ as 5000×10^{-8} cm.

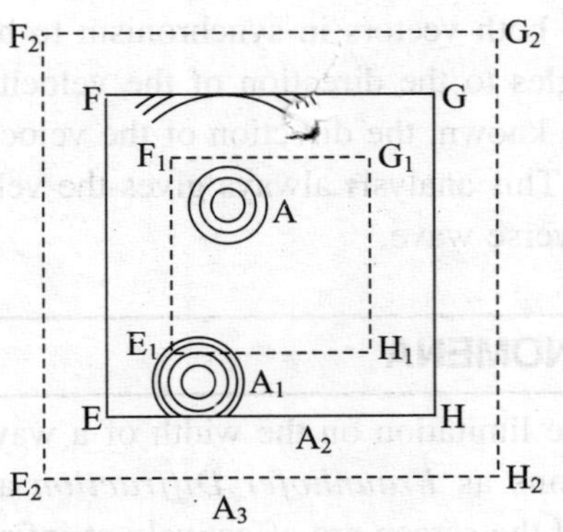

Fig. 13.4 (*a*). Explanation of rectilinear (approximate) propagation

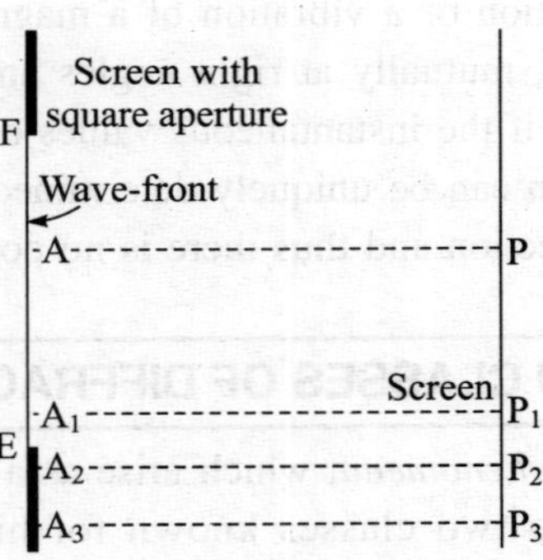

Fig. 13.4 (*b*)

$$\text{Radius of first zone} = \sqrt{b\lambda} = \sqrt{20\times5000\times10^{-8}} = 0.0316 \text{ cm}$$

and Radius of 10,000th zone $= \sqrt{10000\, b\lambda} = 3.16$ cm.

The diameter of the first zone is therefore less than a mm and that of 10,000th zone is less than 10 cm, which is quite the normal width of the wave-front. Therefore, the outer most zone for this case will be of an order even greater than 10000 and its amplitude R_n at *P* will be quite negligible. The resultant amplitude at *P* is simply $R_1/2$ *i.e.* of the portion of the wave-front within about 0.3 mm (radius of first zone) round the point *O*, the pole of the wave-front with respect to *P*. In other words only secondary wavelets from the first half-period zone are effective at *P*, the remainder of the wave-front simply destroys at *P* half the amplitude of the first zone; so that light travels to *P* from a region 0.3 mm in radius around *O*. Thus it travels approximately in straight line.

13.2 ELIMINATION OF REVERSE WAVE IN HUYGENS' PRINCIPLE

An early objection to Huygens' principle of wave propagation, it will be recalled, lies in the fact that it also gave the propagation of wave in the backward direction towards the source. This principle, of course, correctly postulated the propagation of wave in the forward direction. By the help of Fresnel's idea of interference of Huygens' secondary wavelets in conjunction with the inclination factor $(1 + \cos\theta)$ governing the amplitudes of secondary wavelets which was introduced *ad hoc* by Fresnel, it is possible to give a simple explanation of the absence of backward wave. The method of dividing the wave-front into Fresnel half-period zones with respect to a point ahead of it may be also applied to divide the wave-front into zones with respect to a point behind it. An analysis on the lines similar to that given in *Sec.* 13.1 shows that the amplitude of reverse wave at any given point is half of the amplitude which would be produced by the secondary wavelets from Fresnel's first zone alone. The inclination θ for the wave travelling in the backward direction for the first zone is π and accordingly the inclination factor $(1 + \cos \theta)$ of the amplitude for this wave is zero. Hence the amplitude of the reverse wave is zero.

We may also explain the absence of reverse wave on the basis of electromagnetic theory of light. According to this theory, light is a propagation of vibration of an electric vector **E** coupled with the propagation of a vibration of a magnetic vector **H** both vectors in synchronism to be always in equal phase, mutually at right angles and at right angles to the direction of the velocity vector **V**. Conversely, if the instantaneous values of **E** and **H** are known, the direction of the velocity **V** of the wave motion can be uniquely determined analytically. This analysis always gives the velocity in the forward direction and thus there is no possibility of reverse wave.

13.3 TWO CLASSES OF DIFFRACTION PHENOMENA

Diffraction phenomena, which arise as a result of some limitation on the width of a wave-front, are divided into two classes known for historical reasons as *Fraunhofer Diffraction* and *Fresnel Diffraction*. In the former class, the source of light and the screen are effectively at infinite distance from the obstacle of aperture causing the diffraction. This means that the wave-front incident on the obstacle or aperture is plane and the secondary wavelets, which we consider as originating from the unblocked portion of the wave-front at the moment it just touches the aperture, all start at the same time. In other words, the phase of secondary wavelets is the same at every point of the aperture. The screen is at infinity. Therefore, we have to consider the interference as taking place between parallel diffracted rays which can be brought into focus by placing a convergent lens behind the aperture. Thus it is very conveniently observed by employing two convergent lenses: one to render the light from the source parallel before its incidence on the aperture and the other to unite parallel diffracted rays in focus on the screen—an arrangement which effectively removes the source and screen at infinity.

In the *Fresnel Class* of diffraction, the source of light or the screen or both are at finite distances from the obstacle or aperture but no lenses are employed for rendering the rays parallel or convergent. Therefore, the incident wave-front is spherical or cylindrical instead of being plane. As a consequence of this the phase of secondary wavelets is not the same at all points in the plane of aperture causing the diffraction. The resultant amplitude at any point of the screen, however, is obtained by the mutual interference of secondary wavelets from different elements of unblocked portion of the wave-front.

Only diffraction of the *Fresnel type* will be considered in this chapter.

13.4 DIVISION OF A CYLINDRICAL WAVE-FRONT INTO HALF-PERIOD STRIPS

The envelope of secondary wavelets diverging from a long narrow slit SS', illuminated by monochromatic light of wavelength λ, is a cylindrical surface WW' (Fig. 13.5). We shall discuss in Sec. 13.5 Fresnel diffraction produced due to cylindrical wave-front incident on a straight edge. Therefore it is essential first to find its effect at a point P situated ahead of the wave-front, by dividing it into a number of elementary half-period strips by a construction closely similar to that of the division of plane wave-front into zones. We draw a perpendicular PQ from P on SS'. The point of intersection, M_0 in Fig. 13.5, of PQ with the wave-front, is the pole of wave with respect to P.

Let us now consider an equatorial section AA' through the line PM_0Q of the wave surface. Let $QM = a$ and $M_0P = b$. With P as centre and radii

$$b+\frac{\lambda}{2},\ b+\frac{2\lambda}{2},\ b+\frac{3\lambda}{2},$$

etc. we draw arcs so as to cut the section AA' in points $(M_1, M_1'), (M_2, M_2'), (M_3, M_3')$..... respectively.

Let $PM_n = b + \frac{1}{2} n\lambda$, then in triangle PM_nQ, we have by the *law of cosine*,

$$PM_n^2 = QM_n^2 + PQ^2 - 2PQ \cdot QM_n \cos\theta_n$$

$$\left(b+\frac{1}{2}n\lambda\right)^2 = a^2 + (a+b)^2 - 2(a+b)a\cos\theta_n \qquad ...(13.4\ a)$$

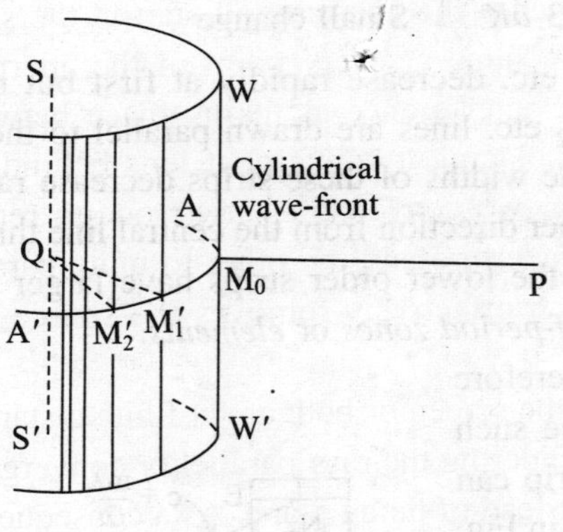

Fig. 13.5. Half-period strips on a cylindrical wave-front

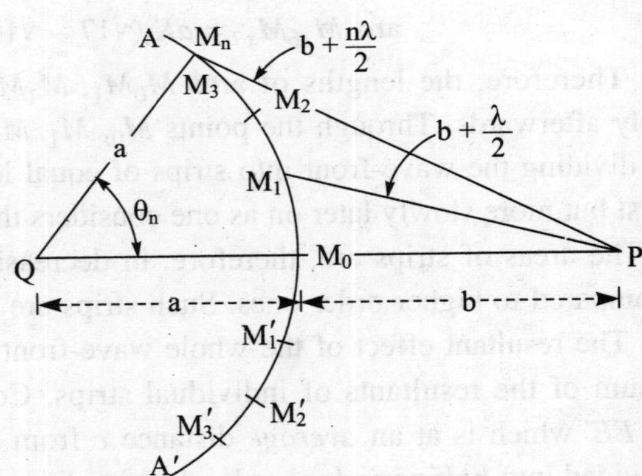

Fig. 13.6. Equatorial section of cylindrical wave-front

As λ is small, the term $(n^2\lambda^2/4)$ in the above equation can be neglected in comparison with other quantities. Also if θ_n is small, we consider only the first two terms in the expression

$$\cos\theta_n = 1 - \frac{\theta_n^2}{2!} + \frac{\theta_n^4}{4!} - \frac{\theta_n^6}{6!} + ...$$

With these considerations, Eqn. (13.4 *a*) reduces to

$$bn\lambda = a(a+b)\theta_n^2$$

$$\theta_n = \sqrt{\frac{nb\lambda}{a(a+b)}} = K\sqrt{n}$$

where K is a constant, as a, b and λ are constants for a given arrangement. Hence

$$\text{arc } M_0M_n = a\theta_n = aK\sqrt{n}$$

Replacing n by 1, 2, 3, 4 etc. we have respectively,

$$\text{arc } M_0M_1 = aK$$

$$\text{arc } M_0M_2 = aK\sqrt{2}$$

$$\text{arc } M_0M_3 = aK\sqrt{3}$$

etc. etc.

Hence

$$\text{arc } M_0M_1 = aK$$

$$\left.\begin{aligned} \text{arc } M_1M_2 &= aK(\sqrt{2}-1) = 0.414\,aK \\ \text{arc } M_2M_3 &= aK(\sqrt{3}-\sqrt{2}) = 0.318\,aK \\ \text{arc } M_3M_4 &= aK(\sqrt{4}-\sqrt{3}) = 0.268\,aK \end{aligned}\right\}\text{Rapid change}$$

$$\left.\begin{aligned} \text{arc } M_{15}M_{16} &= aK(\sqrt{16}-\sqrt{15}) = 0.127\,aK \\ \text{arc } M_{16}M_{17} &= aK(\sqrt{17}-\sqrt{16}) = 0.123\,aK \end{aligned}\right\}\text{Small change}$$

Therefore, the lengths of arcs M_0M_1, M_1M_2, M_2M_3, etc. decrease rapidly at first but rather slowly afterwards. Through the points M_0, M_1, M'_1, M_2, M'_2 etc. lines are drawn parallel to the slit, thus dividing the wave-front into strips of equal lengths. The widths of these strips decrease rapidly at first but more slowly later on as one considers these in either direction from the central line through M_0. The areas of strips are, therefore, in decreasing order, the lower order strips have larger areas as compared to higher order ones. Such strips are called *half-period zones* or *elements*.

The resultant effect of the whole wave-front at P is therefore the sum of the resultants of individual strips. Consider one such strip EE' which is at an *average* distance c from P. This strip can be divided into *half period zones* by drawing lines, as shown in Fig. 13.7 so that the middle points N_1, N_2, N_3, etc. of these lines are respectively at distances

$$c + \frac{1}{2}\lambda,\ c + \lambda,\ c + 3\lambda/2 \text{ etc.}$$

from P. In the right triangle PON_m we have $(c + \frac{1}{2} m\lambda)^2 = c^2 + ON_m^2$ and neglecting $\frac{1}{4} m^2\lambda^2$, the above equation reduces to $ON_m^2 = cm\lambda$

or
$$ON_m = \sqrt{cm\lambda} = K'\sqrt{m}$$

Fig. 13.7. Half-period elements on a single strip

where K' is a constant. The same analysis, as before, will show that the widths and, therefore, the areas of these half-period elements decrease at first rapidly but more slowly afterwards. Consequently the resultants at P of the higher order elements, being alternately in and out of phase with the first element, annual each other's effect. Thus the resultant effect of the whole strip at P is due to a few half-period elements of lower order. These conclusions can be applied to all strips of the wave-front. Therefore the resultant effect at P of the entire cylindrical wave-front is merely due to a narrow equatorial band surrounding the section AM_0A', which is divided into half-period elements by the strips. This is shown in the Fig. 13.8 which represents the equatorial band spread out in a plane.

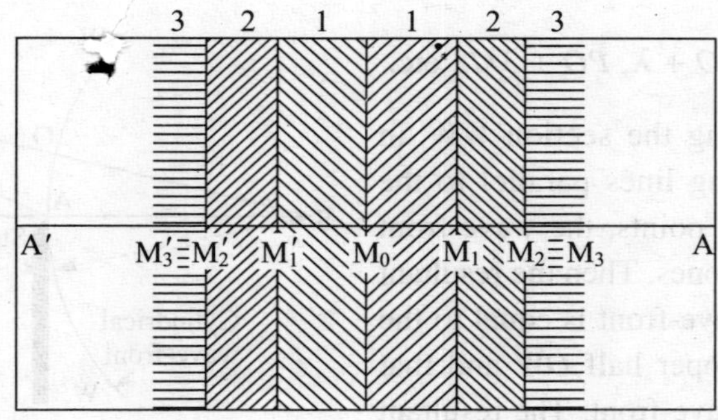

Fig. 13.8. Equatorial band of cylindrical wave-front spread out in a plane

Let $R_1, R_2, R_3, \ldots\ldots\ldots, R_n$ be the resultant amplitudes at the point P due to secondary wavelets reaching P from first, second, third, ... and nth zone on either side of the central line through M_0. Therefore the resultant amplitude R at P due to either half of the wave, by the principle of superposition, is

$$R = R_1 - R_2 + R_3 - R_4 + \ldots\ldots\ldots + (-1)^{n-1} R_n$$

where the alternate plus and minus signs occur because the resultants of successive zones are alternately out of and in phase. The summation of this series is given in Sec. 13.1, the result is

$$R = R_1/2 \pm R_n/2$$

the plus sign corresponds to n odd while negative sign corresponds to n even. And when n is large, then

$$R = R_1/2$$

Obviously the resultant amplitude at P of the entire wavefront is R_1.

13.5 DIFFRACTION AT A STRAIGHT EDGE

Let A be the sharp straight edge of an obstacle placed perpendicular to paper and exactly parallel to the long narrow slit S, which is illuminated with a monochromatic light source. A train of cylindrical waves is incident on the edge A from the left. A careful examination of the screen, reveals that immediately above C there are series of maxima and minima regions, parallel to the straight edge, called diffraction fringes. The intensity of maxima gradually decreases while that of minima increases as we move above the point C along the screen. At a short distance from this point the fringes merge into a uniform illumination. It should be emphasised that minima regions are not absolute. The fringes, moreover, are not equidistant but gradually become closer as we move away from C. Some of the light also penetrates within the geometrical shadow and the illumination falls off continuously and rapidly to zero, without showing any maxima and minima.

To explain these characteristics of diffraction pattern due to a straight edge, we divide the incident cylindrical wave-front into Fresnel's half-period elements. In the Fig. 13.9 WW' represents the section of the wave-front, when it just touches the edge A, by the plane of paper. Let P be any point on the screen. A straight line joining P to S intersects the wave-front at O, its pole with respect to P. With F as centre and radii

$$PO + \frac{1}{2}\lambda,\ PO + \lambda,\ PO + 3\lambda/2 \text{ etc.}$$

we draw circles intersecting the section WW' in different points. By drawing lines parallel to the straight edge through these points, the wave-front is divided into half-period zones. Then the resultant effect at P due to whole wave-front is equal to the sum of the effect of the upper half OW and that of lower half OW' of the wave-front. The resultant amplitude at P due to each half of the wave, as discussed in *Sec.* 13.4, is given by

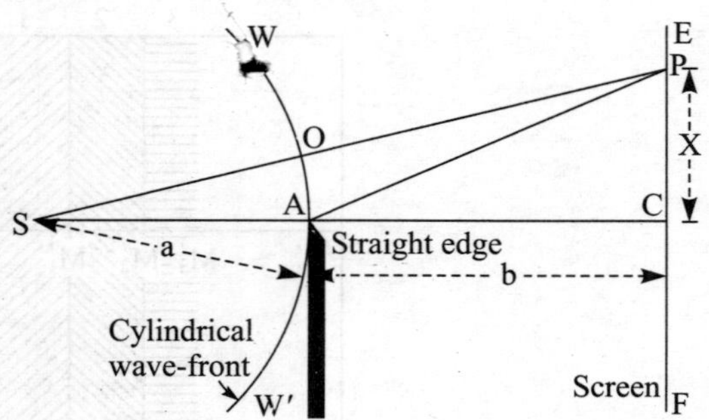

Fig. 13.9. Fresnel diffraction of a cylindrical wave-front by a straight edge

$$R = R_1 - R_2 + R_3 - R_4 + \ldots = \frac{1}{2} R_1$$

where R_1 is the amplitude at P due to first zone in either half of the wave and so on.

Uniform Illumination at Points far above *C*

Let us suppose that the point P is sufficiently far above C so that the portion OA contains all the effective half period zones of the lower half of the wave. In other words, the obstacle AB imposes no limitation on the zones effectively contributing to the resultant amplitude at P. The intensity at P will be, therefore,

$$\frac{1}{2}R_1 + \frac{1}{2}R_1 = R_1$$

i.e., as if the entire wave is unobstructed. This explains the uniform illumination on the screen beyond the diffraction fringes.

Maxima and Minima diffraction Fringes on the Screen

Let us consider a point P_1 on the screen for which the pole of the wave-front is O_1 as shown, in such a position that the following relation.

$$P_1A - P_1O_1 = \frac{1}{2}\lambda$$

is satisfied. O_1A contains only first-half period zone of the lower half of the wave. Therefore, the resultant amplitude D_1 at P_1 will be simply equal to the sum of the resultant amplitude due to upper half of the wave and that of first zone of the lower half that is,

$$D_1 = (R_1 - R_2 + R_3 - R_4 + \ldots) + R_1$$

$$= \frac{1}{2}R_1 + R_1 = \frac{3}{2}R_1$$

and the intensity at P_1 is proportional to $\frac{9}{4}R_1^2$ or 2.25 R_1^2 which is 2.25 times as great as that which would be produced by the unobstructed wave.

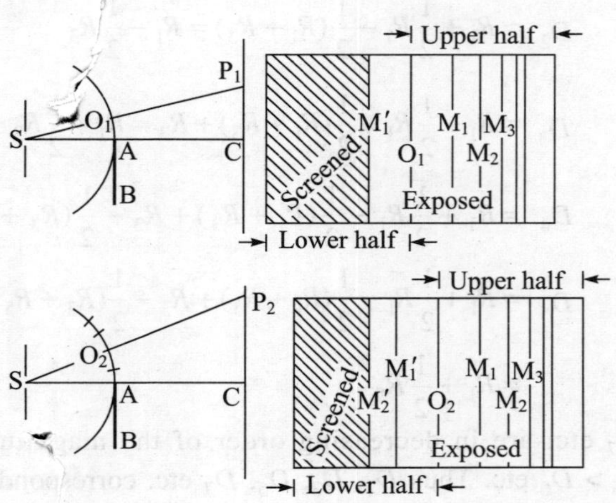

Fig. 13.10

As we move above P_1, we come across a point P_2 so that

$$P_2A - P_2O_2 = \lambda$$

Then O_2A contains, for this point, the first and the second half-period zones of the lower half of the wave. The resultant amplitude at P_2 is therefore,

$$D_2 = \frac{1}{2}R_1 + R_1 - R_2 = \frac{3}{2}R_1 - R_2$$

This is less than the magnitude of R_1, so that bright band just outside the shadow is followed by a darker one. However, the minima is not absolute as D_2 has some positive value.

Further above, we come across a point P_3 so that

$$P_3A - P_3O_3 = 3\lambda/2$$

Then O_3A exposes first three half-period zones. Therefore, the resultant amplitude at P_3 is

$$D_3 = \frac{1}{2}R_1 + R_1 - R_2 + R_3$$

For a point P_4 such that $P_4A - P_4O_4 = 2\lambda$, O_4A exposes first four half-period zones and the resultant amplitude is

$$D_4 = \frac{1}{2}R_1 + R_1 - R_2 + R_3 - R_4$$

We can form some idea about the relative magnitudes of D_1, D_2, D_3, D_4 etc. by writing in the above expressions the approximate relations

$$R_2 = \frac{1}{2}(R_1 + R_3), \quad R_4 = \frac{1}{2}(R_3 + R_5) \text{ etc.}$$

Then we get

$$D_1 = R_1 + \frac{1}{2}R_1$$

$$D_2 = R_1 + \frac{1}{2}R_1 - \frac{1}{2}(R_1 + R_3) = R_1 - \frac{1}{2}R_3$$

$$D_3 = R_1 + \frac{1}{2}R_1 - \frac{1}{2}(R_1 + R_3) + R_3 = R_1 + \frac{1}{2}R_3$$

$$D_4 = R_1 + \frac{1}{2}R_1 - \frac{1}{2}(R_1 + R_3) + R_3 - \frac{1}{2}(R_3 + R_5) = R_1 - \frac{1}{2}R_5$$

$$D_5 = R_1 + \frac{1}{2}R_1 - \frac{1}{2}(R_1 + R_3) + R_3 - \frac{1}{2}(R_3 + R_5) + R_5$$

$$= R_1 + \frac{1}{2}R_5$$

Since R_1, R_3, R_5, R_7 etc. are in decreasing order of the magnitude, hence $D_3 > D_2$ and $D_3 > D_4$; $D_5 > D_4$ and $D_5 > D_6$ etc. Thus D_1, D_3, D_5, D_7 etc. correspond to regions of maximum illumination while D_2, D_4, D_6 etc. correspond to regions of minima illumination of the screen. Moreover it is obvious that $D_1 > D_3 > D_5 > D_7$ and $D_2 < D_4 < D_6$ etc. Thus the amplitude and therefore the intensity of maxima decreases gradually while that of minimum increases gradually as we move along the screen above the point C, until finally the fringes merge into a uniform illumination.

From the above analysis it should be obvious that the intensity at any point P of the screen is maximum or minimum according as OA exposes odd or even number of zones of the lower half of the wave. Mathematically the condition for maxima and minima can be expressed as

$$PA - PO = (2n+1)\lambda/2 \quad \text{[Max. brightness]} \qquad ...(13.5\ a)$$

and

$$PA - PO = 2n\lambda/2 \quad \text{[Min. brightness]} \qquad ...(13.5\ b)$$

The positions of the maxima and minima on the screen can be easily calculated. Let $SA = a$, $AC = b$ and $CP = x$, then

$$PA^2 = b^2 + x^2$$

$$PA = b(1 + x^2/b^2)^{1/2} = b + \frac{1}{2}(x^2/b) \text{ (Approx.)}$$

Similarly $$PS = \sqrt{(a+b)^2 + x^2} = (a+b) + \frac{1}{2}\frac{x^2}{(a+b)} \text{(Approx.)}$$

Hence $$PO = PS - SO$$

$$= (a+b) + \frac{1}{2}\frac{x^2}{(a+b)} - a = b + \frac{x^2}{2(a+b)}$$

Hence, $$PA - PO = \frac{x^2}{2b} - \frac{x^2}{2(a+b)} = \frac{ax^2}{2b(a+b)} \qquad ...(13.5\ c)$$

From Eqns. (13.5 *a*) and (13.5 *c*), we have for maxima,

$$\frac{ax_{max}^2}{2b(a+b)} = (2n+1)\lambda/2$$

or

$$x_{max}^2 = (2n+1)\frac{b(a+b)\lambda}{a}$$

$$x_{max} = K\sqrt{2n+1} \quad ...(13.5\ d)$$

where $K = \sqrt{b(a+b)\lambda/a}$, a constant for a given experimental arrangement and wavelength. Similarly, from Eqns. (13.5 *b*) and (13.5 *c*), we have for minima,

$$x_{min} = \sqrt{b(a+b)2n\,\lambda/a} = K'\sqrt{n} \quad ...(13.5\ e)$$

where

$$K' = \sqrt{b(a+b)2\,\lambda/a} \quad \text{a constant}$$

For 1st Max at P_1, $n = 0$ in Eqn. (13.5 *d*) $\therefore\ x_1 = K$

For 2nd Max at P_3, $n = 1$ in Eqn. (13.5 *d*) $\therefore\ x_2 = K\sqrt{3}$

For 3rd Max at P_5, $n = 2$ in Eqn. (13.5 *d*) $\therefore\ x_3 = K\sqrt{5}$

But x_1, x_2, x_3, x_4 etc. are the distances from the point C of the first, second, third, fourth maxima. Hence the separation between first and second maxima, between second and third maxima etc. is as follows

$$x_2 - x_1 = K\sqrt{3} - K = 0.732\ K$$

$$x_3 - x_2 = K\sqrt{5} - K\sqrt{3} = 0.504\ K$$

$$x_4 - x_3 = K\sqrt{7} - K\sqrt{5} = 0.430\ K$$

Thus the diffraction fringes are not equidistant as in the case of two small coherent sources of light, but get closer and closer as we move above the edge of geometrical shadow. Also, these diffraction fringes are parallel to the edge of the geometrical shadow.

Intensity at the Edge of Geometrical Shadow

The edge A of the obstacle itself is the pole of the wave for the point C. Thus the upper half of the wave is only exposed while the lower half is completely obstructed by the obstacle. The resultant amplitude at C is, therefore,

$$= R_1 - R_2 + R_3 - R_4 \ldots = \frac{1}{2}R_1$$

and the intensity at C is proportional to $R_1^2/4$ which is simply one fourth of the intensity produced by the full wave. Also the intensity at C is less than that at the first maxima point P_1.

Penetration of light within the Geometrical Shadow

Having explained the diffraction pattern above the point C, let us now explain the penetration of light to a certain distance within the Geometrical shadow of the obstacle. Let us suppose a point P_1' on the screen below C so that

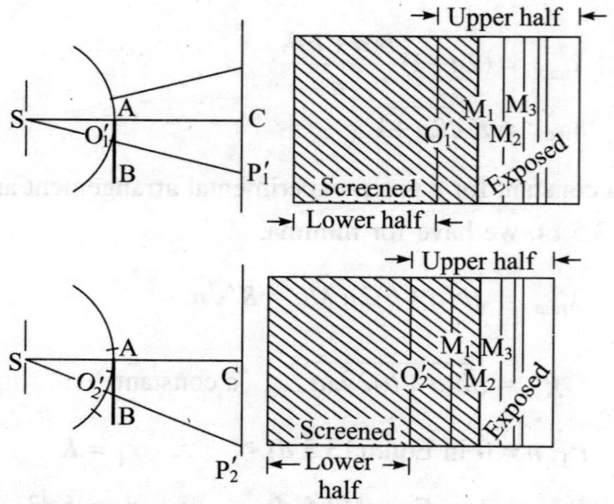

Fig. 13.11. Penetration of light within the geometrical shadow

$$P_1'A - P_1'O_1' = \lambda/2$$

Then for this point, the obstacle not only cuts off completely the lower half of the wave but also the first zone of the upper half. Therefore the resultant amplitude at P_1' due to the exposed portion of the upper half is

$$D_1' = R_2 - R_3 + R_4 - R_5 + \ldots = R_2/2$$

and the intensity at P_1' is proportional to $R_2^2/4$ which is evidently less than the intensity at C.

For the point P_2' below C so that

$$P_2'A - P_2'O_2' = \lambda$$

the obstacle obstructs, in addition to the lower half, the first two zones of the upper half of the wave also. The resultant amplitude at P_2' is

$$D_2' = R_3 - R_4 + R_5 - R_6 + \ldots = R_3/2$$

and the intensity is proportional to $P_3^2/4$ which is evidently less than the intensities at the points P_1' and C. Thus as the position below C are taken to be farther and farther into the geometrical shadow, the smaller will be the number of zones contributing to the resultant amplitude on the screen. Therefore the intensity falls off continuously, but rapidly without showing any maxima or minima, till all the effective zones are cut off and the higher order ones in the upper half AW of the wave-front simply annual the effects of one another. Thus the resultant intensity becomes zero.

The graphs of amplitude and intensity of illumination on the screen against the distance from the edge of the geometrical shadow can be now plotted. These graphs are depicted in the Figs. 13.12 and 13.13 respectively.

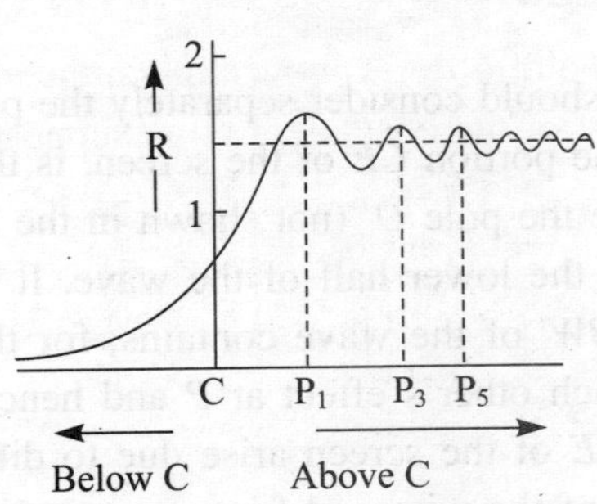

Fig. 13.12. Amplitude contour

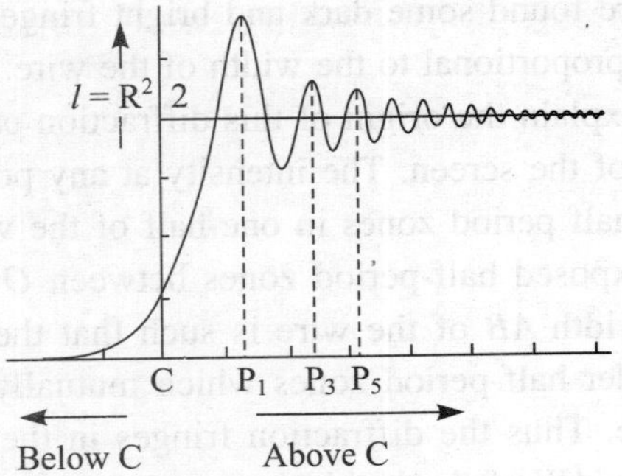

Fig. 13.13. Intensity contour

Measurement of Wavelength

To carry out the measurement of wavelength of light by diffraction at a straight edge, a narrow slit is mounted vertically in one of the uprights of the optical bench. In another upright a tin plate is mounted, with its straight edge parallel to the slit, which is illuminated with monochromatic light. The parallelism of straight edge and the slit is adjusted first with naked eye. The finer adjustment is however, done by observing the diffraction bands through the micrometer travelling eye-piece and rotating the edge in its own plane by the help of tangent screw of the holder until the bands are most distinct. Now the distance between the first or second bright band and one of the most distant that can be seen distinctly is measured by the help of micrometer eye-piece. If the former be *m*th and the latter *n*th, we have by the use of Eqn. (13.5 *d*)

$$x_n = \sqrt{\frac{b\lambda\,(2n-1)(a+b)}{a}}$$

and

$$x_m = \sqrt{\frac{b\lambda\,(2m-1)(a+b)}{a}}$$

Hence

$$x_n - x_m = \sqrt{\frac{b\lambda\,(a+b)}{a}}\,\{\sqrt{2n-1} - \sqrt{2m-1}\}$$

13.6 DIFFRACTION AT A NARROW WIRE

Let *AB* represent, in the Fig. 13.14, a long narrow wire, perpendicular to the plane of the paper and placed exactly parallel to a long narrow slit S_1, which is illuminated by a source of monochromatic light. The cylindrical waves diverging from the slit are incident on the wire from the left. The lines joining the slit *S* with the extremities of the wire when produced, meet the screen *EF* at points *C* and *C'*. Therefore, *CC'* is the region, on the screen, of thé geometrical shadow of the wire. The wire offers some limitation to the incident wave. Therefore what is actually observed on the screen on both sides of the geometrical shadow *CC'* is a diffraction pattern characterised by diffraction fringes exactly similar to those in the case of the straight edge. These fringes gradually get closer and closer, independent of the width of the wire, finally merging into uniform illumination. Within the geometrical shadow are found some dark and bright fringes which are equidistant and narrow, their width being inversely proportional to the width of the wire.

To explain the origin of this diffraction pattern, we should consider separately the portions *CE*, *CC'*, *C'F* of the screen. The intensity at any point *P* in the portion *CE* of the screen, is the resultant effect of half period zones in one-half of the wave above the pole *O'* (not shown in the Fig. 13.14) and the exposed half-period zones between *O'* and *A* of the lower half of the wave. It is assumed that the width *AB* of the wire is such that the portion *BW'* of the wave contains, for the point *P'* higher order half-period zones which mutually cancel each other's effect at *P* and hence are quite ineffective. Thus the diffraction fringes in the portion *CE* of the screen arise due to diffraction of the portion *AW* of the incident wave by the extremity *A* of the wire and fringes are independent of the portion *BW'* of the wave. Exactly in a similar way, the diffraction fringes in the portion *C'F* arise on account of the diffraction of the portion *BW'* by the extremity *B* of the wire and these are independent of the portion *AW*. The explanation of the details of these fringes is, therefore, essentially the same as given in *Sec:* 13.5.

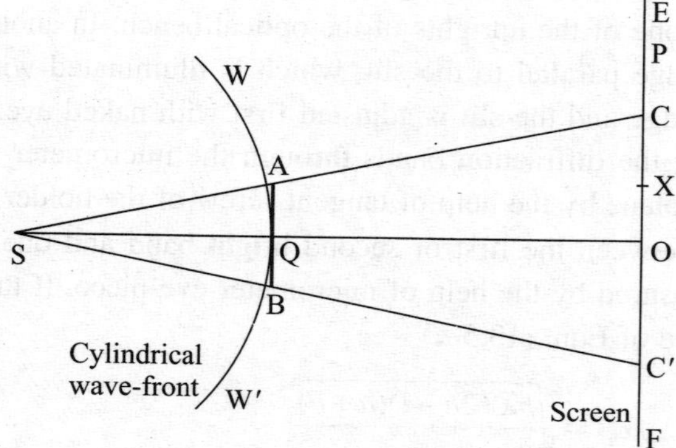

Fig. 13.14. Fresnel diffraction of a cylindrical wave-front by a narrow wire

It now remains to interpret the system of equidistant fringes within the geometrical shadow of the wire. Suppose *X* is a point within the shadow. A straight line (not shown in Fig. 13.14) joining *S* and *X* cuts *BA* at some point between *A* and *B*. The wire cuts off few half period zones from the upper and the lower half of the wave for this point. The resultant intensity at *X* is, therefore, due to combined effect of the resultant of the half period zones contained in *AW* and the resultant of zones in the portion *BW'* of the wave. The resultant amplitude at *X* of the portion *AW* is simply equal to half the amplitude of the zone adjacent to *A*. Therefore the phase of this resultant will be almost the same as that of optical disturbance from the zone adjacent to *A*. Similarly, the phase of the resultant at *X* of the portion *BW'* will be almost the same as that of disturbances from the zone adjacent to *B*. The figure on both sides of the line *SO* is symmetrical in the plane of paper. Hence, it is obvious that the disturbances start from *A* and *B* in the same phase. Therefore on reaching *X* they will mutually interfere constructively or destructively, according as the difference of their paths *BX* and *AX* is even or odd multiple of $\frac{1}{2}\lambda$. This set of fringes is, therefore, simply the interference fringes arising as if waves of light from two coherent sources, situated at *A* and *B*, interfere in the region of the geometrical shadow of the wire. Let the diameter *AB* of the wire be 2*d* and the perpendicular distance between the wire and the screen be *D*. The fringe width of the interference fringes, as obtained in *Sec.* 11.3.4, is given by

$$\bar{X} = \frac{D\lambda}{2d} = \lambda\frac{OQ}{AB}$$

where λ is the wavelength of the light waves incident on the wire. The fringe width is the same for all internal fringes and is inversely proportional to the diameter of the wire.

The explanation of the origin of internal fringes was experimentally verified by *Dr. Young*. He showed that when the light which passed through one edge was intercepted by an opaque screen, fringes within the geometrical shadow as well as diffraction fringes on the side of opaque screen vanished. Therefore it is clear that the internal fringes arise on account of the combined action of the portions *AW* and *BW'*, according to the principles of interference of light.

13.7 DIFFRACTION AT A CIRCULAR APERTURE

Fresnel diffraction at a circular aperture is characterised by alternate concentric circular maxima and minima. The intensity at minima regions, however, is not absolutely zero. Moreover, as the screen is moved towards or away from the aperture, the intensity of the central ring alternately becomes maximum and minimum. The same effect is observable by increasing or decreasing the diameter of the aperture. We shall now explain the origin of this diffraction pattern with the aid of *Fresnel's half-period zones*.

Let *AB* represent a section, in the plane of paper, of the spherical wave-front diverging from a point source *S*, situated on the axis of the circular aperture. Let λ be the wavelength of the light waves. *P* is an axial point on the observing screen. The amplitude at *P* is the resultant of a large number of secondary wavelets which we consider as originating from various points of the spherical wavefront at the instant it just touches the periphery of the circular aperture. To find this resultant, we divide the spherical wave-front into *Fresnel's half-period zones* round about the point *O*, the pole of the wave with respect to the point *P*. The *Fresnel zones* obtained as in *Sec.* 13.1, are shown in the Fig. 13.16 where

$$PO = b,\ PM_1 = b + (\lambda/2),\ PM_2 = b + (2\lambda/2)$$

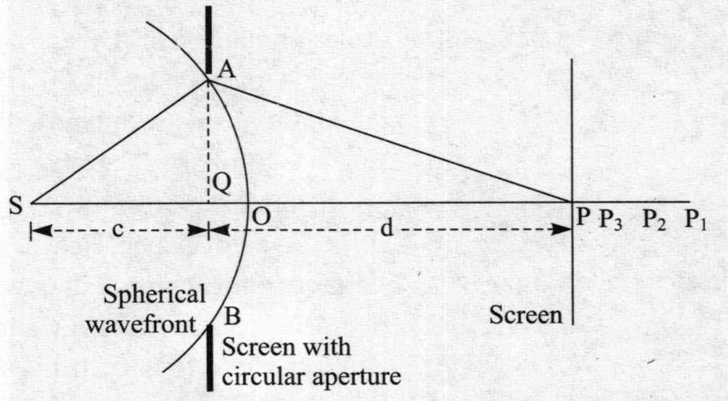

Fig. 13.15. Fresnel diffraction of a spherical wave-front by a circular aperture

The segment of the spherical wave-front by the first circle is first half period zone, the annular space between circles M_1 and M_2 is the second half-period zone and so on. The *area of Fresnel zones i.e.*, of rings between successive circles can be very easily obtained. Let the radius of spherical wave-front at a moment it just touches the periphery of the aperture be a and let us consider a segment of the wave-front which subtends an angle 2α at the point source S. Then from the geometry of the Fig. 13.17, $EG = a \sin \alpha$. The width of the annular ring, shown in Fig. 13.17, is $EF = a\, d\alpha$. Hence the area of the elementary ring is given by

$$dA = 2\pi\, a \sin \alpha \times a\, d\alpha = 2\pi a^2 \sin \alpha\, d\alpha$$

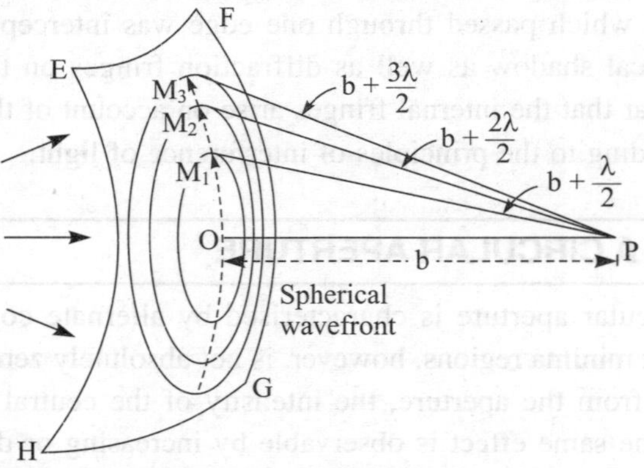

Fig. 13.16. Half period zones on a spherical wave-front

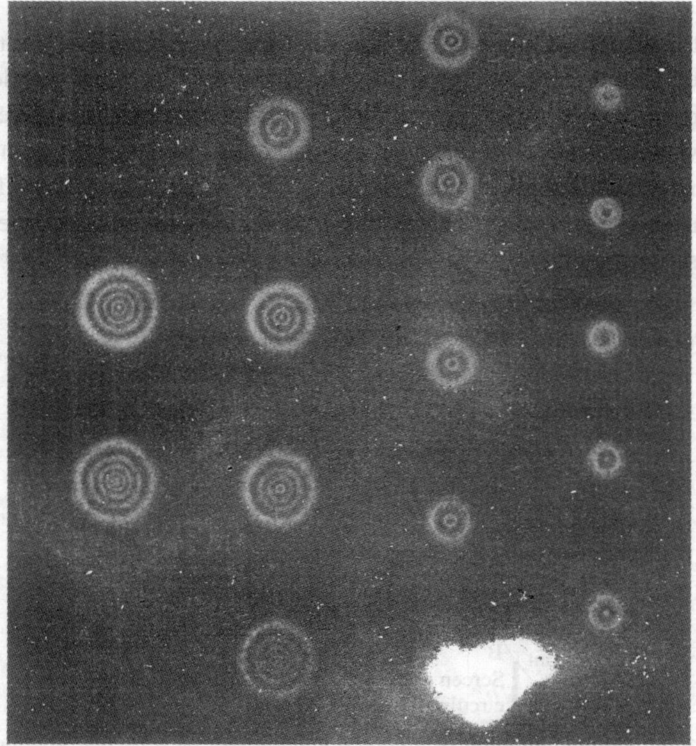

Fresnel diffraction patterns of circular apertures of various sizes

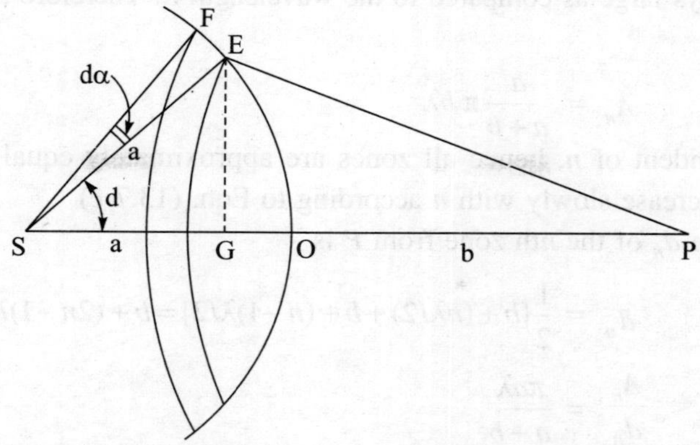

Fig. 13.17. Illustrating the geometry for finding the area of Fresnel zones

Integrating this elementary area between the limits 0 and α, we get the area of the segment which subtends 2α at *S*,

$$A = 2\pi a^2 \int_o^\alpha \sin\alpha \, d\alpha = 2\pi a^2 [-\cos\alpha]_o^\alpha$$

or

$$A = 2\pi a^2 (1 - \cos\alpha) \qquad \text{...(13.7 } a\text{)}$$

The angles α_n for the boundary of the zones are determined by the condition

$$PM_n = b + (n\lambda)/2 \qquad \text{...(13.7 } b\text{)}$$

If the point *E* is on the boundary of *n*th zone, then

$$EP = PM_n = b + n\lambda/2 \text{ and } \alpha = \alpha_n$$

$$\therefore \qquad \cos\alpha_n = \frac{(a+b)^2 + a^2 - (b + n\lambda/2)^2}{2a(a+b)}$$

Substituting this value in Eqn. (13.7 *a*), we get

$$\text{Area of first } n \text{ zones} = 2\pi a^2 - \frac{\pi a}{a+b}\left(2a^2 + 2ab - bn\lambda - \frac{n^2\lambda^2}{4}\right)$$

Replacing *n* by (*n* – 1) in the above expression, we have

$$\text{Area of the first } (n-1) \text{ zones} = 2\pi a^2 - \frac{\pi a}{a+b}\left[2a^2 + 2ab - b(n-1)\lambda - (n-1)^2\frac{\lambda^2}{4}\right]$$

To obtain the area A_n of the *n*th zone, we subtract the area of the first (*n* – 1) zones from that of the first *n* zones. Thus, we get

$$A_n = \frac{\pi a}{(a+b)}\left[b\lambda + (2n-1)\frac{\lambda^2}{4}\right] \qquad \text{...(13.7 } c\text{)}$$

In practice b is always large as compared to the wavelength λ. Therefore the term $(2n - 1)\lambda^2/4$ may be neglected. Hence

$$A_n = \frac{a}{a+b}\pi b\lambda \qquad \text{...(13.7 } d\text{)}$$

This area is independent of n, hence all zones are approximately equal in area. To be more accurate, the areas will increase slowly with n according to Eqn. (13.7 c).

The average distance d_n of the nth zone from P is

$$d_n = \frac{1}{2}[b+(n\lambda/2)+b+(n-1)\lambda/2] = b+(2n-1)\lambda/4$$

Hence, $$\frac{A_n}{d_n} = \frac{\pi a\lambda}{a+b}$$

This ratio, being independent of n, is the same for all zones. The resultant amplitudes of successive zones, starting from the first, owing to obliquity factor $(1 + \cos\theta)$ discussed in *Sec.* 13.1, will decrease slowly at first but more rapidly afterwards as θ increases. The resultant amplitude at P is given by

$$R = R_1 - R_2 + R_3 - R_4 + R_5 \ldots\ldots + (-1)^{n-1} R_n$$

where the alternate plus and minus signs occur because at P the effects of the zones are alternately out of and in phase.

Maxima and Minima Illumination of the Central Ring

Let the screen be at such a distance from the aperture that for the axial point P_1, the aperture exposes only the first Fresnel zone. The amplitude at P_1 is simply R_1, which is twice the amplitude if the entire wave-front were exposed. The intensity at P_1 is, therefore, four times as great as the intensity obtainable if the entire wave were exposed. Hence the point P_1 is extremely bright.

For the light of given wavelength, the area of Fresnel zone

$$A_n = \frac{\pi a\lambda}{1+(a/b)}$$

decreases with decrease in the value of b. Hence, the aperture exposes a greater number of zones as the screen is moved towards it. Let the screen be moved until it arrives at the point P_2 for which the aperture exposes the first and the second *Fresnel zones*. The amplitude at P_2 is therefore $R_1 - R_2$, which since R_1 and R, are nearly equal, is very small. The intensity is approximately zero. Thus the centre of diffraction pattern is practically dark in contrast to *Fraunhofer* diffraction pattern where the central disc is always bright. If the screen is moved still nearer to the aperture, it arrives at P_3, for which the first three zones are exposed. The amplitude at P_3 becomes

$$R_1 - R_2 + R_3 = \frac{1}{2}(R_1 + R_3) \text{ approximately,}$$

which is again large. Thus the axial point alternately becomes bright and dark as the screen is moved towards the aperture.

If the position of the screen be kept fixed while the aperture is gradually increased, it successively exposes one, two, three or four zones or more. Therefore, again, the intensity at the axial point is alternately maximum and minimum. *The illumination at the axial point P of the screen is, therefore, maximum or minimum according as the aperture exposes odd or ever number of Fresnel zones.*

We can now easily calculate the position of the screen for the maximum or minimum brightness of the axial point P. Let $SQ = c$, $QP = d$, $AQ = r$ where Q is the centre of the circular aperture.

$$SA^2 = c^2 + r^2$$

$$\therefore \quad SA = c[1 + r^2/c^2]^{1/2} = c + \frac{1}{2} r^2/c$$

on expanding by binomial theorem and neglecting powers of r^2/c^2, since r is very small in comparison with c and d.

Similarly $\quad AP = d + (r^2/2d)$

$$\therefore \quad SA + AP = (c + d) + \frac{r^2}{2}\left(\frac{1}{c} + \frac{1}{d}\right)$$

Hence, the path difference between the extreme ray SAP relative to the axial ray SQP is

$$SAP - SQP = (c + d) + \frac{r^2}{2}\left(\frac{1}{c} + \frac{1}{d}\right) - (c + d) = \frac{r^2}{2}\left(\frac{1}{c} + \frac{1}{d}\right)$$

For maximum at P, the aperture contains odd number of Fresnel zones. Therefore the path difference ($SAP - SQP$) must be odd multiple of $\lambda/2$, that is,

$$\frac{r^2}{2}\left(\frac{1}{c} + \frac{1}{d}\right) = (2n + 1)\frac{\lambda}{2} \quad \text{[Maxima]} \qquad \ldots(13.7\ e)$$

Similarly for minimum at P, we have

$$\frac{r^2}{2}\left(\frac{1}{c} + \frac{1}{d}\right) = 2n\frac{\lambda}{2} \quad \text{[Minima]} \qquad \ldots(13.7\ f)$$

Equation (13.7 *e*) is a general equation connecting the distance c of the source and d of the axial point P, both measured from Q, the centre of the aperture, when intensity is maximum at P.

Intensity at Non-axial Points

The mathematical treatment for finding the intensity at points a short distance off the axis is more complicated but we can form an idea about the result by a simplified treatment. For concreteness, let us suppose that the aperture is of such a size as to expose three zones for an axial point P_3, so the centre of pattern is bright.

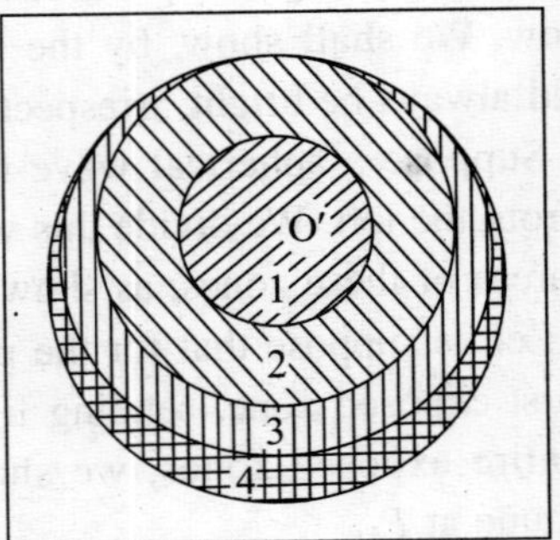

Fig. 13.18. Fresnel zones exposed by a circular aperture for a non-axial point

The Fresnel zones that should be used to find the amplitude at non-axial point P', have their centre on the point O', the intersection of the wave-front with the line joining S and P'. As seen from P', the aperture and the half-period zones appear as in Fig. 13.18, where it would be seen that the first and the second zones are still entirely exposed, together with about half of the third and fourth zones. The resultant amplitude at P' due to exposed zones is equal to

$$R_1 - R_2 + \frac{1}{2}R_3 - \frac{1}{2}R_4 = R_1 - \frac{1}{2}(R_1 + R_3) + \frac{1}{2}R_3 - \frac{1}{2}R_4$$

$$= \frac{1}{2}(R_1 - R_4) \text{ approximately}$$

which is a minimum value as compared to the amplitude at the axial point P_3. The intensity at P' is, therefore, less than that at the centre. The aperture is perfectly symmetrical about its axis. Therefore the central bright spot in this case is surrounded with an approximately dark ring. In a similar way it can be shown that there are other bright and dark rings surrounding the central spot.

Penetration of Light within the Geometrical Shadow

Let us consider a point on the screen just above the edge of the geometrical shadow of the aperture. The exposed zones are shown in Fig. 13.19. About one-third of the first zone, a somewhat larger portion of the second and varying parts of many more zones are exposed. There is, therefore, a small but definite amount of intensity within the geometrical shadow. But if the point is so well within the geometrical shadow that the first few zones and the last few zones are entirely cut off, then the contributions of the exposed portions of different zones mutually cancel, with the result that the intensity is zero.

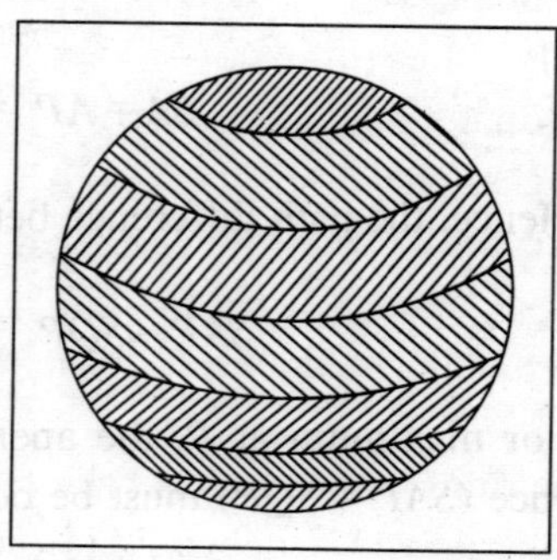

Fig. 13.19. Fresnel zones exposed by a circular aperture for a point within the geometrical shadow

13.8 DIFFRACTION AT AN OPAQUE DISC

A surprising result was deduced by *Poisson* in 1815 *i.e.* bright spot should always be formed at the centre of the shadow of small circular disc. He concluded that this seemingly absurd prediction was so far at variance with the recognised properties of light that it disproved the wave theory. The experiment was tried by *Argo*, who employed a circular disc of 2 mm in diameter. He showed that the geometrical shadow had a central bright spot. The complete diffraction pattern, besides the central spot and faint rings in the shadow, consists of bright circular fringes bordering the outside of the shadow. We shall show, by the help of *Fresnel half period zones*, that the centre of the shadow should always be bright, irrespective of distance of the screen from the disc.

Suppose a spherical wave-front *WW* of monochromatic light is incident on the circular disc *AB*, from the left. We divide this wave-front into *Fresnel zones* for an axial point in the usual manner. The areas of these zones, as shown in *Sec.* 13.7 depend on the distance *b* of the axial point from the disc. Let us suppose that for the point P_1, the disc covers up the first zone. The second zone is then the first exposed zone, sending light to P_1. Therefore, to find the resultant amplitude at P_1, due to the entire exposed zones, we should perform the summation beginning with R_2. We get for the amplitude at P_1,

$$D_1 = R_2 - R_3 + R_4 - R_5 + \ldots = R_2/2$$

and the intensity at P_1 is proportional to $R_2^2/4$, which is approximately the same as $R_1^2/4$, the intensity obtainable if the entire wave were exposed. The centre of shadow therefore, is nearly as bright as when the disc were absent.

When the screen is moved towards the disc, as shown in *Sec.* 13.7, the area of each zone diminishes and the disc, therefore, covers up a large number of zones. Let us suppose that disc covers up the first two zones, when the screen is at P_2, a point the nearer to disc than P_1. The resultant amplitude at P_2 is therefore, given by

$$D_2 = R_3 - R_4 + R_5 - R_6 + \ldots = R_3/2$$

and the intensity at P_2, is proportional to $(R_3/2)^2$. Similarly, the intensity at any point P is proportional to $(R_4/2)^2$, $(R_5/2)^2$, $(R_6/2)^2$ etc., according as the first three, four, five.... zones are intercepted. Thus the centre of geometrical shadow is always bright. The central spot becomes smaller in diameter as the disc becomes large.

The rings in the shadow are similar in origin as the interference fringes within the shadow of the narrow wire (see *Sec.* 13.6) while the bright and comparatively dark fringes *bordering* the outside of the shadow are similar in origin as the diffraction fringes due to a straight edge (*Sec.* 13.5).

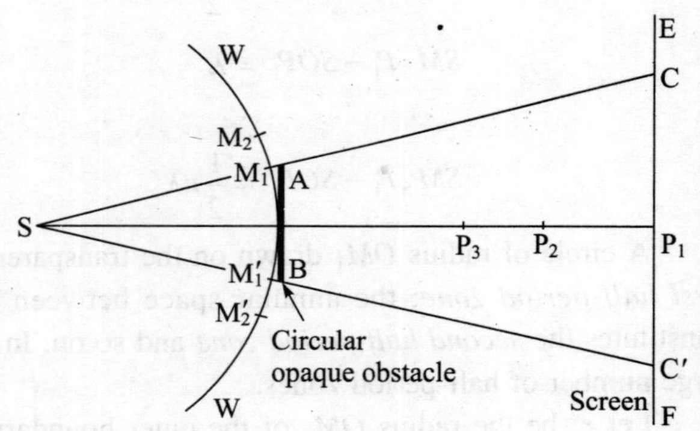

Fig. 13.20. Fresnel diffraction of a spherical wave-front by a circular opaque obstacle

13.9 ZONE PLATE

A remarkable experimental confirmation of *Fresnel's theory of Half-period Zones* is provided by an optical device known as *Zone plate*. The zone plate is so called because it is constructed following the Fresnel zone law. It is simply a plane parallel glass plate, having concentric circles of radii accurately proportional to the square roots of the consecutive natural numbers 1, 2, 3, etc. The even or odd order annular spaces between the circles are completely dark. Let us suppose that a plane monochromatic light wave, for which the plate is constructed, is incident normally on it. On the other side, on its axis we shall encounter a series of maxima points having intensities in the increasing order, as the distance of the points from the plate increases. The zone plate acts, therefore, like a convergent lens but unlike the lens, it has a series of foci

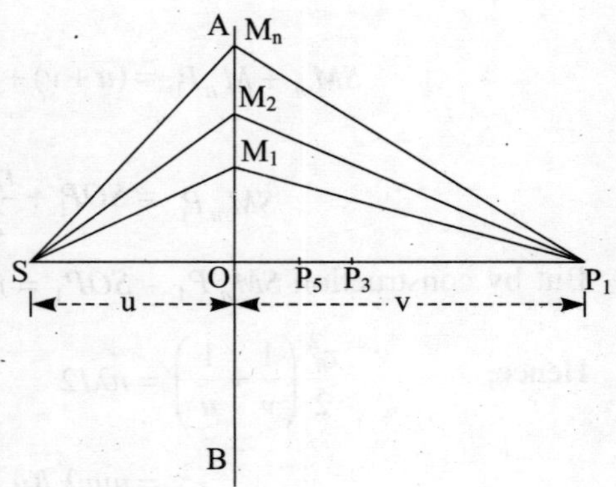

Fig. 13.21. Theory of zone plate

and associated focal lengths. There are also series of virtual foci on the side of the source. In other words, the zone plate is capable of acting as a divergent lens also. The theory of the zone plate and its properties are discussed below.

Let AB in Fig. 13.21 represent a plane parallel transparent screen perpendicular to the plane of paper and S is a point source of monochromatic light of wavelength λ.

We have to find the resultant amplitude at a point P_1 due to secondary wavelets diverging from the various points of the screen under the influence of spherical waves diverging from S. We must, of course, consider that the screen is not a wave surface and so the virtual secondary sources are not in phase. The resultant amplitude can be determined by dividing the entire screen into half-period zones. We mark off points M_1, M_2 etc. on the screen so that

$$SM_1P_1 - SOP_1 = \frac{1}{2}\lambda$$

$$SM_2P_1 - SOP_1 = \lambda$$

$$\dots\dots\dots$$

$$SM_nP_1 - SOP_1 = \frac{1}{2}n\lambda$$

A circle of radius OM_1 drawn on the transparent screen constitutes what is known as *Fresnel's first half-period zone*; the annular space between the concentric circles of radii OM_1 and OM_2 constitutes the *second half-period zone* and so on. In this way the entire screen can be divided into a large number of half-period zones.

Let r_n be the radius OM_n of the outer boundary of the nth zone and let $OS = u$ and $OP_1 = v$. Then from the right angled triangle SOM_n, we have

$$SM_n^2 = SO^2 + OM_n^2 = u^2 + r_n^2$$

$$\therefore \qquad SM_n = u(1 + r_n^2/u^2)^{1/2} = u + \frac{1}{2}r_n^2/u$$

since r_n is small in comparison with u. Similarly

$$P_1M_n = v + \frac{1}{2}r_n^2/v$$

$$SM_n + M_nP_1 = (u+v) + \frac{r_n^2}{2}\left(\frac{1}{v} + \frac{1}{u}\right)$$

$$SM_nP_1 = SOP_1 + \frac{r_n^2}{2}\left(\frac{1}{v} + \frac{1}{u}\right)$$

But by construction $SM_nP_1 - SOP_1 = n\lambda/2$

Hence,
$$\frac{r_n^2}{2}\left(\frac{1}{v} + \frac{1}{u}\right) = n\lambda/2$$

$$r_n^2 = nuv\lambda/(u+v) \qquad \text{...(13.9 a)}$$

$$r_n^2 = nr_1^2 \qquad \text{...(13.9 b)}$$

The value of r for the first, second ..... zones can be obtained by substituting 1, 2, 3, ... for n in Eqn. (13.9 *a*). Thus the external radii of Fresnel's half-period zones, for given values of u, v and λ, are proportional to square roots of natural numbers.

The area of the nth half-period zone is given by

$$A_n = \frac{\pi v \lambda}{1 + v/u}$$

This being independent of n, all zones are equal in area (approximately). If u becomes infinite we shall have plane waves falling on the screen. The area of each half-period zone reduces to $\pi v \lambda$.

Let R_1, R_2, R_3,...... R_n be the resultant amplitudes at P_1 due to secondary wavelets from 1st, 2nd, 3rd,...... and the nth zone. Owing to obliquity factor $(1 + \cos \theta_n)$, already discussed in *Sec.* 13.1, the magnitudes of R_1, R_2, R_3 etc. are in decreasing order. Therefore the resultant amplitude at P_1 due to entire zones is given by

$$D = R_1 - R_2 + R_3 - R_4 + ... + (-1)^{n-1} R_n$$

$$D = \frac{1}{2}(R_1 \pm R_n)$$

where n is finite and small. But when n is very large, then

$$D = ½\, R_1$$

The resultant amplitude at P_1 due to either the odd numbered zones or due to the even numbered zones alone would be clearly much greater than that of the entire wave. This may by tested experimentally by blocking off, by some means, say the even numbered zones. The resultant amplitude at P_1 in this case becomes

$$D_1 = R_1 + R_3 + R_5 + R_7 + ... = \frac{1}{2} N R_1 \text{ (approx.)} \qquad ...(13.9\ c)$$

where N is the total number of zones in the screen and $N/2$ zones are exposed. The intensity at P_1 is, therefore, proportional to ¼ $N^2 R_1^{\,2}$ which is N^2 times greater than the resultant intensity $R_1^{\,2}/4$ due to all zones. For example, if only alternate exposed zones are 10 in number, that is $N = 20$, even then the intensity is 400 times as great as that obtained when all zones are exposed. Under this condition P_1 will by a point of maximum intensity. In other words, most of the light from S will be concentrated at P_1. If odd zones are blocked, the resultant amplitude at P_1 becomes

$$D_1' = R_2 + R_4 + R_6 ......$$

and again P_1 will be a point of maximum intensity. We can, therefore, say that under these conditions P_1 is the image of S and its distance from the zone plate satisfies the relation Eqn. (13.9 *a*) which can be expressed in the form

$$1/v + 1/u = n\lambda / r_n^2 \qquad ...(13.9\ d)$$

a result similar to the lens equation.

A zone plate can be very conveniently prepared by the following method:

On a sheet of white paper we first of all draw concentric circles of radii accurately proportional to the square roots of consecutive integers 1, 2, 3, 4 etc. The alternate zones are then blackened and a very reduced copy of the same is obtained by photographing it on a plane parallel glass negative. When the plate has been fixed, the negative is a zone plate. Because of the difficulty of drawing a large number of circles accurately, another method for constructing the zone plate may be employed. Dark rings of a Newton rings interference pattern it will be recalled, have radii proportional to the square roots of consecutive integers 1, 2, 3, 4... . The dark and bright rings have equal areas when parallel monochromatic light is employed. Thus it is possible to construct a zone plate by photographing Newton's rings as seen by the reflected light.

The zone plate acts as a convergent lens not only for axial points but also for small extended objects. Let us suppose that a small object PQ is situated at a distance u from the plate on its axis. Let $PQ = x$ and $P'Q' = y$. It is easy to see from Fig. 13.22, that

$$PR = [u^2 + (OR - x)^2]^{1/2} = u + \frac{1}{2}(OR - x)^2/u$$

$$P'R = [v^2 + (OR + y)^2]^{1/2} = v + \frac{1}{2}(OR + y)^2/v$$

Hence
$$PRP' = (u + v) + \frac{1}{2}(OR - x)^2/u + \frac{1}{2}(OR + y)^2/v$$

Similarly
$$PR'P' = (u + v) + \frac{1}{2}(OR' - x)^2/u + \frac{1}{2}(OR' + y)^2/v$$

$$PRP' - PR'P' = \frac{1}{2}(OR^2 - OR'^2)\left(\frac{1}{v} + \frac{1}{u}\right) - RR'\left(\frac{x}{u} - \frac{y}{v}\right) \quad ...(13.9\ e)$$

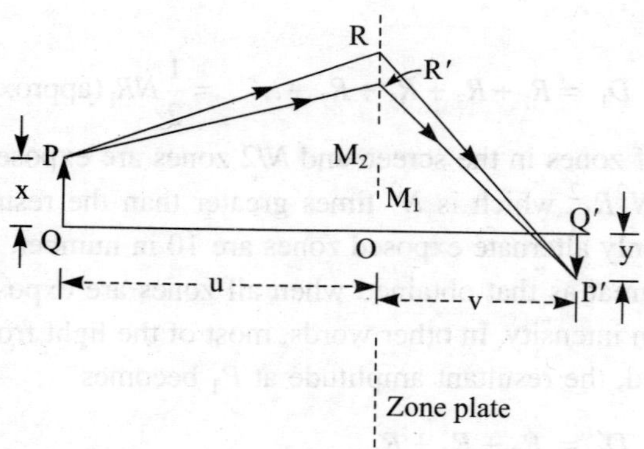

Fig. 13.22. Illustrates image formation by zone plate

Let R and R' be the two corresponding points on two successive transparent zones. For concreteness, let R be on the outer boundary of the nth zone and R' on that of the $(n - 2)$th zone, the $(n - 1)$th zone being opaque. Therefore, by the help of Eqn. (13.9 a), we have

$$OR^2 - OR'^2 = \frac{uv\lambda}{u + v}[n - (n - 2)] = \frac{2uv\lambda}{u + v}$$

Hence Eqn. (13.9 *e*) assumes the form

$$PRP' - PR'P' = \lambda - RR'\left(\frac{x}{u} - \frac{y}{v}\right)$$

If the path difference between the waves through corresponding points of the zone plate be λ the waves reinforce each other at P'. Then the second term of the above equation vanishes and gives the relation

$$y/x = v/u$$

Under these conditions P' is the image of P. The zone plate, therefore, also forms well defined images of small extended objects and the magnification, y/x, follows the same law as in the case of a lens.

Multiple Foci of a Zone Plate

The focal length of the zone plate can be easily obtained by making u infinite in Eqn. (13.9 *d*)

$$f_n = r_n^2/n\lambda \quad \text{...(13.9 } f\text{)}$$

For $n = 1$,

$$f_1 = r_1^2/\lambda \quad \text{...(13.9 } g\text{)}$$

The point P_1 given by the above relation is called the primary focus or the first order focal point and is the most intense. Here one Fresnel zone coincides with each actual zone of the plate. f_1 is known as the primary focal length of the plate.

There are a series of foci, unlike convergent lens, of diminishing intensity as we go along the axis towards the plate. The existence of these foci is simply due to the reason that the area of each zone diminishes as points nearer to the plate are considered. Thus for a point P_3 on the axis at one-third the distance from the plate to the first focus, the actual first zone contains 1st, 2nd and 3rd half-period zones. The second black zone intercepts 4th, 5th and 6th new Fresnel zones while the wavelets from 7th, 8th, 9th new zones, contained in the original 3rd zone, are transmitted and so on. Therefore, the resultant amplitude at P_3 is given by

$$\begin{aligned} D_3 &= (R_1 - R_2 + R_3) + (R_7 - R_8 + R_9) + (R_{13} - R_{14} + R_{15}) + \ldots\ldots \\ &= \left\{R_1 - \frac{1}{2}(R_1 + R_3) + R_3\right\} + \left\{R_7 - \frac{1}{2}(R_7 + R_9) + R_9\right\} + \ldots\ldots\ldots \\ &= \frac{1}{2}(R_1 + R_3 + R_7 + R_9 + \ldots\ldots\ldots) \text{ approximately} \end{aligned}$$

The intensity at P_3 is a maximum which is, therefore, the second focal point but it is much less intense than first one. The light rays transmitted through each successive transparent zone have a path difference of 3λ. Hence, P_3 is also called third order focal point. The first original zone exposes three new zones, all of equal area. Therefore, we have the relation

$$\begin{aligned} 3\pi a_1^2 &= \pi r_1^2 \\ a_1 &= r_1/\sqrt{3} \end{aligned}$$

where a_1 is the radius of the new first zone. Therefore, replacing r_1 by a_1 in Eqn. (13.9 *g*), the focal length associated with the second focal point P_3 is obtained as

$$f_3 = a_1^2/\lambda = r_1^2/3\lambda = (1/3)f_1$$

The fifth-order focal point P_5 occurs at a position in which Fresnel zones are alternately passed and blocked off in sets of five. The resultant amplitude at P_5 is given by

$$D_5 = (R_1 - R_2 + R_3 - R_4 + R_5) + (R_{11} - R_{12} + R_{13} - R_{14} + R_{15}) + \ldots\ldots$$
$$= \frac{1}{2}(R_1 + R_5 + R_{11} + R_{15} + \ldots\ldots) \text{ approximately}$$

Thus P_5 is still less bright focus and the corresponding focal length is given by

$$f_5 = b_1^2/\lambda = r_1^2/5\lambda = (1/5) f_1$$

where b_1 is the radius of the first new zone.

A zone plate has multiple foci at $f_1, \frac{1}{3}f_1, \frac{1}{5}f_1, \frac{1}{7}f_1$ etc. and their intensity is in decreasing order. Generalising the result we can say that, if for any point on the axis odd number of Fresnel zones are alternately passed and blocked off, the point under consideration is a focal point. The focal length associated with various foci can be obtained by putting $m = 0, 1, 2, 3$ etc. in the expression

$$f_{2m+1} = \frac{r_1^2}{(2m+1)\lambda}$$

Similar reasoning shows that there should be no even-order focal points. However, all the zone plates tested had a second order focal point at ½f_1 approximately. This was caused partly due to unequal areas of opaque and transparent zones. Owing to this reason, the two Fresnel zones, which should coincide with each actual zone of the plate at the second order focal point, do not completely cancel each other but a partial reinforcement of light occurs.

The zone plate also acts simultaneously as a concave lens but unlike the lens, it has multiple virtual foci on the side of the source. The explanation is given in *Sec.* 14.14.

SOLVED EXAMPLES

Example 1. *A very narrow vertical slit illuminated with sodium light of wavelength 5.9 × 10⁻⁵cm casts a shadow of a vertical copper wire 1 mm in diameter and 2 metres away, on a screen 3 metres away from the wire. Calculate the distance apart of the bands inside the shadow and the total number of bands which can be seen?*

In Fig. 13.23, *AB* represents the region of geometrical shadow of the wire *O*. As explained in Sec. 13.6, in this region of the screen equidistant bright and dark bands are observed. Their distance apart is expressed by

$$\bar{X} = D\lambda/2d$$

Fig. 13.23

In the given problem, $D = OC = 300$ cm, $2d = S_1S_2 = 0.1$ cm and $\lambda = 5.9 \times 10^{-5}$ cm

Hence, $\bar{X} = 300 \times 5.9 \times 10^{-5}/0.1$

$= \mathbf{0.177}$ **cm**

Also, from similar triangles SS_1S_2 and SAB, we have the relation

$$AB/SC = S_1S_2/SO$$

With the given data, $AB = \dfrac{0.1 \times 500}{200} = 0.25$ cm

At the point C equidistant from S_1 and S_2, a bright fringe will be observed. The separation between the two bright interference fringes, each of the order 1, would be 2×0.177 cm *i.e.*, 0.354 cm (if formed). This separation exceeds the width of the geometrical shadow AB. Hence only **2 dark bands** will be seen within the geometrical shadow.

Example 2. *A strong parallel beam of monochromatic light is incident normally on a thin plate having a small circular hole of diameter 1 mm. If the screen be moved through a distance of 12.5 cm from the first position where the centre is black to the second similar position, what is the wavelength of light used.*

From the theory of Fresnel diffraction due to circular aperture, (*Sec.* 13.7), it follows that for the first position of the screen where the centre of diffraction is black (not absolute), the aperture should expose two Fresnel's half-period zones of the wave-front. Accordingly, the distance of the centre of diffraction pattern from the edge of the aperture must be $(x + \lambda)$, where x is the separation between the screen and the aperture. It is now easy to write the relation

$$(x+\lambda)^2 = x^2 + r^2$$

Neglecting the term λ^2, being extremely small as compared with other terms, the above equation reduces to

$$2x\lambda = r^2$$

The second similar position of the screen, as explained in *Sec.* 13.7, is closer to the aperture. For this setting of screen the aperture should expose four half-period zones of the wave-front. Accordingly, the distance of the centre of diffraction pattern from the edge of the aperture would be $(x - 12.5 + 2\lambda)$. It is now easy to write the relation

$$(x-12.5+2\lambda)^2 = (x-12.5)^2 + r^2$$

Neglecting the term involving λ^2, the above equation reduces to the form

$$4x\lambda - 50\lambda = r^2$$

Hence $\lambda = r^2/50 = 0.05 \times 0.05/50 = \mathbf{5 \times 10^{-5}}$ **cm.**

Example 3. *A zone plate is made by arranging that the radii of the circles which define the zones are the same as the radii of Newton's rings formed between a plane surface and surface whose radii of curvature is 2 metres. Find the primary focal length of the zone plate.*

The square of the radii of dark Newton's rings for the case of air film ($\mu = 1$) reduces to

$$r_n^2 = n\lambda R$$

where n = 1, 2, 3, 4 etc. for the first, second, third..... dark rings respectively and R is the radius of curvature of the curved surface of the film.

According to the condition of the problem, the above equation also expresses the square of the radii of the circles which define the zones of the zone plate. The primary focal length of the zone plate, expressed by Eqn. (13.9 *g*) is

$$f_1 = r_1^2/\lambda$$

which, therefore, reduces to

$$f_1 = R = \mathbf{200\ cm.}$$

PROBLEMS

1. (a) Explain the phenomenon of diffraction of light.
   (b) Explain the difference between Fresnel and Fraunhofer class of diffraction phenomena.
   (c) Distinguish clearly between interference and diffraction.
2. What are half-period zones? Show that the amplitude due to a large wave-front at a point in front is just half that due to the first half-period zone acting alone.
   Hence explain the rectilinear propagation of light on the basis of wave theory.
3. What is a zone-plate? How is it constructed? Show that a zone-plate has multiple foci. Compare the zone-plate with a convex lens.
4. Describe giving necessary theory the Fresnel type of diffraction produced by a straight edge. Explain how you would obtain the wavelength of light in this case, stating precisely what observations you would make.
   Draw a diagram to indicate the intensity distribution of light in the diffraction pattern.
5. Describe and explain the diffraction pattern formed by a narrow wire illuminated by a monochromatic light from a narrow slit parallel to the wire. How would you use the pattern to measure the thickness of the wire.
6. Give Fresnel's theory of diffraction and explain with its help, why the centre of the shadow of a small disc is bright.
7. A narrow circular aperture is held on one side of a fine hole and a screen is placed at some distance on the opposite side. Explain the illumination observed on the screen when a beam of monochromatic light is made to pass normally.
8. What is the radius of the first zone in a zone plate of focal length 20 cm for light of wavelength 5000 A.U.?
9. A zone plate has the radius of the first ring 0.05 cm. If plane waves (λ = 5000 A.U.) fall on the plate, where should the screen be placed so that light is focussed to a bright spot?
10. A zone plate gives a series of images of a point source on its axis. If the strongest and the second strongest images are at 30 cm and 6 cm respectively from the plate, both on the same side remote from the source, calculate the distance of the source from the zone plate.
11. If the diameter of a wire be 0.2 mm, calculate the separation between the fringes formed on a screen placed at 50 cm from the wire. The light source used has a wavelength of 5000 A.U. and is placed at a finite distance from the wire.
12. A narrow slit illuminated by light of wavelength 6.4×10^{-5} cm is placed at a distance of 3 metres from a straight edge. If the distance between the straight edge and the screen is 6 metres, calculate the distance between the first and the fourth dark bands.

13. Light of wavelength 6×10^{-5} cm passes through a narrow circular aperture of radius 0.09 cm. At what distance along the axis will the first maximum intensity be observed?
14. A circular aperture of 1.2 mm diameter is illuminated by plane waves of monochromatic light. The diffracted light is received on a distant screen which is gradually moved towards the aperture. The centre of the circular path of light first becomes dark when the screen is 30 cm from the aperture. Find the wavelength of light used.
15. A monochromatic beam of light on passing through a slit 1.6 mm wide falls on a screen held close to the slit. The screen is then gradually moved away and the middle of the patch of light on it becomes dark when the screen is 50 cm from the slit. Calculate the wavelength of light used.
16. A very narrow vertical slit is illuminated with monochromatic light of wavelength 5461 A.U. A vertical wire of 0.1 cm diameter is placed in front of the slit, parallel to it. A screen is placed 3 metres from the wire. Calculate the band width within the geometrical shadow.
17. The diameter of the central zone of the zone plate is 0.23 cm. If a point source emitting light of wavelength 6000 A.U. is placed 600 cm from the zone plate, find the position of the primary image and that of other weaker images.

DIFFRACTION OF LIGHT
(Fraunhofer Class)

We have considered, in the preceding chapter, *Fresnel diffraction* patterns resulting from the diffraction of spherical or cylindrical wave-fronts at the edges of obstacles while observations are made at a small distance away from obstacles. We now turn our attention to *Fraunhofer diffraction* which arises when the source of light and the screen are effectively at infinite distance from the diffracting aperture. The wave-front incident on the aperture is plane and accordingly the secondary wavelets which we consider as originating from the plane of aperture at this instant originate in equal phase and the diffracted wave is plane. We shall see that owing to this fact the summation of optical disturbances reaching any point is considerably simpler from the mathematical point of view than that in the *Fresnel diffraction* in which the diffracted wave-front is not plane.

14.1 RESULTANT OF *N* SIMPLE HARMONIC VIBRATIONS OF EQUAL AMPLITUDES, SAME PERIOD, AND PHASES INCREASING IN AN ARITHMETICAL PROGRESSION

Let us first of all consider the interference pattern produced due to the superposition of N vibrations of the same period and amplitude and phase varying in arithmetical progression. Such a situation may arise as for example while considering the diffraction pattern due to N identical and uniformly spaced slits. We shall suppose that the distance between the centre lines of two consecutive slits is d and that between the first and the last slits is D. We can thus write

$$D = (N - 1)\, d. \qquad ...(14.1\ a)$$

For the present, we shall suppose that each slit is so fine that the phase shift between contributions from extreme parts of the same slit can be safely neglected.

When a plane wave falls upon this array of fine slits the entire wave-front excepting the parts illuminating the N slits is blocked. Each of the slits then becomes the source of secondary wavelets having the same amplitude and same initial phase. Let us consider the resultant effect of all the secondary wavelets along a direction θ to the axis of the system at a distant field point P.

Since the distance between the slits and field point P is large, the variation of amplitude of successive disturbances reaching the point P can be safely neglected. This is however, not the case with the phases of these disturbances. For example, if we consider, say, the ith slit the phase of the corresponding disturbance on arrival at P will depend upon the distance x_i between the slit and the field point. In this case, the disturbance itself will be the real part of

$$y_i = ae^{-i\omega t} e^{ikx_1} \quad \text{...(14.1 } b)$$

where the symbols have their usual meaning. Also, it is evident from Fig. 14.1 that on arrival at P the phase difference between waves from successive slits is given by

$$\alpha = \phi_{i+1} - \phi_i - \frac{2\pi}{\lambda}(d \sin\theta) \quad \text{...(14.1 } c)$$

Similarly, the phase difference between the waves from extreme slits is given by

$$\phi_N - \phi_1 = \frac{2\pi}{\lambda}(D\sin\theta) \quad \text{...(14.1 } d)$$

Combining Eqns. (14.1 *a*) and (14.1 *d*), we get

$$\phi_N - \phi_1 = (N-1)\,\alpha \quad \text{...(14.1 } e)$$

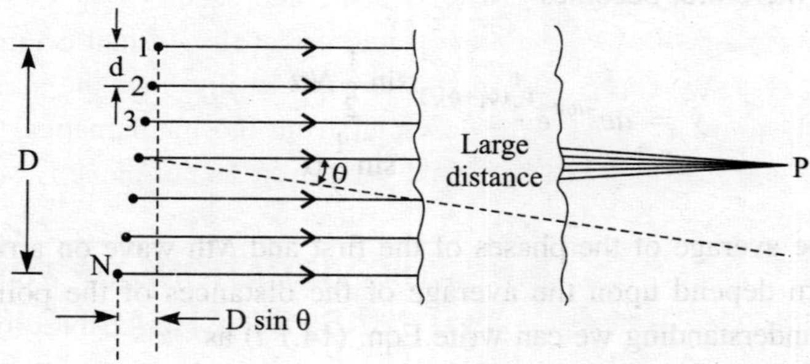

Fig. 14.1. N narrow slits

The problem thus reduces to the calculation of the resultant disturbance at P due to the superposition of N waves of the same amplitude A and phases given by Eqn. (14.1 *c*). This will be the real part of

$$y = ae^{-i\omega t}[e^{i\phi_1} + e^{i\phi_2} + \ldots\ldots + e^{i\phi_N}]$$

$$= ae^{-i\omega t} e^{i\phi_1}[1 + e^{i(\phi_2-\phi_1)} + e^{i(\phi_3-\phi_1)} + \ldots\ldots + e^{i(\phi_N-\phi_1)}] \quad \text{...(14.1 } f)$$

In view of Eqn. (14.1 *c*), we can write Eqn. (14.1 *f*) as

$$y = ae^{-i\omega t} e^{i\phi_1}[1 + e^{i\alpha} + e^{2i\alpha} + \ldots + e^{i(N-1)\alpha}]$$

$$= ae^{-i\omega t} e^{i\phi_1} \cdot \frac{e^{iN\alpha} - 1}{e^{i\alpha} - 1} \quad \text{...(14.1 } g)$$

where the fraction on the RHS of Eqn. (14.1 *g*) is the sum of the *G.P.* of ratio $e^{i\alpha}$ enclosed within the square brackets. We further simplify Eqn. (14.1 *g*) by putting

$$\frac{e^{iN\alpha}-1}{e^{i\alpha}-1} = \frac{e^{iN\alpha/2}-e^{-iN\alpha/2}}{e^{i\alpha/2}-e^{-i\alpha/2}}\cdot\frac{e^{iN\alpha/2}}{e^{i\alpha/2}}$$

thus, we have

$$y = ae^{-i\omega t}\cdot e^{i\phi_1}\cdot e^{i\alpha(N-1)/2}\cdot\frac{e^{iN\alpha/2}-e^{-iN\alpha/2}}{e^{i\alpha/2}-e^{-i\alpha/2}} \qquad \text{...(14.1 } h)$$

We know that

$$\frac{e^{iN\alpha/2}-e^{-iN\alpha/2}}{e^{i\alpha/2}-e^{-i\alpha/2}} = \frac{\sin\frac{1}{2}N\alpha}{\sin\frac{1}{2}\alpha}$$

Also because of Eqn. (14.1 *e*)

$$\phi_1+\frac{1}{2}(N-1)\alpha = \phi_1+\frac{1}{2}(\phi_N-\phi_1) = \frac{1}{2}(\phi_N+\phi_1)$$

Equation (14.1 *h*), therefore, becomes

$$y = ae^{-i\omega t}e^{i\frac{1}{2}(\phi_1+\phi_N)}\frac{\sin\frac{1}{2}N\alpha}{\sin\frac{1}{2}\alpha} \qquad \text{...(14.1 } i)$$

½ $(\phi_1+\phi_N)$ is in fact the average of the phases of the first and *N*th wave on arrival at the point *P*. This average will in turn depend upon the average of the distances of the point *P* from the two extreme slits. With this understanding we can write Eqn. (14.1 *i*) as

$$y = ae^{-i\omega t}e^{i\phi_{Av}}\cdot\frac{\sin\frac{1}{2}N\alpha}{\sin\frac{1}{2}\alpha} \qquad \text{...(14.1 } j)$$

Taking the real part out, we find that the resultant disturbance at *P* is given by

$$y = a\frac{\sin\frac{1}{2}N\alpha}{\sin\frac{1}{2}\alpha}\cos(\omega t-\phi_{Av}) \qquad \text{...(14.1 } k)$$

Substituting the value of α the resultant amplitude is seen to be

$$\text{Amp} = a\frac{\sin\left[\frac{1}{2}N(2\pi/\lambda)d\sin\theta\right]}{\sin\left[\frac{1}{2}(2\pi/\lambda)d\sin\theta\right]}$$

$$= a\frac{\sin[(N\pi/\lambda)d\sin\theta]}{\sin[(\pi/\lambda)d\sin\theta]} \qquad ...(14.1\ l)$$

Also, because $D = (N - 1)\ d$, Eqn. (14.1 l) can be alternatively written as

$$\text{Amp} = a\sin\left(\frac{N\pi}{N-1}\frac{D\sin\theta}{\lambda}\right)\Big/\sin\left(\frac{\pi}{N-1}\frac{D\sin\theta}{\lambda}\right) \qquad ...(14.1\ m)$$

This formulation is of general application for any number of slits starting from $N = 2$. We can also use it for finding the diffraction pattern of a single slit with a neat little trick.

14.2 DIFFRACTION BY A SINGLE SLIT

We first consider, as an example of *Fraunhofer class*, the diffraction of monochromatic light by a narrow slit. Let a beam of parallel monochromatic light be incident normally on an opaque plate in which there is a long narrow slit. According to the laws of *Geometrical Optics*, the transmitted beam should have the same cross-section as the slit and on the screen, there should be uniform illumination over an area of the same size and shape as the slit. Actually, the transmitted beam spreads out perpendicularly to the length of the slit. Therefore, when it is focussed on the screen by a lens, what is observed is a diffraction pattern. It consists of central band, much wider than the slit width, situated directly opposite to the slit and bordered by dark and bright bands of decreasing intensity. The central bright band is extremely intense and its width is twice as great as that of fainter side bands. We shall now theoretically explain these characteristics of the single slit diffraction pattern.

Figure 14.2 represents a section *AE*, by the plane of paper of a slit of width *D*. Let a plane wave-front *WW* of monochromatic light of wavelength λ propagating normally to the slit be incident on it, which obviously limits the size of the wave-front, at the instant it occupies the plane of the slit. According to the principle of *Huygens-Fresnel*, at this instant, every point of the wavefront in the plane of the slit is to be regarded as the origin of secondary spherical wavelets, which spread out to the right in all directions. The parts of each wavelet travelling normally to slit, are brought to focus at P_0 by the lens while the parts, marked by cross, travelling at an angle θ with the normal are brought to focus at P_1. Therefore, we can regard *AB*, *CD*, *EF* etc. as parallel rays, diffracted at an angle θ. Exactly along the same argument it follows that diffracted rays start from every point in every direction.

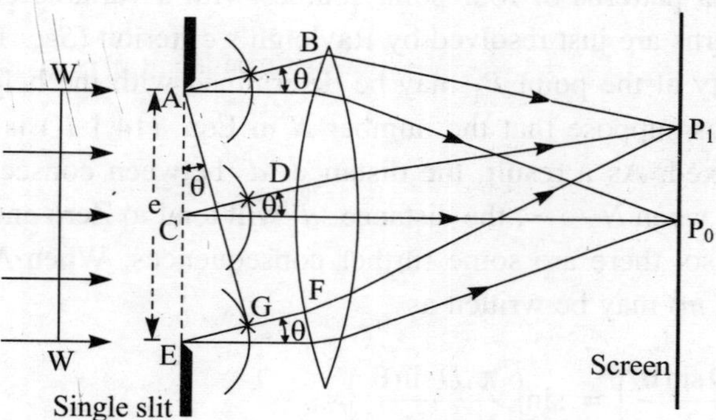

Fig. 14.2. Fraunhofer diffraction of a plane wave-front by a single slit

Fraunhofer diffraction pattern of a single slit

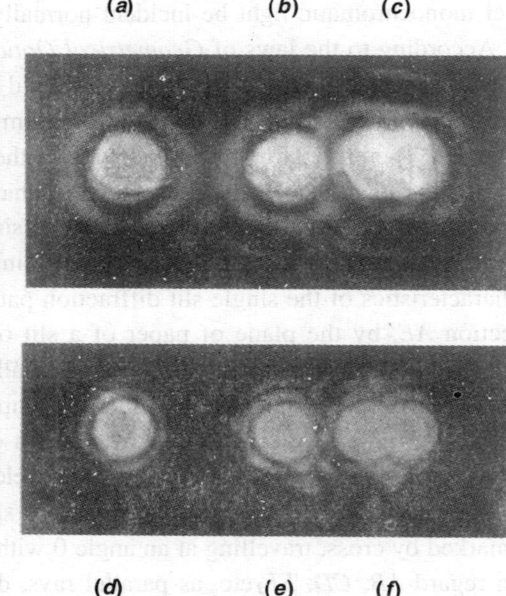

(a) (b) (c)

(d) (e) (f)

Fraunhofer diffraction patterns of four point sources with a variable circular opening in front of the lens. In (*c*) the patterns are just resolved by Rayleigh's criterion (*Sec.* 15.1).

The resultant intensity at the point P_1 may be determined with the help of Eqn. (14.1 *l*). For the sake of argument let us suppose that the number *N* in Eqn. (14.1 *l*) is made very large while keeping the distance *D* fixed. As a result, the distance '*d*' between consecutive slits will become very small and in the limit when $N \to \infty$, the distance '*d*' will tend to Zero and we shall be left with a single slit of width *D*. Also, there are some further consequences. When *N* is infinitely large the denominator of Eqn. (14.1 *m*) may be written as

$$\sin\left(\frac{\pi}{N-1}\frac{D\sin\theta}{\lambda}\right) = \sin\left(\frac{\pi}{N}\frac{D\sin\theta}{\lambda}\right) = \frac{\;}{N} \qquad \text{...(14.2 } a\text{)}$$

At the same time, the factor $N/(N-1)$ in the numerator of Eqn. (14.1 *m*) is effectively equal to unity. With these result, we get

$$\text{Amp }(R) = a\frac{\sin(\pi/\lambda)D\sin\theta}{(\pi/N)(D\sin\pi/\lambda)}$$

$$R = Na\sin\{(\pi/\lambda)D\sin\theta\}/\{(\pi/\lambda)(D\sin\theta)\}$$

$$R = A\sin\alpha/\alpha \qquad ...(14.2\ b)$$

where $A = Na$ and $\alpha = \{(\pi/\lambda)D\sin\theta\}$

The resultant intensity at P_1, being proportional to the square of the resultant amplitude R, is given by

$$I \text{ proportional to } A^2\frac{\sin^2\alpha}{\alpha^2}$$

or

$$I = R_o^2\frac{\sin^2\alpha}{\alpha^2} = I_o\frac{\sin^2\alpha}{\alpha^2} \qquad ...(14.2\ c)$$

where $R_o^1 = I_o$ is interpreted as the intensity of principal maximum. Let us now proceed to analyse the intensity distribution on the screen.

Principal Maximum

The expression for the resultant amplitude R, on writing the expansion of sin α in ascending powers of α, becomes

$$R = \frac{A}{\alpha}\left[\alpha - \frac{\alpha^3}{3!} + \frac{\alpha^5}{5!} - \frac{\alpha^7}{7!} + ...\right]$$

$$= A\left[1 - \frac{\alpha^2}{3!} + \frac{\alpha^4}{5!} - \frac{\alpha^6}{7!} + ...\right]$$

It is now obvious that R will be maximum, if the negative terms vanish. This is only possible when

$$\alpha = \frac{\pi D\sin\theta}{\lambda} = 0 \quad \text{or} \quad \theta = 0$$

The maximum value of R is therefore A and the maximum intensity is proportional to A^2 and we designate this maximum intensity by I_o. The condition $\theta = 0$ simply means that this maximum is formed by parts of secondary wavelets which travel normally to the slit. The position of this maximum is, therefore, directly opposite the slit. This maximum is known as the principal maximum and as explained on the next page, it is bordered symmetrically by the dark and the bright bands.

Minimum Intensity Position

The intensity will be zero in the diffraction pattern at the position where

$$\sin \alpha = 0$$

The values of α which satisfy the above equation are

$$\alpha = \pm \pi, \pm 2\pi, \pm 3\pi, \text{ etc. } = \pm m\pi$$

which gives on substituting the value of α and simplifying

$$D \sin \theta = \pm m\lambda \qquad \text{...(14.2 } d\text{)}$$

where m = 1, 2, 3, 4 etc. give the first, the second, the third etc. points of minimum intensity, on either side of the principal maximum. The value m = 0 is not admissible, because for this value, $\theta = 0$ and this corresponds to the principal maximum. In words, Eqn. (14.2 *d*) simply expresses the condition that *the rays diffracted at an angle θ interfere destructively if the path difference between the extreme diffracted rays is an integral multiple of λ.*

Secondary Maxima

In addition to the principal maximum at $\alpha = 0$, there are weak secondary maxima between the equally spaced minima. The exact positions of these maxima can be obtained, in accordance with the rule of finding maxima and minima of a given function in calculus, by differentiating the expression for I,

$$I = I_o \frac{\sin^2 \alpha}{\alpha^2}$$

with respect to α and equating it to zero. Therefore, for maxima or minima, we have the condition

$$\frac{dI}{d\alpha} = 2 I_o \frac{\sin \alpha}{\alpha} \frac{[\alpha \cos \alpha - \sin \alpha]}{\alpha^2} = 0$$

Hence either $\sin \alpha = 0$ or $\alpha \cos \alpha - \sin \alpha = 0$.

The equation $\sin \alpha = 0$ gives the values of α (except zero) for which the intensity is zero on the screen. Positions of maxima are given by the roots of the equation

$$\alpha \cos \alpha - \sin \alpha = 0 \quad \text{or} \quad \alpha = \tan \alpha \qquad \text{...(14.2 } e\text{)}$$

The values of α satisfying the above equation are obtained graphically, by plotting the curves

$$y = \alpha \quad \text{and} \quad y = \tan \alpha$$

on the same graph. The points of intersection of the two curves give the values of α which satisfy Eqn. (14.2 *e*). Now, $y = \alpha$ is simply a straight line through the origin and inclined at 45° to the axis of α. To plot $y = \tan \alpha$, we derive mathematically the following characteristics of the curve which the equation represents.

(*i*) The curve cuts the axis of α at points where

$$y = \tan \alpha = 0$$

that is at points given by

$$\alpha = 0, \pm\pi, \pm 2\pi, \pm 3\pi \text{ etc.}$$

(*ii*) The inclination of the tangent to the curve with the axis of α is

$$\tan\psi = \frac{dy}{d\alpha} = \sec^2\alpha$$

and at points where the curve cuts the axis of α, we get for the inclination of the tangent

$$\tan\psi = 1$$

that is, it is inclined at 45° to the axis of α.

(*iii*) Asymptotes of the curve are parallel straight lines defined by

$$\alpha = \pm\pi/2, \pm 3\pi/2, \pm 5\pi/2 \text{ etc.}$$

The equation $y = \tan\alpha$, therefore, represents a family of curves, having the above stated characteristics, which can be now very easily drawn on the graph. The abscissae of the points of intersection between the curves $y = \alpha$ and $y = \tan\alpha$ are the require values of α for which the intensity is maximum. By inspection of the Fig. 14.3, it is clear that the first value of α is zero and then there are other values less than and gradually approaching ± 3π/2, ± 5π/2, ± 7π/2 etc. The value α = 0 gives principal maximum while other values, just given, give secondary maxima. They are displaced somewhat towards the middle of two consecutive minima and this displacement increases with the order of the secondary maxima.

We can form a rough idea regarding the relative intensities of maxima by assuming that the secondary maxima occur nearly, though not exactly, at α = ± 3π/2, ± 5π/2, ± 7π/2 etc.

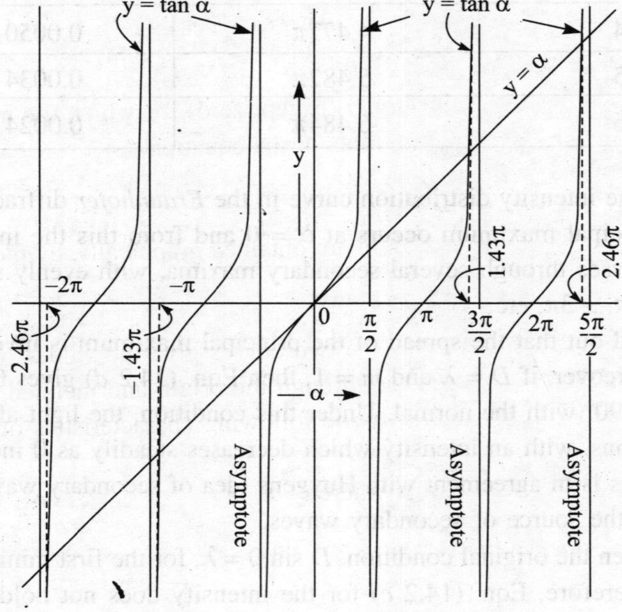

Fig. 14.3. Graphical solution of the equation, α = tan α

Let $R_o = 1$ for simplicity of argument.
For principal maximum $\alpha = 0$, therefore $I_o = 1$
For Ist secondary maximum $\alpha = 3\pi/2$ (approx.), therefore

$$I_1 = R_o^2\left(\frac{\sin 3\pi/2}{3\pi/2}\right)^2 = \frac{4}{9\pi^2} = \frac{1}{22.2}$$

Thus the intensities of the principal, the first, the second,...... Maximum are,

$$1, \frac{4}{9\pi^2}, \frac{4}{25\pi^2}, \frac{4}{49\pi^2}, \ldots\ldots, \text{ that is, } 1, \frac{1}{22.2}, \frac{1}{61.7}, \frac{1}{121} \text{ etc.}$$

The intensity of the principal maximum is over 20 times as great as that of the first secondary maximum, which is also quite obvious, since all the secondary wavelets will arrive in phase at the position of the principal maximum because the path difference is equal to zero. Secondary maxima are very feeble and their intensities diminish very rapidly. Most of the light is, therefore, concentrated in the central maximum. The exact values of α for the principal maximum and the first-six secondary maxima are given below, together with their relative intensities, assuming the intensity of the principal maximum to be unity.

Order of max.	α	I
0	0	1
1	1.430π	0.0469
2	2.459π	0.0168
3	3.471π	0.0083
4	4.477π	0.0050
5	5.482π	0.0034
6	6.484π	0.0024

We can now plot the intensity distribution curve in the *Fraunhofer* diffraction pattern of a slit. In the Fig. 14.4 the principal maximum occurs at $\alpha = 0$ and from this the intensity diminishes to zero at $\alpha = \pm\pi$, then passes through several secondary maxima, with evenly spaced points of zero intensity at $\alpha = \pm\pi, \pm 2\pi, \pm 3\pi$, etc.

It should be pointed out that the spread of the principal maximum is inversely proportional to the width of the slit. Moreover, if $D = \lambda$ and $m = 1$, then Eqn. (14.2 *d*) gives $\theta = \pm\pi/2$, that is, the first minimum occurs at 90º with the normal. Under this condition, the light after traversing the slit spreads out in all directions, with an intensity which decreases steadily as θ increases. There are no maxima and minima. This is in agreement with Huygens idea of secondary wavelets, for the slit has effectively now become the source of secondary waves.

Finally, if $D < \lambda$, then the original condition, $D \sin\theta = \lambda$, for the first minimum gives $\sin\theta > 1$, which is impossible. Therefore, Eqn. (14.2 *c*) for the intensity does not hold when the slit width becomes less than one wavelength. The statement breaks down.

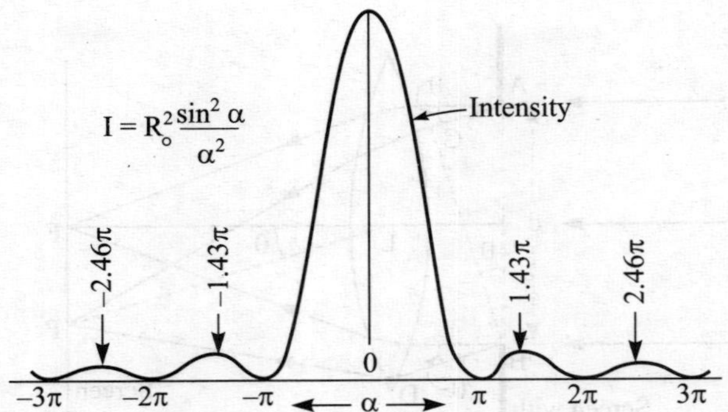

Fig. 14.4. Graph of intensity distribution in Fraunhofer diffraction of a single slit

14.3 DIFFRACTION AT A CIRCULAR APERTURE

The *Fraunhofer* diffraction of plane waves passing through a circular aperture is of fundamental importance in the discussion of the resolving power of telescope and microscope, in view of the circular form of the lenses employed as objectives and eye-pieces in these instruments. The diffraction pattern is characterised by a central bright disc, called *Airy's* disc, surrounded by a number of concentric fainter rings, the common centre being at a point where, in the absence of the diffraction phenomenon, the plane wave-front would have converged by the lens placed on the other side of the aperture. The exact theory was given by *Airy* in 1835 but it is too complicated, involving *Bessel Functions* which are beyond the scope of students for whom the book is meant. We shall content ourselves with a simplified treatment, given below, to explain the diffraction pattern produced by a circular aperture.

Let AB be a section of the circular aperture, of diameter d, by the plane of paper through its diameter, the plane of aperture being perpendicular to the plane of paper. Parallel monochromatic rays are incident normally on the aperture. From every point in the plane of the aperture, waves of light are diffracted in all directions. The rays travelling along the normal to the aperture, are brought to focus by the lens at its principal focus F, while the beam diffracted at an angle θ with the normal is brought to focus at P, distant r from F in the focal plane. Let AC and BD be the extreme rays in this diffracted beam. Then the path difference between them is

$$AC = AB \sin\theta = d \sin\theta$$

By the same reasonings as in the single slit diffraction pattern, discussed in *Sec.* 14.2, if the difference of paths between the extreme diffracted rays AC and BD is integral multiple of λ, there will be zero intensity at P. On the other hand, if the difference of paths is odd multiple of $\lambda/2$, there will be maximum intensity at P. That is, in symbols the conditions for minimum and maximum are

$$d \sin\theta = m\lambda \quad \text{[Minima]} \qquad ...(14.3\ a)$$

$$d \sin\theta = (2m+1)\lambda/2 \quad \text{[Maxima]} \qquad ...(14.3\ b)$$

where m = 1, 2, 3, 4, etc., of course for m = 0, we have central maximum at F.

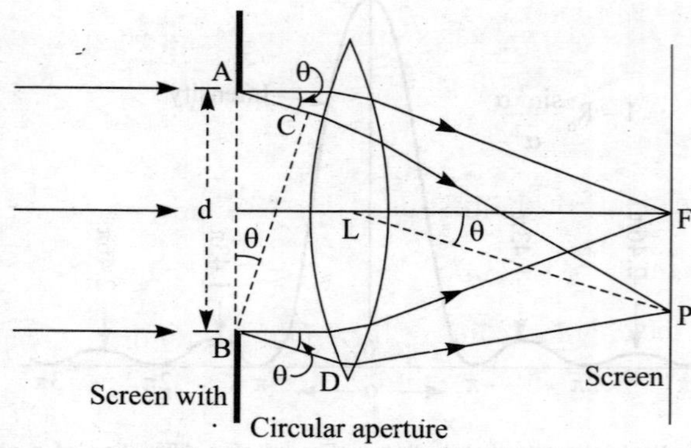

Fig. 14.5. Fraunhofer diffraction of a plane wave-front by a circular aperture

It is obvious from the figure that $\angle PLF = \theta$. Then

$$\sin\theta = PF/LP = PF/LF \text{ approx.} = r/f'$$

where f' is the focal length of the lens. Therefore, the condition for minima given by Eqn. (14.3 *a*) becomes

$$d\frac{r_m}{f'} = m\lambda$$

$$r_m = m\frac{f'\lambda}{d} \quad \text{[Dark rings]} \qquad ...(14.3\ c)$$

Similarly for maxima, we have

$$r_m = (2m+1)\frac{f'\lambda}{2d} \quad \text{[Bright rings]} \qquad ...(14.3\ d)$$

The circular aperture is perfectly symmetrical about the central axis passing through F. Therefore, the resulting diffraction pattern is obtained by considering the Fig. 14.5 rotated about the same axis. The point P traces out a circular ring of uniform illumination, whether zero or maximum depending upon the conditions given by Eqns. (14.3 *c*) and (14.3 *d*). Thus, there is formed a diffraction pattern characterised by a central bright disc surrounded by alternate dark and bright concentric rings. The radius of the first dark ring is obtained by putting $m = 1$ in Eqn. (14.3 *c*), we get

$$r_1 = f'\lambda/d$$

and this dark ring bounds the central maximum. Therefore r_1 may be also taken as the radius of the central bright disc.

Now if the lens is sufficiently close to the aperture or we may consider the diffraction from the periphery of the lens itself, instead of separate circular aperture, then the angle subtended at the centre of the aperture by the radius of the first dark ring is

$$\theta_1 = r_1/f' = \lambda/d$$

The more exact theory by *Airy* indicates that the first dark ring is given by the condition that the difference of paths between the extreme diffracted rays should be 1.22λ that is

$$d \sin \theta_1 = 1.22\lambda$$

Therefore, the angular radius of the first-dark ring becomes

$$\theta_1 = \sin \theta_1 = 1.22\lambda/d$$

which may also be taken as the angular radius of the bright central disc. About 84% of the transmitted light energy is concentrated in the central bright disc. It is now obvious that increasing the diameter of the circular aperture results in the decrement of the radius of the central disc. This fact is utilised in the construction of telescope objective for the astronomical work of high precision.

It should be borne in mind that in the *Fraunhofer diffraction* the central disc is always bright in contrast to *Fresnel diffraction* where the central disc may be bright or comparatively dark depending upon whether the circular aperture exposes odd or even number of *Fresnel zones* for an axial point on the screen.

14.4 DIFFRACTION AT TWO PARALLEL SLITS

The phenomenon of interference of light was first demonstrated by Dr. Young by the help of his double slit experiment. In the discussion of this experiment, it was assumed that the slit width was comparable with the wavelength of the light and they were close together. We now see that these conditions were essential, so that the central maximum in the diffraction patterns of each slit was wide enough to occupy the same large area of the screen (see discussion near the end of *Sec*. 14.2). Under these conditions, it was possible to consider the resultant displacement at any point of the screen, as simply due to superposition of the two interfering beams of light. Diffraction of light at each slit begins to play part in modifying the interference pattern due to two slits, when the width of each slit is increased to such an extent that it becomes comparable with the width of the opaque space separating them. We get on the screen what is now called *Fraunhofer diffraction* by a double slit, which is characterised by a number of equally spaced interference maxima and minima in the region normally occupied by the central maximum in the single slit diffraction. The central interference maximum, having intensity four times that of the central maximum in the single slit diffraction, is bordered by other interference maxima of gradually decreasing intensity. In the region of the secondary maxima (of single slit diffraction), faint interference maxima are now observed. We shall now explain theoretically the characteristics of the diffraction pattern.

Let us consider two parallel slits *AB* and *CD*, each of width *e*, separated by an opaque space *BC* of width *b*. Let a plane wave-front of light of wavelength λ be incident normally on the plane of slits. According to the principle of *Huygens-Fresnel*, every point on the wave-front is to be regarded as an origin of secondary wavelets at the instant it occupies the plane of the slits. Thus from every point in the slits, the rays of light are radiated out in all directions. The resultant amplitude due to rays, from each slit, diffracted in a direction θ to normal, as in *Sec*. 14.2, is $A \sin \dfrac{\sin \alpha}{\alpha}$ and the resultant phase in this diffraction is $\alpha = (\pi\, e \sin \theta)/\lambda$. But the phase difference between the secondary wavelets from A and S_1, the middle point of AB, on reaching the focal point P_1 is $(\pi\ e \sin\theta)/\lambda'$. We can therefore, replace, for the purpose of theoretical discussion, all the secondary wavelets in AB and CD by single waves of amplitude $(A \sin\alpha)/\alpha$ and initial phase α originating from the middle points S_1 and S_2 of AB and CD travelling in a direction θ to normal. Consequently the resultant amplitude at P_1 is simply equal to the resultant due to the interference of two waves of equal amplitude $\dfrac{A\sin\alpha}{\alpha}$ originating in same phase α; on reaching P_1, however these waves develop a phase difference depending upon the difference in their optical paths. We drop perpendicular S_2K_1 from S_2 on S_1K_1. Accordingly the difference in paths of two waves under consideration on reaching the focal point P_1 is $S_1K_1 = (e + b) \sin\theta$. Therefore the phase difference at P_1 between them is

$$\delta = \frac{2\pi}{\lambda}(e+b)\sin\theta = 2\beta$$

The resultant amplitude R at P_1 is obtained with the help of Eqn. (11.3 h) where

$$a = A\sin\alpha/\alpha \quad \text{and} \quad \delta = 2\beta$$

we therefore get,

$$R^2 = 4A^2\frac{\sin^2\alpha}{\alpha^2}\cos^2\left\{\frac{\pi}{\lambda}(e+b)\sin\theta\right\} = 4A^2\frac{\sin^2\alpha}{\alpha^2}\cos^2\beta$$

The intensity at P_1 is proportional to the square of the amplitude.

Hence $$I \propto 4A^2\frac{\sin^2\alpha}{\alpha^2}\cos^2\beta$$

or $$I = 4R_o^2\frac{\sin^2\alpha}{\alpha^2}\cos^2\beta \quad \text{...(14.4 a)}$$

The intensity in the resultant pattern depends on two factors: (1) $R_o^2(\sin^2\alpha/\alpha^2)$ which, as discussed in *Sec.* 14.2, gives diffraction pattern of a single slit and (2) $\cos^2\beta$ which give, interference pattern due to light waves from the two slit. Thus each slit produces a diffracted beam in which the intensity distribution depends upon the slit width; these diffracted beams then interfere with each other to produce the final diffraction pattern.

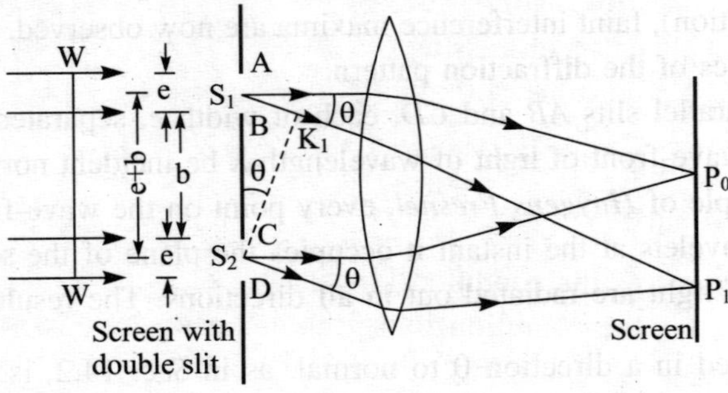

Fig. 14.6. Fraunhofer diffraction of a plane wave-front by a double slit

Dependence of Intensity on $R_o^2 \dfrac{\sin^2 \alpha}{\alpha^2}$

The intensity I is zero, whenever this function vanishes, that is, when

$$\sin \alpha = 0.$$

The values of α satisfying the above equation are $\alpha = \pm m\pi$

or $$e \sin \theta = \pm m\lambda \qquad ...(14.4\ b)$$

where m = 1, 2, 3, 4 etc. The positions of maxima, due to this factor as discussed in *Sec.* 14.2, are at $\alpha = 0$ and at values approaching $\pm 3\pi/2$, $\pm 5\pi/2$, $\pm 7\pi/2$ etc.

Dependence of Intensity on $\cos^2\beta$

The positions of maximum intensity due to this factor are given by values of β satisfying $\cos^2 \beta = 1$ and this equation gives $\beta = \pm n\pi$.

or $$(e + b) \sin \theta = \pm n\lambda \quad \text{[Maxima]} \qquad ...(14.4\ c)$$

where n, = 0, 1, 2, 3, 4 etc. *Points separated by a distance* $(e + b)$ *in the two slits are known as corresponding points*. The condition for maxima, expressed by Eqn. (14.4 *c*), therefore, simply states that for maxima, the path difference between the parallel diffracted rays from any pair of corresponding points in the two slits should be even multiple of $\lambda/2$. The whole number n represents the order of interference maximum.

The positions of zero intensity in the interference pattern are given by values of β satisfying $\cos^2\beta = 0$, that is $\beta = \pm (2n + 1)\ \pi/2$. Thus minima occur at angles θ, such that

$$(e + b) \sin \theta = \pm (2n + 1)\lambda/2 \quad \text{[Minima]} \qquad ...(14.4\ d)$$

where again n = 0, 1, 2, 3, 4 etc. This equation simply expresses the condition that for minima, the path difference between the parallel rays diffracted from any pair of corresponding points in the two slits should be odd multiple of $\lambda/2$ on reaching the focal plane of the focussing lens.

Missing Order of Interference Maxima

If the slit width e is kept constant and the separation b between them is varied, then it is obvious from Eqn. (14.4 *c*) that the distance between two consecutive interference maxima varies, and it follows from Eqn. (14.4 *b*) that the size of the diffraction pattern due to single slit remains constant. A study of the resultant diffraction pattern so obtained reveals that certain orders of interference maxima are missing. The explanation of missing orders is as follows:

Let us suppose that for some value of θ, the following relations simultaneously hold:

$$e \sin \theta = \pm m\lambda \qquad ...(14.4\ b)$$

$$(e + b) \sin \theta = \pm n\lambda \qquad ...(14.4\ c)$$

The first equation expresses the condition for zero intensity in the single slit diffraction while the second equation expresses the condition for a maximum in the interference pattern. Therefore the interference maxima will be absent in this direction θ.

From the above equations, we have

$$(e+b)/e = n/m \qquad ...(14.4\ e)$$

Since n and m are integral numbers, the ratio $(e + b)/e$ should be the ratio of two integers for missing orders. The ratio $(e + b)/e$ determines the orders of the missing interference maxima. For example,

(*i*) Let $(e + b)/e = 2$, that is $e = b$, then $n = 2m$. Giving m integral values, we get the corresponding values of missing orders n. $m = 1, 2, 3, 4, 5$ etc. $n = 2, 4, 6, 8, 10$ etc.

Thus, there will be three interference maxima within the central diffraction maximum (Fig. 14.8).

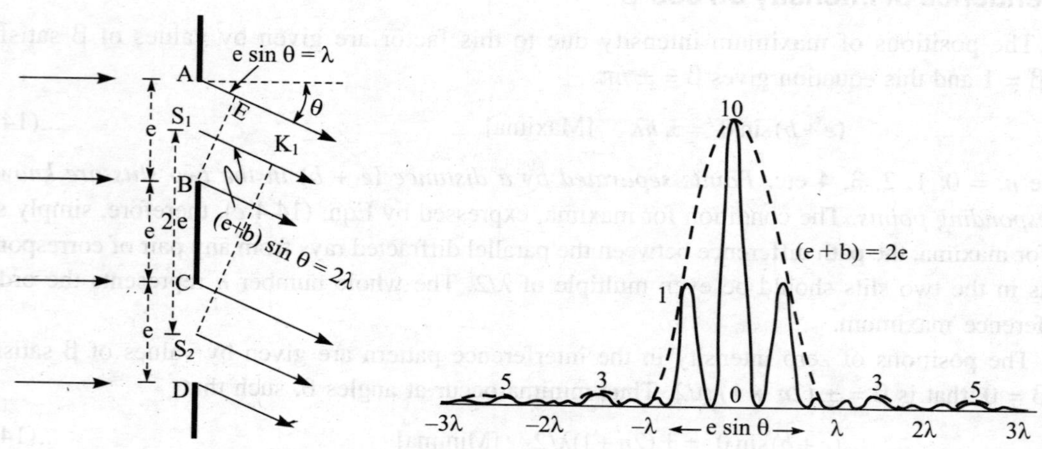

Fig. 14.7. Diffracted rays for missing interference maximum of order 2.

Fig. 14.8. Illustrates missing orders 2, 4, 6 etc

(*ii*) Let $(e + b)/e = 3$, that is, $b = 2e$, then $n = 3m$.

Therefore when $m = 1, 2, 3, 4, 5$ etc.

Missing orders are $n = 3, 6, 9, 12, 15$ etc.

Thus there will be five interference maxima within the central diffraction maximum (Fig. 14.9). In this way, as the ratio $(e + b)/e$ increases, the number of interference maxima within the central diffraction maximum also increases.

The physical explanation of the missing orders is as follows: Consider for an example the missing order $n = 2$. The point on the screen, where we expect this order is just 2λ further from the centre of one slit than from the centre of the other. But since $AB = BC$, this point is one wavelength further from one edge than the other of the same slit. Consequently the secondary wavelets from this slit produce zero intensity at this point. The same is true for the second slit. Thus it is obvious that the resultant intensity at this point, due to resultants of each slit will be zero, that is, the second order maximum will be absent.

Intensity Curve

We are now in position to plot the intensity curve for the double slit diffraction pattern. The upper curve in Fig. 14.9 is a plot of $\cos^2\beta$ against β and the values of the order are marked for the various maxima. We suppose $(e + b) = 3e$, then orders $n = 3, 6, 9$ etc. of interference maxima will be absent in the diffraction pattern. Accordingly, in the plot of $\sin^2 \alpha/\alpha^2$, we mark the first point of zero intensity ($m = 1$) below the third order interference maximum and so on. The complete double slit pattern as given by Eqn. (14.4 *a*) is obtained by multiplying the ordinates of corresponding points of two curves and the constant $4R_o^{\,2}$; the product is then plotted as ordinate at that point. This is shown in the lowest curve of Fig. 14.9.

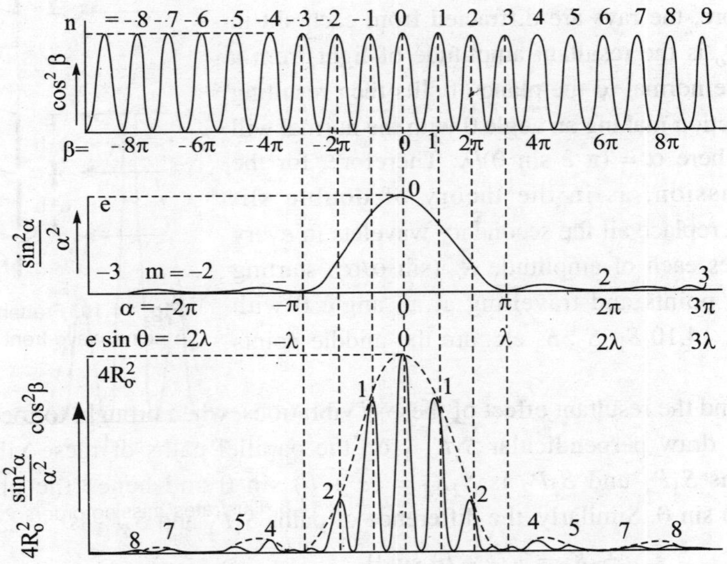

Fig. 14.9. Illustrates missing orders 3, 6, 9 etc

14.5 THE PLANE DIFFRACTION GRATING (DIFFRACTION AT *N* PARALLEL SLITS)

An arrangement consisting of large number of parallel slits of the same width and separated by equal opaque spaces is known as diffraction grating. This was first constructed by *Joseph Fraunhofer* and it was due to his study of diffraction of parallel beams of light by grating in 1819 that his name became associated with this class of diffraction phenomenon.

The diffraction pattern suffers a modification as the number of slits is increased beyond two. In the adjoining plate are shown diffraction patterns from one to five equidistant slits. It would be observed that the maxima occupy the same position in each figure, the only effect of increasing the number of slits consists in narrowing and increasing the sharpness of the principal interference maxima. Moreover, there appear what are called weak secondary maxima between any two adjacent central principal maxima. For three slits there is only one, for four slits there are two and for five slits there are three weak secondary maxima. With very large number of slits, the principal interference maxima become extremely sharp, practically narrow lines, while the intensity of secondary maxima becomes negligibly small so that these are not visible in the diffraction pattern. We shall now discuss theoretically the *Fraunhofer diffraction* pattern to N slits.

Let Fig. 14.10, represent N similar parallel slits, each of width e and separated by an opaque space of width b. Let a plane wave-front of light of wavelength λ be incident normally on the slits. According to the principle of *Huygens-Fresnel*, at the instant the incident wave-front occupies the plane of the slits, every point in each slit is regarded as the origin of secondary spherical wavelets, which spread out in all directions. Therefore, the rays are diffracted from each slit in all directions. If R_o is the resultant amplitude of light from a single slit along the normal to the plane of slits then resulting amplitude in a direction making an angle θ with the normal will be R_o ($\sin \alpha/\alpha$) where $\alpha = (\pi\ e \sin \theta)/\lambda$. Therefore, for the purpose of discussion, as in the theory of double slit diffraction, we can replace all the secondary wavelets in every slit by single waves each of amplitude R_o ($\sin \alpha/\alpha$), starting from their middle points and travelling at an angle θ with normal. In the Fig. 14.10 S_1, S_2, S_3 etc. are the middle points of the slits.

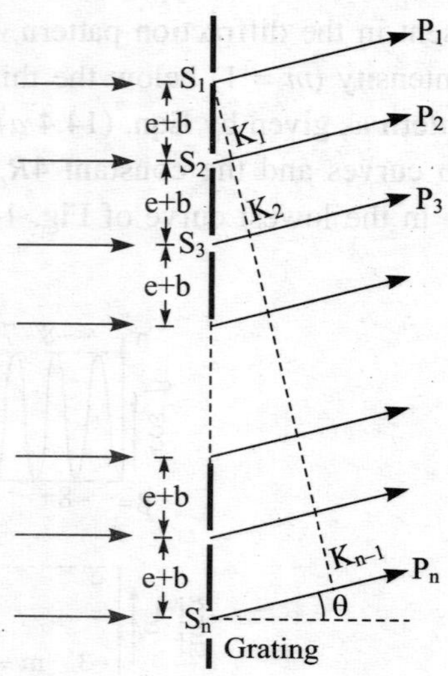

Fig. 14.10. Fraunhofer diffraction of a plane wave-front by N parallel slits

Let us now find the resultant effect of these N vibrations, when brought to focus by a convergent lens. From S_1, we draw perpendicular S_1K_{n-1} on the parallel paths of these vibrations. Then the difference of paths S_1P_1 and S_2P_2 is $S_2K_1 = (e + b) \sin \theta$ and hence the phase difference is $2\beta = (2\pi/\lambda)\ (e + b) \sin \theta$. Similarly, the difference of paths S_1P_1 and S_3P_3 is

$$S_2K_2 = 2\ (e + b) \sin \theta$$

and the phase difference is 4β and so on. Thus, if at the instant the waves under consideration arrive in focus, we consider the phase of the wave from S_1 to be zero, then the phases of waves from S_2, S_3, S_4 etc. should be 2β, 4β, 6β, 8β etc. Thus the phases increase in arithmetic progression. Therefore, the problem reduces to finding the resultant amplitude of N waves of equal amplitude R_o ($\sin \alpha/\alpha$) and period but the phases increasing in the arithmetic progression, the common phase difference being 2β. Therefore, the results of Sec. 14.1 can be applied here. It has been shown that

$$R = \frac{a \sin\left(\frac{1}{2} N\varepsilon\right)}{\left(\sin \frac{1}{2}\varepsilon\right)}$$

In the present case, $a = R_o$ ($\sin \alpha/\alpha$) and

$$\varepsilon = 2\beta = (2\pi/\lambda)(e + b)\sin\theta$$

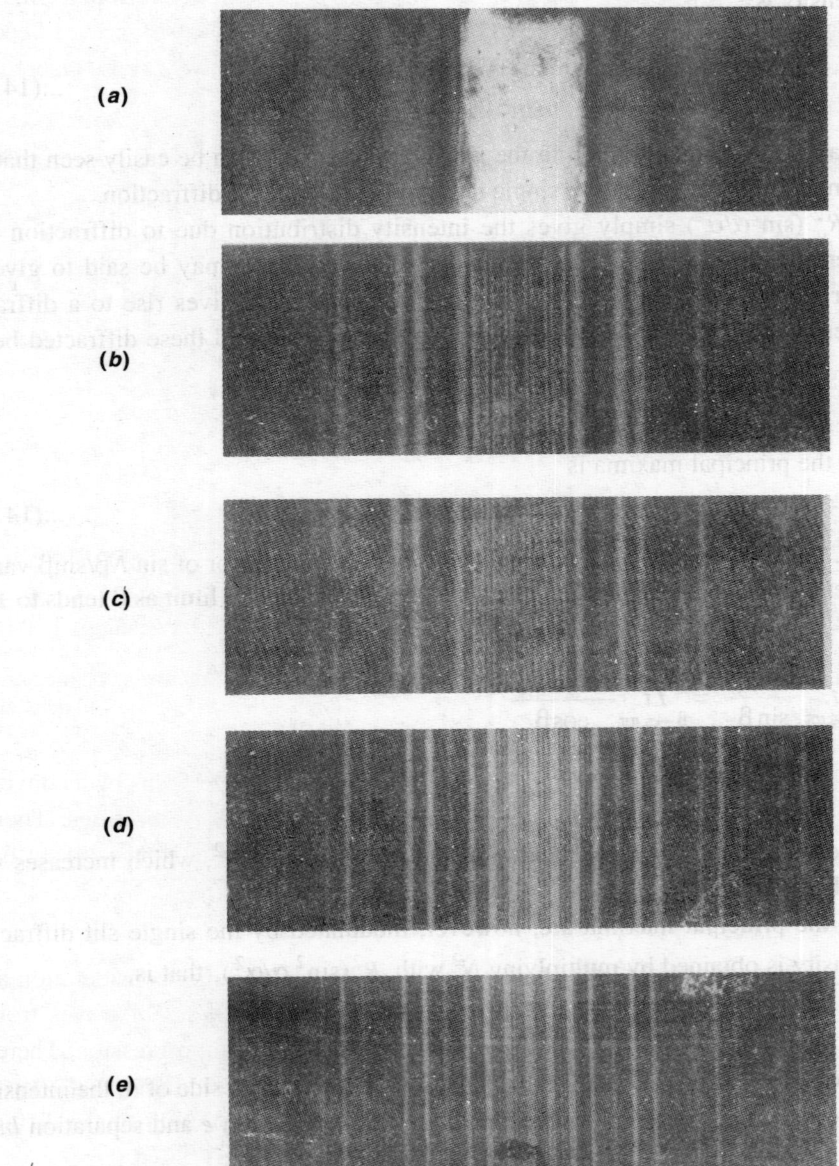

Fig. 14.11(*a*). Fraunhofer diffraction pattern of a single slit (*b*), (*c*), (*d*), (*e*) Interference patterns of 2, 3, 4 and 5 slits respectively

Hence the resultant amplitude is given by

$$R = R_o \frac{\sin \alpha}{\alpha} \frac{\sin N\beta}{\sin \beta} \qquad ...(14.5\ a)$$

The resultant intensity is

$$I = R_o^2 \frac{\sin^2\alpha}{\alpha^2}\frac{\sin^2 N\beta}{\sin^2\beta} \quad ...(14.5\ b)$$

If we put $N = 1$, and $N = 2$ in succession in the above expression, it can be easily seen that this reduces to corresponding expressions from the single slit and the double slit diffraction.

The first factor $R_o^2\ (\sin^2\alpha/\alpha^2)$ simply gives the intensity distribution due to diffraction by a single slit and has been discussed before. The second factor $\sin^2 N\beta/\sin^2\beta$ may be said to give the interference pattern for N slits. Thus, we can say that each of the N slits gives rise to a diffracted beam in which the intensity distribution depends on the width of the slit and these diffracted beams then interfere with one another to produce the final diffraction pattern.

Principal Maxima

The condition for the principal maxima is

$$\sin\beta = 0 \quad \text{that is,} \quad \beta = \pm\, n\pi \quad ...(14.5\ c)$$

where $n = 0, 1, 2, 3$ etc. For these values of β, both numerator and denominator of $\sin N\beta/\sin\beta$ vanish, that is, the term becomes indeterminate. But it may be evaluated by finding its limit as β tends to $\pm\, n\pi$, by the rules of calculus. We have for the limit

$$\underset{\beta\to\pm n\pi}{Lt.} \frac{\sin N\beta}{\sin\beta} = \underset{\beta\to\pm n\pi}{Lt.} \frac{N\cos N\beta}{\cos\beta}$$

$$= \frac{N\cos Nn\pi}{\cos n\pi} = \pm\, N.$$

Therefore, the intensity of the principal maximum is proportional to N^2, which increases with the increase in the number of slits.

The intensities of the principal maxima are, however, modulated by the single slit diffraction pattern. The actual intensity is obtained by multiplying N^2 with $R_o^2(\sin^2\alpha/\alpha^2)$, that is,

$$I = N^2 R_o^2 (\sin^2\alpha/\alpha^2)$$

The intensity of the central principal maximum is greatest while on either side of it, the intensities of other maxima go on decreasing. For the double slit, having the same width e and separation b, we have

$$I = 2^2 R_o^2 (\sin^2\alpha/\alpha^2)$$

We thus find that the relative values of the intensities of the principal maxima for N slits are the same as those for the double slit pattern but they are more intense in the ratio of the square of the number of slits.

Substituting $\beta = (\pi/\lambda)(e+b)\sin\theta$ in Eqn. (14.5 *c*) and simplifying, we get

$$(e+b)\sin\theta = \pm\, n\lambda \qquad \text{[Principal maxima]} \quad ...(14.5\ d)$$

This equation simply expresses the condition that if the difference of paths of the parallel diffracted rays from any pair of corresponding points of grating is an integral multiple of λ, all the parallel diffracted rays in that direction reinforce each other, so as to form the principal maximum in the focal plane of the lens. Eqn. (14.5 *d*) is exactly identical with Eqn. (14.4 *c*) obtained in the double slit diffraction. Therefore, the condition for missing orders [Eqn. 14.4 (*e*)] is applicable in the same form here. Moreover Eqn. (14.5 *d*) also indicates that the principal maxima are formed in the same position as in the double slit pattern. The whole number n in Eqn. (14.5 *d*) represents the order of the interference maximum, and physically it gives the number of wavelengths in the difference of paths of the diffracted rays from any pair of corresponding points of the grating, as shown in the Fig. 14.12 for the first and the second order maxima. The *distance* $(e + b)$ *is called the grating element.* Eqn. (14.5 *d*) does not contain N; so that for a given slit separation, the positions of the principal maxima are independent of the number of rulings in the grating.

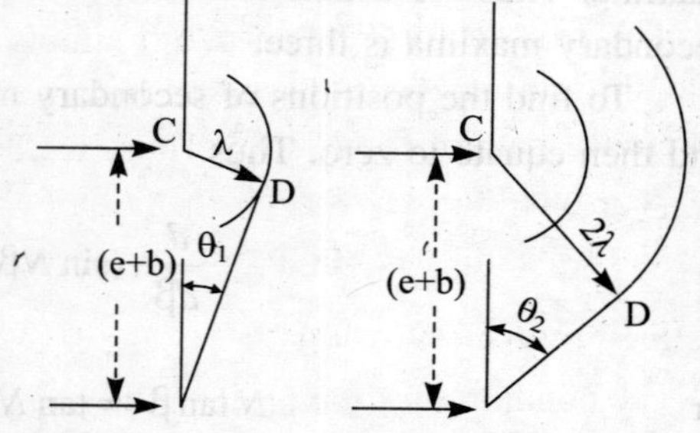

Fig. 14.12. First order maximum when CD = λ, Second order maximum when CD = 2λ

Minima

The values of β for which only the numerator of the term $\sin N\beta/\sin\beta$ vanishes give points of zero intensity in diffraction pattern. That is, for minima

$$\sin N\beta = 0 \quad \text{but} \quad \sin\beta \neq 0$$

or

$$N\beta = \pm m\pi \qquad \text{...(14.5 } e\text{)}$$

where m has any integral value with the exception of 0, N, $2N$, $3N$, etc. because for these values of m, $\beta = 0 \pm \pi, \pm 2\pi, \pm 3\pi$, etc. and therefore, the denominator of the term $(\sin N\beta/\sin\beta)$ also vanishes. These correspond to the positions of the principal maxima discussed above.

Substituting the value of β in Eqn. (14.5 *e*), we get the condition for the minima in terms of the path difference between the parallel diffracted rays from any pair of corresponding points of the grating.

$$(e+b)\sin\theta = \pm m\lambda/N \quad \text{but} \quad m \neq nN$$

where n = 0, 1, 2, 3, 4 etc., being the order of the principal maximum. Thus

$$(e+b)\sin\theta = \pm\frac{\lambda}{N}, \pm\frac{2\lambda}{N}, \pm\frac{3\lambda}{N} \ldots\ldots \frac{(N-1)\lambda}{N}, \frac{(N+1)\lambda}{N}, \frac{(N+2)\lambda}{N} \quad \text{Minima}$$

...(14.5 *f*)

omitting the values

$$(e+b)\sin\theta = 0, \pm\frac{N\lambda}{N}, \pm\frac{2N\lambda}{N}, \frac{3N\lambda}{N} \ldots \text{Principal max.} \qquad \text{...(14.5 } g\text{)}$$

Secondary Maxima

A study of the two expressions given above clearly indicates that between the principal maximum of zero order and the principal maxima of order one there are $(N - 1)$ equally spaced points of zero intensity i.e. minima. This is also true for minima points between any two adjacent principal maxima. Hence there are $(N - 2)$ other maxima known as secondary maxima between any two adjacent principal maxima. Thus for 3 slits there is only one, for 4 slits there are two and for 5 slits the number of secondary maxima is three.

To find the positions of secondary maxima, we first differentiate $\sin N\beta/\sin\beta$ with respect to β and then equate to zero. Thus

$$\frac{d}{d\beta}(\sin N\beta \operatorname{cosec}\beta) = 0$$

or

$$N\tan\beta = \tan N\beta$$

The roots of this equation other than $\beta = \pm n\pi$ give the positions of secondary maxima. They are not of equal intensity but decrease gradually as we go out on either side of principal maxima. Moreover, they are slightly shifted, as in the case of single slit pattern, towards the adjacent principal maxima. These characteristics can at once become evident on plotting the function $\sin^2 N\beta/\sin^2\beta$ for $N = 6$ slit.

Plot of $\frac{\sin^2 N\beta}{\sin^2\beta}$ **for** $N = 6$. Let us plot the function $\sin^2 N\beta/\sin^2\beta$ for 6 slits to bring out clearly the existence of secondary maxima. We first plot separately the functions $\sin^2\beta$ and $\sin^2 N\beta$ against β. The function $\sin^2\beta$ is zero for $\beta = 0, \pi, 2\pi, 3\pi$, etc. and it is maximum (equal to 1) for $\beta = \pi/2, 3\pi/2, 5\pi/2$ etc. Hence the plot of $\sin^2\beta$ is of the form shown in Fig. 14.13. Similarly $\sin^2 N\beta$ is zero for $N\beta = 0, \pi, 2\pi$ etc. and is maximum (equal to 1) for $N\beta = \pi/2, 3\pi/2, 5\pi/2$ etc. The scale of β is the same in the two curves. The function $(\sin^2 N\beta/\sin^2\beta)$ is plotted by taking the quotient of ordinates at corresponding points of two curves and plotting this quotient separately as shown in the bottom curve (Fig. 14.13). It would be observed that quotient at points $\beta = 0, \pi, 2\pi$ etc. is 0/0 which is indeterminate but its limit for these values of β, as obtained above, is N^2 and in this case equal to 36. Thus the principal maxima in the plot $(\sin^2 N\beta)/(\sin^2\beta)$ occur at $\beta = 0, \pi, 2\pi$ etc. In this plot, there are four secondary maxima between any two adjacent principal maxima. It should be borne in mind that the intensities of principal maxima are modulated by the single slit diffraction, that is, the plot of $(\sin^2 N\beta/\sin^2\beta)$ is to be multiplied by $R_o^2(\sin^2\alpha/\alpha)$, the single slit diffraction term, to get the actual intensity-distribution curve of the resultant diffraction pattern.

For the sake of comparison, we give in Figs. 14.14 (*a*), (*b*), (*c*) and (*d*) the plot of diffraction patterns (not in scale) due to 1, 2, 3 and 4 slits respectively. The full line curves in (*b*), (*c*) and (*d*) represent the respective interference patterns while the dotted curves showing several maxima represent the modulated interference patterns but their vertical scales are to be increased by the factor 4, 9 and 16 in (*b*), (*c*) and (*d*) respectively to get the actual intensity distribution in the resultant pattern in each case. It will be seen that although the principal maxima occur at the same positions with 3 and

4 slits as with two but they become brighter and narrower while the secondary maxima become weaker and faint with the increase in the number of slits. The extension to a very large number of slits, as in actual grating (say 15000 lines to an inch), should be now quite obvious. The maxima become very narrow and of sharp intensity while the secondary maxima are of negligible intensity, invisible in the pattern.

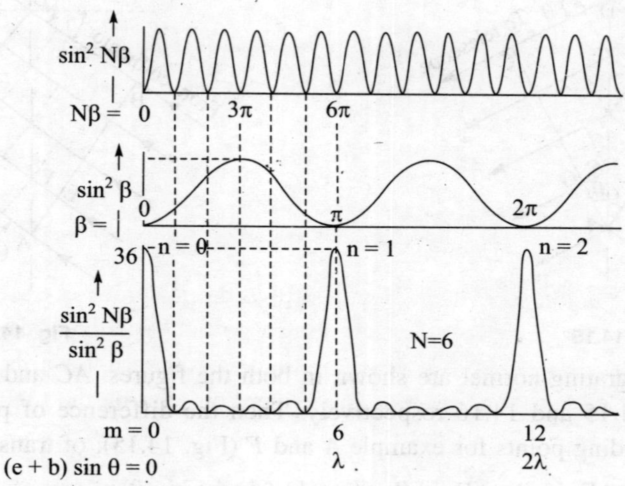

Fig. 14.13. Details of the intensity pattern in Fraunhofer diffraction by a grating of six slits

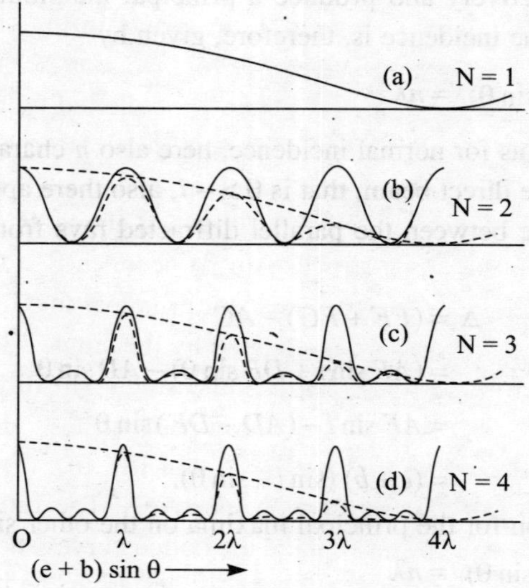

Fig. 14.14. Diffraction and interference patterns of 1, 2, 3, 4 slits

14.6 PLANE GRATING AT OBLIQUE INCIDENCE

Let in the Figs. 14.15 and 14.16, *AB* represent a plane wave-front of light of wavelength λ incident obliquely on the grating, the inclination being i with the grating surface. A set of parallel rays diffracted

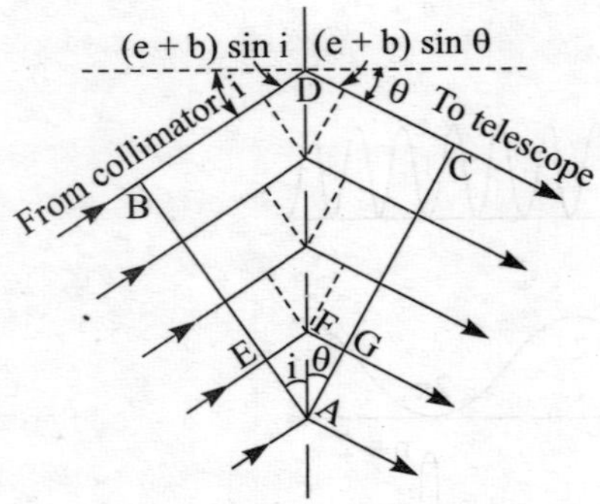

Fig. 14.15

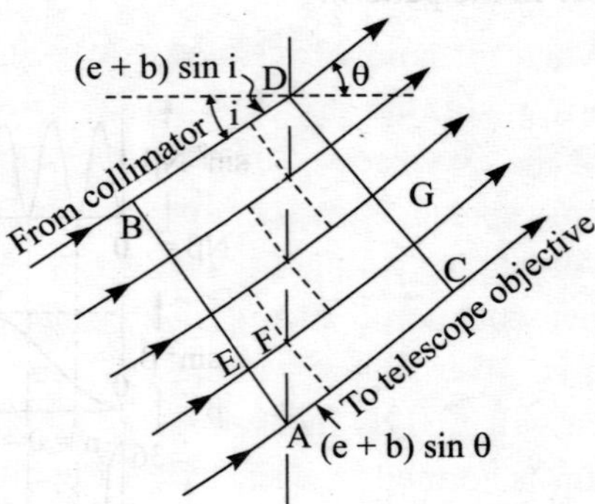

Fig. 14.16

at an angle θ with the grating-normal are shown in both the figures. *AC* and *DC* are the diffracted wave-fronts in Figs. 14.15 and 14.16 respectively. Then the difference of paths of the diffracted rays from the corresponding points for example *A* and *F* (Fig. 14.15), of transparent spaces is

$$\Delta = (EF + FG) = AF \sin i + AF \sin \theta = (e + b) (\sin i + \sin \theta)$$

where $AF = (e + b)$ is the grating element. If the parallel diffracted rays are brought to focus by the objective lens, and if, for light of wavelength λ this path difference is an integral multiple of λ, then these rays interfere constructively and produce a principal maximum. The condition for principal maxima in the case of oblique incidence is, therefore, given by

$$(e+b)(\sin i+\sin\theta) = n\lambda \qquad \text{...(14.6 } a)$$

where $n = 0, 1, 2, 3$ etc. and as for normal incidence, here also n characterises the order of maximum.

On the other side of the direct beam, that is $\theta > -i$, also there appear succession of the principal maxima. The path difference between the parallel diffracted rays from the corresponding points say *A* and *F* in Fig. 14.16 is

$$\begin{aligned}\Delta &= (FF' + FG) - AC \\ &= (AF \sin i + DF \sin\theta) - AD \sin\theta \\ &= AF \sin i - (AD - DF)\sin\theta \\ &= (e+b)(\sin i - \sin\theta).\end{aligned}$$

Therefore, the condition for the principal maxima on the other side of the direct beam is

$$(e+b)(\sin i-\sin\theta) = n\lambda \qquad \text{...(14.6 } b)$$

where $n = 0, 1, 2, 3$ etc.

14.7 FORMATION OF SPECTRA BY A GRATING

We have seen in *Sec.* 14.5 that the principal maxima in the diffraction pattern of plane grating, when light is incident normally on it, are given by Eqn. (14.5 *d*) which is

$$(e+b)\sin\theta = \pm n\lambda$$

When the number of rulings is very large say 14000 to an inch and the primary source of light is a narrow slit, the principal maxima are observed as sharp bright lines (*Sec.* 14.12) in the focal plane of the eye-piece of the telescope adjusted for parallel rays. Consequently principal maxima are commonly spoken of as spectrum lines. If the slit is parallel to the rulings of the grating, the spectrum lines will be also parallel to the rulings. These lines are simply the images of the slit. It is clear from the above equation that in a given order n, the angle of diffraction θ depends upon λ. Therefore, if the source gives light of number of wavelengths e.g. λ_1, λ_2, λ_3, then in the given order n (except $n = 0$) we will find as many lines as there are different wavelengths in the light source. Thus a line spectrum is formed in each order. But for the central image ($n = 0$) the path difference is zero for every wavelength. Hence the central maxima of different wavelengths coincide forming the central image, obviously of the same colour as that of the primary source of the light. On the other side of the central image a similar set of spectra are formed. In each order, the spectrum line corresponding to shortest wavelength is formed on the side towards the central image.

14.8 PRODUCTION OF RULED GRATING

Actual gratings are made by ruling equidistant, extremely close fine grooves with a diamond point on an optically plane parallel glass plate, thus forming a *transmission grating*. If the rulings are made on the polished metal surface the grating is of *reflection type*. The rulings serve as opaque spaces leaving transparent spaces in between. Gratings having 30,000 lines to an inch for 6 in. are ruled for high resolution work. Commercial gratings are made by taking a cast of the ruled surface of the actual grating with some transparent material like *cellulose acetate*. Properly diluted cellulose acetate is poured on actual grating and allowed to dry to a thin strong film which is then easily detatched from the parent grating. The impressions of the rulings so obtained are preserved by mounting the film between two glass plates.

14.9 EXPERIMENTAL DETERMINATION OF WAVELENGTH OF LIGHT

The plane diffraction grating can be very conveniently employed for the experimental determination of wavelength of light. Several experimental arrangements for its determination were described in the chapters on *Interference of Light*. But these arrangements employ only monochromatic light and therefore if light is heterogeneous, previous spectral analysis by some auxiliary device is essential before employing any one of the interference methods. The grating has an advantage over these methods in the sense that we can use heterogeneous light directly, the grating itself produces dispersion due to its spectra forming property.

(*a*) **Normal Incidence.** The fundamental equation of the plane grating for normal incidence of light is

$$(e+b)\sin\theta_n = \pm n\lambda$$

when n characterises the order of the spectrum. The spectrum consists of number of spectral lines, in separate positions, corresponding to each of wavelengths present in the incident light. In the first order, the angle of diffraction corresponding to spectral line of light of wavelength λ is given by the relation

$$(e+b)\sin\theta_1 = \pm \lambda$$

and in the second order, we have

$$(e+b)\sin\theta_2 = \pm 2\lambda$$

where the grating element $(e + b) = 1/N$, N being the number of lines per cm in the grating. Therefore, if n is noted and the corresponding angle of diffraction is measured for a definite spectral line, we can easily calculate its wavelength.

The experiment is performed on a spectrometer which has been adjusted for parallel rays. The grating has to be adjusted on the turn table so that its lines are vertical, that is, parallel to the axis of rotation of the table. Moreover, the light from the collimator should fall normally on the grating. To achieve these adjustments the following procedure is adopted.

The telescope is brought directly in line with the collimator so that the centre of the direct image of the slit falls on the intersection of the cross-wires. In this setting of the telescope, its vernier reading is taken; let it be ϕ. The telescope is now turned through 90° from this position in either direction so that the reading of the vernier becomes $(\phi + 90)$ or $(\phi - 90)$. Now the axis of telescope is at right angles to the direction of rays of light emerging from the collimator. The telescope is clamped in this position. The grating is then properly mounted on the turn table such that its ruled surface lies over the centre of table and is perpendicular to the line joining two of the levelling screws. The table is now rotated in the proper direction till the reflected image of the slit from the grating surface coincides with the intersection of the cross-wires of the telescope. By the help of two levelling screws perpendicular to which grating is fixed on the table the image is adjusted to be symmetrical on the cross-wires. The plane of grating, in this setting, makes an angle of 45° with the incident rays as well as with the telescope axis. The reading of table vernier is taken and by its help the turn table is rotated through 45° from this position so that the ruled surface becomes exactly normal to the incident rays and on the side of telescope. The table is now firmly clamped.

The final adjustment is to set the lines of the grating exactly parallel to the axis of rotation of the telescope. The telescope is rotated and adjusted to view the first order diffracted image. The third levelling screw of the prism table is now worked to get the image in the centre of the field of view. If this adjustment is perfect, the centres of all the diffracted images on either side of the direct one, will be found to lie on the intersection of the cross-wires as the telescope is turned to view them one after another. The rulings on the grating are now parallel to the axis of rotation of the telescope.

Having completed all these adjustments, telescope is turned to get the first order spectral line, corresponding to unknown wavelength, on the intersection of the cross-wires and the two verniers are read. The telescope is then turned to observe the same spectral line in the first order on the other side of the direct image and when it is exactly on the cross-wires, the two verniers are again read. The difference between the readings of the same vernier for the above two settings of the telescope gives twice the angle of diffraction for diffracted rays of light of unknown wavelength in the first order. Half of this difference gives the angle of diffraction θ_1, which may be now substituted in the equation

$$\lambda = (e+b)\sin\theta_1 = (1/N)\sin\theta_1 \qquad ...(14.9\ a)$$

to calculate the value of λ. The same observations may be repeated in the second order but the spectra beyond this are not usually employed, the spectral lines being too faint. The unit in which λ is expressed is Angstrom ($1A. = 10^{-8}$ cm).

(*b*) **Oblique Incidence – Minimum Deviation Method.** In the grating experiment for the determination of λ, outlined above, it is absolutely essential to adjust the ruled surface normally to the path of rays from the collimator. This is rather a troublesome adjustment. If, however, there is a slight departure from this adjustment, light is incident obliquely on the grating and the angle of diffraction measured as in the previous experiment when substituted in Eqn. (14.9 *a*) would lead to considerable error in the computed value of λ. Therefore, we employ an alternative method called *Minimum Deviation Method* for evaluating λ. In this experiment, the grating is adjusted in relation to the collimator and observing telescope such that in a given order *n*, the deviation of incident rays of λ, viz. $D = i + \theta$, is a minimum. The condition for minimum deviation can be easily obtained by writing Eqn. (14.6 *a*), which is

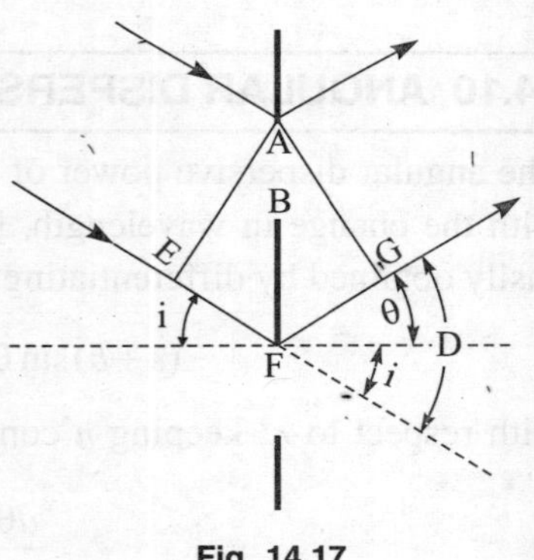

Fig. 14.17

$$n\lambda = (e+b)(\sin i + \sin\theta)$$

in the form

$$n\lambda = 2(e+b)\sin\frac{1}{2}(i+\theta)\cos\frac{1}{2}(i-\theta) = 2(e+b)\sin\frac{1}{2}D\cos\frac{1}{2}(i-\theta)$$

From this equation, it follows that sin 1/2 *D*, in other words deviation *D*, is a minimum when cos 1/2 $(i - \theta)$, is a maximum, which can be easily seen to be true when the relation $i = \theta$ holds good. Then the wavelength is found from

$$2(e+b)\sin\frac{1}{2}D = n\lambda \qquad ...(14.9\ b)$$

To employ the grating in the position of minimum deviation, we proceed as follows: The grating is mounted vertically, as in the previous experiment, with its ruled surface passing through the axis of rotation of the turn table of the spectrometer. The telescope is turned so as to receive the spectral line say in the first order. The grating is rotated by rotating the turn table such that the deviation of incident ray decreases which obviously results in the motion of the spectrum line in the field of view. Ultimately a stage will arrive when the spectrum line begins to retrace its path from a certain point; although rotation of the grating is continued in the same direction. The position of grating at which spectrum line becomes momentarily at rest corresponds to the minimum deviation of incident rays. By trial, the telescope is adjusted to a position such that the spectrum line during its motion just touches the intersection of cross-wires and then retraces its path. This setting of telescope is recorded by the help of telescope-verniers. The table is clamped, grating removed and telescope is rotated to a position such that the direct image of slit is received on the inter-section of cross-wires. This setting of telescope is again recorded by the same vernier. The difference of these two readings is the angle of minimum deviation *D*. This measurement may be repeated by working with the same order spectrum on the other side of the direct image. The mean value of *D* may be substituted in Eqn. (14.9 *b*) and λ evaluated.

It should be emphasised that in the neighbourhood of minimum deviation the variation in the angle of diffraction measured as half the angle of deviation, due to slight rotation of grating, is negligibly small. So a slight departure of the grating from the minimum deviation position would not effect the measured angle of minimum deviation. Thus, employing the grating in the minimum deviation position is more advantageous as compared with its setting for normal incidence of the light.

14.10 ANGULAR DISPERSIVE POWER OF THE GRATING

The angular dispersive power of the grating is defined as the rate of change of the angle of diffraction with the change in wavelength. Mathematically, it is expressed by $(d\theta/d\lambda)$ and its value can be very easily obtained by differentiating the equation,

$$(e+b)\sin\theta = n\lambda \quad \text{[Principal maxima]}$$

with respect to λ, keeping n constant. We get

$$\frac{d\theta}{d\lambda} = \frac{n}{(e+b)\cos\theta} \qquad \text{...(14.10)}$$

A study of the expression for the angular dispersive power, obtained above, leads to the following conclusions:

(*i*) The angular dispersion is directly proportional to *n*, the order of the spectrum. This means that the angular separation $d\theta$ between the principal maxima corresponding to wavelengths λ and $\lambda + d\lambda$ increases with the order of the spectra. Therefore, the second order spectrum is twice as wide as the first and so on.

(*ii*) The angular separation $d\theta$, for a given $d\lambda$ is, inversely proportional to $(e + b)$, the width of the grating element and therefore directly proportional to the number of lines per cm on the grating. The smaller the grating element, the greater is the spread of spectrum of given order.

(*iii*) The angular dispersion increases with θ and therefore with the wavelength, since θ increases with increase in λ. Therefore grating spectra are spread much more at the red than at the blue end of the spectrum.

14.11 LINEAR DISPERSIVE POWER OF THE GRATING

Equation (14.10) expresses the angular separation per unit wavelength in a given order of spectrum. Sometimes we want to know the linear separation of the two spectral lines in a given order spectrum.

Let dl (= l) denote the linear distance AB along the screen between two lines differing in wavelength by $d\lambda$. Then from ΔOAB, since $\angle AOB = d\theta$ we have

$$dl = f' d\theta$$

Hence
$$\frac{dl}{d\lambda} = f' \frac{d\theta}{d\lambda} = \frac{nf'}{(e+b)\cos\theta} \quad \text{...(14.11 a)}$$

where f' is the focal length of the lens employed to focus the spectra. $dl/d\lambda$ is defined as the linear dispersive power of the grating. An increased linear dispersion can be obtained by working with lenses of longer focal lengths. The plate or screen should be curved into an arc of a circle, of radius f' and the lens should be at the centre of curvature O of the screen.

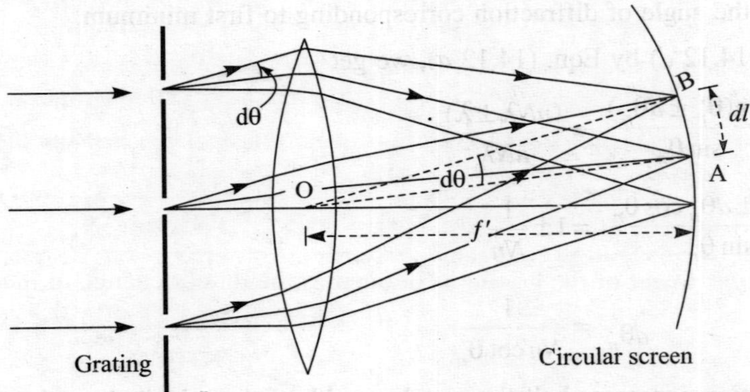

Fig. 14.18

14.12 ANGULAR WIDTH OF A PRINCIPAL MAXIMUM

We can easily deduce the angular separation between the first minima on either side of the principal maxima in the diffraction pattern of any order say n. This angular separation is called the *angular* width of the principal maximum in the nth order. Let θ_n be angle of diffraction corresponding to the principal maximum of wavelength λ in the nth order. This direction θ_n is given by

$$(e + b) \sin \theta_n = n\lambda$$

or
$$N(e+b)\sin\theta_n = nN\lambda \quad \text{...(14.12 a)}$$

It will be recalled that the positions of minimum of intensity in the diffraction pattern due to N slits are given by the condition,

$$(e+b)\sin\theta = \frac{\lambda}{N}, \frac{2\lambda}{N}, \ldots \frac{(N-1)\lambda}{N}, \frac{(N+1)\lambda}{N}, \frac{(N+2)\lambda}{N}$$

$$\ldots\ldots \frac{(2N-1)\lambda}{N}, \frac{(2N+1)\lambda}{N} \ldots\ldots$$

or
$$N(e+b)\sin\theta = \lambda, 2\lambda, \ldots\ldots (N-1)\lambda, (N+1)\lambda, (N+2)\lambda.$$

$$\ldots\ldots (2N-1)\lambda, \ (2N+1)\lambda, \ldots\ldots \quad \text{...(14.12 b)}$$

But for the values

$$N(e+b)\sin\theta = 0, N\lambda, 2N\lambda, 3N\lambda, \ldots \quad \text{...(14.12 c)}$$

we get respectively 0, 1st, 2nd, 3rd .... order principal maximum. It is clear from Eqns. (14.12 *b*) and (14.12 *c*) that the first minimum on either side of the principal maximum of the first order is given by the equation

$$N(e+b)\sin(\theta_1 \pm d\theta_1) = (N\lambda \pm \lambda)$$

where the angle of diffraction θ_1 corresponds to principal maximum of first order and $(\theta_1 \pm d\theta_1)$ corresponds to first minimum on either side of this maximum. Now it should be obvious that the first minimum on either side of the principal maximum of the *n*th order should be given by

$$N(e+b)\sin(\theta_n \pm d\theta_n) = (nN\lambda \pm \lambda) \quad (14.12\ d)$$

where $(\theta_n \pm d\theta_n)$ is the angle of diffraction corresponding to first minimum.

Dividing Eqn. (14.12 *d*) by Eqn. (14.12 *a*), we get

$$\frac{\sin(\theta_n \pm d\theta_n)}{\sin\theta_n} = \frac{(nN\lambda \pm \lambda)}{nN\lambda}$$

or

$$\frac{\sin\theta_n \pm d\theta_n \cos\theta_n}{\sin\theta_n} = 1 \pm \frac{1}{Nn}$$

or

$$d\theta_n = \frac{1}{Nn\cot\theta_n} \quad ...(14.12\ e)$$

The above equation expresses half the angular width of the principal maximum of order *n*. For $n = 1$, $\theta_n = 20°$, cot 20 = 2.745 and N = 14000 lines per inch, the value of $d\theta_1$ = 1/38430 and therefore it is of order of 10^{-4} radian. The principal maxima in actual grating spectra, consequently, appear simply as narrow lines.

The separation of the first and last slits in the plane grating is $(N - 1)(e + b)$ which when N is large, can be approximately written as $N(e + b)$. Now, substituting the value of n from Eqn. (14.12 *a*) in (14.12 *e*), we get

$$d\theta_n = \frac{\lambda}{N(e+b)\cos\theta_n}$$

This expression for the half angular width clearly indicates that the sharpness of the principal maxima depends upon the total width, $N(e + b)$, of the ruled surface of grating and not upon the number of lines per centimeter.

14.13 OVERLAPPING OF ORDERS

When we examine a large range of wavelengths by the help of diffraction grating, then some overlapping of spectral lines occur in the higher order of the spectra. This is due to the reason that for a given value of θ, the fundamental equation of the grating, viz.,

$$(e + b)\sin\theta = n\lambda$$

may be satisfied by several values of λ with appropriate values of n. Thus, for example

$$(e+b)\sin\theta = \lambda_1 = 2\lambda_2 = 3\lambda_3 \text{ etc.}$$

This equation clearly indicates that the wavelengths $\lambda_1, \lambda_2, \lambda_3$........ occurring in the first, the second and the third order respectively will coincide in position in the spectra.

The *visible spectrum* extends from 3900 A to 7700 A. The second order of violet 3900 A, would coincide with 7800 A, which lies beyond the visible region and hence it just fails to coincide with the first order of the red 7700 A. The first order spectrum, therefore, just fails in the visible region to overlap the second order spectrum. The red of the second order, however, overlaps the violet of the third order, as shown in the Fig. 14.19 which depicts to scale, the first, second and third order visible spectra to the left of the zero order. When photographic plates sensitive to wide range of wavelengths are employed the first-two orders may extend down to 2000 A. in the *ultra violet* and therefore these two orders overlap in the *ultra-violet* region. The overlapping of different wavelengths can be avoided by the use of suitable filter to remove the undesired overlapping wavelengths. This overlapping of wavelengths, on the other hand, is often advantageous in photographing different orders simultaneously and in the determination of the unknown wavelength by the so-called *method of coincidence*. If in a spectrum, a line of known wavelength λ, in the order n, coincides in position with a line of unknown wavelength λ' in the order n', then λ' may be determined by the equation

$$(e+b)\sin\theta = n\lambda = n'\lambda'$$

that is,

$$\lambda' = n\lambda/n'$$

In the method of coincidence, it is not essential to find spectrum lines which actually coincide in various orders — they should only fall near each other. Interpolation can be applied for evaluating the unknown wavelength. This method was used with great accuracy by *Rowland*.

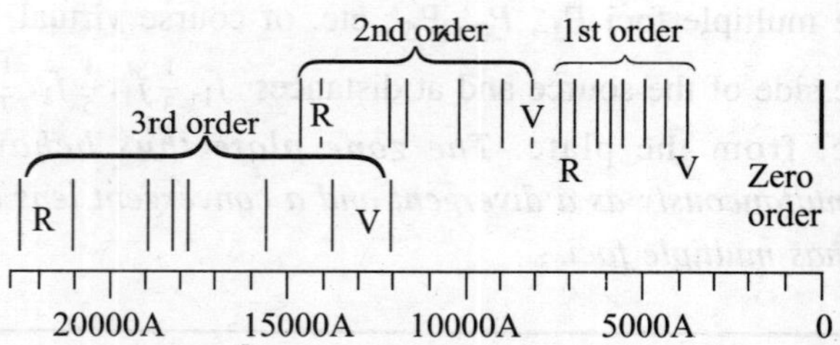

Fig. 14.19. Several orders of the visible spectrum, showing overlapping

14.14 ZONE PLATE AS A DIVERGENT LENS

The fundamental equation of the plane diffraction grating for normal incidence of light is

$$(e+b)\sin\theta = \pm n\lambda$$

where $(e + b)$ is the grating element. It follows from this equation, that if the spacing of clear spaces, that is, $(e + b)$ varies systematically from point to point, then for the formation of intensity maximum corresponding to λ in a given order n, the angle of diffraction from different elements must also vary, in a systematic manner, so as to keep the product, $(e + b)\sin\theta$, a constant. Under this condition, the diffracted rays from different elements would themselves come in focus, at a point, without the aid of lenses. Thus they would form a maximum of intensity.

We can regard the *zone plate as circular diffraction grating*, in which the spacing of clear spaces (transparent zones) decreases systematically from the centre towards its boundary. Now if we regard the primary focus P_1, as the principal maximum of the first order, it follows that the diffraction angles for this order must be, for each clear zone of the plate, such as to cause the diffracted rays to meet at P_1. At points A and B, the zones are closer as compared with the zones at C and D. Therefore, the diffraction angles are smaller at C and D as compared to their values at A and B in such a systematic manner that the diffracted rays, under consideration, from A, B, C and D come in focus at P_1. We may now regard the second focus P_3 as principal maximum of the second order and so on. The diffracted rays, shown in Fig. 14.20, by full lines and marked 1st order and 2nd order, from different zones come in focus at P_1 (not shown in Fig. 14.20) and P_3 respectively, and so on.

Just as in the formation of spectra by grating, in the case of zone plate also, we shall get diffracted rays, shown by dotted lines, on the other side of the normal. But the angles of diffraction would increase in the same systematic manner as for diffracted rays shown by full lines, when we go from the central zone to the outer zones. The diffracted rays shown by dotted lines, therefore, meet only if they are produced backwards behind the zone plate. Those marked 1st order, 2nd order meet at P_1' (not shown) and P_3' respectively. Thus there are multiple foci P_1', P_3', P_5', etc. of course virtual, on the side of the source and at distances $f_1, \frac{1}{3}f_1, \frac{1}{5}f_1, \frac{1}{7}f_1$ etc. from the plate. *The zone plate thus behaves simultaneously as a divergent and a convergent lens but it has multiple foci.*

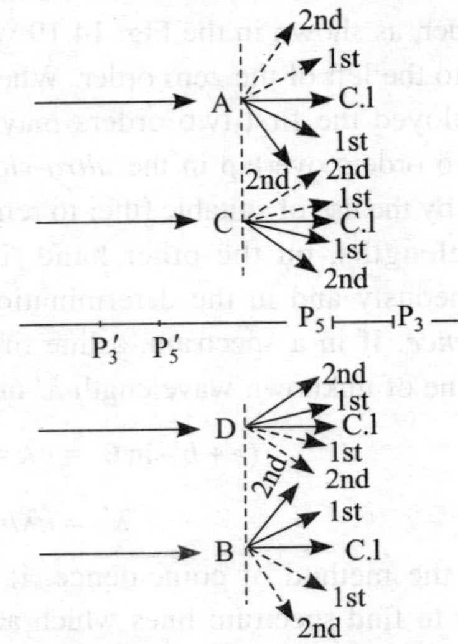

Fig. 14.20. Zone plate as circular grating of variable grating element

14.15 CONCAVE GRATING

The *plane grating* requires the use of two convergent lenses for the formation of spectra, one for rendering the light incident on the grating parallel and the other for bringing the parallel diffracted rays in focus. But the glass lenses are unsuitable in the investigation of *ultra-violet region*, being not transparent outside the visible spectrum. Moreover, the lenses cannot be made truly achromatic, so *chromatic aberration* comes in the spectra, so that lines to be examined cannot be exactly in focus simultaneously. The brilliant discovery was announced by *Henry Rowland* in 1882 that both the lenses may be very easily dispensed with by employing gratings ruled upon polished concave spherical metal surfaces. The ruled surface acts simultaneously as a grating and a concave mirror in producing and focussing the spectra, excelling in sharpness and brightness than those formed by plane grating-lens combination. This discovery is regarded most brilliant in the sense that it made possible the investigations of ultra-violet and extreme ultra-violet regions with an accuracy never before attained. Moreover, the problem of chromatic aberration disappears and superposed spectra of different orders are all in focus but the *astigmatic* defect associated with every concave mirror comes in. But the greatest advantage of this grating over plane grating is the fact that when properly mounted (see Rowland's mounting), the concave grating produces spectra which are truly normal. *A normal spectrum is that in which the distance between two lines is proportional to the wavelength difference.*

Concave gratings are usually made by ruling the polished concave spherical surface of *speculum metal* (an *alloy* of 32% *tin* and 68% *copper*) with short lines equally spaced along the chord *AB* drawn across the surface but not equally spaced along the arc of the surface. The positions of rulings in the Fig. 14.21 are indicated by dots. This mode of ruling the surface was theoretically derived by *Rowland* and also by *Runge* in the theory of *concave grating* for the formation of well focussed spectra. *The polished strips between the rulings, on account of their extremely small width, offer limitation to the incident beam, and therefore causing it, after reflection, to be diffracted in all*

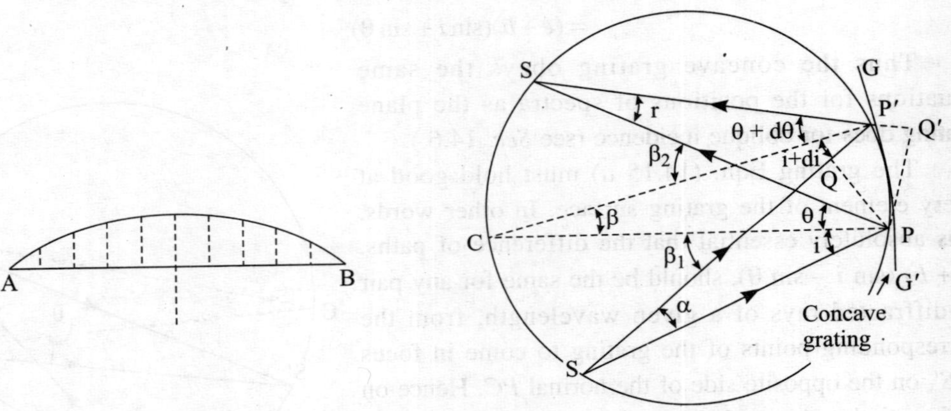

Fig. 14.21. Rulings are at equidistant intervals along a chord

Fig. 14.22. Diffraction by concave grating

directions. The focal properties of the mirror bring the diffracted rays in focus. We shall now give an elementary theory of this grating.

Let P and P' be corresponding points in consecutive polished spaces (greatly enlarged) of the concave grating GG', so that their distance apart is $(e + b)$, the grating element. C is the centre of curvature of the grating surface and S is a vertical slit illuminated by monochromatic light of wavelength λ. Light rays SP and SP' incident at angles i and $i + di$ after diffraction at the grating at angles θ and $\theta + d\theta$ respectively, are brought to focus at S'. Then the difference of their paths is

$$\Delta = (SP' + P'S') - (SP + PS') = (SP' - SP) - (PS' - P'S')$$

$$= P'Q - P'Q' = PP' \sin i - PP' \sin \theta$$

$$= (e + b)(\sin i - \sin \theta)$$

Owing to the fact that the phase change π produced by reflection of SP at P is exactly compensated by a similar phase change at P', the difference in phase between the waves arriving at S' merely depends upon the difference in their paths SPS' and $SP'S'$. Hence the condition for rays PS' and $P'S'$ to interfere constructively so as to form a principal maximum at S' is that the difference in their paths should be integral multiple of λ, that is,

$$(e + b)(\sin i - \sin \theta) = n\lambda \qquad \text{...(14.15 } a\text{)}$$

where n is an integral number and characterises the order of the spectrum.

If S' is taken on the same side of the normal PC as S (Fig. 14.23), the condition for maximum at S' becomes

$$\begin{aligned} n\lambda &= (SP' + P'S') - (SP + PS') \\ &= (SP' - SP) + (P'S' - PS') \\ &= PQ' + PQ \\ &= PP' \cos P'PQ' + PP' \cos P'PQ \\ &= PP' \cos (90 - i) + PP' \cos (90 - \theta) \\ &= (e + b)(\sin i + \sin \theta) \end{aligned} \quad \text{...(14.15 } b)$$

Thus the concave grating obeys the same equations for the positions of spectra as the plane grating does for oblique incidence (see *Sec.* 14.6).

The grating Eqn. (14.15 *a*) must hold good at every element of the grating surface. In other words, it is absolutely essential that the difference of paths, $(e + b)(\sin i - \sin \theta)$, should be the same for any pair of diffracted rays of a given wavelength, from the corresponding points of the grating to come in focus at S', on the opposite side of the normal PC. Hence on differentiating the grating equation, we get

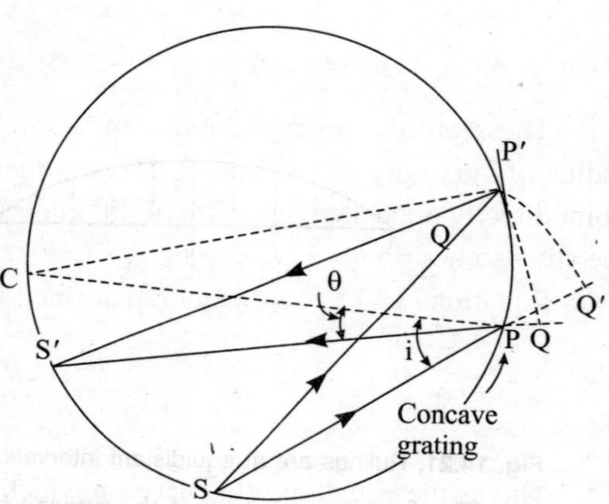

Fig. 14.23. Diffraction by a concave grating

$$d(n\lambda) = d[(e + b)(\sin i - \sin \theta)] = 0$$

and since grating element $(e + b)$ is to be regarded as a constant, we have

$$\cos i \, di - \cos \theta \, d\theta = 0 \quad \text{...(14.15 } c)$$

We shall now derive an equation which relates the position of diffracted image of order n and wavelength λ with the position of the source of light.

From the geometry of Fig. 14.22, we have the following relations between the angles marked:

$$i + \alpha = \beta_1 = \beta + i + di$$

and

$$\theta + \beta = \beta_2 = \gamma + \theta + d\theta$$

whence

$$di = \alpha - \beta$$

and

$$d\theta = \beta - \gamma$$

Hence Eqn. (14.15 *c*) becomes

$$(\alpha - \beta) \cos i - (\beta - \gamma) \cos \theta = 0 \quad \text{...(14.15 } d)$$

Let R be the radius of curvature, PC, of the grating and let $PS = r_1$ and $PS' = r_2$. Then

$$r_1 \cdot \alpha = PQ = (e + b) \cos i$$

$$R \cdot \beta = PP' = (e + b)$$

$$r_2 \cdot \gamma = PQ' = (e+b)\cos\theta$$

Substituting the values of α, β and γ in Eqn. (14.15 *d*) and dropping the common factor (*e* + *b*), we get

$$\left[\frac{\cos i}{r_1} - \frac{1}{R}\right]\cos i - \left[\frac{1}{R} - \frac{\cos\theta}{r_2}\right]\cos\theta = 0 \qquad \text{...(14.15 } e\text{)}$$

This is the required equation. It is obvious that if Eqn. (14.15 *e*) is to be satisfied for all values of *i* and θ the bracketed terms must be zero, since cos *i* and cos θ are neither always in a constant ratio nor always zero. Hence, we get

$$\left.\begin{aligned} r_1 &= R\cos i \\ r_2 &= R\cos\theta \end{aligned}\right\} \qquad \text{...(14.15 } f\text{)}$$

These are the polar equations for a circle such that *S* and *S'* lie on a circle of diameter *R*, the radius of curvature of the grating. This is then the *Rowland circle* to which grating is a tangent at a point directly opposite *C*. Spectra of all orders are in focus on the circumference of this circle, provided the slit also lies on it.

Equation (14.15 *e*) may be transformed by simple algebraic steps into the following form:

$$r_2 = \frac{R\, r_1 \cos^2\theta}{r_1(\cos i + \cos\theta) - R\cos^2 i}$$

This is the equation of a curve on which series of diffracted images are formed when the source of light, for example the slit illuminated by external source, is not on the Rowland circle.

14.16 CONCAVE GRATING MOUNTINGS

There are three important mountings for concave grating, viz.

(*i*) *Rowland mounting*,

(*ii*) *Runge-Paschen mounting*, and

(*iii*) *Eagle's mounting*.

Each of these employ mechanical devices for keeping the slit, spectra and grating, in accordance with the general principle discovered by Rowland, on the Rowland's circle which just touches the ruled surface at one point.

Rowland's Mounting

The mounting designed by Rowland for his own use satisfies in an ingenious manner the requirements mentioned above. Two strong iron rails *SA* and *SB* are mounted, in a horizontal plane, on a heavy framework so as to intersect perpendicularly at *S*. Another metal bar *R*, of length exactly equal to the radius of curvature of the grating surface, is mounted at its two ends on carriages which slide freely in tracks along iron rails. At one end of this bar is mounted the grating with its lines vertical and axis along the length of the bar while at the other end is an eye-piece or photographic plate bent into an arc of a circle of which the bar R is the diameter. In all positions of this bar, the intersection S of the rails also lies on this circle simply due to the reason that the bar always subtends right angle at S. Therefore by mounting a vertical slit at S, the *Rowland circle* is completed and the condition of focussing is satisfied. In order to change from one spectral region to another, the end P of the bar, carrying the plate holder, is moved let us say away from the slit to P', whereupon the grating G moves towards the slit say to G'. The slit S and source remain fixed in position and the incident beam remains fixed in direction. When grating moves the Rowland circle may be thought of moving with it in such a manner that, under given mechanical adjustments, it always slides through the slit. The instrument, therefore, once adjusted, always remains in focus.

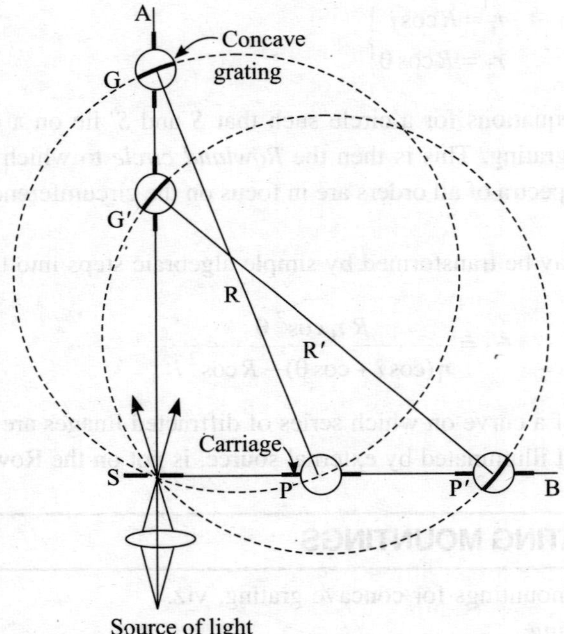

Fig. 14.24. Rowland mounting for a concave grating

Monochromatic rays diffracted from the grating along GP will have $\theta = 0$ and $i = \angle SGP$. Hence, the condition for formation of bright spectral line at P corresponding to wavelength λ, by Eqn. (14.15 *a*), becomes

$$(e+b)\sin i = n\lambda \quad \text{or} \quad (e+b)SP/PC = n\lambda$$

Since PG and $(e + b)$ are constant, as the point P moves away from S along SB, 1st, 2nd, 3rd .... orders of spectral lines corresponding to wavelength λ are formed separated by equal intervals.

The plate covers only a range of 10°–20° of Rowland circle on both sides of the normal. Therefore, the spectra may be taken as nearly the normal spectrum over this region.

The *Rowland mounting* never remains, in practice, in accurate adjustment when the spectral region is changed, as carriages are not extremely rigid. The adjustments for refocussing are so troublesome that this mounting is not in common use.

Paschen Mounting

Rowland mounting described above is rarely used. The most satisfactory mounting for general work was devised by *Paschen* and *Runge*. In this arrangement, the slit is mounted in a hole of the grating room which is painted black, the source of light being in the adjacent room. The position of the slit is fixed relative to grating so as to give a desired angle of incidence, determined by the problem to be investigated. Once the grating and slit are fixed, the Rowland circle is accurately located for the problem. The light from the slit strikes the grating, which diffracts it into well-focussed spectra of various orders on the *Rowland circle*. A semi-circular steel frame coincides with the region of *Rowland circle*. The rim plays the role of photographic plate-holder. This has a horizontal slot in it against which photographic plates can be pressed at suitable positions. The plates are bent to the curvature of the *Rowland circle* by means of clamps, so as to photograph simultaneously spectra of all orders, without altering the adjustments of spectrograph. This is not possible with any other mounting. Therefore, *Paschen's* mounting has gained much popularity over others.

Fig. 14.25. Paschen mounting for a concave grating

Normal Dispersion is obtained only in the region which falls at or near grating normal. The spectra is, therefore, in general not normal in the whole region. But the facility of photographing spectra of all orders outweighs the disadvantage of non-linear dispersion.

Eagle Mounting

Eagle, in 1910, designed a mounting for concave grating, which is so compact that it can be shifted very easily from one part of the laboratory to another. Grating *G* is capable of rotation about the vertical axis and can be moved along a track by external mechanical devices. The whole of this is enclosed in a light tight box at the other end of which a plate holder, capable of rotation about the vertical axis, is mounted. The slit is mounted in one side of the box in a tube, at such a position that its virtual image in a small total reflection prism is formed at a point *O*, just below the centre of the photographic plate.

In this mounting only that part of the spectrum is observed for which the angle of diffraction is nearly equal to the angle of incidence. Diffracted rays, therefore, nearly retrace their path. By slight tilt of grating, these rays pass over the prism and are incident on the plate. To form sharply focussed spectra, the grating is moved along the track and also given rotation about the vertical axis until

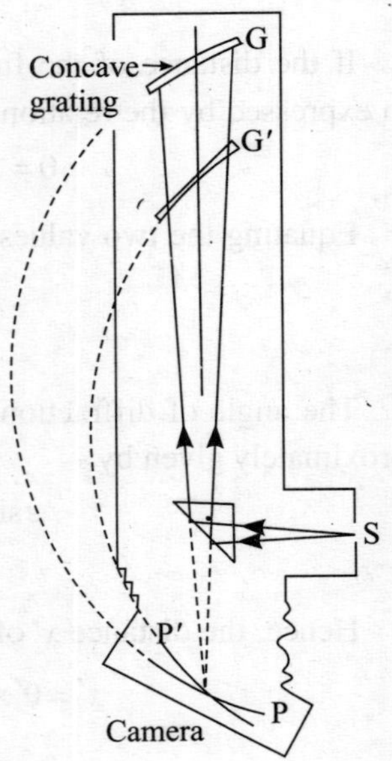

Fig. 14.26. Eagle mounting for a concave grating

the Rowland circle, which we must think as moving with the grating, passes through O, the image of the slit formed by total reflection prism. The plate holder, whose axis of rotation passes through O, is now rotated by proper amount such that the surface of the plate also lies on *Rowland circle*. Of course, some focussing by trial is necessary. To pass from one spectral region to another, the grating is rotated by proper amount to bring the desired region on the plate and the above adjustments are repeated for focussing it. Two positions of grating, corresponding settings of plate and the positions of Rowland circle are marked in Fig. 14.26 for two different spectral regions to be in focus, in succession, on the plate.

SOLVED EXAMPLES

Example 1. *A lens whose focal length is 40 cm forms a Fraunhofer diffraction pattern of a slit 0.3 mm wide. Calculate the distance of the first dark band and of the next bright band from the axis. (Wavelength of light used in 5890 A.)*

In the Fraunhofer diffraction pattern due to a single slit, the angle of diffraction corresponding to zero intensity are given by

$$e \sin\theta = m\lambda, \quad \text{where} \quad m = 1, 2, 3 \text{ etc.}$$

For the first dark band $m = 1$, hence

$$\sin\theta = \lambda / e = 5890 \times 10^{-8} / 0.03$$

or

$$\theta = \frac{1}{3} \times 589 \times 10^{-5} \text{ radian}$$

If the distance of the first dark band from the axis be x cm then the angle of diffraction θ, is also expressed by the relation

$$\theta = x/f' = x/40 \text{ radian}$$

Equating the two values of θ, we get

$$x = \frac{1}{3} \times 40 \times 589 \times 10^{-5} \text{ cm} = \mathbf{0.785\ mm}$$

The angle of diffraction corresponding to first bright band on either side of the central band is approximately given by

$$e \sin\theta' = (3/2)\lambda$$

$$\theta' = 3\lambda / 2e$$

Hence, the distance x' of the first bright band from the axis is

$$x' = \theta' \times 40 = 3 \times 5890 \times 10^{-8} \times 40 / 2 \times 0.03 \text{ cm}$$

$$x' = \mathbf{1.178\ mm}$$

Example 2. *The limits of the visible spectrum are approximately 4000 A.U. and 7000 A.U. Find the angular breadth of the first order visible spectrum produced by a plane diffraction grating having 15000 lines per inch when light is incident normally on the grating. Prove that the violet of the third order visible spectrum overlaps the red of the second order.*

The grating element, $(e + b)$ is

$$(e+b) = 2.54/15000 = 1.69\times10^{-4} \text{ cm}$$

The angular deviation of the violet in the first order is

$$\sin\theta = \frac{4000\times10^{-8}}{1.69\times10^{-4}} = 0.237.$$

$$\theta = 13°40'$$

Similarly, the angular deviation of red is 24°30′. Hence, the angular breadth of the first order visible spectrum is

$$24°30' - 13°40' = \mathbf{10°50'}.$$

The angular deviation of the third order violet is

$$\sin\theta = 3\times 4000\times10^{-8}/(e+b)$$

and of the second order red is

$$\sin\theta' = 2\times 7000\times10^{-8}/(e+b)$$

The first angle is smaller than the second, whatever may be the value of $(e + b)$. Therefore, the third order will always overlap the second order.

Example 3. *What is the maximum width, e, of any clear space in a grating having 15000 lines per inch, if complete second order spectrum is to be formed?*

Grating element, $(e + b) = 2.54/15000 = 1.69 \times 10^{-4}$ cm.

The maximum angle θ in the second order visible spectrum is given by

$$\sin\theta = \frac{2\times7000\times10^{-8}}{1.69\times10^{-4}}$$

For practical purposes light diffracted by one clear space (single slit) can be considered to be confined only in the central diffraction band. The half angular breadth of the central diffraction band must be at least as great as the maximum angle of diffraction in the second order visible spectrum. (In reality, it must be slightly greater, since at the edges of central diffraction band, intensity falls to zero).

The half angular breadth of the central diffraction band is

$$\sin\alpha = \lambda/e = 7000\times10^{-8}/e$$

Hence putting $\alpha = \theta$, we get

$$\frac{2\times7000\times10^{-8}}{1.69\times10^{-4}} = \frac{7000\times10^{-8}}{e}$$

$$e = \mathbf{0.845 \times 10^{-4}\ cm.}$$

This is just half the grating element.

Example 4. *How many orders will be visible if the wavelength of incident radiation be 5000 A.U. and the number of lines on the grating be 2620 to an inch?*

Grating element $(e + b) = 2.54/2620$ cm

The maximum observable angle of diffraction is 90°.

Hence, $$n = \frac{(e+b)\sin 90}{\lambda} = \frac{2.54}{2620 \times 5000 \times 10^{-8}} = 19.4$$

19 orders will be visible in the grating spectra.

Example 5. *Sodium light of wavelength 5890 A. U. is incident normally upon a plane transmission diffraction grating of 10,000 lines per inch. Calculate the angular separation of the sodium lines in the spectrum of the first order as observed in a telescope of which the objective has a focal length of 24 cm and eye-piece 2 cm.*

Grating element, $(e + b) = 2.54/10000 = 2.50 \times 10^{-4}$ cm

In the first order ($n = 1$), grating equation becomes

$$(e+b)\sin\theta = \lambda$$

Differentiating the above equation and simplifying it is easy to get

$$d\theta = \frac{d\lambda}{\sqrt{(e+b)^2 - \lambda^2}} = \frac{d\lambda}{(e+b)} \text{ approximately}$$

From the given data,

$$d\theta = (5896 - 5890) \times 10^{-8}/2.54 \times 10^{-4} = 3/12700 \text{ radian.}$$

The angular magnifying power of the telescope, expressed by Eqn. (14.6 *b*) for the given problem becomes

$$d\theta'/d\theta = -f'_o/f'_e$$

where f'_o is the focal length of the objective and f'_e is the focal length of the eye-piece and $d\theta'$ is the angular separation of images of sodium lines in the telescope. With the given data,

$$d\theta' = -\frac{3}{12700} \times \frac{24}{2} = -\mathbf{0.0028\ radian.}$$

The angular separation of sodium lines as seen by telescope is **0.0028 radian.**

Example 6. *A diffraction grating used at normal incidence gives a green line, λ = 5400 A. U. in a certain order superimposed on the violet line, λ = 4050 A. U. of the next higher order. If the angle of diffraction is 30°, how many lines are there to the centimeter in the grating?*

Suppose the green line, $\lambda = 5400$ A.U., in the *n*th order is superimposed on the violet line, $\lambda = 4050$ A. U., of $(n + 1)$th order. Hence, we write the equation.

$$(e + b)\sin\theta = n \times 5400 \times 10^{-8} = (n + 1) \times 4050 \times 10^{-8}$$

Hence $$n = 3$$

and $$(e+b) = \frac{3 \times 5400 \times 10^{-8}}{\sin 30°} = 324 \times 10^{-6}$$

Suppose, there are x lines per cm in the grating, then $(e + b) = 1/x$.

Hence $$x = 1/(e + b) = 1/(324 \times 10^{-6}) = \textbf{3086 lines/cm.}$$

Example 7. *A Rowland grating has 15000 lines to the inch. Calculate the angular separation of the two mercury lines 5770 A. and 5461 A. in the spectrum of first order.*

The fundamental equation of the concave grating is

$$(e + b)(\sin \theta \pm \sin i) = n\lambda$$

Differentiating the above equation, the angular separation between the spectral lines of wavelengths λ and $\lambda + d\lambda$ is given by

$$d\theta = n d\lambda / (e+b) \cos \theta$$

For Rowland grating, $\theta = 0$ and with the given data

$$d\lambda = (5770 - 5461) \times 10^{-8} \text{cm} = 309 \times 10^{-8} \text{cm}$$

$$n = 1 \text{ and } (e + b) = 2.54/15000 \text{ cm}$$

Hence, $$d\theta = 309 \times 10^{-8} \times 15000/2.54 = \mathbf{1.83 \times 10^{-2}} \textbf{ radian}$$

Example 8. *Light of wavelength 4000 A. U. from a distant point source is incident on an opaque plate in which there is a small circular opening. As a screen is moved from a large distance toward the plate, the Fresnel diffraction pattern on the screen first has a dark centre when the distance of the screen from the plate is 80 cm. Find the radius of the central disc in the Fraunhofer diffraction pattern, if a lens of focal length 80 cm is placed just to the right of circular opening.*

From the theory of Fresnel diffraction at a circular aperture, it follows that as a screen is moved towards the aperture, the centre of diffraction pattern is first dark (not absolutely dark) when the screen is at a position for which the circular aperture exposes two Fresnel's half period zones of the plane wave-front. In this setting of the screen, the distance of the centre of the diffraction pattern to the edge of aperture must be $(80 + \lambda)$, where the separation of screen from the aperture is 80 cm. If a be the radius of the circular aperture, we have the relation

$$(80 + \lambda)^2 = 80^2 + a^2$$

or $$a^2 = 2 \times 80 \times \lambda + \lambda^2$$

On neglecting λ^2 being very small as compared with $2 \times 80 \times \lambda$ the radius of circular aperture is given by

$$a = \sqrt{2 \times 80 \times \lambda} = \sqrt{2 \times 80 \times 4000 \times 10^{-8}} = 0.08 \text{ cm.}$$

As explained in *Sec.* 14.3, the radius of the central disc in the Fraunhofer diffraction pattern is given by

$$r_1 = \frac{1.22\lambda}{d} \times f' = \frac{1.22\lambda f'}{2a}$$

With the given data,

$$r_1 = \frac{1.22 \times 4000 \times 10^{-8} \times 80}{2 \times 0.08} = \mathbf{0.0244\ cm.}$$

PROBLEMS

1. (a) Discuss the Fraunhofer diffraction at a single slit.
   (b) Distinguish between the diffraction fringes due to a single slit of width *e* and the interference fringes produced with two narrow slits of separation *e*.
   (c) Draw separately the amplitude distribution and the intensity distribution for the Fraunhofer diffraction pattern.
2. Explain clearly the difference between the Fresnel and Fraunhofer classes of diffraction phenomena. Discuss Fraunhofer diffraction of monochromatic light from a single slit and show that the intensity of the first subsidiary maximum is roughly 4.5% of that of the principal maximum.
3. Describe the features of a double-slit Fraunhofer's diffraction pattern. How does it change if the distance between the slit centres is varied, keeping the slit width constant?
4. Give the construction and theory of a plane transmission grating and explain the formation of spectra.
5. Give the theory of plane transmission grating and show how you would use to find the wavelength of light.
6. For a plane diffraction grating with 5000 lines per cm, used at normal incidence, answer the following:
   (a) What is the longest wavelength of light for which a spectrum can be observed?
   (b) For the wavelength 6000 A.U. what is the highest order of spectrum which may be observed?
   (c) If the opaque spaces are twice as wide as the transparent spaces, which orders of the spectra will be absent?
   (d) If the number of lines per cm remains the same but the shape of the grooves is changed, what will be the effect on (i) the dispersive power of the grating and (ii) the relative intensities of the spectra in different orders?
7. Two spectral lines have wavelengths λ and $\lambda + d\lambda$ respectively where $d\lambda << \lambda$. Show that their angular separation $d\theta$ in a grating spectrometer is given by

$$d\theta = d\lambda / \sqrt{[(e+d)/n]^2 - \lambda^2}$$

where $(e + d)$ is the grating element and n the order at which lines are observed.

8. (a) Define dispersive power of a grating and obtain an expression for it.

   (b) Find the maximum number of orders available with a grating and show that only first order is possible if the width of the grating element is less than twice the wavelength of light.

9. (a) In what respects does a prism spectrum differ from a grating spectrum?

   (b) Explain the effect of closeness of the rulings and the width of ruled space on the grating spectrum.

10. Give an account of the theory and use of concave grating. What is normal spectrum?

11. Give the theory of concave reflection grating. Describe the various methods of mounting the same and point out their relative merits and demerits.

12. Light of wavelength 5000 A.U. is incident normally on a slit. The first minimum of the diffraction pattern is observed to lie at a distance of 5 mm from the central maximum on a screen placed at a distance of 2 meters from the slit. Calculate the width of the slit.

13. A slit of width 0.002 cm is placed immediately in front of a lens of focal length 60 cm. It is illuminated normally by a parallel beam of light of wavelength 5×10^{-5} cm. Calculate the distance between the centre and the first dark band of the diffraction pattern on a screen placed at 60 cm from the lens.

14. When plane waves from a monochromatic source fall normally on a plane grating having 500 lines per mm. it is observed that the second order spectral line is deviated through 30°. Find the wavelength of the source of light.

15. A diffraction grating used at normal incidence gives a green line (5400 A.U.) in a certain order superimposed on the violet line (4050 A.U.) of the next higher order. If the angle of diffraction is 30°, how many lines per cm are there in the grating?

16. The wavelength of the C-line of hydrogen is 6563 A.U. What is the highest order of the line that can be observed with a spectroscope using a grating with 6000 lines per cm?

17. A plane transmission grating having 6000 lines per cm is used to obtain a spectrum from a sodium lamp in the second order. Calculate the angular separation between the D_1 and D_2 lines.

18. Sodium light of mean wavelength 5893 A.U. is incident normally on a plane transmission grating of 10,000 lines per inch. Calculate the angular separation of the D lines in the first order spectrum as observed in a telescope whoes objective and eye-piece have respective focal lengths of 24 cm and 2 cm.

19. Find the angular separation between two yellow lines of neon (5882 A.U. and 5852 A.U.) in the first order spectrum of a concave grating, having 6000 lines per cm, the spectrum being observed normally.

RESOLVING POWER

The *magnifying power* of a microscope or a telescope depends (except for certain other numerical factors), only on the focal lengths of the lenses forming the objective and the eye-piece of the instrument. By the proper choice of these focal lengths, it appears as if any desired magnification might be attained. Unless the instrument is properly designed, however, while the magnification increases (*i.e.*, the image becomes larger), in reality the image does not reveal new details even if the lens were free from all aberrations. The upper limit to the useful magnification is set by the fact that the image of the point source formed by the lens, even in absence of all aberrations is not a point image. But in reality it is a circular diffraction pattern formed by the limited portion of the wavefront transmitted by the lens system. This diffraction pattern is characterised by the central bright disc bordered by the dark and the bright concentric circular rings of rapidly decreasing intensity. An optical system is said to be able to resolve two point objects if the corresponding diffraction patterns are sufficiently small or sufficiently separated to be distinguished as separate image patterns. With the analysing instruments one is chiefly concerned with its ability to just separate in the diffraction spectrum two spectral lines corresponding to light of slightly different wavelength λ and $\lambda + d\lambda$. We shall now discuss and derive expressions for the resolving power or the limit of resolution, of some of the image forming instruments like telescope and microscope, and of the analysing spectrographs involving prism or grating.

15.1 THE RAYLEIGH CRITERION FOR LIMIT OF RESOLUTION

From a detailed study of the resultant intensity distribution in the diffraction patterns of closely spaced point sources *Lord Rayleigh* arrived at the conclusion that two equally bright point sources could be just resolved by any optical system, when their distance apart is such that the central maximum in the diffraction pattern due to one source coincided exactly with the first minimum in the diffraction pattern due to the other. This is known as *Rayleigh's criterion of resolution* and it is also applicable to the resolution of spectral lines of equal intensity, formed by grating or prism spectrographs. When applied to spectral lines, it is equivalent to the condition that in the diffraction pattern, *the angular separation between the principal maxima of two spectral lines in a given order, should be equal to half angular width of either principal maximum.*

In symbols, this means that if the principal maximum of wavelength λ is formed by an optical instrument at an emergent angle θ, and the principal maximum of $\lambda + d\lambda$, in the same order, is formed at an emergent angle $\theta + d\theta$, then the first minimum in the diffraction pattern due to λ should also be formed at an angle $\theta + d\theta$.

The Physical significance of *Rayleigh's criterion for resolution* can be explained by considering the resultant intensity distribution in the diffraction patterns due to two wavelengths λ and λ_1. Let the wavelength difference be such that the principal maximum of one falls exactly over second point of minimum intensity in the diffraction pattern of the other. The two wavelengths must have been emitted by different sources atoms so that the disturbances are incoherent. Consequently, the total intensity at any point in the spectrum is simply the sum of the intensities for each wavelength. In this case, as depicted in Fig. 15.1 there is a distinct point of zero intensity in the middle of resultant intensity curve which has been obtained by summing the intensities due to separate patterns. Thus, the two spectral lines are distinctly separated. However, if the wavelength difference is smaller, the two spectral lines with their principal maxima are closer. There is one limiting value of λ_1 ($= \lambda + d\lambda$), for which the angular separation between their principal maxima is such that the principal maximum of one falls exactly over the first point of minimum intensity in the diffraction pattern due to the other.

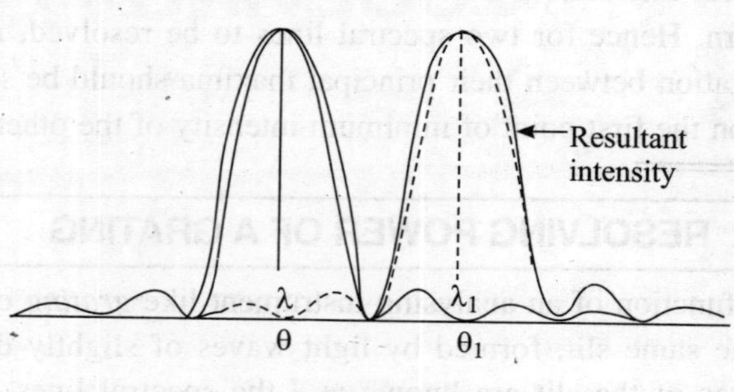

Fig. 15.1

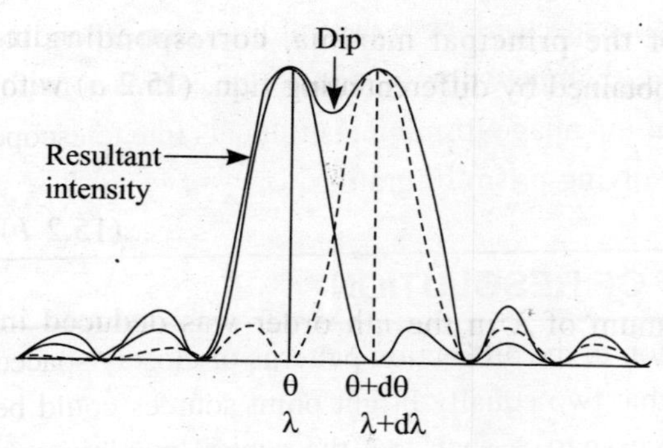

Fig. 15.2. Illustrates Rayleigh's criterion for resolution of spectral lines

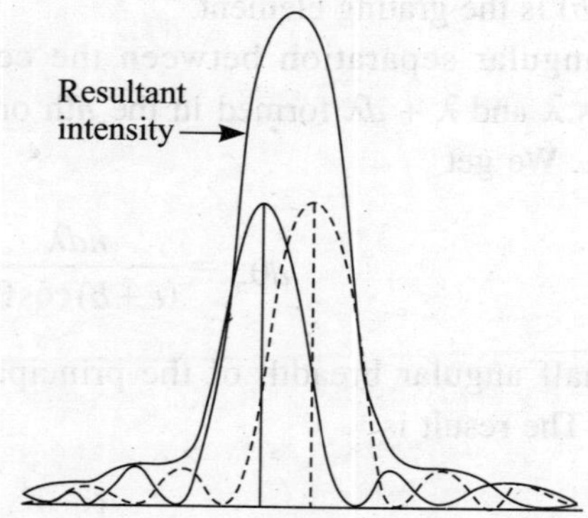

Fig. 15.3

Therefore, the intensities of maxima in the resultant intensity patterns are equal to those of original maxima. Thus, the resultant intensity curve, in Fig. 15.2, is a *double humped* curve with a distinct *dip* at a point half way between the two principal maxima. This change of intensity is easily visible to the eye and enables us to recognise that in the resultant pattern there are two spectral lines corresponding to λ and $\lambda + d\lambda$. The two spectral lines, under this condition are said to be resolved. But if $\lambda_1 < \lambda + d\lambda$, the two spectral lines come still closer and the intensity curves of the principal maxima show considerable overlapping. The resultant intensity curve, shown in Fig. 15.3 exhibits no dip but only one maximum in centre, thereby indicating as, if it is a diffraction pattern of only one spectral line. Thus, under this condition, the two spectral lines are not resolved in their diffraction pattern. Hence for two spectral lines to be resolved, it is absolutely essential that the least angular separation between their principal maxima should be such that the principal maximum of one should fall on the first point of minimum intensity of the other, as depicted in Fig. 15.2.

15.2 RESOLVING POWER OF A GRATING

The function of an analysing instrument like *grating* or *prism spectrograph* is to resolve two images of the same slit, formed by light waves of slightly different wavelengths, say λ and $\lambda + d\lambda$. The images of the slit are known as *l* the spectral lines corresponding to wavelengths λ and $\lambda + d\lambda$. Accordingly, *the limit of resolution of a grating or a prism is defined as the smallest wavelength difference $d\lambda$ for which the spectral lines can be just resolved at the wavelength λ.* This limit of resolution is mathematically expressed by the ratio $\lambda/d\lambda$. We shall apply *Rayleigh's* criterion for resolution to derive an expression for the resolving power of the plane grating.

Let a plane wavefront of light to be analysed be incident normally on the grating, which forms the diffraction pattern in the focal plane of the telescope lens, corresponding to each of the wavelengths, say λ and $\lambda + d\lambda$, present in the incident light. The angular deviation θ_n of the centre of principal maximum formed in the *n*th order, by light of wavelength λ, is given by

$$(e + b) \sin \theta_n = n\lambda \qquad \text{...(15.2 } a)$$

where $(e + b)$ is the grating element.

The angular separation between the centres of the principal maxima, corresponding to wavelengths λ and $\lambda + d\lambda$ formed in the *n*th order, is obtained by differentiating Eqn. (15.2 *a*) with respect to λ. We get

$$d\theta_n = \frac{n d\lambda}{(e+b)\cos\theta_n} \qquad \text{...(15.2 } b)$$

The half angular breadth of the principal maximum of λ in the *n*th order was deduced in *Sec.* 14.12. The result is

$$d\theta_n = \frac{1}{Nn\cot\theta_n} \qquad \text{...(15.2 } c)$$

where N is the total number of lines in the grating.

According to *Rayleigh's criterion for resolution*, the two spectral lines will be just resolved, when the principal maximum of one falls exactly over the first point of minimum intensity of the other line. This condition will be realised if the angular separation, $d\theta_n$, between the two principal maximum is exactly equal to half-angular breadth of the principal maximum of λ in the same order. Hence equating the value of $d\theta_n$ as given by Eqns. (15.2 *b*) and (15.2 *c*), we have for just resolution of spectral lines

$$\frac{1}{Nn\cot\theta_n} = \frac{nd\lambda}{(e+b)\cos\theta_n}$$

or

$$\frac{1}{Nn} = \frac{nd\lambda}{(e+b)\sin\theta_n}$$

and making use of Eqn. (15.2 *a*), we have for chromatic resolving power.

$$\lambda / d\lambda = Nn \qquad \text{...(15.2 } d\text{)}$$

The resolving power is, therefore, directly proportional to the order *n*, and in the given order it is proportional to the total number of lines effective in the formation of the diffraction pattern. It would be observed that the resolving power is independent of the spacing $(e + b)$ of the lines. Of course all of the lines will be effective, only if the diameters of the collimating and telescope lenses are great enough so as to fill the entire ruled surface of the grating with the incident beam. Moreover the apparent dependence of resolving power on *N* is illusory, for in a given direction θ_n and for a given wavelength it simply depends on the width of the ruled surface. This can be seen simply by substituting the value of *n* in Eqn. (15.2 *d*), we get

$$\frac{\lambda}{d\lambda} = \frac{N(e+b)\sin\theta_n}{\lambda} = \frac{B\sin\theta_n}{\lambda} \qquad \text{...(15.2 } e\text{)}$$

where $B = (e + b)\,N$, is the total width of the ruled surface of grating. Therefore a change in the number of lines *N* in a given width of ruled surface, will not effect the resolving power. However, a change in *N* would change the order *n* of spectrum in a given direction.

On putting $\theta_n = 90°$ in Eqn. (15.2 *e*), it will be seen that the resolving power of a grating cannot exceed B/λ, the width of the grating measured in wavelength. In practice the intensity would be extremely small under such conditions. But for oblique incidence, it can be easily seen that the resolving power of a grating cannot exceed twice the width of the grating measured in wavelengths.

15.3 RESOLVING POWER OF A PRISM

The resolving power of a prism, as in the case of a grating, is expressed by the ratio $\lambda/d\lambda$, where $d\lambda$ is the smallest wavelength difference that can be just resolved at the wavelength λ. Neighbouring wavelengths will be only resolved if the angular separation between their spectral lines exceeds the limit of resolution of the telescope. We shall give *Rayleigh's* treatment of deriving an expression for the resolving power of a prism, set at minimum deviation.

The essential parts of a prism spectrograph are illustrated in Fig. 15.4. Heterogeneous light, consisting of wavelengths λ and $\lambda + d\lambda$ diverging from a slit *S* passes through a collimating lens L_1 and the plane wave with surfaces parallel to *AD* is incident on the prism *ABC*. In traversing the prism, the rays are refracted by different amounts and the emergent wavefronts are *BE* for the wavelength λ and *BF* for $\lambda + d\lambda$. The telescope lens L_2 focusses the wavefront *BE* at I_1 and *BF* at

I_2. If the wavelength difference $d\lambda$ is small, the setting of the prism is essentially at minimum deviation for both, dotted as well as the solid paths. By *Fermat Principle*, the total path from S to I_1 is the same for all rays of wavelength λ and the focal property of the lens L_1 implies that the paths S to A and S to D are equal. Similarly the paths from B to I_1 and E to I_1 are equal. Hence we have for the wavelength λ

$$DC + CE = \mu AB \quad ...(15.3\ a)$$

and similarly for $\lambda + d\lambda$, we have

$$DC + CE = (\mu - d\mu)AB \quad ...(15.3\ b)$$

where the refractive index of the material of the prism is μ for λ and $(\mu - d\mu)$ for $(\lambda + d\lambda)$. It should be clearly borne in mind that the refractive index decreases with increase of wavelength.

On subtraction of Eqn. (15.3 *b*) from Eqn. (15.3 *a*), we get

$$CE - CF = d\mu\ AB \quad ...(15.3\ c)$$

but from the geometry of the Fig. 15.4,

$$CE - CF = EF = BE\ d\theta = a\ d\theta$$

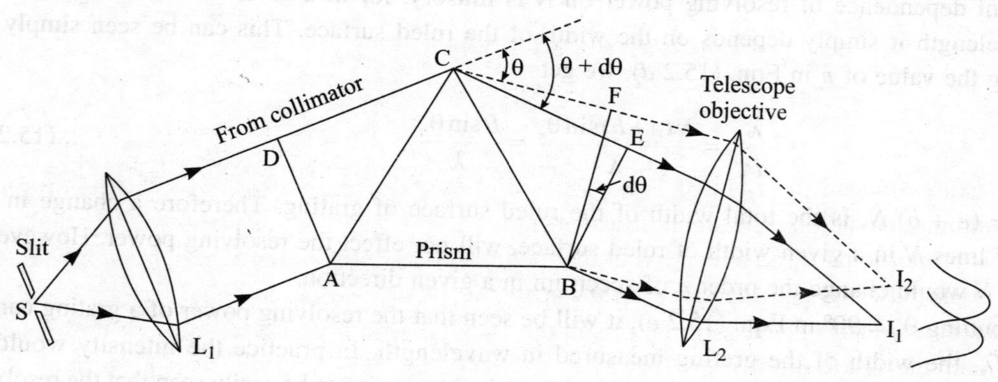

Fig. 15.4. Limit of resolution of a prism

where $BF = a$, is the width of the emergent beam. θ and $\theta + d\theta$ are the angles of deviation of beams of light of $\lambda + d\lambda$ and λ respectively. Therefore Eqn. (15.3 *c*) simplifies to the form

$$d\theta = (t/a)d\mu \quad ...(15.3\ d)$$

where the thickness of the base of the prism has been denoted by t.

The face BC of the prism limits the refracted beam, incident upon it, to a rectangular section. Therefore, we can consider the rays of light, corresponding to λ and $\lambda + d\lambda$, to be diffracted in all directions from every point of the face BC. *Rayleigh's criterion of resolution* is then applied to the diffraction pattern due to the rectangular aperture. The diffracted rays corresponding to λ, which are perpendicular to BE are in the same phase on this wavefront. Therefore, the principal maximum of λ is formed at I_1, the position where BE comes in focus. Similarly the principal maximum of $\lambda + d\lambda$ is formed at I_2, the position where BE comes in focus. Hence the angular separation between the two principal maxima is the angle $d\theta$ between BE and BF, which is given by the expression on the right hand side of Eqn. (15.3 *d*). If these wavelengths are to be just resolved in the focal plane of the telescope lens then according to *Rayleigh's* criterion for resolution, the principal maximum corresponding to $\lambda + d\lambda$ must exactly fall over the first point of minimum intensity in the diffraction pattern of λ and vice versa. This simply means that for just resolution of λ and $\lambda + d\lambda$, the diffracted rays of wavelength λ, which are perpendicular to BF, must give rise to first minimum at I_2 in the diffraction pattern. The condition for this is that the difference in paths of the extreme diffracted rays of λ, which are normal to BF, must be equal to λ, that is,

$$\mu AB - (DC + CF) = \lambda$$

$$(DC + CE) - (DC + CF) = \lambda \qquad \text{[using Eqn. 15.3 (a)]}$$

or

$$CE - CF = \lambda$$

or

$$d\theta = \lambda/a \qquad \text{...(15.3 e)}$$

This is simply half angular breadth of the principal maximum of λ. Therefore for just resolution of λ and $\lambda + d\lambda$, we equate the values of $d\theta$ as given by Eqns. (15.3*d*) and (15.3 *e*). We get

$$\lambda/a = (t/a)d\mu$$

The desired result is now obtained on dividing the above equation by $d\lambda$.

$$\lambda/d\lambda = t(d\mu/d\lambda) \qquad \text{...(15.3 f)}$$

or

$$d\lambda = \frac{\lambda}{t(d\mu/d\lambda)} \qquad \text{...(15.3 g)}$$

The expression on the right of Eqn. (15.3 *f*) gives resolving power only when the prism is used at minimum deviation. The smallest wavelength difference $d\lambda$ that can be resolved at minimum deviation is proportional to wavelength λ at that point in the spectra and inversely proportional to the width of prism base AB and the slope $d\mu/d\lambda$ at the wavelength λ of the index versus wavelength curve. Therefore greater the base of the prism, the smaller the resolvable wavelength difference $d\lambda$, hence greater the resolving power. However, the resolving power is limited due to inability of constructing homogeneous prism with very large base. The diameters of both the lenses must, of course, be great enough so that the full height of the prism may be effective, otherwise, t must be replaced by $(t_1 - t_2)$ where t_1 and t_2 are the thickness of the material through which two extreme rays of the refracted beam pass. It would be observed that the resolving power of a prism is independent of its refracting angle.

It is sometimes convenient to express the resolving power of the prism in terms of angular dispersion D, which is defined as the ratio of the change in the angle of deviation with wavelength. Thus, if $d\theta$ be the change in the angle of deviation when wavelength is changed by $d\lambda$, then

$$D = d\theta/d\lambda$$

Hence using Eqn. (15.3 *e*), it is easy to see that the resolving power is expressed by the relation

$$R.P. = \text{width of the beam} \times \text{angular dispersion}$$

In the end, it should be pointed out that *the resolving power of a train of prisms is equal to the algebraic sum of the resolving powers of individual prism*. This statement can be very easily proved by deriving an expression for the resultant resolving power of two prisms, of unequal bases and constructed from different materials placed as shown in Fig. 15.5. Then, since the optical distance between the wavefronts measured along any ray is the same, we have for the wavelength λ,

$$GC + \mu' CD = \mu AB + BE$$

and for the wavelength $\lambda + d\lambda$ we have

$$GC + (\mu' - d\mu') CD = (\mu - d\mu) AB + BF$$

On subtraction and rearranging, we get

$$BE - BF = CDd\mu' - ABd\mu$$

or

$$ad\theta = t'd\mu' - td\mu$$

or

$$d\theta = (t'd\mu' - td\mu)/a$$

where μ and μ' are the indices of refraction of material of the prisms ABC and BCD, for light of wavelength λ and $AB = t$, $CD = t'$, are the width of the bases of the prisms.

The half angular width of the principal maximum corresponding to λ is given by Eqn. (15.3 *e*), viz.

$$d\theta = \lambda/a$$

Therefore, for just resolution of spectral lines by this combination of prisms, we must have

$$\lambda/a = \frac{t'd\mu' - td\mu}{a}$$

$$\lambda/d\lambda = t'\frac{d\mu'}{d\lambda} - t\frac{d\mu}{d\lambda}$$

$$= R.P. \text{ of prism } BCD - R.P. \text{ of prism } ABC.$$

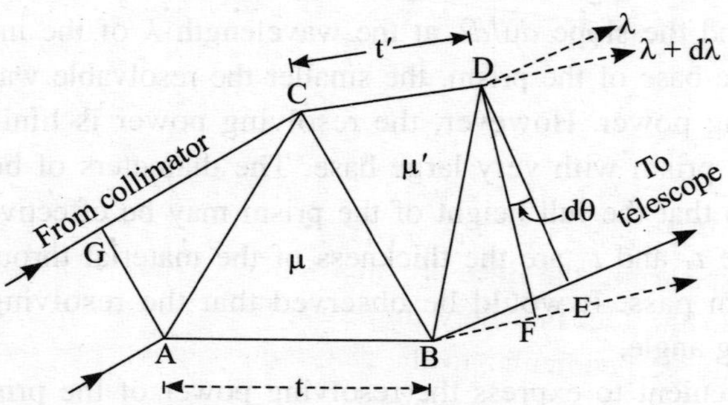

Fig. 15.5. Limit of resolution of two prisms

15.4 DISTINGUISHING FEATURES BETWEEN THE GRATING AND THE PRISM SPECTRUM

The diffraction grating is commonly used instead of a prism as an appliance for dispersing a beam of light into spectrum. The distinguishing features between the grating and the prism spectra may be now given.

(1) A prism produces only one spectrum but as explained in *Secs*. 13.5 and 13.7, the grating gives a number of spectra on either side of the zero order (direct) image of the slit.

(2) Higher order grating spectra partially overlap but in prism, as there is only one spectrum, there is no overlapping of orders. For certain type of experiments employing prism this is an important advantage.

(3) The light incident on the prism is distributed simply in one spectrum which it produces. But in the case of a grating most of the light goes into zero order image of the slit where no spectrum is formed and rest of the incident light is distributed amongst spectra of various orders. Consequently the prismatic spectrum is extremely bright as compared to the grating spectra.

(4) The refractive index of the material of the prism increases towards the violet end of the spectrum. As a result of this violet colour has greatest deviation while the deviation of red is least. The grating, on the other hand, deviates the violet colour least and red the most. This action of grating is just reverse of the prismatic action in dispersing the light. Moreover, the grating produces spectra on either side of the direct image of the slit and in both the sets the violet is nearest to the direct image. Therefore if the *prismatic spectrum*, depicted in Fig. 15.6 and the *grating spectra*, depicted in Fig. 15.7, are photographed side by side on the same plate, then the order of spectral colours in the prismatic spectrum will agree only with the grating spectra formed on one side of the central image (in Fig. 15.7 shown above *C.I.*).

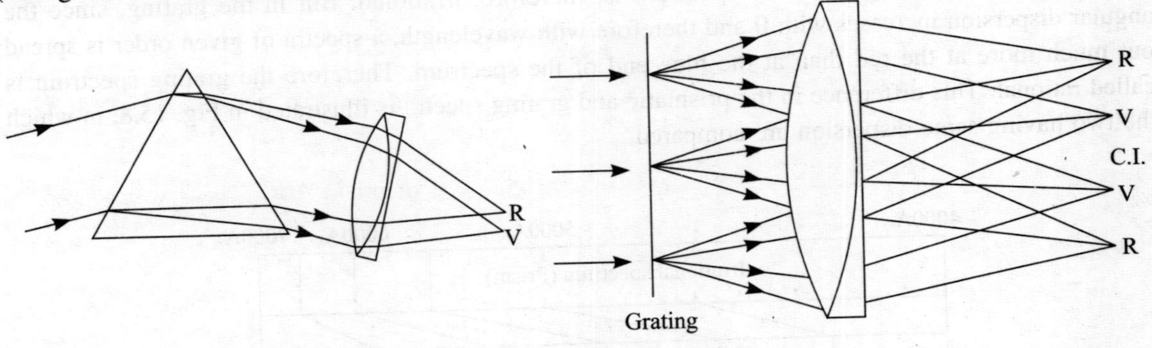

Fig.15.6

Fig. 15.7

(5) In prismatic spectrum, the lines are slightly curved, being concave towards the violet end. This curvature is due to the fact that only the rays of light from the middle of the slit traverse the prism in a plane exactly perpendicular to the refracting edge, while the rays from the ends do not traverse the prism in such a plane. Consequently the section of the prism containing such rays has an effective refracting angle of the prism greater than the actual one. These rays are, therefore, deviated through greater angle than those from the middle of the slit. This accounts for the concave nature of lines towards the violet end. The lines of a grating spectrum, on the other hand, are much more nearly straight but the concave side is towards the red end of the spectrum.

(6) The angular dispersion of the prism, defined as the rate of change of the angle of deviation with wavelength, by Eqn. (15.3 *d*), is expressed as

$$\frac{d\theta}{d\lambda} = \frac{t}{a} \times \frac{d\mu}{d\lambda}$$

This indicates that the dispersion, that is, the prismatic spectrum depends on the characteristics of the material of the prism. The relative spacing of the spectral lines, therefore, varies with the material. But the dispersion produced by the grating is given by Eqn. (15.10 *a*), which is

$$\frac{d\theta}{d\lambda} = \frac{n}{(e+b)\cos\theta}$$

Therefore the grating spectra is independent of the material of the ruled surface. Consequently the relative spacing of the lines is the same, provided n and $(e + b)$ are constant in the reflection as well as transmission type of gratings.

(7) The dependence of the refractive index of the material of the prism on wavelength is roughly expressed by Cauchy's dispersion formula, viz.

$$\mu = A + (B/\lambda^2)$$

where A and B are constants. Hence differentiating the above equation with respect to λ. We get

$$d\mu / d\lambda = -2B/\lambda^3 \quad \text{whence} \quad d\theta / d\lambda = -2Bt / a\lambda^3$$

Thus the angular dispersion of the prism varies inversely as the cube of the wavelength. As a consequence of this, the prismatic spectrum is spread out much more in proportion at the violet than it is at the red end. The prismatic spectrum is, therefore, irrational. But in the grating, since the angular dispersion increases with θ and therefore with wavelength, a spectra of given order is spread out much more at the red than at the blue end of the spectrum. Therefore the grating spectrum is called rational. This difference in the prismatic and grating spectra is illustrated in Fig. 15.8, in which the two having same dispersion are compared.

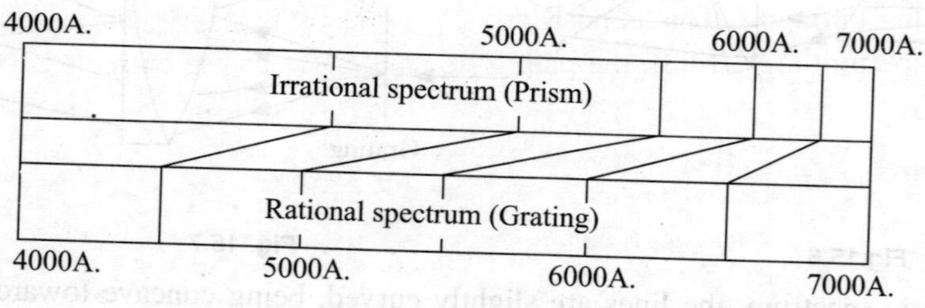

Fig. 15.8. Comparison of dispersion by prism and grating

(8) By employing concave grating with Rowland mounting, the spectrum is nearly normal. But the prismatic spectrum is never normal. A normal spectrum is one in which the wavelength interval is proportional to the distance between the spectrum lines along the spectrum.

(9) The resolution of spectral lines is much greater in the grating spectrum than in the prismatic one. Moreover, the resolving power of grating can be increased by ruling large number of lines and increasing the width of the ruled surface or employing spectra of higher orders. But in the case of a prism, since it depends upon the thickness of the base when full face of the prism is utilised, it cannot be increased beyond a certain limit owing to the inability of constructing homogeneous prisms with large bases.

(10) Sometimes, owing to periodic error in ruling additional spectra are produced by gratings. Such false spectra are known as *Ghosts*. Prismatic spectrum is free from such *Ghosts*.

15.5 RESOLVING POWER OF A MICROSCOPE

The primary function of a microscope is not to magnify an object but to reveal those finer details in the object which are invisible to unaided eye. The extent to which finer details are revealed depends not on the magnifying power but on the resolving power of the microscope. It is conveniently expressed in terms of the smallest linear separation between two point objects which are just resolved, that is, their diffraction pattern formed by the microscope objective are just distinguishable. The smaller is this distance, the greater is said to be resolving power. We shall now derive this least separation between two point objects for two distinct cases. (*i*) *Objects are self-luminous having no phase relations and (ii) objects are non-luminous but their illumination is coherent.*

15.5.1 Resolution with Non-coherent Self-luminous Objects

The objective of the microscope is represented in Fig. 15.9 by a simple convex lens. The waves of light, assumed to be of the same wavelength λ, starting from self-luminous point objects O and O' are diffracted by the circular periphery of the lens. A diffraction pattern corresponding to each object, is formed in the focal plane of the objective. Each image is therefore, a diffraction pattern, consisting of a central bright disc bordered by alternate concentric circular dark and bright rings. The centre of the bright disc in the diffraction pattern of O is at I, while for O' it is at I'. If these objects are to be resolved in their images then, according to Rayleigh's criterion of resolution, the distance II' between the centres of two central maxima should be at least equai to the radius of either one. This simply means that the waves of light from O' after diffraction by the lens, should from first dark ring passing through I. This is possible only if, according to the theory of Fraunhofer diffraction at a circular aperture (*Sec.* 14.3), the path difference between the extreme rays $O'BI$ and $O'AI$ is equal to λ_o, that is

$$O'BI - O'AI = 1.22\lambda_o \qquad ...(15.5\ a)$$

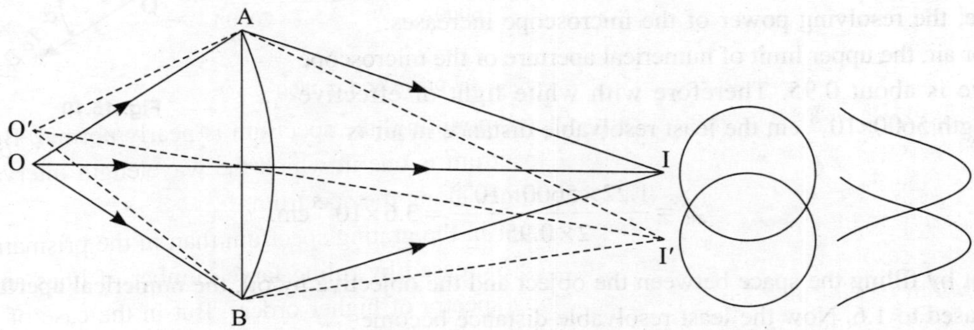

Fig. 15.9. Limit of resolution of a microscope

where λ_o is the wavelength of light waves in a medium of refractive index μ, which is supposed to be between the objects and the objective. Eqn. (15.5 *a*) can be written as:

$$(O'B + BI) - (O'A + AI) = 1.22\lambda_o$$

and taking O to be symmetrical with respect to A and B, the point I also becomes symmetrical, that is $BI = AI$. Hence the above equation reduces to

$$O'B - O'A = 1.22\lambda_o$$

$$(O'B - OB) + (OA - O'A) = 1.22\lambda_o$$

The objects O and O' are, in practice, so close together that we can consider $O'A$ essentially parallel to OA and OB' parallel to OB. Therefore, drawing line OC perpendicular to both OB and $O'B$, and OD perpendicular to OA and $O'A$ (see Fig. 15.10), the above equation gives

$$O'C + O'D = 1.22\lambda_o$$

$$2s \sin\alpha = 1.22\lambda_o$$

$$s = \frac{1.22\lambda_o}{2\sin\alpha} = \frac{1.22\lambda}{2\mu\sin\alpha} \qquad \text{...(15.5 } b\text{)}$$

where s is the linear distance OO' between the point objects and α is half the angle subtended at the axial point object O by the rim of the microscope objective and λ is the wavelength of light in vacuum. It would be observed that the entire difference in paths under consideration, that is, ($O'B$ – $O'A$) is in the medium between the object and the objective.

Equation (15.5 *b*), therefore, gives the linear distance between the two just resolvable objects. The product of the index of refraction of medium in which the object is situated and the sine of half the angle of cone of rays admitted by the objective, *i.e.* μ sin α was called by *Abbe the numerical aperture of the objective.* Equation (15.5 *b*) indicates that smaller the wavelength, smaller the distance between two just resolvable points; the greater the numerical aperture, again the distance decreases. Under both the circumstances, therefore, the resolving power of the microscope increases.

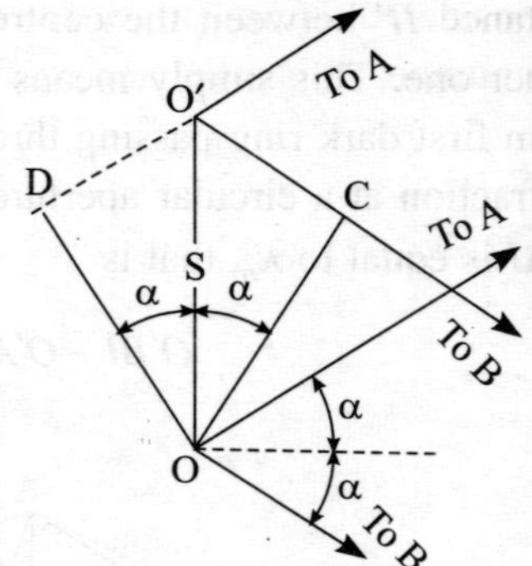

Fig. 15.10

For air, the upper limit of numerical aperture of the microscope objective is about 0.95. Therefore with white light of effective wavelength 5600×10^{-8} cm the least resolvable distance in air is

$$s_{\text{air}} = \frac{1.22 \times 5600 \times 10^{-8}}{2 \times 0.95} = 3.6 \times 10^{-5}\ \text{cm.}$$

But by filling the space between the object and the objective by oil, the numerical aperture may be increased to 1.6. Now the least resolvable distance become

$$s_{\text{oil}} = \frac{1.22 \times 5600 \times 10^{-8}}{2 \times 1.6} = 2.14 \times 10^{-5}\ \text{cm.}$$

Thus the resolving power of the oil immersion type objective is approximately twice that of the ordinary objective.

Mathematically, the reason of employing oil immersion objective is to increase its numerical aperture, $\mu \sin \alpha$. Physically the function of the medium can be clear by considering the Eqn. (15.5 *b*) which indicates that smaller the value of λ_o, the smaller is the separation of two just resolvable objects. This simply means that the wavelength of light is the only tool controlling the resolving power of the microscope. Therefore by employing quartz lenses and ultraviolet light, with its smaller value of λ the resolving power can be increased much further. Of course, it is essential in this case to take photographs for examining the diffraction pattern.

In order to take full advantage of the high resolving power of the microscope it is absolutely essential that its magnifying power should be at least normal. When the radius of the objective lens and the focal length of the eye-piece are so related that the radius of the exit pupil is just equal to the radius of the pupil of the eye, the overall magnification is said to be normal. A higher magnification does not reveal more details while the lower magnification does not take full advantage of the available resolving power of the microscope objective. The proof of this statement will be given in *Sec.* 15.7.

15.5.2 Abbe's Theory of Resolution with Coherent Illumination

We have discussed in *Sec.* 15.5.1 the theory of resolution of microscope for self-luminous objects. But the self-luminous objects are rare in microscopy, which practically always has to examine nonluminous objects, illuminated by light from some external sources. *Abbe* was first to consider the resolution of microscope for these dark objects. As a preliminary to the discussion of resolving power, he considered the image formation by a lens for an object having grating like structure, consisting of large number of parallel and equidistant lines.

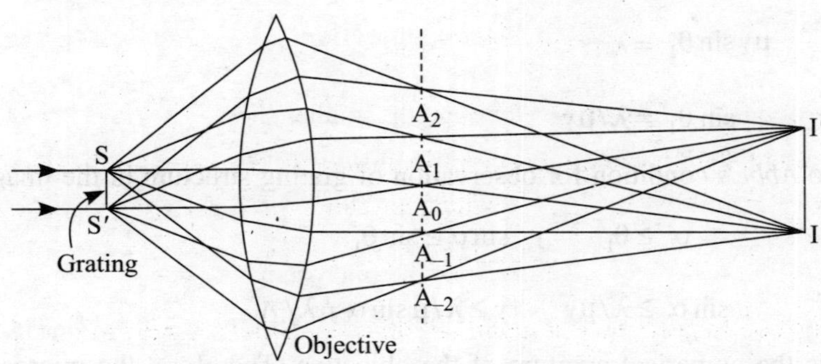

Fig. 15.11. Abbe's diffraction theory

Let a parallel beam of monochromatic light of wavelength λ be incident normally upon the grating *SS′*. The light is partly stopped by the grating and partly transmitted. The light from different transparent spaces is *coherent* and is diffracted in all directions. The diffraction angles θ_n for principal maxima in the Fraunhofer diffraction are determined by the grating equation, viz.,

$$\mu y \sin \theta_n = n\lambda$$

where $n = 0, \pm 1, \pm 2, \pm 3$ etc. In the above equation μ is the index of refraction of the medium on the right of the grating and y is the grating element. It is obvious from the above equation that the diffracted beams deviate further from the normal, the finer the grating structure, that is, smaller the grating element y.

The microscope objective focusses the direct beam and some of the diffracted beams corresponding to foci A_{-2}, A_{-1}, A_0, A_1, A_2 etc. forming so called diffraction images, in its second focal plane. The question now is what light effect will this diffraction figure produce in the image plane II' which is conjugate to the object plane SS'. We can regard the various diffraction images as new sources of light. The image is, therefore, formed by the combined effect of the beams of light coming from various points A_{-2}, A_{-1}, A_0, A_1, A_2, etc. *Abbe* proved that an exactly true image will be formed only if all diffraction beams starting from the object act together. However as the microscope objective has small aperture, all diffraction beams cannot pass through it and consequently the true image is impossible. *Abbe* satisfied himself that if only some of the diffracted beams are admitted, for example corresponding to 0, 1st and 2nd order diffraction maxima, even then the image still very like the grating structure will result. Even with only one beam on either side of the central beam, a grating like structure will result in the image. In the extreme case, however, in which only the central beam corresponding to zero order diffraction image A_0 is transmitted, any cooperation with other diffracted beams is impossible. The image plane is, therefore, simply uniformly illuminated by the light passing via A_0 and no image of grating structure is observed. *In order to observe grating structure, it is therefore absolutely essential that the first order diffracted beams on each side of the central beam must also pass through the microscope objective*. This condition simply demands that the angular aperture 2α of the objective should be at least as large as to admit these beams, that is α should be at least equal to the angle of diffraction θ_1 corresponding to the first order principal maximum. Now, θ_1 satisfies the grating equation

$$\mu y \sin\theta_1 = \lambda$$

$$\sin\theta_1 = \lambda/\mu y$$

But according to *Abbe's* condition for observation of grating structure in the image, we have

$$\alpha \geq \theta_1 \qquad \sin\alpha \geq \sin\theta_1$$

$$\sin\alpha \geq \lambda/\mu y \qquad y \geq \lambda/\mu\sin\alpha = \lambda/A$$

where $A = \mu \sin\alpha$, is the numerical aperture of the objective. Therefore, the microscope objective with numerical aperture A can resolve only points separated at least by λ/A.

The least resolvable distance can be made still smaller with the same objective, by making the light to be incident obliquely on the grating structure in such a way that it enters the microscope objective at one edge and at the other edge diffracted beam corresponding to the first order image is transmitted (see Fig. 15.12). Even in this case the impression of grating like structure can be obtained in the image. For oblique incidence, the fundamental equation of grating is

$$\mu y'(\sin i + \sin\theta) = n\lambda$$

For first order, in the present case, the above equation becomes

$$\mu y'\left(\sin\frac{1}{2}\theta_1+\sin\frac{1}{2}\theta_1\right)=\lambda$$

$$\sin\frac{1}{2}\theta_1 = \lambda/2\mu y'$$

For oblique incidence, in the limiting case, *Abbe*'s condition gives

$$2\alpha \geq \theta_1 \qquad \alpha\geq\theta_1/2$$

$$\sin\alpha \geq \sin\theta_1/2 \qquad \sin\alpha\geq\lambda/2\mu y'$$

$$y'\geq\lambda/2\mu\ \sin\alpha\ =\ \lambda/2A$$

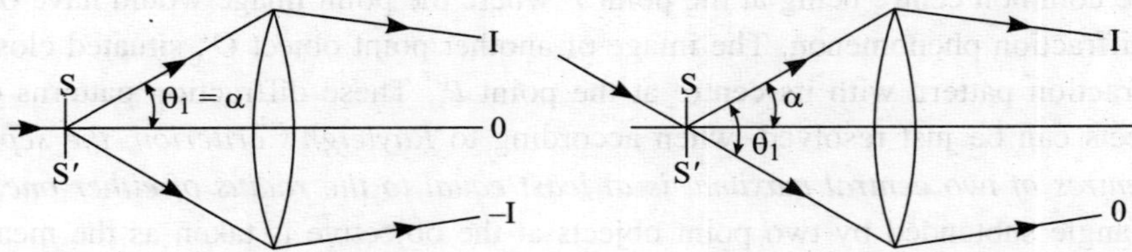

Fig. 15.12. A grating structure can only be observed as resolved when at least two beams take part in forming the image

Therefore, the grating structure can be twice as fine as in the first case, that is, the resolving power of the microscope is doubled when oblique illuminations is substituted for axial illumination.

Abbe was able to prove convincingly by series of experiments that the various diffraction images do really play an important role in the image formation. He made an image of a grating and then he removed odd diffraction images by placing diaphragm at suitable places, so that only the even diffraction images were effective. By this contrivance, the original second order diffraction image, formed at the angle of diffraction θ_2 given by

$$y\sin\theta_2 = 2\lambda$$

apparently becomes the first order diffraction at the same angle of diffraction θ_2, which now satisfies the above equation written after some rearrangement

$$\frac{1}{2}y\sin\theta_2 = \lambda$$

for the first order. Thus the effect of cutting odd order diffraction images, apparently, is halving the grating element. What *Abbe* observed in the image was, therefore, not the real grating structure, but structure in which lines were twice as close as in the original grating. In the same way when he arranged the cooperation of the central beam and the third diffracted beam, the observed grating structure was three times as fine as in the original one.

15.6 RESOLVING POWER OF A TELESCOPE

The telescope is employed to view distant objects and therefore the amount of grain or detail which it reveals depends on the angle the two point objects subtend at the objective rather than on the linear separation between them. *The resolving power of a telescope is therefore defined as the inverse of the least angle subtended at the objective by two distant point objects which can be just distinguished as separate in its focal plane.*

Let O be a monochromatic point source of light of wavelength λ, situated at a great distance from the telescope objective AB. A beam of parallel rays, shown by full lines, from this object is incident normally on the objective. The ring supporting the lens as well as lens itself serves as a circular aperture and therefore Fraunhofer diffraction pattern is produced in the focal plane of the lens. The diffraction pattern of O consists of a central bright disc surrounded by concentric dark and bright rings, the common centre being at the point P where the point image would have occurred in the absence of diffraction phenomenon. The image of another point object O', situated close to O, is also similar diffraction pattern with its centre at the point P'. These diffraction patterns overlap and two point objects can be just resolved, when according to *Rayleigh's criterion, the separation PP' between the centres of two central maxima, is at least equal to the radius of either one.* Under this condition, the angle subtended by two point objects at the objective is taken as the measure of resolving power of the telescope; actually resolving power is inverse of this angle.

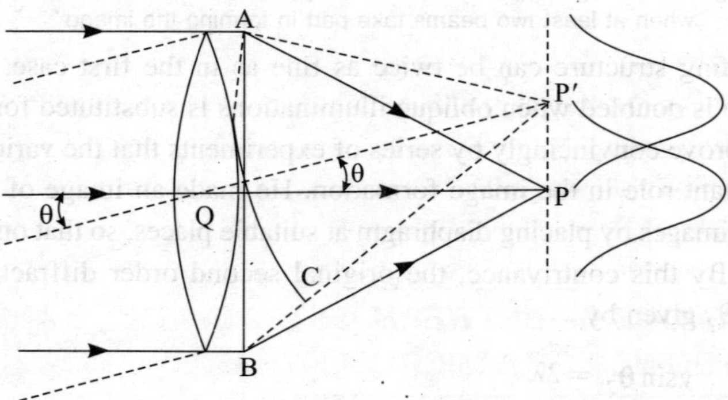

Fig. 15.13. Limit of resolution of a telescope

With P' as centre and radius $P'A$ we draw an arc cutting $P'B$ in the point C. This arc AC, therefore, represents the transmitted wavefront, curved due to the focal property of the lens, which gives rise to the central maximum at P' in the diffraction pattern of O', the second point object near to the first. For just resolution of O and O', the first dark ring in the diffraction pattern of O should pass through the point P'. Therefore AC should also be a diffracted wavefront for O, so as to form in the diffraction pattern the first dark ring passing through P'. Therefore, the path difference (BP'–AP') between the extreme diffracted rays reaching P', according to theory of Fraunhofer diffraction at a circular aperture, should be equal to 1.22 λ. Thus, we have

$$BP' - AP' = BC = 1.22\lambda$$

Hence

$$\angle BAC = \theta = BC/AB = 1.22\lambda / D$$

where D is the diameter of the objective. Since the rays QP' and QP are normal to the respective wavefronts which form central maxima at P' and P, hence

$$\angle PQP' = \angle BAC$$

But $\angle PQP'$ = the angle subtended at the objective by the two point objects.

Hence the minimum angular separation which two point objects should have so that they can be just resolved is given by the relation

$$\theta = 1.22\lambda/D$$

This angle is, therefore, the measure of angular resolving power of the telescope objective. The smaller the value of θ, the greater is said to be resolving power. Therefore telescopes with larger objective have higher resolving power.

We may now calculate the value of limiting angle θ which the two distant stars should subtend on the objective of one inch diameter so as to be just resolved by it.[1″ = 2.54 cm]

$$\theta = \frac{1.22\times5500\times10^{-8}}{2.54}\text{ radian}$$

$$= \frac{180}{\pi}\times\frac{1.22\times5500\times10^{-8}}{2.54}\text{ degree of arc}$$

$$= 5.52 \text{ seconds of arc}$$

where 5500 A. is the effective wavelength of light. The limiting angle θ for 1″ telescope is of the order 5 sec. Therefore a telescope with a 200″ objective will resolve them even if their separation be only 1/40th of a second of arc.

In order to take full advantage of the high resolving power of the telescope, it is essential that its magnification should be at least equal to the normal magnification.

15.7 INTER-RELATIONSHIP BETWEEN THE MAGNIFYING POWER AND RESOLVING POWER OF A MICROSCOPE

From the standpoint of the resolving power of microscope objective, we can express its overall magnification M in a slightly different form as the product of the linear magnification m of the objective and the angular magnification β of the eye-piece.

$$M = m\,\beta \qquad ...(15.7\ a)$$

Let μ represents the refractive index of the medium between the object and the objective but in actual microscope the medium between the eye-piece and objective is air of index μ′ = 1.00. Let a ray of light incident on the objective with inclination α to the axis. After refraction by the objective be inclined at α′ with the axis (Fig. 15.14). Then *Abbe's sine condition*

$$s\mu\sin\alpha = s'\mu'\sin\alpha'$$

applied to the objective gives

$$m = \frac{s'}{s} = \frac{\mu\sin\alpha}{\sin\alpha'} \qquad ...(15.7\ b)$$

The angle α' is always small and therefore from the geometry of Fig. 15.14 it is clear

$$\sin \alpha' = R/f \text{ approx} \qquad ...(15.7\ c)$$

where f is the focal length of the eye-piece and R is the radius of the exit pupil.

The angular magnification produced by the eye-piece is

$$\beta = 25/f \qquad ...(15.7\ d)$$

where f is in centimeters.

When Eqns. (15.7 *b*), (15.7 *c*), (15.7 *d*) are combined with Eqn. (15.7 *a*), we get

$$M = \frac{\mu \sin \alpha}{R/25} \qquad [R \text{ in cm.}]$$

At *normal magnification*, the full apertures of both the eye and the objective are filled with light (Fig. 15.14). Then $R/25$ is the maximum *numerical aperture* of the eye and $\mu \sin \alpha$ is the maximum *numerical aperture* of the objective. Hence

$$\text{Normal Magnification} = \frac{\text{Max. N.A. of objective}}{\text{Max. N.A. of eye}} \qquad ...(15.7\ e)$$

Now the least resolvable distance by the compound microscope objective is

$$s = \frac{1.22\lambda}{2\mu \sin \alpha} = \frac{0.61\lambda}{\text{Max. N.A. of objective}}$$

and the least resolvable distance by an unaided eye is

$$s_1 = \frac{1.22\lambda}{2 \sin \alpha_1} = \frac{0.61\lambda}{\text{Max. N.A. of eye}}$$

whence

$$\frac{s_1}{s} = \frac{\text{Max. N.A. of objective}}{\text{Max. N.A. of eye}} \qquad ...(15.7\ f)$$

Combining Eqns. (15.7 *e*) and (15.7 *f*), we get

Normal magnification of microscope $= s_1/s$ ...(15.7 *g*)

Thus we have an important result that *the limit of resolution of a microscope multiplied by its normal magnifying power, is equal to the limit of resolution of the eye.*

The maximum numerical aperture of an unaided eye, for a pupillary [Eye-lens] radius 1 mm and an object in air atleast distance of distinct vision 25 cm, is 0.004. The maximum numerical aperture of the microscope objective is 1.6, which is 1.6/0.004 = 400 times as great as that of the unaided eye. Hence the least distance between two point objects that can be just resolved by the microscope objective of N.A. = 1.6, from Eqn. (15.7 *f*) is given by

$$s = s_1/400 \qquad ...(15.7\ h)$$

that is, it is 1/400th as great as for the unaided eye. Actually, the resolving power of the microscope objective is defined as the reciprocal of the least distance between two just resolvable point object and similar definition is for the resolving power of the eye. From Eqn. (15.7 *h*), therefore, it follows that the microscope gives 400 times the resolving power of the unaided eye. By the combination of Eqns. (15.7 *g*) and (15.7 *h*), it follows that the magnifying power of the microscope should be at least 400 inorder to take full advantage of the available resolving power of objective of numerical aperture 1.6. In other words, the magnification should be *normal*, then the diffractional discs formed on the retina of the eye would be also resolved. A magnification higher than the *normal* one does not increase detail while a lower magnification does not take full advantage of the available resolving power.

To prove the statement given above, we draw Fig. 15.14, not in scale, representing the optical system of a compound microscope for *normal magnification*. O_1 and O_2 are two point objects, just resolved by the objective in its focal plane, and O_1' and O_2' are the centres of their diffraction discs, also just resolved on the retina. The appearance of central discs is shown in Fig. 15.15 (*a*). The focal length of the eye-piece and the diameter of the objective are so related that the diameter of the exit pupil of the instrument is simply equal to diameter of the pupil of eye [Eye-lens]. This represents the condition that the overall magnification *M* of the microscope is normal.

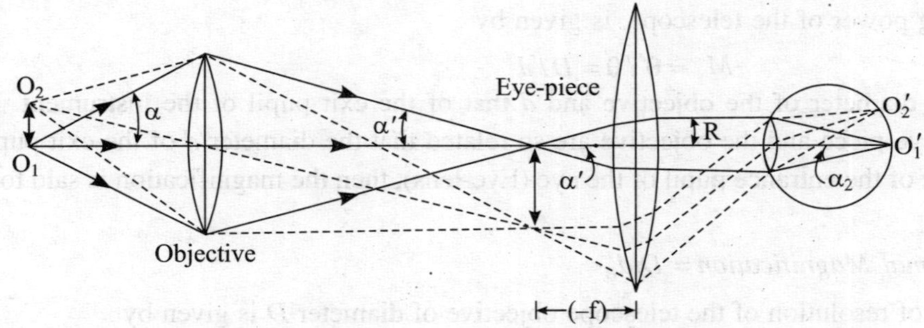

Fig. 15.14. The compound microscope—normal magnification

Now let us suppose that the eye-piece of Fig. 15.14 is replaced by one of shorter focal length. The magnification of the microscope would increase beyond normal magnification and the radius of exit pupil of the instrument would be now smaller than the radius of pupil of the eye. This would decrease the half angle, α_2, of the cone of light converging towards the retina. The radius of the central disc is given by

$$s_1' = \frac{1.22\lambda}{2\mu_2 \sin \alpha_2}$$

Therefore, a decrement in α_2 would increase the radii of diffraction discs formed on the retina. But due to greater magnification, the distance between the centres of diffraction discs would also increase in the same proportion. This simply means that two point objects, which were just resolved when the overall magnification was *normal* (Fig. 15.14), would still be just resolvable when the magnification has been increased by employing a high power eye-piece. The only effect of increased magnification would be a large image, but no more details could be observed as shown in Fig. 15.15 (*b*). Also, since the same amount of light would be now distributed over a much magnified image, a loss in its brightness would result.

Now let us suppose that the eye-piece of Fig. 15.14 is replaced by one of longer focal length. This would result in decrement of magnification below *normal*. The diameter of the exit pupil would be now greater than that of the pupil of eye. But the eye cannot admit a wider beam. The half angle α_2 would be, therefore, the same as in Fig. 15.14. Thus the radii of diffraction discs would be the same as in Fig. 15.15 (*a*). But due to the decrement in the magnification, the centre to centre spacing of diffraction discs on the retina would be decreased, resulting in greater overlapping of diffraction discs as shown in Fig. 15.15 (*c*). Consequently the two point objects, which were just resolved under *normal magnification*, would not be now resolved by the eye when the magnification decreases, although they are just resolved by the microscope objective in its focal plane.

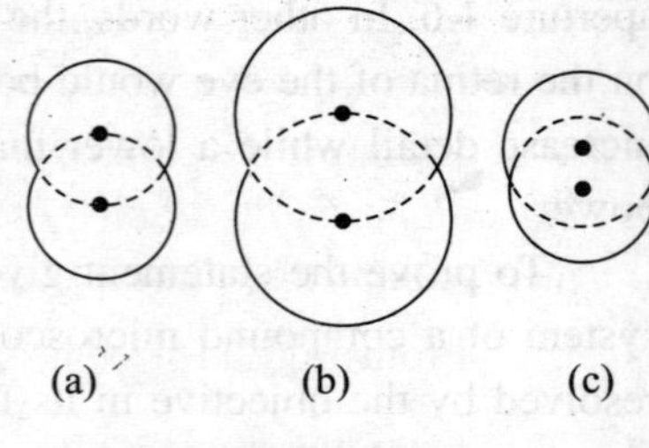

Fig. 15.15

15.8 INTER-RELATIONSHIP BETWEEN THE MAGNIFYING POWER AND RESOLVING POWER OF A TELESCOPE

The magnifying power of the telescope, is given by

$$M = \theta'/\theta = D/d \quad \text{...(15.8 } a\text{)}$$

where D is the diameter of the objective and d that of the exit pupil of the instrument. If the focal lengths of the eye-piece and the objective are so related that the diameter d of the exit pupil becomes d_θ, the diameter of the entrance pupil of the eye (Eye-lens), then the magnification is said to be *normal*.

Hence

$$\textit{Normal Magnification} = D/d_\theta \quad \text{...(15.8 } b\text{)}$$

The limit of resolution of the telescope objective of diameter D is given by

$$\theta = 1.22\ \lambda/D \quad \text{...(15.8 } c\text{)}$$

and the limit of resolution of the unaided eye is given by

$$\theta_1 = 1.22\lambda/d_\theta \quad \text{...(15.8 } d\text{)}$$

When Eqns. (15.8 *c*) and (15.8 *d*) are combined with Eqn. (15.8 *b*), we get

$$\text{Normal magnification} = \theta_1/\theta \quad \text{...(15.8 } e\text{)}$$

Thus we arrive at the result that *the limit of resolution of telescope, multiplied by its normal magnifying power is equal to the limit of resolution of the eye*—a relation similar to that derived for microscope.

Angle θ_1 between two point objects which can be just resolved by the unaided eye with the pupil of diameter d_θ = 2 mm is 1 minute of arc, assuming λ = 5500 A.U. The angle θ between two distant stars, whose diffraction discs are on the limit of resolution in the focal plane of the telescope objective of 200 in. diameter, is 1/40th of a second of arc. The magnified images of these diffraction discs, formed by the telescope, must subtend at the eye an angle θ_1 = 1 minute = 60 second of arc, in order to be resolved by the eye. Therefore the magnifying power of the 200 in. telescope should be at least

$$M = \frac{60}{1/40} = 2400$$

in order to take full advantage of the high resolving power of the objective. Thus the magnification should be at least *normal*.

SOLVED EXAMPLES

Example 1. *Calculate the least width that a grating must have to resolve two components of the sodium D line in the second order, the grating having 800 lines per cm. (Two wavelengths of the two components of the sodium D line are 5896 and 5890 A.U. respectively.)*

From Eqn. (15.2 *d*), $N = \lambda/nd\lambda$

From the given data, $d\lambda = 5896 - 5890 = 6$ A.U., $n = 2$

and $\lambda = 1/2(5896 + 5890) = 5893$ A.U.

Hence, $N = 5893/(2 \times 6) = 491\frac{1}{12}$ = lines

Thus to resolve two components of the sodium *D* line in the second order, the grating must have at least 492 lines. Accordingly, the least width of ruled surface of the grating

= 492/800 = **0.615 cm.**

Example 2. *The index of refraction of a certain glass is given by $\mu_C = 1.6545$, $\mu_E = 1.6635$. The wavelength of the C and E lines in the solar spectrum are 6563 and 5270 A.U. respectively. Calculate the length of the base of a 60° prism made of this glass which is just capable of resolving the sodium lines of wavelength 5896 and 5890 A.U.*

The smallest wavelength difference $d\lambda$ that can be resolved by the prism when set in minimum deviation for light of wavelength λ is expressed by

$$d\lambda = \lambda/t\frac{d\mu}{d\lambda}$$

where t is the length of the base of the prism.

From the given data, $d\lambda = 5896 - 5890 = 6$ A.U.

$$\lambda = \frac{1}{2}(5896 + 5890) = 5893 \text{ A.U.}$$

and $$\frac{d\mu}{d\lambda} = \frac{\mu_E - \mu_C}{\lambda_C - \lambda_E} = \frac{1.6635 - 1.6545}{(6563 - 5270) \times 10^{-8}} = \frac{0.009}{1293 \times 10^{-8}}$$

Substituting these values in the expression for resolving power, it is easy to get t = **1.4 cm** approximately.

Example 3. *What is the smallest wavelength difference that can be resolved in the neighbourhood of 5890 A. by a prism spectrometer using a 60° prism in a minimum deviation position. The width of the light beam passing through the prism is 3 cm. The prism is made of glass whose dispersion curve is represented by Cauchy's equation $\mu = A + (B/\lambda^2)$ where A= 1.47 and B = 1.6×10^6 (angstrom)^2.*

$$\mu = A + (B/\lambda^2)$$

Hence $$\frac{d\mu}{d\lambda} = -\frac{2B}{\lambda^3} = -\frac{2\times1.6\times10^6\times10^{-16}}{(5890)^3\times10^{24}} = -\frac{32\times10^{10}}{(589)^3}$$

When the full height of the prism is not effective, the smallest wavelength difference $d\lambda$ that can be resolved is evaluated from

$$\frac{\lambda}{d\lambda} = (t_1 - t_2)\frac{d\mu}{d\lambda}$$

where t_1 and t_2 are the thickness of the material parallel to the base of the prism through which two extreme rays of the refracted beam pass.

Now, $$t_1 - t_2 = 2\times3\times\tan 30° = 6/\sqrt{3}$$

Hence $$d\lambda = -\frac{5890\times10^{-8}\times\sqrt{3}\times(589)^3}{6\times32\times10^{10}}\ \text{cm}$$

$$= \mathbf{-\ 1.08\ A.U.}$$

The smallest wavelength difference that can be resolved is 1.08 A.U.

Example 4. *The ruled space on a grating is as wide as the base of a 60° rocksalt prism. It has the same resolving power in the second order as the prism has at 6000 A.U. Calculate the size of.the grating interval, given the following data:*

λ	μ
6708 A.U.	1.5400
6438 A.U.	1.5412

From Eqn. (15.2 *d*) $\quad d\lambda = \lambda/nN$

and in the second order, $\quad d\lambda = \lambda/2N$

The smallest wavelength difference that can be resolved by the prism when set at minimum deviation for λ is

$$d\lambda = \frac{\lambda}{t(d\mu/d\lambda)}$$

Since the grating has the same resolving power in the second order as the prism has at 6000 A.U., hence on equating the above two expressions for $d\lambda$ and simplifying we get

$$2N = t(d\mu/d\lambda)$$

The ruled space on the grating is as wide as the base of the prism. Hence, the grating interval is expressed by

$$(e + b) = t/N$$

With the help of the above two equations it is easy to get

$$(e + b) = 2(d\lambda / d\mu)$$

From the given data,

$$\frac{d\lambda}{d\mu} = \frac{(6708 - 6438) \times 10^{-8}}{1.5412 - 1.5400} = \frac{270 \times 10^{-8}}{0.0012}$$

Hence $\qquad (e + b) = \dfrac{2 \times 270 \times 10^{-8}}{0.0012} = \mathbf{4.5 \times 10^{-3}\ cm.}$

Example 5. *Calculate the minimum magnifying power required for a telescope whose objective has an aperture of five metres, in order to make full use of resolving power.*

Aperture of eye = 5 mm. Wavelength of light used = 5000× 10^{-8} cm.

As explained in *Sec.* 15.8, in order to make full use of the resolving power of the telescope, the minimum magnifying power required is given by the relation,

R.P. of telescope × Magnifying power = *R.P.* of the Eye

$$\frac{1.22\lambda}{D} \times \text{Magnifying power} = \frac{1.22\lambda}{d_e}$$

$$\text{Magnifying power} = \frac{D}{} = \frac{5 \times 100}{0.5} = \mathbf{1000}$$

Example 6. *Calculate the least value of the angular separation of two stars which can be resolved by a telescope of 200 cm aperture. If the aperture of human eye be 2 mm and if the focal length of the eye-piece be 1 inch, what must be the minimum focal length of the objective if the full resolving power of the telescope is to be utilised.*

The least value of the angular separation of two stars which can be resolved by the telescope is given by

$$\theta = 1.22\lambda/D$$

With the given data and assuming the mean wavelength of white light to be 5000 × 10^{-8} cm, the least angular separation is

$$\theta = \frac{1.22 \times 5000 \times 10^{-8}}{200} = \mathbf{3.05 \times 10^{-8}\ radian}$$

As explained in Example 5, here also the minimum magnifying power of telescope required to make full use of its resolving power is expressed by

$$\text{Magnifying power} = \frac{D}{d_e} = \frac{200}{0.2} = 1000$$

Magnifying power of a telescope from sec. 5.6 is

$$\gamma = -\frac{f_0'}{f_e'} = -\frac{f_0'}{1}$$

Hence $f_O'/1 = 1000$ (Neglecting negative sign)

or $f_O' = \mathbf{1000\ inches}$

The focal length of the objective is **1000 inches.**

Example 7. *Calculate the number of lines per inch on a plane transmission grating which when used on a spectrometer with telescope and collimator of aperture 3/4 inch, will just resolve the two D lines in the sodium spectrum in the second order spectrum. $D_1 = 5896$ A. and $D_2 = 5890$ A.*

Diameter of telescope objective = 3/4 inch = 3/4 × 2.54 cm.

The resolving power of the telescope is

$$d\theta = \frac{1.22\lambda}{D} = \frac{1.22 \times 5893 \times 10^{-8}}{\frac{3}{4} \times 2.54}$$

Let, the requisite number of lines in the grating be x per inch or $x/2.54$ per cm for just resolving the D lines in the second order spectrum.

Hence $(e + b) = 2.54/x$ cm

Now $(e + b) \sin\theta = n\lambda$

Hence $(e + b) \cos\theta\, d\theta = nd\lambda$

$$\frac{nd\lambda}{(e+b)\sqrt{1-\sin^2\theta}} = \frac{nd\lambda}{(e+b)\sqrt{1-[n^2\lambda^2/(e+b)^2]}}$$

$$= \frac{nxd\lambda}{2.54\sqrt{1-[n^2x^2\lambda^2/(2.54)^2]}}$$

From the given data, $n = 2$, $d\lambda = 6 \times 10^{-8}$ cm

Hence $d\theta = 2x \times 6 \times 10^{-8}/2.54$ approx.

For just resolution of D lines of sodium in the second order this angular separation between the diffracted rays corresponding to $\lambda_1 = 5896$ A.U. and $\lambda_2 = 5890$ A.U. must be equal to the limit of resolution of telescope of the spectrometer. Hence we get

$$\frac{2x \times 6 \times 10^{-8}}{2.54} = \frac{1.22 \times 5893 \times 10^{-8}}{\frac{3}{4} \times 2.54}$$

$$\mathbf{x = 799\ lines\ per\ inch.}$$

PROBLEMS

1. (a) Discuss the usefulness of Rayleigh's criterion for determining the resolving power of optical experiments.

   (b) Explain clearly what is meant by resolving power of an optical instrument.

2. (a) Derive an expression for the chromatic resolving power of a prism spectroscope.

   (b) Show that the chromatic resolving power of a spectroscopic instrument is given by

   Resolving Power = Dispersive Power × Width of the dispersed beam

3. What is meant by the resolving power of a microscope? How is it related to the numerical aperture of its objective?

4. Discuss Abbe's theory of resolution of a microscope. In what respects does the criterion of this theory differ from that of Rayleigh? Obtain expressions for resolving power of a microscope on the basis of the two theories and show how it is possible to approach the limit of resolution as obtained by Rayleigh's criterion.

5. Obtain in expression for the magnifying power of a telescope and show that for full resolution the magnifying power must not be less than the ratio of the diameter of the objective to the diameter of the pupil of the eye.

6. Explain the board features of the diffraction pattern due to a circular aperture. Derive an expression for the limit of resolution of a telescope.

   Explain on the basis of diffraction theory the need of large apertures for telescopes used for astronomical purposes.

7. Explain what is meant by magnifying power and resolving power of an optical instrument. Show that the product of the magnifying and resolving powers of a telescope is equal to the resolving power of eye.

8. Calculate the magnifying and resolving powers of microscope from the following data:

   Numerical Aperture = 0.12

   Focal length of the objective as well as of the eye-piece = 2.5 cm

   Distance of the crosswires from the objective = 15 cm

   Wavelength of light = 6×10^{-5} cm

9. An oil immersion microscope will just resolve test lines drawn 112,000 per inch, using monochromatic light of wavelength 4200 A.U. Calculate numerical aperture of the objective.

10. What will be the diameter of a telescope objective required to resolve two stars separated by an angle 10^{-3} degrees, of light 5000 A.U.?

11. Two stars at a distance of 10 light years are viewed through a telescope having a lens of 20 cm diameter. What is the minimum separation of these stars for which they would still be distinguishable as separate objects?

12. The telescope at an observatory has a diameter of 100 inches. Estimate the smallest angle between two stars that can be separated by it. Assume the wavelength of light to be 5080 A.U.

13. Calculate the minimum magnifying power required for a telescope whose objective has an aperture of 5 meters, in order to make full use of the resolving power.

    Aperture of the eye = 5 mm
    Wavelength of light = 5000 A.U.

14. What must be the base of a flint glass prism ($d\mu/d\lambda$ = –982 cm^{-1}) necessary to resolve the sodium D lines?

$$D_1 = 5896 \text{ A.U. and } D_2 = 5890 \text{ A.U.}$$

15. Calculate the minimum thickness of the base of a prism which will be able to resolve D_1 and D_2 lines of sodium. Given μ for 6563 A.U. = 1.6545 and for 5270 A.U. = 1.6635.

16. Find the minimum number of lines in a plane grating so that it may just resolve the sodium doublet in the first order.

17. Find if a one inch grating of 300 lines per cm can resolve the sodium D lines in the first order spectrum.

18. A plane transmission grating has 16000 lines per inch over a length of 5 inches. Find the wavelength region of 6000 A.U. in the second order:

(a) the resolving power of the grating.

(b) the smallest wavelength difference that can be resolved.

19. In the second order spectrum of a grating a spectral line appears at 10° another of wavelength 5×10^{-9} greater appears at 3″ farther. Find the wavelengths of the lines and the minimum grating width required to resolve them.

Chapter 16

POLARISATION OF LIGHT

The successful interpretation of the phenomena of *interference* and *diffraction* of light on the basis of *wave theory* established conclusively that like sound, light is also a wave motion. It was, however, not at all essential at that stage to raise the question whether light waves are longitudinal or transverse, owing to the fact that these phenomena occur for all types of waves. With the help of experiments illustrating these phenomena we are able to measure accurately the wavelength of light, but fail to obtain any information on the nature of light waves. In this connection we may ask whether light waves are transverse or longitudinal, or whether the vibrations are linear, circular, elliptical or torsional. In order to get correct answers to these questions we have to take recourse to another group of experiments which bring into evidence another property of light called the *polarisation of light*. This phenomenon demands for its explanation that light must be a transverse wave motion in *contradistinction to Huygens'* conception of longitudinal wave motion.

Before discussion the phenomenon of the polarisation of light we will first try to explain the meaning of the phrase '*polarisation of wave*'.

16.1 POLARISATION OF A WAVE

In a longitudinal wave, the vibrations of the medium are parallel to the direction of the propagation. As a consequence, vibrations are exactly similar in all planes drawn through the direction of propagation. To be more precise, the appearance of longitudinal wave is the same when viewed from different angles around this direction. This is expressed by saying that the longitudinal wave is perfectly symmetrical about the direction of propagation. On the other hand, in the case of a transverse wave, particles of medium execute simple periodic vibrations in a direction at right angles to the direction of propagation. For example in a transverse wave, sketched in Fig. 16.1, progressing along *X*-axis, the vibrations of the medium are confined to the *XY* plane. There are no vibrations in the *XZ* plane. As a consequence, the transverse wave would appear quite differently when viewed from different angles around the direction of propagation *i.e.*, this wave is not symmetrical around the axis of *X*. This want of symmetry is the distinguishing feature of every transverse wave and is spoken of as the polarisation of wave. However, it is only so if we observe a limited portion of the wave or observe it momentarily, for it may happen that the displacement of the medium may be changing its direction continuously, although always remaining perpendicular to the direction of propagation to preserve its transverse character. Besides this, these changes may become too rapid to be perceived by the eye or photographic plate, both of which are capable of registering only the average effect. As a consequence, the resultant effect perceived by the eye may show it to be perfectly symmetrical about its direction of propagation, although there is no doubt about its transverse character. Such a wave is said to be unpolarised.

Ordinary or natural light behaves in this manner and is said to be unpolarised. To elaborate this statement we may add that the electric vector, transverse to the direction of propagation, in the ordinary light wave undergoes so rapid and random changes of direction that if it were possible to view the vibrations physically, then due to the persistence of vision the eye would observe the wave to be quite symmetrical about the direction of propagation. As we have discussed in chapter 9, this is due to the fact that ordinary light contains millions of transverse waves, about a metre in length and confined in all possible planes through the direction of propagation. Each wave is radiated out for 10^{-8} sec. from one of the millions of excited. Atoms or molecules, in random orientations in the source of light, it is impossible for the electric vector to have all possible orientations simultaneously at any moment, it is assumed that the constituent transverse waves follow one another in such rapid succession that, during the minimum period required by the eye or photographic plate to register the effect, all possible orientations of electric vector are equally represented. In other words, vibrations in any one direction cannot be isolated in the analysis of ordinary light. This accounts for the symmetrical character of ordinary or natural light around the direction of propagation.

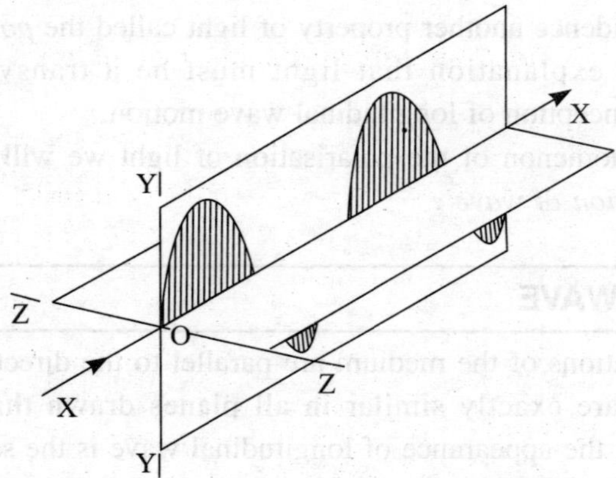

Fig. 16.1. Linearly polarised wave

To this change in the direction of vibration in the transverse waves we add another possibility of change in the character of vibration viz., it may somehow undergo many random changes of shape, say from linear to circular and from circular to elliptical. Accordingly, we define the polarised wave as follows: *A transverse wave is said to be polarised when the vibrations of the medium do not change in direction or form.*

By suitable devices described later, the electric vector may be constrained to describe the same locus *linear*, *circular* or *elliptical* of fixed orientation transverse to the direction of propagation of the wave. Accordingly, we shall have *linearly polarised, circularly polarised or elliptically polarised wave*.

16.2 MECHANICAL EXPERIMENT TO DEMONSTRATE POLARISATION OF A WAVE

A very simple mechanical experiment may be arranged to demonstrate the polarisation of a wave. Suppose a narrow loosely stretched string AB passes through two narrow rectangular slits N_1 and N_2, each a little wider than the diameter of the string. Its one end B is fixed while transverse waves are generated in it by shaking its free end A and changing the direction of shaking in a random fashion from one instant to the next. In the portion AN_1, the vibration of every element of string is changing arbitrarily; accordingly in this portion we have unpolarised wave. The slit N_1, being only slightly wider than the string, will allow only those transverse waves to pass through it freely whose vibration planes are parallel to its length. The transmitted vibrations are thus confined to a single fixed plane; and therefore the wave between two slits is a symmetrical and is spoken of as plane polarised or linearly polarised wave, while the slit N_1 is spoken of as the polariser in this experiment. This polarised wave will be transmitted by the second slit N_2 only if the lengths of N_1 and N_2 are mutually parallel [Fig. 16.2 (*a*)], but it will be completely stopped if N_2 is rotated to a position perpendicular to the slit N_1 [Fig. 16.2 (*b*)]. The string beyond N_2 in the latter case will show no vibrations at all. The slit N_2 which functions as the analyser or detector of polarisation of incoming waves is spoken of as analyser.

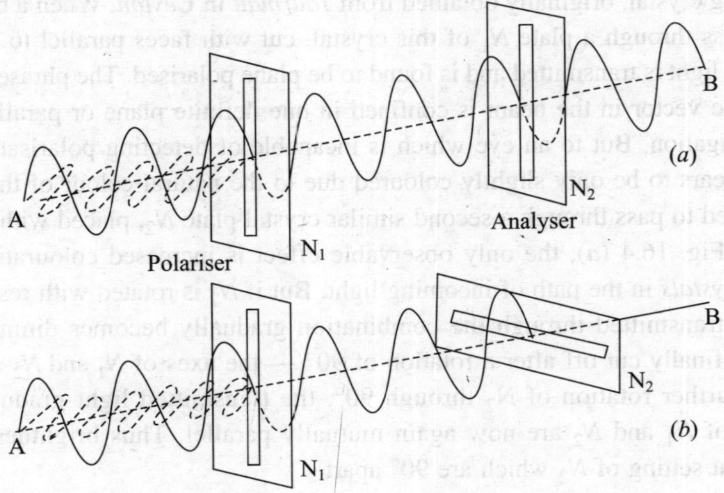

Fig. 16.2. Mechanical experiment to demonstrate polarisation of a wave

If, however, the strin were replaced by a spiral spring longitudinal waves created in it would pass freely from one end to the other, unaffected by the presence of slits or their relative orientation. Polarisation is thus the characteristic of transverse waves only.

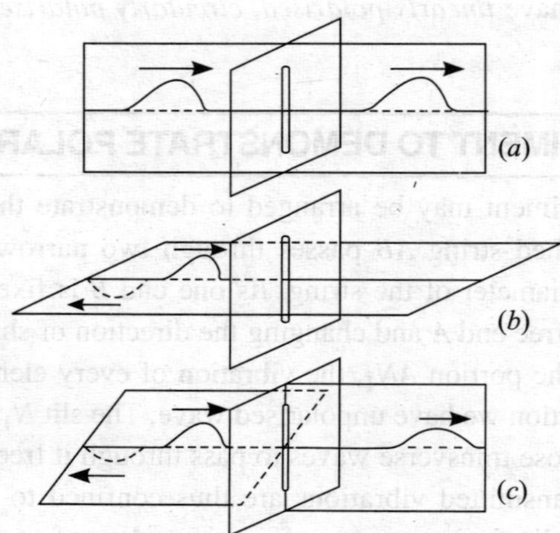

Fig. 16.3. (*a*) A vertical disturbance in a rubber tube is transmitted freely through a vertical slit; (*b*) a horizontal disturbance is not transmitted; (*c*) a disturbance in an intermediate plane is partially transmitted

16.3 OPTICAL EXPERIMENT TO ILLUSTRATE THE POLARISATION OF A WAVE

A similar experiment in optics throws considerable light on the nature of vibrations in light. *Tourmaline* is a naturally occurring crystal, originally obtained from *Tourmali* in *Ceylon*. When a beam of ordinary light is allowed to pass through a plate N_1 of this crystal, cut with faces parallel to its vertical axis, only a part of incident light is transmitted and is found to be plane polarised. The phrase *plane polarised* means that the electric vector in the beam is confined in one definite plane or parallel to it, through the direction of propagation. But to an eye which is incapable of detecting polarisation of light, the transmitted beam appears to be only slightly coloured due to the natural colour of the crystal. If this beam is further allowed to pass through a second similar crystal plate N_2, placed with its axis parallel to N_1 as depicted in Fig. 16.4 (*a*), the only observable effect is increased colouration of emergent light, owing to *two crystals* in the path of incoming light. But if N_2 is rotated with respect to N_1 in its own plane, the light transmitted through the combination gradually becomes dimmer and dimmer [Fig. 16.4 (*b*)] and is finally cut off after a rotation of 90° — the axes of N_1 and N_2 are now crossed [Fig. 16.4 (*c*)]. On further rotation of N_2 through 90°, the transmitted light gradually attains full brightness; the axes of N_1 and N_2 are now again mutually parallel. Thus brightness and darkness continue to alternate at setting of N_2 which are 90° apart.

This experiment clearly shows that light wave emerging from N_1 is asymmetric, for had it been symmetrical it would have passed freely through N_2 in all its orientations. Obviously, *tourmaline crystals* possess some property analogous to that of slits described in *Sec.* 16.2. This simple experiment, which illustrates the *polarisation of light*, demands for its explanation as first advanced by Fresnel, that the light wave must be transverse. For had it been a longitudinal wave it would have passed freely through the crystal plates irrespective of their mutual orientation. Accordingly, the intensity of transmitted light would have remained constant. After traversing N_1 the vibrations in the light wave are executed in one single direction, parallel to the *crystallographic axis* of N_1. Thus, the emergent light is plane polarised. The crystal plate N_1 is spoken of as polariser and N_2 as the analyser because it enables us to observe polarization of light.

The plane of polarisation is defined as the plane perpendicular to the plane of vibration in the light wave.

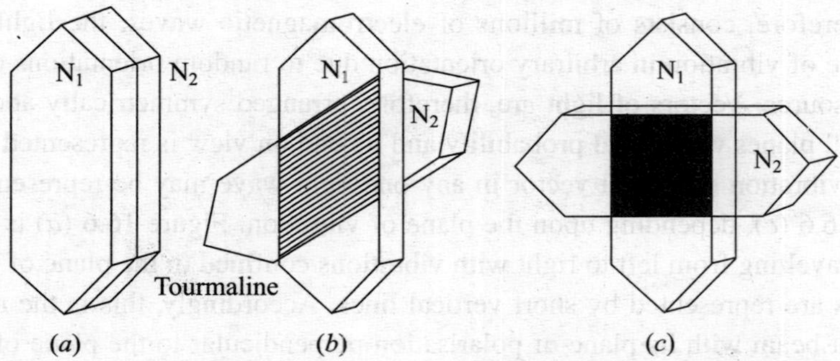

Fig. 16.4. Optical experiment to demonstrate polarisation of light

16.4 DISCOVERY OF POLARISATION OF LIGHT

The history of polarisation began in 1669 with the discovery of the phenomenon of double refraction by a *Dutch* philosopher, *Erasmus Bertholinus* while experimenting with a *calcite* crystal. This phenomenon was further studied by *Huygens* in 1690 who found that calcite forms two images of equal brightness, except when light traversed it in a direction parallel to the crystallographic axis. Furthermore, he discovered that each of the emergent beams from the calcite can be further subdivided, by its passage through a second crystal, into two beams of equal or unequal intensities or not decomposed at all, depending upon the orientation of the second crystal with respect to the first. This experiment clearly established that the natural light is perfectly symmetrical while the emergent beams from calcite are asymmetrical and therefore polarised. Although *Huygens* became the discoverer of polarisation of light yet he remained ignorant of its true nature and he could not explain this phenomenon according to the prevalent longitudinal conception of light waves. *Newton* recognised the essential features of polarisation when he said that a ray of light might have sides.

In 1808 *Malus* discovered that light reflected from glass possessed partly the characteristics (asymmetry) of beams of light emerging from calcite, but at one particular angle of reflection, it is completely polarised. He could not explain this asymmetry on the basis of *Young's* longitudinal vibration theory. Accordingly, he suggested that this phenomenon must be caused by some induced property of light corpuscles like that of magnetic pole or electric charge, giving rise to definite bias or *polarity* in the light beam. This suggestion gave birth to the title *Polarisation* for this new phenomenon. The interpretation of polarisation in terms of transverse nature of light waves was, however, given by *Fresnel*.

16.5 PICTORIAL REPRESENTATION OF LIGHT

For further study of polarisation of light we shall have to represent light waves on paper. Accordingly, we should adopt some convention for their proper representation. According to *electromagnetic theory of light*, it will be recalled that light is nothing but the propagation of mutually perpendicular vibrating electric and magnetic vectors and the electric vector functions as the light vector. A beam of ordinary or natural light, therefore, consists of millions of electromagnetic waves, the light vector of each having its own plane of vibration in arbitrary orientation due to random orientations of excited atoms or molecule in the source. Vectors of light are, therefore, arranged symmetrically about the direction of propagation, in all planes with equal probability and its end on view is represented by Fig. 16.5.

The mode of vibration of a light vector in any one light wave may be represented as shown in Figs. 16.6 (*a*) and 16.6 (*c*), depending upon the plane of vibration. Figure 16.6 (*a*) is a representation of a wave of light travelling from left to right with vibrations confined to the plane of paper or parallel to it. The vibrations are represented by short vertical lines. Accordingly, this is the representation of plane polarised light beam with its plane of polarisation perpendicular to the plane of page. Fig. 16.6 (*c*) is a representation of light wave in which vibrations are executed perpendicular to the plane of the page; accordingly they are represented by dots on the direction of propagation. End on views of the same waves are shown respectively in Figs. 16.6 (*b*) and 16.6 (*d*).

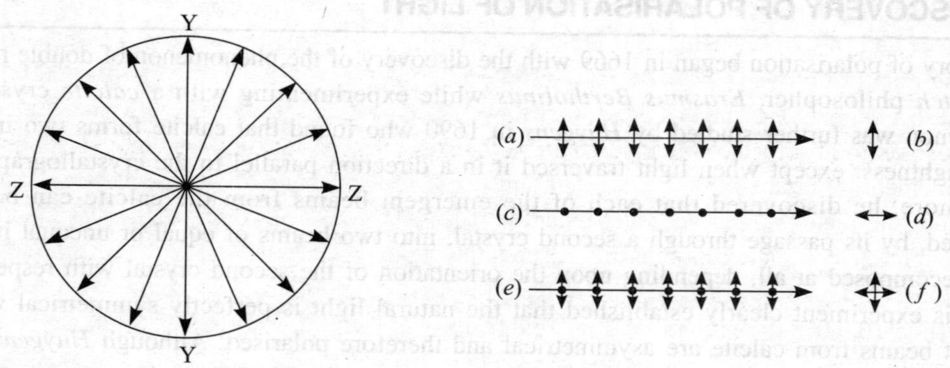

Fig. 16.5. End on view of unpolarised beam of light

Fig. 16.6. Pictorial representations of plane polarised and ordinary light beams

In ordinary light, light vectors are in random orientations and all the amplitudes of light vectors may be resolved into components along any two transverse and mutually perpendicular directions, say in Y and Z direction. Figure 16.7, shows components OA_Y and OA_Z of the light vector OA of one constituent wave, where

$$OA_Z = OA\cos\theta \quad \text{and} \quad OA_Y = OA\sin\theta$$

If this resolution of amplitude is continued for all light vectors, then owing to all vibration directions being equally probable, the average of the amplitude components along the axis of Y will be just equal to the average along Z axis. Consequently, for theoretical discussion, we may replace an ordinary beam of light by two waves of equal amplitudes,

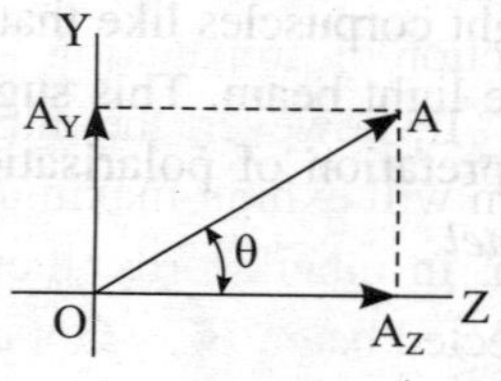

Fig. 16.7

vibrating in mutually perpendicular planes, of course, having no permanent phase relations to each other, due to the arbitrary phase changes in waves constituting the ordinary light. Figures 16.6 (*e*) and 16.6 (*f*) may be taken as the theoretical representation of ordinary light.

16.6 PRODUCTION OF PLANE POLARISED LIGHT

The extreme importance of the discovery of polarisation of light lies in the fact that the transverse nature of the waves constituting the light was thereby established, in contradistinction to *Huygens*' conception of longitudinal waves. The common methods for the production and analysis of plane polarised light may be grouped under the following heads: Polarisation by (*a*) reflection, (*b*) refraction, (*c*) double refraction, (*d*) selective absorption and (*e*) scattering.

16.7 POLARISATION BY REFLECTION

In 1808, the *French* Physicist *Malus* discovered that when a beam of ordinary light is incident at particular angle about 57°, on a glass plate, the reflected light possesses properties similar to those found in light obtained by transmission of natural light through *tourmaline* or *calcite i.e.*, reflected light is plane polarised light. The phrase *plane polarised* means that the light vector in the reflected light is vibrating transversely to the direction of transmission, in a fixed plane through this direction, pursuing the enquiry further, he found that the same phenomenon occurs when light is reflected from water and other transparent substances.

The polarisation of reflected light can be easily demonstrated by an experimental arrangement sketched in Fig. 16.8. in which N_1 and N_2 are two plane parallel glass plates blackened on their back surfaces so as to absorb the refracted beam. A beam of unpolarised light, *AB*, consisting of parallel rays is incident at an angle 57° on N_1 at *B* and the reflected beam *BC* is again reflected at the same angle from the plate N_2, placed exactly parallel to N_1. The plate N_2 is now rotated about *BC* as an axis through one complete rotation; this ensures the angle of incidence on N_2 to be the same in all its orientations. The intensity of twice the reflected beam is found to be maximum when N_1 and N_2 are parallel [Fig. 16.8 (*a*)] or antiparallel [Fig. 16.8 (*c*)] that is the planes of incidence are mutually parallel. The intensity is zero when they are crossed [Fig. 16.8 (*b*)] *i.e.*, the planes of incidence are mutually perpendicular; and in passing from parallel to crossed position the intensity diminishes from maximum to zero value. The variation in intensity of twice the reflected beam due to rotation of N_2 clearly proves that when light is reflected at 57°, the reflected beam consists of waves in which vibrations are confined to a certain definite direction transverse to the ray *i.e.*, it is a plane polarised beam. The lower plate N_1 is called polariser while the upper plate N_2 is called analyser.

If we rotate the lower plate N_1 about the incident beam as an axis, we get the reflected beam *BC* of the same intensity, thereby showing that the incident light is perfectly symmetrical about the direction of propagation.

If, however, the angle of incidence on the upper or lower plate is not 57°, the twice-reflected beam will exhibit maximum and minimum intensities as before but the minimum intensity will not be zero. In other words, there will always be a reflected beam from N_2. At angle other than 57°, the reflected beam *BC* is not completely plane polarised, while at 57° the polarisation is most complete.

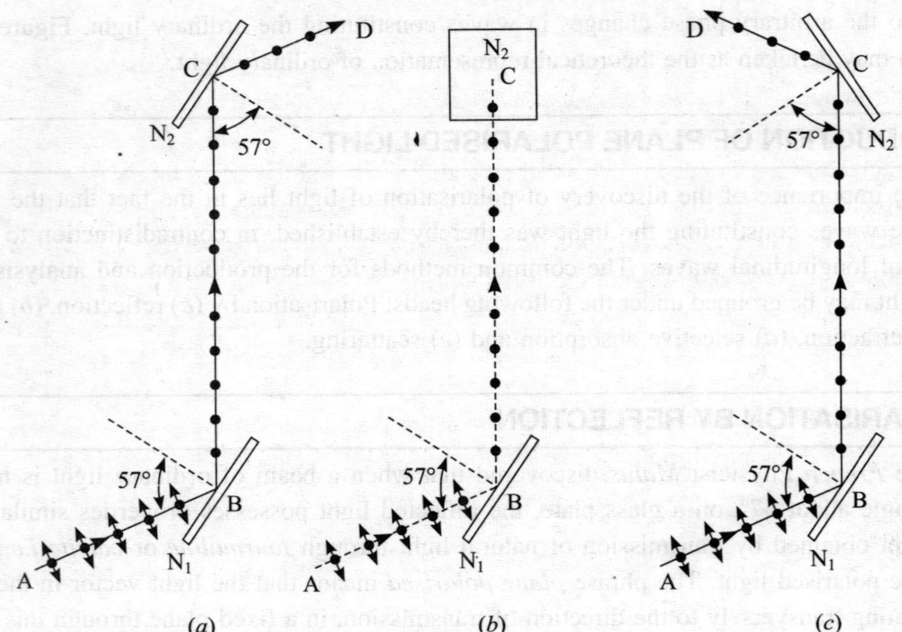

Fig. 16.8. Polarisation by reflection from glass surface

The angle of reflection, i_p, at which the polarisation of reflected light is most complete is called the polarising angle and it varies with the nature of glass and also with the wavelength of light. *Malus* defined the plane of polarisation as the plane of incidence of polarised light when the reflected light was of maximum intensity. We Shall explain in *Sec.* 16.9 the fact that the vibration in the reflected plane polarised light *BC* are perpendicular to this plane.

16.8 BREWSTER'S LAW

Brewster made a remarkable discovery that at polarising angle the reflected and refracted rays are just 90° apart. The angle or refraction r_p is, therefore, equal $(90 - i_p)$ and we have for the refractive index,

$$_a\mu_g = \frac{\sin i_p}{\sin r_p} = \tan i_p$$

The tangent of the angle of polarisation is equal to the refractive index of the reflecting substance; this is called *Brewster's law*. As a consequence of this law it follows that the angle of polarisation depends upon the wavelength λ of light employed.

Brewster's law also applies to reflection at the second surface of glass plate in the denser medium. In this case, it states that the refractive index is equal to the cotangent of polarising angle. For a polished metallic reflector this law does not hold good.

16.9 EXPLANATION OF POLARISATION BY REFLECTION

We may now explain the action of reflecting surface in producing plane polarised light. The light vectors in the unpolarised light may be resolved at the point of incidence *B*, into two components, one perpendicular and the other parallel to the plane of incidence. At polarising angle since the reflected and the refracted rays are just 90° apart, therefore, at the point of incidence *B* in Fig. 16.9, the vibrations which are in the plane of incidence become parallel to the direction of reflected ray *BC* and so they can generate only longitudinal waves along *BC*. Since the waves of light are transverse and not longitudinal, we conclude that, the wave with vibrations in the plane of incidence is not at all reflected at polarising angle but totally (100%) transmitted as a refracted beam. The other wave in which vibrations are perpendicular to the plane of incidence, is partly reflected (15%) and partly transmitted (85%). The refracted beam is, therefore, strong but only partially polarised as it always contains both kinds of vibrations, while the reflected beam is weak but completely plane polarised, as it contains only vibrations which are executed perpendicular to the plane of incidence.

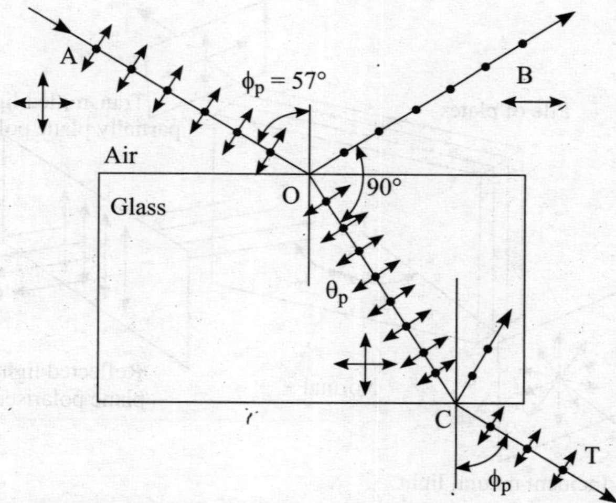

Fig. 16.9. Illustrates Brewster's law at the polarising angle

We may now explain the variation of the intensity of twice the reflected beam at the polarising angle owing to the rotation of glass plate N_2 in the experiment described in *Sec.* 16.7. The plane polarised reflected beam *BC* contains only those vibrations which are perpendicular to the plane of page, which for the lower plate N_1 is the plane of incidence. When N_1 and N_2 are crossed the vibrations in *BC* which are perpendicular to the plane of incidence for N_1 will be in the plane of incidence of the ray incident on N_2 and as explained earlier, they are totally transmitted within N_2, thereby making the reflected intensity as zero. But when N_1 and N_2 are parallel the vibrations in *BC* are also perpendicular to the plane of incidence of N_2, and so partly transmitted within N_2 and partly reflected along *CD*. The intensity of the reflected beam is maximum. For intermediate orientation of N_2 the intensity of twice the reflected beam varies as the square of cosine of angle between the two planes of reflections and so the intensity is intermediate of maximum ($\cos\theta = 1$) and minimum ($\cos\theta = 0$) values.

16.10 POLARISATION BY REFRACTION

Brewster's law applies not only for reflection in rarer medium but also for reflection in denser medium. In other words, it can be easily shown that if ordinary light is incident at polarising angle on the upper surface of a plane parallel glass plate, the refracted ray will be incident also at polarising angle at the lower surface. This remarkable fact is used in producing polarisation by refraction through a pile of large number of glass plates as shown in Fig. 16.10. At each reflection, the reflected ray contains 15% of vibrations which are perpendicular to plane of incidence while the refracted ray contains vibrations which are parallel (100%), and perpendicular (85%) to the plane of incidence, assuming μ of glass as 1.5. The larger the number of plates in the pile, the more nearly plane polarised is the transmitted beam and so in the limit we expect that vibrations which are perpendicular to the plane of incidence will be completely quenched from the transmitted beam and the vibrations of light in it will be only parallel to the plane of incidence.

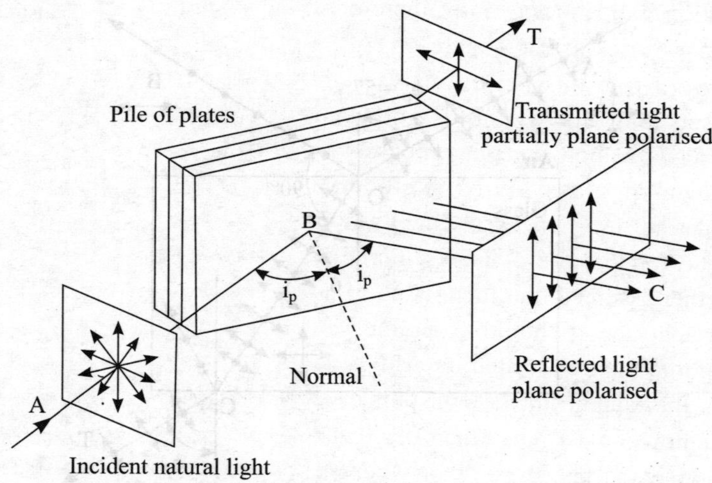

Fig. 16.10. Polarisation of light by a pile of glass plates

The pile of plates when used as polariser is mounted in a tube in such a way that the unpolarised light, along the axis of the tube, is incident at polarising angle i_p, on the pile. The second pile of plates, placed parallel to the first as shown in Fig. 16.11 functions as analyser.

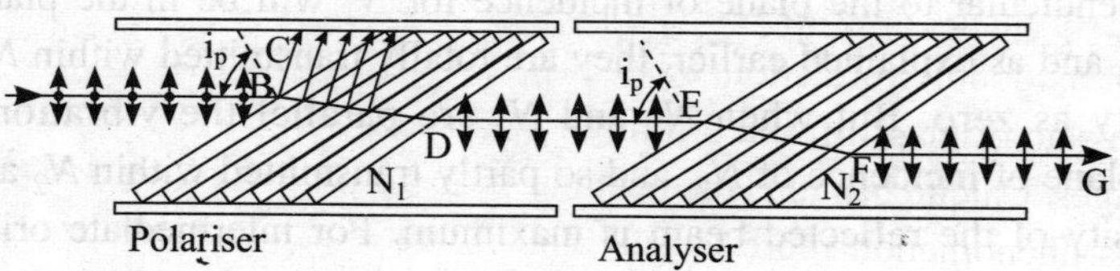

Fig. 16.11. Pile of glass plates as polariser and analyser

16.11 LAW OF MALUS

In 1809 *Etienne Louis Malus* experimentally discovered a relation which indicates that the intensity of light transmitted by the analyser depends upon the inclination of its plane of transmission with that of the polariser. To derive the relation let us suppose, that the angle between the two planes of transmission is θ at any instant. The light vector, $AP = a$, in the plane polarised light emerging from the polariser may be resolved into two components, as shown in Fig. 16.12,

$$AE = a \cos \theta \text{ and } AO = a \sin \theta$$

respectively, along and perpendicular to the plane of transmission of the analyser. The perpendicular component is eliminated in the analyser while the parallel component is freely transmitted through it. Therefore, the intensity I of light that emerges from the analyser is given by

$$I = a^2 \cos^2 \theta = I_o \cos^2 \theta \qquad ...(16.11\ a)$$

where I_o is the intensity of the plane polarised light incident on the analyser. This relation, called *Malus' Law*, therefore, states that the transmitted intensity varies as the square of cosine of the angle between the planes of transmission of analyser and polariser. It may be emphasised that this law holds good only when light incident on analyser is completely plane polarised, which is true of the combination of glass plates sketched in Fig. 16.8, combination of polaroids (*Sec.* 16.27) or Nicol's prism (*Sec.* 16.24), but is partially true for the pile of plates. For Eqn. (16.11 *a*) to hold true, it is further assumed that there is no loss of light due to absorption in the analyser. It should be pointed out that even in the case of absorption, the transmitted intensity depends on $\cos^2\theta$, the only effect is the change in the constant in Eqn. (16.11 *a*). The transmitted light is a maximum when $\theta = 0$ and is zero (if the polarisation is complete) when $\theta = 90°$ or when the polariser and analyser are crossed.

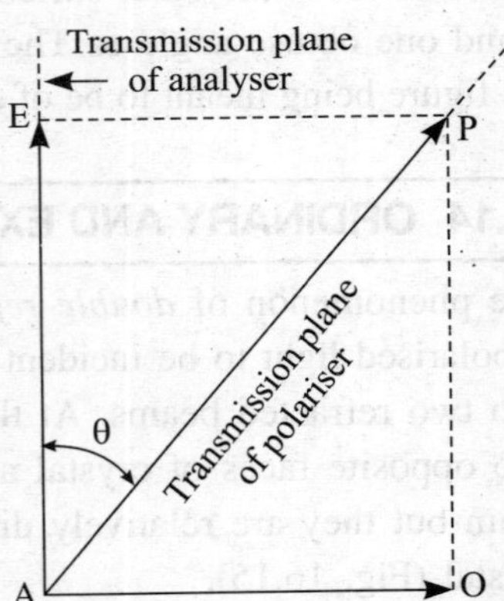

Fig. 16.12. Resolution of the amplitude of plane-polarised light

16.12 DOUBLE REFRACTION

In the year 1669 a Dutch philosopher, *Erasmus Bartholinous* discovered that when a ray of ordinary light is incident on a crystal, called Iceland spar or calcite, it splits into two refracted rays, one of which always obeys ordinary laws of refraction and the other, in general, may not obey them. This phenomenon known as double refraction was called by Huygens as 'strange refraction' of calcite. A slab of calcite and a slab of glass lift the printing underneath; but while the glass lifts only one image, calcite lifts two images. Calcite has thus the peculiar property of exhibiting double refraction. Huygens in 1690 studied this phenomenon more closely and found that both the refracted rays were linearly polarised in mutually perpendicular planes. This phenomenon is also exhibited by other crystals like quartz, except those belonging to cubic system which are optically anistropic *i.e.*, their optical properties are the same in all directions like glass.

We shall chiefly study this phenomenon in Iceland spar, in which double refraction is very marked and easy to study experimentally.

16.13 GEOMETRY OF CALCITE CRYSTAL

Iceland spar, which chemically is hydrated *calcium carbonate,* $CaCO_3$ was found in large quantities in times gone by in *Iceland.* $CaCO_3$ crystalises in a large colourless clear hexagonal prism with a blunt pyramid at each end. When struck, it easily cleaves obliquely in three definite planes forming a *rhombohedral* body whose six rhombic sides make an angle of 45°23′ with the lines joining the vertices of the pyramids, the *crystallographic axis* of the crystal. A spar rhombohedran of this kind is bounded by six faces each of which is a parallelogram whose angles are $\alpha = 101°55'$ and $\beta = 78°5'$. At the two opposite corners *A* and *H*, as shown in Fig. 16.14, there are three obtuse angles α but at the other six corners there are two acute angles each equal to β and one obtuse angle α. The length of the edges is not important, those in the figure being meant to be of equal length.

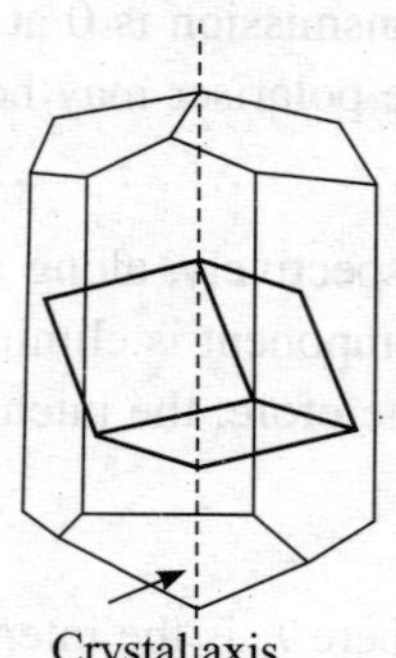

Fig. 16.13. Iceland spar crystal

16.14 ORDINARY AND EXTRAORDINARY RAYS

The phenomenon of *double refraction* can be easily demonstrated by allowing a narrow beam of unpolarised light to be incident normally on calcite crystal. The incident beam (in general) splits up into two refracted beams. At the second face double refraction does not at all occur and since the two opposite faces of crystal are parallel, the two refracted beams emerge parallel to the incident beam but they are relatively displaced by a distance, which is proportional to the thickness of the crystal (Fig. 16.15).

On a screen, two images, *O* and *E*, of the pinhole are thus obtained. One image, *O* lies in the direction of incident beam.

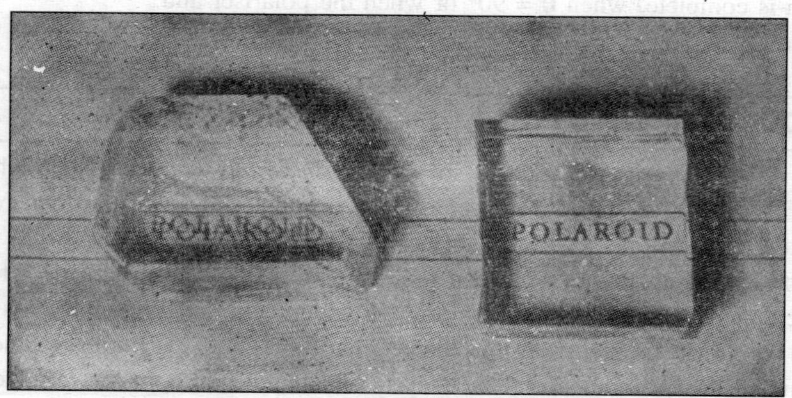

Calcite Glass

Figure illustrates double refraction in calcite and ordinary refraction in glass. Glass lifts one image while calcite lifts two images. (*Courtesy Polaroid Corporation*)

Therefore it must have been formed in accordance with the ordinary laws of refraction. The other image, E is found to be separated from the O image despite perpendicular incidence, in contrast to the ordinary phenomenon. The corresponding refracted beam, therefore, does not, in general, obey the ordinary laws of refraction. On rotating the crystal about the incident beam as an axis it is observed that the O image remains stationary while the E image revolves in a circular path with its centre at the O image but *the line joining the two images is always parallel to the shorter diagonal of the end face of the crystal*.

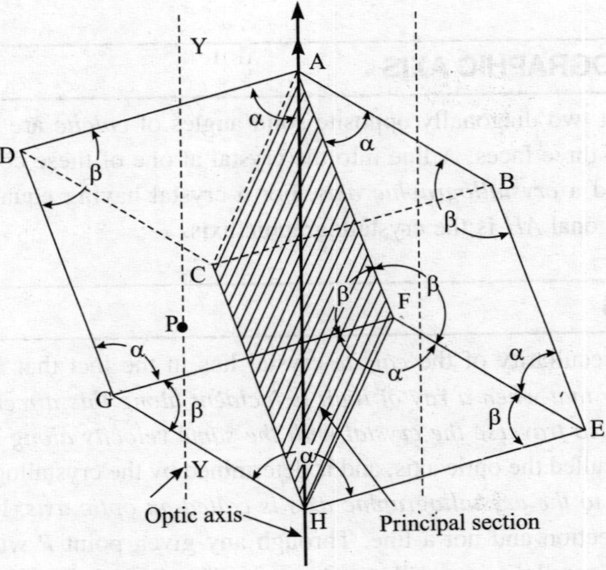

Fig. 16.14. Rhombohedran of calcite

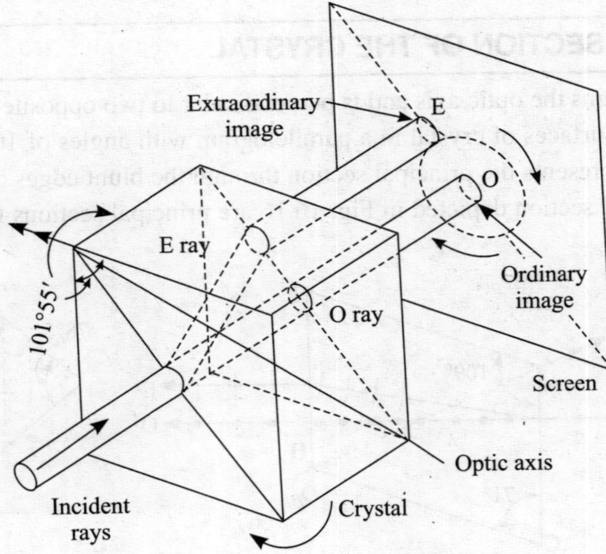

Fig. 16.15. A narrow beam of natural light can be split into two beams by a doubly refracting crystal

Of the two refracted rays, the one which always obeys ordinary laws of refraction is called the *ordinary ray* or the *O ray* while the other, which behaves in quite extraordinary manner is called the *extraordinary ray* or the *E ray*. The ratio sin i/sin r physically represents the ratio of the velocity of light within vacuum to that in the refracting medium. It is, therefore, obvious that the velocity of *O ray* is the same in all directions within the crystal. In the case of an extraordinary ray, since the ratio sin i/sin r varies with the angle of incidence, its velocity is different in different directions within the crystal. *O* ray is always in the plane of incidence but in general this is not true of the *E* ray.

16.15 CRYSTALLOGRAPHIC AXIS

It will be recalled that two diagonally opposite solid angles of *calcite* are formed by the junction of three obtuse angles of three faces. A line into the crystal at one of these corners and equally inclined to three faces is called a *crystallographic axis*. For a crystal having equal edges, as shown in Fig. 16.14, its shortest diagonal *AH* is the crystallographic axis.

16.16 OPTIC AXIS

One most important peculiarity of the *calcite* crystal lies in the fact that *there is one and only one direction through it so that when a ray of light is incident along this direction, the O ray and E ray do not separate and also traverse the crystal with the same velocity along this direction*. Hence, this particular direction is called the optic axis, and is determined by the crystallographic axis. *Any direction in the crystal parallel to the crystallographic axis is called an optic axis*. It may be emphasised that the optic axis is a direction and not a line. Through any given point *P* within the crystal, only one line *YY*, can be drawn parallel to crystallographic axis. This line gives the direction of the optic axis for this point and for all points lying on it.

16.17 PRINCIPAL SECTION OF THE CRYSTAL

Any plane which contains the optic axis and is perpendicular to two opposite faces is called a principal section. This cuts the surfaces of crystal in a parallelogram with angles of 109° and 71°, as illustrated in Fig. 16.16 which represents the principal section through the blunt edges of the crystal. All sections parallel to the principal section depicted in Fig. 16.16 are principal sections for points lying on it.

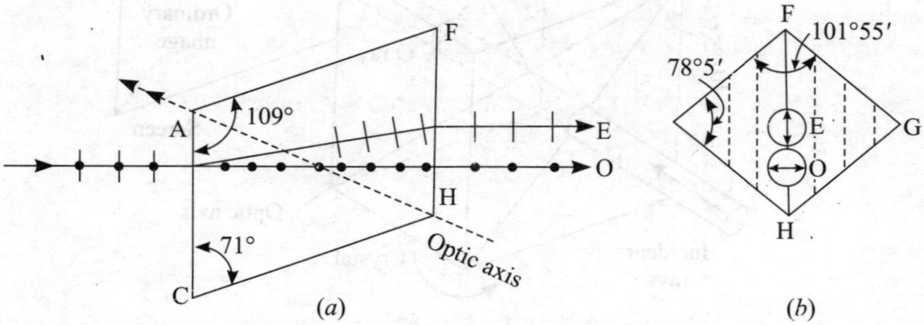

Fig. 16.16. Principal section in a calcite crystal

Through every point three principal sections can be drawn corresponding to three pairs of opposite crystal surfaces. An end on view of any principal section is a straight line (shown dotted) in the crystal surface parallel to its shorter diagonal *FH*, which is the end on view of the principal section through the blunt edges.

16.18 PRINCIPAL PLANE OF O RAY AND E RAY

The principal plane of the ordinary ray is defined as a plane in the crystal drawn so as to contain the optic axis and the ordinary ray. *The principal plane of the extraordinary ray* is defined as a plane in the crystal drawn through the optic axis and extraordinary ray. The principal planes of the two refracted rays, in general do not coincide. They do coincide, however, under special circumstances when the plane of incidence is the principal section of the crystal. In a case like this, as illustrated in Fig. 16.16, the plane of incidence, the principal planes are all coincident. The line joining the *O* image and the *E* image of the same point is in the principal section. This accounts for the rotation of the *E* image in a circle around *O* image as the crystal is rotated about the ordinary ray as an axis.

16.19 POLARISATION BY DOUBLE REFRACTION

When the *O* ray and the *E* ray from the calcite crystal, in the experimental arrangement sketched in Fig. 16.15, are examined by a polaroid or tourmaline, it is found that as the analyser is rotated in its own plane, the intensity of one of the two images, say, the *O* image gradually increases while that of the *E* image gradually diminishes. It is possible to get a position of tourmaline in which the *O* image acquires maximum intensity while the *E* image completely disappears. In this setting of tourmaline it will be observed that its longer axis (*crystallographic axis*) which determines the plane of transmission in it, is exactly parallel to the longer diagonal of the end face of the crystal [Fig. 16.17 (*a*)]. A rotation of tourmaline from this setting to a position obtained on 90° rotation will render the intensity of the *E* image maximum while the *O* image will completely vanish. In this setting of tourmaline, it will be observed, its longer axis is parallel to the shorter diagonal of the crystal face [Fig. 16.17 (*b*)], which determines the trace of the principal section in the end on view. We, therefore, conclude that the rays by which the ordinary and extraordinary images are seen must be polarised. Since the *E* ray is intercepted by the tourmaline when its plane of transmission is perpendicular to the principal section of crystal while the *O* ray is intercepted on 90° rotation, it follows that the vibrations in the *E* ray and the *O* ray are executed respectively along and at right angels to the principal section *i.e., the O ray is polarised in the principal section while the E ray is polarised perpendicularly to the principal section.*

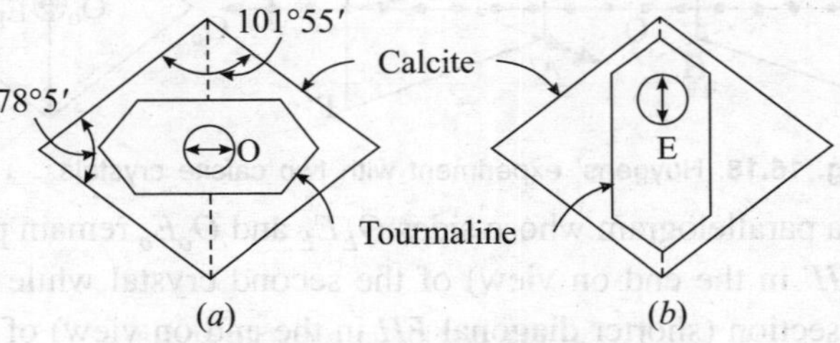

Fig. 16.17. Polarisation by double refraction

In general, when the plane of incidence is not parallel to the principal section, the vibrations in the wavefront of the E ray are executed in the principal plane of the E ray while those in the wavefront of the O ray are executed at right angels to the principal plane of the O ray. Fig. 16.26. fully illustrates this case.

Huygens' Experiment

The polarisation of light by double refraction in calcite was most convincingly demonstrated by *Huygens* in his experiment with two calcite crystals. He found that if a beam of light be sent normally through two calcite crystals, their principal sections being exactly parallel and symmetrical as shown in Fig. 16.18, the ordinary ray from the first traverses the second crystal without any deviation while the extraordinary ray traverses the second one as E ray along a path exactly parallel to the path inside the first crystal. Therefore, if the crystals are of equal thickness, as we may suppose for the present, the two rays represented by O_o and E_E emerge exactly parallel but with a separation twice as much as with one crystal. Now, if the second crystal be rotated, with respect to the first, through a small angle about the ordinary ray as an axis, then each of the two rays O and E from the first undergoes double refraction in the second, giving rise to four images. The old images O_o and E_E are slightly fainter and in between them we may notice two new faint images O_E and E_o. On rotating the second crystal further it is found that the two images, O_o and O_E remain stationary while E_o and E_E rotate in a circular path around O_o and O_E, respectively. Furthermore, the intensity of the old images O_o and E_o goes on diminishing while that of the new images, O_E and E_o goes on increasing until at a rotation of θ = 45°, all the four images are of equal intensity.

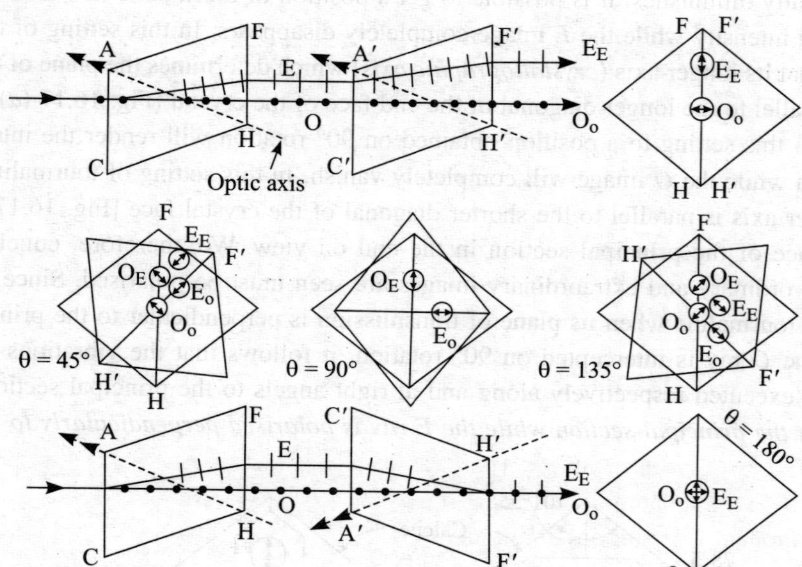

Fig. 16.18. Huygens' experiment with two calcite crystals

The four images form a parallelogram whose sides $O_E E_E$ and $O_o E_o$ remain parallel to the principal section (shorter diagonal $F'H'$ in the end on view) of the second crystal while sides $E_o E_E$ and $O_o O_E$ are parallel to the principal section (shorter diagonal FH in the end on view) of the first crystal.

At $\theta = 90°$ rotation, the old images, O_o and E_E disappear while the new images O_E and E reach maximum intensity. On continuing the rotation, O_o and E_E reappear and increase in intensity while O_E and E_o correspondingly begin to decrease in intensity and at $\theta = 135°$, the four images are again of equal intensity. At $\theta = 180°$, the principal sections are again parallel but the optic axis of crystals are oriented in opposite directions, as shown in the bottom left hand diagram of Fig. 16.18. Therefore, the images O_E and E_o disappear while the O_o ray and the E_E ray coalesce, as the two crystals are supposed to be of equal thickness, on emerging from the second crystal, thereby forming a single image. As the rotation is continued from 180° to 360° the same changes take place in the reverse order.

From the above observations, we can easily conclude that the characteristic of O and E rays is quite different from ordinary light. For, while the ordinary light, on double refraction in calcite always gives two images of the same equal intensity, except when incident along the optic axis, the double refraction of O and E rays in the second calcite depends upon its orientation with respect to the first. This shows that the wave displacements in O and E ray lack symmetry about the direction of propagation in contrast to symmetrical vibrations in the ordinary light. Therefore, the O ray and the E ray are said to be polarised.

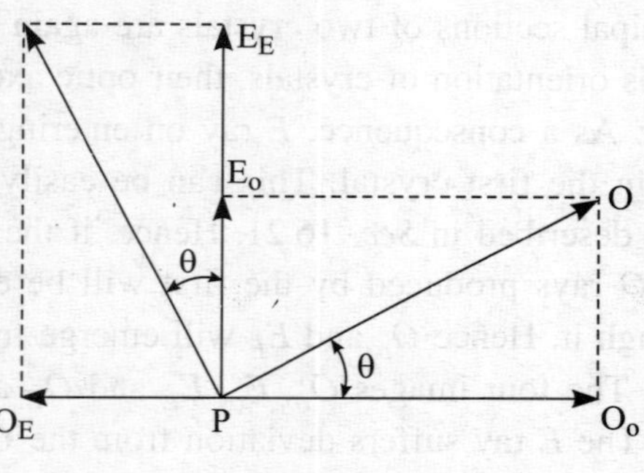

Fig. 16.19

Explanation

Suppose PX and PY represent in Fig. 16.19 in the end on view the principal sections of the first and the second crystal respectively, mutually inclined at an angle θ. Ordinary light undergoes double refraction within the first crystal forming the O ray with vibrations perpendicular to its principal plane, which in this case is coincident with the principal section PX, and the E ray with vibrations in its principal plane, here the principal section PX. So in Fig. 16.19, $PO = a$ represents the amplitude of the O ray, and $PE = a$, represents the amplitude of the E ray. The second crystal like the first transmits light with vibrations in its principal section along one path while light with vibrations at right angles to it is transmitted along another path. Each of the rays O and E, on just entering the second crystal, is refracted into two components. The O ray is resolved into two components viz. O_o with an amplitude $PO_Q = a \cos \theta$, which is perpendicular to the principal section PY of the second crystal, giving rise to ordinary image, while the other component E_o, with an amplitude $PE_o = a \sin \theta$ along PY, gives rise to extra-ordinary image. The E ray from the first on entering the second is resolved into the component E_E with an amplitude, $PE_E = a \cos \theta$ along PY, which gives rise to extra-ordinary image, and the component O_E with an amplitude, $PO_E = a \sin \theta$, at right angles to PY gives rise to ordinary image. We, therefore, observe four images according to the following scheme:

$$O\begin{cases} O_o = a\cos\theta & \textit{Old ordinary image} \\ E_o = a\sin\theta & \textit{New extraordinary image} \end{cases}$$

$$E\begin{cases} E_E = a\cos\theta & \textit{Old extraordinary image} \\ O_E = a\sin\theta & \textit{New ordinary image} \end{cases}$$

The intensities of old images O_o and E_E are each $a^2 \cos^2\theta$ while that of the new images O_E and E_o are each $a^2 \sin^2\theta$. Thus, for $\theta = 0$, new images O_E and E_o vanish. When the angle of rotation is 45° or 135°, all the four images are of equal intensity viz. $a^2/2$. But when θ = 90°, old images vanish while new images O_E and E_o acquire maximum intensity each equal to a^2. When θ = 180°, principal sections of two crystals are again parallel and so the new images O_E and E_o disappear. But in this orientation of crystals, their optic axes are oriented in opposite directions with respect to each other. As a consequence, E ray on entering the second crystal is deviated in a direction opposite to that in the first crystal. This can be easily verified by Huygens' construction for the O and the E rays, described in *Sec.* 16.21. Hence, if the crystals are of equal thickness, the separation between E and O rays produced by the first will be exactly neutralised by the second crystal in transmission through it. Hence O_o and E_E will emerge from the same point of the second crystal.

The four images O_o, E_o, E_E and O_E always form a parallelogram. It is not difficult to explain this. The E ray suffers deviation from the O_o ray (which pursues straight path in the direction of the O ray since $\angle i = 0$) in the principal section of the second crystal. Consequently, both rays emerge from respective points E_o and O_o, situated in the principal section passing through the point of incidence of the O ray. Similarly the O_E and the E_E rays emerge from respective points O_E and E_E situated in the principal section through the point of incidence of the E ray on the second crystal. Since the thickness of the second crystal is the same for both E_o and E rays, hence $O_oE_o = O_E\, E_E$. Now it is clear that the figure $O_o\, E\, E_EO_E$ is a parallelogram.

16.20 HUYGENS' THEORY OF DOUBLE REFRACTION

Soon after the discovery of the phenomenon of double refraction in uniaxial crystals like *calcite* by Bertholinus, Huygens, who had already put forward the wave theory and explained ordinary reflection and refraction in the light of his theory of secondary wavelets, was naturally desirous to explain the peculiar properties of *calcite* according to the same theory. Although, with his conception of longitudinal nature of light waves, he could not explain the polarisation of the O and the E rays, yet he gave a suitable explanation of their origin by extending his principle of secondary wavelets in a suitable manner to *uniaxial crystals*. Accordingly he put forward the following postulates:

1. Every point of the crystal disturbed by the incident wavefront must become the origin of two wavelets— one *the ordinary and the other extraordinary wavelet* — in order to account for the ordinary and extraordinary refracted ray within the crystal.

2. It is found that the ordinary ray always obeys both the laws of refraction. The refractive index of calcite for the ordinary ray,

$$\mu_o = \sin i/\sin r_o = V/V_o$$

is, therefore, independent of the direction of its propagation within the crystal. This is only possible when the velocity V_o of the ordinary wave is the same in all directions. Hence *the wavelet for the ordinary wave must be spherical in form just* as in *isotropic media*, like glass, due to the equality of speed of light waves in all directions, the secondary wavelets were drawn spherical. Thus, according to *Huygens*, the ordinary refracted wavefront in calcite must be formed as a result of mutual reinforcement of spherical wavelets originating at the various points of the refracting surface of the crystal, successively or simultaneously disturbed by the incident wavefront.

3. The *E* ray does not always obey ordinary laws of refraction. The ratio sin i/sin r_E for *E* ray has a value which varies with the direction of its propagation relative to optic axis. It is the same as the ordinary refractive index along the optic axis but differs most from this value in a direction perpendicular to the optic axis. The velocity of the propagation of the *E* ray is, therefore, different in different directions. The extraordinary wavelet cannot be spherical in form. *Huygens assumed the form of this wavelet to be ellipsoid of revolution*—a surface generated by the revolution of an ellipse about one of its principal axes (major or minor axis) just as a sphere can be obtained by revolving a circle about any diameter.

4. The properties of uniaxial crystals are perfectly symmetrical about the optic axis. Hence, the *E* ray is propagated with equal velocity in directions which are equally inclined to optic axis. Consequently, the ellipsoidal wave surface must be symmetrical about the optic axis. To account for this, Huygens put forward the hypothesis that *the axis of revolution* (*i.e.*, the axes of symmetry) *of the ellipsoid must be coincident with the direction of optic axis through the point of origin of the wavelets*.

5. The velocities of the *O* and *E* rays are equal along the optic axis. As a consequence, uniaxial crystals do not exhibit double refraction along the optic axis. To account for this, Huygens postulated that *the spherical wavelet and the ellipsoidal wavelet touch each other along the extremities of the axis of revolution* (optic axis) *of the ellipse generating the ellipsoidal wave surface.* To be more precise, the two wavelets touch each other at points where the optic axis through the point of origin of wavelets intersects them.

Thus, according to Huygens, every point of the doubly refracting crystalline substance, disturbed by the incident wave, is to be considered as the source of two secondary wavelets sphere for the ordinary wave and ellipsoid of revolution about the optic axis for the extraordinary wave, the two wave surfaces touching each other along the extremities of the axis of revolution of the generating ellipse.

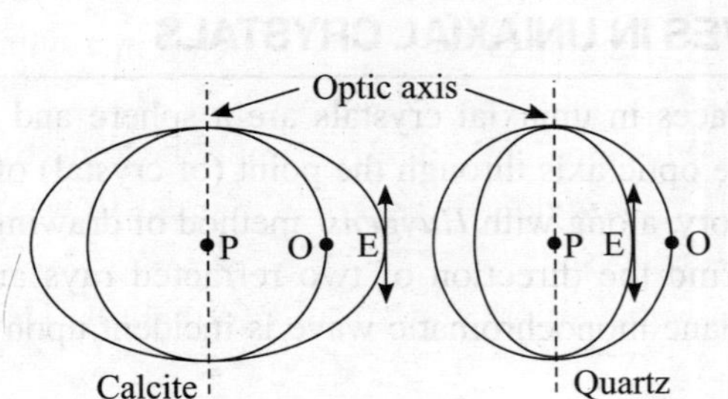

Fig. 16.20. Wavesurface diagrams in calcite and quartz

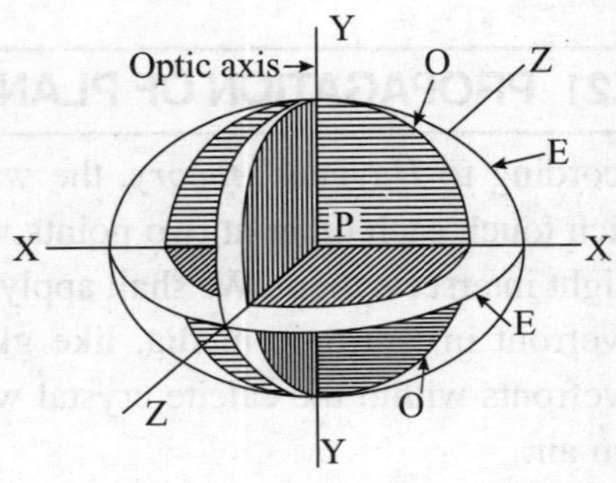

Fig. 16.21

Now, imagine a point source P of monochromatic light within the crystal; then Fig. 16.20 represents the O and E wavefronts in two dimensions, at any instant surrounding the point source P. The wave surfaces in three dimensions are obtained by rotating the figure about the optic axis – the circle generating the sphere, the wave surface for the O wave and the ellipse generating the ellipsoid, the wave surface for the E wave. These are shown in Fig. 16.21 for calcite while in Fig. 16.22 are shown the sections of two surfaces by three principal planes viz. XY, YZ and ZX planes. The origin of three dimensional coordinate system is at P and in the Y axis is along the optic axis. The velocity of the E ray varies as the radius vector of ellipsoid and in crystals like calcite, its value is maximum along a direction perpendicular to the optic axis and least, equal to the velocity of the O ray, along the optic axis. In calcite, the sphere is, therefore, enclosed within the ellipsoid, the latter being formed by rotating the ellipse about its minor axis. Crystals belonging to this class are known as *negative crystals*. In quartz, ZnO, TiO_2 the velocity of the E ray is maximum, equal to the velocity of the O ray, along the optic axis but it is least along a direction perpendicular to optic axis. Consequently, in quartz, the ellipsoid is enclosed within the sphere, the former being obtained by the rotation of an ellipse about its major axis, which is equal to the diameter of the spherical wave. Such crystals are known as *positive* crystals.

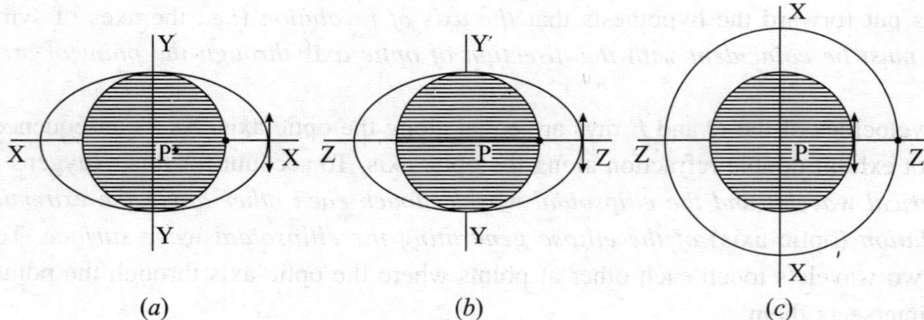

Fig. 16.22. Cross-sections of the wave surfaces for calcite

In quartz, the sphere and the enclosed ellipsoid do not quite touch each other at points where the optic axis through P intersects these surfaces. Owing to this reason, quartz exhibits a new phenomenon called *optical activity*, which is discussed in detail in Chapter 18.

16.21 PROPAGATION OF PLANE WAVES IN UNIAXIAL CRYSTALS

According to *Huygens' theory*, the wave surfaces in uniaxial crystals are a sphere and an ellipsoid which touch each other at two points where the optic axis through the point (of crystal) of origination of light intersects them. We shall apply this theory, along with *Huygens*' method of drawing a refracted wavefront in isotropic media, like glass, to find the direction of two refracted rays and refracted wavefronts within the calcite crystal when a plane monochromatic wave is incident upon it obliquely from air.

Suppose AB represents a section of the plane wavefront of monochromatic light incident obliquely on the plane surface AC of calcite crystal, both planes being perpendicular to the plane of Fig. 16.23. The plane of incidence is the plane of paper. The optic axis of the crystal, however, is inclined at some arbitrary angle with the crystal surface. At the instant when the incident wavefront touches A, this crystal point becomes the origin of two disturbances within the crystal and hence of two secondary wavelets.

One of them spreads in spherical form and the other in a ellipsoidal form. As the incident wavefront AB advances, every point along AC successively becomes the origin of two secondary wavelets and all these would have advanced different distances till the instant the point C of the crystal is just being disturbed by the advancing wavefront. Let us now find the position of the secondary wavelets from A at the moment when B reaches C. Let V, V_o and V_E be respectively the velocity of light in air, the velocity of O ray, and velocity of E ray within the crystal (V_E in a direction perpendicular to the optic axis). Therefore the time taken by the incident wavefront to travel from B to C in air is

$$t = BC/V$$

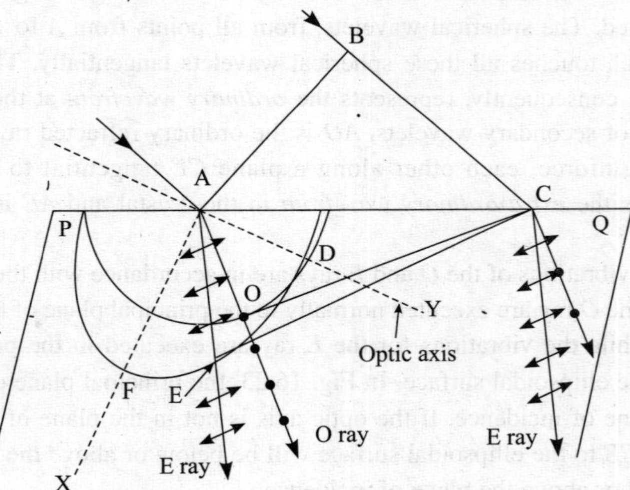

Fig. 16.23

In this period, the secondary spherical wavelet from A would acquire a radius AD within the crystal, given by

$$AD = V_o t = \frac{V_o}{V} BC = \frac{BC}{\mu_o}$$

where μ_o is the refractive index of calcite for ordinary ray. The section of this spherical wavelet by any plane through A is a circle of a radius AD. Therefore, in the Fig. 16.23 we draw a circle of radius AD and this represents in two dimensions, the position of the secondary spherical wavelet from A at the moment when C just becomes the origin of secondary wavelets.

The semi-major axis of the ellipsoidal wavelets from A, at the moment when C just becomes the origin of wavelets, is given by

$$a = AF = V_E t = \frac{V_E}{V} BC = \frac{BC}{\mu_E}$$

where μ_E is the refractive index of calcite for the propagation of extraordinary ray in a direction normal to optic axis. In calcite the sphere is enclosed within the ellipsoid, hence the semi-minor axis of ellipsoidal wavelet is given by

$$b = AG = AD = BC/\mu_o$$

The third principal axis of ellipsoid is given by

$$2c = 2a$$

simply due to the fact that the ellipsoid is obtained by the rotation of an ellipse about its minor axis. The section of the ellipsoid in this particular case, by the plane of incidence is an ellipse. Therefore, with the above values for a and b, we draw an ellipse touching the circle at the point of its intersection with optic axis through A. this ellipse represents, in two dimensions, the position of the ellipsoidal wavelet from A, at the moment when C has just become the origin of secondary wavelets.

Exactly in a similar way, the positions of secondary wavelets, originating from any point between A and C, can be obtained. The spherical wavelets, from all points from A to C, reinforce each other along a plane CO which touches all these spherical wavelets tangentially. This plane CO which is perpendicular to paper, consequently, represents the *ordinary wavefront* at the moment when C has just become the origin of secondary wavelets. AO is the ordinary refracted ray from A, similarly, the ellipsoidal wavelets reinforce, each other along a plane CE tangential to these wavelets which consequently represents the *extra-ordinary wavefront* in the crystal and AE is the extraordinary ray from A.

The directions of vibrations of the O and E rays are in accordance with their state of polarisation, viz., the vibrations for the O ray are executed normally to the principal plane of the O ray and tangential to spherical surface, while the vibrations for the E ray are executed in the principal plane of the E ray and tangential to the ellipsoidal surface. In Fig. 16.23, the principal plane of the O and E rays are coincident with the plane of incidence. If the optic axis is not in the plane of incidence, the point of contact of the tangent CE to the ellipsoidal surface will be below or above the plane of the figure and the E ray will be below or above the plane of incidence.

16.22 PROPAGATION OF PLANE WAVES IN CALCITE CRYSTAL CUT WITH OPTIC AXIS HAVING SPECIAL DIRECTIONS WITH RESPECT TO CRYSTAL SURFACE

The principles of *Huygens*' construction will now be applied to find the O and the E rays in the following three cases in each of which crystal plane is so cut that the optic axis occupies special direction with respect to the refracting surface:

Case I. *Optic axis parallel to the refracting face of the crystal and situated in the plane of incidence.* Suppose AB represents a section of the plane wavefront of monochromatic light incident obliquely on the plane surface AC of calcite crystal plate, both planes being perpendicular to the plane of paper. The plane of incidence is the plane of paper. According to conditions imposed on the orientation of optic axis, a line drawn through any crystal point parallel to AC and the plane of paper represent optic axis for that point. In Fig. 16.24 dotted lines drawn parallel to PQ represent the direction of optic axis. The O and E wavelets spread to a position shown by the time the edge B of the incident wavefront reaches C. Since the section PQ of the refracting surface by the plane of paper is the optic axis for the point A, hence the minor axis of the ellipsoidal wavelet (E wavelet) lies in the section PQ and its semi-major axis is perpendicular to PQ. Thus, the section of the E wavelet by the plane of incidence is a semi-ellipse, while the section of the spherical wavelet is a semi-circle, which touches the semi-ellipse at the points of intersection of optic axis (through A) with the semi-ellipse. This construction can be further verified by taking the axis of Y along PQ, the optic axis, and the axis of X perpendicular to it and in the plane of incidence. Then, the portion of Fig. 16.22 (*a*) to the right of YY will represent, for the case under consideration, the section of the O and E wavelets in the plane of incidence. The O and E rays lie in the plane of incidence but for different values of the angle of incidence the ratio $\sin i/\sin r_E$ is not constant for the E ray.

Figure 16.24 (*b*) illustrates a similar construction for normal incidence of the wavefront. Here, the O and E rays are perpendicular to the ordinary wavefront OO' and the extraordinary wavefront EE', the two wavefronts being parallel to the crystal surface. The O and E rays pursue the same path but since in this case the velocity of E wave is greater than that of the O wave, these emerge through the crystal with a certain path difference.

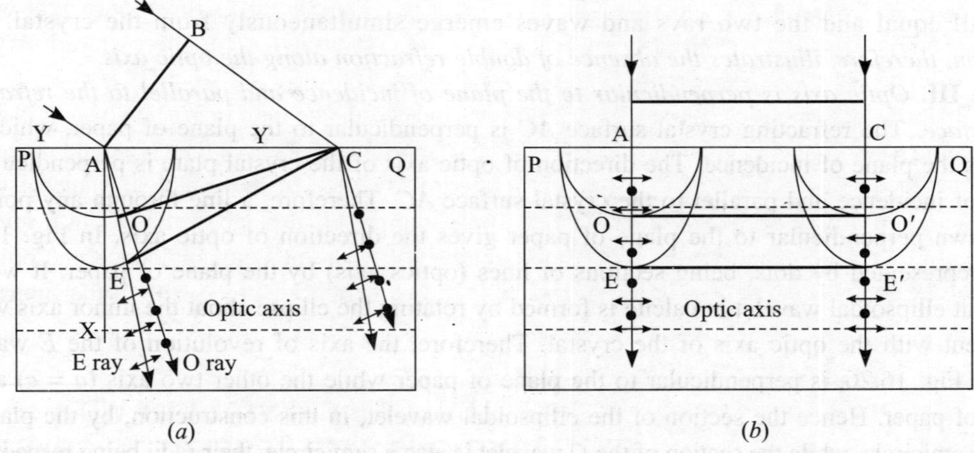

Fig. 16.24. Double refraction of plane waves in calcite crystal cut parallel to the optic axis

Case II. *Optic axis lies in the plane of incidence and perpendicular to the refracting surface.* In Fig. 16.25, AC is the trace of the boundary of plane refracting crystal surface, supposed perpendicular to the plane of paper, which we consider as the plane of incidence. A line through any point of the crystal drawn parallel to the plane of paper and perpendicular to AC, therefore, gives the direction of optic axis in the crystal plate. As before, the O and E wavelets from A spread to positions shown by the time the edge B of the incident wavefront reaches C. The semi-minor axis of the ellipsoidal wavelet, which is along the optic axis, is perpendicular to the crystal surface while the major axis, lies in AC. Thus, the section of the E wavelet by the plane of incidence (plane of paper) is a semi-ellipse as shown, while the section of O wavelet is a semicircle which touches semi-ellipse at the end of minor axis.

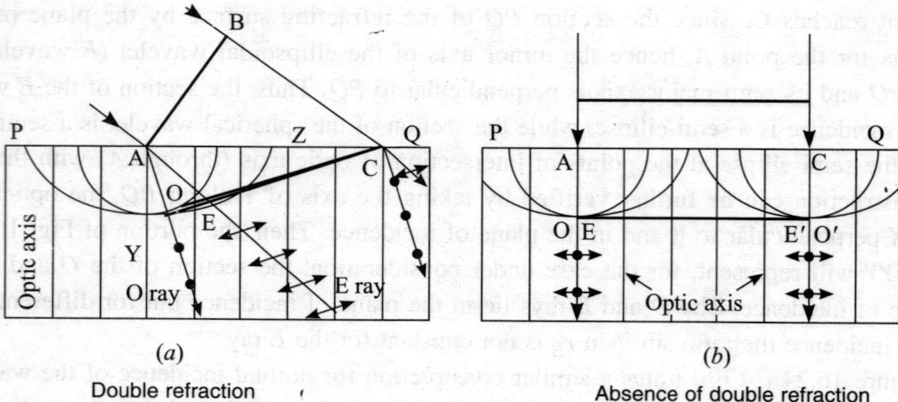

(a)
Double refraction

(b)
Absence of double refraction

Fig. 16.25

AO is the ordinary ray while *AE* is the extraordinary ray. In this construction also, for different values of the angle of incidence, the ratio sin i/sin r_E is not constant for extra-ordinary ray.

Figure 16.25 (*b*) illustrates the similar construction for normal incidence of the wavefront on the crystal surface *AC*. It is obvious from Fig. 16.25 (*b*) that ray and wave velocities of the *O* and *E* rays are all equal and the two rays and waves emerge simultaneously from the crystal. *This construction, therefore, illustrates the absence of double refraction along the optic axis.*

Case III. *Optic axis is perpendicular to the plane of incidence and parallel to the refracting crystal surface.* The refracting crystal surface *AC* is perpendicular to the plane of paper, which we consider as the plane of incidence. The direction of optic axis of the crystal plate is perpendicular to the plane of incidence and parallel to the crystal surface *AC*. Therefore, a line through any point of crystal drawn perpendicular to the plane of paper gives the direction of optic axis. In Fig. 16.26, these are represented by dots, being sections of lines (optics axis) by the plane of paper. It will be recalled that ellipsoidal wavelet in calcite is formed by rotating the ellipse about the minor axis which is coincident with the optic axis of the crystal. Therefore, the axis of revolution of the *E* wavelet from *A*, in Fig. 16.26, is perpendicular to the plane of paper while the other two axis ($a = c$) are in the plane of paper. Hence the section of the ellipsoidal wavelet, in this construction, by the plane of paper, is a semicircle, while the section of the *O* wavelet is also a semicircle, their radii being respectively the semi-major axis and semi-minor axis of the ellipse which generates the *E* wavelet. The points of tangency of the *O* and *E* wavelets lie below and above the plane of paper. The *E* and *O* wavelets spread to positions shown at the instant the end *B* of the incident wavefront reaches *C*. the ordinary wavefront, *CO*, is the tangent plane from *C* to *O* wavelet and similarly the extraordinary wavefront *CE* is the tangent plane from *C* to the *E* wavelet. *AO* and *AE* are respectively the ordinary and extra-ordinary rays from *A*.

This construction reveals that in the plane of incidence the speed of the *E* ray is the same in all directions and the extraordinary ray pursues the path normally to the extraordinary wavefront. Therefore, calcite has a constant refractive index for the *E* ray in this particular construction. Thus, both the laws of refraction are obeyed by the *E* ray in this construction. Furthermore, it will be observed that the plane of incidence, the principal plane of the *E* ray and the principal plane of the *O* ray do not coincide each other. The principal plane of the *E* ray is the plane containing *AE* and the direction of optic axis; here it is perpendicular to the plane of incidence. The vibrations of the *E* ray are tangential to its wave surface and in the principal plane of the *E* ray. Hence the direction of vibrations of the *E* ray are perpendicular to the plane of incidence (plane of paper) and are, therefore, represented in Fig. 16.26 by dots on *AE*—sections of vibrations by the plane of incidence. The principal plane of the *O* ray is defined in an exactly similar way as that of the *E* ray. In this case, it is also perpendicular to the plane of incidence. The vibrations of the *O* ray are executed perpendicular to principal plane and tangential to the *O* wave surface. They are shown by lines in the plane of paper and perpendicular to the *O* ray.

Figure 16.25 (*b*) illustrates a similar construction for the normal incidence of the wavefront on the crystal surface. It will be observed that for normal incidence, in all three cases of special orientation of optic axis, the ordinary as well as extraordinary light waves are not refracted, though only in case II double refraction does not occur and in other cases it does.

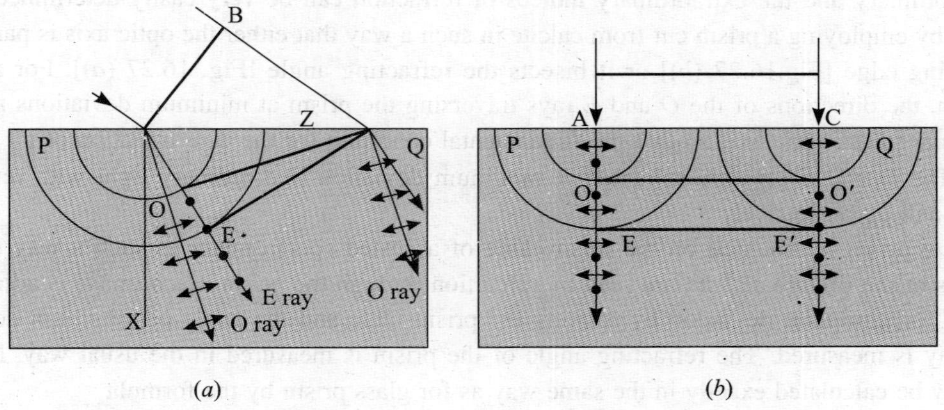

Fig. 16.26. Illustrates double refraction of plane wave such that both the laws of refraction are obeyed by the *E* ray

16.23 INDICES OF REFRACTION FOR UNIAXIAL CRYSTALS

The index of refraction of a medium is usually defined as the ratio of the velocity of light in vacuum to the velocity in the refracting medium under consideration. In uniaxial crystals like calcite, there are two principal indices of refraction, one corresponding to the propagation of the *O* ray and the other corresponding to the propagation of the *E* ray in a direction perpendicular to the optic axis. The two indices are respectively known as *ordinary and extraordinary indices of refraction.* The refractive index μ_o of the crystal for the *O* ray is defined as,

$$\mu_o = \frac{\text{Velocity in vacuum}}{\text{Velocity of the } O \text{ wave}}$$

The *E* ray obeys ordinary laws of refraction in a plane perpendicular to the optic axis of the crystal and, as shown in Fig. 16.26, in this plane the velocity of *E* ray is the same as that of the *E* wave, which is maximum for negative crystals and least for positive crystals. For *negative crystals like calcite*, the principal index of refraction for the *E* ray is, therefore, defined as

$$\mu_E = \frac{\text{Velocity in vacuum}}{\text{Maximum velocity of the } E \text{ wave}}$$

For *positive uniaxial crystals*, like quartz, the principal index of refraction for the *O* ray is defined exactly in a similar way as for uniaxial negative crystals but the principal index of refraction for the *E* ray is defined as

$$\mu_E = \frac{\text{Velocity in vacuum}}{\text{Minimum velocity of the } E \text{ wave}}$$

μ_o and μ_E vary with the wavelength of light.

Experimental Determination of μ_o and μ_E

The ordinary and the extraordinary indices of refraction can be very easily determined in the laboratory by employing a prism cut from calcite in such a way that either the optic axis is parallel to the refracting edge [Fig.16.27 (*b*)] or it bisects the refracting angle [Fig. 16.27 (*a*)]. For a given wavelength, the directions of the *O* and *E* rays traversing the prism at minimum deviations must be perpendicular to the optic axis so that the fundamental condition for the determination of μ_E may be satisfied. The *O* and *E* rays are refracted at minimum deviation like ordinary light with refractive indices μ_o and μ_E respectively.

Calcite prism is mounted on the prism-table of adjusted spectrometer in such a way that the two images of the illuminated slit are seen by refraction through the prism. Each image is adjusted in succession for minimum deviation by rotating the prism table and the angle of minimum deviation for each ray is measured. The refracting angle of the prism is measured in the usual way. Now μ_o and μ_E may be calculated exactly in the same way as for glass prism by the formula

$$\mu_o = \frac{\sin\frac{1}{2}(A + \delta_{mO})}{\sin\frac{1}{2}A}; \quad \mu_E = \frac{\sin\frac{1}{2}(A + \delta_{mE})}{\sin\frac{1}{2}A},$$

where δ_m is the angle of minimum deviation for the *E* ray and *A* is the refracting angle of the prism.

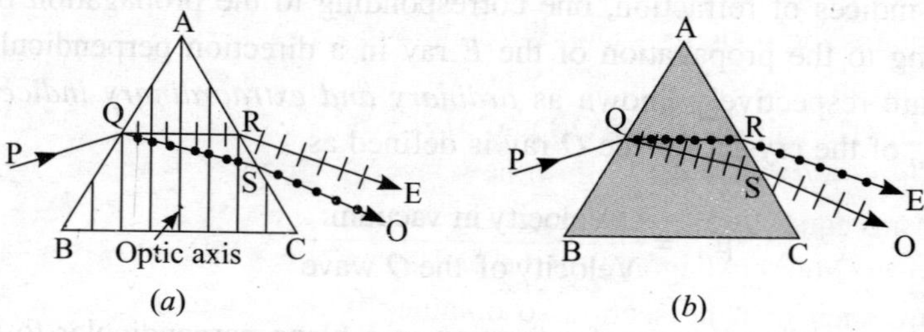

Fig. 16.27. Double refraction in prisms cut from uniaxial crystals

16.24 THE NICOL PRISM

Nicol's prism is an ingenious optical device, designed from a crystal of calcite, and until recently much employed in many optical instruments for producing and analysing plane polarised light, but now largely replaced by polaroids. It will be recalled that when a beam of ordinary light is transmitted through a crystal of calcite, in general, two completely plane polarised beams, the *O* wave and the *E* wave, with vibrations in two mutually perpendicular planes are obtained. Consequently, if some means can be found to eliminate one beam, calcite may be used to obtain plane polarised light from natural light. The elimination of the ordinary wave was accomplished in an ingenious manner by *William Nicol* in 1828 by utilising the phenomenon of total reflection at a thin film of Canada balsam separating the two pieces of calcite. His device is known after his name as Nicol's prism (which is really not a prism but a parallelopipedon).

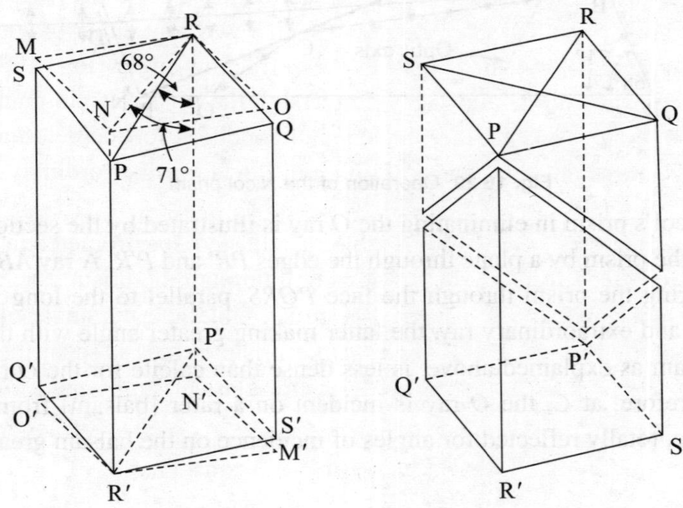

Fig. 16.28. The Nicol prism

To construct a Nicol prism, a calcite rhombohedron of length three times its width is produced by cleavage from a clear natural crystal. At the corners *N* and *N'* (Fig. 16.28) of the crystal there are three obtuse angles of the three faces meeting at these ends. An imaginary plane through the edges *NR'* and *N'R* contains the optic axis and this is, therefore, the principal section of the crystal (*NRN'R'* in Fig. 16.28). The angles *NRN'* and *N'R'N* are equal to 71° in the crystal produced by cleavage from the natural crystal. The end faces *MNOR* and *M'N'O'R'* are cut artificially by planes perpendicular to the principal section *NRN'R'* so as to form new faces (*PQRS* and *P'Q'R'S'*) in such a way that angles *PRP'* and *P'R'P* are equal to 68°. The crystal is then cut along a plane passing through *P* and *P'* and perpendicular to the shorter diagonal of the end faces (*PR* and *P'R'* in Fig. 16.28). The dimensions of the crystal are such that the section so obtained is perpendicular to the principal section of the crytstal. The cut surfaces are polished to optical flatness and cemented together with a thin layer of *Canada balsam* which is a transparent non-doubly refracting plaster with an index of refraction intermediate between the indices of refraction of calcite for the *O* and *E* rays. For sodium yellow light at mean wave-length λ = 5893 A.U., the values of indices are:

Index of calcite for *O* ray, $\mu_o = 1.65836$

Index of Canada balsam, $\mu = 1.55$

Index of calcite for *E* ray, $\mu_E = 1.48641$

Since $\mu_B > \mu_E$, Canada balsam is optically more dense than calcite for *E* ray; and since $\mu_B < \mu_o$ it is optically less dense than calcite for *O* ray. The end faces of the prism are left clear while the remaining faces are painted black.

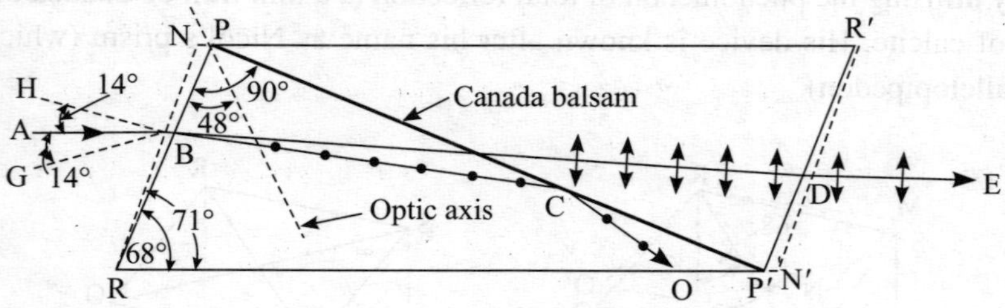

Fig. 16.29. Operation of the Nicol prism

The action of Nicol's prism in eliminating the *O* ray is illustrated by the sectional diagram where *PRP'R'* is a section of the prism by a plane through the edges *PR'* and *P'R*. A ray *AB* of monochromatic unpolarised light entering the prism through the face *PQRS*, parallel to the long sides of the prism, splits into an ordinary and extraordinary ray, the latter making greater angle with the optic axis (paths *BC*, *BD*). Canada balsam as explained above, is less dense than calcite for the *O* ray and denser than calcite for *E* ray. Therefore, at *C*, the *O* ray is incident on a rarer [balsam] from a denser [calcite] medium. It is, therefore, totally reflected for angles of incidence on the balsam greater than the critical angle, viz.

$$\theta = \sin^{-1}\frac{\mu_B}{\mu_b} = \sin^{-1}\frac{1.55}{1.65} = 69°,$$

for the two media. This is accomplished simply due to the mode of construction and dimensions of the prism. After total reflection it pursues the path *CO* and finally effectively absorbed by the blackened side of the prism. The *E* ray, on the other hand, is passing from rarer (calcite) into a denser (balsam) medium and consequently it is transmitted through the balsam, simply, suffering a slight deviation in its path. Finally the *E* ray emerges from the prism parallel to the incident ray. The prism, therefore, only transmits the *E* ray which is plane polarised with vibrations in the principal section. In the end on view, these vibrations are parallel to the shorter diagonal of the end face *P'Q'R'S'* (Fig. 16.30).

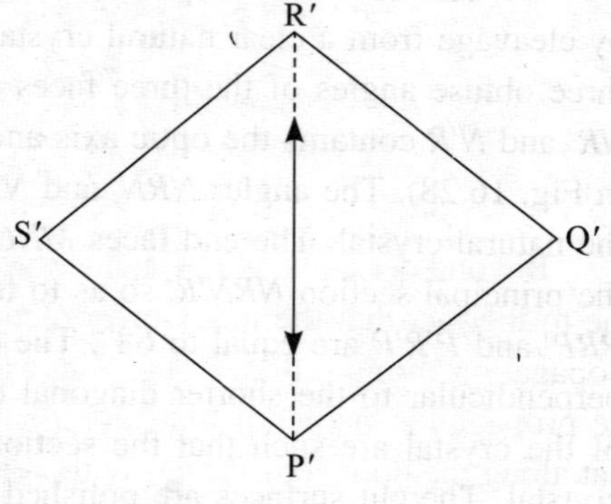

Fig. 16.30

Although Nicol prisms are very good polarisers throughout the region of the visible spectrum yet it should be emphasised that they fail to be effective polarisers with too convergent or too divergent incident beam. On one hand the angular limit of the incident ray is limited to $\angle ABG$ of about 14° (Fig. 16.29), for at angles greater than this some of the *O* ray will also be transmitted through Canada balsam; and on the other side direction of incident light is limited to $\angle ABH$ also nearly equal to 14°, for at angles greater than this, the corresponding *E* ray is also totally reflected, as explained in the sequence.

Referring to Fig. 16.31 it is clear that for ray incident along *AB*, the angle of incidence is 22°; the corresponding angle of refraction for *O* ray may be calculated.

$$\sin r_o = \frac{\sin i}{\mu_o} = \frac{\sin 22°}{1.65} = \frac{0.3746}{1.65} = 0.2270$$

$$r_o = \sin^{-1}(0.2270) = 13° \text{ approximately}$$

Hence $$\angle BCF = 90 - r_o = 77°$$

which is greater than the critical angle 69° for the *O* ray. But if the angle of incidence be increased by 15°, then the corresponding angle of refraction r'_o is given by

$$\sin r'_o = \frac{\sin i'}{\mu_o} = \frac{\sin(22+15)}{1.65} = \frac{0.6018}{1.65} = 0.3647$$

$\therefore$ $$r'_o = \sin^{-1}(0.3647) = 21.4°$$

and now the corresponding $\angle BCF = 90 - r_o' = 68°.6$, which is less than the critical angle 69°. Therefore some of the *O* ray is transmitted through balsam along with the *E* ray. Thus, we can say that the maximum value of $\angle ABA'$ is about 14°, beyond which as explained, some of the *O* ray is also transmitted.

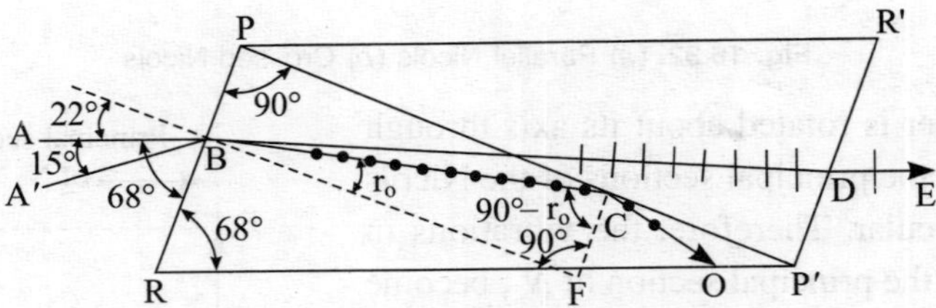

Fig. 16.31

Beyond certain angle of incidence the *E* ray will be totally reflected by the balsam. This is simply due to the fact that the index of refraction of calcite varies for *E* ray, depending upon its direction of propagation through the crystal. The extreme values of refractive index are 1.48641 and 1.65836 for the propagation perpendicular and along the optic axis. The latter value is due to the simple reason that along the optic axis the *E* ray travels with the same speed as the *O* ray. For intermediate direction, the effective index lies between these two limits viz. 1.48641 and 1.65836. There will, therefore, be a maximum angle of incidence *ABH* of the original light beyond which the balsam will be optically less dense than calcite for the E ray and thus there will be total reflection of E vibrations also. The dimensions of Nicol's prism are such that this angle also lies in the neighbourhood of 14°.

Thus to avoid the total reflection of the E ray and the transmission of the O ray, the angle between the extreme rays of light incident on Nicol is limited to about 28°.

16.25 PARALLEL AND CROSSED NICOLS

Nicol's prism can be used both as polariser and as analyser. When two Nicols are mounted co-axially, as shown in Fig. 16.32, then the first Nicol N_1, which produces plane polarised light, is called the polariser while the second Nicol N_2, which analyses the incoming light is called the analyser. When the principal sections of the two Nicols are parallel, then the vibrations in the E ray, which are in the principal section of the polariser, are also in the principal section of the analyser. Consequently the E ray from N_1 is freely transmitted by N_2 just as it was freely transmitted by the first. [Fig. 16.32 (*a*)]. In this setting of the combination—technically known as Parallel Nicols the intensity of the field of view is maximum.

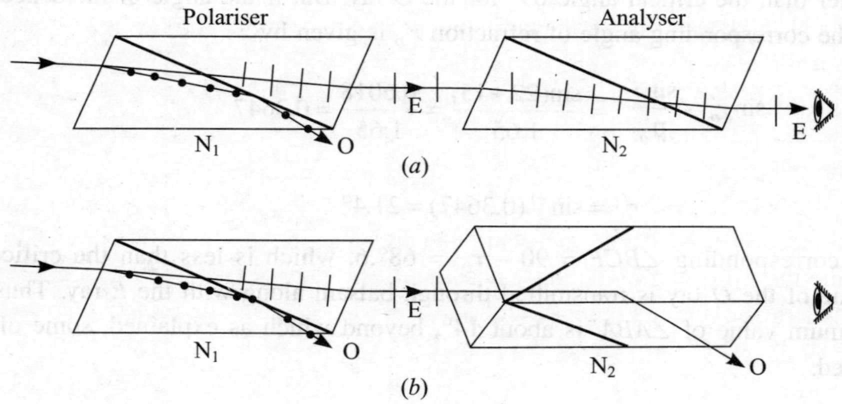

Fig. 16.32. (*a*) Parallel Nicols (*b*) Crossed Nicols

When the analyser is rotated about its axis through 90° from this position, the principal sections of the Nicols are mutually perpendicular. Therefore, the vibrations in the E ray, which are in the principal section of N_1, become perpendicular to the principal section of N_2. The E ray from the polariser, consequently, enters the analyser as O ray and therefore, it is totally reflected from the balsam, exactly in the same way as O vibrations were totally reflected in the polariser. Thus no light is transmitted by this setting of the combination, which is technically known as *Crossed Nicols*.

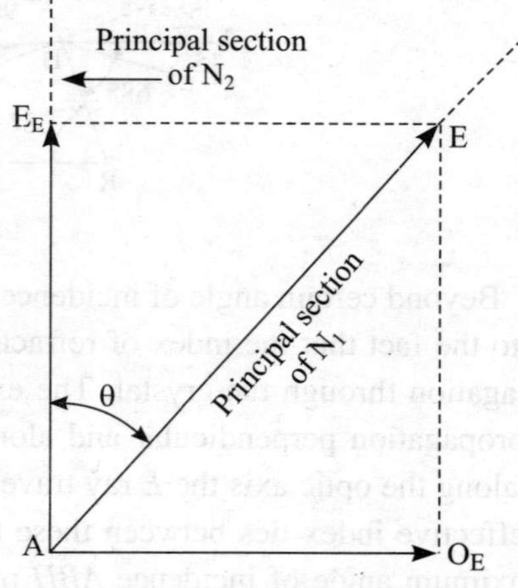

Fig. 16.33

After a rotation of 180° from the first position of the analyser the principal sections are again parallel and E vibrations are again transmitted by the analyser. After a rotation of 270°, Nicols are again crossed and no light is transmitted by the combination.

In the intermediate position between the crossed and parallel settings of the combination, some light is transmitted through the analyser. The E vibrations from the polariser are resolved, just on entering the analyser, into two component vibrations—the ordinary component $a \sin \theta$, which is perpendicular to principal section of the analyser, is totally reflected while the extraordinary component $a \cos \theta$, which is in the principal section is freely transmitted by the analyser. Thus for intermediate positions, the intensity of light is proportional to the square of the cosine of the angle θ between the principal section of the analyser and the polariser.

16.26 POLARISATION BY SELECTIVE ABSORPTION

The production of plane polarised light by selective absorption is exhibited by certain class of doubly refracting crystals, which not only produce two internal beams polarised at right angles to each other but also absorb one of the polarised components much more strongly than the other. Hence if crystal is cut of the proper thickness, one of the components is completely extinguished by absorption while the other is transmitted in appreciable amount as shown in Fig. 16.34. This phenomenon is known by the name *dichroism* and crystals exhibiting this property are said to be *dichroic*. The best known dichroic mineral crystal is tourmaline and the Fig. 16.34 illustrates its action. A beam of ordinary light incident normally on a thin crystal of tourmaline, cut with its faces parallel to the optic axis is broken up into O and E rays, traversing the crystal in the same direction but with vibrations respectively in and perpendicular to the plane of incidence. The O beam is completely absorbed in the crystal, while the E beam, only slightly absorbed, is transmitted, as far as the green region of the spectrum is concerned. The emergent polarised light can be analysed by the second tourmaline as shown in Fig. 16.35, which obviously explains the experiment with two tourmalines described in *Sec.* 16.3

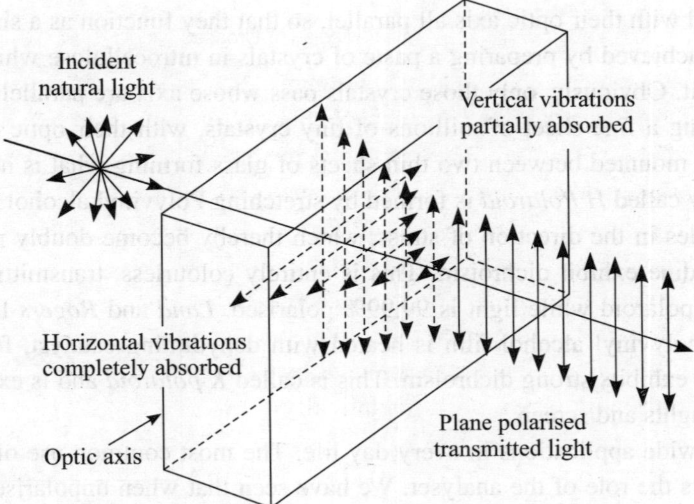

Fig. 16.34. Plane polarised light transmitted by a dichroic crystal

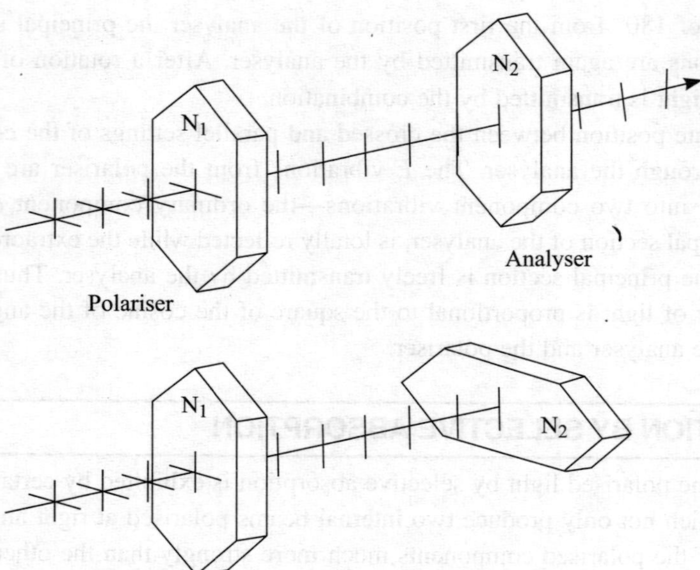

Fig. 16.35 Tourmaline as polariser and analyser

The use of tourmaline as polariser is limited, as the plane polarised transmitted light is coloured due to unequal absorption of E rays of various wavelengths.

16.27 POLAROID

In 1852, the English Physician, *W.H. Herapath* succeeded in producing small crystals of *idosulphate of quinine*, called *herapathite* after its discoverer, which exhibit strong dichroism—they absorb completely one component of polarisation while transmitting the other with little loss. These crystals are ultra-microscopic, pulverising easily when subjected to relatively small stress and so they are quite useless as such. In 1932, *E.H. Land* developed a process which arranges herapathite crystals side by side, oriented with their optic axis all parallel, so that they function as a single crystal of large dimensions. This is achieved by preparing a paste of crystals in nitrocellulose which is then squeezed out through a fine slit. Obviously, only those crystals pass whose axis are parallel to the length of the slit, thereby producing a fine sheet of millions of tiny crystals, with their optic axis parallel to one another. This is then mounted between two thin sheets of glass forming what is now called *Polaroid*.

Another variety called *H Polaroid* is formed by stretching Polyvinyl alcohol films so as to orient the complex molecules in the direction of stress, which thereby become doubly refracting and when impregnated with iodine exhibit dichroism. This is entirely colourless, transmitting 33% more light than the herapathite polaroid while light is 99.99% polarised. *Land* and *Rogers* later discovered that when the stretched polyvinyl alcohol film is heated with dehydrating catalyst, for example, HCl, it slightly darkness but exhibits strong dichroism. This is called *K polaroid* and is extensively employed in automobile head lights and screen.

Polaroids find wide applications in every day life. The most common use of polaroid is in *sun-glasses* where it plays the role of the analyser. We have seen that when unpolarised light is reflected, vibrations perpendicular to plane of incidence are predominantly reflected. Accordingly, when sunlight is reflected from horizontal surfaces such as wet pavements, the plane of incidence is vertical and in the reflected light horizontal vibrations predominate the vertical vibrations. The direction of transmission of polaroid is, therefore, kept vertical to cut off completely the horizontal reflected vibrations, while objects are seen by diffusely reflected light, which is, however, unpolarised.

Polaroids are fitted in motor car head-lights and wind screen, with their planes of transmission inclined at 45° to vertical. The driver can see the light of his own head-light while that from oncoming automobile is cut off. However, road and other objects are visible by light scattered by them, which is unpolarised and a part of it can pass through the wind screen polaroid.

Polaroids are used in windows of railways, aeroplanes etc. in which there is a fixed outer and rotable inner polaroid disc for controlling the amount of light entering it from outside.

Polaroids are employed to view stereoscopic motion picture in which the three dimensional effect is obtained by first taking two pictures of the same view from slightly different angles—one showing the scene exactly as viewed by the right eye and the other as viewed by the left eye. These pictures are projected side by side on the screen but light in the two pictures is polarised at right angles to each other. Pictures are now viewed by the polarised glasses so that each eye views only the picture which had been taken from its own angle.

Polaroid has now largely replaced Nicol prism for producing and analysing plane polarised light in the laboratory.

16.28 POLARISATION BY SCATTERING

The sky is blue while sunset and sunrise are red. Sky light is partially plane polarised. Maximum polarisation is observed in a direction perpendicular to the incident light. This can be easily verified by looking the part of the sky overhead through a Nicol prism or a polaroid. All the three effects are due to scattering of sunlight by obstacles, such as dust particles, free water molecules and even by oxygen molecules, present in the atmosphere. The process of scattering is simply the absorption and reradiation of energy.

The process of scattering of light can be easily explained on the basis of electromagnetic theory according to which in a light wave periodic electric and magnetic vibrations are executed in mutually perpendicular planes. Furthermore, according to Wiener's experiment, the electric vector is responsible for all optical phenomena. Referring to Fig. 16.36, suppose the natural sunlight (unpolarised) is incident on one of the molecules of earth's atmosphere, situated at O just over the observer. Every molecule consists of aggregate of electrically charged particles of opposite signs. Owing to electric field in the incident light beam, positive charges experience a force in the direction of field while on negative charges force is in the opposite direction. Since charges are not rigidly bound in a molecule, a relative displacement of the charges occurs first in one direction and then periodically reversed, owing to periodical reversal of electric field in the light wave. In Fig. 16.36 these directions necessarily lie in YZ plane. An oscillating electric dipole, according to electromagnetic theory, itself radiates electromagnetic waves, which form the scattered light. As in *Sec.* 16.4, we may here also employ the equivalent components E_Y and E_Z for all the E vectors in the incident beam. The result is that the light incident on the molecule at O produces the equivalent two oscillating molecules along the Y axis and the Z axis, the frequency of oscillation is the same as that of the incident light, since the oscillations of electric charges in a molecule are (nearly) in unison with the oscillating electric field. The observer looking along the *Z* axis receives light only from the *Y* component of incident vibration. This light is, therefore, plane polarised with vibrations parallel to *Y* axis. In fact, the light scattered in any direction in a plane transverse to the direction of propagation of incident light is plane polarised. In all other directions scattered light will be only partially polarised.

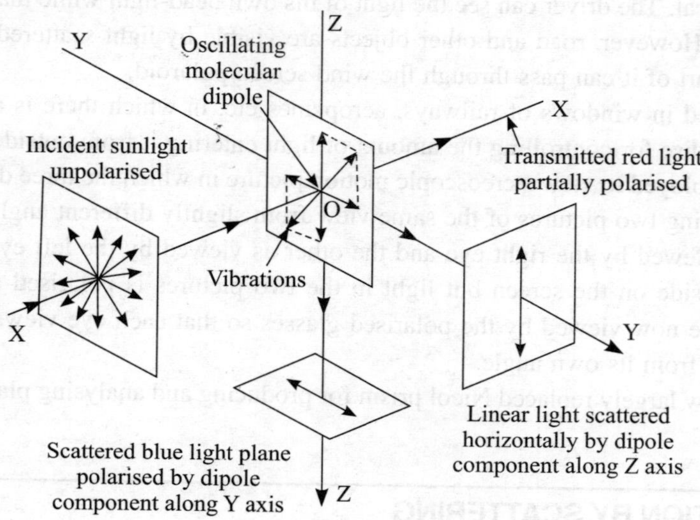

Fig. 16.36. Polarisation by scattering of light by an oscillating molecular dipole

16.29 BLUE COLOUR OF SKY

To explain the blue colour of sky we observe that the vibration of electric charges in a molecule, under the influence of electric field in the incident light, is a forced vibration. Its amplitude is greater and as consequence intensity of scattered light is greater, the closer is the frequency of the incident light to the natural frequency of oscillation of electric charges, which however, is the same as that of a wavelength in the ultraviolet region of the spectrum. The frequencies of light waves in the visible region of spectrum are less than the natural frequency of molecule but the frequency of blue light is very close to it. Therefore, the blue colour in sunlight is scattered more than the red by the molecules in the atmosphere.

Exact theory of scattering process has been worked out by Lord Rayleigh who showed that the percentage of light scattered is inversely proportional to the 4th power of the wavelength when the size of the particles is of the order of magnitude of the wavelength of a light wave. This is almost 16 times greater for the violet end of the spectrum as compared with the scattering of red light. Consequently, the blue contained in the sunlight is scattered to a much greater extent and more intensely than the red.

In reality, looking the sky overhead, the blue light due to scattering of incident sunlight from molecules in the neighbourhood of *O* will be only partially plane polarised, with maximum vibrations perpendicular to the plane *XOZ*. This is due to the fact that the incident light waves are scattered several times by the molecules of the atmosphere before reaching the observer, thus accounting for the blue colour of unclouded sky in all directions. Blue colour is more brilliant when the atmosphere is free from dust particles and water molecules, so scattering from molecules of gases is to a considerable extent responsible for the blue colour of the sky.

If the earth had no atmosphere, we could receive no sky light at earth's surface. The sky would be perfectly black even in day. This may be proved by flying at high altitudes, where there is less atmosphere above the observer.

The sun and the neighbouring sky appear red at the sunrise and at the sunset. We observe them directly, referring to Fig. 16.37 by placing our eye at *E*. The rays coming from the sun have to traverse a greater distance approximately horizontally through the earth's atmosphere before reaching the observer. A large proportion of blue is scattered to side and lost by fine dust and smoke particles near the earth's surface. White light when blue is removed is yellow or red in hue. When such light is incident on the cloud near sunset or sunrise, the light reflected from the cloud has the yellow or red hue.

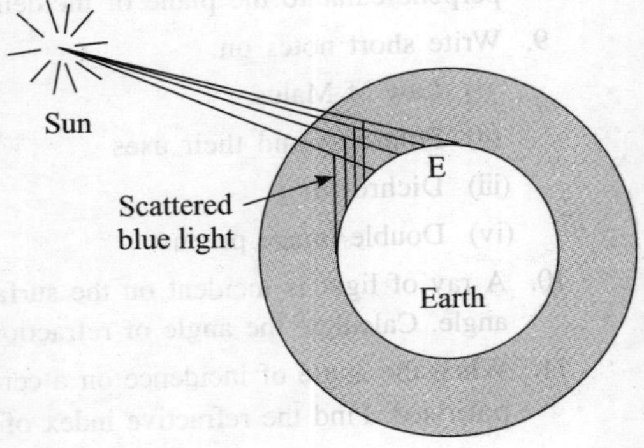

Fig. 16.37

Red light is only partially plane polarised.

PROBLEMS

1. (a) What is meant by the term 'Polarisation of light'?
   (b) Comment on the statement 'Polarisation requires that vibrations are transverse'.
2. (a) What is Brewster's law? Show that at Brewster's angle the reflected ray is perpendicular to the refracted ray.
   (b) Explain how you would obtain and detect a beam of plane polarised light by reflection. How does this prove the transverse character of light vibrations?
3. Describe the construction and use of a Nicol's prism and explain how it produces plane polarised light. Would a similar prism prepared from quartz serve a similar purpose?
4. Describe Polarisation of light by scattering. Explain:
   (a) Blue of the sky.
   (b) Red colour of sunset.
5. Describe different methods for producing plane polarised light.
6. Describe the phenomenon of double refraction in uniaxial crystals. How is double refraction explained by Huygen's theory?

7. Thin plates are cut from a negative crystal possessing the property of double refraction with positions of the optic axis as described below. Explain giving diagrams, the nature of refraction observed in each case.
   (i) Optic axis parallel to the refracting face and lying in the plane of incidence.
   (ii) Optic axis parallel to the refracting face but perpendicular to the plane of incidence.
   (iii) Optic axis perpendicular to the refracting face and lying in the plane of incidence.
8. Explain what is meant by a negative uniaxial crystal. Draw the necessary diagram to show the path of light through the crystal when the optic axis is in the surface of the crystal and perpendicular to the plane of incidence.
9. Write short notes on:
   (i) Law of Malus
   (ii) Polaroids and their uses
   (iii) Dichroism
   (iv) Double image prisms
10. A ray of light is incident on the surface of a glass plate of refractive index 1.5 at the polarising angle. Calculate the angle of refraction of the ray.
11. When the angle of incidence on a certain material is 60°, the reflected light is almost completely polarised. Find the refractive index of the material.

PRODUCTION AND ANALYSIS OF POLARISED LIGHT

Plane polarised light represents a special and relatively simple type of polarisation. We have studied the production and analysis of this kind of light in the preceding chapter. For such light, the light vector varies simple harmonically along a fixed straight line perpendicularly to the direction of propagation. By the superposition of two plane polarised beams of monochromatic light under suitable conditions (see *Sec.* 17.1), the resultant light vector so produced rotates in a plane perpendicular to the direction of propagation and if its magnitude remains constant, the resulting light is spoken as *circularly polarised*. If, however, the magnitude of the resultant light vector varies periodically between its maximum and minimum values during its rotation, its tip would trace out an ellipse and the resulting light is spoken as *elliptically polarised*.

We shall first consider the conditions under which circular and elliptical motions may be accomplished.

17.1 SUPERPOSITION OF TWO PLANE POLARISED WAVES VIBRATING IN TWO MUTUALLY PERPENDICULAR PLANES

Let a beam of plane polarised monochromatic light from a polariser be incident normally on a calcite plate, cut with faces parallel to the optic axis, but oriented in such a way that the linear vibrations

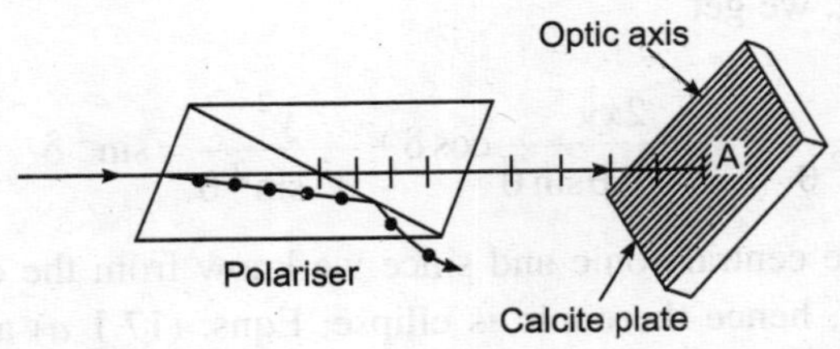

Fig. 17.1. Plane-polarised light incident normally on a crystal plate cut parallel to optic axis

make an arbitrary angle θ with the optic axis. The amplitude a of the incident wave is resolved by the doubly refracting calcite, as shown in Fig. 17.2, into two components (i) $AE = a \cos \theta$ along the optic axis forming the extra-ordinary vibrations, and (ii) $AO = a \sin \theta$ perpendicular to optic axis forming the ordinary vibrations. As they enter the plate, the incident wave and its two components have the same phase. Therefore, they may be expressed by equations.

$$S = a \sin \omega t; \quad x = a \cos \theta \sin \omega t; \quad y = a \sin \theta \sin \omega t$$

From the wave surface diagram for normal incidence of plane polarised light [Fig. 15.24 (*b*)], it follows that the *O* and *E* vibrations traverse the crystal plate without separation *i.e.*, they traverse the same path but in calcite, the *E* wave travels faster than the *O* wave ($\mu_o > \mu_E$). Therefore, the wavelength of the *E* wave $\lambda_E = \lambda/\mu_E$, in the crystal is greater than that of the *O* wave, $\lambda_o = \lambda/\mu_o$, where λ is the wavelength in vacuum. Due to this inequality in their wavelengths, the *O* wave is retarded behind the *E* wave. The two waves, therefore, emerge from the crystal with a certain relative phase difference say δ, for a given wavelength, depending upon the thickness of the plate and the magnitudes of μ_o and μ_E.

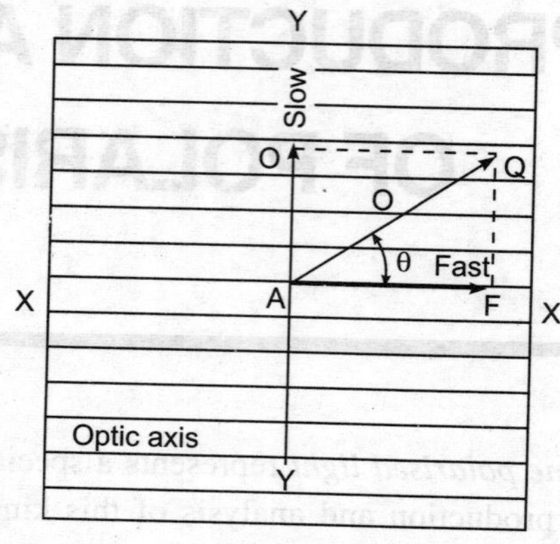

Fig. 17.2

Therefore, as the two waves leave the plate they may be represented by the following equations:

$$x = a \cos \theta \sin(\omega t + \delta) \quad [E \text{ Wave}] \qquad \text{...(17.1 } a)$$

$$y = a \sin \theta \sin \omega t \quad [O \text{ Wave}] \qquad \text{...(17.1 } b)$$

The resultant vibration may be readily found by eliminating *t* between the above two equations. For this purpose we proceed as follows:

Equation (17.1 *a*) may be written in the form

$$x/a \cos \theta = \sin \omega t \cos \theta + \sqrt{1 - \sin^2 \omega t} \sin \delta$$

Substituting the value of sin ω*t* from Eqn. (17.1 *b*) in the above and rearranging, we get

$$\frac{x}{a \cos \theta} - \frac{y}{a \sin \theta} \cos \delta = \sqrt{1 - \frac{y}{a^2 \sin^2 \theta}} \sin \delta$$

Squaring and simplifying, we get

$$\frac{x^2}{a^2 \cos^2 \theta} - \frac{2xy}{a^2 \cos \theta \sin \theta} \cos \delta + \frac{y^2}{a^2 \sin^2 \theta} = \sin^2 \delta \qquad \text{...(17.1 } c)$$

This is an equation of the central conic and since we know from the condition of the problem that *x* and *y* are never infinite, hence the conic is ellipse. Eqns. (17.1 *a*) and (17.1 *b*) show that *x* varies from $+a \cos \theta$ to $-a \cos \theta$ and that *y* varies from $+a \sin \theta$ to $-a \sin \theta$. Hence the ellipse represented by these equations or Eqn. (17.1 *c*) is inscribed in a rectangle with sides of length $2a \cos \theta$ and $2a \sin \theta$ and whose diagonal represents the rectilinear vibrations of the incident wave. Thus, during the passage of the wave, the vibrating particle describes, in general, an ellipse about its position of rest in the plane perpendicular to the direction of propagation. The exact nature of the resultant motion, and therefore of the light emerging from calcite, however, depends upon the value of δ. We consider the following special cases:

(1) *Resultant Vibration Rectilinear—Emergent Light Plane Polarised.* (*a*) If $\delta = 2n\pi$ where $n = 0, 1, 2, 3$ etc. Eqn. (17.1 *c*) reduces to

$$\frac{x^2}{a^2\cos^2\theta} - \frac{2xy}{a^2\cos\theta\sin\theta} + \frac{y^2}{a^2\sin^2\theta} = 0$$

which easily simplifies to the form

$$y = (\tan\theta)x$$

Thus, when the two mutually perpendicular superposed plane polarised waves are in phase, the ellipse degenerates into a straight segment, inclined at an angle θ to the axis of x, coinciding with the diagonal of the rectangle that lies in the first and third quadrants [Fig. 17.3 (*a*)]. We express this fact by saying that the light emerging from the crystal is plane polarised, vibrating in the same plane as that of the plane polarised light [*QAQ* in Fig. 17.3 (*a*)] incident on the plate.

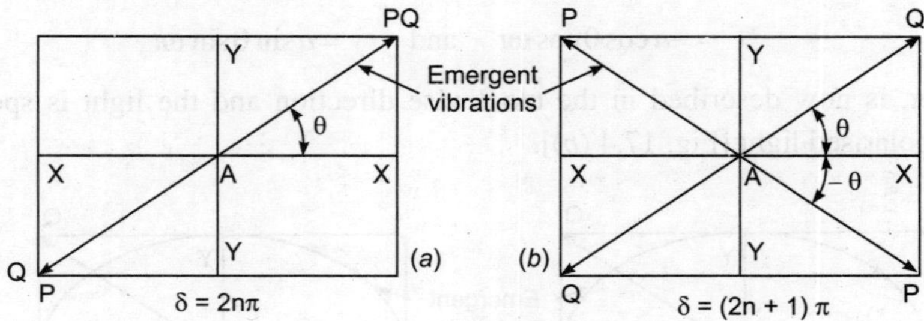

Fig. 17.3. Emergent light plane polarised corresponding to $\delta = 2n\pi$ and $\delta = (2n + 1)\pi$

(*b*) If $\delta = (2n + 1)\pi$ where $n = 0, 1, 2, 3$ etc. Eqn. (17.1 *c*) easily reduces to the form

$$y = -\tan\theta \cdot x = \tan(-\theta)x$$

The light emerging from the crystal plate is again linearly polarised, but now the vibration *PAP* is parallel to the diagonal of the rectangle that lies in the second and fourth quadrants [Fig. 17.3 (*b*)]. It is, therefore, inclined at an angle 2θ with the plane of vibration [*QAQ* in Fig. 17.3 (*b*)] in the plane polarised light incident on calcite plate.

(2) *Resultant Motion Elliptical —Emergent Light Elliptically Polarised.* If δ is an odd multiple of $\pi/2$ *i.e.*, $\delta = (2n + 1)\pi/2$, the product term in Eqn. (17.1 *c*) vanishes and the equation reduces to

$$\frac{x^2}{a^2\cos^2\theta} + \frac{y^2}{a^2\sin^2\theta} = 1.$$

This is an equation to ellipse, its principal axes $2a \cos \theta$ and $2a \sin \theta$ are parallel and perpendicular to the optic axis of the crystal plate [*XAX* in Fig. 17.4 (*a*)]. The light emerging from the crystal plate is, therefore, elliptically polarised; the tip of the resulting light vector continuously sweeps out an ellipse in a plane perpendicular to the direction of propagation. At any one time the direction of the resultant varies from point to point along the direction of propagation. It has a space-periodicity λ.

If $\delta = \pi/2, 5\pi/2, 9\pi/2$ etc. the components of the elliptic motion along the major and the minor axis are obtained by putting this value in Eqns. (17.1 *a*) and (17.1 *b*).

$$x = a\cos\theta\cos\omega t \text{ and } y = a\sin\theta\sin\omega t$$

The ellipse is described counter-clockwise with respect to an observer towards whom the wave travels and the light is spoken as left-handed elliptically polarised light [Fig. 17.4 (*a*)]. This statement can be easily verified by plotting the above equation on a graph paper. For example, we proceed as follows:

(*i*) When $\omega t = 0$, $x = a \cos \theta$ and $y = 0$
(*ii*) When $\omega t = \pi/4$, $x = a \cos \theta/\sqrt{2}$ and $y = a \sin \theta/\sqrt{2}$
(*iii*) When $\omega t = \pi/2$, $x = 0$ and $y = a \sin \theta$

The reader can draw his own graph.

But, if $\delta = 3\pi/2, 7\pi/2, 11\pi/2$ etc. the components of elliptical motion along the major and minor axis are

$$x = -a\cos\theta\cos\omega t \quad \text{and} \quad y = a\sin\theta\sin\omega t$$

the ellipse, however, is now described in the clockwise direction and the light is spoken as right-handed elliptically polarised light [Fig. 17.4 (*b*)].

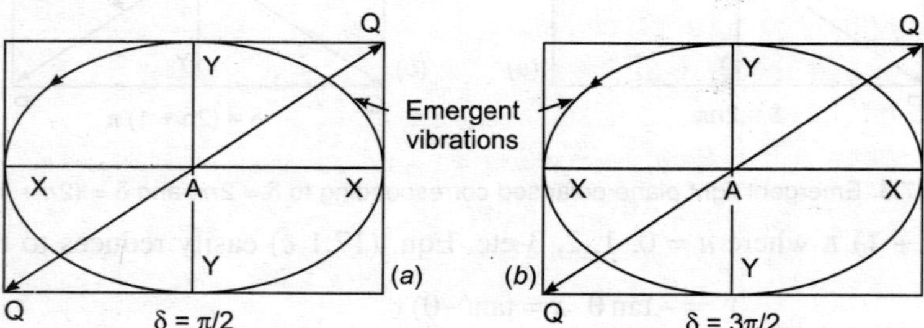

Fig. 17.4. Emergent light elliptically polarised with principal axes along and perpendicular to optic axis

For every value of δ, other than $n\pi$ and $(2n + 1)\pi/2$, the principal axes of the ellipse are inclined to x and y axis.

(3) *Resultant Motion Circular — Emergent Light Circularly Polarised.* In the special case when $\theta = 45°$ in addition to $\delta = (2n + 1)\pi/2$, the ellipse reduces to a circle,

$$x^2 + y^2 = a^2/2$$

and the emergent light is said to be circularly polarised. In this case, the optical vector representing the optical disturbance at a given point in space rotates with uniform angular speed without change of magnitude.

When $\delta = \pi/2$, $5\pi/2$, $9\pi/2$ etc. the circle is described counter-clockwise with respect to an observer towards whom the wave travels [Fig. 17.5 (*a*)]. But if $\delta = 3\pi/2$, $7\pi/2$, $11\pi/2$, etc. circle is described in the clockwise direction [Fig. 17.5 (*b*)].

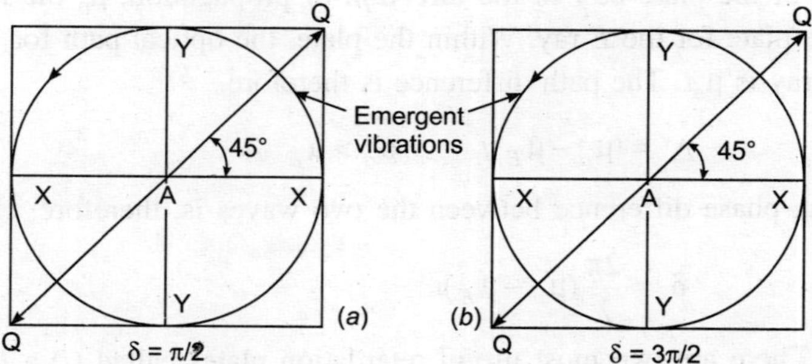

Fig. 17.5 Emergent light circularly polarised when $\theta = 45°$ and $\delta = (2n + 1)\pi/2$

It is now easy to discuss the change in the polarisation of the plane polarised beam as it penetrates deeper and deeper normally to optic axis within the crystal plate, cut with faces parallel to the optic axis. This is illustrated in Fig. 17.6 for a special case when the plane of vibration of the incident linear light is inclined at 45° to the principal plane of the crystal. At the point of incidence ($t = 0$), the components E and O of the incident vibration are in phase ($\delta = 0$), hence equivalent to the original linear vibration. In the 2nd figure (Fig. 17.6) the thickness traversed by the incident beam is such that a phase difference of $\delta = \pi/4$ is introduced between the two components. Hence at this point of the crystal, the two components combine to form an elliptically polarised light. In the third figure, the phase difference advances to $\pi/2$ and since $\theta = 45°$, hence at this point of the crystal the light is circularly polarised. Thus, as the beam penetrates deeper and deeper δ gradually increases until in the fifth figure $\delta = \pi$ is reached and again we get plane polarised light but the plane of vibration is coincident with the diagonal of the square that lies in the second and fourth quadrants. As the beam penetrates further; the resultant vibrations go through the same cycle of figures but the figures are now described in the clockwise direction with respect to an observer towards whom the light travels, and ultimately when $\delta = 2\pi$ the resultant vibration is linear in the same plane as that of the incident beam. Beyond this thickness, the figures go through the same cycle and again the first figure is repeated when $\delta = 4\pi$ radians and so on.

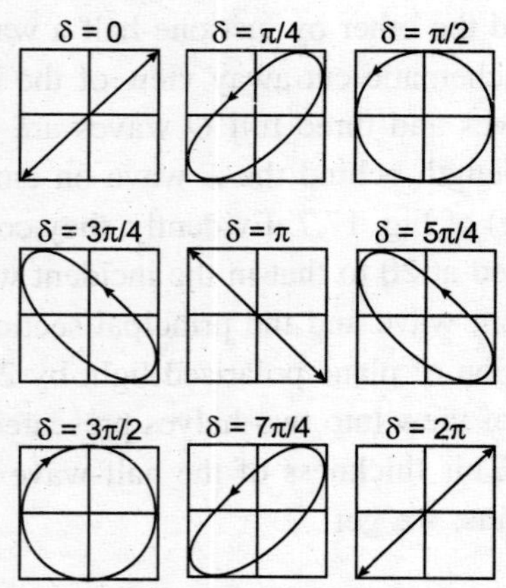

Fig. 17.6. States of polarisation corresponding to different values of the phase difference from 0 to 2π and $\theta = 45°$

In concluding our discussion of the superposition of two plane polarised coherent light waves vibrating in two mutually perpendicular planes, it should be emphasised that they never produce destructive interference, no matter whatever the phase difference.

17.2 RETARDATION PLATES

A plate cut from a doubly refracting crystal by sections parallel to the optic axis and employed to introduce a given phase difference between the ordinary and extraordinary waves in transmission normally through it, is called a *retardation plate*. This phase difference may be deduced as follows:

Let the thickness of the plate be t in the direction of propagation, μ_o the index for the O ray and μ_E the index of the plate for the E ray. Within the plate, the optical path for the E ray is simply $\mu_E t$ and that for the O ray is $\mu_o t$. The path difference is therefore.

$$\Delta = (\mu_o - \mu_E)t \qquad \mu_o > \mu_E$$

The corresponding phase difference between the two waves is, therefore, expressed by

$$\delta = \frac{2\pi}{\lambda}(\mu_o - \mu_E)t \qquad \text{...(17.2 } a\text{)}$$

Half-wave plate. There are two most useful retardation plates called (*i*) a *half-wave plate* and (*ii*) a *quarter-wave plate*. The former is of such a thickness that in traveling through the plate a relative phase difference of π comes in between the O and E waves and accordingly one wave drops behind the other by just one-half a wavelength. This action is illustrated in more details in Fig. 17.7. In a schematic cut away view of the half-wave plate, shown in part (*c*) of Fig. 17.7, two and a half E waves and three full O waves are marked, thereby indicating that the O wave is just one-half a wavelength behind the E wave on emergence. The two waves on emergence are shown in parts (*d*) and (*e*) of Fig. 17.7. Evidently, they combine to form a linearly polarised light with vibration direction inclined at 2θ to that in the incident wave, where θ is the angle between the plane of vibration of the incident wave and the principal section of the plate. Owing to this property of rotating the plane of vibration of plane polarised light by 2θ, this plate is employed in *Laurent's polarimeter* to divide the field of view into two-halves presented side by side.

This thickness of the half-wave plate may be computed by substituting $\delta = \pi$ in Eqn. (17.2 *a*) and thus, we get

$$t = \frac{\lambda}{2(\mu_o - \mu_E)} \qquad \text{(Calcite plate)} \qquad \text{...(17.2 } b\text{)}$$

In particular, a plate whose thickness satisfies the equation $t(\mu_o - \mu_E) = (2n + 1)\lambda/2$ behaves like a *half-wave plate*.

Quarter-wave plate. It is of such a thickness that a relative phase difference of $\pi/2$ comes in between the O and E waves in transmission through its full length. The physical conception of this is that one wave lags behind the other by just one quarter of a wavelength on emergence. The quarter-wave plate is obviously only half as thick as the half-wave plate. For a given wavelength, thickness of this plate may be computed from the relation

$$t = \frac{\lambda}{4(\mu_o - \mu_E)} \qquad \text{(Calcite plate)} \qquad \text{...(17.2 } c\text{)}$$

In particular, a plate whose thickness satisfies the equation $t(\mu_o - \mu_E) = (4n + 1)\lambda/4$ behaves like a quarter-wave plate.

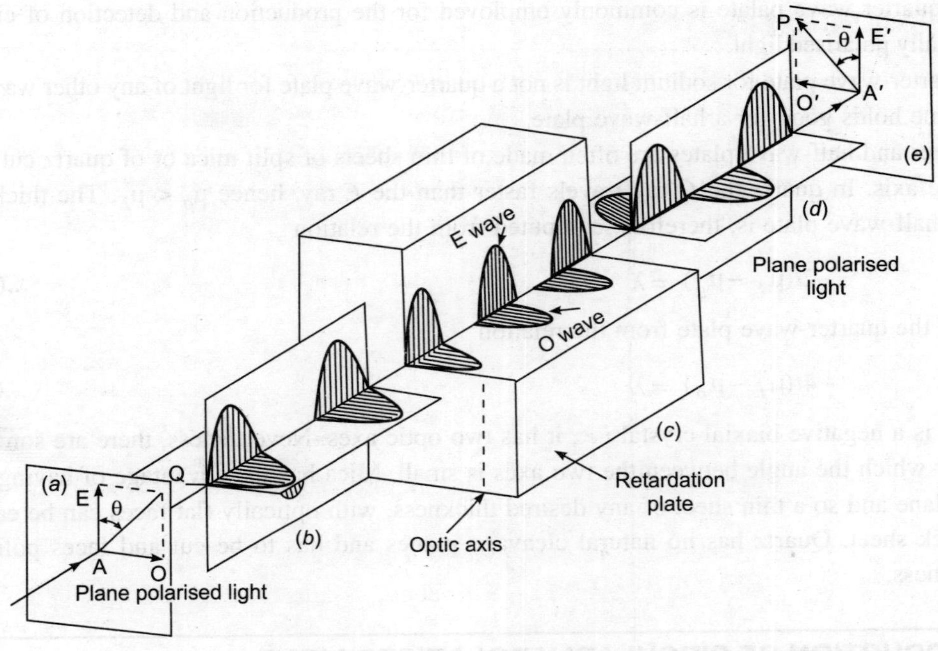

Fig. 17.7. A half-wave plate rotates the direction of polarisation of linear light by 2θ, when light is incident on the plate at an angle θ with the optic axis

The ordinary and extraordinary waves on emergence from this plate are shown in Fig. 17.8 and 17.9 while the action within the plate may be imagined by considering the retardation plate in Fig. 17.7 only half as thick as that actually shown in Fig. 17.7. The *O* wave, therefore, lags behind by just one-quarter of a wavelength. The nature of resultant wave is circular as shown in Fig. 17.9 for a special case *viz.* θ = 45°; for other angles it is an ellipse.

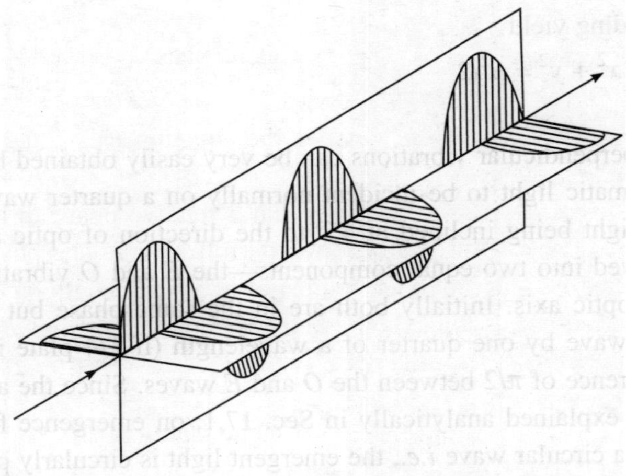

Fig. 17.8. Emergent waves from λ/4 plate

The quarter wave palate is commonly employed for the production and detection of circularly and elliptically polarised light.

A quarter-wave plate for sodium light is not a quarter-wave plate for light of any other wavelength and the same holds good for a half-wave plate.

Quarter and half-wave plates are often made of thin sheets of split mica or of quartz cut parallel to the optic axis. In quartz the *O* ray travels faster than the *E* ray, hence $\mu_o < \mu_E$. The thickness of the quartz half-wave plate is, therefore, computed from the relation

$$2t(\mu_E - \mu_o) = \lambda \qquad \text{...(17.2 } d\text{)}$$

and that of the quarter-wave plate from the relation

$$4t(\mu_E - \mu_o) = \lambda \qquad \text{...(17.2 } e\text{)}$$

Mica is a negative biaxial crystal *i.e.*, it has two optic axes. Nevertheless, there are some forms of mica for which the angle between the two axes is small. Mica has the advantage of having natural cleavage plane and so a thin sheet of any desired thickness, with optically flat faces can be easily cut from a thick sheet. Quartz has no natural cleavage planes and has to be cut and faces polished to optical flatness.

17.3 PRODUCTION OF CIRCULARLY POLARISED LIGHT

We are now in a position to describe the experimental technique employed for the production of circularly polarised light. It would be recalled that a *circular motion results when two mutually perpendicular coherent linear vibrations of equal amplitude and period but differing in phase by π/2 are compounded together*. For example, consider two mutually perpendicular coherent linear vibrations,

$$x = (a/\sqrt{2})\sin(\omega t + \pi/2) = (a/\sqrt{2})\cos\omega t$$

$$y = (a/\sqrt{2})\sin\omega t$$

which, on squaring and adding yield

$$x^2 + y^2 = a^2/2$$

an equation to a circle.

Two such mutually perpendicular vibrations can be very easily obtained by allowing a beam of plane polarised monochromatic light to be incident normally on a quarter wave plate, the vibration direction in the incident light being inclined at 45° to the direction of optic axis of the plate. The incident vibration is resolved into two equal components—the *E* and *O* vibration-respectively along and perpendicular to the optic axis. Initially both are in the same phase but on emergence, the *O* wave drops behind the *E* wave by one quarter of a wavelength (if λ/4 plate is of calcite) which is equivalent to a phase difference of π/2 between the *O* and *E* waves. Since the amplitudes of *E* and *O* waves are equal, hence as explained analytically in Sec. 17.1, on emergence from the quarter-wave plate, they compound into a circular wave *i.e.*, the emergent light is circularly polarised.

The physical picture of the formation of circularly polarised wave can be very easily seen by referring to Fig. 17.9, in which *E* and *O* waves are shown on emergence from the plate for the sake of explaining the rotation of the resultant light vector representing the optical disturbance at a given point. At points *A, C, E, G, I* and *K,* the magnitude of the horizontal component is a positive or negative maximum while that of vertical component is a zero. The resultant light vector at these points is, therefore, shown by horizontal arrow and its magnitude is equal to the maximum of the *O* wave. At points *B, D, F, H* and *J*, the magnitude of horizontal component is zero, hence the resultant light vector is vertical, equal in magnitude to the maximum of *E* wave and hence equal in magnitude to the maximum of the *O* wave ($\theta = 45°$). At a point midway between the points *A* and *B*, the magnitude of each component is 0.707 times the maximum of each wave *i.e.*, it is equal to $0.707a/\sqrt{2}$ where *a* is the amplitude of incident vibration. Obviously, the resultant light vector at this point is inclined at 45° to the horizontal and vertical planes and its magnitude is $a/\sqrt{2}$, the maximum of the *E* and *O* waves. Hence it is equal to the resultant at the points *A* and *B*. Exactly in a similar way, it is easy to see that the resultant light vector at a point midway between *B* and *C* is equal in magnitude to that at *A, B* and *C* but inclined at 45° to the vertical plane on the opposite side of it. It is now obvious that the magnitude of the resultant light vector is the same at all points and it rotates continuously around the direction of propagation, completing one revolution in the interval the wave advances one wavelength. In a plane perpendicular to the direction of propagation, through any point on this direction, the light vector rotates with uniform angular velocity as the wave advances, the tip of the vector describing a circular path. The light is, therefore, said to be circularly polarised.

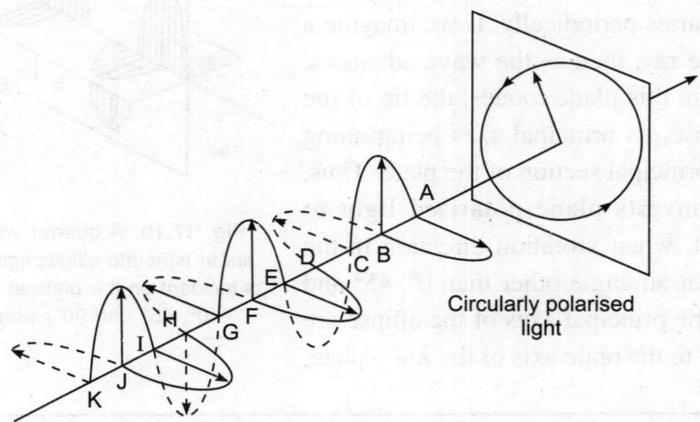

Fig. 17.9. A quarter-wave plate converts linear light to circular light, when linear light is incident on the plate at an angle of 45º with the optic axis

17.4 PRODUCTION OF ELLIPTICALLY POLARISED LIGHT

It will be recalled that *the elliptically polarized light arises when two mutually perpendicular plane polarised coherent light waves of unequal amplitudes but differing in phase by the value other than $n\pi$ are compounded together*. We are, however, more interested in the special case arising when $\delta = \pi/2$.

For example, consider coherent linear vibrations,

$$x = a\cos\theta\sin(\omega t + \pi/2) \quad \text{or} \quad x/a\cos\theta = \cos\omega t$$

$$y = a\sin\theta\sin\omega t \quad \text{or} \quad y/a\sin\theta = \sin\omega t$$

which on squaring and adding yield

$$\frac{x^2}{a^2\cos^2\theta} + \frac{y^2}{a^2\sin^2\theta} = 1$$

an equation to ellipse.

Elliptically polarised light can be, therefore, conveniently produced by passing a beam of plane polarised monochromatic light through a quarter wave plate, its optic axis being inclined at about $\theta = 30°$ to the plane of vibration of the incident light.

The physical picture of the formation of elliptically polarized wave can be obtained by referring to Fig. 17.10 in which the E and O waves of unequal amplitudes are shown on emergence from a quarter wave plate. When the light vectors at every point in the E and O waves are added up it will found, as with circular light, while the resultant light vector rotates around the direction of the propagation of the ray, it magnitude varies periodically. If we imagine a plane perpendicular to the ray, then as the wave advances, the resultant light vector in this plane rotates, the tip of the vector describing an ellipse, its principal axes being along and perpendicular to the principal section of the plate. Thus, a quarter-wave plate converts plane polarised light to elliptically polarised light, when vibration direction in the incident light is inclined at an angle other than 0°, 45° and 90° with the optic axis. The principal axes of the ellipse are parallel and perpendicular to the optic axis of the $\lambda/4$ – plate.

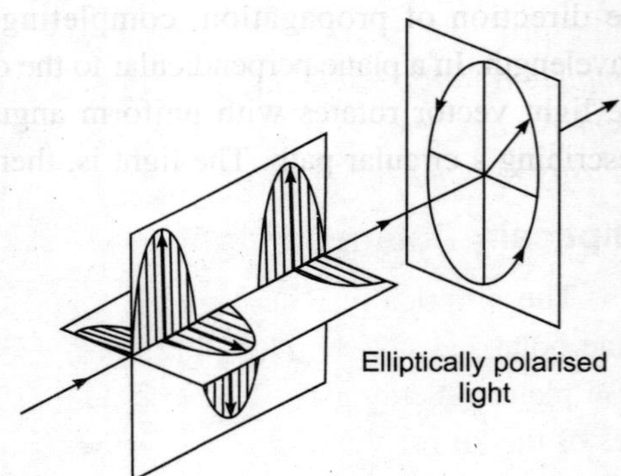

Fig. 17.10. A quarter wave plate-converts linear light into elliptic light, when linear light is incident on the plate at an angle (excluding 0°, 45° and 90°) with the optic axis

17.5 ANALYSIS OF LIGHT

Let us suppose that as a result of some experiment we have a beam of light and it is desired to ascertain its nature from the following seven possibilities.

(*i*) *Unpolarised light*

(*ii*) *Plane Polarised light*

(*iii*) *Elliptically polarised light*

(*iv*) *Circularly polarised light*

(*v*) *Mixture of unpolarised and plane polarised light*

(*vi*) *Mixture of unpolarised and elliptically polarised light*

(*vii*) *Mixture of unpolarised and circularly polarised light.*

Before giving a systematic qualitative investigation to ascertain the nature of given beam, we shall first give the characteristics exhibited by every one of the above listed kind when examined by the analyser, if one knows beforehand the nature of light.

Unpolarised Light

The unpolarised light on entering the Nicol prism is decomposed into ordinary and extraordinary beams of equal intensity. The E vibration is freely transmitted, the O vibration is totally reflected at the balsam surface, whatever may be the orientation of Nicol, which is, however, rotated about the direction of the incoming beam as an axis. *Thus, whatever may be the orientation of Nicol, the transmitted intensity is the same.*

Plane Polarised Light

The plane polarised light when examined by the Nicol can be completely extinguished for one setting of the analyser. In this setting, the direction of vibration of the incident light is perpendicular to the principal section of the analyser—that is to the shorter diagonal of its end face. *For each complete rotation of the analyser, there are two positions (180º from each other) at which the incoming beam is completely extinguished and two positions (90º from the former) at which the transmitted intensity is a maximum.*

Elliptically Polarised Light

The elliptically polarised wave may be regarded as the resultant of two mutually orthogonal plane polarised waves. The amplitudes and the phase difference, however, depend on the choice of these planes of vibration. In the special case when the phase difference is $\pi/2$ the major and minor axes of the ellipse are along the planes of vibration of the component waves (Fig. 17.11). Conversely, a given elliptically polarised wave can be resolved into two linearly polarised waves of unequal amplitudes, vibrating in planes parallel to the major and minor axes of the ellipse and differing in phase by $\pi/2$. The amplitude of one wave is equal to the semi-minor axis b and that of the other equal to semi-major axis a of the elliptic vibration. Let the principal section NN of the analysing Nicol make an angle θ with the major axis of the ellipse (Fig. 17.11). The resolved components along NN of the amplitudes of component waves, *viz*, $AB = a \cos \theta$ and $AC = b \sin \theta$ are freely transmitted as extraordinary vibrations, but they differ in phase by $\pi/2$. The transmitted intensity by Nicol is, therefore, given by

$$I \propto R^2 = a^2 \cos^2 \theta + b^2 \sin^2 \theta + 2ab \cos \theta \sin \theta \cos \pi/2$$

$$= (a^2 - b^2) \cos^2 \theta + b^2$$

when $\theta = 0, I_{max} = a^2$ and when $\theta = \pi/2, I_{min} = b^2$

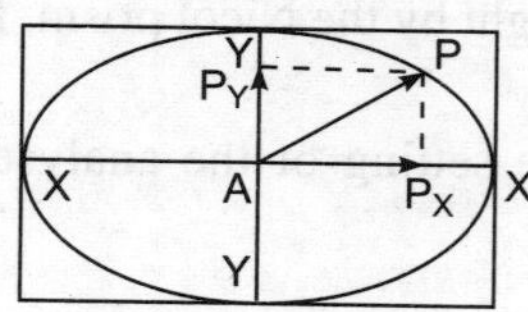

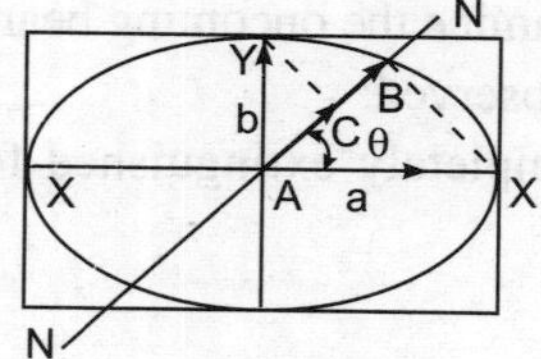

Fig. 17.11. Analysis of elliptically polarised light

Thus, for each complete rotation of the analyser about an axis parallel to the incident beam, there are two positions of NN (180° from each other) *at which the transmitted intensity is a minimum and two positions (at 90° from the former) at which the transmitted intensity is maximum*; the beam is never completely extinguished as with plane polarised light.

Circularly Polarised Light

Circular polarisation is a special case of elliptic polarisation *viz*, when $a = b$, the ellipse degeneration into a circle. Therefore, when circularly polarised light is examined by a Nicol, *the transmitted intensity is*

$$I = a^2$$

which is the same whatever may be the orientation of the analyser. In this respect the circularly polarised light resembles the unpolarised light.

It should be remarked that the circularly polarised wave is equivalent to two mutually perpendicular plane polarised waves of equal amplitude and differing in phase by $\pi/2$. Because of symmetry of the circular wave about the direction of propagation, all mutually perpendicular directions of resolution are equivalent.

Mixture of Unpolarised and Plane Polarised Light

(*Partially plane polarised light*). We have explained above that when unpolarised light is examined by Nicol, the transmitted intensity is the same for all orientations of the analyser but linear light is extinguished for two settings (180° from each other) of the analyser. Consequently, when a mixture of these light is examined by the analyser, the transmitted intensity will never become zero but *it will periodically fluctuate between a maximum and a minimum.* In this respect this mixture resembles elliptically polarised light.

Mixture of Unpolarised and Circularly Polarised Light

It will be recalled that unpolarised and circularly polarised light when examined separately by the analysing Nicol exhibit no variation in intensity. Obviously, when the mixture of these light is examined by the Nicol, *the transmitted intensity is the same in all orientations of the analyser.*

Mixture of Unpolarised and Elliptically Polarised Light

For one complete rotation of the analyser, *there are two settings at which the transmitted intensity is a maximum and two settings at which the transmitted intensity is a minimum.* In this respect this mixture resembles with elliptically polarised light and also with a mixture of unpolarised and plane polarised light.

Systematic Analysis of Light

We are now in a position to give a scheme for the systematic analysis of the nature of given light. The first step is to examine the oncoming beam of light by the Nicol prism. Either of the following three phenomena will be observed:

(*a*) Light can be completely extinguished for one setting of the analyser. This indicates that light is *plane polarised.*

(*b*) The transmitted intensity is the same for all settings of the analyser. This indicates that light is either (*i*) *unpolarised* or (*ii*) *circularly polarised* or (*iii*) *mixture of circularly polarised and unpolarised light*. To distinguish between these light we make use of quarter wave plate in a manner given later.

(*c*) The transmitted intensity varies as the Nicol is rotated and for each complete rotation of the analyser we get two positions (180° from each other) at which the transmitted intensity is a maximum and two positions at which transmitted intensity is a minimum. This indicates that the light is either (*i*) *elliptically polarised or* (*ii*) *mixture of unpolarised and plane polarised light* or (*iii*) *a mixture of unpolarised and elliptically polarised light*. To distinguish between these light we employ a quarter-wave plate in the manner described later.

Distinction between Unpolarised, Circularly Polarised and a Mixture of Unpolarised and Circularly Polarised Light

To accomplish this identification a quarter-wave plate is inserted in the path of oncoming beam of light and the light emerging from the plate is examined by the Nicol prism, rotating it about the direction of propagation of the beam. Either of the following three phenomena will be observed:

(*i*) The light emerging from the $\lambda/4$-plate can be completely extinguished for two settings (180° from each other) of the analyser in its one complete rotation. This indicates that the original light is *circularly polarised*. The reason for this conclusion is as follows: Circularly polarised light, just on entering the $\lambda/4$-plate, can be resolved into two components of equal amplitudes—one along and other perpendicular to the optic axis—differing in phase by $\pi/2$ radians. In transmission through the $\lambda/4$-plate a further phase difference of $\pi/2$ in introduced so that on emergence the resultant phase difference between the phases of two components becomes $\pi/2 + \pi/2 = \pi$ if in the original light circle is described in the anticlockwise direction [Fig. 17.12 (*a*)] and if the circle is described in the clockwise direction with respect to an observer toward whom the wave travels the resultant phase difference will be $\pi/2 - \pi/2 = 0$. Consequently, as explained in Sec. 17.1, in either case they compound into rectilinear vibrations *PAP i.e.,* the emergent light is plane polarised. This may be completely extinguished by the Nicol prism when its principal section *NN* is at 45° with the optic axis of the quarter wave plate [Fig. 17.12 (*a*)].

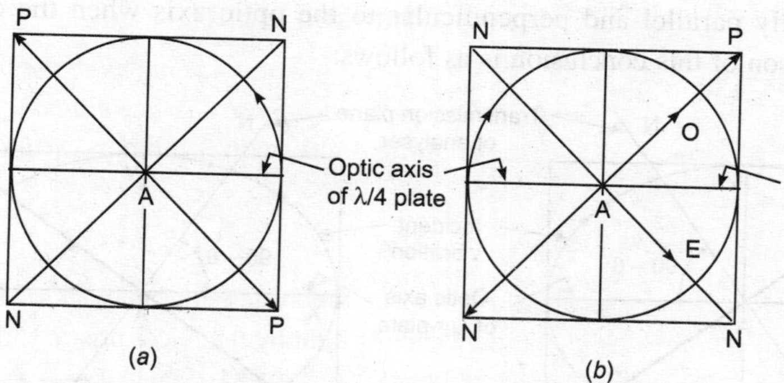

Fig. 17.12. (*a*) Analysis of circularly polarised light (*b*) Analysis of mixture of circularly and unpolarised light

(*ii*) The light emerging from the quarter wave plate exhibits no variation in intensity as the analysing Nicol is rotated about its axis. This indicates that the *original light is unpolarised*. The reason for this conclusion is as follows: Unpolarised light is theoretically equivalent to two mutually perpendicular linear light having no constant phase relation to each other. That is, incoherent *O* and *E* disturbances are propagated through the crystal. Hence $\lambda/4$-plate cannot introduce any constant phase difference between the two components of unpolarised light. (No matter what phase difference is introduced between the two components of unpolarised light, if the components recombined, the resultant will always be unpolarised light. One can, in fact, be said to have unpolarised light at all points through the plate). Hence, we conclude that unpolarised light is not at all changed in transmission through the quarter wave plate. Hence we observe no variation in the intensity of emergent light when examined by the analyser.

(*iii*) The emergent light exhibits two maxima and two minima in one complete rotation of the analyser. This indicates that the original beam is a *mixture of unpolarised and circularly polarized light*. The reason for this conclusion is as follows: The circularly polarised part of the original beam on transmission through $\lambda/4$-plate, as explained earlier, is converted into plane polarised light *PAP* [Fig. 17.12 (*b*)] while the unpolarised part is transmitted without any modification. Thus, the light emerging from the $\lambda/4$-plate is a mixture of unpolarised and plane polarised light. Obviously, when the principal section *NN* of the Nicol prism is at right angles to *PAP*, shown in Fig. 17.12 (*b*), the transmitted intensity is a minimum. For, while the linear vibration *PAP* is completely extinguished but the unpolarised part gives rise to *E* vibrations [*AE* in Fig. 17.12 (*b*)] in transmission through the analyser. Hence, we observe the variation in the intensity as stated earlier.

Distinction between Elliptically Polarised Light, Mixture of Elliptically Polarised and Unpolarised Light, and Mixture of Plane Polarised and Unpolarised Light

We insert a quarter wave plate between the analyser and the incoming beam of light. The quarter wave plate is rotated gradually in steps of 1° or so and for each setting of the plate the analyser is given one rotation. Then either of the following three phenomena will be observed:

(*a*) It is possible to get one setting of the $\lambda/4$-plate with respect to the original light so that the emergent light can be completely extinguished for one setting of the analyser. The principal section of the analyser (in other words, the shorter diagonal of the end face of Nicol) is inclined to the optic axis of the plate. This indicates that the original beam is *elliptically polarised*, and the axes of the ellipse are respectively parallel and perpendicular to the optic axis when the emergent light is extinguished. The reason for this conclusion is as follows:

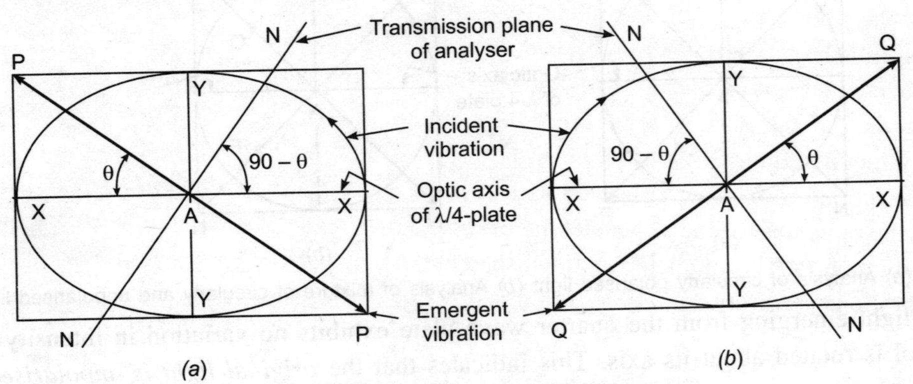

Fig. 17.13. Detection of elliptically polarised light

Suppose the optical vector in the elliptically polarised wave rotates in the anti-clockwise direction with respect to an observer towards whom the wave travels and that the optic axis (say *x*-axis) of the plate is parallel to the major axis of the ellipse [Fig. 17.13 (*a*)]. The incident wave, just on entering the plate, is resolved into two linearly polarised waves, vibrating in the *x* (major axis) and *y* (minor axis) directions *i.e.*, along and perpendicular to the optic axis of the plate, differing in phase by $\pi/2$ radians. The equations of component vibrations are respectively,

$$x = a\cos\omega t = a\sin(\omega t + \pi/2) \quad [E \text{ wave}]$$

$$y = b\sin\omega t \quad [O \text{ wave}]$$

There is a relative phase advancement of $\pi/2$ of the first wave with respect to the second. If we assume that $\mu_E < \mu_o$ *i.e.*, $V_E > V_o$, the quarter waveplate produces an additional relative phase advancement of $\pi/2$ of the first wave with respect to the second. Therefore the waves, as the emerge from the plate, have a relative phase difference of $\pi/2 + \pi/2 = \pi$ and so they combine into a plane polarised wave. Therefore, the light emerging from the plate can be completely extinguished when the principal section of the analyser is perpendicular to the piane of vibration of the emergent light.

If in the original beam, the ellipse in described in the clockwise direction, then the resolved plane polarised vibrations along major and minor axis are

$$x = -a\cos\omega t = a\sin(\omega t - \pi/2) \quad [E \text{ wave}]$$

$$y = b\sin\omega t \quad [O \text{ wave}]$$

The first wave lags behind the second by a phase angle $\pi/2$. If we assume, as before that $V_E > V_o$, the $\lambda/4$-plate produces a relative phase advance of $\pi/2$ of the first wave with respect to the second. Therefore, the two waves, as they emerge from the plate have a phase difference of $1/2\pi - 1/2\pi = 0$. Consequently they combine into plane polarised wave, the plane of vibration being parallel to the diagonal of the rectangle that lies in the first and the third quadrants and so the emergent light can be completely extinguished for the setting of the analyser marked in Fig. 17.13 (*b*).

(*b*) For one complete rotation of the analyser there are two settings (180º from each other) at which the transmitted intensity is minimum (not equal to zero) and *in this setting its principal section* (transmission plane) *is inclined to the optic axis of the* $\lambda/4$*-plate* [Fig. 17.13 (*a*)]. This indicates that the original light is a mixture of *unpolarised light and elliptically polarised light*. The reason of this conclusion is as follows:

When the optic axis of the $\lambda/4$-plate is parallel to the major or minor axis of the elliptically polarised part of the mixture, then on transmission through the plate, as explained earlier, the elliptic vibration degenerates into a linear vibration. The plane of vibration [*PAP* in Fig. 17.14 (*a*)] is inclined to the optic axis of the plate. On the other hand the unpolarised part of the mixture is transmitted without any modification. Obviously, when the principal section of the analyser is at right angles to the plane of vibration of plane polarised part of the mixture emerging from $\lambda/4$-plate and therefore inclined to the optic axis of the $\lambda/4$-plate, the transmitted intensity is a minimum. For, while the plane polarised vibration *PAP* is completely extinguished but the unpolarised part gives rise to plane polarised light [*E* vibrations, *AE* in Fig. 17.14 (*a*)] in transmission through the analyser.

(c) The principal section of the analyser in the setting for a minimum of transmitted intensity (not equal to zero) is either parallel to or perpendicular [Fig. 17.14 (b)] to the optic axis of the quarter wave plate. This indicates that the original light is a mixture of unpolarised light and plane polarized light. The reason of this conclusion is as follows: The plane polarised part [*QAQ* in Fig. 17.14 (*b*)] of the mixture is converted into elliptically polarized wave on emergence from the $\lambda/4$-plate while the unpolarised part is transmitted without any modification. The principal axes of the ellipse are respectively parallel and perpendicular to the optic axis of the $\lambda/4$-plate. It will be recalled that when the elliptically polarised light is separately examined by the Nicol prism the transmitted intensity is a minimum when the principal section is parallel to the minor axis of the ellipse while the unpolarised light exhibits no variation in intensity. It is now obvious that the mixture emerging from the $\lambda/4$-plate will exhibit the variation in the intensity in a manner stated earlier.

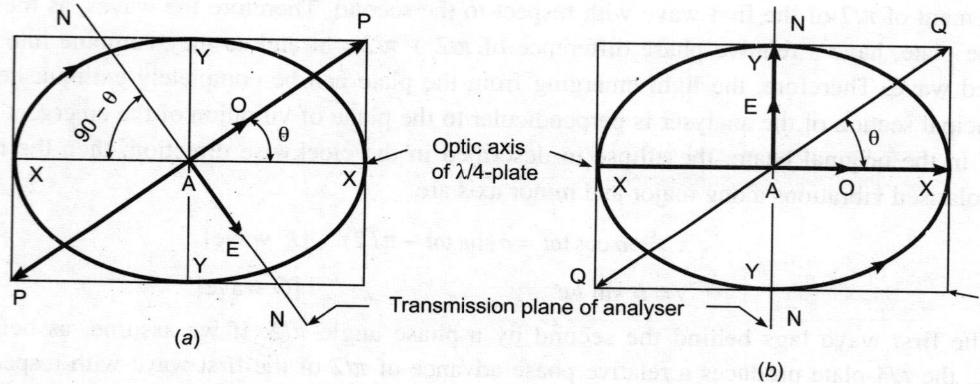

Fig. 17.14. (*a*) Detection of mixture of unpolarised light and elliptically polarised light (*b*) Detection of mixture of unpolarised light and plane polarised light

The analysis given above may be summarised as follows:

Detection

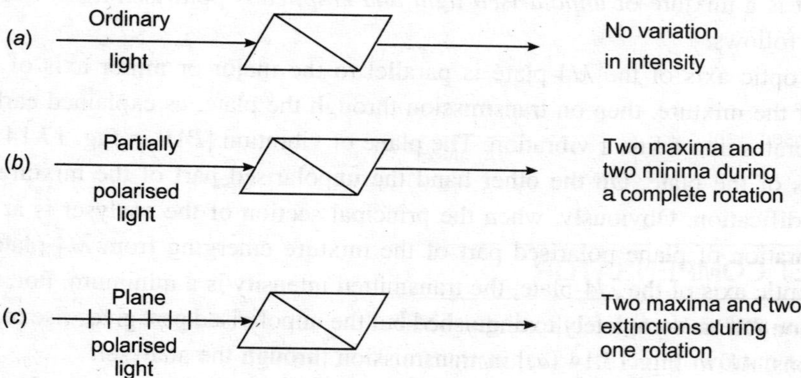

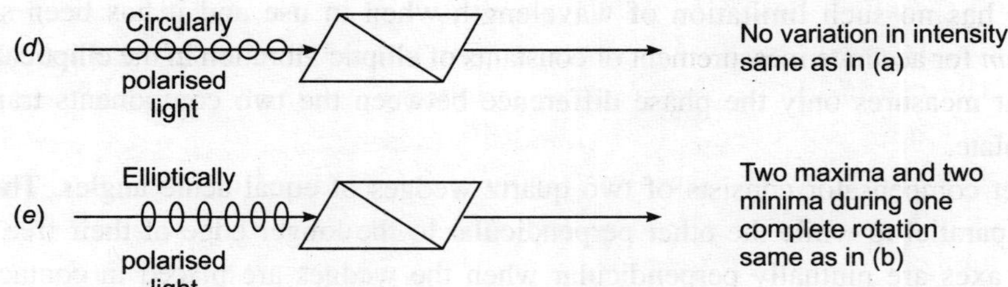

Distinction

(*i*) Between circularly polarised and unpolarised light.

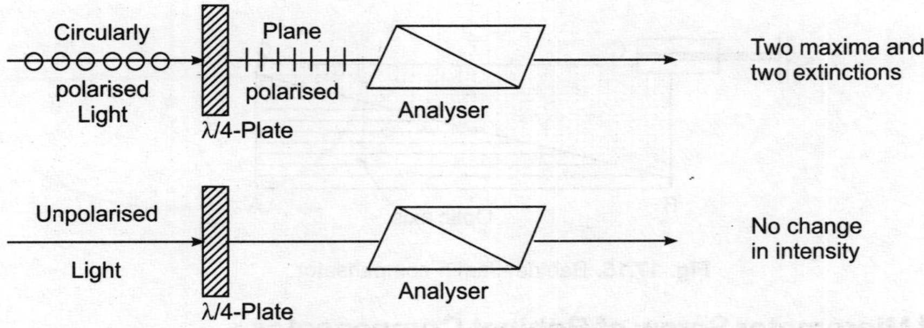

(*ii*) Between elliptically polarised and partially polarised light.

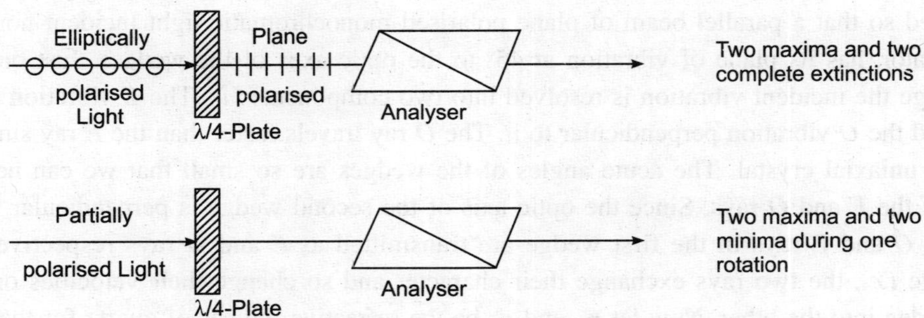

17.6 BABINET COMPENSATOR

In the production and analysis of elliptically polarised light the use of given *quarter wave plate* is limited only to a narrow range of wavelengths for which the path difference between *E* and *O* rays on transmission through it is $(4m + 1)\lambda/4$ where $m = 0, 1, 2, 3$ etc. *The Babinet compensator* is an apparatus which has no such limitation of wavelength when in use and it has been successfully employed by *Jamin* for accurate measurement of constants of elliptic vibration in the elliptically polarised light. In reality it measures only the phase difference between the two components transmitted by any anisotropic plate.

The Babinet compensator consists of two quartz wedges of equal acute angles. The optic axis in one wedge is parallel to while the other perpendicular to the longer edge of their free rectangular faces. The optic axes are mutually perpendicular when the wedges are placed in contact with each other as shown in Fig. 17.15 so as to form a thin plate of rectangular cross-section *ABCD*. The angles of wedge are much exaggerated in Fig. 17.15 for the sake of convenience in drawing the figure. As modified by *Jamin*, the instrument has fixed cross-wires in front of the upper wedge and a micrometer screw to displace the lower wedge relative to the upper fixed wedge along their plane of contact, thus forming a plane parallel plate whose thickness can be varied.

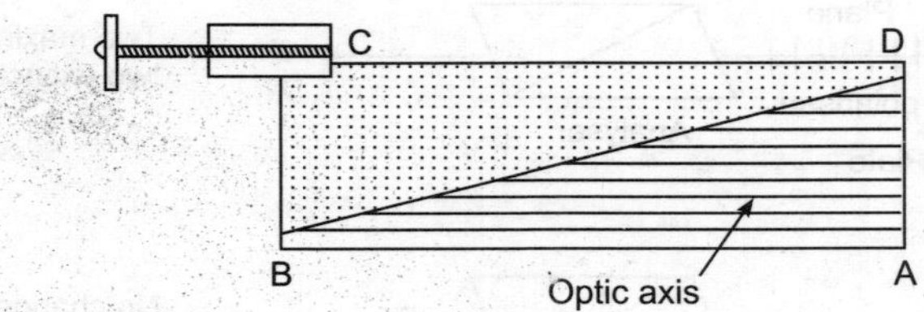

Fig. 17.15. Babinet-Jamin compensator

Calibration of Micrometer Screw of Babinet Compensator

It is first essential to calibrate the micrometer screw in terms of phase difference or path difference between the *E* and *O* rays originating from the monochromatic light which is to be employed in the experiment. The analyser and polariser are crossed and the compensator is placed between them, oriented so that a parallel beam of plane polarised monochromatic light incident normally on the compensator, has its plane of vibration at 45° to the optic axes of the wedges. Just on entering the first wedge the incident vibration is resolved into two components *viz*. The *E* vibration along the optic axis and the *O* vibration perpendicular to it. The *O* ray travels faster than the *E* ray since quartz is a positive uniaxial crystal. The acute angles of the wedges are so small that we can neglect the separation of the *E* and *O* rays. Since the optic axis of the second wedge is perpendicular to that of the first, the *O* and *E* rays in the first wedge are transmitted as *E* and *O* rays respectively in the second wedge *i.e.*, the two rays exchange their character and so change their velocities on passing from one wedge into the other. Now let n_E and n_o be the refractive indices of quartz for the *E* and *O* rays respectively. Let us consider a point on the compensator where the upper wedge is of thickness h_1 and the lower wedge of thickness h_2. The optical path difference introduced between the *E* and *O* rays in the transmission through the first wedge at this point is $h_1(n_E - n_o)$ and that introduced by the second wedge is $-h_2(n_E - n_o)$. The negative sign arises because the rays that are *E* rays and *O* rays in the first wedge become, respectively, *O* rays and *E* rays in the second wedge. Hence, the total optical path difference is

$$\Delta = (h_1 - h_2)(n_E - n_o) \quad ...(17.6\ a)$$

and the resultant phase difference is

$$\delta = (2\pi/\lambda)(h_1 - h_2)(n_E - n_o) \quad ...(17.6\ b)$$

Due to increasing thickness of one component wedge and decreasing thickness of the other Δ and δ increase linearly on one side of the centre ($h_1 = h_2$) and decrease on the other side *i.e.*, the retardation is of opposite signs on the two sides of the zero point.

The compensator has constant thickness along lines perpendicular to the longer edge of the wedge. For the central ray $h_1 = h_2$ and $\delta = 0$ *i.e.*, in effect the plate is of zero optical thickness at the centre. The emergent light at the central line is plane polarised with its plane of vibration parallel to that of the incident light—in effect the original vibration is transmitted. This is also true for the emergent light at equidistant parallel lines, on either side of the central ray, where $\delta = \pm 2\pi, \pm 4\pi, \pm 6\pi, ... = \pm 2m\pi$. Therefore the light is extinguished as if no birefringence medium were present between the crossed analyser and polariser. At intermediate parallel lines where $\delta = \pm(2m + 1)\pi$ the emergent light is plane polarised but its plane of vibration is inclined at $\pi/2$ to that of the incident light *i.e.*, the emergent vibrations are along the principal section of the analyser and so they are freely transmitted through it. At other parallel lines the emergent light is either circularly polarised or elliptically polarised. The polarisation of emergent light varies along the length of the compensator. When the compensator is viewed through the analyser (crossed) by the help of low power microscope focussed on the fixed cross-wires, a series of equally spaced alternate dark and bright bands, perpendicular to the long edge, are seen in the field of view. If the plane of vibration of the incident light is not exactly at 45° to the optic axes, the emergent vibration at points where $\delta = \pm 2m\pi$ will not be exactly perpendicular to the principal section of the analysing nicol. The fringes will be dark but not black and the contrast in the fringes will be reduced.

When the lower wedge is displaced with respect to the other by rotating the micrometer screw the position of zero optical thickness moves across the wedge. Therefore the fringes move laterally across the field of view. Thus, any desired dark fringe can be brought under the wire. The movable wedge is adjusted until one of these dark bands in on the cross-wires. The movement $2c$ of the quartz wedge as read on the micrometer screw or its angular rotation ϕ necessary to produce a shift of one fringe spacing is that which corresponds to a change of phase difference by 2π or a change of path difference of one wavelength. The screw is thus calibrated in terms of phase difference or path difference. Now, the central difference of path can be altered by a known amount by displacing the wedge by the known amount.

In order to locate the central black band corresponding to $\delta = 0$, it is only necessary to illuminate the compensator with plane polarised white light. Only the central fringe at $h_1 = h_2$ is dark and on either side of it the Newtonian colours of thin plate appear.

Measurement of the Birefringence ($n_E - n_o$)

The experimental arrangement is the same as sketched in Fig. 17.16 and monochromatic light is employed. By moving the lower wedge a dark fringe is brought under the cross-wires. A doubly refracting plate, cut from the given crystal with its faces parallel to the optic axis, is introduced between the compensator and one of the nicols. The plate is oriented so that its optic axis is inclined at 45° with the principal sections of nicols. Thus, its optic axis is parallel and perpendicular to the optic axis of the wedges. An extra phase difference

$$\delta' = (2\pi/\lambda)(n_E - n_o)h \qquad \text{...(17.6 } c)$$

is introduced by the plate between the E and O rays at every point thereby producing a lateral shift of fringes. The shifted fringes can be brought back to their original position if this extra phase difference is neutralised. This is accomplished by displacing the lower wedge so as to diminish effective thickness of quartz wedges at every point by an amount necessary to produce a reduction of δ' in the phase difference produced by the compensator. Thus, the phase difference δ' introduced by the doubly refracting plate has been compensated by the Babinet compensator, hence its name the compensator. The difference of two micrometer readings gives δ' and on substituting this in Eqn. (17.6 c) one can determine ($n_E - n_o$).

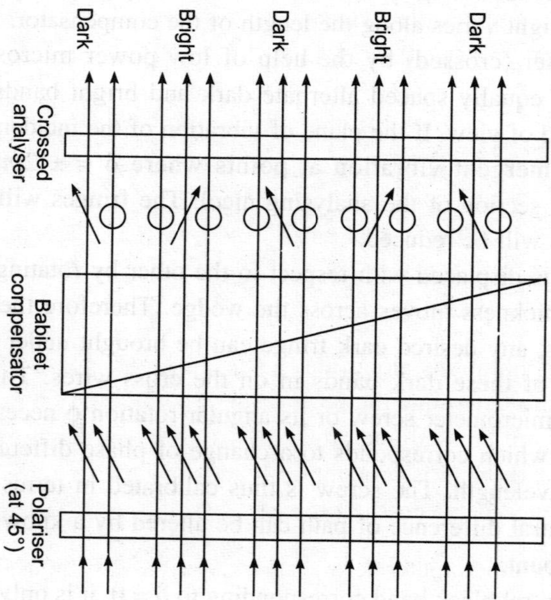

Fig. 17.16. Calibration experiment

Determination of Constants of Elliptic Vibration in Elliptically Polarised light

We shall now describe the use of the Babinet compensator to determine the constants of elliptic vibration.

Phase Difference

When an elliptically polarised monochromatic light is incident normally on the compensator we may consider the elliptic vibration as resolved at the surface into two components *viz.*

$$x = A\sin(\omega t + \alpha); \quad y = B\sin(\omega t + \beta)$$

along and perpendicular to the optic axis of the first quartz wedge. The phase difference between the two components is ($\alpha - \beta$). Transmission through the compensator changes this phase difference by

$$\delta = (2\pi/\lambda)(h_1 - h_2)(n_E - n_o)$$

The total phase difference $\alpha - \beta + \delta$ will be integral multiple of π along a system of equally spaced lines. The light emerging along these lines will be plane polarised. Therefore when the compensator is observed through a properly oriented nicol, bright and dark bands will be seen in the field of view and the central band again corresponds to $\alpha - \beta + \delta = 0$.

To begin with we employ a plane polarized light and the wedge is moved to bring the central band at the cross-wires. The phase difference between the E and O rays at this point is zero. On substituting the elliptically polarised light the central band shifts to a point where the initial phase difference $\alpha - \beta$ between the components of elliptic vibration is exactly neutralised by the phase difference δ introduced by the compensator. The lower wedge is now displaced by turning the micrometer screw until the central dark band is again under the cross-wires. If the displacement be x, we have

$$\frac{\alpha - \beta}{\pi} = \frac{x}{c} \quad \text{or} \quad \alpha - \beta = \pi x / c$$

where $2c$ is the distance through which the wedge must be moved to introduce a phase difference of 2π at the cross-wires. The value of $2c$ has been previously determined in the calibration experiment.

Position and Ratio of Axes

The components of elliptic vibrations along the major and minor axes differ in phase by $\pi/2$. With this knowledge we proceed to determine the position and the ratio of the axes as follows:

The analyser and polariser are crossed and adjusted with their principal sections at 45° to the optic axes of quartz wedges. A dark band is brought at the cross-wires by moving the lower wedge. In this setting, the phase difference between the ordinary ray and the extraordinary ray, emerging under the cross-wires is $2m\pi$. The micrometer screw is turned through $\frac{\phi}{4}$ turn to introduce a phase difference of $-\pi/2$ at the cross-wires so that the total phase difference becomes $2m\pi - 1/4\pi$ and therefore there results a displacement of fringes in the field of view. The polarising nicol is removed and the elliptically polarised light under examination is allowed to fall normally on the compensator, which is now rotated in its own plane about the line of vision until the same dark band again moves on to the cross-wires. This will happen when the phase difference at the cross-wires again becomes $2m\pi$. The analyser must be adjusted to make this dark band as black as possible *i.e.*, to achieve the extinction perfect at the cross-wires. The situation is now shown in Fig. 21.58. The optic axes of the two quartz wedges now give the directions of the axes of the elliptic vibration. The reason of this conclusion is as follows: When the major and minor axes of the elliptic vibration are along the optic axes of the quartz wedges the components of the elliptic vibration along the optic axes differ in phase by $\pi/2$ all over the surface of the compensator. In traversing the compensator a further phase difference $2m\pi - 1/2\pi$ is introduced between these components at cross-wires so that the resultant phase difference becomes $2m\pi$. Thus, the light emerging under the cross-wires is plane polarised. Hence the same dark fringe should appear under the cross-wires.

Suppose the analysing nicol is in the setting of complete extinction of the dark fringe under the cross-wires. Its principal section is perpendicular to the direction of the resultant of the (emerging)

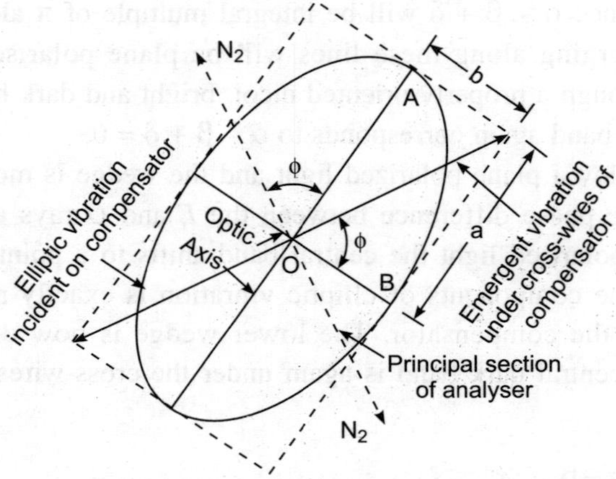

Fig. 17.17. Analyser of elliptically polarised light

two components of the elliptic vibration whose phase difference has been compensated by the compensator at the cross-wires. The tangent of the angle which the principal section by the analyser makes with the optic axis of the quartz wedge gives the ratio of the axis of the elliptic vibration.

$$\tan\phi = a/b$$

17.7 DETERMINATION OF DIRECTIONS AND RATIO OF THE MAJOR AND MINOR AXIS OF THE ELLIPTICALLY POLARISED LIGHT

The elliptically polarised light is examined by a Nicol prism, which is rotated until the transmitted intensity is a maximum. In this setting of the analyser, as explained in Sec. 17.5 its principal section is parallel to the major axis of the ellipse. Accordingly, the shorter diagonal of the end face of the Nicol is parallel to the major axis of the ellipse. Obviously, its minor axis is parallel to the longer diagonal of the end face. This setting of the analyser is left undisturbed and a quarter waveplate is placed in front of it in such a way that the optic axis is parallel to the shorter diagonal of the end face of Nicol and therefore parallel to the major axis of the ellipse. Suppose, for example, that the optical vector of the elliptically polarised light rotates in the anti-clockwise direction. Then, as explained in *Sec.* 18.5, the light emerging from the quarter-wave plate is a plane polarised wave. The plane of vibration of this wave is parallel to the diagonal of the rectangle that lies in the second and fourth quadrants [Fig. 17.13 (*a*)]. The shorter diagonal of the end face of Nicol (which determines the transmission axis of the analyser) is initially parallel to the optic axis of quarter wave plate. The transmitted light is extinguished by rotating the analyser in the anti-clockwise direction through an angle equal to $(90 - \theta)$ where θ is the angle which the plane of vibration of plane polarised emergent wave makes with the optic axis of the plate. If the rotation of Nicol is β as measured by the help of graduated circle and vernier, then

$$\beta = 90 - \theta.$$

$$\theta = 90 - \beta$$

The tangent of the angle θ so determined gives the ratio of the minor axis to the major axis of the ellipse.

$$\tan \theta = b/a$$

17.8 CHARACTER OF NATURAL LIGHT

We have explained in *Sec.* 16.5 that vibrations in unpolarised light can be resolved into transverse linear components in any two arbitrary mutually perpendicular directions. The average amplitudes of two components are equal and these components have no definite mutual phase relationship. So they must be considered to be incoherent—the phase difference between them is constantly changing in quite random fashion. Therefore, at any given instant, these components correspond to a particular elliptically polarised light. Due to incoherent components of unpolarised light we can regard it as elliptically polarised light, in which the shape of the ellipse as well as positions of its axes, change rapidly and irregularly with time, circularly polarised and plane polarised light being included as special cases.

SOLVED EXAMPLES

Example 1. *A beam of linearly polarised light is changed into circularly polarised light by passing it through a slice of crystal 0.003 cm thick. Calculate the difference in the refractive index of the two rays in the crystal assuming this to the minimum thickness that will produce the effect and that the wavelength is* 6×10^{-5} *cm.*

It will be recalled that when the least phase difference between the *O* and *E* waves on emergence from the doubly refracting plate is $\pi/2$ and incident linear vibrations are inclined at 45° to the optic axis, then the *O* and *E* waves recombine to form circularly polarised light. Obviously, the crystal plate is a quarter waveplate and its thickness is computed from Eqn. (17.2 *c*) which when solved for $(\mu_o - \mu_E)$ gives

$$\mu_o - \mu_E = \lambda / 4t = 6 \times 10^{-5} / 4 \times 0.003 = \mathbf{0.005.}$$

Example 2. *Plane polarised light is incident on a piece of quartz cut parallel to the axis. Find the least thickness for which the ordinary and extraordinary rays combine to form plane polarised light given that*

$$\mu_o = 1.5442, \mu_E = 1.5533,\ \lambda = 5\times 10^{-5}\ cm$$

It will be recalled that if the least phase difference between the ordinary and extraordinary waves in transmission through the quartz plate is π, they combine to form plane polarised light. Obviously, the quartz plate is a half-wave plate and its thickness is computed from Eqn. (17.2 *d*).

$$t = \frac{\lambda}{2(\mu_E - \mu_o)} = \frac{5\times 10^{-5}}{2(1.5533 - 1.5442)} = \mathbf{0.00274\ cm.}$$

Example 3. *Quartz has a refractive index of 1.554 for the ordinary ray and 1.553 for the extra-ordinary ray when measured with sodium light, What thickness of quartz between a crossed polariser and analyser will produce annulment of the light, the quartz being cut parallel to the optic axis.*

It will be recalled that no light is transmitted when the principal section of the analyser is perpendicular to that of the polariser. If the quartz is of such a thickness that the phase difference between the *O* and *E* waves is 2π in transmission through the plate, then on emergence they combine to form linearly polarised light, the plane of vibration coinciding with that of the incident linearly polarised light. Obviously, this light will be again extinguished by the analyser. To compute the thickness of quartz plate we substitute $\delta = 2\pi$ in the equation

$$\delta = (2\pi/\lambda)(\mu_E - \mu_o)t$$

and solve it for *t*.

$$t = \frac{\lambda}{\mu_E - \mu_o} = \frac{5896\times 10^{-8}}{1.553 - 1.544} = \mathbf{0.0065\ cm.}$$

Example 4. *A quartz plate is a half-wave plate for light whose wavelength is λ. Assuming that the variations in the indices of refraction with wavelength can be neglected, how would this behave with respect to light of wavelength $\lambda_1 = 2\lambda$.*

The thickness of half-wave plate is computed from Eqn. (17.2 *b*)

$$t = \frac{\lambda}{2(\mu_o - \mu_E)}$$

which can be easily re-written as

$$t = \frac{2\lambda}{4(\mu_o - \mu_E)} = \frac{\lambda_1}{4(\mu_o - \mu_E)}$$

which is identical with Eqn. (17.2 *c*). The given plate, therefore, behaves like a **quarter wave plate** for light of wavelength λ_1.

PROBLEMS

1. Discuss the phenomenon of superposition of two rectangular S.H. Vibrations of the same period and show how it may be used for the production of linearly, circularly and elliptically polarised lights.
2. (a) What is meant by plane polarised, circularly polarised and elliptically polarised lights?
   (b) Show that plane and circularly polarised lights are special cases of elliptically polarised light.
3. Describe, giving relevant theory, how linearly, circularly and elliptically polarised lights are produced and detected.
4. How will you find whether a beam of light is plane polarised, circularly polarised or elliptically polarised?
5. What is circularly polarised light? How is it produced? How can you distinguish between circularly polarised and unpolarised lights?
6. What is elliptically-polarised light? How is it produced? How will you distinguish between elliptically polarised light and a mixture of plane and unpolarised lights?
7. (a) Explain the construction and use of a quarter-wave plate and a half-wave plate.
   (b) If a quarter-wave plate and a half-wave plate be given to you, how would you proceed to distinguish them from each other?
8. Give the construction of the Babinet's compensator and explain how you would use it to analyse elliptically polarised light.
9. Calculate the thickness of quarter-wave plate of quartz for sodium light (λ = 5893 A.U.). The refractive indices of quartz for the *E* and *O*-waves are respectively 1.5533 and 1.5442.
10. Calculate the thickness of a doubly-refracting crystal required to introduce a path difference of $\lambda/2$ between the ordinary and extraordinary rays when λ-6000 A.U., μ_o = 1.55 and μ_o = 1.54.
11. Calculate the thickness of a calcite plate which would convert plane-polarised light into circularly polarised light. Given μ_o = 1.658, μ_e = 1.486 and λ = 5890 A.U.
12. (a) Plane polarised light falls normally on a quarter wave plate. Explain what will be the nature of the emergent light if the plane of polarisaton of the incident light makes the following angles with the principal plane of the quarter wave plate:

    $$0°, 30°, 45°, 90°$$

    (b) Elliptically polarised light falls normally on a quarter wave plate. Explain the nature of the emergent light if the major axis of the ellipse makes the following angles with the principal plane of the quarter wave plate:

    $$0°, 30°, 90°$$

13. Show that the resultant of two coherent beams of elliptically polarised light is in general another beam of elliptically polarised light.
14. How may a quarter wave plate by made to produce
    (a) right-handed and (b) left-handed circularly polarised light.
    How may a Nicol prism and a quarter wave plate be used to distinguish between right-handed and left-handed circularly polarised light?

15. How may a doubly refracting plate be made to change (a) a right-handed elliptically polarised beam into a left-handed beam (b) a right-handed circularly polarised light into left handed beam?
16. (a) How will you produce right-handed circularly polarised light?
    (b) How will you distinguish between elliptically polarised light and partially polarised light?
    (c) Two nicols are oriented with their principal planes making an angle of 30°. What percentage of incident polarised light will pass through the system?
17. A parallel beam of plane polarised light of wavelength 5890 A.U. (in vacuum) is incident on a quartz crystal. Find the wavelengths of the ordinary and extraordinary waves in the crystal. [$\mu_o = 1.5418$, $\mu_E = 1.5508$].

OPTICAL ACTIVITY

In the year 1811 Arago, while experimenting with a crystal of quartz (placed between two Nicols), discovered that when a beam of plane polarised light is directed through this crystal along its optic axis, the plane of vibration gradually undergoes rotation about the optic axis, as the beam penetrates deeper and deeper into the crystal. Finally, it emerges with its plane of vibration, in general, inclined to the plane of the incident vibrations. Obviously, the plane of polarisation also undergoes an equal rotation in the same direction–the phenomenon being termed as *rotatory polarisation* or more commonly, as *optical activity*. It can be illustrated schematically by Fig. 18.1 where the plane of vibration of the plane polarised light before entering the crystal is in the plane *QQ* and on emergence the direction of vibration is in the plane *PP*. The angle of rotation of the plane of polarisation is obviously equal to that between *QQ* and *PP*.

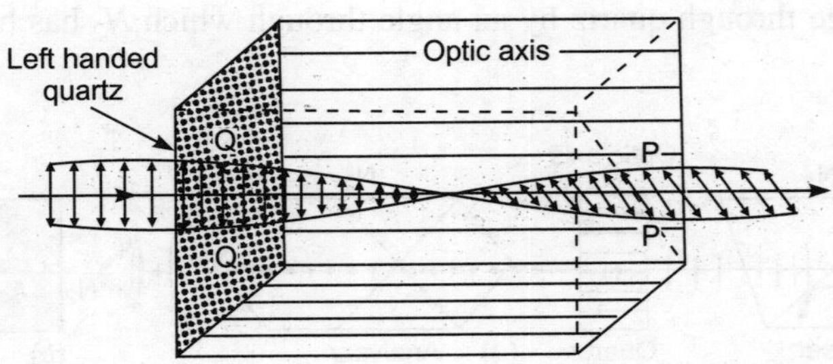

Fig. 18.1. Rotation of plane of vibration in an optically active substance like quartz

Many other crystals like sodium chlorate, sodium bromate and cinnabar exhibit the phenomenon of optical activity and such substances are said to be optically active. The angle of rotation of the plane of polarisation is found experimentally to be directly proportional to the thickness of the crystal section traversed by the plane polarised light. This fact brings into light the fact that the action occurs within the optically active medium and not at the surfaces. The angle of rotation is also found to depend upon the wavelength of the light employed and the nature of the optically active substance.

In addition to solids mentioned above, there are certain liquids *e.g.*, turpentine, sugar solution, alcoholic solution of camphor etc. which are optically active. The optical rotation produced by liquids is however very small as compared to that produced by solids. Thus one millimetre thick quartz plate produces a rotation of about 18° in the red region of the spectrum while the same thickness of turpentine produces a rotation of only twenty five minutes. The amount of rotation produced by liquids is proportional to the following factors:

(*i*) The length of path into the liquid.

(*ii*) Inverse square of wavelength (The law is approximate).

It was discovered by Biot that optically active liquid retains its property when it is dissolved in a solvent which does not react with it. The rotation produced by such solutions is proportional to the concentration. This property makes it possible to determine the concentration of an optically active solution by measuring the rotation produced by a standard length of the solution.

An optically active liquid *e.g.*, turpentine retains its property when it is transformed into vapour.

While in solids the optical activity is attributed to their crystalline structure, in the case of liquids it is attributable to the molecular structure itself.

18.1 EXPERIMENTAL DEMONSTRATION OF ROTATORY POLARISATION

The phenomenon of rotatory polarisation can be experimentally demonstrated by an arrangement sketched in Fig. 18.2 where N_1 and N_2 are Nicol prisms mounted at a small distance apart and are capable of rotation about the common axis. In the crossed settings of the Nicols, the field of view is dark – the *E* vibrations from the polariser enter the analyser as *O* vibrations in this particular setting and hence are totally reflected within the analyser. However, on inserting a plate of quartz (cut perpendicularly to the optic axis) between the crossed. Nicols the field of view brightens *i.e.*, some light is now transmitted by the combination. If the light is monochromatic (of single wavelength) then on rotating N_2 through a certain angle the light emerging from quartz can be once more completely extinguished. This shows that it is still plane polarised but the plane of vibration of the incident light has been rotated in passage through quartz by an angle through which N_2 has been rotated.

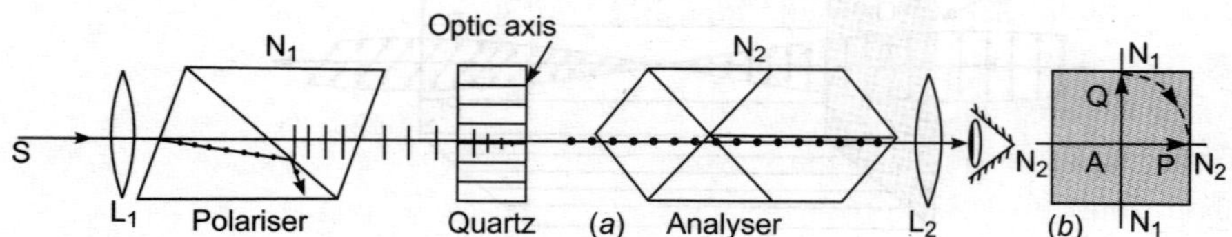

Fig. 18.2. Demonstrating the special case of rotation of plane of vibration through 90°

All this is more closely explained schematically in Fig. 18.3 where $N_1 N_1$ represents the principal section of the polarising Nicol and therefore *AQ* represents the vibration incident on the quartz. This vibration is rotated from the plane *AQ* to *AP* in passing through the quartz and therefore *AP* represents the vibration emergent from quartz and incident on the analysing Nicol N_2, whose principal section is represented by $N_2 N_2$ [Fig. 18.3 (*a*)]. On entering the analyser, its component $AE = AP \sin \theta$ along $N_2 N_2$ (plane of transmission in analyser) is transmitted as *E* vibration. On the other hand, the perpendicular component $AO = AP \cos\theta$ is totally reflected as O vibration within the analyser. This explains the brightening of the field of view on the insertion of quartz between the crossed Nicols. If now N_2 is rotated, then $N_2' N_2'$ is the required new position of its principal section, being perpendicular to AP, for complete extinction of AP. The angle of rotation of the plane of vibration due to quartz plate is, therefore, equal to the angle between $N_2 N_2$ and $N_2' N_2'$ *i.e.*, the angle through which the analysing Nicol has been rotated. Unfortunately, this is not the most sensitive and accurate method for measuring the angle of rotation.

If, however, a plate of calcite, cut perpendicularly to the optic axis, be similarly inserted between the crossed Nicols, the field of view remains dark–the plane polarised light from the first Nicol is transmitted by the calcite without any modification and, therefore, it is extinguished by the analyser. Thus, *calcite is an optically inactive crystal.*

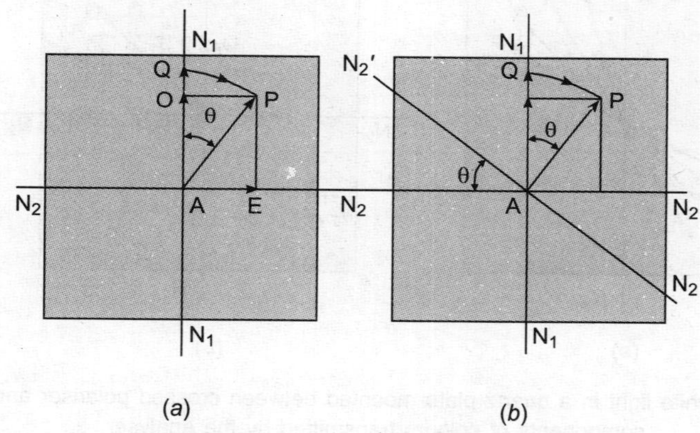

Fig. 18.3. Rotation of monochromatic light in a quartz plate mounted between crossed polariser and analyser

18.2 ROTATORY DISPERSION

A careful study of optical activity would show that the planes of vibration of light of different wavelengths are rotated by different amounts in traversing the same thickness of quartz. The first accurate results in this direction are due to Biot who found that the angle of rotation is very nearly inversely proportional to the square of the wavelength—the light of wavelength in the violet region being rotated about four times more than the light of wavelength in the red region of the spectrum. Therefore, if a plane polarised beam through quartz along its optic axis, the effect can be illustrated diagrammatically as in Fig. 18.4 where QQ represents the direction of vibration in the polarised white light incident on the quartz.

The waves of different wavelengths in the emergent beam will have different planes of vibration as shown in Fig. 18.4 (*a*), where AR is the plane of vibration of the wave of some particular wavelength in the red region and so on. Consequently, when emergent vibrations from quartz are analysed by the second Nicol with its principal section $N_2 N_2$ (plane of transmission) as shown in Fig. 18.4 (*b*), their components along $N_2 N_2$ – from AE_R to AE_v – are transmitted as E vibration while the perpendicular components–from AO_v to AO_R – are totally reflected as O vibrations within the analyser. Since $AE_v > AE_R$ the resultant transmitted colour will have violet tinge. What happens is that more of the red colour is totally reflected as O vibrations within the analyser. In fact, the resultant colour of the transmitted light is exactly complementary to the resultant colour of the totally reflected light. The field of view, therefore, appears coloured, the colours obviously changing as the analysing Nicol is rotated about its own axis, light never disappearing entirely for anyone setting of the analyser.

The decomposition of white light into its constituent colours in the transmission of plane polarised white light through quartz, (along its optic axis) is termed as the phenomenon of *rotatory dispersion.*

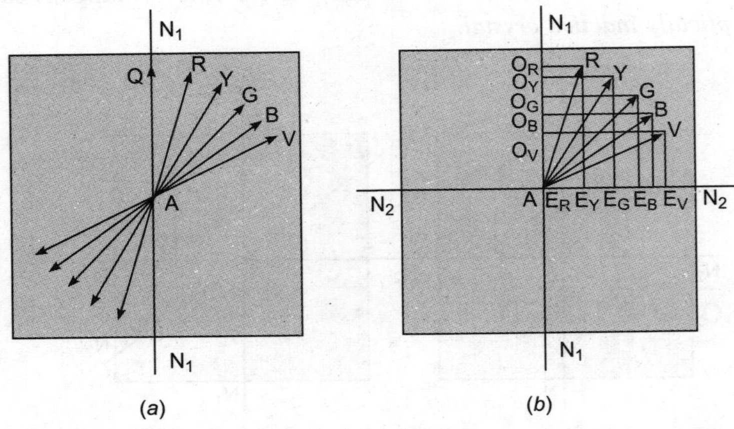

Fig. 18.4. Rotation of white light in a quartz plate mounted between crossed polariser and analyser showing components of colours transmitted by the analyser

18.3 TWO SAMPLES OF QUARTZ

There exist two kinds of quartz which are mirror images of each other even in their crystalline structure and form. The most peculiar thing about them is that their optical properties are also mirror images of each other. Thus, if one rotates the plane of vibration to the right, the other would rotate it to left. Crystals belonging to the former class are called *dextrorotatory* or simply *right-handed*, while those belonging to the latter class are called *levorotatory* or simply *left-handed.* The right-handed rotation means that the rotation of the plane of vibration occurs in the clockwise direction with respect to an observer looking towards the oncoming beam of light. A left-handed crystal rotates the plane of vibration in the anti-clockwise direction.

A simple inspection of a quartz crystal enables one to ascertain its nature conclusively. The character of the crystal is determined by the relative position of S and B with respect to K, the surface of the hexagon which both touch. If S and B lie to the right of K, the crystal is a right-handed one; but if they are to the left it is a left-handed variety.

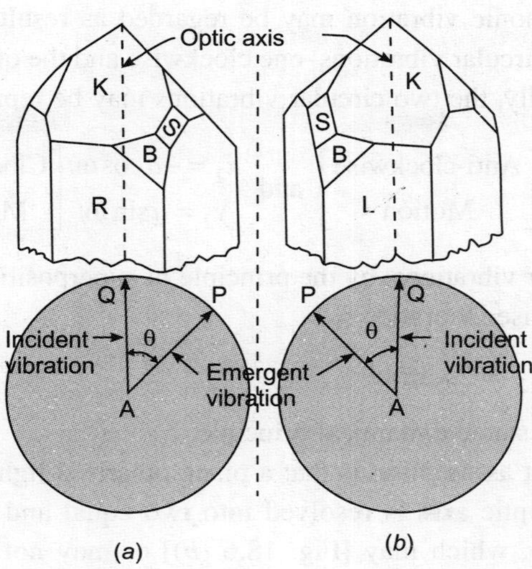

Fig. 18.5. Each quartz is a mirror image of the other

18.4 LAWS OF ROTATION OF THE PLANE OF POLARISATION

Shortly after the discovery of the phenomenon of rotatory polarisation by *Arago Biot*, in about 1815 made a detailed study of the phenomenon of rotatory polarisation with right-handed as well as left-handed crystals and employing plane polarised light of different wavelengths. His conclusions are best expressed in the form of the following three laws:

1. The angle of rotation of the plane vibration, for any given wavelength, is directly proportional to the thickness of the optically active substance traversed by the incident light. In other words, the greater the thickness, the greater is the angle of rotation.
2. The rotation produced in transmission through two or more optically active substances is the algebraic sum of rotations which would be produced by these substances separately. In mathematical language we may express the resultant rotation θ due to number of substances arranged one after another as

$$\theta = \theta_1 - \theta_2 + \theta_3 - \theta_4 + \theta_5 - \theta_6 + \ldots\ldots$$

   when substance alternately right-handed and left-handed varieties. *Therefore, a solution containing an equal number of right-handed and left-handed molecules is optically inactive.*
3. The angle of rotation is approximately inversely proportional to the square of the wavelength for a given thickness of the optically active substance.

18.5 FRESNEL'S EXPLANATION OF ROTATORY POLARISATION

The first explanation of the phenomenon of rotatory polarisation in optically active crystals like quartz was given by Fresnel on the supposition that the circularly polarised light is propagated along the optic axis without any modification. The theory is based on the elementary principle in dynamics that any rectilinear simple harmonic vibration may be regarded as resultant of the superposition of equal and oppositely described circular vibrations–one clockwise and the other anti-clockwise–rotating with equal frequency. Analytically, the two circular vibrations may be represented by

$$\begin{matrix} x_1 = a\cos\omega t \\ y_1 = a\sin\omega t \end{matrix} \begin{bmatrix} \text{Anti-clockwise} \\ \text{Motion} \end{bmatrix} \quad \text{and} \quad \begin{matrix} x_2 = -a\cos\omega t \\ y_2 = a\sin\omega t \end{matrix} \begin{bmatrix} \text{Clockwise} \\ \text{Motion} \end{bmatrix}$$

The x components of the circular vibrations, by the principle of superposition of wave motions, cancel each other, leaving a linear polarised vibration

$$y = y_1 + y_2 = 2a\sin\omega t$$

This establishes the above stated dynamical principle.

Accordingly, Fresnel's first assumption is that a plane polarised light (linear vibrations) just on entering any crystal along its optic axis is resolved into two equal and opposite circular motions, described with equal frequency, which may [Fig. 18.6 (*b*)] or may not [Fig. 18.6 (*e*) and (*h*)] be propagated with the same speed, depending upon the nature of the crystal. In crystals like calcite, since the spherical and the ellipsoidal wave surfaces actually touch each other at points where optic axis through the point of origin of light intersects them, the two circular components travel with the same speed (V_L=V_R) along the optic axis. Consequently, both arrive simultaneously at any point B along their path in the crystal; and as the wave travels on, due to equality of frequencies, the R and the L vectors rotate through equal angles at equal intervals of time—the R vector in the clockwise and the L vector in the anti-clockwise direction—both initially starting from the same point. For every simultaneous position of the R and the L vectors, as shown in Fig. 18.6 (*c*), their resultant lies along the direction BP_o, which is coincident with AQ, the vibration plane of the incident wave. Thus the light at the point B is still plane polarised and the plane of vibration is the same as that of the incident wave. Calcite, thus, does not exhibit the phenomenon of rotatory polarisation.

In an optically active crystal the two wave surfaces do not actually touch each other at points where the optic axis through the point of origin of the light intersects them. What would normally be the spherical surface, it is distorted outwards by a small amount in the region of the axis, and the ellipsoidal surface is distorted inwards. Hence, circularly polarised components travel along the optic axis with slightly unequal speeds. In the right-handed quartz the clockwise component travels *faster* [Fig. 18.6 (*g*)] while in the left-handed quartz anti-clockwise component travels faster [18.6 (*d*)]. Due to this inequality in their speeds a certain phase difference will be introduced between the two components in transmission through the quartz. As a result of this, as explained in the sequence their resultant vibration will be inclined to the plane of vibration of the incident light and thus the rotation of the plane of polarisation will be accounted for.

Let us now consider more closely the rotation of the plane of vibration in the left-handed quartz. Let the light vector AQ in the plane polarised light, incident normally on the crystal face, be initially vertical. Since $V_L > V_R$. The L component arrives first at any point B along the path of propagation of light in the crystal, and as the wave travels on, the L vector continues to rotate (at B) in the anti-clockwise direction, having turned through an angle δ, say, before the R component arrives at B.

This angle δ, therefore, is the phase difference between the L and the R components on reaching the point B. At this instant, which we call $t = 0$, both circular components rotate in opposite direction the one starting at R_o and the other at L_o, due to equality of frequencies, tracing out equal angles in equal intervals of time. For every simultaneous positions of R and L vectors, as shown in Fig. 18.6 (f), their resultant lies along the plane BP, which it is easy to see makes an angle $1/2\delta$ with the vibration plane of the incident wave. We, therefore, conclude that at the point B light is still plane polarised but the plane of vibration has been rotated through an angle $1/2\delta$ in the anti-clockwise direction in traversing the thickness AB of the crystal.

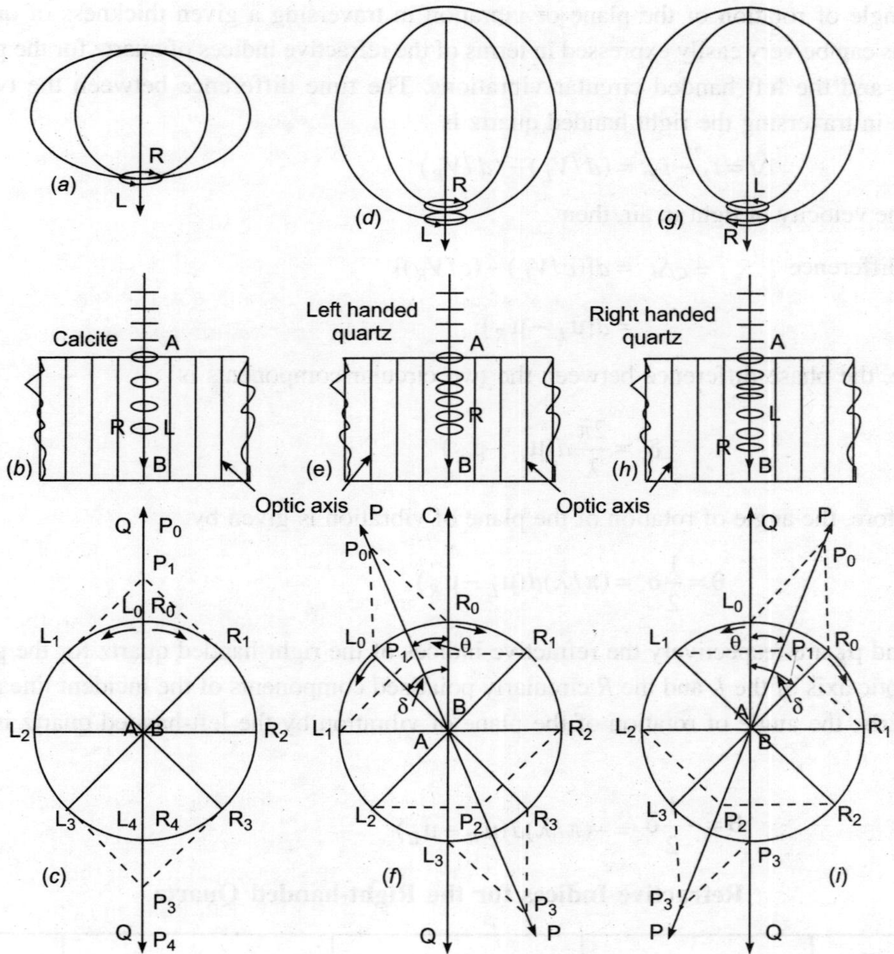

Fig. 18.6. Fresnel's explanation of optical rotation

In the right-handed quartz, the R component arrives first at B and when its vector has described an angle δ in the clockwise direction, the L component arrives at B. At this instant, $t = 0$, both circular components rotate in opposite direction with equal frequency, the R vector starting at R_o and the L

vector at L_o. As shown in Fig 18.6 (*i*), the point *P* executes linear vibrations along *BP*. Thus, in travelling from *A* to *B* along the optic axis, the plane of vibration has been rotated through an angle $1/2\delta$ in the clockwise direction. It is now obvious that the plane of vibration would rotate gradually as the light penetrated deeper and deeper into the crystal.

It may be remarked that Fresnel's explanation is quite satisfactory as far as it goes, but it fails to give any plausible reason for the curious resolution of the plane polarised light into equal and opposite circularly polarised beams.

Calculation of the Angle of Rotation

The angle of rotation of the plane of vibration in traversing a given thickness of quartz along the optic axis can be very easily expressed in terms of the refractive indices of quartz for the propagation of the right and the left handed circular vibrations. The time difference between the two circular components in traversing the right handed quartz is

$$\Delta t = t_L - t_R = (d/V_L) - (d/V_R)$$

and if *c* is the velocity of light in air, then

$$\text{Path difference} = c\Delta t = d[(c/V_L) - (c/V_R)]$$

$$= d[\mu_L - \mu_R]$$

Hence, the phase difference between the two circular components is

$$\delta = \frac{2\pi}{\lambda} d(\mu_L - \mu_R)$$

Therefore, the angle of rotation of the plane of vibration is given by

$$\theta = \frac{1}{2}\delta = (\pi/\lambda)d(\mu_L - \mu_R)$$

where μ_L and μ_R are respectively the refractive indices of the right-handed quartz for the propagation along the optic axis of the *L* and the *R* circularly polarised components of the incident linear vibration.

Similarly, the angle of rotation of the plane of vibration by the left-handed quartz is expressed by

$$\theta = -\frac{1}{2}\delta = -(\pi/\lambda)d(\mu_R - \mu_L)$$

Refractive Indices for the Right-handed Quartz

Wavelength	μ_R	μ_L	μ_E	μ_o
7620 A	1.53914	1.53920	1.54811	1.53917
3968 A	1.55810	1.55821	1.56771	1.55815

18.6 ANALYTIC TREATMENT OF FRESNEL'S THEORY OF ROTATORY POLARISATION

We may also explain, in keeping with Fresnel's principles, the phenomenon of rotatory polarisation by considering the equations of various vibrations. Let the vibrations in the plane polarised light incident normally on a quartz plate, cut with faces perpendicularly to the optic axis, be represented by the equation of simple periodic motion, *viz*,

$$y = 2a \sin \frac{2\pi}{T} t$$

According to Fresnel's idea, this linear vibration is resolved, just on entering the crystal, into two equal and opposite circular motions, which may be represented by the equations of their components along two mutually perpendicular directions.

$$x_1 = -a\cos(2\pi/T)t \qquad \text{[Right-handed}$$

$$y_1 = a\sin(2\pi/T)t \qquad \text{circular motion]}$$

and

$$x_2 = a\cos(2\pi/T)t \qquad \text{[Left-handed}$$

$$y_2 = a\sin(2\pi/T)t \qquad \text{circular motion]}$$

These circular motions travel along the optic axis of the quartz with slightly unequal velocities, say V_R and V_L, the velocities of the *right* and the *left* motions respectively. The right-handed motion, therefore, traverses the thickness d of the crystal in the time interval

$$t_R = d/V_R$$

Hence in traversing the thickness d of the crystal, in each component of the R motion, the phase difference will have the value $\frac{2\pi}{T} t_R$ or $\frac{2\pi}{T} \times \frac{d}{V_R}$ (since in complete period T, the phase of the wave changes by 2π). The components of the right-handed motion just on emerging from the crystal of thickness d along the optic axis, may, therefore, be represented by equations

$$x_1 = -a\cos\frac{2\pi}{T}\left(t - \frac{d}{V_R}\right) \qquad y_1 = a\sin\frac{2\pi}{T}\left(t - \frac{d}{V_R}\right)$$

Similarly, the left-handed motion on emergence from the crystal can be represented by the equations

$$x_2 = a\cos\frac{2\pi}{T}\left(t - \frac{d}{V_L}\right) \qquad y_2 = a\sin\frac{2\pi}{T}\left(t - \frac{d}{V_L}\right)$$

The resultant of the two emergent circular motions can be obtained, according to the principle of superposition of wave motions, by the algebraic summation of X and Y components individually.

Then we get

$$x = x_1 + x_2$$

$$= a\left[\cos\frac{2\pi}{T}\left(t-\frac{d}{V_L}\right)-\cos\frac{2\pi}{T}\left(t-\frac{d}{V_R}\right)\right]$$

$$= 2a\sin\frac{\pi d}{T}\left(\frac{1}{V_L}-\frac{1}{V_R}\right)\sin\frac{2\pi}{T}\left[t-\frac{d}{2}\left(\frac{1}{V_R}+\frac{1}{V_L}\right)\right]$$

and $$y = y_1 + y_2$$

$$= a\left[\sin\frac{2\pi}{T}\left(t-\frac{d}{V_L}\right)+\sin\frac{2\pi}{T}\left(t-\frac{d}{V_R}\right)\right]$$

$$= 2a\cos\frac{\pi d}{T}\left(\frac{1}{V_L}-\frac{1}{V_R}\right)\sin\frac{2\pi}{T}\left[t-\frac{d}{2}\left(\frac{1}{V_R}+\frac{1}{V_L}\right)\right]$$

where, we have assumed $V_R > V_L$ *i.e.*, the crystal is of the right-handed variety.

The resultant x and y motions, being in the same phase, compound into a rectilinear simple harmonic motion, *viz*,

$$s = \sqrt{x^2 + y^2}$$

or $$s = 2a\sin\frac{2\pi}{T}\left[t-\frac{d}{2}\left(\frac{1}{V_R}+\frac{1}{V_L}\right)\right]$$

Thus, the light emerging from the crystal is still plane polarised but the direction of vibrations is inclined to Y axis *i.e.*, to the direction of vibrations in the incident light by an angle given by

$$\tan\theta = \frac{x}{y} = \tan\frac{\pi d}{T}\left(\frac{1}{V_L}-\frac{1}{V_R}\right)$$

Hence, $$\theta = \frac{\pi d}{T}\left(\frac{1}{V_L}-\frac{1}{V_R}\right) = \pi d\frac{c}{\lambda}\left(\frac{1}{V_L}-\frac{1}{V_R}\right)$$

or $$\theta = \frac{\pi d}{\pi}(\mu_L - \mu_R) \text{ radians}$$

This angle is independent of t, which simply means that at each point of the quartz the two circular vibrations combine into linear vibrations, However, θ is proportional to d, indicating that the plane of vibration rotates gradually in the clockwise direction as the plane polarised wave penetrates deeper and deeper into the right-handed quartz along its optic axis.

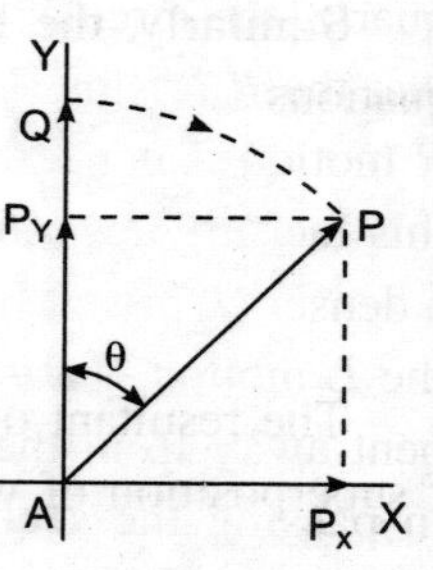

Fig. 18.7

In the left-handed quartz; since $V_L > V_R$, the angle of rotation of the plane of vibration is given by

$$\theta = \frac{\pi d}{\lambda}(\mu_R - \mu_L)$$

On the other hand, if the two circularly polarised components travel with the same speed, as in calcite crystal, then $\mu_L = \mu_R$ and hence $\theta = 0$. The plane polarised light is, therefore, propagated along the optic axis of such crystals with direction of vibrations unaffected. Thus, calcite does not exhibit the phenomenon of rotatory polarisation.

18.7 EXPERIMENTAL PROOF OF FRESNEL'S THEORY

The experimental proof of the curious resolution of plane polarised light into two circularly polarised components, in the manner described above, was also given by *Fresnel*. He reasoned that if there existed in reality two circular components, which travelled with unequal velocities, they should be refracted by different amounts into the air from an oblique surface of a quartz prism. But he failed to observe this separation, as the angular separation $(i_L - i_R)$ is too small to be observed. The increase in the angular separation between the two beams was, however, accomplished by him with the help of the ingenious device of employing a train of alternate right and left-handed quartz prisms, cut and arranged as shown in Fig. 18.8—the optic axis of each prism being in the same direction and perpendicular to the end faces.

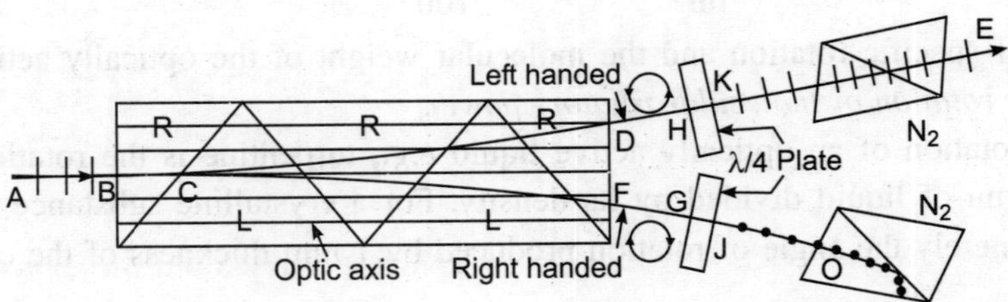

Fig. 18.8. Fresnel multiple prism for demonstrating the circularly polarised components of plane polarised light

The reason that the angular separation between the two beams, is increased at each oblique surface may be very easily explained. Since $V_R > V_L$ in the right-handed quartz and in the left-handed quartz, the circular components simply interchange their velocities *i.e.*, $V_L > V_R$, it therefore follows that the *R* prisms act as the rarer media as compared to the *L* prisms so far as the propagation of the *R* motion is concerned. The reverse of this holds good for the propagation of the *L* motion. Keeping this fact in view, on the first oblique surface, since the *R* motion is passing from rarer (*R* prism) into a denser (*L* prism) medium, it will be bent towards the boundary normal and hence turned downwards; the *L* motion is passing from a denser (*R* prism) into a rarer (*L* prism) medium; it will therefore be bent away from the normal and hence turned upwards. The reasoning along these lines shows that in passing the second oblique surface, the angular separation between the two beams is further increased. The net result is that the angular separation of the two rays increases at each successive refraction, the two rays finally emerging separated from each other.

The two images of the same source formed by the compound prism are examined by a Nicol prism. It would be observed that their intensities remain independent of the rotation of Nicol. This is also the characteristic of the ordinary unpolarised light when examined by the Nicol. But on introducing a $\lambda/4$ plate of mica, as shown in the figure, it is found that images are extinguished in succession as the Nicol is rotated—the $\lambda/4$ plate transforming the circularly polarised light into plane polarised light, the plane of vibration in the two beams being mutually perpendicular. This conclusively proves that the light emerging from the rectangular quartz block is circularly polarised in opposite directions.

18.8 SPECIFIC AND MOLECULAR ROTATION

The dependence of the rotation produced by an optically active substance dissolved in a non-active solvent upon the concentration of the solution has led to a large number of applications *e.g.*, in the estimation and analysis of sugars. For a comparison of optical activity, therefore, a unit is adopted. This is called *specific rotation* and is defined by the following equation

$$\text{Specific rotation} = \frac{\text{Rotation produced by 10 cm length of the solution}}{\text{Density of the solution in gms per c.c.}}$$

Thus if θ be the rotation produced by 'l' decimeters of the solution and ω be the amount of the optically active substance dissolved in one hundred c.cs of the solvent then the specific rotation or the rotatory power α is given by

$$\alpha = \frac{100\,\theta}{l\omega} \quad \text{or} \quad \theta = \frac{\omega\alpha l}{100}$$

The product of specific rotation and the molecular weight of the optically active substance is termed as *molecular rotation or molecular rotatory power.*

The specific rotation of an optically active liquid *e.g.*, turpentine is the rotation produced by one decimetre column of liquid divided by its density. For a crystalline substance like quartz, the specific rotation is merely the angle of rotation produced by 1 mm thickness of the crystal along the optic axis.

To distinguish between dextro and levo-rotatory substances the specific rotation of the former is written positive while for the latter it is written negative.

The specific rotations of a large number of substances dissolved in various non-active solvents have been measured. It has been found from these experiments that the specific rotation varies not only with the nature of the solvent but also with the concentration of the active substance present. Thus if P, Q and R are constants and C be the concentration of the active substance, the specific rotation is given by the equation

$$\alpha = P + QC + RC^2$$

The specific rotation of a substance also depends upon the temperature and the wavelength of light used and is usually written in the form $[\alpha]_t^D$ where 't' stands for the temperature and D stands for the D lines of sodium (λ_1 = 5890 Å, λ_2 = 5896 Å). Thus for quartz the specific rotation at any temperature T is given by

$$[\alpha]_t^D = 21.72\{1 + 0.000147(T - 20)\}$$

In the case of sugar the variation of specific rotation with temperature is represented by

$$[\alpha]_t^D = 66.5 - 0.0184\ (T - 20)$$

Thus the specific rotation of sugar at 20ºC is 66.5º. The rotatory powers of a few substances are given in the following table:

Active substance	**Solvent**	$[\alpha]_D^{20°}$
Cane Sugar	Water	+ 66.5°
Glucose	Water	+ 52.5°
Fructose	Water	– 91.5°
Turpentine	Pure	– 37°
Nicotine	Pure	– 162°
Camphor	Alcohol	+ 41°

Specific rotations of certain substances
λ = 5893 Å and T = 20°C

18.9 POLARIMETERS

The famous scientist Biot discovered that a mixture of an optically active substance with an optically inactive one exhibits a rotatory power proportional to the concentration of the optically active compound present. This discovery has paved the way for the application of this phenomenon in commerce and industry, for the estimation of the quantity of sugar in the presence of an optically inactive impurity. Consequently, a number of instruments called *polarimeters* have been designed for the accurate measurement of the angle of rotation. A polarimeter calibrated to read directly in percentage of cane sugar in the solution is known as *saccharimeter*.

The simplest form of polarimeter consists of two Nicols N_1 and N_2, capable of rotation about a common axis and mounted a little distance apart. This simple arrangement, first employed by *Mitcherlich* and described in *Sec.* 18.1, fails to help us in finding accurately the angle of rotation, owing to the difficulty in ascertaining one position of the analyser for complete extinction of the transmitted plane polarised light. This is due to the fact that the analyser can be rotated through an appreciable angle from its setting for complete extinction without allowing any appreciable amount of light to pass through it. Thus the field of view remains totally dark not for a single position but for a considerable range of the rotation of the analyser.

In order to increase the sensitiveness of the pair of crossed Nicols, polarimeters have been designed on what is known as *half-shadow principle*. In these instruments, the field of view is divided into two halves, presented side by side. Consequently, the analyser can be set accurately for the equality of brightness of the two halves. It should be remarked that the eye is a better judge of equality of brightness than either the point of complete extinction or maximum brightness of the whole field of view.

18.10 LAURENT HALF SHADE

Laurent removed the drawbacks of the simple polarimeter by interposing between the two Nicols a device known as *Laurent half-shade*. It consists of a semicircular piece of quartz Q cut parallel to the optic axis (which is in turn parallel to CD). The thickness of this quartz plate is such that a phase difference of π is introduced between the E and O waves during their passage through it. In other words this semi-circular piece CAD is essentially a half waveplate made of quartz. To this piece another semi-circular piece CBD of glass is joined along CD so as to make a composite circular plate $CADB$. The thickness of this glass half G is so adjusted that it transmits and absorbs the same amount of light as the quartz half Q does (Fig. 18.9).

This circular plate is known as a half shade and is placed just behind the polariser in such a manner that the quartz and glass halves both cover a semi-circle each of the field of view. Its action may be explained as follows:

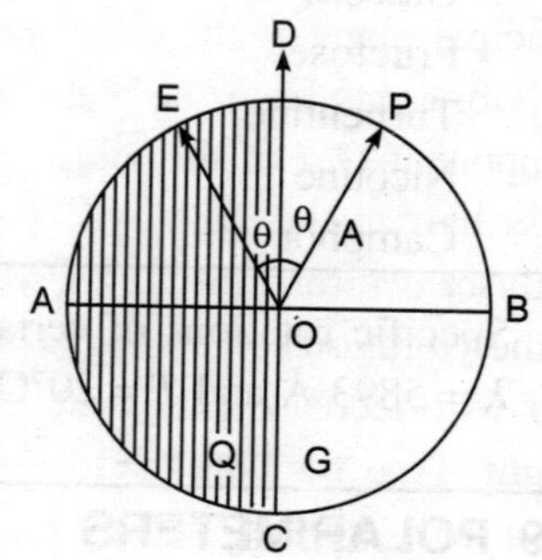

Fig. 18.9. Laurent's plate

Let the vibration plane of the light incident on this glass quartz combination be parallel to OP inclined at an angle θ to the optic axis CD of the quartz half. So far as the glass half is concerned the vibrations of light transmitted through it will remain in the same plane. But a change occurs in the quartz half.

To understand this change suppose that the vibrations incident upon the first face of the half shade are represented by

$$P_i = A \sin \omega t$$

where A is their amplitude and the subscript 'i' stands for the incident wave. This amplitude can be resolved along the two principal directions of the quartz half *i.e.*, along the optic axis (Y direction) and perpendicular to the optic axis (the X direction). The components of A along the X and Y directions are respectively $A \sin \theta$ and $A \cos \theta$. Hence the incident plane polarised vibrations contained in the plane OP can be written as

$$x_i = A \sin \theta \sin \omega t \qquad [O\text{-wave}]$$

$$y_i = A \cos \theta \sin \omega t \qquad [E\text{-wave}]$$

These two mutually perpendicular vibrations travel with unequal speeds along the same direction perpendicular to both the X and Y directions defined above. Since the quartz half is essentially a half waveplate, the two vibrations which emerge out of the quartz half will be

$$x_e = A \sin \theta \sin(\omega t + \pi)$$

$$= -A \sin \theta \sin \omega t$$

$$y_e = A \cos \theta \sin \omega t$$

where the subscript 'e' stands for the emergent beam. The resultant vibration will be given by

$$R = \sqrt{x_e^2 + y_e^2} = A \sin \omega t$$

and the angle ϕ which the plane of R makes with the axis of X is given by

$$\tan\phi = y_e / x_e = -\cot\theta$$

$$\therefore \qquad \phi = (\pi/2 + \theta)$$

It is, therefore, clear that the vibration plane of light emerging from glass half is *OP* while that of the light emerging from the quartz half is *OE* making the same angle θ with the optic axis as does *OP*.

The working of the half shade can now be understood with the help of Fig. 18.10. If the principal plane of the analyser is parallel to *AOB*, the components *OL* and *OM* of the vibrations emerging out of the two halves will be equal and, therefore, they will appear equally bright [Fig. 18.10 (*i*)]. If the principal plane of the analyser is rotated by a small angle in the clockwise direction [Fig. 18.10 (*ii*)] the vibrations from the glass half will be completely cut off (provided *OP* is normal to *A'B'*) while a component *OM* of the vibrations emerging out of the quartz half will be transmitted. In general, the glass half in this case will be less bright as compared to the quartz half. If on the other hand the analyser is turned through a small angle in the anti-clockwise direction [Fig 18.10 (*iii*)] the component of the vibrations emerging from glass half will increase while the component emerging from the quartz half will decrease and if *OQ* is normal to *A'B'* the quartz half will be dark and the glass half will be bright. Thus the half shade serves the purpose of dividing the field of view in two halves. The device is so accurate that if the principal plane of the analyser is rotated through even a small angle with respect to *AOB* a marked change in the intensity of the two halves is observed.

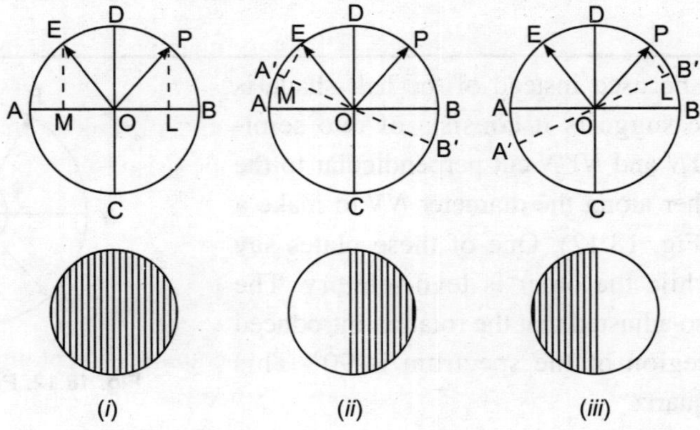

Fig. 18.10. Principle of half-shade polarimeter

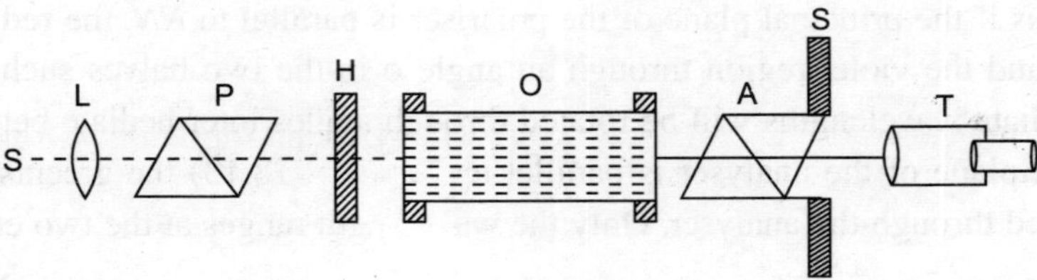

Fig. 18.11. Laurent saccharimeter

The scheme of the polarimeter employing a half shade is shown in Fig. 18.11. Here S is a source of monochromatic light usually a sodium lamp. The lens L is employed to condense light on to the Nicol prism P which serves as the polariser. Immediately behind the polariser the half-shade plate H is kept. The optically active liquid is in the tube O provided with plane transparent glass windows at the ends. The analyser A is mounted coaxially and at the same height as the polariser and the tube O. It can be rotated and its orientation read on the graduated scale S (The polariser is also fitted with a similar device in order to change the half shadow angle θ). Lastly there is a short focus telescope T focused on H and mounted in a manner similar to others.

To measure the optical rotation of any solution, the tube is first filled with the pure solvent and introduced at its position. The analyser A is turned to obtain the condition of equal brightness of the two halves of the field of view. This reading of the analyser is noted.

The solution having a known concentration of the solute is now filled in the tube and the process repeated thereby obtaining a second reading for the analyser. The difference in the two readings gives the rotation produced by the solution. Knowing the length of the tube the specific rotation can be calculated.

To obtain maximum accuracy with the instrument the half shadow angle θ (Fig. 18.9) is set at the proper angle by rotating the polariser. The accuracy depends upon this angle because of the simple reason that the components of the vibrations (emerging out of the two halves) transmitted through the analyser will depend upon it.

The device has, however, a disadvantage for it can be used only with a light source having the wavelength for which the quartz half serves as a half wave plate.

18.11 BIQUARTZ

Another device which can be used instead of the half-shade is the biquartz. As the name suggests it consists of two semi-circular pieces of quartz NYN and $NY'N$ cut perpendicular to the optic axis and joined together along the diameter NN to make a composite circular plate (Fig. 18.12). One of these plates say $NY'N$ is dextro-rotatory while the other is levo-rotatory. The thickness of these plates is so adjusted that the rotation introduced for the greenish yellow region of the spectrum is 90°. This thickness is 3.75 mm for quartz.

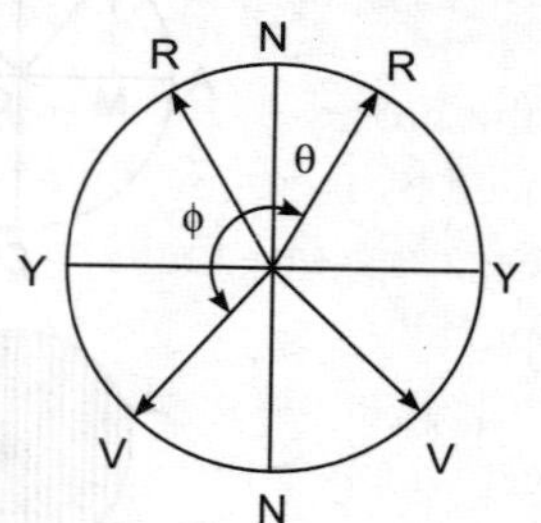

Fig. 18.12. Principle of biquartz

Since the plates are cut perpendicular to the optic axis light travels along the optic axis through these plates and, therefore, the phenomenon of rotatory dispersion takes place. As a result of this the longer wavelengths have their planes of vibrations rotated through a smaller angle as compared to the shorter ones. Thus if the principal plane of the polariser is parallel to NN, the red region is rotated through an angle θ and the violet region through an angle ϕ in the two halves such that ϕ is greater than θ. The intermediate wavelengths will be rotated through angles intermediate between ϕ and θ.

If the principal plane of the analyser is parallel to [illegible]. 18.13) the greenish yellow portion will not be transmitted through the analyser. Only the wa[illegible]gth ranges at the two ends of the visible spectrum can be transmitted equally from the two halves. Their resultant colour in this case will be a mixture of red and blue namely greyish-violet which is called the tint of passage.

If the analyser is turned through a small angle in the clockwise direction [Fig. 18.13 (*b*)] the longer wavelengths from the right half will be transmitted to a greater degree than the shorter ones. For the left half, however, just the reverse is the case–the shorter wavelengths are transmitted to a greater degree than the longer ones. In this particular setting, therefore, the right half will appear red while the left half will appear blue. If on the other hand the analyser is turned in the anti-clockwise direction [Fig. 18.13 (*c*)] the left half will appear red while the right half will appear blue.

It is, therefore, obvious that the tint of passage is a very sensitive position. A slight rotation of the analyser with respect to this position makes one half of the field of view blue and the other half red and *vice versa*. This position is, therefore, employed for the accurate measurement of the angle of rotation. For this purpose, the biquartz is kept in the position occupied by the half shade (Fig. 18.11). The tube is filled with the solvent and introduced in its position. The reading of the analyser for the tint of passage setting is noted. The process is repeated with the solution. The difference between the two readings of the analyser gives the rotation produced by the solution.

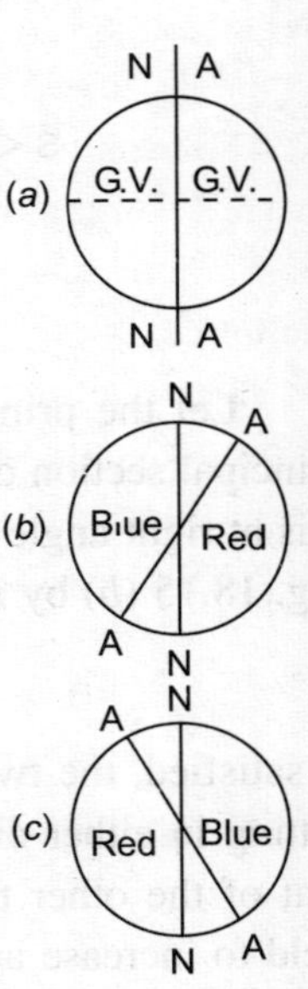

Fig. 18.13

The biquartz possesses a great advantage over a half shade. It can be used with white light but the rotation is obtained for the wavelength which is quenched when the two Nicols are parallel. This is usually the wavelength of sodium light $\lambda = 5893$Å.

The disadvantages of biquartz polarimeter are as follows:

(1) Even if colourless optically active substance are examined, it is practically impossible to achieve the accurate setting of the analyser for which the tint in both the halves of the prism is the same. This is due to the rotatory dispersion produced by the optically active substance itself, and so some error is introduced in measuring the angle of rotation. This error is considerably increased when the optically active substance is coloured.

(2) It is impossible for a colour blind person to use this method. Even man with normal vision will get widely different readings for the angle of rotation, as the relative sensibility curves of individual eyes (to light of different wavelengths) are not identical but show considerable variation.

18.12 LIPPICH'S TWO PRISM POLARISER

Lippich's polarimeter is also based on the half shadow principle and can be used with light of any wavelength. The essential parts of this polarimeter are sketched in Fig. 18.14. The polarising system consists of a bigger Nicol prism N_1 followed by a smaller Nicol N_2 covering half the field. The principal sections of N_1 and N_2 are mutually inclined at a small angle 2ϕ, called the half shadow angle. The field of view as seen through the telescope and the analysing Nicol is, therefore, divided into two halves separated by a sharp line—the image of the near edge T of the prism N_2. One half of the field is illuminated by plane polarised light produced by the light which passes through N_1 and N_2 and hence the emerging light has vibrations in the principal section of N_2. The other half is illuminated by light which passes through N_1 only and hence the emerging light has vibrations in the principal section of N_1. Furthermore, the amplitude of vibration in the plane polarised light from N_2 is slightly less than the amplitude of vibration of light emerging from N_1 only, because a small fraction of the later amplitude in the upper half of the beam emerging from N_1 is totally reflected as ordinary vibrations within the smaller Nicol N_2, while the amplitude of the lower half of the beam emerging from N_1 remains unchanged. (*O* rays are not shown in N_1.)

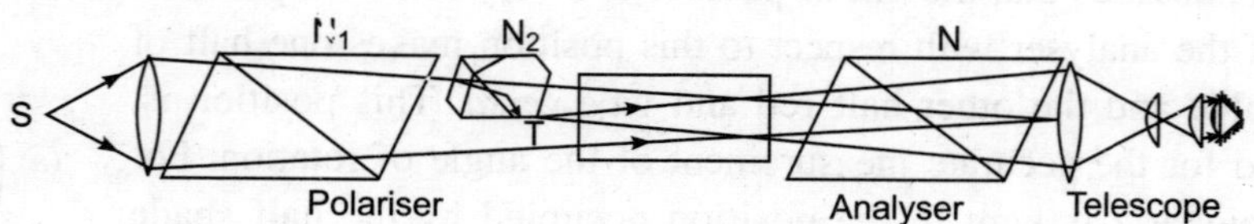

Fig 18.14. Lippich's (two prism polariser) polarimeter

Let the principal section of N_1 and N_2 be represented by AN_1 and AN_2. Accordingly, if the principal section of the analyser is at right angle to AN_2, one half of the field is obviously dark; but if it is at right angle to AN_1, the other half is dark. Consequently, for an intermediate position, shown in Fig. 18.15 (*b*) by *NN*, so that the relation

$$r_1 \cos \theta_1 = r_2 \cos \theta_2$$

is satisfied, the two halves of the field are equally bright. A slight rotation of the analyser from this setting in either direction causes the transmitted component of one vibration to increase rapidly and that of the other to decrease rapidly, thereby causing the intensities of corresponding halves of the field to increase and decrease rapidly. The analyser can be, therefore, very easily set for the position of quality of brightness of the two halves of the field. Thus, Lippich's polarimeter employing two Nicol prisms as polariser is exactly similar in action to Laurent's half-shade polarimeter. It has a distinct advantage over the latter in as much as light of any wavelength can be employed while working with it which is, however, not possible in using Laurent's polarimeter.

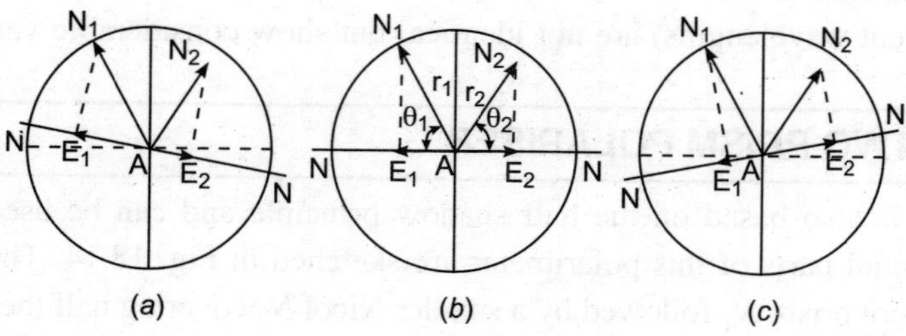

Fig. 18.15. Principle of Lippich's polarimeter

18.13 LIPPICH'S THREE PRISM POLARISER

In modern polarimeters, the accuracy in the measurement of the angle of rotation is increased by dividing the field of view into three parts. The two outer parts are equally illuminated for every setting of the analyser while the central portion which may be illuminated differently from the adjacent parts, has to be adjusted for the same intensity (as that of outer parts) by the help of the analysing Nicol. In Nicol modern polarimeter this is achieved by employing a polarising system consisting of three Nicols N_1, N_2 and N_3, arranged as shown in Fig. 18.16. The principal section of N_2 and N_3 are mutually parallel but slightly inclined to that of the bigger Nicol N_1. Accordingly, the central parts of the field are illuminated by light emerging from N_1 while the outer parts are illuminated by the light which passes through N_1 and one of the prisms N_2 and N_3. Let AN_1, AN_2 and AN_3 respectively represent the principal sections of N_1, N_2 and N_3. Consequently, AN_1, AN_2 and AN_3 also represent the vibrations incident on the analysing Nicol. If the analyser is set in a position so that its principal section occupies the position NN satisfying the relation

$$AN_1 \cos \theta_1 = AN_2 \cos \theta_2 = AN_3 \cos \theta_3$$

the three parts of the field will be uniformly illuminated. A slight rotation of the analyser from this setting in either direction causes the intensity of the central portion of the field to be markedly different from that of the two outer portions and so the accurate setting of the analyser for uniform illumination of the field can be very easily accomplished.

This polarimeter is used for most accurate work. In the half shadow instrument, a slight displacement of the eye from the axis of the instrument may match the two halves at a wrong setting of the analyser. This error is not possible with the polarimeter employing three prism polarising system, for a slight displacement of the eye from the axis would cause the side towards which the eye is displaced, to appear too bright as compared to the other side. Hence the experimenter would immediately adjust his eye on this axis. Owing to the same reason, the experimenter can correct any slight error in the alignment of the light source with the optical axis of the polarimeter.

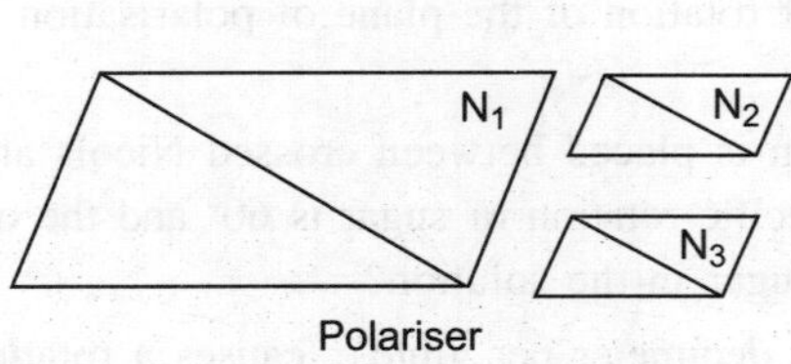

Fig. 18.16. Lippich's three prism polariser

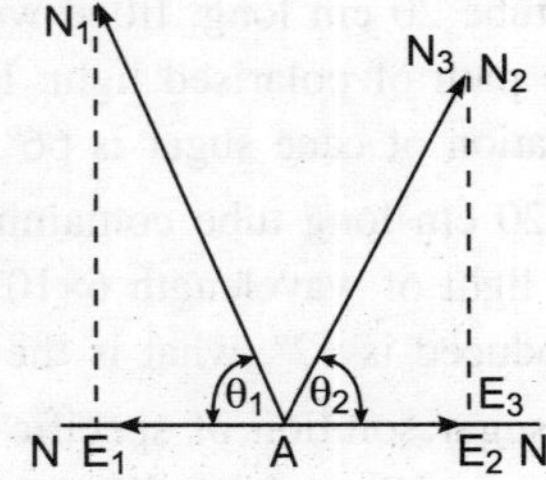

Fig. 18.17

PROBLEMS

1. State the laws of rotary polarisation. Give Fresnel's hypothesis of rotary polarisation and derive the formula for the rotation produced by quartz. Give the experimental verification of this formula.
2. Discuss the phenomenon of rotation of the plane of polarisation of light by optically active materials. Give the necessary theory.

   Show that the rotation of the plane of vibration is given by

   $$(\pi d/\lambda)(n_A - n_C)$$

   where n_A and n_C are respectively the refractive indices of the crystal in the direction of the optic axis for anti-clockwise and clockwise circularly polarised light and 'd' is the thickness of the crystal plate.
3. What is specific rotation?

   Give with full details the construction and working of a Laurent's half-shade polarimeter. Explain how you would use it to determine the specific rotation of sugar.
4. Describe briefly the various methods that may be used to increase the sensitively of a pair of crossed Nicols in polarimetry.
5. What is optical rotation? Describe a polarimeter using a biquartz plate and explain how you would use it to find the specific rotation of an optically active substance.
6. (a) What do you understand by rotary polarisation and rotary dispersion?

   Describe either the half-shade or the biquartz polarimeter explaining the function of each part. How will you use it to measure the rotation of plane of polarisation?

   (b) Discuss the relative merits of biquartz and half shade polarimeter.
7. Calculate the specific rotation of sugar solution from the following data:

Length of tube containing solution	= 11 cm.
Volume of solution	= 44 c.c.
Amount of sugar in solution	= 3 g
Angle of rotation	= 4°57′

8. A tube 20 cm long, filled with a solution of 15 g of cane sugar in 100 c.c. of water is placed in the path of polarised light. Find the angle of rotation of the plane of polarisation if the specific rotation of cane sugar is 66°.
9. A 20 cm long tube containing sugar solution is placed between crossed Nicols and illuminated by light of wavelength 6×10^{-5} cm. If the specific rotation of sugar is 60° and the optical rotation produced is 12°, what is the percentage of sugar in the solution?
10. A sugar solution of specific rotation 52° per decimeter per gm/c.c. causes a rotation of 12° in a column 10 cm long. What is the concentration of the solution?
11. A solution of camphor in alcohol in a tube 20 cm long is found to rotate the plane of vibration of light by 27°. What is the mass of camphor in unit volume of the solution? The specific rotation of camphor is + 54°.

Appendices

FABRY-PEROT ETALON

FABRY-PEROT INTERFEROMETER AND ETALON

A multiple beam, high resolution interferometer was designed by C. Fabry and A. Perot in 1899. This yields bright fringes excelling the Michelson's fringes in sharpness. Extremely narrow bright fringes impart to this instrument very high resolving power in the sense that the presence of two close components in the incident light give rise to two sets of clearly separated bright fringes in the interference pattern. Thus, one can observe directly the fine structure of spectral lines which Michelson could only infer from the visibility curves. Fabry-Perot instrument is, therefore, most suitable for the study of hyperfine structure of the spectral lines.

The Fabry-Perot interferometer consists essentially of an air space bounded by two high quality glass or quartz plates with inner surfaces optically plane and parallel to less than $\lambda/20$. These faces are uniformly coated with partially transparent highly reflecting thin film of silver or aluminium. Each plate is slightly prismatic; the outer unsilvered surfaces are at an angle 1 to 10 min. of arc away from the silvered surfaces. This is essential in order to throw to one side the false interference pattern formed by reflections between a silvered and the outer unsilvered surfaces which is thus not superposed on the pattern formed due to multiple reflections between the inner silvered surfaces. The slight prismatic dispersion due to wedge shaped plates does not materially affect the observed fringe system. In the interferometer the plate facing the observer is fixed to the frame of the instrument and is provided with screws pressing on spring for achieving accurate parallelism of the silvered surfaces. The other plate is mounted on a carriage which can be moved along precision machined straight and optically parallel tracks, in a direction perpendicular to the reflecting surfaces. This slow and accurately controlled motion is accomplished by means of a calibrated accurate screw provided with a wheel and a graduated circle. However, due to difficulties of mechanical construction this arrangement is now obsolete.

The *etalon* is now invariably used for research. In the etalon two semi-silvered plates are mounted in a frame work. Their coated faces are kept in constant parallel position a fixed distance apart by a fixed spacer, which is commonly a hollow cylinder of invar or silica with three projecting studs at each end. The plates are kept in place slightly pressed against the spacers by the pressure of adjustable springs attached in a proper way to the etalon housing. Etalons are supplied with a variety of spacers of lengths ranging from 1 to 200 mm for use in the investigation of hyperfine structure of spectral lines.

Theory of Fabry-Perot Instrument

Let *EF* and *GH* represent two plane parallel lightly silvered surfaces of the Fabry-Perot instrument. Light from a broad source of monochromatic light is incident on the plates at all angles. Consider a plane wave of amplitude A_i travelling along *AB* and incident on *EF* at an angle θ to the normal. The series of parallel transmitted waves C_1T_1, C_2T_2, C_3T_3... arise from the same incident wave by its internal multiple partial reflections and refractions between *EF* and *GH*. These waves being coherent interfere when brought to a focus at *P* in the focal plane of an achromatic lens *L*. The problem here is one of superposition of a very large number of waves of falling amplitudes. For facilitating its discussion we shall assume that the medium bounded by the silvered surfaces *EF* and *GH* is uniform and continuous and has a refractive index μ. In other words, the problem may be taken as that of interference due to a long thick air plate of uniform thickness.

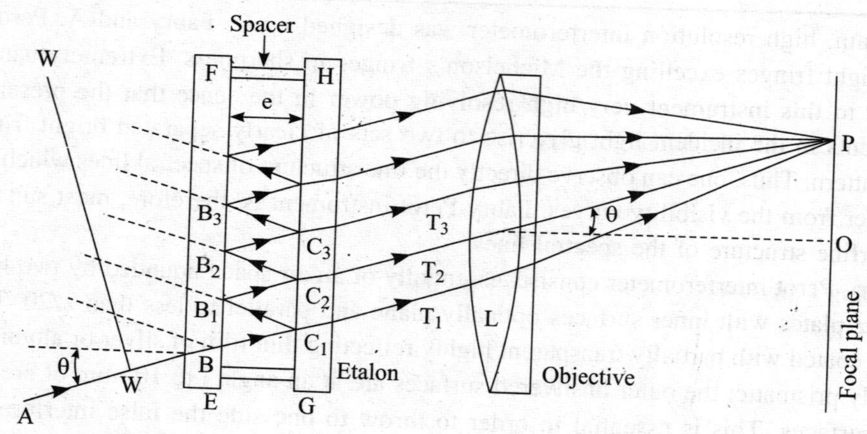

Fig. A-1. Formation of multiple reflection fringes by Fabry-Perot etalon

Suppose that for the wave *travelling from the surrounding medium into the plate* the reflection coefficient (defined as the ratio of the reflected to the incident amplitude) is *r* and the transmission coefficient (defined as the ratio of transmitted and incident amplitude) is *t*. Also, suppose that the corresponding values for the wave travelling from the plate into the surrounding medium are respectively *r'* and *t'*. The amplitudes of the successive transmitted waves may then be written as

$$tt'A_i, tt'r'^2 A_i e^{i\delta}, tt'r'^4 A_i e^{2i\delta}, ..., tt'r'^{2(n-1)} A_i e^{i(n-1)\delta}$$

where δ is the phase difference between any two consecutive waves introduced due to a path difference $2\mu e \cos\theta$ on reaching *P*. Thus

$$\delta = \left(\frac{2\pi}{\lambda}\right) 2\mu e \cos\theta \qquad ...(A\text{-}1)$$

Similarly the amplitude of the successive reflected waves may be written as

$$rA_i, tt'r'A_ie^{i\delta}, tt'r'^2A_ie^{2i\delta} ....., tt'r'^{(2n-3)}A_i^{ei(n-1)\delta}$$

It has been seen in *Sec.* (12.1) that

$$tt' = T \qquad ...(A\text{-}2)$$

and

$$r = -r' \qquad ...(A\text{-}3)$$

so that

$$r^2 = r'^2 = R \qquad ...(A\text{-}4)$$

where T and R are respectively the transmitivity and reflectivity of the assembly *EFGH*. Such that

$$R + T = 1 \qquad ...(A\text{-}5)$$

The resultant amplitude due to the superposition of 'n' transmitted waves is therefore given by

$$A_t = tt'[1 + r'^2e^{i\delta} + r'^4e^{2i\delta} + ... + r'^{2(n-1)}e^{i(n-1)\delta}]A_i$$

$$= tt'\left(\frac{1 - r'^{2n}e^{in\delta}}{1 - r'^2e^{i\delta}}\right)A_i$$

If there are an infinite number of waves reaching P, $n \to \infty$ and therefore

$$A_t = \frac{tt'}{1 - r'^2e^{i\delta}}A_i \qquad ...(A\text{-}6)$$

Combining Eqns. (A-2), (A-4) and (A-6) we get

$$A_t = \frac{T}{1 - Re^{i\delta}}A_i \qquad ...(A\text{-}7)$$

The intensity of the resultant illumination at P then we can be determined by multiplying Eqn. (A-7) with its complex conjugate. Thus

$$I_t = \frac{T}{1 - Re^{i\delta}} \cdot \frac{T}{1 - Re^{-i\delta}}I_i$$

$$\frac{I_t}{I_i} = \frac{T^2}{1 + R^2 - 2R\cos\delta} = \frac{T^2}{(1-R)^2 + 4R\sin^2\delta/2}$$

$$= \frac{T^2}{(1-R)^2}\frac{1}{[1 + 4R/(1-R)^2\sin^2\delta/2]} \qquad ...(A\text{-}8)$$

Similarly the resultant intensity due to the superposition of an infinite number of reflected waves may be shown as

Fabry-Perot Interferometer

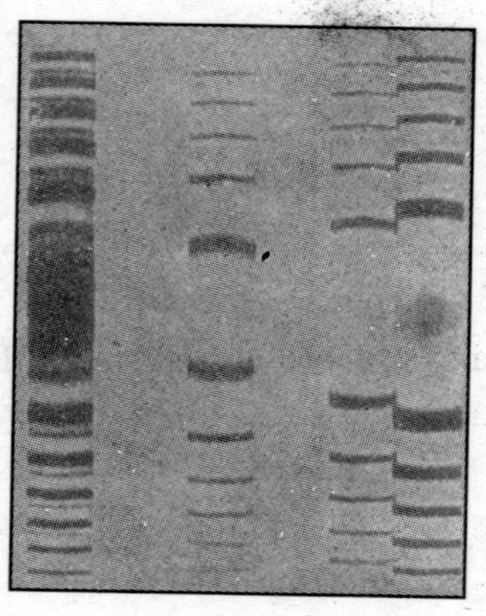

Helium spectrum, 3 mm etalon (negative)

Fabry-Perot etalon

$$I_r = \frac{4R\sin^2 \delta/2}{(1-R)^2 + 4R\sin^2 \delta/2} I_i \qquad \text{...(A-9)}$$

The resultant transmitted intensity varies with the phase δ and with the film properties. It is maximum at points where $\delta = 2n\pi$ and minimum when $\delta = (2n + 1)\pi$ where n is an integer. In terms of the optical path difference, the conditions are

$$\delta = n\lambda \qquad \text{[Max.]} \qquad \text{...(A-10)}$$

$$\delta = \left(n + \frac{1}{2}\right)\lambda \qquad \text{[Min.]}$$

The integer n is called the order of interference. Inserting $\delta = 2n\pi$ and $\delta = (2n + 1)\pi$ successively in Eqn. (A-8) we have

$$I_{max} = T^2/(1-R)^2 \quad \text{and} \quad I_{min} = T^2/(1+R)^2$$

When there is no absorption at each reflecting film, $T = (1 - R)$ then

$$I_{max} = 1 \quad \text{and} \quad I_{min} = (1-R)^2/(1+R)^2$$

No matter what the reflecting power may be I_{max} is equal to the intensity of the incident light when there is no absorption in the film.

The minima are situated half way between the two maxima. Moreover I_{min} never reaches zero at any point in the fringe system. I_{min} is a decreasing function of R and so when R is near to unity the minima are very weak.

Visibility of Fringes

The visibility of fringes is given by

$$V = \frac{I_{max} - I_{min}}{I_{max} + I_{min}} = \frac{2R}{1+R^2}$$

which is independent of the transparency of the film. It depends on the reflecting power only but the visibility increases from 0.8 to 1 only as the reflecting power increases from 50% to 100%.

Sharpness of Fringes

Following *Fabry* the intensity distribution within the fringes may be written as

$$I = I_{max} \Big/ \left(1 + F \sin^2 \frac{1}{2}\delta\right) \qquad \text{...(A-11)}$$

where $F = 4R/(1 - R)^2$ and was called by Fabry the coefficient of finesse, since, as explained in the sequence, it determines the sharpness of the fringes.

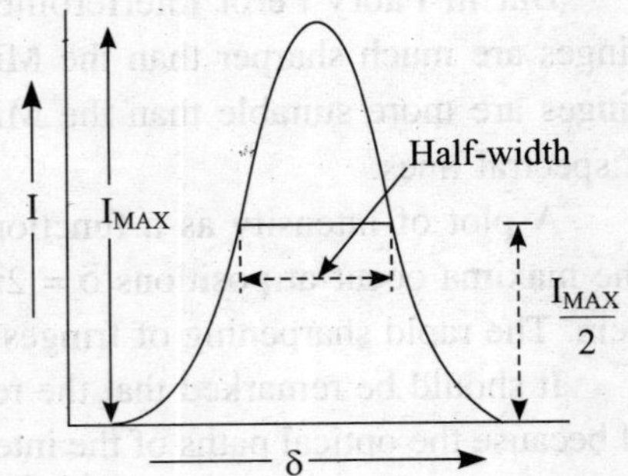

Fig. A-2. Half-width of a fringe

A measure of sharpness of fringes *i.e.*, how rapidly the intensity diminishes on either side of the maximum, is the fringe half-width, illustrated in Fig. A-2. The fringe half width is the total width of a fringe at points where the intensity has fallen to half the intensity maximum. If this point on the intensity curve differs in phase by δ' from the maximum, the equation for determining δ' can be easily obtained by substituting $I = 1/2\, I_{max}$ in Eqn. (A-11) and we get

$$\sin\frac{1}{2}\delta' = 1/\sqrt{F}$$

For values of R greater than 1/2, $1/\sqrt{F}$ is so small that we may write $\sin 1/2\delta' = 1/2\delta'$ radian without any serious error. Then we get

$$\delta' = 2/\sqrt{F}$$

But the phase difference between successive maxima is 2π, hence δ' expressed as fraction of inter-order distance is $1(\pi\sqrt{F})$. The total half width, measured in order numbers is, therefore, given by

$$W = 2\delta'/2\pi = 2/(\pi\sqrt{F})$$

$$W = (1-R)/(\pi\sqrt{R}) \qquad \text{...(A-12)}$$

The half-fringe width and therefore, the sharpness of the interference fringes depends upon the coefficient of finesse $F = 4R/(1-R)^2$. For getting sharp lines, obviously this factor has to be increased upto a certain value by increasing the thickness of the metal film. The values of F and W (as a fraction of inter-order distance) for different values of R are tabulated below.

R	0.5	0.7	0.8	0.9	0.94	0.98
F	8	31	80	360	1044	9800
W	0.225	0.114	0.071	0.033	0.020	0.006
N_{eff}	4.3	8.4	13.6	28.8	49	150

The bright fringes are extremely narrow for higher reflectivities.

In the Michelson interferometer fringes are formed due to interference of two waves of equal amplitudes and so the intensity distribution in the fringes is given by

$$I = I_{max}\cos^2\frac{1}{2}\delta$$

and when $I = 1/2\ I_{max}$, we get the corresponding δ',

$$\delta' = 2\cos^{-1}(1/\sqrt{2}) = \pi/2 = 1.57 \text{ radians}$$

But in Fabry-Perot interferometer if $R = 0.9$, then $\delta' = 0.105$ radian. Thus, the Fabry-Perot fringes are much sharper than the Michelson fringes. Due to their extreme narrowness, Fabry-Perot fringes are more suitable than the Michelson fringes for the investigation of the hyperfine structure of spectral lines.

A plot of intensity as a function of δ for selected values of reflectivity is shown in Fig. A-3. The maxima occur at positions $\delta = 2\pi, 4\pi, 6\pi$ etc. with relatively broad regions of minima between them. The rapid sharpening of fringes with increasing R is very striking in Fig. A-3.

It should be remarked that the refraction in the glass plate GH does not affect the final result at all because the optical paths of the interfering waves in the plate are all equal to each other. Neglecting the loss of light by the internal reflections in the glass plates, this interferometer acts as if we are dealing only with the plane parallel air plate whose faces are characterised by a high reflecting power. Therefore, this interferometer is often called *the silvered plane parallel air plate.*

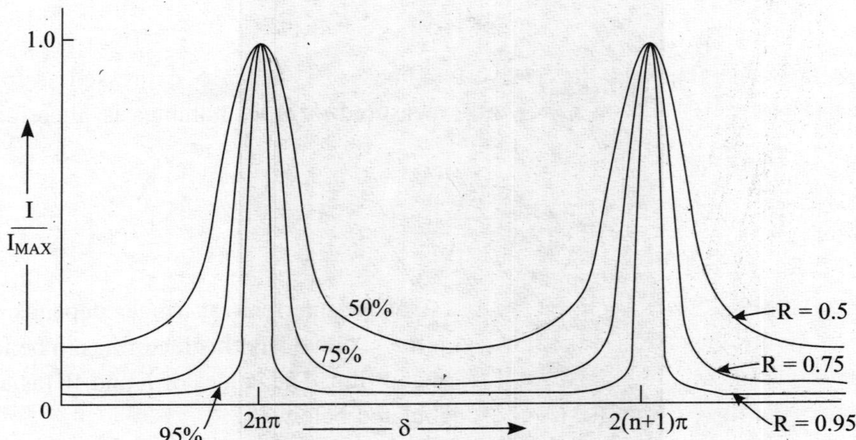

Fig. A-3. Intensity distribution in multiple beam fringes

Shape of Fringes

The condition for reinforcement of the transmitted waves given by Eqn. (A-10) may be written as

$$2\, e \cos \theta = n\lambda \qquad \text{...(A-13)}$$

for an air film at atmospheric pressure. For a given value of e, λ and n, θ must be a constant, which is obviously true for all points on a circle through P, of radius OP and centre at O, the principal focus of the lens L, provided the silvered surfaces are accurately parallel.

A given fringe corresponds to a constant value of θ. In reality, light is not only incident on the silvered surfaces in one direction θ but in various directions. Therefore, for any wavelength there will be several values of θ for which n is an integer. Hence, the maxima are a series of exceptionally sharp concentric rings with relatively wide dark regions between them, with the principal focus O as their common centre.

(*a*)

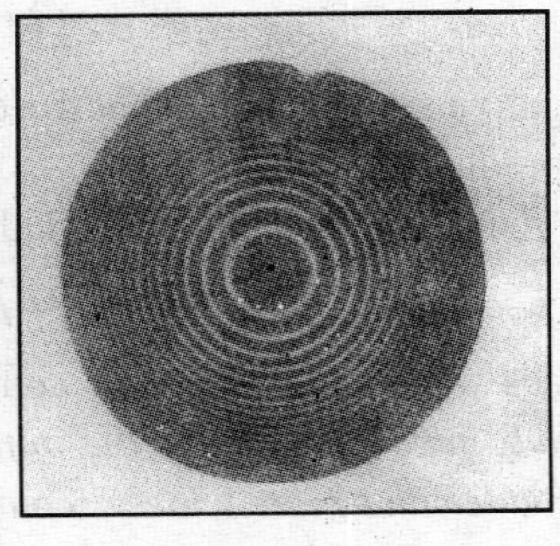

(*b*)

Comparison of the fringes produced with (*a*) the Michelson interferometer and (*b*) the Fabry-Perot Interferometer

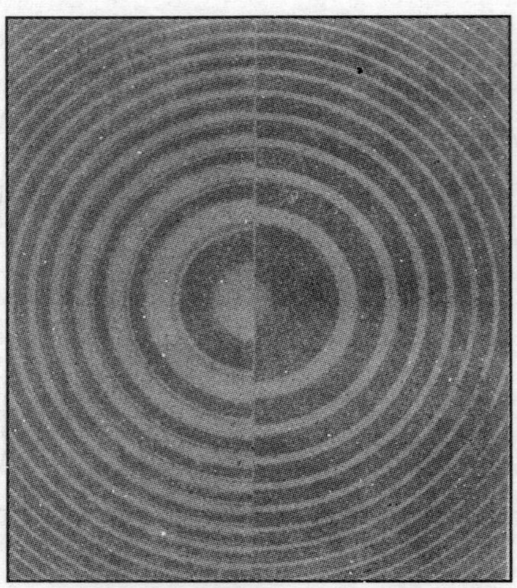

Fabry-Perot patterns with green light 5461 A from (*a*) natural mercury and (*b*) from artificial mercury 198. (By courtesy – Dr. William F. Meggers, National Bureau of Standards)

From Fig. A-4, we can see that

$$\cos\theta = \frac{f}{\sqrt{f^2 + R^2}} = \left(1 + \frac{R^2}{f^2}\right)^{\frac{1}{2}} = \left(1 - \frac{1}{2}\frac{R^2}{f^2}\right) = \left(1 - \frac{D^2}{8f^2}\right)$$

where D is the diameter of the bright ring. Introducing this value of cos θ in Eqn. (A-13) and solving for n, we get

$$n = (2e/\lambda)\cos\theta = (2e/\lambda)(1 - R^2/2f^2) \quad \text{...(A-14)}$$

Differentiating this equation, we get

$$dn = -(2e/\lambda)\sin\theta d\theta = -(2e/\lambda)(R/f^2)dR$$

The negative sign signifies that the order of fringes decreases with increasing angle of incidence or increasing radius R.

For $dn = -1$ the above formula gives

$$\Delta R = \lambda f^2/2eR$$

which indicates that *for larger radii consecutive circles are closer together. Near to the centre the rings are widely separated but in the outer field rings are closer together. The closer the two etalon plates, the broader and more widely separated will be the fringes.*

Furthermore when e and f are constant *i.e.*, in any given arrangement, in the neighbourhood of certain value of R, bright rings of shorter wavelength are closer together as compared to those of the longer wavelength.

From Eqn. (A-14), it is clear that to equal increments of n there correspond unequal increments of R but equal increments of R^2. *Hence, square of the diameters represent an arithmetical progression, the common difference being*

$$\Delta D^2 = D_{p+1}^2 - D_p^2 = 4\lambda f^2 2e$$

The maximum value of the order is at the centre of the ring system. If the order of the first ring from the centre is given by Eqn. (A-14), the order of the second ring will be given by

$$(n-1) = 2e\cos\theta'/\lambda$$

and so on. But the order at the centre of the interference pattern, given by $n_o = 2e/\lambda$, is not necessarily an integer and it may be written as $n_o = n + \varepsilon$, where ε is a small fraction less than 1. The Eqn. (A-13) is exactly identical with Eqn. (11.6 *e*) for the Michelson interferometer. Therefore, the spacing of Fabry-Perot rings is the same as that of the circular fringes in that instrument, provided e has the same value in both the instruments.

The Fabry-Perot fringes arising due to interference of infinite number of light waves of constant inclination to the axis are called *Haidinger fringes*. Path difference of several centimeters may be used without loss of visibility of fringes, and hence very high orders of rings may be examined.

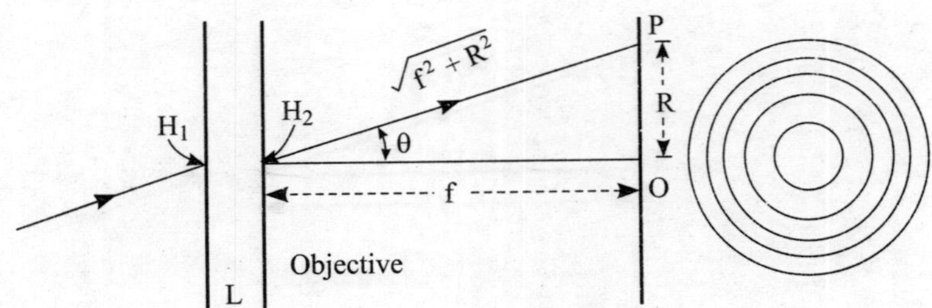

Fig. A-4. Formation of circular fringes in the focal plane of a lens

Spectral Range without Overlap

The wavelength difference $(\Delta\lambda)_{S.R.}$ corresponding to a displacement of one order $(dn = 1)$ is called the spectral range of the interferometer. Near normal incidence Eqn. (A-13) gives

$$n\lambda = 2e$$

Hence $\qquad \lambda\, dn + n\, d\lambda = 0$

Inserting $dn = 1$, we get

$$(\Delta\lambda)_{S.R.} = -\lambda/n = -\lambda^2/2e \qquad \text{...(A-15)}$$

The spectral range $(\Delta\lambda)_{S.R.}$ is the maximum wavelength difference that can be examined without overlapping of orders with this instrument alone.

The spectral range can be expressed in wave numbers also because of the relation $\bar{v} = 1/\lambda$.

$$(\Delta\bar{v})_{S.R.} = -\frac{(\Delta\lambda)_{S.R.}}{\lambda^2} = \frac{1}{2e} \qquad \text{...(A-16)}$$

This indicates that the range in wave numbers is a constant for a given etalon, independent of the order and wavelength as well. It varies inversely as the separation between the plates. For $e = 0.5$ cm, the spectral range is 1 cm^{-1} and for $e = 20$ cm it is 0.025 cm^{-1}. Thus, with the help of different spacers the free spectral range can be varied at will. Consequently, the variable gap etalon is superior to any other single instrument. The increased spectral range without overlapping of order is an important asset in the study of hyperfine structure of spectral lines.

Uses

The Fabry-Perot etalon is generally used for:

1. Determination of the wavelength of light.
2. Determination of the difference between two close wavelengths.
3. Study of the hyperfine structure of spectral lines. (Of all the spectroscopic devices the etalon possesses the highest resolving power).

ELECTROMAGNETIC NATURE OF LIGHT

It has been seen in earlier chapters that Electromagnetic theory suffices to explain the main features of the phenomena associated with the propagation of light. The first definite indication of the electromagnetic character of light was however, obtained in 1845 by *Faraday* who observed the rotation of plane of polarisation of light through glass when a strong magnetic field was applied parallel to the beam of light. Faraday's ideas were translated into mathematical expression by *Maxwell*. The theory received direct experimental confirmation in 1888 when *Hertz*, produced electromagnetic waves from an oscillatory electric current. He also demonstrated that these waves could be reflected, refracted, focussed by a lens, polarised and so on. In fact, they exhibited properties similar to those of light. In addition, it was experimentally determined that electromagnetic waves are propagated in free space with the same velocity as that of light in vacuum. The evidence thus went far to show that *light consisted of electromagnetic waves*.

19.1 THE ELECTROMAGNETIC SPECTRUM

The solar spectrum as seen by human eye extends from violet to red regions of wavelengths.

However, it was discovered by *Herschel* that on the longer wavelength side it extends *beyond* the red region. Although invisible to the human eye these radiations can be detected if the proper device is used. Thus Herschel observed that a thermometer with a blackened bulb when moved beyond the red region continued to show a rise in the mercury column. Also, *Ritter* discovered that a photographic plate is affected when exposed to radiations of wavelength shorter than the shortest visible radiations. These discoveries led to the belief that the solar spectrum actually extends far beyond on either side of violet and red regions and it is only a very small region of this spectrum that is visible to man. Gradually,

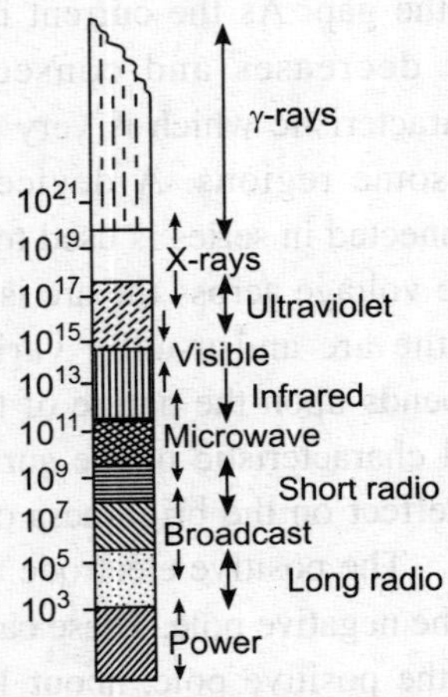

Fig. 19.1. Electromagnetic spectrum

this was seen to be actually true — only the proper detecting devices were to be employed. At present radiations extending from *Cosmic rays* on one end to *Radio waves* on the other are known to exist. Although these radiations have different physical properties, all of them move with the same velocity in free space. This velocity is equal to the velocity of transverse waves in free space. This whole family of wavelengths goes under the collective name of *Electromagnetic spectrum* (Fig. 19.1).

The so-called light which produces to sensation of vision extends from about 4000 A.U. on the shorter wavelength side to about 7000 A.U. on the longer wavelength side. This range is a very insignificant part of the whole spectrum. A short account of the different methods of generation, detection and properties of some of these radiations is given below.

19.2 THE VISIBLE REGION

Practically all modern sources of light operate on electrical energy and, therefore, the study of light sources is a very specialised branch of engineering. In optical studies, however, we are not very much concerned about these technical questions. We are nevertheless concerned about laboratory sources of light, their shape, size and brightness. We therefore, give below short descriptions of the more important sources of light in laboratory use.

Carbon Arc

Before the advent of incandescent lamps the carbon arc devised by *Davy* in 1801 was the most popular light source. The equipment required is very simple. The only things needed are two carbon electrodes with insulating holders, a D.C. source of about 100 volts and ballast resistance. These are connected in series.

When the carbon rods are brought into contact with each other and then drawn apart an arc is struck. The heat generated in the column of the incandescent gas with a rising air current causes it to take the form of an arc and hence the name.

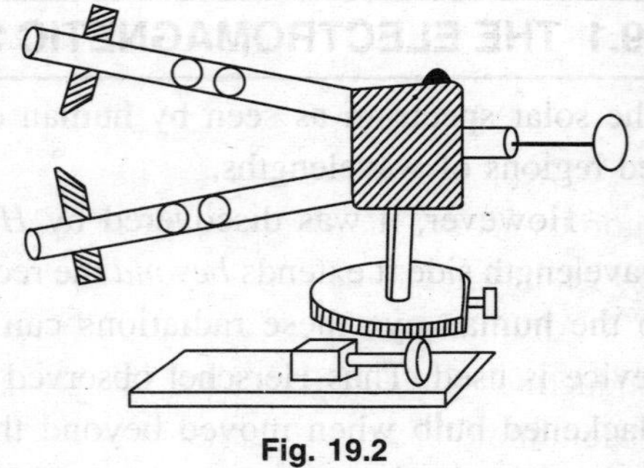

Fig. 19.2

The carbon electrodes are placed at an angle and are provided with special control mechanisms for the adjustment of the gap. As the current increases, the resistance of the arc decreases and consequently the voltage current characteristic which is very steep may have negative slope in some regions. A device usually a ballast resistance connected in series is used to limit and stabilise the current. The voltage across the arc is a linear function of the length of the arc and usually varies from 40–70 volts. It also depends upon the nature of the electrodes and the pressure and characteristic of the surrounding gas. Increasing the current through the arc has got practically no effect on the brightness of the arc.

The positive electrode at a temperature of about 4000ºC repels out carbon particles which travel to the negative pole. These carbon particles radiate light. Nearly 85% of the luminous flux is contributed by the positive pole, about 10% is contributed by the negative pole at temperature of 2000ºC while the remaining 5% is contributed by the arc itself. The continuous "repulsion" of the carbon particles from the positive pole causes the latter to wear and in course of time a "crater" is produced. This cup-shaped depression produces most of the light. An increase in arc current only causes this crater to increase in area.

The contribution of the arc itself is very small. This can be increased by drilling the electrodes lengthwise and filling them with cores of inorganic salts. An arc having a core consisting of the fluorides of rare earths is used for photography. A blue flame arc having a core of iron salts emitting the vital rays (290–310 mμ) copiously is used for therapeutic purpose. Yet another arc for the same purpose has a nickel wire in its core.

The Electric Spark

Whenever a higher excitation is required, the electric spark is invariably used. It is produced by connecting the secondary of a transformer giving about 10 to 50 kilovolts across an insulated electrode holder carrying the electrodes. Invariably a condenser which increases the discharge across the gap is also used. A small amount of inductance is also necessary to prevent the excitation of the line and band spectra of air molecules. These spectra if present cause lot of confusion and background in the spectrum. If a larger inductance is used it decreases the initial current and also the degree of excitation of the spark. The degree of excitation has been found to be higher with a condenser of larger capacity, larger initial current and rate of energy dissipation.

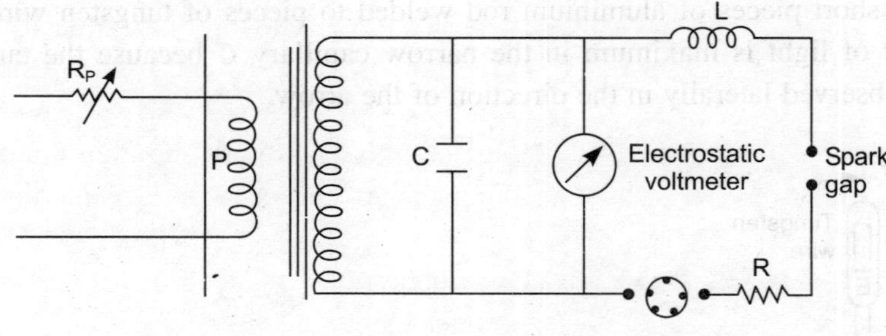

Fig. 19.3

The temperature of the spark discharges is always higher than that of arcs. In most of the vigorous sparks the temperature may be as high as 10,000ºC. The higher the temperature the lower will be the excitation.

Any conducting electrodes can be directly used in spark. When the substances are rare, non-conducting, powdered or otherwise unsuitable for use as electrodes, they are packed in hollow carbon or aluminium electrodes, alloyed with other material. They can also be packed and sintered with metal powders into a solid electrode. A *vacuum spark* operating in vacuum has also been devised and is used for the excitation of lines in the extreme ultraviolet.

Geissler Tubes

Gases at ordinary atmospheric pressure do not conduct electricity but if the pressure be reduced, their resistance assumes quite a low value to permit flow of electricity if a high voltage is applied. The high voltage ionises the gas and makes it conducting.

Ionisation is a mild and superficial form of atom splitting. Every atom of the gas (or any other form of matter) consists of a nucleus having a positive charge and electrons revolving round the nucleus in fixed orbits. If by some means, an electron is detached from the atom the electrical equilibrium existing inside the atom is disturbed. The removal of the electron from the atom causes it to attain an excess positive charge. The gas gets ionised.

When a gas contained in a tube containing two electrodes is ionised the electron carrying a negative charge is attracted towards the anode while the "ion" will be attracted towards the cathode. The electron being lighter of the two acquires a large amount of kinetic energy. Several atoms of the gas may be simultaneously ionised. The electrons moving with great energy encounter other atoms of the gas contained in the tube. An electron however may again fall back into the residual part of an already ionised atom under the mutual force of attraction between the two thereby making it a complete atom.

While some of the electrons do combine with the ions (residual atom) and convert them into complete atoms, there are others which collide with ionised atoms of the gas. If the kinetic energies of these electrons are quite large, they may dislodge electrons on collision with other atoms. These detached electrons would then move towards the anode. If however the energy of the colliding electron is not sufficiently high the detached electron is not removed from the sphere of influence of the atom concerned. It is only displaced. These displaced electrons when they return to their normal position give rise to electromagnetic radiations.

This principle is made use of in construction of *Geissler's tubes* for laboratory purposes. The electrodes *E*, *E* are short pieces of aluminium rod welded to pieces of tungsten wire fused through glass. The intensity of light is maximum in the narrow capillary *C* because the current density is greatest here. It is observed laterally in the direction of the arrow.

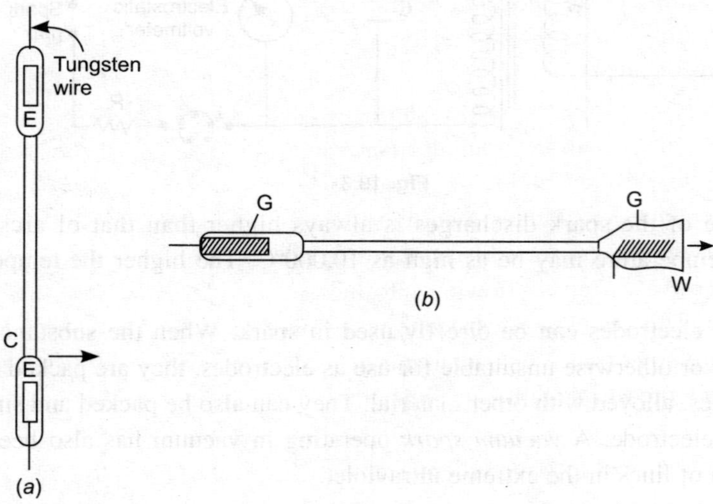

Fig. 19.4

If greater intensities are required the "end on" type (Fig. 19.4) is used. The electrodes in this case are sheets of aluminium which are rolled and then slipped into two loosely fitting glass tubes. They are fastened to tungsten leads. Because of the larger area of the electrodes greater currents can be used without causing any undue heating of the electrodes. The light is observed longitudinally through the glass window *W*:

The pressure at which the tube must be sealed off varies according to the nature of the gas and may be between 0.5 mm to 1 cm of Hg. Only a few gases, can be used for long periods in such tubes. They are the inert gases, helium, argon and neon. Other gases *e.g.*, hydrogen, nitrogen and carbon dioxide last only for a comparatively small time.

Sodium Lamp

It consists of a tubular hairpin shaped glass envelope G.G. An electrode *E* is sealed into each end of the tube. The lead in wires are connected to contacts in the lamp base. One of the electrode is connected to a narrow metal strip *S* called the starting strip. This strip is located between the two legs of the tube and is anchored at the end by a fork *V*. An inert gas mainly *neon* is filled into the tube at a low pressure. Some quantity of pure sodium is also introduced. This arrangement is enclosed in a double walled glass tube evacuated to provide thermal insulation.

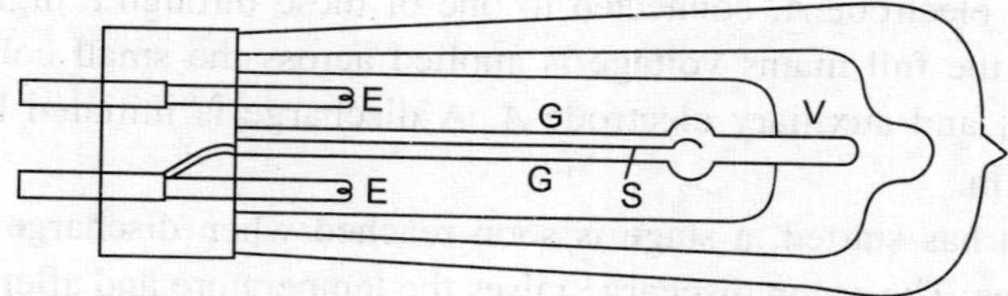

Fig. 19.5

When an A.C. is applied the polarity of the electrodes changes. The voltage of the starting strip *S* changes in step with that of the electrode with which it is connected. This strip may even touch one of the legs of the tube containing an electrode of the opposite sign. The combination of the electrode and the strip together with the gas, sodium and glass forms a condenser of a very small capacity. A feeble current, therefore, passes through this condenser and causes the gas to ionise. The resistance between the electrodes thus decreases and the electrons and the ions set is motion. This motion further ionises other atoms of the gas by collision. The process goes on till the gas becomes sufficiently conductive to allow a current to pass between the electrodes themselves.

The ionisation of the gas causes a discharge to be initiated and the lamp begins to glow with a reddish tinge. Since now a current is flowing into the tube it will cause a further ionisation which reduces the resistance. This decreased resistance will cause a further increase in current. In a short time due to this negative resistance characteristic of the lamp the current attains such a value that unless measures are adopted to stabilise it, either the lamp breaks down or the fuse in the supply burns out.

Let us assume that current into the tube is stabilised immediately after it has established. Since the resistance of the gas is lower when the gas is running, the voltage applied at start (starting voltage) is greater than the running voltage. Due to the passage of current a large amount of heat is generated. This causes the sodium to evaporate and soon the tube is filled with sodium vapours which promote further ionisation. With the advent of sodium vapour a yellow colour soon dominates the red light of neon. The emitted radiations contain two wavelengths 5890 Å and 5896 Å.

Mercury Vapour Lamps

The basic principal of mercury lamps is the same as that of the sodium lamp but there are the following differences between the two:

(1) The light radiation from the secondary discharge ceases immediately after striking.

(2) The vapour pressure and the temperature developed are very high as compared to sodium lamp.

(3) The discharge forms a thin luminous column of the gas between the electrodes. It does not fill the whole of the tube.

There are two types of mercury vapour lamps:

(*i*) High pressure lamps.

(*ii*) Very high pressure lamp.

A high pressure mercury lamp is shown is Fig. 19.6. In addition to the two main electrodes there is another auxiliary electrode *A*, connected to one of these through a high resistance *R*. When the lamp is switched on the full mains voltage is applied across the small column of gas included between the electrode E_1 and auxiliary electrode *A*. A discharge is initiated between the two and ionisation of the gas sets in.

Once the ionisation has started, a stage is soon reached when discharge begins to take place between the two electrodes. The argon discharge raises the temperature and after a short time mercury is vaporised. As soon as vapours of mercury are formed the discharge fills the whole of the tube but gradually contracts into a thin luminous column between the electrodes. The argon discharge becomes ineffective when the lamp is fully run up. As soon as the lamp is running the resistance of the starting circuit becomes too high as compared to that of the parallel circuit and hence it plays a very minor part in subsequent working of the lamp.

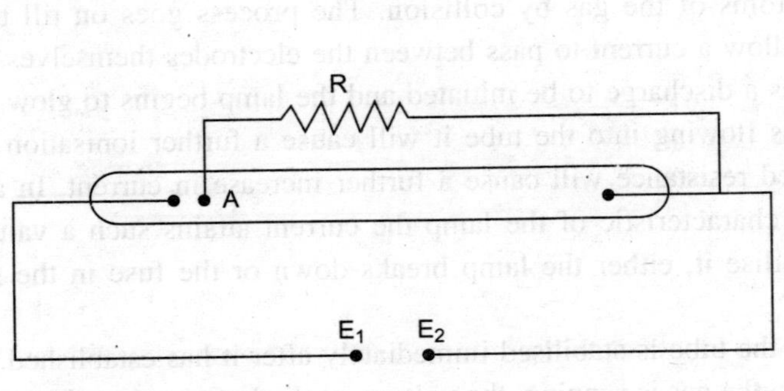

Fig. 19.6

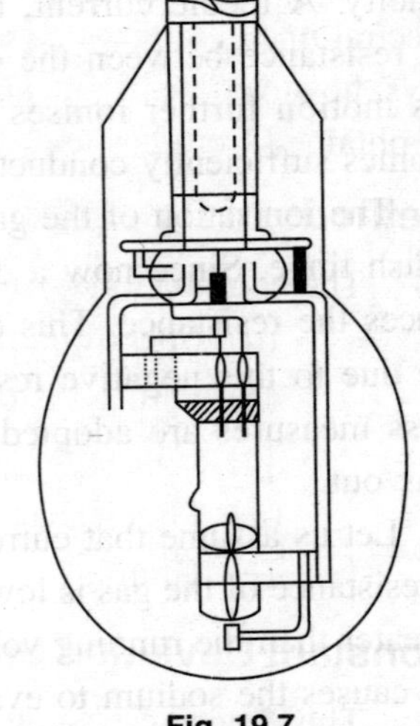

Fig. 19.7

The very high pressure mercury lamp is naturally the bigger of the two. Moreover because of the high temperature produced the discharge tube is made of quartz. The shape of the outer envelope is the same as that of an electric bulb (Fig. 19.7). Such types of lamps produce some red although it is insufficient. This deficiency can be corrected by blending mercury vapour light with incandescent light. Moreover by connecting a filament in series with the discharge tube the ballast can be dispensed with. Consequently such type of mercury lamps can be directly operated from 220 volts mains.

Monochromatic Light

In laboratory many optical measurements are made with monochromatic light. For this purpose the simplest source frequently used is the *sodium lamp*. Although it emits two lines 5890 and 5896 A.U. They are so close together that for all practical purposes the source may be regarded as monochromatic.

The intensity of sodium light is however small. Hence wherever a source of greater brightness is required a *mercury arc* is used. There are only a few lines in the visible region of the spectrum of mercury. These lines are widely separated and as such they can be easily isolated. The green line of mercury $\lambda = 5461$ A.U. which is also the strongest is usually used. It can be isolated by using a combination of two glass filters, one of *didymium* glass (absorbing the yellow lines) and one of *bichromate* glass or other yellow glass (absorbing the blue and violet lines).

Red line of cadmium $\lambda = 6438.46$ A.U. is also sometimes taken as the primary standard of wavelength. But for certain practical difficulties the cadmium arc is not used much. Since the lamp is operated at a high temperature a tube of quartz must be used. Moreover on cooling cadmium solidifies on the walls of the hot quartz tube thereby breaking it. This difficulty can however be avoided by adding a trace of gallium to cadmium. This addition has got no effect on the cadmium radiations.

Although very narrow spectral regions can be isolated by using special monochromators, the resultant intensity of the radiation is very poor. The technique therefore, has very limited practical application.

The best available monochromatic radiation, obtained from conventional light sources, is the spectral line of approximate wavelength 6056 Å emitted by *krypton isotope* of mass number 86. These days, however, nearly ideal monochromatic light of negligible band-width and very high intensity can be obtained from gas and solid state lasers.

Methods of Detection

The chief methods of detection of the visible region are the following:

(*a*) Spectrometer

(*b*) Direct Vision Spectroscope

(*c*) Constant Deviation Spectrograph

(*d*) Photoelectric Cell

(*e*) Gratings.

Constant Deviation Spectrograph

This device essentially consists of the following three parts:

(*i*) Collimator.

(*ii*) Camera.

(*iii*) Constant deviation prism.

The collimator and the camera are mounted at right angles to each other on a heavy stand. A prism of flint glass — the constant deviation prism (Fig. 19.8) is mounted on a table, (in the same horizontal plane as the collimator and the camera), in such a way that the mean ray enters the camera, at right angles to its original direction, after suffering minimum deviation during its passage through the prism. The prism table can be rotated by a tangent screw. This screw carries a drum which is calibrated in Angstrom units. Sometimes it is possible to detach the camera and introduce a telescope in its place. The instrument in this case is called a constant deviation spectroscope. It is possible with the help of such a spectroscope to know the wavelength of spectral lines directly. The prism is set in the minimum deviation position for a particular wavelength. The drum is then rotated till the other wavelengths successively coincide with the vertical cross-wire of the eye-piece. When such is the case, the wavelengths of spectral lines can be read on the scale. The accuracy of this measurement is ± 2 Å.

If the camera is used in place of the telescope the results obtained are very accurate. To determine the wavelengths of the spectral lines of a certain source a photograph of its spectrum is taken in the middle. Both above and below this spectrum the lines of a known source are also photographed for comparison purposes. This is accomplished with the help of a *Hartmann Slit* (Fig. 19.9). This is simply a metallic piece carrying three holes. This piece can be slipped on the slit of the collimator.

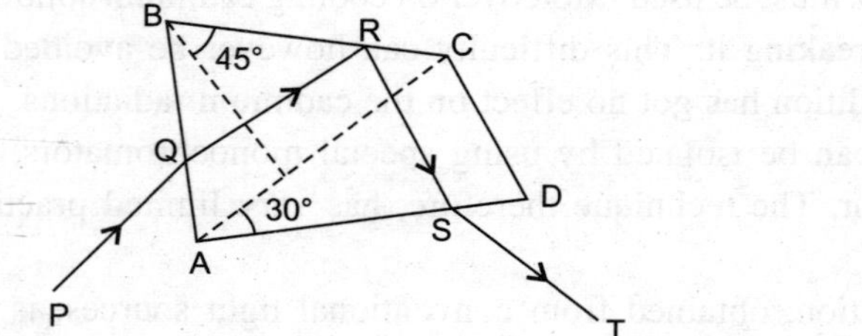

Fig. 19.8. The constant deviation prism

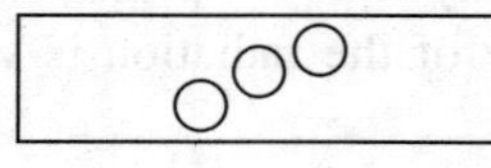

Fig. 19.9. Hartmann slit

The three holes allow three different portions of the slit to be exposed. Having obtained a photograph the separation of the spectral lines is measured by a comparator (which is nothing but a travelling microscope having a least count of the order of 10^{-4} cm). The Hartmann's formula is then applied to determine the wavelengths of the unknown lines.

The constant deviation spectrograph is suitable for use with a Lummer plate, Fabry-Perot Etalon and Echelon grating for determining the fine structure of spectral lines. It is also sometimes employed as a monochromator.

19.3 ULTRAVIOLET REGION

These radiations were discovered in the year 1801 by *Ritter*. He found that silver chloride turns black when exposed to the invisible radiations beyond the violet end of the solar spectrum. Later on, in 1804 *Thomas Young* discovered that these invisible radiations exhibit the phenomenon of interference. He projected the Newton's rings pattern formed by these radiations on a paper treated with silver chloride and found that they had a wavelength shorter than those of visible light.

The ultraviolet radiations are usually defined as the electromagnetic radiations having wavelengths lying between 40 Å and 4000 Å. They bridge the gap between the longest wavelength *X*-rays and the shortest of wavelengths of visible light. The ultraviolet radiations are usually divided into three regions:

(*a*) Near ultraviolet 4000 A.U. – 3000 A.U.
(*b*) Far ultraviolet 3000 A.U. – 200 A.U.
(*c*) Extreme ultraviolet 200 A.U. – 20 A.U.

This division is rather loose and is done on the basis of the experimental methods employed for their study.

The sources of ultraviolet radiation are essentially of two types (*i*) natural and (*ii*) artificial. The natural source is obviously the sun while the artificial sources includes a wide variety of arc and incandescent lamps. These are as follows:

1. Electric spark
2. Mercury arc
3. Argon filled glow lamps
4. Zirconium arc
5. Hydrogen discharge lamp in quartz envelope
6. Carbon arc
7. Photo flash lamps.

Most of the ultraviolet radiation from the sun are absorbed by the earth's atmosphere. Cornu using a photographic method employing a reflection grating found that radiations beyond 3000 Å are absorbed. Quartz, water and fluorspar were found by *Miller* to be transparent to these radiations. Much of the work on transparency of various substances to the ultraviolet was done by Stoke. The phenomenon of fluorescence was employed as a tool for this. Stoke found that many substances such as paraffin oil, alcoholic solution of chlorophyll, quinine sulphate solution, uranium glass and fluorspar exhibited the phenomenon when exposed to ultraviolet radiations. It was during this study that Stoke found that quartz is more transparent to ultraviolet than glass. It is because of this that lenses and prisms used in ultraviolet work are mostly made of quartz. The transparency of various substances to ultraviolet radiations has been shown in the following table:

It will be clear from the table below that lenses and prisms of quartz or fluorite (if available) should be used for ultraviolet work. Moreover air should be evacuated from the apparatus. These are the essential features of all works on ultraviolet radiations.

Substance	*Wavelength absorbed*
Flint glass	Up to 3900 Å
Crown glass (2 mm thick)	Below 3100 Å
Air	Below 2970 Å
Quartz	Below 1850 Å
Photographic plates (Gelatin)	Below 2200 Å
Iceland spar	Below 2150 Å
Fluorite	Below 1000 Å

The chief methods of detection are the following:

(*a*) Photoelectric cell.

(*b*) Soret's fluorescent eyepiece.

(*c*) Spectroscopes and spectrographs.

The fluorescent eyepiece consists of a thin plate of a fluorescent substance such as uranium glass placed at the position usually occupied by cross-wires such that it is normal to the axis of the telescope. The eyepiece however is inclined to it. Ultraviolet radiations falling upon the plate produce a fluorescent image which can be seen with the help of the eyepiece.

The spectroscopes and spectrographs used in the visible region can be adapted for use in the ultraviolet provided the glass lenses and prisms are replaced by quartz ones. The following are the essential features of these instruments.

1. Fluorescent eyepieces are used.
2. The prisms used are of Cornu type.
3. The objectives of the telescope and the collimator are made of quartz cut parallel to the optic axis. Moreover, while one is made out of a piece of *d*-quartz the other is cut from a piece of *l*-quartz.

Studies in the Extreme Ultraviolet

It will be clear from the above discussion that the use of gelatin and glass as also air is to be avoided for studying ultraviolet radiations below the wavelength 1850 Å. Using Fluorite prisms and lenses *Schuman* was able to extend his observations up to 1000 Å. The photographic plates used by him had a very small quantity of gelatin. The lower limit was pushed still further down by *Lyman* who employed a vacuum reflection grating spectrograph. *Millikan* and his co-workers then used a very high vacuum grating spectrograph coupled with high tension spark to get up to 100 Å. *Seighahn* using concave grating spectrographs with very high vacuum has been able to bridge the gap between the ultraviolet and *X*-rays.

Uses of Ultraviolet Radiations

The ultraviolet radiations are used in the treatment of various kinds of tuberculosis, promoting the healing of wounds, in the prevention and cure of rickets and in the prevention of airborne disease. Germicidal tubes are used to purify air in closed spaces. Various foodstuffs are irradiated by ultraviolet to increase their vitamin *D* potency. Small doses are used for the sterilisation of milk.

19.4 INFRARED RADIATIONS

The credit of discovering the Infrared radiations goes to *Sir William Herschel* who while studying the heating effect of the different regions of solar spectrum found that it goes on increasing as one moves towards the longer wavelength side. He observed that the heating effect continued to increase even in the invisible region beyond the red end of the solar spectrum. But it was *Ampere* who proved for the first time that these radiations belong to the same class as ordinary light. *Sir J. Herschel* later on detected Fraunhofer lines in this invisible region of the spectrum.

The wavelength range between 8000 Å and 1 mm is called Infrared region. This is usually divided into two parts:

(1) The near infrared – 7800 Å $\rightarrow$ 20 μ ($1\mu = 10^{-4}$cm)

(2) The far infrared – 20 μ $\rightarrow$ 1 mm.

The chief sources of infrared radiations can be divided into two classes (*a*) natural and (*b*) artificial. The sun is of course the most powerful natural source of infrared. Radiations having wavelength as large as 18 μ have been detected by *Langley* in the solar spectrum although these are rather faint in this region. The chief artificial sources are the following:

1. Nernst glower.
2. Globar heater.
3. Welsbach mantle.
4. Powdered crystals of certain materials.

Most of these sources give continuous infrared radiations with different emissivity in different regions. When heated to incandescence these sources emit a characteristic band of strong intensity (*see* curves of black body radiation Fig. 20.1), the intensity of other wavelengths which are also simultaneously emitted out being too small.

The Nernst glower is essentially a filament made of oxides of certain rare earths which is a non-conductor at ordinary temperatures. However, if it is heated externally by some flame or other source it becomes a conductor. A current sufficient to heat it to incandescence can then be passed through it. The Welsbach mantle radiates strongly in the visible and in the infrared region beyond 8 μ but not in the near infrared. The Globar heater which is simply a rod of silicon carbide gives out radiations beyond 20 μ.

Powdered crystals of certain substances on heating give out bands of wavelengths. The wavelength region of these bands depends upon the material used. Thus, quartz emits out bands near 8.5 μ, 9.0 μ and 20.75μ; fluorspar does so near 24.0 μ and 31.6 μ while sodium bromide has emission bands at 50–55μ.

Particular regions of the infrared radiations can also be obtained by using the focal isolation method. The method consists in eliminating all but the desired range with the help of a quartz lens. The principle is illustrated in Fig. 19.10.

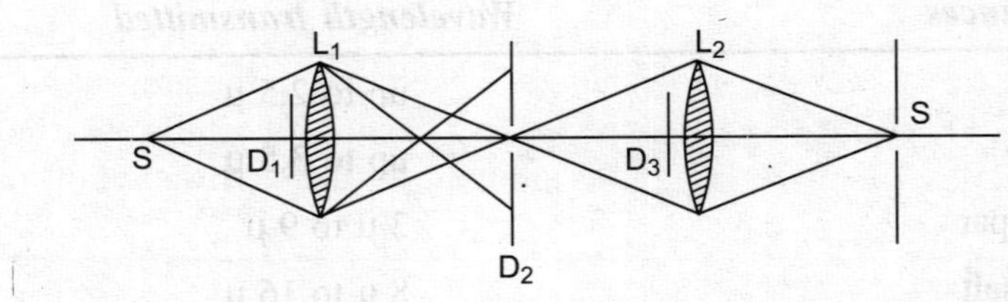

Fig 19.10. Focal isolation method

Light from a source *S* of continuous infrared radiations is refracted through a quartz lens L_1. The central portion of L_1 is shadowed by the diaphragm D_1. The desired wavelength range is sorted out by placing another diaphragm D_2 at the focal plane of that particular region. The radiations are further purified by passing them through another lens L_2 whose action is similar to that of L_1.

The chief methods employed for the detection of infrared radiations fall under two groups: (*a*) Non-selective (*b*) Selective. To the first class belong the devices whose response is proportional to energy of radiations just as thermopiles, bolometer, radiometer and pneumatic cell. Other devices such as photographic plate, photo cells, photo conductive cells etc. give spectral response to the radiations.

The usual thermopiles have now been replaced by vacuum thermocouple formed by evaporating bismuth and antimony on a thin film of nitrocellulose. The radiations falling upon one junction of Bi-Sb produce an electric current which is amplified and measured.

The bolometer originally devised by Langley is essentially a resistance which forms one arm of a balanced Wheatstone network. It is usually a thin film of gold, nickel or platinum formed by evaporating the metal on a thin film of nitrocellulose. When the radiations fall upon this film, it is heated and its resistance increases. The change in resistance disturbs the balance of the bridge which can be restored in the usual manner and the change measured.

The Crooke's radiometer consists of very light mica vanes mounted in an evacuated vessel on a spindle. In a modification by *Nicols* the vanes are suspended by a thin quartz fibre. When the radiations fall upon the vanes, the latter are heated and tend to rotate. The quartz fibre exerts a restoring force and the resulting deflection can be measured by the usual lamp and scale arrangement.

The pneumatic cell employs the principle of the change in pressure of the enclosed gas to detect the infrared radiations. This change in pressure produces movement of a mirror which reflects light on a photo cell.

Infrared Spectroscopy

The spectrometers usually employed in the study of the infrared are either of the prism type or grating type or a combination of both. They, however, have one thing in common. All the types use mirrors instead of lenses to get a parallel beam or to converge the radiations at a particular point.

The material of the prism used in the prism spectrometers depends upon the region of wavelength in which the investigations are desired to be carried out. The range of wavelengths transmitted by various substances is given in the following table:

Substances	*Wavelength transmitted*
Glass	up to 2.5 μ
Quartz	up to 3.5 μ
Fluorspar	3 μ to 9 μ
Rock salt	8 μ to 16 μ
Sylvine	15 μ to 20 μ
Potassium bromide	19 μ to 28 μ
Thallium bromide	25 μ to 40 μ

The prism spectrometers are usually used in the near infrared. They are preferred on account of the sharpness of spectra. Also, because of their low resolving power they are best suited for the study of simple spectra. Two types of mountings are usually employed (*i*) *Wadsworth mounting* (*ii*) *Littrow mounting*.

A prism spectrograph of sufficiently good resolving power designed by *McAllister* is shown in Fig. 19.11.

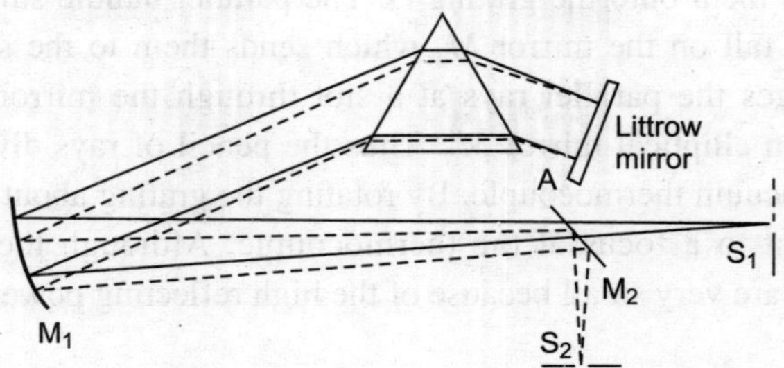

Fig. 19.11

The infrared radiations from the entrance slit fall upon the parabolic reflector M_1 which in turn collimates them onto a prism. The rays travel through the prism and are reflected by the Littrow mirror. They pass through the prism a second time and are focused by the mirror on a vacuum thermocouple. Other wavelengths are brought to a focus on the thermocouple by rotating the prism and the Littrow mirror as a unit about the axis.

For greater resolving power and higher dispersion a grating spectrograph is used. The gratings employed for infrared work are of three types: (*a*) Transmission gratings (*b*) Echelette gratings, (*c*) Laminary gratings. The transmission gratings are of the wire type. They are seldom used because of the weak spectra produced by them. They have been replaced by the other two types which are capable of concentrating energy in the first order. These gratings have been shown in Figs. 19.12 and 19.13.

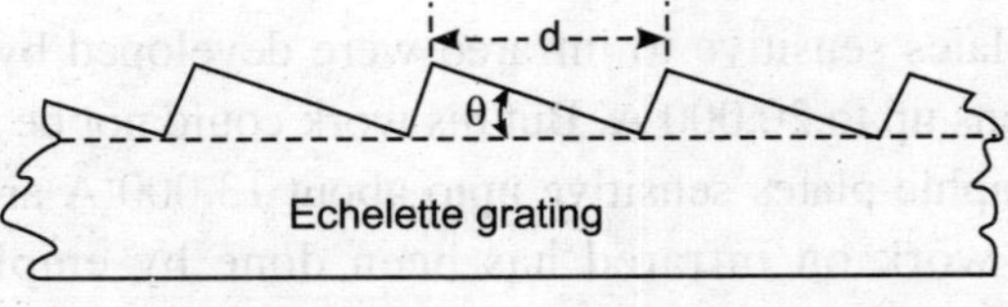

Fig. 19.12

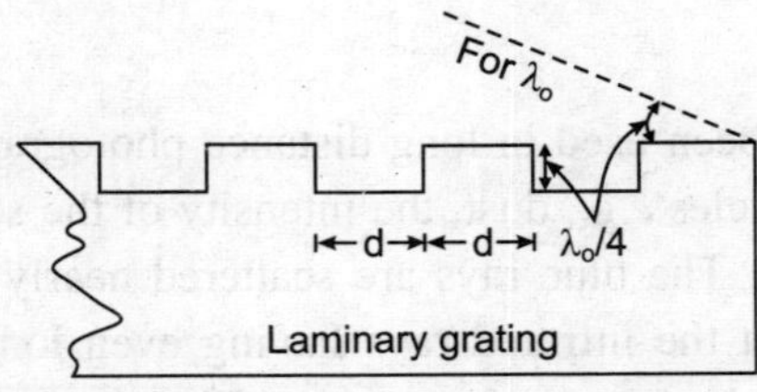

Fig. 19.13

The simplest grating spectrograph is obtained by replacing the prism and mirror combination of Figs. 19.12 and 19.13 by a grating. For greater precision however several modifications have been used. A typical grating spectrograph designed by *Hardy* has been shown in Fig. 19.14. The infra region which is required to be investigated is made to pass through an entrance slit S_1. This slit may even be the exit slit of a prism spectrometer. The rays entering through S_1 and a slot in the plane mirror M_1 are rendered parallel on suffering reflection at the parabolic mirror M_2.

The mirror M_1 reflects them onto the grating G. The parallel bundle suffers a diffraction at the grating. The diffracted rays fall on the mirror M_3 which sends them to the second parabolic mirror M_4. The mirror M_4 converges the parallel rays at a slot through the mirror M_3. The rays passing through this slot fall upon an elliptical mirror M_5. Thus the pencil of rays diverging from the slot in M_3 is focussed by M_5 at a vacuum thermocouple. By rotating the grating about a vertical axis different wavelengths can be brought to a focus at the thermocouple. Although the method involves five reflections the energy losses are very small because of the high reflecting power of the metallic mirrors used.

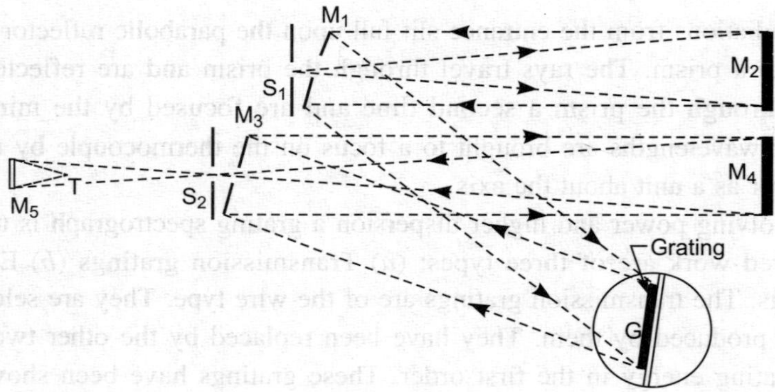

Fig. 19.14. Grating spectrograph for infrared

The first photographic plates sensitive to infrared were developed by *Abney*. With these plates he could extend this observations up to 20,000 Å. But his work could not be repeated by other workers. These days, however, photographic plates sensitive upto about 13,000 Å are available.

Most of the important work on infrared has been done by employing non-photographic techniques just as bolometer and the method of residual rays. The work of *Paschen, Rubens* and their co-workers has extended the infrared into the region in which it overlaps the shortest radio waves.

Uses of Infrared Radiations

The infrared radiations have been used in long distance photography. It has been observed that when light is scattered by tiny particles *e.g.*, dust, the intensity of the scattered beam varies inversely as the fourth power of wavelength. The blue rays are scattered nearly sixteen times as much as the red ones. It, therefore, follows that the infrared rays having even longer wavelengths than the red region of the visible spectrum will be scattered even less. Thus because of their greater penetrating power the infrared rays are generally used to photograph objects at a distance of say fifty miles. For this very reason these rays are best suited for photographing objects enveloped in fog. Only a proper filter and suitable photographic plates are needed.

They are also used for scientific research and medicinal purposes.

19.5 OTHER ELECTROMAGNETIC RADIATIONS

The discovery, methods of production and important properties of the other radiations have been given in the following table:

Radiation	*Discoverer*	*Wavelength range*	*Methods of production*	*Properties and methods of detection*
Radio Waves	Hertz Bose (Micro waves)	10^6 Å →	Spark gap discharges, valve oscillators. Klystrons and Magnetrons. (For shortest waves).	Reflected, refracted and diffracted; coherer, receiving circuits.
X-rays	Rontgen	0.06 Å → 100 Å	Sudden stoppage of fast moving cathode rays.	Diffracted by crystals. Produce phosphore scence, chemical action, ionisation and exhibit photoelectric phenomenon. These properties are also employed as the methods of detection.
γ-rays	Becquerel	0.01 → 1.4 Å	Disintegration of nuclei.	Same as in the case of X-rays except that they have a greater penetrating power.
Cosmic rays	Hess	→0.0004 Å	Not yet certain	Highly penetrating rays reaching the earth day and night. Geiger-Muller counters and Cosmic ray telescopes. Photographic plates. Ionising action.

19.6 SOME MAGNETO AND ELECTRO-OPTIC EFFECTS

There is a group of experiments which to start with were performed to test the electromagnetic nature of light. These experiments have however been responsible for the development of quantum theory of radiations. Some of these experiments which demonstrate the interactions between light and matter depend for their action on magnetic field while there are certain others which depend upon an electric field. To the first group belong the phenomena of Zeeman effect, Voigt effect, Faraday effect, Cotton-Mouton effect and the Kerr-magneto-optical-effect. Stark effect, Electric-double refraction and Kerr-electro-optical effect belong to the second group. A short description of some of these is given on next page.

Zeeman Effect

The first attempt to observe the effect of magnetic field on spectral lines was made by *Faraday* in 1862. But because of the weak magnetic fields used and the low resolving power of the optical instruments employed, he was unsuccessful in observing any effect. *Zeeman* in 1932 could observe a broadening of sodium lines when a sodium flame was placed between the pole pieces of a strong electromagnet. On improving the apparatus Zeeman and later on *Preston* found that the spectral lines were split into many components which were polarised.

The Zeeman effect can be observed (*i*) transversely and (*ii*) longitudinally. In the transverse arrangement the spectral line is viewed at right angles to the applied magnetic field. In this case each spectral line is split up into three components. The central line has the same frequency as that of the original line while the other two have got different frequencies. The central line is plane polarised with its plane of vibration parallel to the direction of the applied magnetic field while the other two lines are also plane polarised but the vibrations are contained in a plane perpendicular to the direction of the magnetic field (Fig. 19.15).

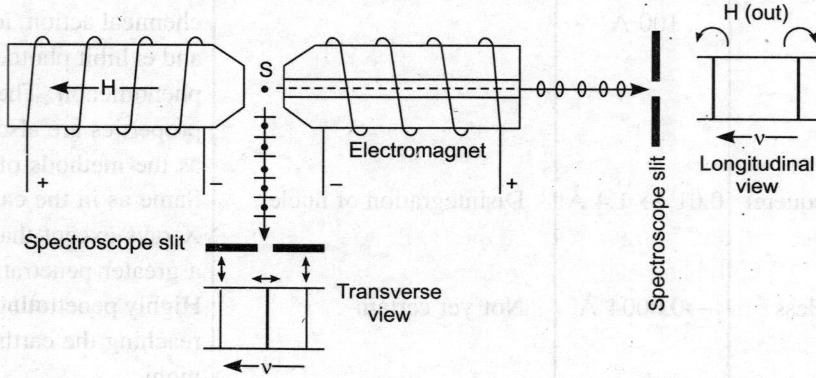

Fig. 19.15. Experimental arrangement for observing the Zeeman effect

In the longitudinal arrangement a hole is drilled in the pole pieces of the electromagnet and the phenomenon is viewed along a direction parallel to the field. Each spectral line when thus observed will be seen to split up into two components. Each of these has a frequency different from that of the original line–one having a greater and the other a smaller frequency. Also both of them are circularly polarised having opposite senses of rotation.

The frequencies of the new components which appear on the application of the magnetic field are given by

$$\nu_1 = \nu + \frac{eH}{4\pi mc} \quad \text{(anti-clockwise)}$$

$$\nu_2 = \nu - \frac{eH}{4\pi mc} \quad \text{(clockwise)}$$

where ν is the frequency of the original line, e/m the specific charge of the electron, c the velocity of light and H the magnetic field in Oersteds.

It is found that the change of frequency $\frac{eH}{4\pi mc}$ is very small. Consequently a spectroscope having a very high resolving power should be used to observe the splitting. In practice a *Lummer plate* is used.

An experimental arrangement for the study of Zeeman splitting has been shown in Fig. 19.16.

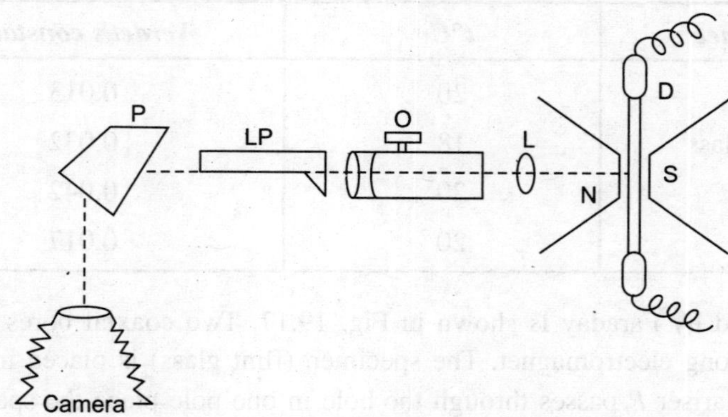

Fig. 19.16

A neon tube is placed between the pole pieces of strong electromagnet. Light from this lamp is focused on the slit of the collimator *C* of the constant deviation spectrograph. The parallel beam emerging out of the collimator falls on the Lummer plate and is then deviated through a right angle by the constant deviation prism. The splitting can be seen through the telescope *T* or photographed with the help of a camera.

By measuring the change in frequency it is possible to determine the specific charge (*e*/*m*) of the electron by this method.

If the original spectral line is a doublet or a triplet the splitting is much more complex. Thus the sodium D_1 line gives six and the D_2 line gives four components. This phenomenon is called *Anomalous Zeeman Effect.*

Paschen Back Effect

It has been seen that alkali (Li, Na, K) doublets or triplets exhibit the phenomenon of anomalous Zeeman effect on the application of a weak magnetic field. However if the magnetic field is increased, so that the Zeeman separation becomes larger than the separation of the doublet or triplet, the anomalous pattern becomes a normal triplet and the lines exhibit normal Zeeman effect. This phenomenon, originally observed with lithium lines on the application of a field of 10,000 oersteds, is called *Paschen Back Effect.*

Faraday Effect

It was observed by *Faraday* in 1845 that dense flint glass acquires the property of rotating the plane of polarisation of light when placed in a strong magnetic field. As a result of subsequent researches many more substances solids, liquids and gases have been found to exhibit this property. The angle of rotation θ is seen to vary (*i*) directly with the strength of the field '*H*',

(*ii*) directly with the length of the path traversed by light into the field *l*.

Thus $\theta = VHl$

where *V* is a constant called Verdet's constant. The values of this constant for a few substances have been listed below.

Substance	*t°C*	*Verdet's constant*
Water	20	0.013
Flint Glass	18	0.032
CS_2	20	0.042
Quartz	20	0.017

The apparatus used by Faraday is shown in Fig. 19.17. Two coaxial bores are bored into the two pole pieces of a strong electromagnet. The specimen (flint glass) is placed in between the pole pieces. Light from a polariser *P* passes through the hole in one pole piece the specimen, the hole in the second pole piece and then enters the analyser.

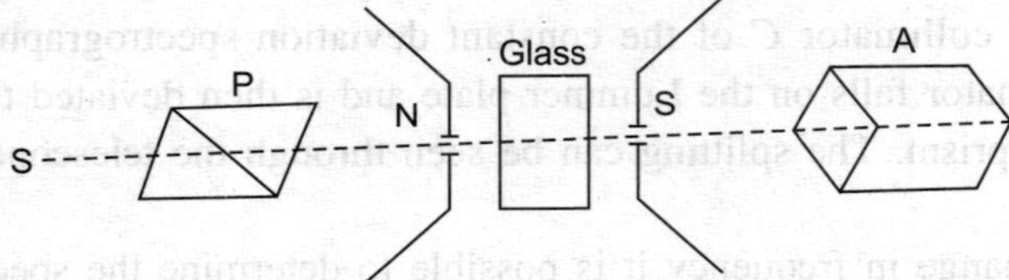

Fig. 19.17

The polariser and the analyser are first adjusted in the crossed position. As soon as the magnetic field is switched on light at once appears in the field of view of the analyser. This can be extinguished by rotating the analyser. If the direction of the magnetic field is reversed the direction of rotation of the plane of polarisation is also reversed. If, for an observer looking along the direction of the magnetic field the rotation is clockwise it is said to be *positive,* if anti-clockwise it is *negative*.

It is, therefore, seen that there is a fundamental difference between the rotation produced by an optically active substance and that produced by a substance under the influence of a magnetic field. In the case of an optically active substance the rotation is always in the same sense whatever be the direction of the beam. Consequently a rotation produced (by the substance) when a beam travels in one direction will be annulled when it travels back in the opposite direction. But in the case of substances exhibiting Faraday effect the rotation will go on adding up if the beam is reflected back and forth parallel to the lines of force for the simple reason that in such substances if the direction of light is reversed the sense of rotation is also reversed. This property is made use of in measuring the rotation produced by substances which exhibit Faraday effect very feebly.

The explanation offered for Faraday effect is similar to Fresnel's explanation of natural rotation. The two circularly polarised waves—one left handed the other right handed, which when combined together give a plane polarised beam, travel with unequal speeds in the direction of the magnetic field. It has been shown that the left handed circularly polarised wave travels faster along the direction of the magnetic field than the right handed one. If the direction of propagation of light is reversed the left handed vibration proceeds slower than the right handed one. Because of these unequal speeds a rotation of the plane of polarisation is observed.

Kerr Effect

It was discovered by Kerr in 1875 that a plate of glass becomes doubly refracting when placed in a strong electric field. The phenomenon is also exhibited by liquids and gases.

A liquid placed in an electric field behaves like a uniaxial crystal with its optic axis parallel to the direction of the field. When a beam of plane polarised light enters the liquid the two rectangular vibrations, (one along the optic, the other perpendicular to it) of which it can be supposed to be made-up, are initially in the same phase. But as these travel into the liquid a relative phase difference is introduced between the two because of the unequal speeds with which they are propagated. The amount of this phase difference is directly proportional to the length of liquid placed in the electric field. It also varies directly as the square of the field strength. Thus if δ be the path difference introduced due to a length l of the liquid column placed in an electric field of intensity E, then

$$\delta = jlE^2\lambda$$

where λ is the wavelength of the incident plane polarised radiations. The light emerging out of the cell will be linearly, circularly or elliptically polarised according to this phase difference δ.

If V is the potential difference applied between the two electrodes each of length l and separated by a distance d then

$$\delta = jl\frac{V^2}{d^2}\lambda$$

Here j is called *Kerr constant*. The constant has been given for certain liquids in the following table:

Substance	*Kerr constant*
Water	4.7
CS_2	3.2
Nitrobenzene	220
Nitrotoluene	124

It will be seen that nitrobenzene shows this phenomenon to a very marked degree.

An arrangement for observing this effect has been shown in Fig. 19.18. P and A are two nicols adjusted in the crossed position with the help of the monochromatic source. A glass tube C containing a liquid say nitrobenzene is introduced in between the crossed nicols. It will be seen that the field of view remains dark. As soon as a large potential difference is applied between the two electrodes some light will at once appear in the analyser. The tube C together with the immersed electrodes is called a *Kerr-cell.*

The Kerr-cell can be conveniently used as an electro-optic shutter for in the absence of the field there is no light in the analyser, on switching on the electric field light at once appears.

The Kerr-electro-optical effect is caused by natural or induced anisotropy of the molecules which line up on the application of the electric field. It is, therefore, dependent upon the temperature of the liquid.

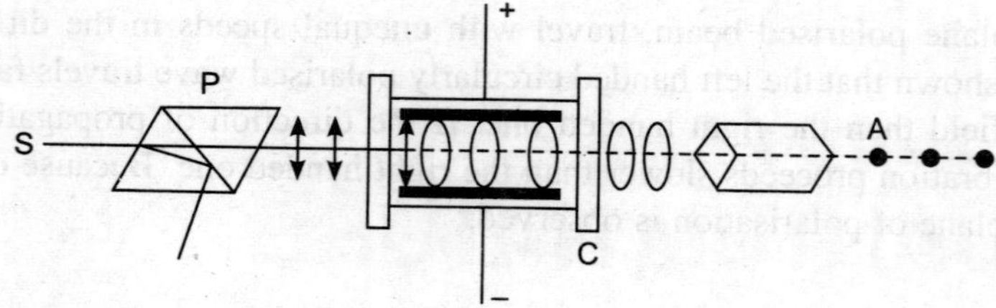

Fig. 19.18

PROBLEMS

1. What are invisible spectra? Give the arrangements for studying infrared and ultraviolet regions of spectra.
2. Write short notes on:
   (a) Ultraviolet rays
   (b) Infrared radiations
   (c) Faraday Effect
   (d) Zeeman Effect
3. What is a Kerr-cell? Discuss its principle of working and give its uses.

QUANTUM NATURE OF LIGHT

The sources of light described in the preceding chapter when excited give rise to what is called *emission spectra*. The spectra can be divided into two general classes—(*i*) the continuous spectra and (*ii*) the discontinuous spectra. The continuous spectra are emitted by incandescent solids and under special circumstances by fairly dense luminous gases. These spectra consist of a wide range of unseparated wavelengths and have a maximum intensity in the *greenish yellow* region of the spectrum.

The discontinuous spectra can be further subdivided into two classes—(*a*) *line spectra* and (*b*) *band spectra*. The line spectra are emitted by substances in their atomic state. They consist of a large number of sharply defined isolated wavelengths which are characteristic of every substance. Nearly all the laboratory sources of light give line spectra. Substances in their molecular state, on the other hand, give rise to band or fluted spectra. These consist of a number of bands separated from each other. In each band there are a large number of lines whose separation goes on decreasing and intensity goes on increasing as one moves from one end of the band to another. The lines become so close together and intense at one end that it is impossible to see them separately by unaided eye. This end is called the *band head*. If the band head is examined under a powerful optical instrument it is seen to consist of a large number of lines very close together.

20.1 CONTINUOUS SPECTRA

The continuous spectra emitted by different solids depend upon the nature of surfaces. Thus, if a piece of glazed china with red design on it is thrown into a furnace and then withdrawn after sometime it will emit a bluish light. The reason for this lies in the fact that the red design absorbs the shorter wavelengths more as compared to the longer ones when it is thrown in the furnace. On taking out, the solid begins to emit the absorbed radiations.

To compare the continuous spectra emitted out by substances reference is usually made to a black body spectrum. The distribution of energy in a black body radiation has been shown in Fig. 20.1. It will be noticed from these curves that at lower temperatures up to 2000° K most of the emitted radiations lie in the infrared regions. At a temperature of 3000° K there is some emission of energy in the visible region (shaded in Fig. 20.1) but the maximum emission still takes place in the infrared region. At a temperature of 4000° K—the temperature of the carbon arc nearly the whole of the visible region is emitted out and the maximum shifts to the red region of the spectrum. Finally, at 6000° K—the temperature of the sun—the intensity of all the wavelengths is quite sufficient while the maximum emission shifts even more towards the shorter wavelength side and lies in the greenish yellow portion of the spectrum. Moreover the ultraviolet radiations are also emitted out.

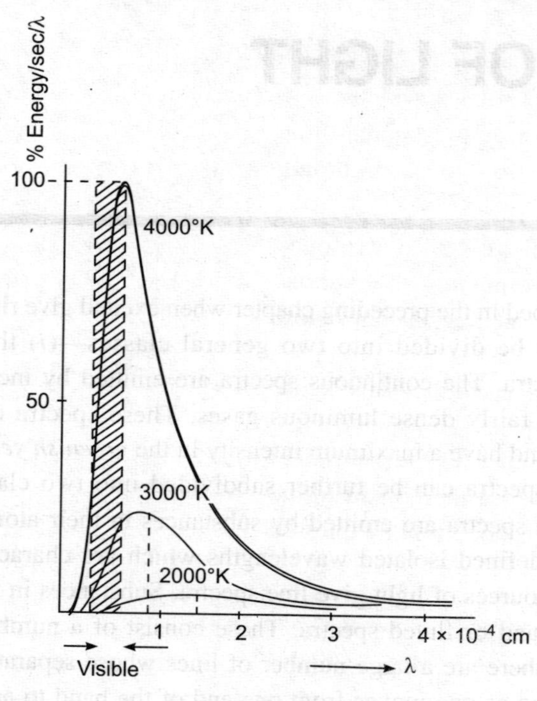

Fig. 20.1

The curves depicting distribution of energy in the emission spectrum of a black body exhibit some important features which played a vital role in the development of some of the most notable concepts of Modern Physics. These are:

1. A fairly sharp lower limit to the emitted wavelengths at each temperature.
2. A distinct wavelength at which the emitted energy has a maximum value corresponding to every temperature.
3. An asymptotic extension to infinity of the upper limit of the emitted wavelengths.

A satisfactory explanation of the energy distribution curve of the black body radiation along with its features was therefore thought to be the key to an understanding of spectral emission by physicists at the turn of the century. All such explanations had to be based upon the state of related knowledge at that time. This included the following firmly established experimental facts:

(*a*) The generation of electromagnetic waves by oscillating charges.

(*b*) The inter-relationship between heating of a body and the kinetic energy of its molecules.

Another notion that somehow seemed relevant was the concept that all bodies are made-up of atoms which in turn contain electrons. Assembling these pieces of information *Wien* proposed in

1896 that thermal emission of energy resulted from the oscillations of charges attached to the atoms which in turn were excited by virtue of the kinetic energy imparted to them by heating. Thus he came forward with the suggestion that the distribution of the emitted frequencies was related to the Maxwellian distribution of molecular velocities. Wien also calculated the distribution of spectral frequencies and obtained a curve which agreed well with the experimental results up to a wavelength of about 3 μ. Beyond this, the disagreement became progressively more pronounced.

Another curve for the spectral distribution of energy was obtained by *Rayleigh* and *Jeans* by postulating:

(*a*) The existence of harmonic frequencies of the oscillating particles, and

(*b*) The application of law of equipartition of energy to each of these harmonic frequencies.

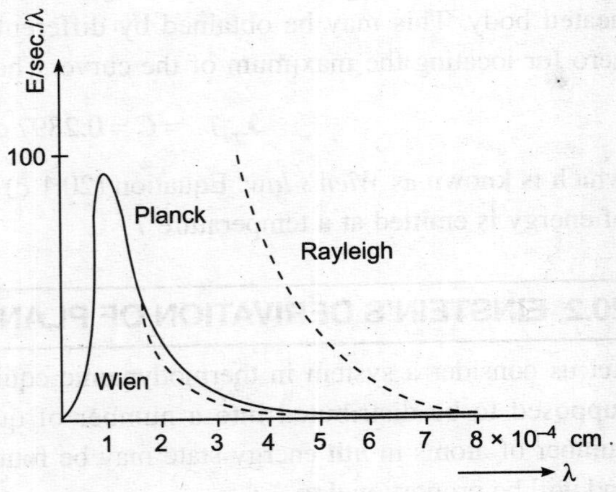

Fig. 20.2

The theoretical curve thus obtained was found to be in agreement with experimental observations beyond a wavelength of about 7 μ but not below (Fig. 20.2).

It was left to *Planck* to put forth a satisfactory theoretical interpretation of the experimental curves. He started by obtaining empirically a theoretical equation which on plotting gave the best fit to the experimental curves. This equation may be expressed as

$$E\lambda = \frac{k_1}{\lambda^5 (e^{k_2/\lambda T} - 1)} = \frac{8\pi hv}{\lambda^4} \frac{1}{(e^{hc/k} - 1)} \qquad ...(20.1\ a)$$

where

$E\lambda$ = energy emitted per unit area per second per unit λ.

k_1 and k_2 = constants

T = Absolute temperature

λ = Wavelength

He further noted that Wien's and Rayleigh Jeans laws were just special cases of his more general expression. But this equation defined all attempts at its theoretical deduction based on concepts of Physics, known at that time. It was then that Planck introduced a new, and what later turned out to be a revolutionary concept which marked a complete break with the previous notions of Physics. He discarded the concept of continuous and smooth emission of energy by the oscillators. He also dropped the idea that energy was emitted in amounts that could be cut off at any point by stopping the oscillations. Instead, Planck postulated that energy was given out abruptly in *lumps* or *quantas* at given instants. Once this energy was liberated it proceeded as a wave of frequency v given by

$$E = hv \qquad ...(20.1\ b)$$

where the correlating constant h has since been known as Planck's constant (6.62×10^{-27} ergs sec). Applied to black body radiation this quantum concept meant that higher frequencies and hence shorter wavelengths could be given out only by those atoms which possessed it and their number is generally much smaller according to Maxwellian distribution of velocities. Similar considerations governed the emission at lower temperatures. In this manner Planck succeeded in giving a theoretical basis to black body spectrum.

It may be mentioned that Planck's law also succeeded in explaining the existence of a preferred wavelength at which most of the energy is emitted corresponding to various temperatures of the heated body. This may be obtained by differentiating Eqn. (20.1 *a*) and setting the result equal to zero for locating the maximum of the curve. The result comes out as

$$\lambda_m T = C = 0.2897 \text{ cm°K} \qquad \text{...(20.1 } c\text{)}$$

which is known as *Wien's law*. Equation (20.1 *c*) gives the wavelength λ_m at which maximum amount of energy is emitted at a temperature T.

20.2 EINSTEIN'S DERIVATION OF PLANCK'S LAW

Let us consider a system in thermodynamic equilibrium at a temperature T. Then its atoms may be supposed to be distributed into a number of quantum energy states E_1, E_2, E_3......, E_n etc. The number of atoms in nth energy state may be found with the help of Maxwell Boltzmann distribution and will be proportional to

$$e^{-E_n/kT}$$

where k is Boltzmann's constant.

An atom in the nth energy state has a finite probability A_{nm} of making a spontaneous transition to a lower energy state m in unit time. The excess energy $E_n - E_m$ will be radiated as monochromatic radiation. On the other hand, if the system is flooded with quantised radiation of energy $E_n - E_m$, two simultaneous and competing processes may be set in motion. A number of atoms in the state m may absorb the energy quantum $E_n - E_m$ and jump to the state n. At the same time, some atoms in the state n may be stimulated, by resonance process, to emit radiations and descend to the lower state m.

If σ be the density of the incident radiation of the required resonant energy, the probability of a stimulated transition of an atom from state n to m in unit time is

$$\sigma B_{nm}$$

Similarly the probability of the reverse process in unit time is

$$\sigma B_{mn}$$

If the exposed system attains thermodynamic equilibrium corresponding to the temperature T, the number of atoms going from state m to n must equal the number experiencing the reverse transition. Hence, by *weighting* the transition probabilities by the number of atoms in the corresponding state, we can write

$$e^{-E_m/kT}\sigma B_{mn} = e^{-E_n/kT}(\sigma B_{nm} + A_{nm}) \qquad \text{...(20.2 } a\text{)}$$

As T tends to infinity both the exponential factors in Eqn. (20.2. *a*) tend to unity and we get

$$B_{mn} = B_{nm} + \frac{A_{nm}}{\sigma}$$

However, as T tends to infinity it is logical to expect the energy density σ to approach infinity. Hence we find that

$$B_{mn} = B_{nm} \qquad \text{...(20.2 } b\text{)}$$

Combining Eqns. (20.2 *a*) and (20.2 *b*), we get

$$\sigma = \frac{A_{nm}}{B_{nm}}\left[e^{(E_n - E_m)/kT} - 1\right]^{-1}$$

which is identical with Planck's law provided

$$E_n - E_m = h\nu \qquad \text{and} \qquad \frac{A_{nm}}{B_{nm}} = \frac{8\pi h\nu}{\lambda^4}.$$

20.3 LINE SPECTRA

The line spectra emitted by substances in their atomic state depend upon the method of excitation of the source and upon the source itself. Thus if a little common salt is introduced into a non-luminous sodium flame only a single yellow line is observed through a spectroscope. This is actually a doublet which can only be seen through an optical instrument having a good resolving power. But the sodium arc spectrum contains a large number of spectral lines. In the spark spectrum of sodium, on the other hand, some of the lines of the arc spectrum are absent while some new lines make their appearance. Thus sodium gives three types of line spectra under different states of excitation. In addition the intensities of the lines common to three classes sometimes undergo a change.

The simplest line spectrum is due to *atomic hydrogen*. The first half a dozen lines or so are visible to the unaided eye. Many more lines are found to occur in the ultraviolet and infrared regions if suitable detecting instruments are employed. The spectra of other atoms are much more complex and exhibit greater number of lines.

Looking at these line spectra with Planck's law in mind we would immediately notice two things.

(1) Most of the wavelengths of the continuous spectra of incandescent substances are absent. Only a discrete number of lines are obtained.

(2) These lines appear to show a certain regularity, at least in the simpler atoms. For example, in the case of hydrogen the difference in wavelengths of consecutive lines diminishes rapidly as we go to the shorter wavelengths.

Historically, the initial attempts were made to explain this regularity. In 1885 *Balmer* succeeded in giving an empirical formula which gave the wavelengths of the lines of hydrogen known in his time to a fair degree of accuracy. This may be written as

$$\lambda = 3645.6\frac{m^2}{m^2 - 4} \qquad \text{...(20.3 } a\text{)}$$

where λ is the wavelength in Angstrom units and m is an integral number 3, 4, 5, 6,...... etc. This empirical result is not valid for $m = 1$ or 2 because in both cases the wavelengths obtained with its help are physically impossible.

The success of Balmer's formula can be judged on comparing the calculated and experimental values of the wavelengths of a few hydrogen lines. The results are summarised in the following table:

m	*Observed* λ	*Calculated* λ
3	6562.8 A.U.	6562.8 A.U.
4	4861.5 A.U.	4861.3 A.U.
5	4340.6 A.U	4340.5 A.U.
6	4101.8 A.U	4101.7 A.U.

Balmer's formula if expressed in wave numbers (the number of wavelengths in one cm) may be reduced to a more familiar form. Thus

$$\frac{1}{\lambda} = \frac{1}{3645.6}\left(\frac{m^2-4}{m^2}\right) = \frac{4}{3645.6}\left(\frac{1}{2^2} - \frac{1}{m^2}\right)$$

or
$$\frac{1}{\lambda} = \bar{\nu} = R_H\left(\frac{1}{2^2} - \frac{1}{m^2}\right)$$

where $R_H = 109678 \text{ cm}^{-1}$ is a constant known after *Rydberg* who gave a general law expressing the wavelengths of spectral lines of other elements as well. For Hydrogen this law is

$$\bar{\nu} = R_H\left(\frac{1}{m^2} - \frac{1}{n^2}\right)$$

where m and n are integral numbers and n is greater than m. As seen above, Rydberg's formula yields the Balmer series of Hydrogen with $m = 2$.

Soon after Rydberg's formulation of the series relationship, new series of lines were discovered in the spectrum of Hydrogen. These were the *Lyman* ($m = 1$), *Paschen* ($m = 3$), *Brackett* ($m = 4$) and *Pfund* series ($m = 5$). These series can be expressed as:

$$\bar{\nu} = R_H\left(\frac{1}{1^2} - \frac{1}{n^2}\right); \; n > 1 \text{ Lyman}$$

$$\bar{\nu} = R_H\left(\frac{1}{2^2} - \frac{1}{n^2}\right); \; n > 2 \text{ Balmer}$$

$$\bar{\nu} = R_H\left(\frac{1}{3^2} - \frac{1}{n^2}\right); \; n > 3 \text{ Paschen}$$

$$\bar{\nu} = R_H\left(\frac{1}{4^2} - \frac{1}{n^2}\right); \; n > 4 \text{ Brackett}$$

$$\bar{\nu} = R_H\left(\frac{1}{5^2}-\frac{1}{n^2}\right);\ n > 5 \text{ Pfund}$$

Thus, as far as the spectrum of Hydrogen is concerned, its spectrum can be resolved into a number of series. In each series, the wave-numbers, of various members can be expressed, in terms of simple integral numbers, as difference of two terms—one of these is constant term and the other is called the current or variable term. As $n \to \infty$, the current term tends to zero and the expression for the wave-numbers approaches the so called series limit. Thus the constant term of a series is in fact the limit to *which the wave-numbers tend as* $n \to \infty$. The series limits of the different series of hydrogen are therefore given by the expression

$$\bar{\nu} = R_H / m^2$$

where $m = 1, 2, 3, 4, 5$ etc. for the Lyman, Balmer, Paschen, Brackett and Pfund series respectively.

These elementary facts regarding the spectrum of Hydrogen point to the important result that *line spectra comprise of only those wavelengths which are functions of integral numbers*. Thus unlike the continuous spectra, all wavelengths are *not* possible. This is the first hint that energy in these spectra also is radiated in discrete amounts or in other words, the emission of line spectra is somehow quantised. Carrying the idea further, we can express the emitted wave-numbers in terms of energy with the help of Planck's Law. Thus

$$1/\lambda = \bar{\nu} = \nu/c = R_H\left(\frac{1}{m^2}-\frac{1}{n^2}\right) \quad \text{or} \quad \nu = c\,R_H\left(\frac{1}{m^2}-\frac{1}{n^2}\right)$$

According to Planck's law

$$E = h\nu$$

Hence the energy E associated with the emission of a frequency ν is given by

$$E = h\,c\,R_H\left(\frac{1}{m^2}-\frac{1}{n^2}\right) \qquad ...(20.3\ b)$$

This important result expresses the fact that the emitted radiation carries an energy which is the *difference between two energy states of Hydrogen atom.* Since in each of these states, the numbers m and n take only integral values we further infer that in each state of the atom the associated energy is quantised.

20.4 BAND SPECTRA

As mentioned before substances in their molecular state give rise to band or fluted spectra. The most common sources are *cored* carbon arcs, gas discharge tubes and flame. Usually solid compounds are used with arc and flame while the gases are most suitable for discharge tubes. An essential difference between the line and band spectra is that *while line spectrum is independent of the nature of compound and is only characteristic of the atom, the band spectrum is different for different compounds of the same element.* Thus the three halides of calcium give different band spectra. In addition, band spectra are emitted by substances which have been subjected to a less violent treatment as compared to that given for the emission of line spectra.

20.5 ABSORPTION SPECTRA

If a transparent substance is introduced between a source of continuous spectrum and a direct vision spectroscope some dark regions are observed. These dark regions of the spectrum are always *characteristic of the substance*, that is introduced in the path of light. The absorption spectra are also of three kinds (*a*) continuous absorption spectrum, (*b*) line absorption spectrum and (*c*) band absorption spectrum.

Blue glass which absorbs all the wavelengths except the shorter ones gives a continuous absorption spectrum.

Dilute solutions of blood, chlorophyll and a large number of other substances when introduced between the source and the detector show a large number of separate dark bands. These bands are characteristic of the substance and, therefore, are used in the analysis of solutions.

If, however, the absorbing substance is a *gas* or a *vapour* the absorption spectrum consists of *a number of sharp black lines isolated from each other*. Thus the spectrum of sun is seen to be crossed by a very large number (20,000) of dark lines called the *Fraunhofer lines* named after their discoverer. The wavelength of some of these lines is given in the table below:

Name	*Origin*	*Wavelength Å*	*Name*	*Origin*	*Wavelength (Å)*
A	O_4	7594 - 7621	F	H	4861
B	O_2	6867 - 6884	G	Fe	4308
C	H	6563	G	Ca	4308
D_1	Na	5896	H	Ca^+	3968
D_2	Na	5890	K	Ca^+	3934
E_2	Fe	5269			

It can be seen from the above table that some of the Fraunhofer lines have the same wavelengths as the lines emitted by certain substances *e.g.*, the two Fraunhofer lines D_1 and D_2 have the same wavelength as those of the two yellow lines of sodium.

These lines remained unexplained for a long time until *Kirchoff* gave a thermodynamic proof of a law that goes after his name. Kirchoff's law states that *at a given temperature the ratio of the emissive and absorptive powers is the same for all bodies and is equal to the emissive power of a perfectly black body.* It follows from this law that if a substance gives out certain radiations in an excited state, it will absorb those very wavelengths when a continuous spectrum falls on it. The occurrence of the Fraunhofer lines is now very easily explained. The sun is a very hot body which emits out a continuous spectrum. These radiations pass through the surrounding atmosphere at a lower temperature. This atmosphere contains hydrogen, oxygen, nitrogen etc., and also practically all elements in a gaseous form. During its passage through this envelope of gases the solar continuous spectrum is robbed of those wavelengths which the gases would have emitted had they been in an excited state. It was, therefore, with the help of *Fraunhofer lines* that the existence of various gases and elements in the sun's atmosphere was established. A large number of experiments have proved this explanation to be correct. We thus arrive at the important result that *energy is not only radiated but also absorbed in quanta.*

20.6 PHOTOELECTRIC EFFECT

The evidence presented so far regarding the quantum nature of the emitted and absorbed radiations has been indirect. We shall now discuss certain important discoveries which placed the quantum theory of light on a firm footing. One of these is the *Photoelectric Effect.*

It was observed by *Hallwachs* (1888) that a carefully insulated metallic plate carrying a negative charge lost the charge completely after exposure to ultraviolet radiations. Since such loss of negative charge can be brought about only through emission of electrons, Hallwachs observation constituted a direct observation of Photoelectric effect. This important discovery was followed by some very accurate and painstaking experimental study by *Lenard, Hughes, Millikan* and ultimately *Einstein.* The apparatus, in principle is sketched in Fig. 20.3.

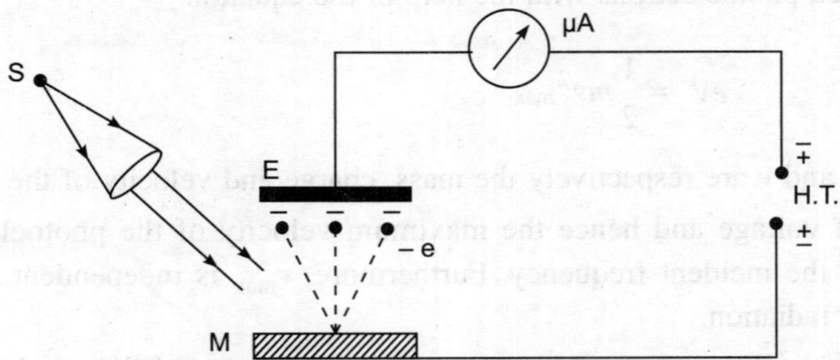

Fig. 20.3

Light from a source *S* is allowed to fall on a metallic plate *M*. An electrode *E* connected to the positive pole of a battery is kept near *M*. The negative pole of the battery is connected to the metal *M*. The source of light is so chosen that the intensity and frequency of the incident radiation may be independently controlled. A meter in the circuit is used for measuring the photocurrent and hence the number of photoelectrons reaching *E*.

In fact the apparatus is much more elaborate *e.g.,* the photo-cathode and the anode (*M* and *E*) are enclosed in a highly evacuated glass envelope, the meter is highly sensitive and arrangements are made for applying in small steps both negative and positive potentials to the anode. Moreover, great care is needed in the preparation of the surface of the photo-cathode. When all these precautions are taken, the following general features of photoelectric emission can be observed:

1. The photoelectrons are not emitted at all unless the frequency of the incident light is equal to or greater than a critical value called the *threshold frequency*. If the frequency of the

incident radiation is less than the threshold value, any amount of increase in the intensity of light is of no avail. The threshold frequency is found to be different for different metals.

2. Keeping the incident frequency constant (and above the threshold value) the strength of the photocurrent and hence the number of photoelectrons is found to increase on increasing the intensity of the incident radiation.

3. With a given frequency, if the battery connections are reversed (so that E is connected to the negative and M to the positive pole) the photoelectrons continue to reach the electrode for small differences of potential between E and M. This shows that the emitted photoelectrons have a kinetic energy which is sufficient to overcome the retarding potential. But as E is made more negative with respect to M, the number of photoelectrons reaching it is reduced and finally falls to zero. (Fig. 20.4) This cut-off voltage, which succeeds in stopping the flow of photoelectrons, has a sharp value and may be used for determining the maximum velocity of the ejected photoelectrons with the help of the equation

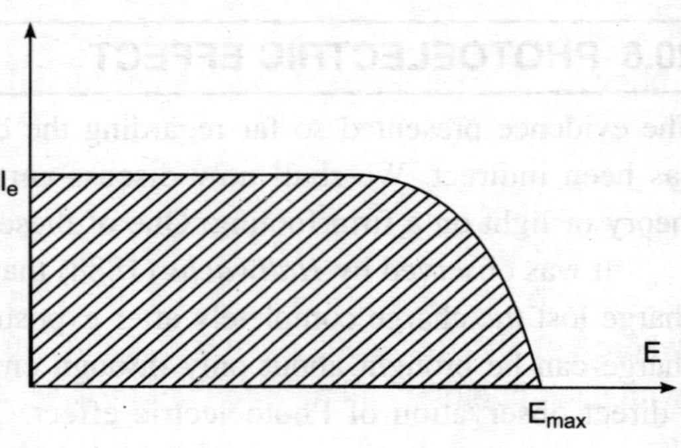

Fig. 20.4

$$eV = \frac{1}{2} m v^2_{max} \qquad \text{...(20.6 } a\text{)}$$

where m, e and v are respectively the mass, charge and velocity of the electron.

4. The cut-off voltage and hence the maximum velocity of the photoelectrons is a linear function of the incident frequency. Furthermore, v_{max} is independent of the intensity of the incident radiation.

5. There is practically no time delay between the incidence of light and emission of photo-electrons. It has been discovered that if there is any time delay, it is less than 10^{-9} secs.

6. Photoelectric emission may also take place in a gas provided that the incident frequency is high enough (higher than that in the case of metals).

The reigning electromagnetic theory of light could offer only a qualitative explanation of the phenomenon. It completely failed in accounting for the special features of photoelectric effect. According to this theory the electron in the surface of the metal begins to oscillate under the influence of the incident electromagnetic radiation. When it accumulates an amount of energy sufficient to overcome the energy barrier binding it to the surface, the electron escapes from the surface. But if this is true, the energy of the photoelectron should depend upon the intensity rather than frequency of the incident light. In that case there should be no E_{max}, the upper limit determining the maximum energy of the photoelectrons. Moreover, since the electron, according to the classical theory, can absorb energy from a wave of any frequency, it should be possible to observe photoelectric effect with light of any frequency. But we have seen above that the experimental facts are quite to the contrary. In addition there are several more and equally serious objections to applicability of electromagnetic theory for explaining photoelectric phenomenon. At the end of the nineteenth century, the physicists, therefore, started looking for fresh ideas to explain these experimental facts.

20.7 EINSTEIN'S PHOTOELECTRIC EQUATION

It was Einstein, who succeeded in giving a satisfactory theoretical explanation of the phenomenon in 1905. By this time Planck's quantum hypothesis had gathered some ground. *Einstein* took it a step further. He asserted that radiation is not only absorbed and emitted in quanta but *also travels in space as a stream of discrete localised particles of energy*. These "particles" were later on named as photons. The energy associated with a photon is given as before by the relation

$$E = h\nu$$

where ν is the frequency of light, as understood in the classical theory. The momentum p associated with each photon is therefore given by

$$p = E/c = h\nu/c \qquad \text{...(20.7 } a\text{)}$$

It may be mentioned that Einstein's assumption of photons in motion as a stream of particles having the wave like properties of wavelength and frequency contained an apparent contradiction. It was however, resolved when the mechanics of these quanta (Quantum Mechanics) was properly understood. But the theory also revealed that it is not possible to describe a photon of discrete energy $h\nu$ as *a sharply* localised particle in space.

Einstein's assumption, nevertheless, provided a very elegant explanation of the quantitative aspects of photoelectric effect. Thus, when a beam of monochromatic light falls on a metallic surface, the photons undergo collisions with the electrons of the metal. During these impacts, both energy and momentum is conserved just like in the case of mechanical collisions. Each electron therefore receives energy from a collision with a photon. (The probability of an electron absorbing energy from two photons is negligible). If this energy $h\nu_o$ (ν_o is the frequency of the incident radiation) is equal to energy binding the electron to the metal surface, it will just be in a position to escape from the surface. If the incident frequency is ν ($\nu > \nu_o$), the electron which is just dislodged in the first case is also able to pick up some kinetic energy which is equal in magnitude to the excess energy available $h(\nu - \nu_o)$ after overcoming the surface barrier. It will then be ejected with a velocity given by the equation

$$\frac{1}{2}mv^2{}_{\max} = h(\nu - \nu_o) = h\nu - \phi \qquad \text{...(20.7 } b\text{)}$$

where the *LHS* denotes the maximum kinetic energy of the ejected photoelectron. If $\nu = 0$, Eqn. (20.7 *b*) gives

$$\phi = h\nu_o = hc/\lambda_o$$

ϕ is called the *work function of the metal* under reference and ν_o is known as threshold frequency.

Einstein's photoelectric equation therefore, postulates the existence of a threshold frequency of the incident radiation, below which no photoelectric emission can take place at all. If the intensity of the incident radiation is increased, only the number of photons falling upon the metal surface becomes more but the energy of each photon remains unaltered. Increasing the number of photons causes an increase in the number of ejected photoelectrons provided the incident radiation has a frequency greater than the threshold value. In this case, the strength of the photoelectric current will increase.*

According to Einstein's equation, for a given metal and radiation of a definite frequency (both ν and ϕ fixed), there is a well defined value for the kinetic energy of the ejected photoelectron. This is of course the upper limit mentioned in Eqn. (20.7 *b*). Other photoelectrons may have a value less than this maximum if they experience secondary collisions as they emerge from the metallic surface.

Again, combining Eqns. (20.6 *a*) and (20.7 *b*), we find that

$$eV = \frac{1}{2} mv^2{}_{max} = h(\nu - \nu_o)$$

Therefore, the cut-off voltage as well as the maximum energy of the photoelectrons is a linear function of ν. If therefore, a graph is plotted between the retarding voltage V and the frequency of incident radiation a straight line is always obtained irrespective of the material of the photoelectric surface. The slope of all these straight lines is the same and gives the value of Planck's constant; the intercepts on the ν axis give the work functions of the material in each case.

20.8 PHOTOELECTRIC CELLS

The metals commonly used in the construction of photocells along with their threshold wavelengths are given in the following table:

Element	*Threshold wavelength*
Caesium	6560 A.U.
Sodium	5400 A.U.
Lithium	5200 A.U.
Potassium	5650 A.U.

The photoelectric cell essentially consists of a glass bulb containing alkali metals, their alloys or amalgams which is evacuated and then sealed off from pumps. An auxiliary electrode which is given a positive potential is also introduced. The photoelectrons which are emitted out when the cell is exposed to light are collected by this anode.

These cells are generally of two types: (1) *gas-filled type* and (2) *vacuum type*. The gases used in the first type of cells are generally argon, or helium at such a pressure as to give the biggest magnification of the original photoelectric current. This magnification is brought about by ionisation due to collision. In the gas-filled type of cells, the photoelectrons ejected out from the cathode collide with the gas molecules on their way to the anode. If the potential difference across the cell is

*There is however no certainty that every photon ejects an electron. In fact, the efficiency is quite low; it is found that roughly one photon out of every hundred produces photoelectric emission.

sufficiently high this collision will ionise the gas molecules. Due to this ionisation the gas-filled cells are more sensitive than vacuum type. The sensitivity however depends upon the impressed voltage and reaches its limit when the voltage is high enough to cause cumulative ionisation. Since argon has got the lowest ionisation potential among the inert gases, the ionisation by collision is greatest in argon filled cells. Hence wherever a high magnification of the photocurrent is required usually argon-filled cells are used.

The simplest types of cells have got plane electrodes (Fig. 20.5). The cathode which is usually an alkali metal is run into position in a molten state. The anode is generally a single wire, a grid or a plate with suitable provision for passage of the incident light.

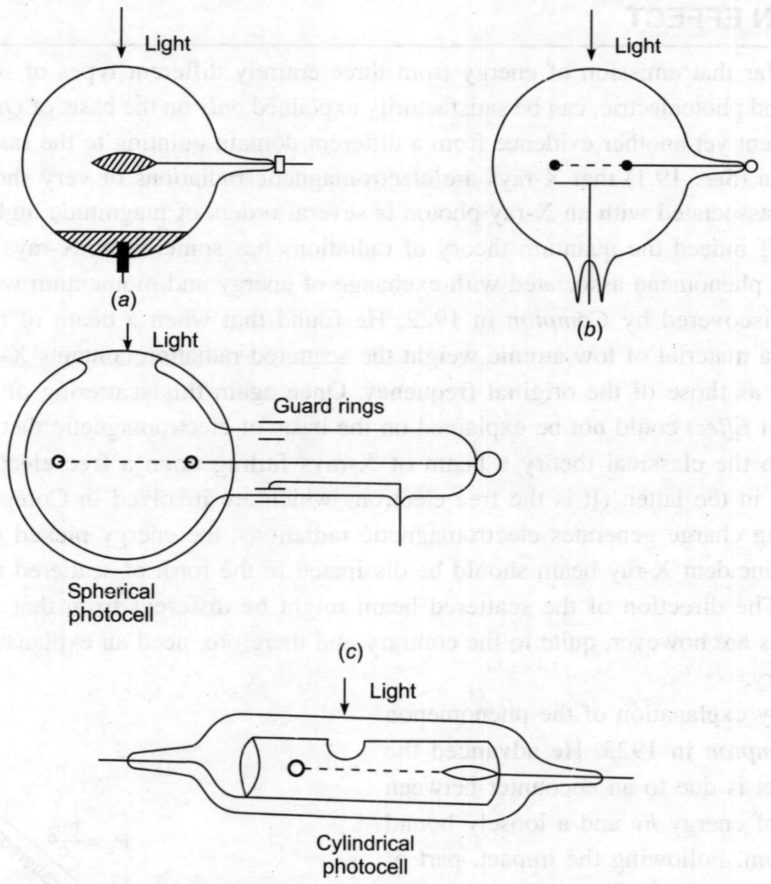

Fig. 20.5

In other types of photocell one electrode surrounds the other. The cathode consists of a layer of alkali metal which is deposited on the inside by condensation. A small window is made by applying the flame to remove the deposit over a small patch. This type of the cell is known as the *black body type* because all the light incident on the surface is employed in producing photoelectrons.

The choice of the alkali metal depends upon the spectral region in which the cell is proposed to be used. Cathode surface of pure potassium and sodium are used for light of wavelength shorter than 5000 A.U. If the incident wavelength is in the range 3500–4500 A.U. sodium is always preferred. The photoelectric characteristics of these cells are permanent and reproducible but the yield measured in μ amp/lumens small.

Commercial photoelectric cells contain a thick layer of potassium sensitized by a glow discharge in hydrogen. Such surfaces give the highest yield in region 4000–5000 A.U. If observations are to be made with incident light whose wavelengths lie over the range 4000–7000 A.U., *i.e.*, the visible region, photocells having thin surfaces films of cesium or its oxide on silver give a better performance. Such cells give excellent performance in the near infrared region.

20.9 COMPTON EFFECT

We have seen so far that emission of energy from three entirely different types of sources, namely thermal, spectral and photoelectric, can be satisfactorily explained *only* on the basis of *Quantum Theory*. We shall now present yet another evidence from a different domain pointing to the same conclusion.

We have seen (*Sec.* 19.1) that X-rays are electromagnetic radiations of very short wavelength. Hence the energy associated with an X-ray photon is several orders of magnitude higher than that of visible radiation. If indeed the quantum theory of radiations has some basis, X-rays should exhibit very markedly the phenomena associated with exchange of energy and momentum with matter. One such effect was discovered by *Compton* in 1922. He found that when a beam of monochromatic X-rays falls upon a material of low atomic weight the scattered radiation contains X-rays of smaller frequency as well as those of the original frequency. Once again this scattering of X-ray photons known as *Compton Effect* could not be explained on the basis of electromagnetic theory.

According to the classical theory a beam of X-rays falling upon a free electron excited an oscillatory motion in the latter. (It is the free electrons which are involved in Compton scattering.) Since an oscillating charge generates electromagnetic radiations, the energy picked up by the free-electron from the incident X-ray beam should be dissipated in the form of scattered radiation of the same frequency. The direction of the scattered beam might be different from that of the incident radiation. The facts are however, quite to the contrary and therefore, need an explanation in the light of Quantum Theory.

A satisfactory explanation of the phenomenon was given by *Compton* in 1923. He advanced the view that the effect is due to an encounter between an X-ray photon of energy *hv* and a loosely bound electron of the atom. Following the impact, part of the incident energy is taken up by the electron which therefore recoils with a definite velocity. The incident radiation having been deprived of a part of its energy at impact is then scattered along a direction governed by the laws of conservation of energy and momentum. The scattered beam along this direction comprises of a radiation of frequency smaller than the incident.

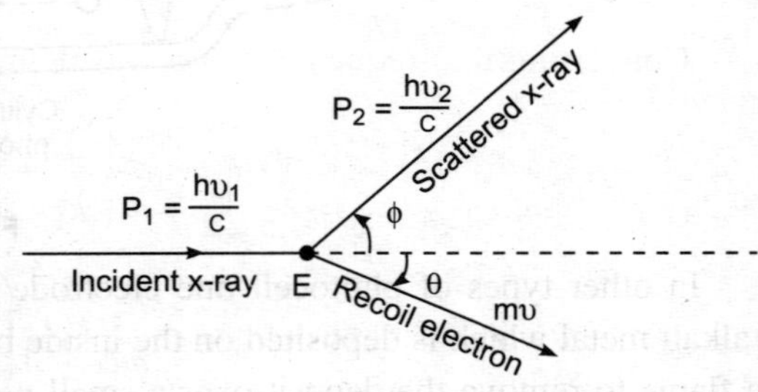

Fig. 20.6. Compton effect

We shall now give an elementary theory of the phenomenon in the light of Compton's explanation.

Consider a free electron E (Fig. 20.6) exposed to monochromatic X-rays of frequency v. This electron may be taken to be practically at rest. According to Quantum Theory, the incident radiation carries an energy

$$E = hv_1 \qquad ...(20.9\ a)$$

and momentum

$$p_1 = hv_1/c \qquad ...(20.9\ b)$$

Due to the impact, the electron will acquire a certain velocity v and recoil along a direction θ to that of the incident beam. Let us now consider the change in energy and hence frequency of the scattered X-rays along a direction ϕ.

Applying the principles of conservation of energy and momentum, we get

$$hv_1 = hv_2 + \frac{1}{2}mv^2 \qquad ...(20.9\ c)$$

$$\frac{hv_1}{c} = \frac{hv_2}{c}\cos\phi + mv\cos\theta \qquad ...(20.9\ d)$$

$$0 = \frac{hv_2}{c}\sin\phi - mv\sin\theta \qquad ...(20.9\ e)$$

where m is the mass of the electron. Eqns. (20.9 d) and (20.9 e) respectively express the conservation of momentum along and perpendicular to the direction of incident beam. Squaring and adding these we get

$$m^2v^2 = \frac{h^2}{c^2}(v_1^2 + v_2^2 - 2v_1v_2\cos\phi)$$

or

$$\frac{1}{2}mv^2 = \frac{1}{2}\cdot\frac{h^2}{mc^2}(v_1^2 + v_2^2 - 2v_1v_2\cos\phi) \qquad ...(20.9\ f)$$

Combining Eqns. (20.9 c) and (20.9 f) we can therefore write

$$h(v_1 - v_2) = \frac{h^2}{2mc^2}(v_1^2 + v_2^2 - 2v_1v_2\cos\phi)$$

Converting the frequencies into wavelengths, we get

$$hc\left(\frac{1}{\lambda_1} - \frac{1}{\lambda_2}\right) = \frac{h^2}{2m}\left(\frac{1}{\lambda_1^2} + \frac{1}{\lambda_2^2} - \frac{2}{\lambda_1\lambda_2}\cos\phi\right)$$

or

$$\lambda_2 - \lambda_1 = \frac{h}{2mc}\lambda_1\lambda_2\left(\frac{1}{\lambda_1^2} + \frac{1}{\lambda_2^2} - \frac{2}{\lambda_1\lambda_2}\cos\phi\right)$$

$$= \frac{h}{2mc}\left(\frac{\lambda_2}{\lambda_1} + \frac{\lambda_1}{\lambda_2} - 2\cos\phi\right) \qquad ...(20.9\ g)$$

Let us assume that

$$\lambda_2 = \lambda_1 + \Delta\lambda$$

where $\Delta\lambda$ is the increase in wavelength of the scattered *X*-ray photon. The quantity inside the brackets on the *RHS* of Eqn. (20.9 *g*) may therefore be written as

$$\frac{\lambda_1 + \Delta\lambda}{\lambda_1} + \frac{\lambda_1}{\lambda_1 + \Delta\lambda} - 2\cos\phi$$

which on simplification gives

$$2(1 - \cos\phi) + \frac{(\Delta\lambda)^2}{\lambda_1(\lambda_1 + \Delta\lambda)}$$

Further the change in wavelength being much smaller than the incident wavelength, the second factor in the above expression may be neglected. We can therefore write

$$\Delta\lambda = \lambda_2 - \lambda_1 = \frac{h}{mc}(1 - \cos\phi) \qquad \text{...(20.9 } h\text{)}$$

as the expression for the increase in wavelength of the scattered radiation in Compton scattering.

In may be recalled that the energy of the incident photon involved in Compton scattering is many orders of magnitude higher than that of a light quantum. Consequently the recoil electron may sometimes possess velocities approaching the velocity of light. Under these circumstances the mass of the electron, *m*, can no longer be assumed to remain the same as its rest mass m_o. However, the two values are related to each other by Einstein's expression

$$m = m_o\left(1 - \frac{v^2}{c^2}\right)^{-1/2}$$

At the same time the kinetic energy of the recoil electron cannot be taken equal to $\frac{1}{2}mv^2$. If the relativistic variation of electronic mass is taken into account, the kinetic energy of the electron is written as difference in energies associated with the rest mass m_o and mass *m* of the electron in motion. These energies are given by Einstein's expression for mass-energy equivalence

$$E = mc^2$$

Thus

$$E_{\text{kin}} = c^2(m - m_o)$$

where E_{kin} now stands for the kinetic energy of the recoil electron.

A rigorous theory of Compton Effect should therefore, take these factors also into account. Once this is done there is no necessity to neglect any factor in the expression for the change in wavelength. Thus the apparent inaccuracy creeping in the simple theory discussed above (neglect of factor involving square of $\Delta\lambda$) has got nothing to do with the observed effect but has been introduced due to our simpler approach. The more exact theory yields the same expression for $\Delta\lambda$ as before. Thus we can write

$$\Delta\lambda = \frac{h}{m_o c}(1-\cos\phi)$$

where h, m_o and c are universal constants. Substituting their values, we get

$$\frac{h}{m_o c} = \frac{6.62\times10^{-27}}{3\times10^{10}\times9.108\times10^{-28}} = .0242 \text{ A.U.}$$

This factor is known *Compton wavelength* λ_o. Hence we can write

$$\Delta\lambda = \lambda_o(1-\cos\phi)$$

We thus find that the change in wavelength of the scattered X-ray photon is:

1. Independent of the incident wavelength.
2. Depends only upon the scattering angle ϕ. As ϕ increases $\Delta\lambda$ increases till it becomes maximum at $\phi = \pi$. In this case the electron recoils in the forward direction parallel to that of the incident radiation and the scattered X-rays double back on their path.

The occurrence of the original frequency along with one of a lower magnitude in the scattered radiation can also be explained on the basis of the above theory. In this case the beam of primary (incident) X-rays, on encountering an atom instead of a free electron, does not suffer any appreciable loss of energy. This becomes evident by calculating $\Delta\lambda$ from Eqn. (20.9 h) by replacing the mass m_o of the electron with the mass of the atom. The change in wavelength thus found is seen to be negligible.

The expression (20.9 h) has been verified experimentally and found to be in excellent agreement with theoretical facts for a large number of elements.

PROBLEMS

1. What is meant by spectrum? Describe various types of spectra.
2. Discuss the evidence on which the Quantum Theory of Light is based.
3. What is Photoelectric Effect? Derive Einstein's photoelectric equation and discuss how it explains the experimental facts of the phenomenon.
4. What is Compton Effect? Derive an expression for the change in wavelength of an X-ray beam involved in Compton scattering. How will you account for the presence of primary radiations in the scattered beam?
5. Give Einstein's derivation of Planck's radiation formula and point out the difference between spontaneous and stimulated emission.
6. Write short notes on:
   (a) Spectral series of Hydrogen
   (b) Photons
   (c) Photoelectric cells
   (d) Einstein's A and B co-efficients
   (e) Quantum Theory of Light.

ORIGIN OF SPECTRA

From the studies of emission and absorption spectra of various substances two very important conclusions emerge. These are:

1. The atoms of a substance are responsible for the emission of its line spectrum. This follows from Kirchoff's law discussed in the previous chapter.
2. The radiant energy is always quantised. It is absorbed, emitted and propagated in quanta.

It was therefore apparent after the first decade of the current century that any successful theory of spectral emission would have to take these factors into account. Also, it became evident by that time that 'atom' which had hitherto been regarded as indivisible has in fact some 'structure'. This view was inspired by Thomson's work on electrons which established that electrons are a fundamental constituent of all matter. In one of the early theories (proposed by Thomson himself) an atom was regarded as a smooth, hard, rigid sphere of dimensions determined by the results of kinetic theory. These spheres were supposed to be made up of a diffuse distribution of positively charged matter, (carrying most of the atomic mass) in which a number of electrons, sufficient to make the atom electrically neutral were imbedded. These electrons were supposed to have the freedom to oscillate about their mean positions and thus generate electromagnetic radiations. But his views received a serious setback when it was discovered (*Lenard* 1903) that swift cathode rays could penetrate sheets of aluminium and other metals. It was then evident that a large portion of the atom was in fact, an empty space and could not be regarded as containing positively charged matter. Almost at the same time a Japanese physicist, *Nagaoka* (1904), came forward with a theory which was quite at variance with the Thomson model. He postulated a planetary model of the atom and likened it to the planet *Saturn*. It was known at that time that Saturn had a heavy central body surrounded by rings containing lighter particle and the stability of the system was sustained by forces of attraction between the planet and its satellites. According to Nagaoka, "The atom will evidently be approximately realised if we replace these satellites by negative electrons and the attracting centre by a positively charged particle".

The existence of a central positively charged particle was however, discovered by *Rutherford* and his associates following their researches on radioactivity. Their apparatus, in principle, consisted of a source of α-particles (which may be taken as helium atoms deprived of both their planetary electrons.) This source, a small quantity of radium which spontaneously emits α-particles, is kept within a lead chamber provided with only a small aperture at one end. The chamber absorbs all α-particles except those emerging out of this aperture in fine pencil. The velocities of the particles moving in this stream may be as high as 2×10^{10} cm/sec. The beam is then allowed to bombard a thin metallic foil *M*. The transmitted beam is scanned by a suitable detecting device which may be, for example, a scintillation counter *C*.

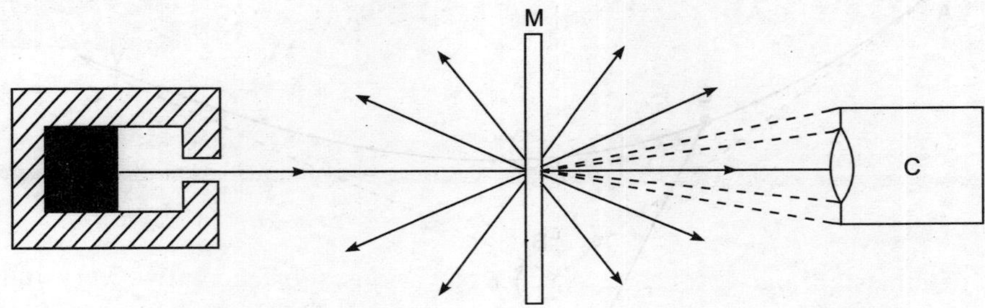

Fig. 21.1

The beam of highly energetic α-particles shooting through the foil will naturally come across the distribution on the concentration of positive charge of the metallic atoms. Since the particles themselves carry a positive charge, they are expected to experience a force of repulsion on passing through the foil. If the Thomson model of the atom is considered, the α-particles will be expected to undergo only a slight deviation in their trajectory during their passage through a continuous distribution of positive charge with embedded electrons. The beams of the emergent α-particles will therefore be only slightly deviated from their original direction as shown by dotted lines in Fig. 21.1.

The experimental results are however quite to the contrary. It was noticed by Rutherford and his associates that most of the α-particles are not deviated *at all* from their original direction. Those which do get deviated are deflected through large angles so much so that a few may even be turned back towards the source.

It thus became evident that since most of the α-particles go undeviated, the space inside atoms is mostly empty and devoid of any positive charge. The particles which are deviated through large angles, therefore, do so only on encountering a concentration of positive charge in each case.

21.1 SCATTERING OF α-PARTICLES

We shall now give an outline of the classical theory of scattering of α-particles. It is assumed that the scattering is due to elastic impact between two point charges — the α-particle approaching a heavy concentration of positive charge *Ze* at rest. If there were no Coulombian interaction between the particles, the α-particle would have followed the linear path *AB* (Fig. 21.2) and the distance of closest approach between the two particles would have been *b*. But on account of the force of repulsion, the α-particle follows the path *AEC*. However, the path of the α-particle at infinite distances from the charge *Ze* is linear because no forces of repulsion act at such distances. The straight lines *AB* and BC would therefore be the asymptotes to the trajectory of the α-particles. The bisector BD of the angle ABC would therefore pass through the positive charge Ze at rest. Let the polar coordinates of any point E, with the charge Ze at the origin be (r, θ).

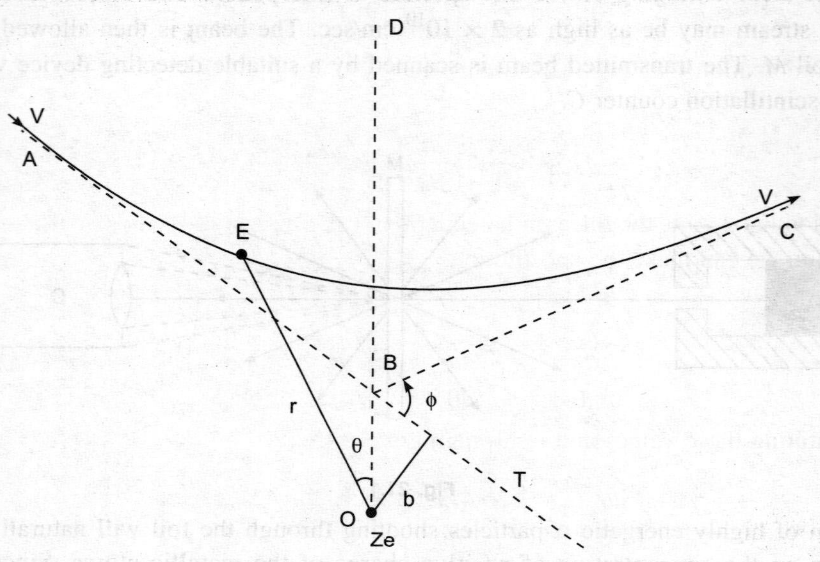

Fig. 21.2

For the problem outlined above it may be shown that:

1. The areal velocity, that is, the area traced by the radius vector r in one second is constant. We can, therefore, write

$$\frac{1}{2}r^2(d\theta/dt) = c \quad \text{(constant)}$$

2. The angular momentum of the α-particle at very large distances from Ze is mvb where m is the mass of the particle.
3. The angular momentum of the systems is conserved.

Combining these factors we get

$$mvb = m.r(d\theta/dt).r$$

where $r(d\theta/dt)$ is the tangential velocity of the α-particle at the point E. We thus get

$$r^2\frac{d\theta}{dt} = vb \qquad \text{...(21.1 } a)$$

Applying the law of conservation of energy we can write

$$\frac{1}{2}mv^2 = \frac{1}{2}m\left[\left(\frac{dr}{dt}\right)^2 + r^2\left(\frac{d\theta}{dt}\right)^2\right] + \frac{2Ze^2}{r} \qquad \text{...(21.1 } b)$$

where the first term on the RHS represents the kinetic energy of the α-particle at any point E and the second gives its potential energy while moving in the Coulomb field of the charge Ze.

We can also write Eqn. (21.1 *b*) as

$$v^2 = \left(\frac{d\theta}{dt}\right)^2\left[\left(\frac{dr}{d\theta}\right)^2 + r^2\right] + \frac{4Ze^2}{mr}$$

which on combination with Eqn. (21.1 *a*) gives

$$v^2 = \frac{v^2 b^2}{r^4}\left[\left(\frac{dr}{d\theta}\right)^2 + r^2\right] + \frac{4Ze^2}{mr} \qquad \text{...(21.1 } c)$$

Equation (21.1 *c*) is the differential equation of the trajectory of the α-particle. For obtaining its polar equation we can make the substitution

$$r = 1/u$$

so that

$$\frac{dr}{d\theta} = -\frac{1}{u^2}\frac{du}{d\theta}$$

Substituting these values and re-arranging we get

$$\left(\frac{du}{d\theta}\right)^2 = \frac{1}{b^2} - \frac{4Ze^2}{m}\frac{u}{v^2 b^2} - u^2$$

$$= \frac{1}{b^2} - \frac{2u}{l} - u^2 \qquad \text{...(21.1 } d)$$

where

$$l = \frac{mv^2 b^2}{2Ze^2} \qquad \text{...(21.1 } e)$$

Equation (21.1 *d*) can be reduced to a standard form by adding and subtracting $1/l^2$ to the RHS. It can then be solved by separation of variables. In this manner we get

$$\cos\theta = \frac{u + 1/l}{(1/p^2 + 1/l^2)^{1/2}} = \frac{lu + 1}{l(1/p^2 + 1/l^2)^{1/2}}$$

Substituting the value of '*u*' we find that

$$l/r = l(1/p^2 + 1/l^2)^{1/2}\cos\theta - 1$$

or

$$l/r = e\cos\theta - 1 \qquad \text{...(21.1 } f)$$

where

$$e = l(1/p^2 + 1/l^2)^{1/2}$$

Hence we find from Eqn. (21.1 *f*) that the path of the α-particle in the neighbourhood of a heavy concentration of positive charge is a hyperbola.

Impact Parameter

We can derive an expression for the impact parameter for α-scattering from the following considerations.

The momentum *p* transferred to the α-particle during an elastic collision is given by (See Fig. 21.2)

$$p = 2mv\cos\left(\frac{\pi}{2} - \frac{\phi}{2}\right) \qquad ...(21.1\ g)$$

This momentum change may also be evaluated from first principles with the help of Newton's second law.

$$p = dp/dt$$

where P is the external force causing the change in momentum.

When the α-particle is at a distance r from Ze, the component of the Coulomb force of repulsion along the direction BD will be

$$\frac{2Ze^2}{r^2}\cos\theta$$

and hence the change in momentum along this direction

$$p = \int_{-\infty}^{+\infty} \frac{2Ze^2}{r^2}\cos\theta\, dt \qquad ...(21.1\ h)$$

where the limits of the integral have been inserted so as to yield an expression alternative to Eqn. (21.1 *g*). But from Eqn. (21.1 *a*) we have

$$dt = \frac{r^2}{vb}d\theta$$

and hence the above integral may be written as

$$p = \int_{-(\pi/2-\phi/2)}^{+(\pi/2-\phi/2)} \frac{2Ze^2}{r^2}\cos\theta(r^2/vb)d\theta$$

which gives

$$p = \frac{4Ze^2}{vb}\cos\frac{\phi}{2} \qquad ...(21.1\ i)$$

Combining Eqns. (21.1 *g*) and (21.1 *i*) we get

$$2mv\cos\left[\frac{\pi}{2} - \frac{\phi}{2}\right] = \frac{4Ze^2}{vb}\cos\frac{\phi}{2}$$

$$\therefore \qquad b = \frac{2Ze^2}{mv^2}\cot\frac{\phi}{2} \qquad ...(21.1\ j)$$

Equation (21.1 *j*) shows that when the distance of the closest approach is large, *i.e.* when the α-particle moves at a large distance from the concentration of a heavy positively charged particle, the angle of scattering ϕ is very small. As b decreases, the scattering angle goes on increasing in accordance with this equation and in the limit when b tends to zero, ϕ tends to π. In this eventuality therefore, the α-particle would have to go back almost on the same trajectory along which it approached Ze.

Scattering Cross-section

Equation (21.1 *j*) may be utilized for calculating the probability of an α-particle passing within a range db at a distance b from the charge Ze. For this we define two concentric circles of radii b and $b + db$ with the location of Ze as centre and normal to the incident direction of the α-particle. Then the annular area enclosed by the two circles is given by $2\pi b\, db$, where db is obtained with the help of Eqn. (21.1 *j*).

If the impact parameter lies within the range defined above the scattering angle will lie between φ and φ + dφ and the scattered α-particles will be between cones enclosing a solid angle

$$d\omega = 2\pi \sin\phi\, d\phi \qquad \text{...(21.1 } k\text{)}$$

If Ze is the only concentration of positive charge in the surrounding unit area of the target material, the probability of an α-particle passing within a range b to $b + db$ is given by

$$\frac{2\pi b\, db}{1} = 2\pi \cdot \frac{2Ze^2}{mv^2} \cot\phi/2 \cdot \frac{Ze^2}{mv^2} \operatorname{cosec}^2\phi/2\, d\phi$$

$$= \pi\left(\frac{2Ze^2}{mv^2}\right)^2 \cot\phi/2 \operatorname{cosec}^2\phi/2\, d\phi \qquad \text{...(21.1 } l\text{)}$$

Combining Eqns. (21.1 *k*) and (21.1 *l*) we can therefore calculate the probability of the α-particle being scattered into a cone of solid angle $d\omega$. If this probability is represented by $d\sigma$ we find that

$$d\sigma = \left(\frac{Ze^2}{mv^2}\right) \operatorname{cosec}^4\phi/2\, d\omega \qquad \text{...(21.1 } m\text{)}$$

The co-efficient $d\sigma/d\omega$ is known as differential scattering cross-section.

If there are n positive charges of the type Ze per unit area of the target material of thickness t, the fraction F of α-particles scattered by this material within the solid angle $d\omega$ will therefore be

$$F = nt\left(\frac{Ze^2}{mv^2}\right) \operatorname{cosec}^4\phi/2\, d\omega$$

The number of scattered α-particles can be determined by counting the scintillations produced on a fluorescent screen at a distance R from the target. This number striking an area dS of the screen will therefore be given by

$$F = nt\left(\frac{Ze^2}{mv^2}\right) \operatorname{cosec}^4\phi/2 \frac{dS}{R^2}$$

This scattering formula was verified experimentally by Geiger and Marsden in Rutherford's laboratory. The dependence of F on different variables, *e.g.* the kinetic energy $\frac{1}{2}mv^2$ of the incident α-particle, the scattering angle φ, the thickness of the target material etc. was also tested. The agreement between the theory and experiments was fairly good.

In addition to providing a sound experimental background to Rutherford's theory, the experiments of Geiger and Marsden brought out the important fact that the number Z occurring in the above expressions was in fact equal to approximately half the atomic weight in the case of all elements examined. Later work by Chadwick firmly established that the number Z is in fact the same as the atomic number assigned to the elements in the periodic table.

21.2 RUTHERFORD MODEL OF THE ATOM

The experiments in Rutherford's laboratory provided the missing link between the observations of Lenard and the postulates of Nagaoka. He was thus quick to realise that the atom consists of a central core or "nucleus" at which all the positive charge of the atom and most of its mass is tightly concentrated. He also estimated the size of the nucleus from the following considerations:

Suppose an α-particle approaches a stationary nucleus along the line joining their centres. Owing to the mutual repulsion, the particle (carrying a positive charge $2e$) possesses a potential energy $2Ze^2/r$ at a distance r from the nucleus. As r goes on diminishing, the potential energy goes on increasing till it becomes maximum and attains a value equal to the kinetic energy $\frac{1}{2}mv^2$ with which it had started at a large distance away from the nucleus. It is at this value of r called the distance of closest approach b, that the α-particle comes instantaneously to rest and is then repelled back along its own previous trajectory. This distance of the closest approach therefore determines an upper limit for the radius of the target nucleus. It may be estimated from the equation

$$\frac{1}{2}mv^2 = \frac{2Ze^2}{b} \quad \text{or} \quad b = \frac{4Ze^2}{mv^2} \qquad ...(21.2\ a)$$

The mass of an α-particle may be taken to be nearly the same as that of helium atom. This in turn can be determined by dividing the atomic weight of helium by Avogadro's number. Thus

$$m = \frac{4.00}{6.02\times10^{23}} = 2/3\times10^{-23} \text{ gms.}$$

Also $$e = 4.802 \times 10^{-10} \text{ E.S.U.}$$

If the average velocity of an α-particle is taken as 1.5×10^9 cm. per sec. and a target of atomic weight 40 is chosen then Z may be approximately put down as 20. In this manner we find that

$$b = \frac{4\times20\times4.8\times4.8\times10^{-20}\times3}{2\times10^{-23}\times1.5\times1.5\times10^{18}}$$

$$= 1.28\times10^{-12} \text{ cms.}$$

Therefore, the radius of the target nucleus which is less than this figure should be of the order of 10^{-12} to 10^{-13} cms. The nucleus is thus 10^{-4} to 10^{-5} smaller than an atom.

According to the ideas of Nagaoka, Rutherford imagined the electrons to be revolving round the nucleus in planetary orbits under the Coulombian centripetal force due to the positive charge at the nucleus. These revolving electrons (Fig. 21.3) by virtue of their constantly accelerated path around the nucleus emit electromagnetic radiations of frequencies equal to their frequencies of rotation.

Inspite of the outstanding success of Rutherford in explaining α-scattering and successfully establishing the existence of an atomic nucleus, his model is open to very serious objections. Two most serious shortcomings are described below.

1. Rutherford postulated the emission of electromagnetic radiations by the constantly accelerated planetary electrons. Now, it follows as a corollary to the electromagnetic theory that an accelerated charge gives rise to electromagnetic radiations. Combining these two concepts we find that the rotating electrons emit electromagnetic radiations at their own expense. On account of the consequent decrease in energy, the revolving electron will continue to spiral down towards the nucleus. Thus the atom, as postulated by Rutherford, is found to be highly unstable.

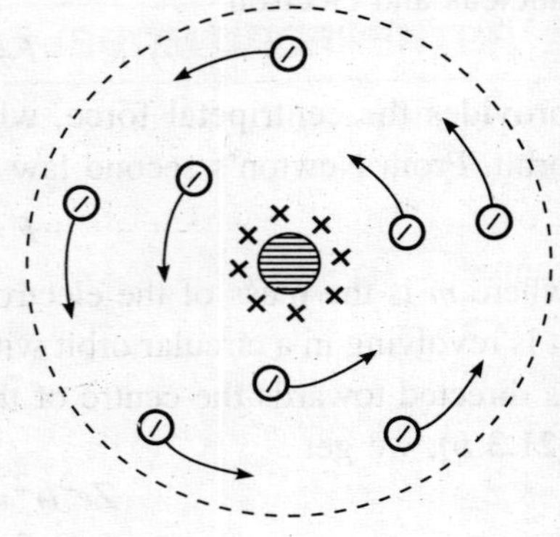

Fig. 21.3

2. The radius of the electron orbit in Rutherford model is determined by the following equation between the centripetal and centrifugal forces

$$\frac{Ze^2}{kR^2} = \frac{mv^2}{R}$$

where Ze is the charge on the nucleus, k, the permittivity of the subatomic space, R, the distance between the nucleus and the electron of mass m moving with a velocity round the nucleus.

Rutherford's model therefore does not place any restrictions upon the radii of the electron orbits and hence upon the frequencies of their rotations. The model therefore, by implication, leads to the conclusion that atomic systems give rise to continuous spectra. The same conclusion is reached on considering the rapid spiralling down of the electrons towards the nucleus as discussed above. This result of theory is known to be at variance with experimental facts, for, it is seen that substances in their atomic state give rise to line rather than continuous spectra.

In view of these and some more objections, it was soon noticed that the model had to be radically revised in order to explain the emission of line spectra by atoms.

21.3 BOHR MODEL OF THE ATOM

The first clues towards the successful quantitative interpretation of the spectra of elements was given by *Niels Bohr* in his theory of hydrogen spectrum. In his explanation he was guided by Rutherford's nuclear model, *viz*., an atom consists of a positively charged nucleus, in which practically all the mass of atom is concentrated, around which, at distances far removed from it, a large number of electrons revolve. In a neutral atom, number of electrons is equal to the atomic number Z of the element. The atomic number of hydrogen is 1. Accordingly, Bohr concluded that hydrogen atom is composed of an electron, of charge $-e$, and

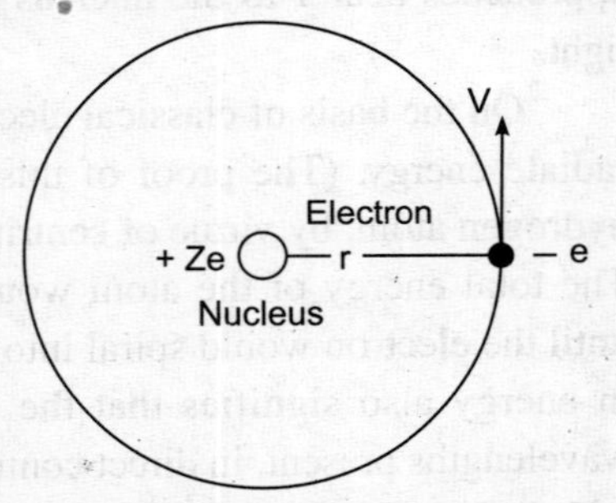

Fig. 21.4. Hydrogen like (ionized) atom

one proton, of charge $+e$ (since $Z = 1$). Since the hydrogen nucleus is nearly 1840 times heavier than the electron, he assumed at first that the nucleus is practically at rest.

Classical Theory

Bohr first considered that under the action of Coulomb's field of force an electron revolves in a circular orbit about the relatively massive nucleus. The electrostatic force of attraction between the nucleus and electron,

$$F = -Ze^2/r^2 \quad \text{...(21.3 } a\text{)}$$

provides the centripetal force, where r the electron-nuclear distance, is the radius of the circular orbit. From Newton's second law of motion.

$$F = m \times (-v^2/r) \quad \text{...(21.3 } b\text{)}$$

where m is the mass of the electron and $(-v^2/r)$ is the centripetal acceleration of the electron, since it is revolving in a circular orbit with a velocity v. The minus sign simply signifies that the acceleration is directed towards the centre of the circle. Equating the values of F as given by Eqns. (21.3 *a*) and (21.3 *b*), we get

$$Ze^2/r^2 = mv^2/r$$

$$mv^2 = Ze^2/r \quad \text{...(21.3 } c\text{)}$$

The kinetic energy of the electron is

$$E_{kin} = \frac{1}{2}mv^2 = Ze^2/2r$$

The potential at any point distant r from the nucleus of charge Ze is Ze/r per unit (positive) charge. Assuming that the potential energy is zero when the electron is at infinity, its potential energy at a distance r from the nucleus is

$$E_{pot} = (Ze/r)(-e) = -Ze^2/r$$

Therefore, the total energy E, which is the sum of the potential and kinetic energies is

$$E = \frac{Ze^2}{2r} - \frac{Ze^2}{r} = -\frac{Ze^2}{2r} \quad \text{...(21.3 } d\text{)}$$

Since Bohr considered nucleus to be stationary, Eqn. (21.3 *d*) represents the total energy of the atom excluding the rest mass energy of the two particles. Obviously, $E = 0$ when $r = \infty$ *i.e.* electron is detached from the atom. The zero level of energy consequently represents the ionized state of the atom. The negative sign in Eqn. (21.3 *d*) signifies that the energy of the atom decreases as the electron approaches nearer to the nucleus and this decrement of energy should be radiated out in the form of light.

On the basis of classical electromagnetic theory, an accelerated electric charge must continuously radiate energy. (The proof of this proposition lies beyond the scope of this book). Accordingly, the hydrogen atom, by virtue of centripetal acceleration of electron, must continuously emit radiant energy. The total energy of the atom would continuously decrease which means r would gradually decrease until the electron would spiral into the nucleus. But hydrogen atom is stable. The continuous decrement in energy also signifies that the hydrogen spectrum should be a continuous one with all possible wavelengths present, in direct contradiction to experimental fact that hydrogen atoms emit line spectrum with well defined wavelengths.

Quantum Theory

These considerations led *Niels Bohr* to the conviction that the classical theory is totally inadequate to account for the origin of the line spectrum. Therefore, he applied *Planck's quantum theory* to the *Rutherford's* model of atom and succeeded in giving a satisfactory explanation of spectral lines of hydrogen spectrum on the basis of following postulates:

1. Only a limited number of stable orbits or stationary states exist with respect to the nucleus, in which electron can continue to revolve without radiating energy, even though it has an acceleration v^2/r. A limited set of orbits obviously suggests the quantum idea.

2. Only those orbits are stable for which the angular momentum $I\omega$ of the electron is an integral multiple of $h/2\pi$, where h is Planck's constant ($h = 6.62 \times 10^{-27}$ ergs./sec.) I the moment of inertia of the electron about the nucleus and ω the angular velocity of revolution of the electron. That is, the angular momentum of the electron in the nth stationary orbit is

$$I\omega_n = mr_n^2\omega_n = nh/2\pi$$

$$mv_nr_n = nh/2\pi \qquad \text{...(21.3 } e)$$

where n is called the quantum number and can take only the whole number values 1, 2, 3...... according as the electron is in 1st, 2nd, 3rd ...... orbit, respectively.

The equation (21.3 *e*) for determining the stationary orbits is the *Bohr's quantum condition*.

3. The radiant energy is absorbed or emitted by the atom in whole quanta of amount hv, where v is the frequency of the radiation, only when the electron passes over or jumps from one stationary state to another possible stationary state.

4. The frequency v of radiation emitted during the transition of electron from the stationary state of energy E_n to another of lower energy E_p is given by the quantum energy formula.

$$hv = E_n - E_p \qquad \text{...(21.3 } f)$$

This formula is called *Bohr's frequency relation*.

5. So long as the electron is in one of the stable orbits, its dynamical equilibrium is governed by the classical law *viz*. Coulomb's force of attraction is equal to the centripetal force.

Suppose the electron is in the nth stationary state. According to 5th postulate, Eqn. (21.3 *c*) is true also in the quantum theory and for the nth stationary orbit it may be written as

$$mv_n^2 = Ze^2/r_n \qquad \text{...(21.3 } g)$$

and the Bohr's quantum condition is

$$mv_nr_n = nh/2\pi \qquad \text{...(21.3 } h)$$

On squaring $\quad m^2v_n^2r_n^2 = n^2h^2/4\pi^2$

Substituting the value of v_n^2 from Eqn. (21.3 *g*) in the above relation and then solving it for r_n, we get,

$$r_n = \frac{n^2h^2}{4\pi^2mZe^2} \qquad \text{...(21.3 } i)$$

Utilising Eqn. (21.3 *d*), the energy of the atom when the electron is in the nth stationary state, is given by

$$E_n = -\frac{Ze^2}{2r_n} = \frac{2\pi^2 mZ^2e^4}{n^2h^2} \quad ...(21.3\ j)$$

According to Bohr's fourth postulate, the frequency v of the radiation emitted as the electron jumps from the stationary state characterised by the quantum number n to another stationary state characterised by the quantum number p is

$$v = \frac{E_n - E_p}{h} = \frac{2\pi^2 mZ^2e^4}{h^3}\left(\frac{1}{p^2} - \frac{1}{n^2}\right) \quad ...(21.3\ k)$$

If c is the velocity of light and λ the wavelength of the spectral line, then since $\lambda v = c$, we have

$$1/\lambda = v/c = \frac{2\pi^2 mZ^2e^4}{ch^3}[1/p^2 - 1/n^2]$$

$$= R_\infty Z^2[1/p^2 - 1/n^2] \quad ...(21.3\ l)$$

where $\bar{v} = l/\lambda$ is called the wave number of the spectral line and $R_\infty = 2\pi^2 me^4/ch^3$ is called the Rydberg's constant and it is expressed in wave number. The wave number is the number of waves in one centimeter.

For the case of hydrogen atom $Z = 1$, hence Eqn. (21.3 l) becomes

$$\bar{v} = R_\infty[1/p^2 - 1/n^2] \quad ...(21.3\ m)$$

Interpretation of the Hydrogen Spectrum

The entire spectrum of hydrogen is explained by Eqn. (21.3 m). For example, if we put $p = 1$ and $n = 2, 3, 4, 5, 6$ etc. successively we get the formula representing the Lyman series. The Balmer series is obtained by setting $p = 2$ in Eqn. (21.3 m) which simply means that all lines of series are produced by electron jumps ending in the second orbit. The Paschen series in the near infrared is given by $p = 3$ and $n = 4, 5, 6$ etc. and so on.

It is now possible to visualise the physical process by which the spectral series of hydrogen are emitted. Eqn. (21.3i) shows that the radii of stable orbits of the electron are proportional to the square of quantum numbers n *i.e.*, proportional to $1^2, 2^2, 3^2$...... or 1, 4, 9...... Some of these orbits are shown pictorially to scale in Fig. 21.5. The orbit marked $n = 1$ is called the ground orbit and represents the stationary state of unexcited atom.

Suppose by processes such as thermal agitation or impact of electrons in a discharge tube the electron of hydrogen atom may be removed from the ground orbit and on equilibrium may again find itself in one of the stable orbits of higher energy say $n = 3$. The atom possessing more energy than the normal is obviously less stable as compared to normal state. If the excited atom is not disturbed, the electron will eventually fall to either ground orbit (from $n = 3$ to $n = 1$) or to intermediate orbit (from $n = 3$ to $n = 2$) and then to the ground orbit (from $n = 2$ to $n = 1$). In the former process frequency corresponding to second member of Lyman series is emitted, while in the other process the first members of Balmer and Lyman series are emitted. The loss of energy in the two processes just visualised is the same and so the frequency of radiation of the second member of Lyman series is equal to the sum of the two other frequencies. Now it is obvious that in a given mass of incandescent hydrogen, with millions of atoms, all possible transitions are occurring, producing the experimentally detectable spectral series (Lyman, Balmer etc.) of hydrogen spectrum.

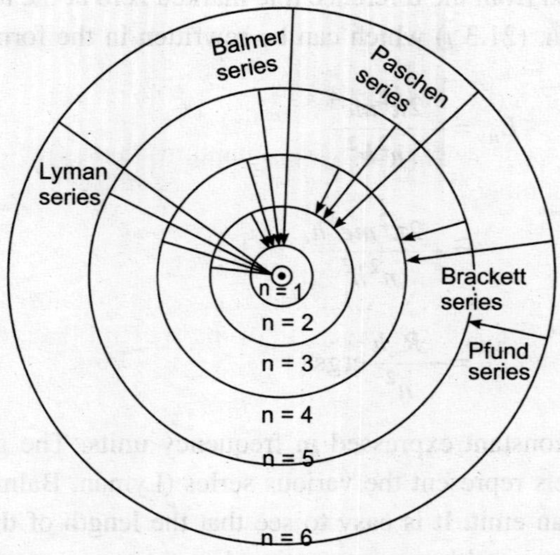

Fig. 21.5. Bohr circular orbits of hydrogen showing quantum jumps of electron

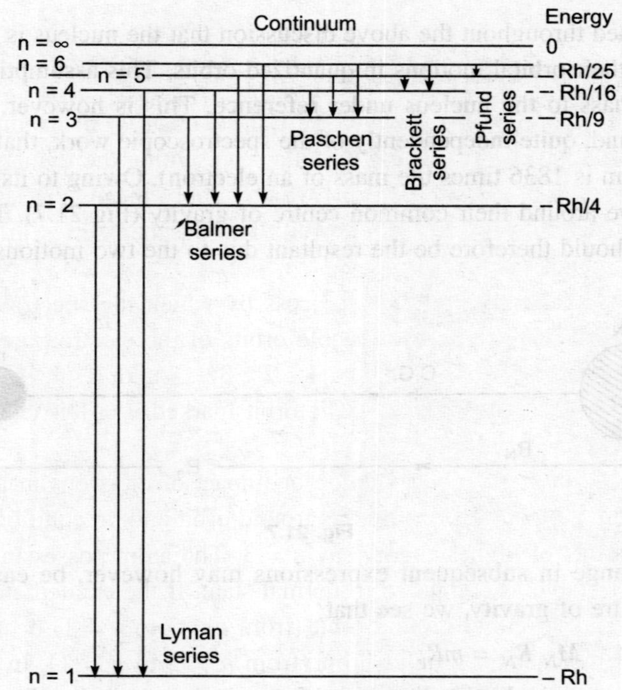

Fig. 21.6. Energy level diagram for hydrogen

Energy Levels

Energy level diagram representing energies of different stationary states and transitions for hydrogen atom are shown in Fig. 21.6. Different energy levels are represented by horizontal lines drawn in positions, measured from the reference line marked zero at the top, corresponding to energies of atom computed from Eqn. (21.3 *j*) which can be rewritten in the form (since $Z = 1$).

$$E_n = -\frac{2\pi^2 me^4}{n^2h^2}$$

$$= -\frac{2\pi^2 me^4 h}{n^2h^3}$$

$$= -\frac{R_\infty h}{n^2}\ \text{ergs.}$$

where R_∞ is the Rydberg constant expressed in frequency units. The group of arrows drawn from higher to lower energy levels represent the various series (Lyman, Balmer, Paschen, etc.) of spectral lines that hydrogen atom can emit. It is easy to see that the length of these lines are proportional to corresponding energy changes and hence proportional to the wave numbers of the resulting spectral line.

Finite Mass of Nucleus

It has been assumed throughout the above discussion that the nucleus is practically at rest while the electrons continue their orbital motions in quantized orbits. This assumption in effect amounts to assigning an infinite mass to the nucleus under reference. This is however, not the actual state of affairs. (It has been found, quite independently of the spectroscopic work, that the mass of the proton – ionized hydrogen atom is 1836 times the mass of an electron). Owing to its finite mass the nucleus and the electron revolve around their common centre of gravity (Fig 21.7). The angular momentum used in Eqn. (21.3 *b*) should therefore be the resultant due to the two motions.

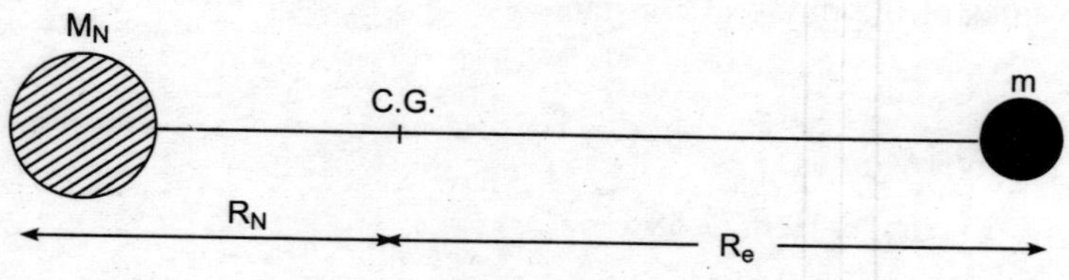

Fig. 21.7

The resulting change in subsequent expressions may however, be easily calculated. Taking moments about the centre of gravity, we see that

$$M_N R_N = mR_e \qquad \text{...(21.3 } n\text{)}$$

where R_N and R_θ are respectively the distances of the nucleus and the electron from the centre of gravity and M_N is the mass of the nucleus. The electron-nucleus distance r is then given by

$$r = R_e + R_N \qquad \text{...(21.3 } o\text{)}$$

Combining Eqns. (21.3 *n*) and (21.3 *o*) we find that

$$\left.\begin{aligned} R_e &= \frac{r}{1+m/M_N} \\ R_N &= \frac{m}{M}r \Big/ \left[1+\frac{m}{M_N}\right] \end{aligned}\right\} \qquad \text{...(21.3 } p)$$

and the resulting moment of inertia of the system as

$$\begin{aligned} I &= mR_e^2 + M_N R_N^2 \\ &= \frac{M_N}{M_N+m} mr^2 \end{aligned} \qquad \text{...(21.3 } q)$$

because of Eqn. (21.3 *p*).

Hence instead of the original $I = mr^2$ we should in fact have I given by Eqn. (21.3 *q*). In other words m, the mass of the electron has to be replaced by its so-called reduced mass μ defined by the equation

$$\frac{1}{\mu} = \frac{1}{M_N} + \frac{1}{m}$$

It therefore follows that the Rydberg's constant

$$R_\infty = \frac{2\pi^2 me^4}{ch^3}$$

should be replaced by

$$R = \frac{2\pi^2 \mu e^4}{ch^3}$$

so that $R = R_\infty (1+m/M)^{-1}$

Substituting the values of the physical constants we find that

$$R_\infty = 109737.3 \text{ cm}^{-1}$$

and Rydberg constant for hydrogen as

$$R_H = 109677.58 \text{ cm}^{-1}$$

The effect of considering this small correction in Rydberg constant is to increase the calculated wavelengths of hydrogen lines by about 3 A.U. The results thus obtained by the theory are found to be in excellent agreement with experimental data (see the following table)

Line	***Observed wavelength***	***Calculated wavelengths***	
		with R_∞	***with*** R_H
$H\alpha$	6562.8	6559.3	6562.8
$H\beta$	4861.3	4858.7	4861.3
$H\gamma$	4340.5	4338.1	4340.5

(*Wavelengths of hydrogen lines in A.U.*)

If there is another atom of mass different from that of the hydrogen but which contains a single electron revolving round the nucleus, the foregoing theoretical considerations can be applied to it. Such an atom is obtained for example, by removing one electron from a helium atom which in the normal state contains two electrons revolving round the nucleus carrying a charge +2*e* and mass approximately 3.98 times the mass of the hydrogen atom. For such an ionized helium atom

$$R_{H_e^+} = R_\infty\left(1+\frac{m}{3.98M_H}\right)^{-1}$$

Also,
$$R_H = R_\infty\left(1+\frac{m}{M_H}\right)^{-1}$$

Hence
$$R_{H_e^+} = R_H\frac{1+m/M_H}{1+m/3.98M_H}$$

Substituting the value of m/M_H we find that

$$R_{H_e^+} = 109722.26\ \text{cm}^{-1}$$

Conversely if the Rydberg constant of singly ionized helium atom is experimentally determined, the value may be used for finding the ratio m/M_H and hence the mass M_H of a proton (hydrogen nucleus).

21.4 EFFECT OF ISOTOPES

In the year 1932, *Urey* and his co-workers working at Princeton University found that the Hα line of hydrogen was accompanied by another line separated by 1.79 A.U. They used a concave grating for observing this separation. These lines were called $H\alpha^1$ and $H\alpha^2$. This provided an experimental evidence of the existence of two types of hydrogen, one having an atomic weight equal to one and the other, two, on the atomic weight scale. The existence of these isotopes, which have the same atomic number and equal number of planetary electrons had already been predicted by *Birge* and *Menzel* in 1929. They had found that the atomic weight of hydrogen determined by chemical methods differed by about two parts in 10,000 with that determined by mass spectrograph. This difference was attributed to the probable existence of two isotopes of hydrogen, which as seen above was confirmed by Urey.

Following Bohr's theory the *H*α is given by the transition

$$\bar{\nu} = \frac{1}{\lambda} = \frac{R_\infty}{1+\dfrac{m}{M_H}}\left[\frac{1}{2^2}-\frac{1}{3^2}\right]$$

the wavelengths λ_1 and λ_2 corresponding to $M_H^1 = 1$ and $M_H^2 = 2$ are then obtained as

$$\lambda_1 = 6562.79\ \text{A.U.}$$

and
$$\lambda_2 = 6561.00\ \text{A.U.}$$

Difference
$$= 1.79\ \text{A.U.}$$

The theory was thus also found to be successful in confirming the presence of a heavier isotope of hydrogen called *deuterium*.

21.5 SOMMERFELD'S EXTENSION

The striking success of Bohr's theory in explaining the spectral emission of mono-electron atoms like H, He^+ and Li^{++} led to its application in understanding some fine details of atomic spectra. When examined with instruments of high resolving power, the line spectrum of even a simple atom like hydrogen is seen to be complex. The lines H_α, H_β etc. are, in fact found to be actually composed of several lines very close together. The simple theory therefore, was totally inadequate for explaining this fine structure and needed a modification.

It was *Sommerfeld* who pointed out the first modification. He regarded Bohr's circular orbits as an unnecessary limitation on the electrons motion round the nucleus. He therefore, proposed that the electron, in fact, moves in an ellipse with the nucleus located at one of its foci; the circular orbits postulated by Bohr, are only a special case of this general treatment. In this manner, Sommerfeld based his atomic model on planetary systems.

Since in Sommerfeld's treatment the position of the electron could be determined by two coordinates (r, ϕ) – radial and azimuthal, two quantum conditions were found necessary for defining the two components of the angular momentum. These were

$$\oint p_r dr = n_r h$$

and

$$\oint p_\phi d\phi = n_\phi h$$

where p_r and p_ϕ are respectively the radial and angular components of electrons momentum and n_r and n_ϕ are integral numbers governing these components. Thus Sommerfeld used two quantum numbers, namely the radial (n_r) and azimuthal (n_ϕ) in the place of Bohr's total quantum number n. It was also found that if

$$n = n_r + n_\phi$$

the expression for energy is identical with that obtained with the help of the simple Bohr's theory. Thus, according to Sommerfeld it is the total quantum number n which governs the energy state of an atom. The two quantum numbers n_r and n_ϕ are found to simply govern the size and shape of the ellipse. It was also found that the radial quantum number n_r could take the values

$$n_r = 0, 1, 2, 3, \ldots\ldots \text{ etc.}$$

The azimuthal* quantum number n_ϕ can however take only the values

$$n_\phi = 1, 2, 3, \ldots\ldots \text{ etc.}$$

Applying these results to the hydrogen atom we find that for $n = 1$, the only possible values of n_r and n_ϕ are respectively zero and one. In the $n = 1$ state of the atom, also called the ground state the electron can therefore revolve in a single orbit which is circular. For $n = 2$, there are two possibilities

$$n_r = 0,\ n_\phi = 2 \text{ or } n_r = 1,\ n_\phi = 1$$

*According to Wave Mechanics n_ϕ should be replaced by the azimuthal quantum number $l = n_\phi - 1$

For these the electron's orbit may either be circular ($n_r = 0$) or elliptic. For $n = 3$ state, one circular and two elliptic orbits are similarly possible. These possibilities have been shown in Fig. 21.8.

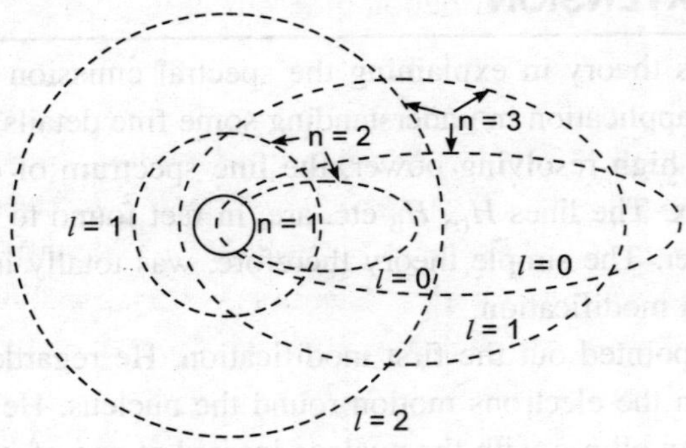

Fig. 21.8

With these modifications, Sommerfeld set out to examine the relativisitic increase in the mass of the electron while moving in elliptic orbits of different eccentricities. He thus found that even for the same n, the energy of the electron is slightly different for orbits of different eccentricities. As a result of this each atomic level is subdivided into as many sub-levels as the possible number of states of the electron. Thus $n = 1$ state remains single, the $n = 2$ state is found to comprise of two sub-levels with a very small difference of energy, and so on account of the multiplicity of levels of different total quantum numbers many more transitions between these levels become possible. This leads to the explanation of the fine structure of spectral lines.

21.6 SHORTCOMINGS OF BOHR-SOMMERFELD THEORY

The success of Bohr-Sommerfeld theory in explaining the spectra of hydrogen and hydrogen-like atoms, the role it played in the discovery of deuterium and in accounting for the fine-structure of spectral lines led to its application to atoms with two and more electrons. In these cases, however, the agreement was less perfect and the discrepancies became more pronounced as more complicated atoms were considered. Moreover, even in the case of hydrogen-like atoms Bohr's quantization of angular momentum, although successful, was nevertheless arbitrary. A similar remark may be made regarding Bohr's postulate concerning the emission of radiations from atoms in the excited state. It could therefore, at best provide a working hypothesis and needed several modifications to account for these and other equally serious objections. It may however, be added that later modifications retained the main features of Bohr's theory and even the new quantum theory owes its birth to Bohr's ideas.

21.7 ATOMIC ENERGY LEVELS

It will be clear from the preceding discussion that mono-electron atoms can exist in a series of excited states. The energy of the atom in each of these states is given by the expression

$$E_n = -\frac{RhcZ^2}{n^2} \qquad ...(21.7\ a)$$

where n is a non-zero integral number. The distribution of atoms in these states has also been seen to be proportional to

$$e^{-E_n/\kappa T} \qquad ...(21.7\ b)$$

At a given temperature therefore the higher excited states are less populated than the ground state $n = 1$. As the temperature is increased, as for example by passing an electric discharge through the gas, the number of atoms in higher excited states also increases. These atoms in higher excited states then lead to the spontaneous emission of monochromatic radiations as postulated by Bohr-Sommerfeld theory. The wave-number of the emitted radiation is then given by the difference in two energy states or energy levels of the atom. Thus

$$\bar{\nu} = RZ^2\left(\frac{1}{n^2} - \frac{1}{m^2}\right) \quad m > n$$

The first empirical evidence of the existence of atomic levels was contained in Ritz's Combination Principle which states, "By additive or subtractive combination, whether of the series formula themselves, or, of the constants that occur in them, formulae are formed that allow us to calculate certain newly discovered lines from those known earlier".

Thus Ritz's Combination Principle when combined with the expressions for wave-number $\bar{\nu}$ leads to the conclusion that each spectral line of the several series emitted by an atom represents an energy state or level of the atom under reference. It therefore also follows that, subject to certain restrictions, called selection rules, the various combinations of these levels lead to different spectral lines of the atom.

Unlike Bohr's theory, Ritz's Combination Principle is applicable in the whole-region of spectroscopy from infrared to X-ray spectra and is an exact physical law. For, as we go to atoms with an electronic structure, different and more complex from the simple hydrogen-like atoms, the expressions for energy of atomic levels no longer retain the simple form derived by Bohr. In particular the quantum numbers n no longer retain their integral values. Nevertheless, the wave numbers of the various spectral series can still be obtained as difference of two terms or energy levels of the atom and the Combination Principle still applies.

21.8 EXISTENCE OF ATOMIC LEVELS

The outstanding success of Bohr's theory of hydrogen like atoms and also the general applicability of Ritz Combination Principle led *Franck* and *Hertz* to a direct determination of atomic energy levels. These experiments, in addition to confirming the existence of discrete energy states of an atom, proved to be very valuable in supplementing spectroscopic data.

In principle, their method consists of accelerating a stream of electrons and allowing them to suffer collisions with the atoms of a gas or vapour. At every impact an exchange of energy (between the bombarding electrons and the target atoms) takes place. This energy exchange is at the cost of the kinetic energy of the electrons given by the expression

$$\frac{1}{2}mv^2 = eV$$

where V is the accelerating voltage experienced by an electron of charge e. The loss in energy at each impact may either appear as an increase in kinetic energy of the target atom or under special circumstances, may bring about a change in state of the atom. In collisions of the former type, known as *elastic collisions*, the loss in energy of the electron may be shown to be negligible* according to the laws of elementary dynamics. In collisions of the latter type, called *inelastic collisions* the loss in energy, will be equal to the difference in the initial and final energy levels of the target atom.

Franck and Hertz aimed at studying the circumstances under which inelastic collisions are brought about. Their experimental set up is outlined in Fig. 21.9.

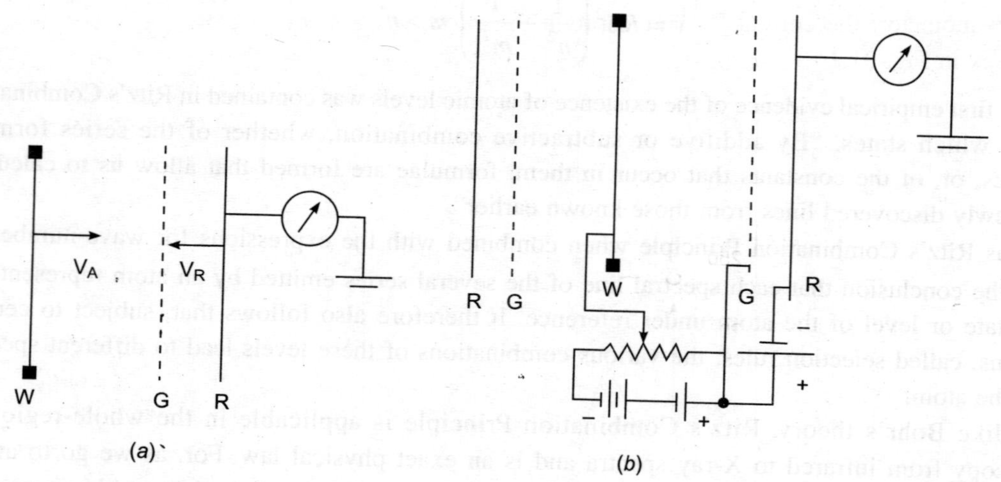

Fig. 21.9

The source of electrons was a thermionic cathode in the form of a hot wire W stretched along the axis of a cylindrical glass envelope, a few inches long. A gauze electrode G and a receiving electrode R both cylindrical in form were mounted co-axially inside the envelope. Arrangements were also made for applying various potential differences between the wire and the gauze and also between the gauze and the receiver. A sensitive galvanometer was used for detecting the strength of current reaching R.

In the actual experiments, mercury vapour, at a low pressure, was exposed to the stream of electrons issuing from the wire W. The initial low velocity of these electrons was increased by applying an accelerating voltage V_A between the cathode and the gauze electrode. A small retarding voltage

*Loss in energy = 2m/M . ½mv^2 where M = mass of the atom. Thus for the sodium atom fractional loss in energy = $(2 \times 9.1 \times 10^{-28})/(1.66 \times 10^{-24} \times 23) = 4.77 \times 10^{-5}$

V_R ($V_R < V_A$) was applied between G and the receiving electrode. Keeping V_R fixed, the accelerating voltage V_A was increased. This brought about the expected increase in current flowing to the galvanometer till V_A attained a critical value $V_C = 4.9$ volts. At this voltage the galvanometer current showed a sudden fall (Fig. 21.10). This sudden fall in current may be explained as follows:

As long as V_A is less than V_C, the electrons accelerating through mercury vapour, suffer only elastic collisions. Since the loss in energy accompanying such collisions is negligible, the electrons carry sufficient kinetic energy to overcome the retarding potential between G and R. Increasing V_A under these circumstances, therefore, leads to an increase in current flow. As soon as V_A reaches the critical value $V_C = 4.9$ volts, the electrons near G which have experienced the full accelerating voltage suffer inelastic collisions accompanied by a significant loss of energy. This energy loss is however compensated by an increase in energy of the atoms which go from the ground to the first excited state as a result of their collision with the electrons. After inelastic collisions, therefore, the electrons considerably slow down and are unable to overcome the retarding potential V_R for reaching the galvanometer; the current flowing through the circuit then suddenly falls. This is shown by the first peak in Fig. 21.10.

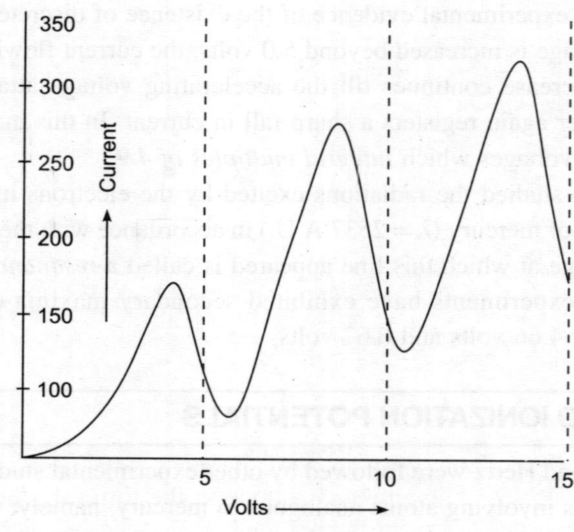

Fig. 21.10

At this stage, we may, as well, consider the significance of this figure 4.9 volts. The energy acquired by an electron under this accelerating voltage is 4.9 electron-volts.* If this energy when imparted to mercury atom raises it to the first excited state, it must be equal to the excess energy possessed by this level over that of the ground state. Thus if the wave-number difference between

*One electron volt is the energy acquired by an electron under an accelerating voltage of one volt. Thus expressing 'e' in e.s.u. of charge we have

$$\text{Energy} = \frac{1}{2} mv^2 = eV = \frac{4.8 \times 10^{-10}}{300} = 1.6 \times 10^{-12} \text{ ergs}$$

the two levels is $\bar{\nu}$, we can write

$$4.9\times1.6\times10^{-12} = hc\bar{\nu}$$

or
$$\bar{\nu} = \frac{4.9\times1.6\times10^{-12}}{6.62\times10^{-27}\times3\times10^{10}}$$

$$\therefore \quad \lambda = 1/\bar{\nu} = \frac{6.62\times3\times10^{-17}}{4.9\times1.6\times10^{-12}}$$

$$= 2520 \text{ A.U. approximately}$$

Hence we find that the inelastic collisions occurring at an accelerating voltage of 4.9 volts involve an energy exchange (between the mercury atoms and the electrons) which is equal to the energy liberated spontaneously when the mercury atoms pass from the first excited state to the ground state emitting the wavelength 2537 A.U. in the process. (The slight difference between the actual and calculated values of the wavelength is within experimental error). The experiments of Franck and Hertz thus provided a direct experimental evidence of the existence of discrete atomic levels.

If the accelerating voltage is increased beyond 5.0 volts, the current flowing to the galvanometer again shows a rise. This increase continues till the accelerating voltage attains a value of 2 × 4.9 volts. At this stage, the meter again registers a sharp fall in current. In this manner, successive peaks are obtained at accelerating voltages which *integral multiples of 4.9.*

Franck and Hertz also studied the radiations excited by the electrons in their experiments and found a single line spectrum of mercury ($\lambda = 2537$ A.U.) in accordance with the theoretical calculations mentioned above. The voltage at which this line appeared is called a *resonance potential of mercury.* More refined and delicate experiments have exhibited secondary maxima corresponding to other resonance potentials namely 4.66 volts and 6.67 volts.

21.9 RESONANCE AND IONIZATION POTENTIALS

The experiments of Franck and Hertz were followed by other experimental studies of the luminescence caused by inelastic collisions involving atoms analogous to mercury, namely; Cd, Zn, Mg, Ca, Sr, Ba etc. In each of these cases, a single spectral line, characteristic of the atom under study made its appearance in the beginning, at well defined voltages. This line is called the *resonance line of the atom* and the voltage at which it flashes forth is known as *resonance voltage*. In the case, of monovalent atoms a single value of resonance voltage is obtained. In the case, of bivalent atoms conditions get more complicated and more than one values of resonance *voltage* are obtained. The reason lies in the multiplicity of first excited states of the valence electrons which are responsible for the optical spectra. This multiplicity also occurs in monovalent atoms but there the energy difference in the multiple levels of the various excited states is very small and therefore does not make its existence felt during inelastic collisions. In all cases, however, a resonance line always involves only the ground and the next higher state of the valence electron and hence of the atom under reference.

As the voltage accelerating the colliding electrons is gradually increased, a limiting value is reached when the complete spectrum of the atom becomes visible. This limiting value, once again a characteristic of the atom under study, is called ionization potential and corresponds to the ionization of the atom. In other words, ionization potential may be considered as a measure of energy required to lift an electron from its ground state to infinity. As soon as this is accomplished all the intervening levels get involved in spontaneous transitions thereby giving rise to the full set of spectral lines allowed by selection rules; a complete spectrum of the atom is thus obtained. For the ionized atom, the quantum state is infinitely large and hence the current term for the series of transitions ending on the level corresponding to the ground state tends to zero for the highest member. We are thus left with only the constant term which is also called the *series limit*. We therefore find that an experimental determination of ionization potential leads to a knowledge of the series limit of the atom for lines involving its ground state.

Let us now see how these potentials called *critical potentials* may be calculated from spectroscopic data.

If λ is the wavelength of the resonance line and V is the resonance potential we have

$$h\nu = h\,(c/\lambda) = \text{eV}$$

This equation reduces to a convenient form by taking the electronic charge in electromagnetic units

$$e = 1.60 \times 10^{-20} \text{ e.m.u.}$$

A potential of V volts is equivalent to a potential $V \times 10^8$ e.m.u. Hence we have

$$hc/\lambda = \text{eV} \times 10^8$$

Taking the wavelength in A.U. (instead of cm.) we find that

$$V(\text{volts}) \times \lambda\ (\text{Angstrom}) = \frac{hc}{e} = 12420 \quad \text{...(21.9 } a)$$

As an example, we shall calculate the resonance potential of sodium. For this we must know the wavelength of the resonance line of sodium. By definition, this wavelength is the first member of the spectral series having the ground state as its final state. This series, always appears as the absorption series of the cold metallic vapour. Its first member, is seen to be the well known *D*-line 5893 A.U. From Eqn. (21.9 *a*) we thus find that

$$V = \frac{12420}{5893} = 2.11 \text{ Volts (approximately)}$$

This value is found to be in good agreement with experimental observations.

Again, the limit of the series containing the sodium D lines expressed in wave numbers is

$$\bar{\nu}_\infty = 41{,}449 \text{ cm.}^{-1}$$

Hence, according to Eqn. (21.9 *a*), the ionization potential of sodium is

$$\bar{\nu} = \frac{12420}{\lambda(\text{A.U.})} = \frac{12420 \times 41449}{10^8}$$

$$= 5.15 \text{ Volts (approximately)}$$

Alternatively, if the ionization potential of an element is determined accurately, the value may be used for calculating the series limit as discussed above. In this manner a knowledge of critical potential is sometimes used for supplementing spectroscopic data.

SOLVED EXAMPLES

Example 1. *Calculate the radius of the Bohr orbit for the H atom in the ground state. Charge of electron = 4.80 × 10^{-10} e.s.u., mass of electron = 9.11 ×10^{-28} gm.*

Velocity of light = 3.00 × 10^{10} cm./sec.

Planck's constant = 6.62 × 10^{-27} erg. sec.

By putting n = 1 in Eqn. (21.3 *i*) we get the radius of Bohr orbit for the *H* atom in the ground state.

$$r = \frac{h^2}{4\pi^2 me^2} = \frac{(6.62\times10^{-27})^2}{4\times(22/7)^2\times9.11\times10^{-28}\times(4.80\times10^{-10})^2} \text{ cm.}$$

$$r = \mathbf{0.528 \times 10^{-8} \text{ cm.} = 0.528 \text{ A.}}$$

Example 2. *The first member of the Balmer series of hydrogen has a wavelength of 6563 A. Calculate the wavelength of the second member.*

The (wavelengths)$^{-1}$ of the spectral lines in the Balmer series are given by

$$\bar{v} = \frac{1}{\lambda} = R\left[\frac{1}{2^2} - \frac{1}{n^2}\right]; \text{ where } n = 3, 4, 5, 6.....$$

By putting n = 3 and n = 4 successively in the above formula we get

$$\textit{First member} \quad \frac{1}{\lambda_1} = R\left[\frac{1}{2^2} - \frac{1}{3^2}\right] = \frac{5R}{36}$$

$$\textit{Second member} \quad \frac{1}{\lambda_2} = R\left[\frac{1}{2^2} - \frac{1}{4^2}\right] = \frac{3R}{16}$$

Whence by eliminating R from the above two equations we get

$$\lambda_2 = \frac{20}{27}\lambda_1 = \frac{20}{27}\times6563 \text{ A.} = \mathbf{4861.5 \text{ A.}}$$

Example 3. *Calculate the ionization potential of hydrogen atom.*

The ionization potential is defined as equal to the amount of energy required to detach the electron completely from the normal atom and the atom is said to be ionized. If the energy required to ionize an atom of a certain gas or metallic vapour is I eV, then we say the ionization potential of that atom is I volts. Obviously, it is the energy with which the electron is bound to the normal atom. The ionization potential of the hydrogen atom is therefore the energy of the ground state orbit of electron. In Eqn. (21.3 *k*) putting p = 1 and $n = \infty$, the ionization energy I in ergs is

$$I = h\nu = \frac{2\pi^2 me^4}{h^2} [\because Z = 1]$$

$$= \frac{2\times\pi^2\times 9.11\times 10^{-28}\times(4.8)^4\times 10^{-40}}{(6.62)^2\times 10^{-54}} \text{ ergs}$$

$$= \frac{2\times\pi^2\times 9.11\times 10^{-28}\times(4.8)^4\times 10^{-40}}{(6.62)^2\times 10^{-54}}\times 0.63\times 10^{12}\text{eV}$$

$$= 13\text{eV}, \quad \text{since} \quad 1\text{erg} = 0.063\times 10^{12}\text{eV}$$

The ionization potential of hydrogen atom is **13 volts.**

PROBLEMS

1. What is an α-particle? Show that its path in the neighbourhood of a nucleus is a hyperbola.
2. Describe Rutherford's experiment on the scattering of α-particles and discuss the significance of the results. How may the size of the nucleus be estimated from such experiments?
3. (a) Obtain an expression for the differential scattering cross-section of an α-particle.

   (b) α-particles of energy 4 million electron volts (MeV) are scattered back from a gold foil $Z = 79$. Calculate the maximum volume in which the positive charge of the atom is likely to be concentrated.

   (c) If the positive charge of the gold atom is supposed to be spread uniformly over a spherical surface of diameter 1 A.U. show that α-particles of energy greater than a certain value E will not be reflected back. Calculate the value of E.

   (d) Discuss the significance of the results of (b) and (c) $h = 6.6 \times 10^{-27}$ ergs. sec., $c = 3 \times 10^{10}$ cm/sec.

   Mass of electron = 9.0×10^{-28} gms., 1 eV = 1.6×10^{-12} ergs.
4. (a) Calculate the orbit radius, velocity, angular velocity and energy of an electron in the first and last orbit in the hydrogen atom according to the Bohr model.

   (b) Calculate the reduced mass of the electron in the helium ion.
5. (a) Describe the various spectral series of hydrogen spectrum. Show how they can be explained on the basis of Bohr's theory.

   (b) Why do the lines of Balmer series not appear in the absorption spectrum?
6. (a) Draw the energy level diagram for the hydrogen atom and indicate the radioactive transitions classified as (i) Lyman series (ii) Balmer series.

   (b) For the Lyman series calculate the wavelength of the first member and the series limit.
7. Discuss Bohr's theory of the hydrogen spectrum and deduce the formula for the Balmer series. If the wavelength of the first line of the Balmer series is 6563 A.U., calculate the wavelength of the first line of the Lyman series.

8. The Rydberg constant for hydrogen is 109678 cm^{-1} and for ionized helium 109722 cm^{-1}. Calculate the ratio of the mass of the proton to that of the electron, assuming the mass of helium nucleus to be four times that of the proton.
9. The wavelength of the H_α line of hydrogen is 6563 A.U. and the isotope shift for heavy hydrogen for the same is 1.791 A.U. Assuming the nucleus of the heavy hydrogen to have twice the mass of the proton, calculate the mass of the electron.
10. (a) Calculate the radius of the second orbit of hydrogen atom from the following data

$$h = 6.6 \times 10^{-27} \text{ ergs. sec.,}$$
$$e = 4.8 \times 10^{-10} \text{ e.s.u.}$$
$$m = 9.1 \times 10^{-28} \text{ gm.}$$

(b) Calculate the ionization energy of the hydrogen atom from this data.
11. What are ionization and resonance potentials? Describe the experiment of Franck and Hertz and discuss its importance.
12. What is the radius of the second Bohr orbit of hydrogen and the velocity of the electron in this orbit?
13. Calculate the critical potentials of mercury from the following data.
Wavelength of resonance line = 2537 A.U.
Series limit corresponding to this line = 84181^{-1} cm.

MATTER WAVES

The preceding pages, while giving the details of a host of physical phenomena, reveal two main undercurrents in the development of physical thought. The first, namely the wave theory—describing radiant energy as the propagation of electromagnetic waves in space and matter held the fort till the year 1900 A.D. On recollection we find that this concept could offer very satisfactory explanations for the phenomena of reflection, refraction, interference, diffraction and polarisation of radiation. The common feature of all these is the interaction of radiation with matter *without any transformation of radiant energy into any other form of energy.* This happy state of affairs received a set-back as soon as the wave theory was applied for explaining black-body radiation. The second stream of thought then took over. It postulated a quantum nature of radiation and sought to substitute the concept of photon in place of electromagnetic waves. This new concept was in turn found very suitable for explaining the phenomena, like photoelectric effect and Compton Effect, *where radiant energy undergoes a transformation on interaction with matter*. This underlying aspect is implied even in Einstein's derivation of Planck's law for black body radiation—the radiation is considered to be in equilibrium with matter and hence the two types of transitions, spontaneous and stimulated are taken into account. Nevertheless, the quantum theory as developed by Planck and Einstein contained an apparent paradox in the expression

$$E = h\nu$$

for the energy associated with a photon. In this expression a basic property of wave motion namely frequency ν has been used for determining the energy of radiation while assuming its quantum nature. Admittedly the wave-concepts could not be discarded in a hurry specially as the state of polarisation of light had been found to have a profound effect on its photoelectric emission properties. Around 1922, the physicists were thus faced with a dilemma. Neither the wave theory could be discarded nor the quantum theory could be given up.

22.1 DUAL NATURE OF LIGHT

The seriousness of the situation can be judged on looking back at some of the phenomena in the light of the above discussion. Let us take the case of interference. For a satisfactory explanation of the phenomenon it is necessary to assume a regular succession of waves extending over a distance at least, equal to the difference of path in experimental set up. This is equivalent to assuming that the source or sources continue to emit these waves for a time equal to that needed for the light to travel through the path difference. It has also been observed that visible interference fringes are obtained even for path differences of the order of 10^6. Taking these two facts into consideration we find that the wave trains emitted by the source extend at least upto $10^6\ \lambda$ or about 50 cm. in the direction of propagation.

If we now apply quantum theory of light for explaining these experimental observations we have to visualise a quantum "extending over a length of 50 cm. in space". This becomes necessary unless we make the highly improbable assumption that a single elementary emitter sends out several successive quanta in phase with one another. The facts of photoelectricity are however, not in conformity with the picture of an extended photon. We know, for instance, that the ejection of photo-electrons is almost instantaneous. This implies an immediate rather than an extended exchange of energy. Moreover, the energy of the photo-electrons is found to depend only upon the frequency of the incident radiation and not on the intensity of light (and hence upon the number of photons falling per second over unit area of the photo-cathode). We are thus faced with a difficult situation in applying the quantum ideas for explaining interference of light.

Such a dilemma is encountered not only with interference but also with diffraction effects. *Taylor*, in 1909 studied the diffraction patterns formed by a needle at low intensities of illumination. He found no change in *sharpness* of the diffraction pattern even when the illumination was so low that an exposure of 2000 hours was needed to photograph it. *Janossy* and *Naray* in 1957 measured the intensity of the interference pattern formed in a Michelson's interferometer with arms nearly 14 metres long. The intensity used by these workers was so low that only one photon at a time was passing through the set up. They found that the interference pattern remains independent of the intensity of light. The obvious conclusion of these studies is that the phenomenon of interference, either between two coherent wavefronts or between various portions of the same wavefront is a property of the individual photon. Increasing the intensity of the light beam, or in other words, increasing the number of photons in the incident light, only increases the density of the record and has no effect on the sharpness and hence on visibility of fringes. If we therefore try to apply quantum theory of light, we are thrown back to the concept of an "extended quantum" which the phenomena of Photo-electric and Compton Effects forbid.

On the contrary, the efforts of using the wave-theory of light to explain both Photo-electric and Compton Effects have been seen (Chapter 20) to be unsuccessful. The difficulties discussed with either theory are not confined to the phenomena discussed in this section. They are of a fundamental nature and many more phenomena, leading to the same impasse, could be cited.

The intriguing situation, however, has granted us one concession. As mentioned in the introduction, the wave theory suffices to explain satisfactorily, all phenomena in which radiant energy does not undergo any transformation on interaction with matter. In case of any such transformation, recourse has to be made to the quantum concept to provide a workable theoretical basis of the observed phenomena. Physicists were thus forced to live with a dual concept, particle as well as wave nature, of radiant energy. The situation in a nutshell is as follows:

According to wave theory light is a wave-motion and has a wavelength and a frequency. This wave-motion involves the propagation of alternating electric as well as magnetic fields through free space as well as matter Fig. (22.1). We can work out the detailed structure of the electromagnetic field at any instant. But at this instant we cannot find out the exact location of light; in fact it is found every where along the direction of propagation.

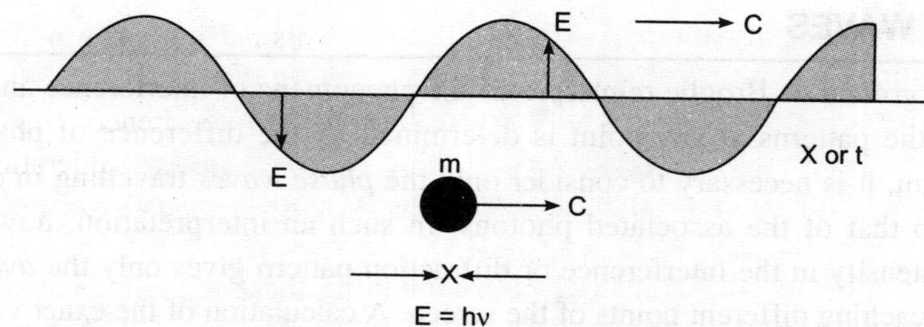

Fig. 22.1. Wave and particle

According to the quantum theory light has an energy and therefore an associated mass. The corpuscle or bundle of energy, called *photon* travels with the velocity of light and has zero rest mass. Viewed as a corpuscle therefore, we know where it is at any given instant, but we do not know its structure or the precise description of the *inside*.

22.2 DE BROGLIE'S HYPOTHESIS

In his effort to reconcile the two seemingly incompatible points of view de Broglie was guided by the infinite though very small interval of time during which the atomic radiators emit radiant energy. The electromagnetic waves, therefore, have a beginning as well as an end (as opposed to a continuous wave of infinite length) and have an appearance like that of Fig. 7.11. For large wavelengths, the two extremes are far apart and the disturbance resembles a wave. At high frequencies, the two extremes are much closer and the disturbance appears more as a corpuscle or bundle of energy even though there is a wave structure within it. At intermediate frequencies the wave is extended but still has the features of a "photon" Fig. (22.1).

According to de Broglie the wave and photon natures of radiation are two manifestations of the same physical phenomenon. In this manner, they are complementary to each other and do not involve any contradiction in the physical picture. Thus on one hand the wave picture gives the relation $c = \nu\lambda$, on the other, the photon concept leads the equation* $E = mc^2$ for the associated energy. The link between these two is provided by Planck's quantisation of energy $E = h\nu$. We can thus write

$$m = h\nu/c^2 \qquad ...(22.2\ a)$$

as the mass of the photon and $h\nu/c$ as its momentum.

Substituting the value of ν in Eqn. (22.2 *a*) we get

$$\lambda = h/mc \qquad ...(22.2\ b)$$

as the wavelength associated with the photon. Eqn. (22.2 *b*) thus combines the wave and particle aspects of radiation.

---

*According to Einstein's special theory of relativity a particle of mass m is associated with an energy $E = mc^2$

This is applicable to particles at rest or in motion. Thus if the rest (intrinsic) mass of a particle is m_o, the equivalent energy is given by $E_o = m_oc^2$. The mass in motion m and the rest mass are related by the equation $m = m_o (1 - \beta^2)^{-1/2}$ where $\beta = v/c$ and v is the velocity of the particle. Thus for a photon moving with the velocity of light $v = c$, the rest mass $m_o = 0$. If this is not so, its energy would be infinite when moving with the velocity of light. Also, for any other particle the kinetic energy, $E_{kin} = c^2 (m - m_o)$.

22.3 MATTER WAVES

Against this background de Broglie reinterpreted the phenomena of interference and diffraction. Since the intensity of the patterns at any point is determined by the difference of phase between waves reaching that point, it is necessary to consider only the *phase waves* travelling in empty space with a velocity equal to that of the associated photons. In such an interpretation, however, the classical distribution of intensity in the interference or diffraction pattern gives only the *average values* of the flux of photons reaching different points of the screen. A calculation of the exact values of the number of photons reaching an unit area of screen in a definite time thus becomes impossible.

Following de Broglie, the electromagnetic waves are seen to lose all their physical content. All that we can rescue from the elaborate classical theory is the concept of *phase waves* or *probability* waves to which the normal laws of superposition apply.

de Broglie went a step further. He postulated that if waves (whose structure we know) sometimes behave as particles (whose structure we do not know) the converse is also true. Thus atoms, molecules and all fundamental particles of matter whose structure we know may sometimes behave like waves whose structure we do not know. In this manner, de Broglie visualised the existence of *matter waves* in 1924.

Let us consider, for instance, a lead shot. Its size shape, and structure are known. When it moves with a velocity v, its kinetic energy is given by

$$E_{\text{k.E.}} = c^2(m - m_o)$$

$$= m_o c^2 \left(\frac{1}{\sqrt{(1-\beta^2)}} - 1 \right)$$

$$= m_o c^2 \left[\frac{1}{2}\beta^2 + \frac{3}{8}\beta^4 + \ldots\ldots \right]$$

$$= \frac{1}{2} m_o v^2 + \frac{3}{8} m_o v^4 / c^2 + \ldots\ldots$$

or

$$E_{\text{k.E.}} = \frac{1}{2} m_o v^2 \quad \text{if} \quad v << c$$

Thus its kinetic energy is $\frac{1}{2}mv^2$ and its total energy is mc^2. This particle possesses all the known attributes of particle mechanics, namely position, momentum etc. de Broglie ascribed an additional property — the associated wavelength. By analogy with Eqn. (22.2 *b*), this wavelength is given by

$$\lambda = h/mv \qquad \ldots(22.3\ a)$$

where the momentum mv of the particle now appears in the denominator [instead of the momentum mc of the photon in Eqn. (22.2 *b*)]. Thus for particles at rest the associated wavelength is infinite.

Equation (22.3 *a*) implies that a particle of momentum $p = mv$ is associated with an infinitely extended wave of the form

$$\xi = Ae^{-i(\omega t - kx)}$$

The planes of constant phase are then propagated with a velocity *V*, called the phase velocity (*Sec.* 8.14) given by

$$V = \omega / k = v\lambda$$

We may note that de Broglie's hypothesis does not mention anything regarding the nature of these waves *i.e.* whether they are electromagnetic waves or of any other type. They were therefore called matter waves. Thus for a particle of mass one milligram moving with a velocity of 10^6 cm/sec., the wavelength of the associated wave will be

$$\lambda = \frac{6.625 \times 10^{-27}}{10^{-3} \times 10^{6}} = 6.625 \times 10^{-24} \text{ cm.}$$

This wavelength is so small that the particle nature of matter predominates. Naturally, therefore, the electron which has the smallest known mass should give rise to the longest wavelength matter waves. For example, an electron on being accelerated by one volt should have a velocity given by

$$\frac{1}{2} mv^2 = Ve$$

or

$$v^2 = \frac{1 \times 10^8 \times 1.60 \times 10^{-20} \times 2}{9 \times 10^{-28}}$$

$$= \frac{3.2}{9} \times 10^{16}$$

$$\therefore \quad v = 0.6 \times 10^8 \text{ cm / sec. approximately.}$$

The de Broglie wavelength of the electron is therefore given by

$$\lambda = \frac{6.62 \times 10^{-27}}{9 \times 10^{-28} \times 0.6 \times 10^{8}}$$

$$= 12.2 \times 10^{-8} \text{ cm. approximately.}$$

This value is of the same order as the wavelength of soft X-rays and also the lattice spacing in crystals. We should therefore, be able to detect such matter waves in experiments on diffraction of electrons by crystal lattices. The evidence thus obtained is presented in Section 21.5.

The essence of de Broglie's theory of matter waves is the recognition of a wave process associated with the transport of energy by a particle of finite mass travelling with a velocity less than that of light. The basic equations governing this process are the following:

$$E = mc^2 = h\nu \qquad \text{...(22.3 } b\text{)}$$

$$\lambda = h/mv \qquad \text{...(22.3 } c\text{)}$$

and

$$m = m_o (1 - \beta^2)^{-1/2} \qquad \text{...(22.3 } d\text{)}$$

With the help of these equations we find that the velocity V of the phase wave travelling with the particle is given by

$$V = \nu\lambda = c^2/v = c/\beta \qquad \text{...(22.3 } e)$$

The phase waves, therefore, do not travel with the velocity of the particle with which they are associated. On the contrary, they travel with a velocity greater than that with which energy may be transported, according to the theory. The phase velocity, however, cannot be experimentally determined and is purely an artificial concept. For its measurement, we need to affix a mark on the infinitely extended wave train and then measure its velocity. By itself, this is physically impossible. We can however, define such a mark by superposing another wave train upon the first thereby creating a noticeable deformation in the original wave profile. If now we measure the velocity of this deformation, we shall in fact obtain the group velocity U rather than the phase velocity.

We have seen [Eqn. (8.14 *e*)] that the group velocity is given by

$$U = d\nu / d(1/\lambda)$$

From Eqns. (22.3 *b*), (22.3 *c*) and (22.3 *d*) we find that

$$h\nu = mc^2 = \frac{m_o c^2}{\sqrt{1-\beta^2}}$$

so that

$$\nu = \frac{m_o c^2}{h} \frac{1}{\sqrt{(1-\beta^2)}} \qquad \text{...(22.3 } f)$$

and

$$\frac{d\nu}{d\beta} = \frac{m_o c^2}{h} \frac{\beta}{(1-\beta^2)^{3/2}} \qquad \text{...(22.3 } g)$$

Also, Eqn. (22.3 *e*) may be re-written as

$$\beta\nu = \frac{m_o c^2}{h} \frac{\beta}{\sqrt{(1-\beta^2)}} = c(1/\lambda)$$

From this we can see that

$$\frac{d(1/\lambda)}{d\beta} = \frac{m_o c}{h} \frac{1}{(1-\beta^2)^{3/2}} \qquad \text{...(22.3 } h)$$

Combining Eqns. (22.3 *g*) and (22.3 *h*) we find that

$$U = \frac{d\nu}{d(1/\lambda)} = \beta c = v \qquad \text{...(22.3 } i)$$

This equation shows that the velocity of the material particle is equal to the group velocity of the associated de Broglie waves. It is in this context that we are tempted to interpret a particle of matter as a wave packet due to the superposition of a number of wave trains. Proceeding along these lines *Schrodinger* developed his famous wave equation for particles of matter. Although his formulae are entirely correct, his interpretation of particles as wave packets had to be subsequently discarded in favour of a more tenable interpretation due to *Max Born*.

22.4 DE BROGLIE'S MODEL OF THE ATOM

In addition to the shortcomings mentioned earlier, the Bohr model of the atom presents certain conceptual difficulties. These are encountered while explaining the emission of monochromatic radiations as a result of transitions between atomic energy levels. During these transitions, the change in energy agrees quite well with that associated with the spectral line. The model, however, does not say anything about the origin of oscillations associated with the spectral frequency. In other words, the model fails to account for the manifestation of the radiated energy in the form of an electromagnetic wave.

These doubts were removed by de Broglie in a very elegant manner, by providing a modification to Bohr's model. It has been seen above that electrons being the lightest of known particles possess the longest wavelength matter waves. In free space these long wavelengths appear localised. But they make their appearance felt in the confines of space of approximately equal dimensions. With this reasoning de Broglie looked at the electron orbits in a different light. He reasoned that if indeed the electrons have associated waves, their motion in closed orbits should *give rise to standing waves* in accordance with similar phenomena with other types of wave motion, as for example, with sound waves in a resonance column. Following this logic, de Broglie verified that the lengths of the various Bohr orbits were in fact simple integral multiples of the wavelengths associated with the electrons moving in those orbits. We have seen that the radii of the Bohr orbits are given by

$$r_n = \frac{n^2 h^2}{4\pi^2 m Z e^2} \qquad \text{...(22.4 } a\text{)}$$

and the velocities of the electrons moving in these orbits, by

$$v_n^2 = \frac{Ze^2}{r_n} = \left(\frac{2\pi Z e^2}{nh}\right)^2 \qquad \text{...(22.4 } b\text{)}$$

Combining Eqns. (22.3 *a*) and (22.4 *b*) we find that wavelengths of the electron waves are given by

$$\lambda_n = \frac{h}{m v_n} = n \cdot \frac{h^2}{2\pi m Z e^2} \qquad \text{...(22.4 } c\text{)}$$

Also, from Eqn. (22.4 *a*) the circumferences of the various orbits are seen to be

$$2\pi r_n = \frac{n^2 h^2}{2\pi m Z e^2} \qquad \text{...(22.4 } d\text{)}$$

Combining Eqns. (22.4 *c*) and (22.4 *d*) we get

$$\frac{2\pi r_n}{\lambda_n} = n \qquad \text{...(22.4 } e\text{)}$$

Equation (22.4 *e*) expresses the important and significant result that the numbers of wavelengths (of the electron waves) contained in the orbits allowed in Bohr atom are always integral.

Let us now try to understand the physical meaning of this result. According to the Bohr model, the electron treated like a particle moves along discrete orbits in which its angular momenta are quantised. According to de Broglie's model, the electron treated as a wave, moves along only those orbits whose lengths allow the associated waves to fold on themselves. We are thus lead to a number of stationary states in which the electron waves resonate with themselves. The situation is depicted in Fig. 22.2. In each of the quantised states, the wave form is seen to close on itself. For cases where this is not possible, as for example, one with a broken part, resonance conditions are not satisfied and the waves get damped very fast. These are the so-called forbidden states.

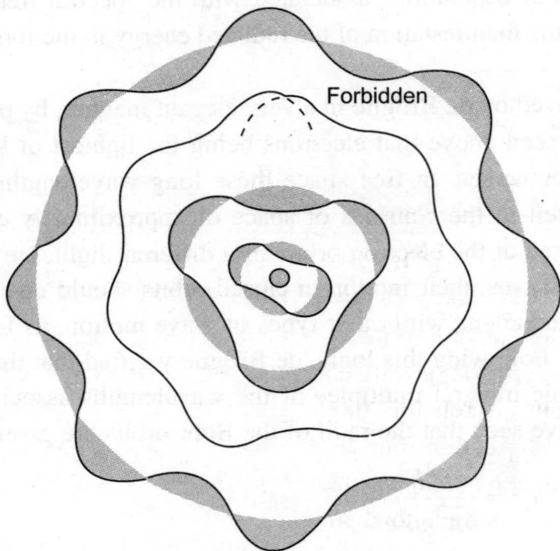

Fig. 22.2

de Broglie's model of the atom, therefore, postulates the existence of standing waves corresponding to the various quantum states and enveloping the concentration of positive charge at the centre. Further, since the standing waves do not allow any transfer of energy, the atom in any one quantum state does not radiate energy. However, when the atom undergoes a transition from a higher to a lower state, the standing wave corresponding to the initial state is replaced by that of the final state. But the decay of the initial standing wave is not instantaneous; it takes an interval of time Δt, although very small. The decay of the initial state is simultaneously accompanied by a growth of the final state. During the interval Δt the two states therefore coexist. (The situation is similar to the one shown in Fig. 22.3). Now, two coexisting waves give rise to the phenomenon of beats. If therefore, the standing wave frequencies in the initial and the final states are respectively ν_i and ν_f, the beat frequency ν_b is given by

$$\nu_b = \nu_i - \nu_f \qquad ...(22.4\ f)$$

where the ν's are obtained from Planck's relation

$$E = h\nu$$

We can thus write Eqn. (22.4 f) as

$$\nu_b = \frac{E_i}{h} - \frac{E_f}{h}$$

or

$$h\nu_b = E_i - E_f \quad \text{...(22.4 } g\text{)}$$

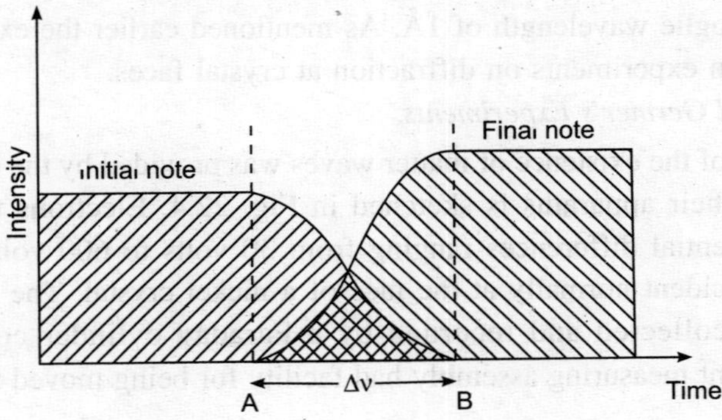

Fig. 22.3. de Broglie's concept of atomic radiations

The beat frequency ν_b, which according to de Broglie's model is the frequency of the emitted radiation, as calculated from Eqn. (22.4 g), thus comes out to be identical with that obtained from the simple Bohr theory. We thus find that this model not only removed the conceptual difficulties mentioned at the beginning of this section, but also provided in some ways, a model of the atom complementary to the one postulated by Bohr. At the same time it succeeded in resolving the apparent contradictions associated with wave-particle dualism of light.

But the theory was not without its own share of drawbacks. To mention just one, the theory while treating electron as wave, is naturally quiet about its charge and mass. If electron is treated as a wave it is pertinent to ask as to how its charge (and mass) is distributed in space. All this led to a still more sophisticated model of the atom-called the *wave-mechanical model* which although an advance, still retains the essential features of de Broglie's model.

22.5 EVIDENCE OF MATTER WAVES

It has been seen in *Sec.* 21.2 that an electron on being accelerated through a potential difference V acquires in vacuum a kinetic energy given by

$$\frac{1}{2}\, mv^2 = Ve$$

It, therefore, follows that the de Broglie wavelength of the accelerated electron is

$$\lambda = \frac{h}{mv} = \frac{h}{\sqrt{(2mVe)}} \quad \text{...(22.5 } a\text{)}$$

Substituting the values of the physical constants we can replace Eqn. (22.5 *a*) by the approximate relation

$$\lambda = \frac{\sqrt{150}}{V} \text{ Å} \quad \text{...(22.5 } b\text{)}$$

This expression is valid so long as the velocity of the electron is such that its relativistic increase of mass may be neglected. This limiting value is reached at an accelerating voltage of 10 kilovolts. Beyond this voltage, the energy and momentum of the electron have to be determined with the help of relativistic mechanics. With this understanding we find that electrons of kinetic energy 150 electron volts (eV) have a de Broglie wavelength of 1Å. As mentioned earlier the existence of these electron waves can be detected in experiments on diffraction at crystal faces.

(*a*) *Davisson and Germer's Experiments.*

The first evidence of the existence of matter waves was provided by the experiments of *Davisson* and *Germer* in 1927. Their apparatus is sketched in Fig. 22.4. Electrons from a filament *F* were accelerated through potential differences ranging from 30 volts to 600 volts in a tube at very low pressure. These were incident normally at the face of a nickel crystal. The electrons scattered from the crystal face were collected and recorded by a Faraday cylinder connected to a sensitive galvanometer. The current measuring assembly had facility for being moved through different angles.

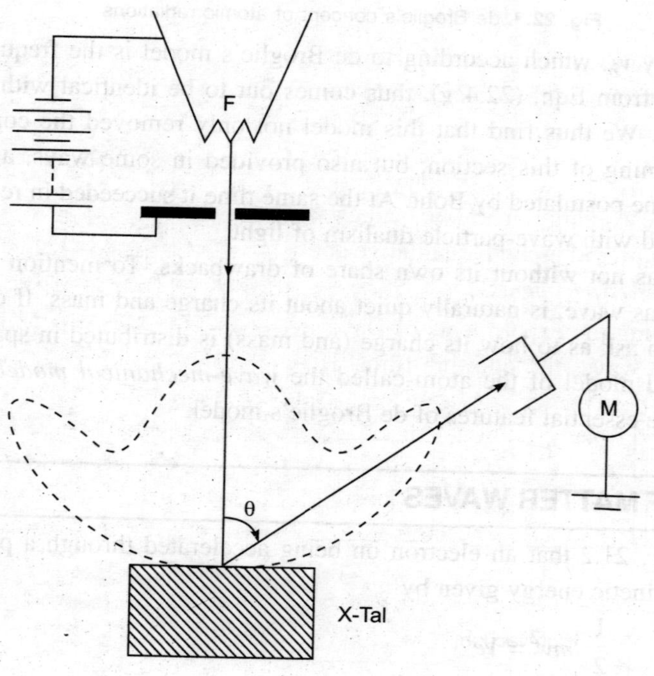

Fig. 22.4. Davisson and Germer's Apparatus

In the experiments, Davisson and Germer measured the strength of the electron current at different angles to the crystal face. The relative intensities when plotted along different directions yielded a curve like the one shown by broken lines (Fig. 22.4). It is seen from this polar diagram that the maximum current density is obtained at zero degrees to the incident beam of electrons. Also, comparatively high currents are observed at $\theta = \pm 50°$ to the incident beam. Further, it was noticed that the magnitude of these maxima was unusually large for electrons of 54 eV energy.

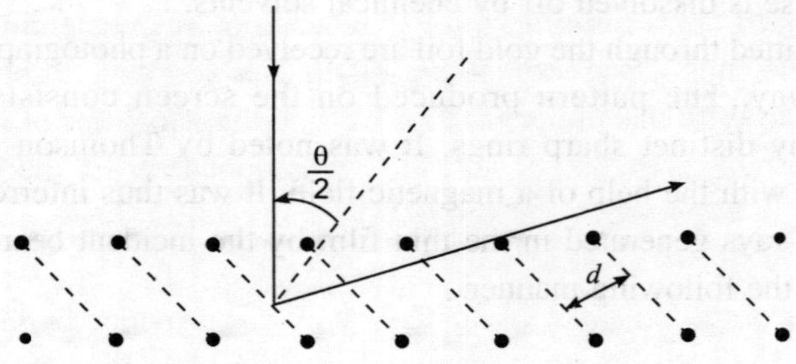

Fig. 22.5

The preferential scattering of electrons was interpreted as being caused by the interference of electron waves undergoing Bragg reflection at different lattice planes of the crystal. If the interatomic distance of nickel is d, the maxima due to Bragg reflection are given by

$$2d \sin \theta = n\lambda$$

where θ is known as the *glancing angle*. For nickel $d = 0.91$Å, so that for $\theta = 50°$, the wavelength of the incident electron waves comes out to be 1.65Å. This wavelength was compared with that calculated from de Broglie equation

$$\lambda = h/mv$$

for electrons of 54 eV energy. The two values were found to agree. In this manner Davisson and Germer succeeded in providing both qualitative and quantitative proof of the existence of matter waves.

(*b*) *Thomson's Experiments.*

Closely following the experiments of Davisson and Germer, *Thomson* studied the diffraction of high energy electron waves through thin films. His apparatus is shown in Fig. 22.6.

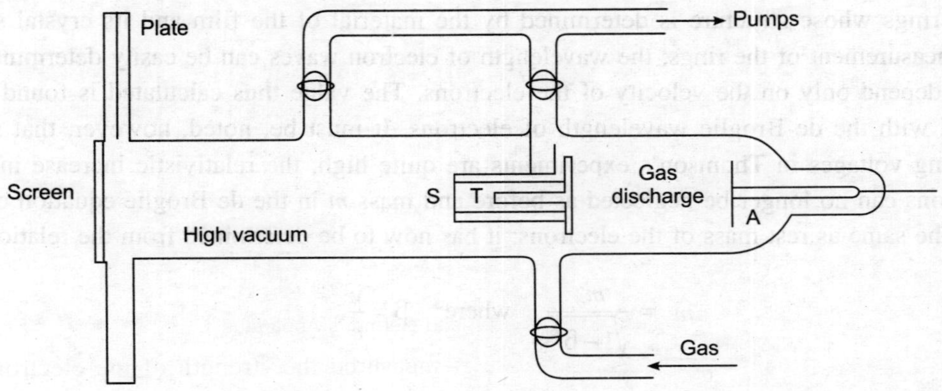

Fig. 22.6

The right hand side is a cold cathode gas discharge tube producing electrons accelerated through potentials upto 60 kV. A fine beam of these highly energetic electrons is isolated by means of a diaphragm tube *T* of diameter 0.25 mm and length 6 cm and is allowed to pass into the high vacuum part of the apparatus through a very thin (10^{-6} cm) film of gold (*S*). Such fine films are produced by special techniques. In one method the metal is deposited on a celluloid acetate base by *cathode sputtering* and then the base is dissolved off by chemical solvents.

The electrons transmitted through the gold foil are received on a photographic plate or fluorescent screen placed 30 cm away. The pattern produced on the screen consists of a central spot *O* (Fig. 22.7), surrounded by distinct sharp rings. It was noted by Thomson that the entire pattern could be shifted sideways with the help of a magnetic field. It was thus inferred that it is caused due to electrons and not by X-rays generated in the thin film by the incident beam of cathode rays. The pattern was interpreted in the following manner.

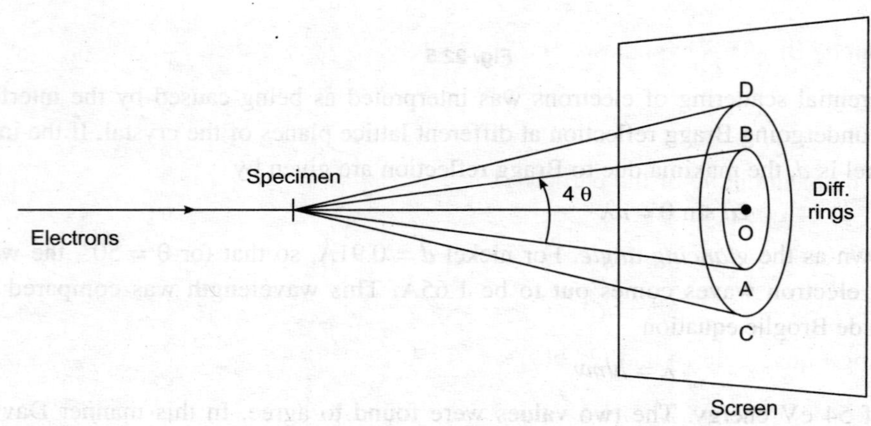

Fig. 22.7. Diffraction of electrons

The metallic film consists of a random distribution of minute crystals. On arrival at the film, therefore, the electrons are scatter in accordance with Bragg's law by the crystals of the right orientation. The cones of diffraction, on intersection with the plane of the screen given rise to a series of rings whose structure is determined by the material of the film and its crystal structure. From a measurement of the rings, the wavelength of electron waves can be easily determined and is found to depend only on the velocity of the electrons. The value thus calculated is found to agree very well with the de Broglie wavelength of electrons. It must be, noted, however, that since the accelerating voltages in Thomson's experiments are quite high, the relativistic increase in mass of the electrons can no longer be neglected as before and mass *m* in the de Broglie equation cannot be taken as the same as rest mass of the electrons: it has now to be determined from the relation.

$$m = \frac{m_o}{\sqrt{1-\beta^2}} \quad \text{where} \quad \beta = \frac{v}{c}$$

22.6 THE UNCERTAINTY PRINCIPLE

The outstanding success of de Broglie's theory of wave-particle dualism naturally leads to the question — "Can we know whether we are dealing with a wave or a particle at a given instant?" For completely defining a particle we must be certain of its exact momentum and also its definite position in space at any instant. Let us take the electron, associated with its wave packet moving with the group velocity U. Since the wave packet has a velocity spread, it is impossible to know simultaneously *where* within this group the electron actually is and what is its exact velocity. In the case of a short wave packet, the position of the particle can be located fairly accurately. But since the velocity spread is large in this case the particle velocity cannot be exactly determined. On the other hand, in the case of long wave packets, the velocity spread being small, the velocity of the particle can be detrmined fairly accurately but not its position. In general, a wave packet has a definite length Δx and the extreme ends have a wave number difference $\Delta \bar{v}$ (see *Sec.* 8.16) such that

$$\Delta x \, \Delta \bar{v} = 1 \quad \text{...(22.6 } a\text{)}$$

Also, from de Broglie's hypothesis

$$\lambda = \frac{h}{mv} = \frac{h}{p}$$

where p is the momentum mv of the particle. Thus we can write

$$p = \frac{h}{\lambda} = h\bar{v}$$

or

$$\Delta p = h \Delta \bar{v} \quad \text{...(22.6 } b\text{)}$$

Combining Eqns. (22.6 *a*) and (22.6 *b*) we find that

$$\Delta x \, \Delta p = h \quad \text{...(22.6 } c\text{)}$$

Δx in Eqn. (22.6 *c*) may be interpreted as the uncertainty in determining the exact location of the particle within the wave packet; similarly Δp represents the uncertainty in evaluating the momentum of the particle. In this manner, we find that the product of the uncertainties in determining the position and momentum of a particle is of the order of the Planck's constant. This is called *Heisenberg's Uncertainty Principle*. According to this principle, we find that if somehow we manage to reduce the uncertainty Δx in determining the position of a particle, the uncertainty Δp in the value of its momentum becomes correspondingly larger so that the product of the two uncertainties remains the same as that given by the relation (22.6 *c*). In other words, we find that it is impossible to determine simultaneously both the position and momentum of the particle accurately no matter what methods we utilise for such determinations. The uncertainty principle therefore sets an optimum limit to the possible accuracy of observations and is of a fundamental character. We shall see how the principle is enforced with help of a few examples.

Let us suppose that we want to determine the position of a particle, as exactly as possible, with the help of a microscope (Fig. 22.8). We have seen in *Sec.* 15.5, that the resolving power of a microscope depends upon the wavelength employed to illuminate it. Suppose we illuminate the particle with the shortest available wavelengths, γ-rays, in order to achieve the highest resolution to facilitate the most accurate determination of its position. Let the frequency of the illuminating radiation be ν and the particle under observation be an electron.

From the theory given in Sec. 15.5, the position of the electron can be determined to an accuracy

$$\Delta x = \lambda/\sin \alpha \quad \text{...(22.6 d)}$$

where α is the angular aperture of the microscope. However, the electron will be *visible* only if the illuminating radiation has first been cattered by it. As a result of the scattering, the electron suffers a Compton recoil and undergoes a change of momentum, of the order of $h\nu/c$, in an indeterminate direction. The uncertainty in the direction of the scattered radiation will therefore be of an equal magnitude. This uncertainty, however, can be estimated as follows:

Since we have assumed that we are in a position to see the electron, it follows that the scattered radiation certainly enters the microscope in use.

Therefore, the uncertainty in the direction of the scattered quantum of radiation is equal to the angular aperture of the microscope. Thus, in this process, the uncertainty Δp creeping in the determination of the electron's momentum in a direction perpendicular to the instruments axis is

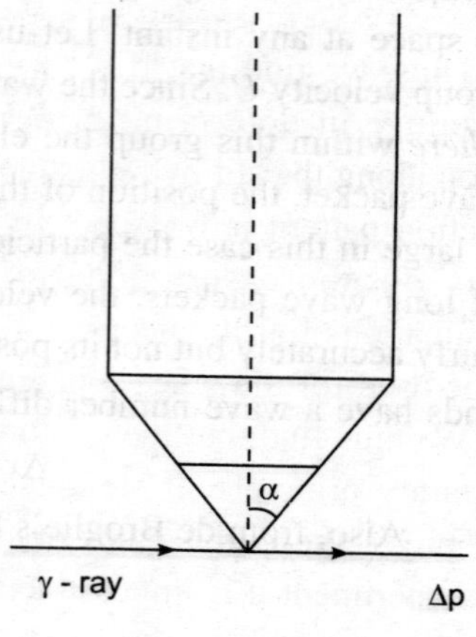

Fig. 22.8

$$\Delta p \sim \frac{h\nu}{c} \sin \alpha \quad \text{...(22.6 e)}$$

Combining Eqns. (22.6 *d*) and (22.6 *e*) we find that

$$\Delta p \cdot \Delta x \sim h$$

22.7 DIFFRACTION OF ELECTRONS AT A SLIT

As another example of the generality of the uncertainty principle, let us consider the diffraction of electrons at a single slit. Viewed as a wave an electron will produce a diffraction pattern Fig. (22.9) in a direction θ defined by the expression

$$\Delta x \sin \theta \sim \lambda \quad \text{...(22.6 a)}$$

where Δx is the slit width and λ is the associated wavelength given by

$$\lambda = h/p \quad \text{...(22.6 b)}$$

The existence of a diffraction pattern, on the other hand, is an evidence that the electron viewed as a particle has passed through the slit. Actually at what point along the slit, the actual passage of the electron takes place, is however indefinite. There is thus an uncertainty of Δx in the determination of the position perpendicular to direction of flight. Also, the diffraction pattern on the screen may be taken to mean that the electron during its passage through the slit undergoes a lateral deflection and thus acquires a component Δp of momentum in a direction perpendicular to its original line of flight.

From the figure, component is given by

$$\Delta p \sim p \sin\theta \sim h/\lambda \sin\theta \qquad ...(22.6\ c)$$

Combining Eqns. (22.6 *a*) and (22.6 *b*), we get

$$\Delta p\, \Delta x \sim h$$

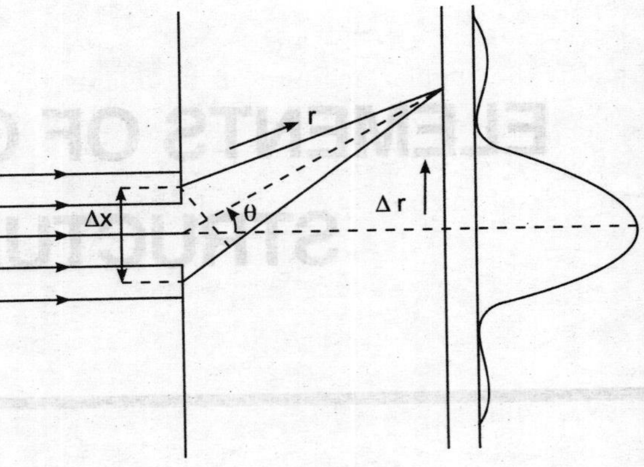

Fig. 22.9

Thus, we find that if we try to minimise the uncertainty in determining the position of the electron along the slit by reducing its width Δx, the diffraction pattern is widened. In other words this causes an increased uncertainty Δp in the determination of its momentum. A similar argument may be used for the other conjugate variable.

We thus find that there is no escape from the uncertainty principle. In general, we observe that if we prove the corpuscular or particle character of an experiment it is impossible to prove its wave character *at the same* time and vice versa. In fact, the two characters are complementary to each other. While dealing with matter waves, therefore, we are forced to abandon the exactness of classical Physics and talk in terms of *probabilities*. In the mechanics of matter waves, called Wave Mechanics, there is a finite probability even of those events which are impossible according to the classical concepts.

PROBLEMS

1. What are matter waves? Describe the evidence supporting the existence of these waves.
2. Write an essay on Heisenberg's Principle of Uncertainty.
3. Write short notes on:
   (a) Wave and Group Velocities of Matter Waves.
   (b) de Broglie's hypothesis.
   (c) Photons and interference of light.
4. Calculate the de Broglie wavelengths of the following particles:
   (a) An electron of kinetic energy 500 MeV.
   (b) An α-particle accelerated through 2000 volts.
   (Mass of an α-particle = 6.576×10^{-24} gm)
   (c) A thermal neutron at 300° K.
   (Mass of Neutron = 1.00898 a.m.u.)

ELEMENTS OF CRYSTAL STRUCTURE

Crystals have held the fascination of man from the earliest of times. Apart from their use as personal ornaments the minerals were used for pigments and also for smelting of metals. *Stensen* was, however, the first person to make a scientific study in the field of crystallography. He made a detailed study of the geometry of quartz crystal. He was followed by *Guglielmini, Rome de l'Isle* and *Haüy* who ultimately placed the subject on a sound footing. Some of his researches on calcite crystals led to the realization of certain characteristic features of crystals in general and thus encouraged systematic studies of their shape and symmetry. These were however confined to only the external appearances and had little to do with the structure of crystals. It was the work of *von Laue* which provided the key for revealing the secrets of crystal structure. Laue's researches led to the discovery that crystals possess an internal structure. Bragg's use of the crystals as three dimensional gratings for X-rays ultimately perfected the tools for their detailed study.

23.1 DEFINITION OF CRYSTALS

In modern usage, the term *crystals* is used for certain types of matter in the solid state. For a solid to be classified as such, it must have:

(*a*) A definite (not necessarily fixed) composition.

(*b*) An ordered or regular atomic arrangement. Under natural conditions free from external constraints or objects the growing crystals are usually bounded by plane surfaces. These plane crystal faces are a manifestation of their regular internal structure and thus they constitute an important aspect of the crystalline state. By implication, therefore, there is an order in the disposition of the crystal faces and symmetry of varying degrees in their structural details. For example, in some cases the same *angular settings* of faces may be observed twice or four times during a complete rotation of the crystal about a fixed axis or axes. At the same time, there may be certain directions, along which, the crystal when cut yields two *halves* which are mirror images of each other. There may be a *centre of symmetry* as well. In view of the variation in shape of natural crystals it is evident that different types will show symmetry of different degrees. This symmetry is, however, different from geometric symmetry. If a crystal is geometrically symmetrical its corresponding faces will be of the same size and shape. In addition the crystal will be crystallographically symmetrical about its axes and centre of symmetry. On the other hand a damaged crystal although crystallographically symmetrical need not be symmetrical geometrically. Crystallographic symmetry therefore implies a more fundamental symmetry and has got nothing to do with *mere* external appearances.

23.2 CONCEPT OF THE BOND

For making workable models of crystalline state the following characteristics of solid state have to be borne in mind:

1. The solids exist in equilibrium at temperatures much lower than those of the corresponding liquids and gases. Their energy configurations are therefore lowest at these temperatures.
2. A random spatial arrangement of atoms necessarily possesses a greater energy of configuration than an ordered and symmetrical arrangement as is present in the case of crystals.
3. The atoms of a solid do not collapse to a configuration of high density.

The first two characteristics lead to the conclusion that crystalline states possess an ordered arrangement of the constituent atoms. The third characteristic implies that while at *large* distances there exist attractive forces holding these atoms together, at *small* distances a repulsive force comes into play which prevents these atoms from piling up on each other. These forces of attraction and repulsion, however, act only between immediate neighbours so that the situation is similar to that existing in chemical bonds of inorganic chemistry. The strength of these bonds is of the order of 0.25 to 0.5 electron volts. It therefore follows that energy has to be supplied for upsetting the symmetry of crystals.

The basic factors governing the arrangement of atoms in crystals can be understood by considering the arrangements of marble balls in a plane. If we have two balls their minimum energy configuration results when the two are in contact with each other (Fig. 23.1 *a*). For three balls the corresponding arrangement is a triangular array (Fig. 23.1 *b*) in which every ball touches the other two. In other words, each ball rests against the depression resulting from the contact of the other two. Proceeding in a similar fashion we find that the minimum energy configuration for seven balls in a plane is a hexagonal array of six balls surrounding the central. Any additional numbers arrange themselves around these central seven (Fig. 23.1 *c*) repeating the endless symmetry of the initial group.

Such two-dimensional arrays may be considered to be stacked upon each other to give the three dimensional structures of crystals. Several such stacking sequences are possible and are discussed in the following sections. Before discussing these arrangements we may note that in two-dimensional arrays of marble balls (Fig. 23.1 *d*) the position of every ball may be designated by translating a rhombus along its sides. Such a rhombus constitutes an *unit cell* for the arrangement. In three dimensional stacking of atoms in a crystal, the unit cells are parallelopipeds. One such unit cell is shown by bold lines in Fig. 23.2. The entire structure may be constructed by translating the edges of this unit cell along their own directions. The edge lengths therefore constitute the three translation vectors.

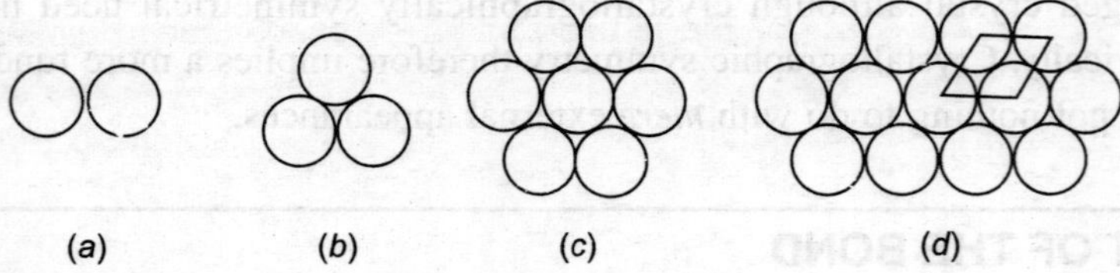

Fig. 23.1. Arrays of balls

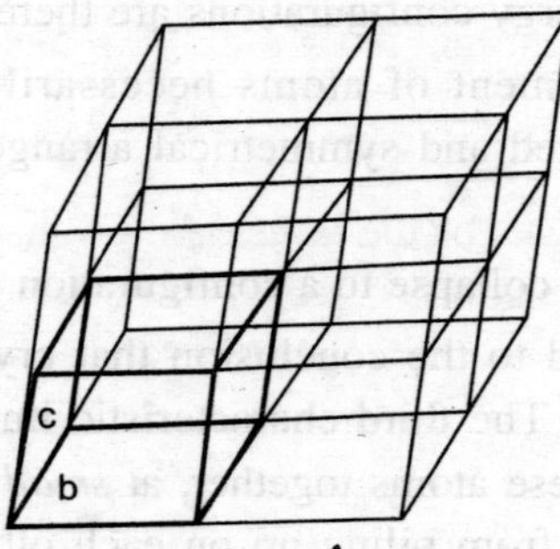

Fig. 23.2. Unit cell

23.3 CRYSTAL LATTICES

The locations of atomic sites in the various stacking sequences are responsible for three dimensional fabrics of crystal structures. These fabrics may be imagined to be formed by joining the atomic sites by straight lines. The resulting network of lines is called crystal lattice. The various possible types of crystal lattice have been theoretically studied and their number calculated. *Bravais* and *Frankenheim* have found that only fourteen lattice types are possible and all crystals conform to one or the other of these so-called *Bravais lattices*. Following their shapes, the fourteen Bravais lattices can be subdivided into seven groups or systems. These are:

1. The Cubic system
2. The Tetragonal system
3. The Orthorhombic system
4. The Monoclinic system
5. The Triclinic system
6. The Hexagonal system
7. The Trigonal system.

In the cubic system, the three edges of the lattice are all of the same length and are at right angles to each other thereby forming a cubic structure. The atomic sites are either (*a*) located at the eight corners or (*b*) located at the corners as well as the centre of the body of the crystal or (*c*) located at the eight corners and centers of the six faces of the lattice. There are thus three types in the cubic system.

In the tetragonal system the edges of the lattice remain perpendicular to each other as in the cubic system but one set (of edges) differs in length from the other two. The atomic sites in these systems are either (*a*) located at the corners or (*b*) located at the corners and at the centre of the body of the lattice.

In the orthorhombic system all the three edges of the lattice, though perpendicular to each other, are of different lengths. The atomic sites in these systems are located either (*a*) at the corners of the lattice or (*b*) at the corners and body-centre of the lattice or (*c*) at the corners and centres of two opposite faces of the lattice or (*d*) at the corners and centers of all the faces of the lattice.

In the monoclinic system there occurs a departure from normality of the three edges. In these systems two sets of edges while remaining normal to the third, are inclined to each other. The atomic sites are located either (*a*) at the corners of the lattice or (*b*) at the corners and centres of two opposite faces of the crystal.

In the triclinic system all the three edges of the lattice are of unequal length and are inclined to each other. The atomic sites in these systems are located at the corners of the lattice.

In the hexagonal system the lattice base is a rhombus with two opposite pairs of edges inclined at 120 degrees to each other. The third edge is normal to both these edges. These systems differ from the orthorhombic lattices where the lattice base is a rectangle and also the monoclinic systems where the lattice base is not a rhombus. The atomic sites are located at each corner of the hexagonal lattice.

In the trigonal system the atomic sites are located at the corners of a rhombohedral lattice.

23.4 THE SIMPLE CUBIC STRUCTURE

As seen above the simplest lattice structure is that of the cubic system where the atomic sites are located only at the corners of a cube (which therefore constitutes a unit cell of the system). Each atom in such systems has only six immediate neighbours (Fig. 23.3) and therefore the structure is loosely packed. Inspite of its simplicity, the simple cubic lattice is rarely encountered in practice. Nevertheless the concept is useful specially as several cubic lattices are directly related to the simple cubic lattice. This in turn simplifies an understanding as well as discussion of their structures.

(*a*) ***The Body-centred Cubic System*:** A slightly less simple structure is that of the body-centred cubic lattice. Here, in addition to the atomic sites at the corners of a cube, there exists an atomic site at the centre of the cube. Each atom, in such an arrangement has now eight neighbours and the structure is thus less loosely packed than that of the simple cubic (Fig. 23.4).

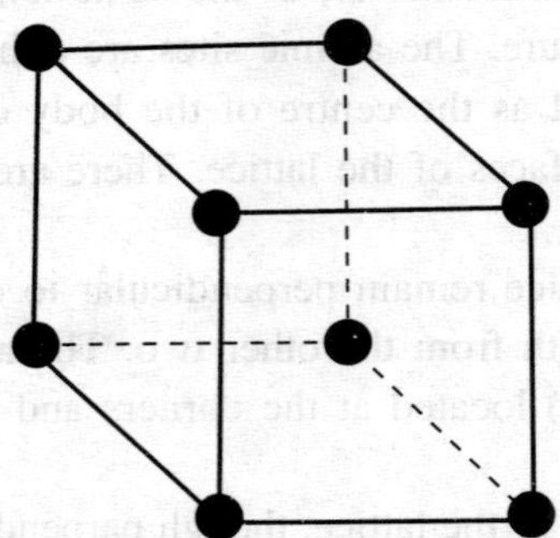
Fig. 23.3. Simple cubic lattice

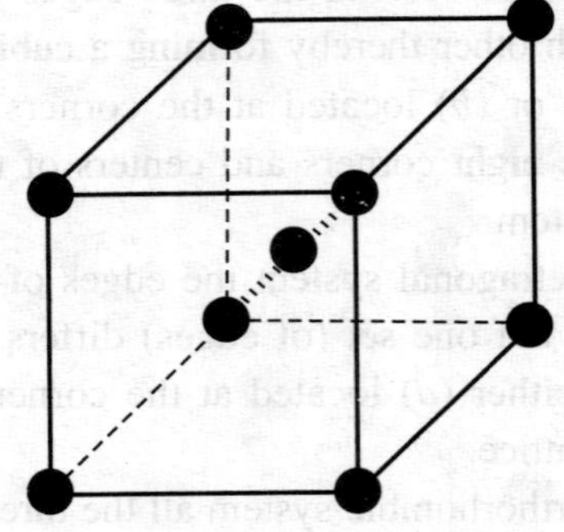
Fig. 23.4. *bcc* lattice

(*b*) ***The Face-centred Cubic System***: Another simple system is the face-centred cubic (fcc) structure. The stacking of atomic sites in these systems can be visualised as follows: Consider four closely packed marble balls arranged in a plane at the corners of a square and a fifth at the centre in contact with all the four in the same plane. Another set of four balls can then be placed in a plane above it by locating each of these balls at the four depressions formed by the five balls in contact in the lower plane. A third set of five balls can now be placed in a third plane such that each ball in the second plane is simultaneously in contact with three balls in the plane above. This is possible if the locations of balls in the first and the third planes are mirror images of each other. The three-dimensional assembly will then have a cubic structure with eight balls at the corners and six at the centres of the faces.

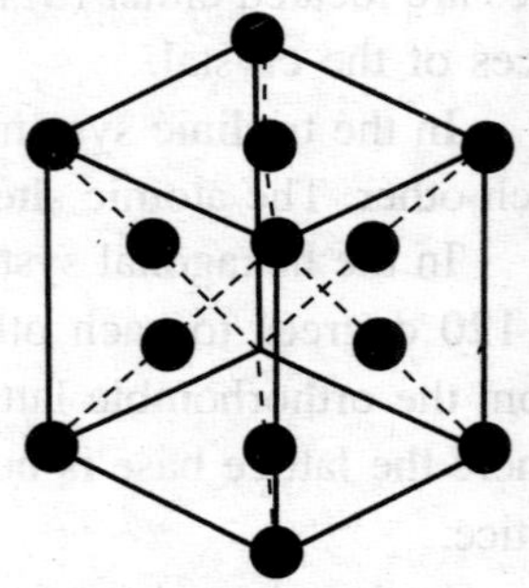
Fig. 23.5. *fcc* lattice

Such an assembly of atomic sites (Fig. 23.5) is called a face-centred cubic lattice. The fcc lattice is a closely packed structure because each atom has twelve immediate neighbours. Further, its structure is symmetrical with respect to a 90° rotation about any of the cube's edges.

(*c*) ***Examples of Cubic Structure***: Rocksalt (NaCl) crystal is an example of the cubic structure (Fig. 23.6). It consists of two face-centred cubic-sub-lattices, one of sodium ions and the other of chlorine ions midway between two sodium atomic sites. Although rocksalt is thus seen to be a combination structure its lattice has a simple cubic structure if the difference in positions of the two ions are ignored.

Another example of the cubic structure is the crystal of cesium chloride (Fig. 23.7). It is in fact made up of two simple cubic lattices interpenetrating each other. The corner of one sub-lattice is the body centre of other sub-lattices; one sub-lattice is formed by the cesium ions and the other is occupied by chloride ions.

Yet another example is the crystal of zinc blende. This crystal has two face-centred cubic sub-lattices (Fig. 23.8) occupied by two different elements —zinc and sulphur.

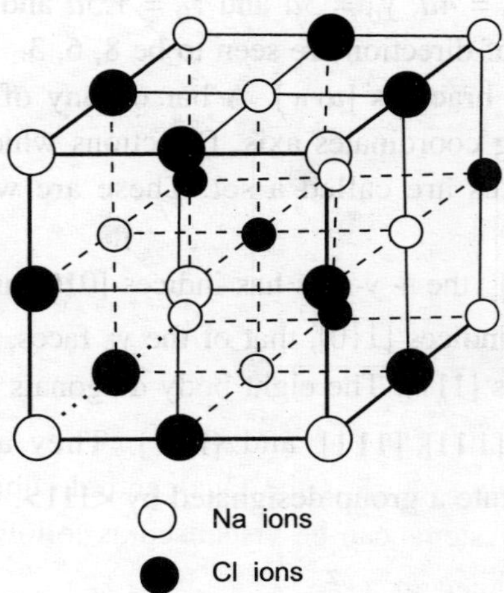

Fig. 23.6. NaCl crystal

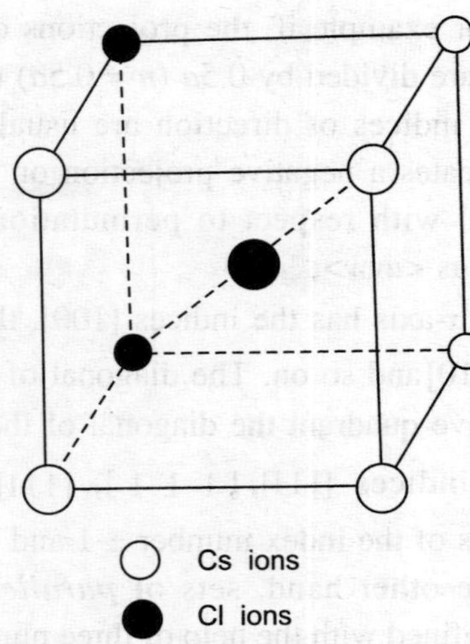

Fig. 23.7. CsCl structure

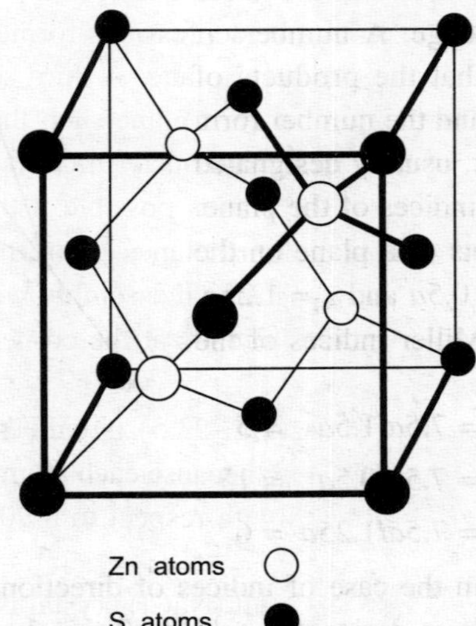

Fig. 23.8. Zinc blende structure

23.5 CRYSTALLOGRAPHIC NOTATION

Since the atomic sites in a crystal produce three dimensional structure it becomes necessary to devise some method of defining rows of atoms in the lattice. This can be done in a simple manner in cubic systems by following the vector notation. Consider a vector whose projections on the coordinates axes are x_0, y_0 and z_0 measured in terms of the cube's edge a. A number n can then be always found such that the quotients x_0/n, y_0/n and z_0/n form the *lowest set of integers* proportional to these projections. These quotients are called the indices of direction and are represented by the symbols u,

v and w. For example if, the projections of a vector are $x_0 = 4a$, $y_0 = 3a$ and $z_0 = 1.5a$ and these projections are divided by $0.5a$ ($n = 0.5a$) the three indices of direction are seen to be 8, 6, 3.

These indices of direction are usually written inside brackets [*uvw*]. A bar on any of these indices indicates a negative projection on the corresponding coordinates axis. Directions which are symmetrical with respect to permutation of the three axes are called a set. These are written collectively as <*uvw*>.

The + *x*-axis has the indices [100], the – *x*-axis, $[\bar{1}00]$, the + *y*-axis has indices [010] and the – *y*-axis, [010] and so on. The diagonal of the *xy* faces has indices [110], that of the *yz* faces, [011]. In the positive quadrant the diagonal of the cube has indices [111]. The eight body diagonals of the cube have indices [111], $[\bar{1}\,\bar{1}\,\bar{1}]$, $[\bar{1}11]$, $[1\bar{1}1]$, $[11\bar{1}]$, $[\bar{1}\bar{1}1]$, $[\bar{1}1\bar{1}]$ and $[1\bar{1}\bar{1}]$. They are all permutations of the index number ± 1 and are said to constitute a group designated by <111>.

On the other hand, sets of *parallel planes* of atoms are defined with the help of three numbers called *Miller indices*. Consider a plane of atoms passing through the origin (Fig. 23.9). Let the parallel plane have intercepts x_1, y_1 and z_1 on the coordinate axes measured in terms of the cube's edge. A numbers r, can then be always found such that the products of the reciprocals of these intercepts and the number form the lowest set of integers. This set, usually designated as (*hkl*) gives the so called Miller indices of the plane. For example, suppose the intercepts of a plane on the coordinate axes are $x_1 = 1.5a$, $y_1 = 0.5a$ and $z_1 = 1.25a$. If now r is taken to be $7.5a$, the Miller indices of the plane are seen to be

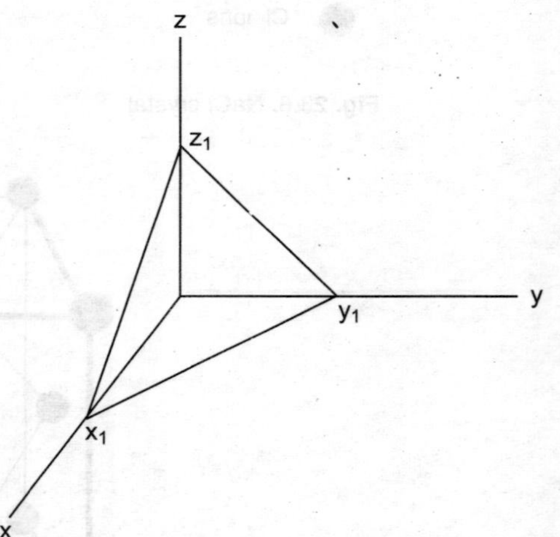

Fig. 23.9. Defining miller indices

$$h = 7.5a/1.5a = 5$$

$$k = 7.5a/0.5a = 15$$

$$l = 7.5a/1.25a = 6$$

The same general rules (as in the case of indices of direction) apply in this case also. Planes which are equivalent in the crystal are designated a {*hkl*}. Typical planes of cubic crystal are shown in (Fig. 23.10). The scheme outlined above has the following features:

(*a*) A plane passing through the origin is designated in terms of a parallel plane having finite intercepts.

(*b*) A plane parallel to one of the coordinate axes has an index number 0 for that direction.

(*c*) All equally spaced parallel planes have the same index numbers (*hkl*).

(*d*) The angle between two crystallographic directions $[u_1\ v_1\ w_1]$ and $[u_2\ v_2\ w_2]$ is given by

$$\cos\theta = \frac{u_1u_2 + v_1v_2 + w_1w_2}{\sqrt{(u_1^2 + v_1^2 + w_1^2)}\sqrt{(u_2^2 + v_2^2 + w_2^2)}}$$

(*e*) The normal to the plane (*hkl*) is the direction [*hkl*].

(*f*) The distance between adjacent planes is given by

$$d = \frac{a}{\sqrt{(h^2 + k^2 + l^2)}}$$

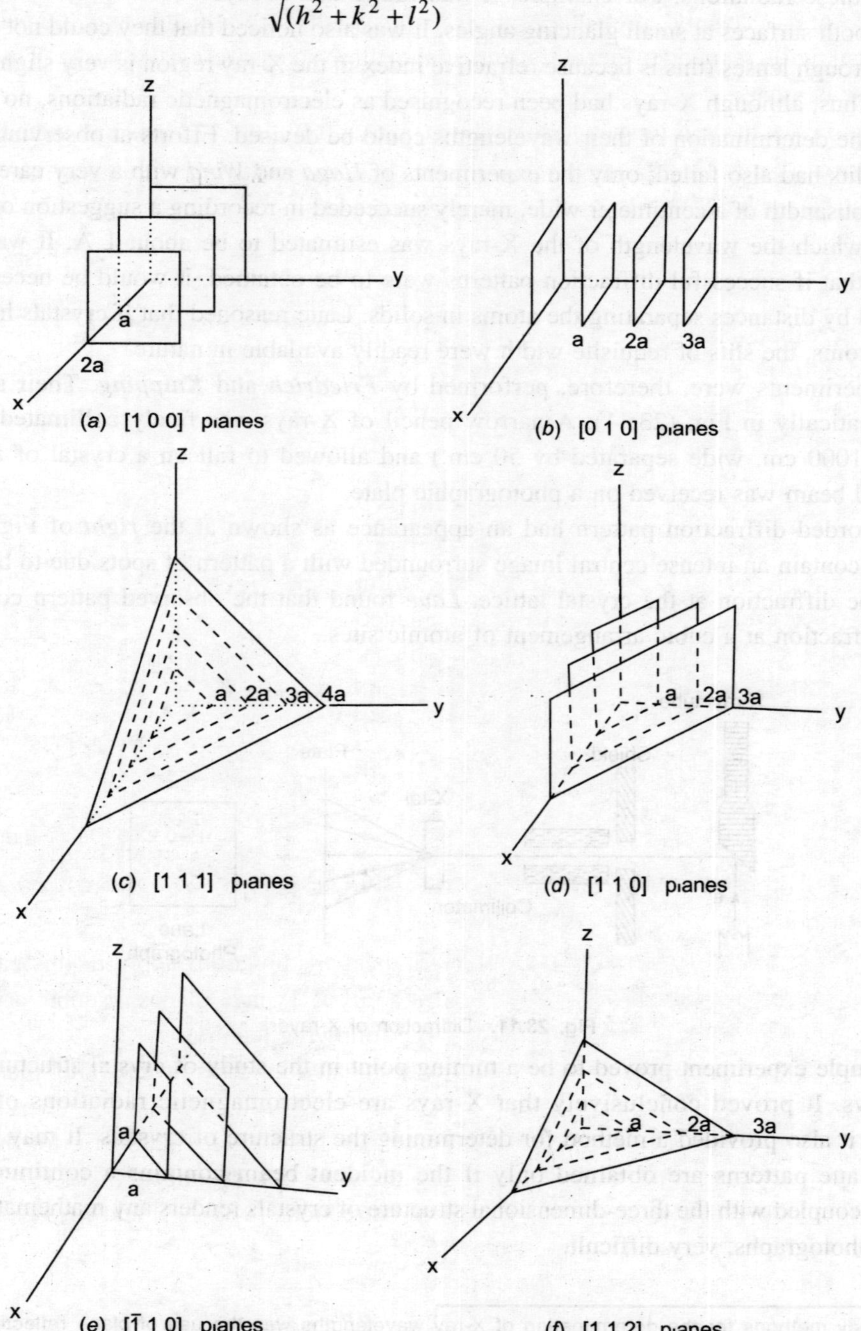

Fig. 23.10. Typical planes of cubic crystal

23.6 CRYSTALS AS THREE-DIMENSIONAL GRATINGS

The use of crystals as diffraction gratings for X-rays was occasioned by the initially unfamiliar properties of these radiations. For example, it was found that X-rays could be *externally reflected* from very smooth surfaces at small glancing angles. It was also noticed that they could not be focussed or refracted through lenses (this is because refractive index in the X-ray region is very slightly different from unity). Thus, although X-rays had been recognised as electromagnetic radiations, no satisfactory method* for the determination of their wavelengths could be devised. Efforts at observing diffraction of X-rays at slits had also failed; only the experiments of *Haga* and *Wind* with a very carefully tooled slit about a thousandth of a centimeter wide, merely succeeded in recording a suggestion of diffraction pattern from which the wavelength of the X-rays was estimated to be about 1 Å. It was *von Laue* who realised that if successful diffraction patterns were to be obtained, it would be necessary to use slits separated by distances separating the atoms in solids. Laue reasoned that if crystals had a regular structure of atoms, the slits of requisite width were readily available in nature.

The experiments were, therefore, performed by *Friedrich* and *Knipping*. Their apparatus is shown schematically in Fig. (23.11). A narrow pencil of X-rays was finely collimated (using two slits about 1/1000 cm. wide separated by 50 cm.) and allowed to fall on a crystal of zinc blende. The diffracted beam was received on a photographic plate.

The recorded diffraction pattern had an appearance as shown at the *right* of Fig. (23.11). It was found to contain an intense central image surrounded with a pattern of spots due to beams which had undergone diffraction at the crystal lattice. *Laue* found that the observed pattern could only be caused by diffraction at a cubic arrangement of atomic sites.

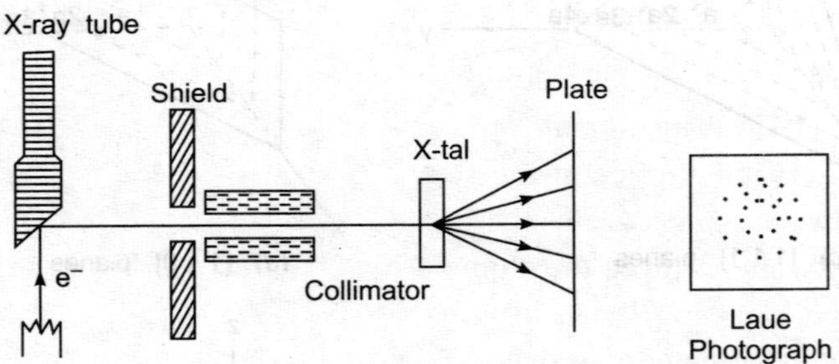

Fig. 23.11. Diffraction of X-rays

This simple experiment proved to be a turning point in the study of crystal structure as well as that of X-rays. It proved conclusively that X-rays are electromagnetic radiations of very short wavelengths; it also provided a method for determining the structure of crystals .It may however, be added that Laue patterns are obtained only if the incident beam contains a continuous band of X-rays. This, coupled with the three-dimensional structure of crystals renders any mathematical analysis of the Laue photographs, very difficult.

*One of the early methods for the determination of X-ray wavelengths was the use of plane reflection gratings at very small glancing angles.

23.7 BRAGG REFLECTION

An insight into the phenomenon of X-ray diffraction was provided by *Bragg* in 1912, soon after Laue's discovery. This was based on *effective reflection* of X-rays from successive lattice planes of a crystal and consequent reinforcement of reflected beams. Consider a beam of *monochromatic* X-rays incident at a glancing angle θ (Fig. 23.12) on a set of parallel planes of a crystal containing rows of atoms, ions or molecules. As the X-rays penetrate the crystal, each atom (or ion or molecule) scatters part of the incident X-rays and in effect behaves like a source of secondary wavelets. These secondary wavelets reinforce each other only if they are in phase.

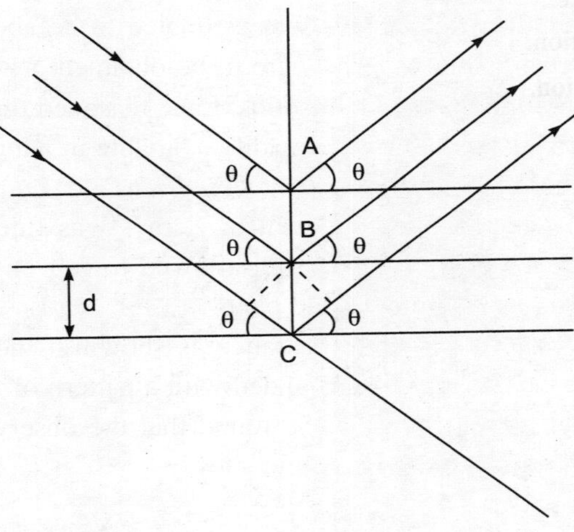

Fig. 23.12. Bragg reflection

The scattered wavelets from the *atomic sites in the same plane* will be in the same phase in a direction inclined to the crystal surface at the same angle as the beam of incident X-rays. If the glancing angle θ and the distance d between lattice planes bears the proper relationship with the incident wavelength, the wavelets scattered from the atomic sites in successive planes of the crystal also will be in the same phase. The X-rays *reflected* from successive planes of the crystal will then reinforce each other producing intense brightness in that direction. The necessary condition can be very easily derived by finding the path difference between rays *reflected* from successive planes of the crystal. This path difference is seen to be $2d \sin \theta$ and hence the condition for reinforcement of the reflected beams is

$$2d \sin \theta = n\lambda \qquad ...(23.7\ a)$$

where n is called the *order of reflection* and has integral values only. This is known as Bragg's law.

If the wavelength λ is known and the angle θ is experimentally determined, the value of d can be calculated with the help of this law. Further, by rotating the crystal and studying the reflections from other sets of planes, the shape and volume of the unit cell can be determined. This knowledge can then be used for calculating the number of atoms of various types in the cell provided the density of the crystal is known.

Although the above derivation of Bragg's law is very elementary the physical background outlined in its derivation may be utilised to get a qualitative picture of the formation of Laue patterns. In view of a variety of orientations of reflection planes in a crystal, intense beams are reflected along as many directions. As a result of this complex Laue patterns are obtained.

PROBLEMS

1. Write an essay on crystal structure.
2. Write short notes on:
   (a) Bravais Lattice.
   (b) Bragg Reflection.
   (c) X-ray diffraction.

PHYSICAL CONSTANTS*

1. Velocity of light in free space $c = 2.99793 \times 10^{10}$ cm./sec.
2. Avogadro's number $N = 6.02486 \times 10^{23}$(g. mole)$^{-1}$
3. Electronic charge $e = 4.8029 \times 10^{-10}$ e. s. u.
   $e = 1.6021 \times 10^{-19}$ Coulombs
4. Planck's constant $h = 6.6252 \times 10^{-27}$ ergs. sec.
5. Atomic Mass Unit a.m.u. = 931.14 MeV
6. Rest Mass of Electron $m_o = 9.1083 \times 10^{-28}$ gm.
7. Specific Charge of Electron $= 1.7589 \times 10^{7}$ e.m.u./gm.
8. Rest Mass of Proton $= 1.6724 \times 10^{-24}$ gm.
9. Ratio of Proton mass to Electron mass = 1836.12
10. Rest Mass of Neutron $= 1.6747 \times 10^{-24}$ gm.
11. Rest mass of Hydrogen atom = 1.00814 a. m. u.
12. Sommerfeld's fine structure constant $\alpha = 137.037$
13. Compton wavelength of electron $= .02426 \times 10^{-8}$ cm.
14. Rydberg constant $R_\infty = 109737.31$ cm.$^{-1}$
15. Rydberg constant for Hydrogen atom $R_H = 109677.58$ cm.$^{-1}$
16. Boltzmann's constant $k = 1.38 \times 10^{-16}$ ergs. °K^{-1}
17. Electron volt $eV = 1.6 \times 10^{-12}$ ergs.

* Based on *The Fundamental Constants of Physics* by Cohen, Crowe and Dumond Interscience, New York.

MATRIX PRESENTATION OF POLARISED LIGHT: JONES' VECTOR AND JONES' MATRIX

24.1 INTRODUCTION

There are many modern mathematical formalism developed to deal with the problems of polarisation, one is due to stokes, another is developed by Jones, still another by Wolf. Of these methods having relative merits and demerits, the formalism due to R. Clark Jones ("A New Calculus for the Treatment of Optical systems", J.O.S.A. 31(1941), 488) is simpler to understand and that is why we are here making use of this method to deal with the problems of polarisation.

In this method a polarised light is represented by a column vector. The first element represents the amplitude and phase of the y-component, the second element does it for z component of the light wave (according to electromagnetic theory, y-component of electric vibration), when the wave is supposed to be travelling along the x-direction.

In Fig. 24.1, the oscillation is contained in horizontal plane, hence it is linearly polarised, horizontally. Mathematically, such a disturbance can be represented as $y = \hat{j} a_2 e^{i(k_1 x - wt)}$,

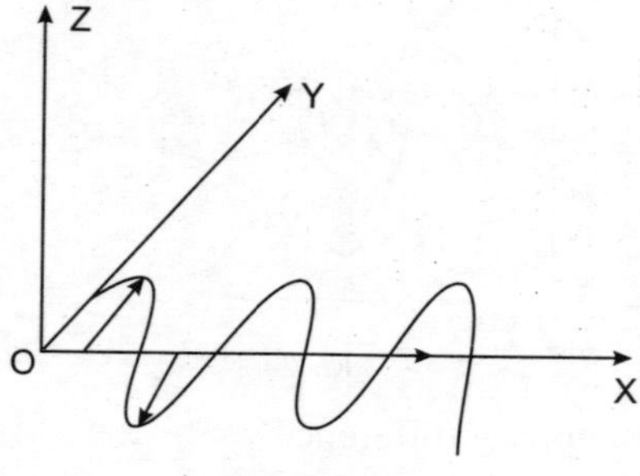

Fig. 24.1

where a_2 is the amplitude of oscillation and $k_1 = \dfrac{2\pi}{\lambda}$. This disturbance has no z-component. So the horizontally polarised light disturbance can be written as

$$y = a_2 e^{i\phi_x}, \ \ \phi_x = kx - wt, \ \ z = 0$$

$$\therefore \quad \begin{bmatrix} y \\ z \end{bmatrix} = \begin{bmatrix} a_2 e^{i\phi_x} \\ 0 \end{bmatrix} \quad \text{...(24.1 a)}$$

Thus, in Eqn. (24.1*a*), the first element of the column matrix represents amplitude and phase of horizontal oscillation. If the intensity is normalised and the phase indicating factor omitted, the relation (1) becomes.

$$\begin{bmatrix} y \\ z \end{bmatrix} = \begin{bmatrix} 1 \\ 0 \end{bmatrix} \quad \text{...(24.1 b)}$$

So, in Jones' formalism, a horizontally polarised light is represented by Jones' vector J_1 given as

$$J_1 = \begin{bmatrix} 1 \\ 0 \end{bmatrix}, \text{ for horizontally polarised light } (\longleftrightarrow)$$

Similarly, the oscillation in the vertical plane is shown in Fig. 24.2. In this figure it shown a disturbance in the vertical plane, represented by an equation

$$z = \hat{k} a_3 e^{i(kx - wt + \phi)}$$

Fig. 24.2

where a_3 is the amplitude, ϕ is the phase difference w.r.t. the horizontal disturbance and $\hat{k}$ is the unit vector along *z*-direction. This polarised wave can be written as

$$y = 0, \; z = a_3 e^{i\phi_x} e^{i\phi}$$

$$\therefore \quad \begin{bmatrix} y \\ z \end{bmatrix} = \begin{bmatrix} 0 \\ a_3 e^{i\phi_x} e^{i\phi} \end{bmatrix}$$

For normalised intensity and if the phase difference factor is omitted, the vertically polarised disturbance can be represented as Jones' vector J_2, given as

$$J_2 = \begin{bmatrix} 0 \\ e^{i\phi} \end{bmatrix} \rightarrow \begin{bmatrix} 0 \\ 1 \end{bmatrix} \text{ if } \phi = 0$$

Thus,

$$J_2 = \begin{bmatrix} 0 \\ 1 \end{bmatrix}, \text{ for vertically polarised light } (\updownarrow)$$

If the light disturbance is polarised at 45° w.r.t. horizontal or vertical direction, then Jones' vector for such polarised light can be written as

$$\begin{bmatrix} y \\ z \end{bmatrix} = \begin{bmatrix} ae^{i\phi_x} \cos\frac{\pi}{4} \\ ae^{i\phi_x} \sin\frac{\pi}{4} \end{bmatrix} \rightarrow \begin{bmatrix} 1/\sqrt{2} \\ 1/\sqrt{2} \end{bmatrix} = \frac{1}{\sqrt{2}}\begin{bmatrix} 1 \\ 1 \end{bmatrix}$$

So a linearly polarised light polarised at an angle of 45° with either of the horizontal or vertical directions is represented by Jones vector $\frac{1}{\sqrt{2}}\begin{bmatrix} 1 \\ 1 \end{bmatrix}$.

This polarised light may be obtained by addition of a horizontally polarised light and vertically polarised light of equal intensity. Thus,

$$\underset{\text{Horizontal}}{\frac{1}{\sqrt{2}}\begin{bmatrix} 1 \\ 0 \end{bmatrix}} + \underset{\text{Vertical}}{\frac{1}{\sqrt{2}}\begin{bmatrix} 0 \\ 1 \end{bmatrix}} = \underset{\text{Linearly polarised light at 45°}}{\frac{1}{\sqrt{2}}\begin{bmatrix} 1 \\ 1 \end{bmatrix}}$$

So, the normalised linearly polarised light at 45° is represented by $J_3 = \frac{1}{\sqrt{2}}\begin{bmatrix} 1 \\ 1 \end{bmatrix}$.

So, if we have a light disturbance represented as

$$\varepsilon = \begin{bmatrix} a \\ b \end{bmatrix}$$

it will mean a disturbance polarised along the direction making an angle θ w.r.t. the y-direction given by $\tan\theta = \left(\frac{b}{a}\right)$, so if $a = b$, $\tan\theta = 1 \Rightarrow \theta = \frac{\pi}{4}$.

We note that $\begin{bmatrix} 1 \\ 0 \end{bmatrix}$ and $\begin{bmatrix} 0 \\ 1 \end{bmatrix}$ are mutually orthogonal and individually normalised. For orthogonality of two vectors represented by matrices $\begin{bmatrix} a \\ b \end{bmatrix}$ and $\begin{bmatrix} a' \\ b' \end{bmatrix}$, we will have

$$[a^* b^*]\begin{bmatrix} a' \\ b' \end{bmatrix} = a^*a' + b^*b' = 0$$

here $[a^* b^*]$ is the transpose of the matrix $\begin{bmatrix} a \\ b \end{bmatrix}$ and a^*, b^* are the complex conjugates.

For the condition of normalisation,

$$[a^* b^*]\begin{bmatrix} a \\ b \end{bmatrix} = a^* a + b^* b = 1$$

with this idea in the background, let us consider the two Jones' vectors as stated above.

$$J_1 = \begin{bmatrix} 1 \\ 0 \end{bmatrix}, \ J_2 = \begin{bmatrix} 0 \\ 1 \end{bmatrix}$$

$$[1 \ 0]\begin{bmatrix} 1 \\ 0 \end{bmatrix} = 1 + 0 = 1$$

hence, $\begin{bmatrix} 1 \\ 0 \end{bmatrix}$ is normalised, so also is $\begin{bmatrix} 0 \\ 1 \end{bmatrix}$.

Again, $[1 \ 0]\begin{bmatrix} 0 \\ 1 \end{bmatrix} = 0$, so it satisfies the condition of orthogonality. Thus the two Jones' vectors J_1 and J_2 are mutually orthogonal and individually normalised, but a combination J_3 of these two linearly polarised light is not normalised, because

$$J_3 = \begin{bmatrix} 1 \\ 1 \end{bmatrix}$$

$$\therefore \qquad [1 \ 1]\begin{bmatrix} 1 \\ 1 \end{bmatrix} = 2$$

hence $\frac{1}{\sqrt{2}}\begin{bmatrix} 1 \\ 1 \end{bmatrix}$ is a normalised vector.

A device which transforms unpolarised light or ordinary light into a linearly polarised light is referred to as a linear polariser. Since the function of the device is to effect transformation, matrix formalism may be conveniently employed. So let us introduce Jones' matrix for linear polariser. Let us consider a matrix.

$$[M_t] = \begin{bmatrix} 1 & 0 \\ 0 & 0 \end{bmatrix}$$

Let it operate on another matrix $\begin{bmatrix} a \\ b \end{bmatrix}$, thus

$$\begin{bmatrix} 1 & 0 \\ 0 & 0 \end{bmatrix}\begin{bmatrix} a \\ b \end{bmatrix} = \begin{bmatrix} a \\ 0 \end{bmatrix} = a\begin{bmatrix} 1 \\ 0 \end{bmatrix}$$

So operating the matrix $[M_t]$ representing the transformation matrix on the matrix $\begin{bmatrix} a \\ b \end{bmatrix}$ representing the incident light, we get light linearly polarised in the horizontal direction with amplitude 'a' times as great. Such a matrix is called Jones' linear polariser matrix.

So a linear polariser with its transmission axis horizontal is given as

$$[L_h] = \begin{bmatrix} 1 & 0 \\ 0 & 0 \end{bmatrix}$$

with this mathematical formalism, the matrix transformation may be utilised to represent the functioning of a device producing linear polarisation. The formalism may be schematically developed as given below:

Let $\begin{bmatrix} A \\ B \end{bmatrix}$ be the incident matrix $[M_i]$ representing the incident light, the transformation matrix $[M_t]$, describing the behaviour of the polarising device that operates on it resulting in the product matrix $[M_f]$ representing the nature of the emergent light.

Thus,

$$\underset{\text{Transformation matrix}}{\begin{bmatrix} a\ b \\ c\ d \end{bmatrix}} \quad \underset{\text{Incident light matrix}}{\begin{bmatrix} A \\ B \end{bmatrix}} = \underset{\text{Emergent light matrix}}{\begin{bmatrix} A' \\ B' \end{bmatrix}}$$

$$\Rightarrow \qquad [M_t]\begin{bmatrix} A \\ B \end{bmatrix} = [M_f]$$

24.2 MATRIX FOR LINEAR POLARISER

Since a polariser transforms state of polarisation, any polarising device can be represented by a matrix, called transformation matrix or polarising element matrix.

Thus, the matrix $\begin{bmatrix} 1\ 0 \\ 0\ 0 \end{bmatrix}$ represents a linear polariser with transmission axis horizontal, the matrix $\begin{bmatrix} 0\ 0 \\ 0\ 1 \end{bmatrix}$ represents a linear polariser with transmission axis vertical, the matrix $\frac{1}{2}\begin{bmatrix} 1\ 1 \\ 1\ 1 \end{bmatrix}$ represents a linear polariser with transmission axis at 45°, because the first matrix operates on an incident light transforming it into a horizontally polarised light, the second into a vertically polarised light etc.

The following Jones matrices represent horizontal, vertical, + 45° polariser.

Horizontal $\rightarrow \begin{bmatrix} 1 & 0 \\ 0 & 0 \end{bmatrix}$

Vertical $\rightarrow \begin{bmatrix} 0 & 0 \\ 0 & 1 \end{bmatrix}$

+ 45° $\rightarrow \frac{1}{2}\begin{bmatrix} 1 & 1 \\ 1 & 1 \end{bmatrix}$

The interpretation of the matrices given on previous page are:

The first matrix represents a polarising element, called a horizontal polariser, because it transforms the incident light into a horizontally linearly polarised light.

Thus,

$$\begin{bmatrix}1 & 0\\ 0 & 0\end{bmatrix}\begin{bmatrix}ae^{i\phi_1}\\ be^{i\phi_2}\end{bmatrix}=\begin{bmatrix}ae^{i\phi_1}\\ 0\end{bmatrix}\rightarrow\begin{bmatrix}1\\ 0\end{bmatrix}\text{ i.e., horizontal polarisation.}$$

Similarly,

$$\begin{bmatrix}0 & 0\\ 0 & 1\end{bmatrix}\begin{bmatrix}ae^{i\phi_1}\\ be^{i\phi_2}\end{bmatrix}=\begin{bmatrix}0\\ be^{i\phi_2}\end{bmatrix}\rightarrow\begin{bmatrix}0\\ 1\end{bmatrix}\text{ i.e., vertically polarised light.}$$

Similarly,

$$\frac{1}{2}\begin{bmatrix}1 & 1\\ 1 & 1\end{bmatrix}\begin{bmatrix}ae^{i\phi_1}\\ be^{i\phi_2}\end{bmatrix}=\frac{1}{2}\begin{bmatrix}(ae^{i\phi_1}+be^{i\phi_2})\\ (ae^{i\phi_1}+be^{i\phi_2})\end{bmatrix}$$

If $\phi_1=\phi_2=0, a=b=1,$ then

$$\frac{1}{2}\begin{bmatrix}1 & 1\\ 1 & 1\end{bmatrix}\begin{bmatrix}ae^{i\phi_1}\\ be^{i\phi_2}\end{bmatrix}=\frac{1}{2}\begin{bmatrix}2\\ 2\end{bmatrix}=\begin{bmatrix}1\\ 1\end{bmatrix}\text{ linearly polarised light at }+45^{\circ}.$$

Q. 1. Linearly polarised light of unit intensity oscillating in the horizontal direction is added to a 16-times as intense vertically polarised beam. Find the nature of the emergent beam.

Ans. The two beams are represented by Jones' vectors $\begin{bmatrix}1\\ 0\end{bmatrix}$ and $\begin{bmatrix}0\\ 4\end{bmatrix}$. So the emergent light has Jones' vector given as

$$J=\begin{bmatrix}1\\ 0\end{bmatrix}+\begin{bmatrix}0\\ 4\end{bmatrix}=\begin{bmatrix}1\\ 4\end{bmatrix}$$

which represents a linearly polarised light at an angle θ given by $\tan\theta=\left(\frac{4}{1}\right)\Rightarrow\theta=76^{\circ}$, the intensity is given by $(1^2+4^2)=17$. So we remember that Jones' vectors represent either the incident or emergent beam but Jones' matrix represents the functioning device or the polariser.

Q. 2. Show that the matrix $\frac{1}{2}\begin{bmatrix}1 & -1\\ -1 & 1\end{bmatrix}$ represents a polariser of – 45°.

Ans. $\frac{1}{2}\begin{bmatrix}1 & -1\\ -1 & 1\end{bmatrix}\begin{bmatrix}ae^{i\phi_1}\\ be^{i\phi_2}\end{bmatrix}$ if $\phi_1=\phi_2=0,\ a=b=1,$ then $\frac{1}{2}\begin{bmatrix}1 & -1\\ -1 & 1\end{bmatrix}\begin{bmatrix}1\\ 1\end{bmatrix}=\frac{1}{2}\begin{bmatrix}0\\ 0\end{bmatrix}$

The matrix operates on a polarised light at +45° producing the emergent light of zero intensity, *i.e.*, light is cut off. So, the matrix represents a polariser of –45°.

Q. 3. Operate $\frac{1}{2}\begin{bmatrix} 1 & -1 \\ -1 & 1 \end{bmatrix}$ on – 45° polarised light. Interprent the emergent light.

Ans. $\frac{1}{2}\begin{bmatrix} 1 & -1 \\ -1 & 1 \end{bmatrix}\begin{bmatrix} -1 \\ +1 \end{bmatrix} = \frac{1}{2}\begin{bmatrix} -2 \\ 2 \end{bmatrix} = \begin{bmatrix} -1 \\ +1 \end{bmatrix}$

The emergent light is linearly polarised at – 45°.

Q. 4. Let light first be horizontally polarised, then pass through a + 45° polariser, then through a – 45° polariser. What is the polarisation of emergent light.

Ans. The complete step by step process is represented by a sequence of matrix operations given as below:

$$\frac{1}{2}\begin{bmatrix} 1 & -1 \\ -1 & 1 \end{bmatrix}\frac{1}{2}\begin{bmatrix} 1 & 1 \\ 1 & 1 \end{bmatrix}\begin{bmatrix} 1 \\ 0 \end{bmatrix} = \frac{1}{2}\begin{bmatrix} 1 & -1 \\ -1 & 1 \end{bmatrix}\frac{1}{2}\begin{bmatrix} 1 \\ 1 \end{bmatrix} = \frac{1}{4}\begin{bmatrix} 0 \\ 0 \end{bmatrix} = \begin{bmatrix} 0 \\ 0 \end{bmatrix}$$

No light is transmitted.

In general terms, we can represent polarising action of each of the optical elements by one matrix and operate with the product on the incident light giving the polarisation of the emergent light. Thus the schematic representation of the action of a set of polarisers is given as

$$\begin{bmatrix} 1 & 0 \\ 0 & e^{i\phi_1} \end{bmatrix}\begin{bmatrix} 1 & 0 \\ 0 & e^{i\phi_2} \end{bmatrix}\ldots\begin{bmatrix} 1 & 0 \\ 0 & e^{i\phi_n} \end{bmatrix}\begin{bmatrix} \text{Incident} \\ \text{polarised} \\ \text{light} \end{bmatrix} = \begin{bmatrix} \text{Emergent} \\ \text{polarised} \\ \text{light} \end{bmatrix}$$

Q. 5. A linear polariser with its transmission axis horizontal is taken. Describe what would happen when

(*i*) linearly polarised light of horizontal orientation is incident on it.

(*ii*) the same light but now with vertical orientation is incident on it.

Ans. (*i*) A linear polariser with its transmission axis horizontal is represented by a matrix $\begin{bmatrix} 1 & 0 \\ 0 & 0 \end{bmatrix}$.

It operates on the matrix $\begin{bmatrix} 1 \\ 0 \end{bmatrix}$ representing linearly polarised light in the horizontal direction to give the state of polarisation of the emergent light.

$$\therefore \quad \begin{bmatrix} 1 & 0 \\ 0 & 0 \end{bmatrix}\begin{bmatrix} 1 \\ 0 \end{bmatrix} = \begin{bmatrix} 1 \\ 0 \end{bmatrix} \text{ Linear horizontally polarised}$$

$$(ii) \quad \begin{bmatrix} 1 & 0 \\ 0 & 0 \end{bmatrix}\begin{bmatrix} 0 \\ 1 \end{bmatrix} = \begin{bmatrix} 0 \\ 0 \end{bmatrix} \text{ No transmission}$$

Q. 6. A linearly polarised light of horizontal orientation is incident on a linear polariser oriented at + 45°. Then

(*i*) the light enters a polariser whose transmission axis is vertical.

(*ii*) the vertical polariser is placed ahead of the 45° polariser. Describe the emergent ray.

Ans. (*i*) $\frac{1}{2}\begin{bmatrix}1 & 1\\1 & 1\end{bmatrix}\begin{bmatrix}1\\0\end{bmatrix} = \frac{1}{2}\begin{bmatrix}1\\1\end{bmatrix}$

So, the emergent light is linearly polarised at 45°, and of the same intensity as that of the incident light. The polariser with axis vertical is represented by a matrix $\begin{bmatrix}0 & 0\\0 & 1\end{bmatrix}$.

$$\therefore \qquad \begin{bmatrix}0 & 0\\0 & 1\end{bmatrix}\begin{bmatrix}1\\1\end{bmatrix} = \begin{bmatrix}0\\1\end{bmatrix}$$

Thus the emergent light is linearly polarised of vertical orientation, the intensity of the emergent light is reduced to half of the original value.

(*ii*) The step by step functioning here is represented by a sequence of matrix operation as shown

$$\begin{bmatrix}0 & 0\\0 & 1\end{bmatrix}\begin{bmatrix}1\\0\end{bmatrix} = \begin{bmatrix}0\\0\end{bmatrix}, \text{ then } \frac{1}{2}\begin{bmatrix}1 & 1\\1 & 1\end{bmatrix}\begin{bmatrix}0\\0\end{bmatrix} = \frac{1}{2}\begin{bmatrix}0\\0\end{bmatrix}$$

There is no transmission of light.

Q. 7. An unpolarised beam falls on a polariser. What is the emergent wave.

Ans. Jones vector of incident light $= \begin{bmatrix}E_x\\E_y\end{bmatrix}$

Jones matrix of polariser $= \begin{bmatrix}1 & 0\\0 & 0\end{bmatrix}$

$\therefore$ Jones vector of emergent light $= \begin{bmatrix}1 & 0\\0 & 0\end{bmatrix}\begin{bmatrix}E_x\\E_y\end{bmatrix} = \begin{bmatrix}E_x\\0\end{bmatrix}$

$\therefore$ The emergent wave is linearly polarized along the *x*-axis.

Only half the average intensity is transmitted.

Q. 8. A linearly polarized light of unit intensity is incident on an analyser at an azimuth *r*. Calculate the fraction transmitted.

Ans. For an azimuth *r*, Jones vector of incident light $= \begin{bmatrix}\cos r\\\sin r\end{bmatrix}$

The Jones matrix of the analyser $= \begin{bmatrix}1 & 0\\0 & 0\end{bmatrix}$

$\therefore$ Jones vector of transmitted light $= \begin{bmatrix}1 & 0\\0 & 0\end{bmatrix}\begin{bmatrix}\cos r\\\sin r\end{bmatrix}$

$$\vec{E} = \begin{bmatrix}\cos r\\0\end{bmatrix}$$

$$\therefore \quad \text{Transmitted intensity} \quad E^+E = [\cos r \; 0]\begin{bmatrix} \cos r \\ 0 \end{bmatrix} = \cos^2 r$$

If a light beam of amplitude a is incident with orientation angle θ with the y-direction and $[(90 - \theta) = \beta$ with the z-direction, then the disturbance is represented as

$$y = a\cos\theta e^{i(kx-wt)} = a\cos\theta e^{i\phi_x}$$

$$z = a\sin\theta e^{i(kx-wt)} = a\sin\theta e^{i\phi_x}$$

$$\therefore \quad \begin{bmatrix} y \\ z \end{bmatrix} = \begin{bmatrix} a\cos\theta e^{i\phi_x} \\ a\cos\beta e^{i\phi_x} \end{bmatrix} \rightarrow \begin{bmatrix} a\cos\theta \\ a\cos\beta \end{bmatrix} \rightarrow \begin{bmatrix} A_1 \\ A_2 \end{bmatrix}$$

Thus we get a linearly polarised light with the orientation axis making an angle ϕ, given by,

$$\tan\phi = \frac{A_2}{A_1} = \frac{a\cos\beta}{a\cos\theta} = \frac{a\sin\theta}{a\cos\theta} = \tan\theta \Rightarrow \phi = \theta$$

So, the linear polariser matrix transforming an incident linearly polarised light into another linearly polarised light with orientation θ with y-direction is written as

$$\begin{bmatrix} A_1 & A_1 \\ A_2 & A_2 \end{bmatrix}$$

Thus,

$$\begin{bmatrix} A_1 & A_1 \\ A_2 & A_2 \end{bmatrix} \begin{bmatrix} 1 \\ 0 \end{bmatrix} = \begin{bmatrix} A_1 \\ A_2 \end{bmatrix}$$

Horizontally polarised — Linearly polarised

$$\begin{bmatrix} A_1 & A_1 \\ A_2 & A_2 \end{bmatrix} \begin{bmatrix} 0 \\ 1 \end{bmatrix} = \begin{bmatrix} A_1 \\ A_2 \end{bmatrix}$$

Vertically polarised — Linerly polarised

$$\begin{bmatrix} A_1 & A_1 \\ A_2 & A_2 \end{bmatrix} \frac{1}{\sqrt{2}} \begin{bmatrix} 1 \\ 1 \end{bmatrix} = \frac{1}{\sqrt{2}} \begin{bmatrix} 2A_1 \\ 2A_2 \end{bmatrix} = \sqrt{2} \begin{bmatrix} A_1 \\ A_2 \end{bmatrix}$$

Writing out the full form in the last case, we get

$$y = \sqrt{2}A_1\cos(kx - wt), \; z = \sqrt{2}A_2\cos(kx - wt)$$

$$\therefore \quad \tan\theta = \frac{z}{y} = \frac{A_2}{A_1}$$

So, here also the emergent light is plane polarised with orientation in the direction making an angle θ with y-direction.

24.3 CIRCULAR AND ELLIPTICAL POLARISATION

We have seen that how to produce linearly polarised light, polarised either vertically or horizontally or polarised at arbitrary orientation w.r.t. vertical or horizontal direction. Now, two such polarised components under certain conditions on superposition produces circularly polarised light where the tip of the displacement vector lies on a circle or an ellipse.

Circularly Polarised Light

In a linearly polarised light with orientation in the horizontal direction, vibration will be in a horizontal plane, if such light disturbance combines with an orthogonal disturbance at some place, the resultant vibration may be linear, circular or elliptical depending on the phase and amplitude relationship of these two orthogonal linear vibrations under consideration. Thus, let the y-direction vibration be represented as

$$\vec{y} = \hat{j}a_1 e^{i(kx-wt)} = \hat{j}a_1 e^{i\phi_x},$$

where $k = \dfrac{2\pi}{\lambda}$, called the propagation constant and $\phi = kx - wt$, which definitely depends on time. This is represented in Fig. 24.3(a)

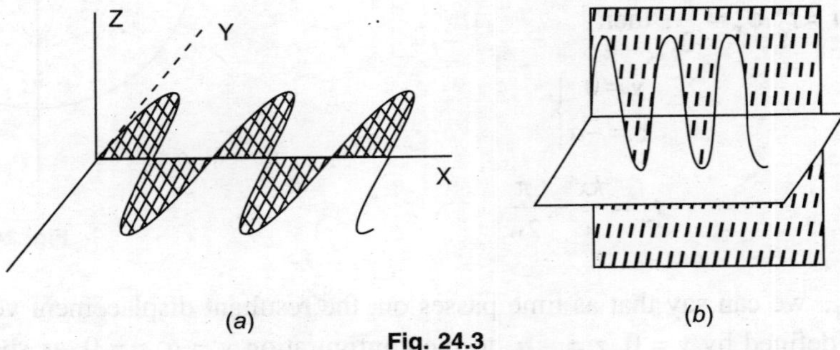

Fig. 24.3

The orthogonal vibration where displacement occurs in the z-direction while propagation direction is along x-axis is shown in Fig. 24.3 (b) and is written as

$$\vec{z} = \hat{k}a_2 e^{i(\phi_x+\phi)}$$

where, ϕ represents the phase difference between the two linear vibrations. If ϕ is positive, z-vibration is in phase lead and if ϕ is negative, z-vibration is phase lag w.r.t. y-vibration. The resultant can be written and shown in Fig. 24.4.

$$\vec{\xi} = \vec{y} + \vec{z} = (\hat{j}a_1 + \hat{k}a_2 e^{i\phi})e^{i\phi_x}$$

Fig. 24.4

Now, if $a_1 = a_2 = a$, $\phi = \dfrac{\pi}{2}$, then $\xi = a(\hat{j} + \hat{k}e^{i\frac{\pi}{2}})e^{i\phi_x}$ the two components fluctuate in different phases, the two vibrations here are said to be in quadrature, at any place, defined by a fixed value of

x, the displacement due to one is maximum when that due to the other is minimum but the resultant amplitude is always constant, hence the tip of the amplitude vector lies on a circle.

When z-component leads in phase by $\frac{\pi}{2}$: Under this case, the two components are represented as

$$y = R.P.\, ae^{i\phi_x} = a\cos\phi_x \qquad \text{...(24.3 } a)$$

$$z = R.P.\, ae^{i(\phi_x+\phi)} = a\cos(\phi_x+\phi)$$

$$= a\cos(\phi_x + \frac{\pi}{2}) = -a\sin\phi_x \qquad \text{...(24.3 } b)$$

where $\phi_x = kx - wt$ and $R.P.$ is real part

Now, at $t = t_1$, let $\phi_x = 0$, then, we get

$$\left.\begin{aligned} y &= a \\ z &= 0 \end{aligned}\right\}$$

corresponding to $t_1 = \frac{kx}{w}$

Also, let at $t = t_2, \phi_x = \frac{\pi}{2}$, then

$$\left.\begin{aligned} y &= 0 \\ z &= -a \end{aligned}\right\}$$

corresponding to $t_2 = \frac{kx}{w} - \frac{\pi}{2w}$

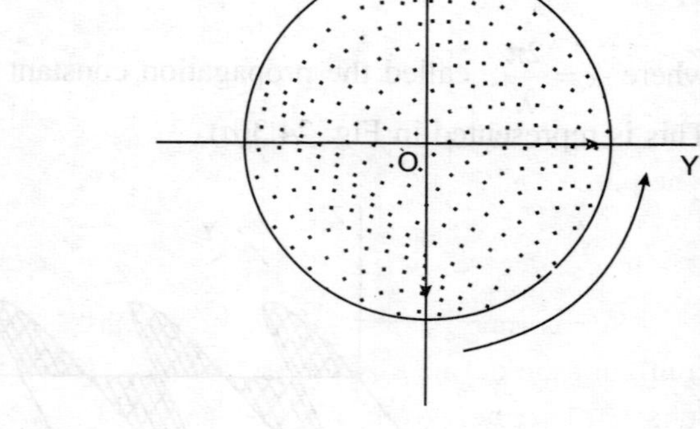

Fig. 24.5

Since $t_1 > t_2$, we can say that as time passes on, the resultant displacement vector passes from the configuration defined by $y = 0$, $z = -a$ to the configuration $y = a$, $z = 0$, as shown in Fig. 24.5. So it represents that the resultant displacement vector rotates in a circle in the anti-clockwise sense.

Also, from Eqns. (24.3 a) and (24.3 b), we get

$$\frac{y^2}{a^2} + \frac{z^2}{a^2} = 1 \qquad \text{...(24.3 c)}$$

Equation (24.3 c) shows that at a certain place, the displacement under the influence of the two vibrations under specified restriction satisfies an equation of a circle traversed anti-clockwise. So light in this case is said to be circularly polarised in the anti-clockwise sense. This is also referred to as Right-handed Circularly Polarised (RCP).

Conclusion : Of the two orthogonal vibrations, one along y-direction, the other along z-direction, if z-vibration leads in phase by $\left(2p + \frac{1}{2}\right)\pi$, where p = 0, 1, 2 etc., the resultant vibration would be of constant amplitude, not fixed in space but rotating anti-clockwise in a circle.

In Jones' matrix formalism, this type of circularly polarised light is represented as

$$[J]_{RCP} = \begin{bmatrix} a \\ ae^{i\frac{\pi}{2}} \end{bmatrix} e^{i\phi_x} \rightarrow \frac{1}{\sqrt{2}}\begin{bmatrix} 1 \\ -i \end{bmatrix}, \quad RCP$$

with intensity normalised. It is the right circularly polarised light.

When the z-Component Lags in Phase by $\frac{\pi}{2}$

In this case, the two component vibrations are represented as

$$y = a\cos\phi_x, \; z = a\cos\left(\phi_x - \frac{\pi}{2}\right) = a\sin\phi_x$$

when $\quad \phi_x = 0 \Rightarrow t_1 = \frac{kx}{w}$, we have

$$y = a, \quad z = 0$$

when $\quad \phi_x = \frac{\pi}{2} \Rightarrow t_2 = \frac{kx}{w} - \frac{\pi}{2w}$, we have

$$y = 0, \quad z = a$$

So, as time passes on, the vector passes from the configuration defined by $y = 0$, $z = a$, to the configuration defined by $y = 0$, $z = a$, as shown in Fig. 24.6. It is said to be circularly polarised vibration in the clockwise sense or Left-handed Circularly Polarised (LCP).

So, in Jones' formalism,

$$[J]_{LCP} = \begin{bmatrix} a \\ ae^{-i\frac{\pi}{2}} \end{bmatrix} e^{i\phi_x} \rightarrow \frac{a}{\sqrt{2}}\begin{bmatrix} 1 \\ -i \end{bmatrix} \rightarrow \frac{1}{\sqrt{2}}\begin{bmatrix} 1 \\ -i \end{bmatrix} \rightarrow LCP$$

with intensity normalised. Thus the two circularly polarised lights are represented by Jones' matrices as

$$[J]_{RCP} = \frac{1}{\sqrt{2}}\begin{bmatrix} 1 \\ i \end{bmatrix}, [J]_{LCP} = \frac{1}{\sqrt{2}}\begin{bmatrix} 1 \\ -i \end{bmatrix}$$

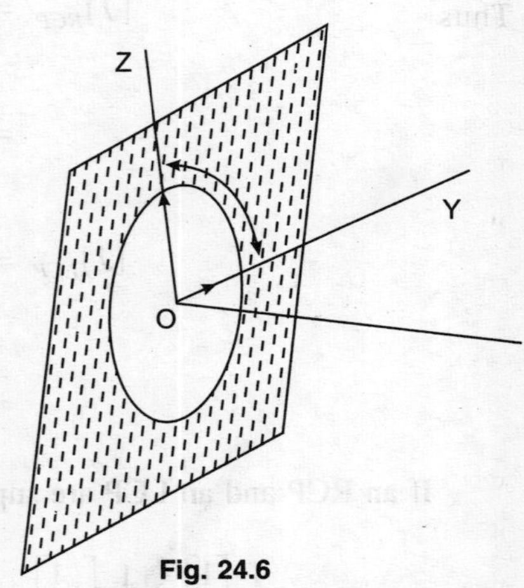

Fig. 24.6

When the y-Component Leads by $\frac{\pi}{2}$

In this case, we can write

$$y = R.P.\, ae^{i(\phi_x + \frac{\pi}{2})} = a\cos\left(\frac{\pi}{2} + \phi_x\right) = -a\sin\phi_x$$

$$z = R.P.\, ae^{i\phi_x} = a\cos\phi_x$$

when $$\phi_x = 0 \Rightarrow t_1 = \frac{kx}{w}, \text{ we get}$$

$$y = 0, \quad z = a$$

when $$\phi_x = \frac{\pi}{2} \Rightarrow t_2 = \frac{kx}{w} - \frac{\pi}{2w}, \text{ we get}$$

$$y = -a, \quad z = 0$$

So, as time passes on, the configuration changes from one defined by $y = -a, z = 0$ to $y = 0, z = a$ as shown in Fig. 24.7. It is a clockwise circularly polarised vibration (LCP).

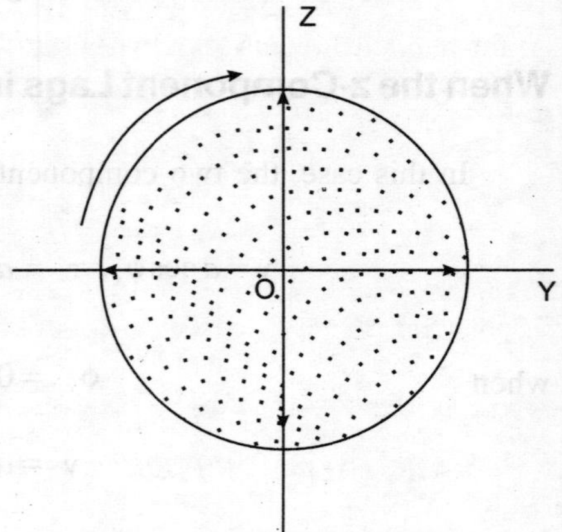

Fig. 24.7

In Jones' formalism,

$$[J]_{LCP} = \frac{1}{\sqrt{2}}\begin{bmatrix} i \\ 1 \end{bmatrix} \rightarrow LCP$$

the y-component leads. Similarly, if y-component lags, then

$$[J]_{RCP} = \frac{1}{\sqrt{2}}\begin{bmatrix} -i \\ 1 \end{bmatrix} \rightarrow RCP$$

Thus $$[J]_{RCP} = \frac{1}{\sqrt{2}}\begin{bmatrix} 1 \\ i \end{bmatrix} \text{ for } RCP \text{ when } z\text{-component leads}$$

$$= \frac{1}{\sqrt{2}}\begin{bmatrix} -i \\ 1 \end{bmatrix} \text{ for } RCP \text{ when } z\text{-component lags}$$

$$[J]_{LCP} = \frac{1}{\sqrt{2}}\begin{bmatrix} 1 \\ -i \end{bmatrix} \text{ for } LCP \text{ when } z\text{-component lags}$$

$$= \frac{1}{\sqrt{2}}\begin{bmatrix} i \\ 1 \end{bmatrix} \text{ for } LCP \text{ when } y\text{-component leads}$$

If an RCP and an LCP are superposed, we would get the emergent light as

$$\frac{1}{\sqrt{2}}\begin{bmatrix} 1 \\ i \end{bmatrix} + \frac{1}{\sqrt{2}}\begin{bmatrix} 1 \\ -i \end{bmatrix} = \frac{1}{\sqrt{2}}\begin{bmatrix} 2 \\ 0 \end{bmatrix} = \begin{bmatrix} \sqrt{2} \\ 0 \end{bmatrix}$$

So the emergent light is linearly polarised, with horizontal orientation and intensity $\sqrt{(\sqrt{2})^2 + (0)^2} = 2$ (sum of the two incident intensities). Similarly,

$$\frac{1}{\sqrt{2}}\begin{bmatrix}-i\\1\end{bmatrix}+\frac{1}{\sqrt{2}}\begin{bmatrix}i\\1\end{bmatrix}=\frac{1}{\sqrt{2}}\begin{bmatrix}0\\2\end{bmatrix}=\begin{bmatrix}0\\\sqrt{2}\end{bmatrix}$$

The emergent light is linearly polarised with orientation in the vertical direction. The above result leads to a very important conclusion:

A linearly polarised light may be obtained by superposing two circularly polarised light, one right-handed (or anti-clockwise) and another, left-handed (or clockwise).

Q. 9. **Indentify the state of polarisation of the following waves:**

(*i*) $E_x = E_1 \cos(wt - kz), E_y = 0$

(*ii*) $E_x = E_2 \cos(wt - kz), E_y = E_2 \cos(wt - kz - \frac{\pi}{2})$

(*iii*) $E_x = E_3 \cos(wt + kz), E_y = E_3 \cos(wt + kz + \frac{\pi}{2})$

(*iv*) $E_x = E_4 \cos(wt - kz), E_y = E_5 \cos(wt + kz)$.

Ans. (*i*) $E_x = R.P.E_1 e^{i\phi_x}, \phi_z = wt - kz, E_y = 0$

$$\Rightarrow \begin{bmatrix}E_x\\E_y\end{bmatrix}=\begin{bmatrix}E_1\\0\end{bmatrix}e^{i\phi_x}\to\begin{bmatrix}1\\0\end{bmatrix}$$

So it is a horizontally polarised with orientation axis along the *x*-direction.

(*ii*) $E_x = R.P.E_2 e^{i\phi_x}, E_y = R.P.E_2 e^{i(\phi_x - \frac{\pi}{2})}$

$$\therefore \begin{bmatrix}E_x\\E_y\end{bmatrix}=\begin{bmatrix}E_2\\E_2 e^{-i\frac{\pi}{2}}\end{bmatrix}e^{i\phi_x}\to\frac{1}{\sqrt{2}}\begin{bmatrix}1\\-i\end{bmatrix}$$

So, it represents an *LCP*.

(*iii*) **It is represented as** $\frac{1}{\sqrt{2}}\begin{bmatrix}1\\i\end{bmatrix}$, **hence it represents an *RCP***

(*iv*) $E_x = R.P.E_4 e^{i\phi_x}$, $E_y = R.P.E_5 e^{i\phi_x}$

$$\begin{bmatrix}E_x\\E_y\end{bmatrix}=\begin{bmatrix}E_4\\E_5\end{bmatrix}e^{i\phi_x}\to\begin{bmatrix}E_4\\E_5\end{bmatrix}$$

So, it is also a plane polarised light with orientation axis making an angle θ with the *x*-direction given by $\tan\theta = \left(\frac{E_5}{E_4}\right)$.

Production of Circularly Polarised Light

From the above discussion it is clear that to obtain a circularly polarised light, we have to take

two linearly polarised lights orthogonal to each other and produce in one a phase difference of $\frac{\pi}{2}$. A phase difference of $\frac{\pi}{2}$ is equivalent to a path difference of

$$\Delta = \frac{\lambda}{2\pi} \cdot \frac{\pi}{2} = \frac{\lambda}{4} \text{ (phase difference } = \frac{2\pi}{\lambda} \text{ path difference)}$$

A device which introduces a phase difference of $\frac{\pi}{2}$ is called a quartz-wave plate or a $\frac{\lambda}{4}$ plate or a $\frac{\pi}{2}$-retarder. Such a retarder can be used to produce and analyse circularly polarised and elliptically polarised light.

Construction of a Quarter-wave Plate

Wave plate, also called retarders are converters of polarisation forms. With the help of appropriate retarder a given form of polarisation can be converted into any other form.

A quarter-wave plate introduces a phase-change of $\frac{\pi}{2}$. It may be made of a thin plate of calcite or quartz whose optic axis lies on the surface. Both of these materials are bi-refringent but calcite is said to be a –ve crystal because in it the extraordinary ray (E-ray) travels faster than the ordinary ray (O-ray), whereas quartz is a +ve crystal where the ordinary ray travels faster than the extraordinary ray.

We know from double refraction phenomenon that when light, ordinary or polarised, is incident on a crystal, in a direction other than the optic axis direction, it breaks up into two waves-ordinary and extraordinary. Both are linearly polarised and they are orthogonal to each other. Depending on the nature of the crystal (negative or positive), extraordinary or ordinary ray will travel faster. The direction along which speed is greater, is called 'fast axis' the other direction is the 'slow axis' of the crystal.

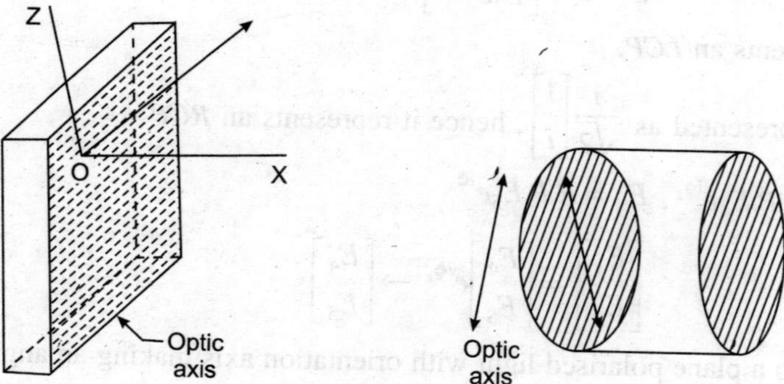

Fig. 24.8

In case of a retarder it is true that both of ordinary ray and extraordinary ray get retarded, but it is the relative phase difference that is of concern to us.

Consider a calcite crystal whose plane faces are parallel to the optic axis and let light fall on the crystal face from left, also suppose that light is linearly polarised with horizontal orientation. The crystal is so oriented that the vibration makes an angel of 45° with the optic axis. The amplitude of vibration is, suppose, *a*. On incidence it breaks up into two vibrations, one along the optic axis or parallel to the optic axis, called E-ray, the other orthogonal to it, called O-ray. Since in a calcite crystal, extraordinary travels faster the optic axis direction is the fast axis, and direction ⊥ to the optic axis, along which ordinary ray travels, is the slow axis. The calcite crystal retarder retards both these components, but due to difference in speeds of the ordinary and extraordinary rays and the thickness of the plate is so chosen that there is a relative change of phase by an amount equal to $\frac{\pi}{2}$ on emergence.

Thickness of a $\frac{\lambda}{4}$ - plate (90°-retarder)

Let us take up once again the equations

$$\vec{y} = \hat{j}ae^{i(k_1x-wt)}, \vec{z} = \hat{k}ae^{i(k_2x-wt)}$$

Let after emergence $(x = L)$, the relative phase change introduced by the plate is $\frac{\pi}{2}$, then,

$$(k_2 - k_1)L = \frac{\pi}{2} \Rightarrow L = \frac{\pi}{2(k_2 - k_1)} \quad \text{...(24.3 } d\text{)}$$

Now, $k = \frac{2\pi}{\lambda}$, hence k_2 = propagation constant for E-ray $= \frac{2\pi}{\lambda_E} = k_e$

Again $\mu_E = \frac{c}{v_e} = \frac{\lambda_{air}}{\lambda_E} = \frac{\lambda}{\lambda_E}$

Similarly, for ordinary ray,

$$k_1 = k_0 = \frac{2\pi}{\lambda_0}, \mu_0 = \frac{c}{v_0} = \frac{\lambda_{air}}{\lambda_0} = \frac{\lambda}{\lambda_0}$$

$$k_2 - k_1 = k_e - k_0 = 2\pi\left(\frac{1}{\lambda_E} - \frac{1}{\lambda_0}\right) = 2\pi\left(\frac{\mu_E}{\lambda} - \frac{\mu_0}{\lambda}\right) = \frac{2\pi}{\lambda}(\mu_E - \mu_0)$$

$$\therefore \quad L = \frac{\pi}{2} \cdot \frac{\lambda}{2\pi(\mu_E - \mu_0)}$$

$$= \frac{\lambda}{4} \cdot \frac{1}{(\mu_E - \mu_0)} \quad \text{...(24.3 } e\text{)}$$

$$\Rightarrow \quad L(\mu_E - \mu_0) = \frac{\lambda}{4}$$

Since $\mu_E = \frac{c}{v_e}, \mu_0 = \frac{c}{v_0}$, for a positive crystal v_0 being greater than v_e, $\mu_E > \mu_0$ and for a negative crystal $v_e > v_0$, so $\mu_E < \mu_0$, so if $(\mu_E - \mu_0) > 0$, the crystal is +ve, if on the other hand, $(\mu_E - \mu_0) < 0$, the crystal is a negative crystal,

In case of calcite crystal,

$$\mu_0 = 1.6584,\ \mu_E = 1.4864$$

If we take Na – light, λ = 5893 × 10^{-8} cm

$$\therefore \quad L|(\mu_E - \mu_0)| = \frac{\lambda}{4}$$

$$\Rightarrow \quad L = \frac{5893\times10^{-8}}{4}\times\frac{1}{[1.4864-1.6584]} = 8565\times10^{-8}\text{ cm} = .00086\text{mm}$$

So it is extremely thin, hence calcite is not suitable quarter-wave plate material, because the difference of $(\mu_E - \mu_0)$, called bi-refringence is large for calcite. If $\mu_E > \mu_0$ in a crystal, it is said to be +ve crystal, if $\mu_E < \mu_0$, it is a –ve crystal.

For mica, the bi-refringence is,

$$\mu_E - \mu_0 = 1.5977 - 1.5936 = 4.1\times10^{-3}$$

hence, $\mu_E > \mu_0$, therefore, mica is a +ve bi-axial crystal and since its bi-refringence is small, it is suitable for making the wave plate. The thickness required is,

$$L = \frac{\lambda}{4(\mu_E - \mu_0)} = \frac{5893\times10^{-8}}{4\times4.1\times10^{-3}} = 3.59\times10^{-3}\text{cm} = .036\text{ mm}$$

Thus in Jones' matrix formalism, a matrix represented as

$$[M] = \begin{bmatrix} 1 & 0 \\ 0 & e^{i\delta} \end{bmatrix},$$

where δ is the phase difference introduced, would represent a linear retarder, either right-handed or left-handed depending on the value of δ.

Jones' matrix written as

$$[J_c] = \begin{bmatrix} 1 & 0 \\ 0 & e^{\pm i\frac{\pi}{2}} \end{bmatrix}$$

will be called quarter-wave plate matrix producing right circularly polarised light with the positive sign, with negative sign, it will be called a quarter-wave plate matrix producing left circularly polarised light. Thus,

$$[J_c] = \begin{bmatrix} 1 & 0 \\ 0 & i \end{bmatrix} \rightarrow \text{Quarter-wave plate matrix producing } RCP$$

$$[J_c] = \begin{bmatrix} 1 & 0 \\ 0 & -i \end{bmatrix} \rightarrow \text{Quarter-wave plate matrix producing } LCP$$

Consider the matrix $\begin{bmatrix} 1 & 0 \\ 0 & i \end{bmatrix}$ and let it operate on a linearly polarised light with orientation 45°, then

$$\begin{bmatrix} 1 & 0 \\ 0 & i \end{bmatrix} \frac{1}{\sqrt{2}} \begin{bmatrix} 1 \\ 1 \end{bmatrix} = \frac{1}{\sqrt{2}} \begin{bmatrix} 1 \\ i \end{bmatrix} RCP$$

$$\begin{bmatrix} 1 & 0 \\ 0 & -i \end{bmatrix} \frac{1}{\sqrt{2}} \begin{bmatrix} 1 \\ 1 \end{bmatrix} = \frac{1}{\sqrt{2}} \begin{bmatrix} 1 \\ -i \end{bmatrix} LCP$$

In general, if the linearly polarised light be represented as $\begin{bmatrix} a \\ b \end{bmatrix}$, then

$$\begin{bmatrix} 1 & 0 \\ 0 & i \end{bmatrix} \begin{bmatrix} a \\ b \end{bmatrix} = \begin{bmatrix} a \\ ib \end{bmatrix} = \begin{bmatrix} a \\ be^{i\frac{\pi}{2}} \end{bmatrix}$$

Writing out the y and z components, we get

$$y = R.P.ae^{i\phi_x} \Rightarrow y = a\cos\phi_x$$

$$z = R.P.be^{i(\phi_x + \frac{\pi}{2})} \Rightarrow -b\sin\phi_x$$

Thus, when $\phi_x = kx - \omega t_1 = 0 \Rightarrow t_1 = \dfrac{kx}{w}$

$$y = a, \quad z = 0$$

When $\phi_x = kx - wt_2 = \dfrac{\pi}{2} \Rightarrow t_2 = \dfrac{kx}{w} - \dfrac{\pi}{2w}$,

$$y = 0, \quad z = -b = -a \text{ when } a = b$$

Thus we get the emergent light as circularly polarised in the anti-clockwise direction (or RCP).

Q. 10. A plane polarised light at +45° to the y-axis enters a $\dfrac{\lambda}{4}$ plate, whose fast axis is vertical. Determine nature of emergent light.

Ans. A plane polarised light at +45° is represented by Jones' vector $\dfrac{1}{\sqrt{2}}\begin{bmatrix} 1 \\ 1 \end{bmatrix}$. A $\dfrac{\lambda}{4}$ plate with fast axis vertical or horizontal is represented by Jones' matrix.

$$\begin{bmatrix} 1 & 0 \\ 0 & i \end{bmatrix}$$

or

$$\begin{bmatrix} 1 & 0 \\ 0 & -i \end{bmatrix}$$

$$\therefore \quad \begin{bmatrix} 1 & 0 \\ 0 & i \end{bmatrix} \frac{1}{\sqrt{2}} \begin{bmatrix} 1 \\ 1 \end{bmatrix} = \frac{1}{\sqrt{2}} \begin{bmatrix} 1 \\ -i \end{bmatrix}$$ Right-handed circularly polarised light

RCP

or $$\begin{bmatrix} 1 & 0 \\ 0 & -i \end{bmatrix} \frac{1}{\sqrt{2}} \begin{bmatrix} 1 \\ 1 \end{bmatrix} = \frac{1}{\sqrt{2}} \begin{bmatrix} 1 \\ -i \end{bmatrix}$$ Left-handed circularly polarised light

LCP

Half-wave Plate or 180° - Retarder

If the relative phase difference introduced by a plate be equal to π or path retardation be $\frac{\lambda}{2}$, then the plate will be termed as a half-wave plate.

Suppose, the incident beam is linearly polarised at +45°. The incident beam is equivalent to two simple harmonic vibrations, vibrations being parallel to y and z-axis and are in same phase on incidence. But, if one component be retarded within the plate by an amount equal to 180°, (say, z-component), it must be represented by a line OB' which is opposite to OB. The resultant lies on a line at –45°, thus a 180°-retarder converts linearly polarised light at +45° to a linearly polarised light at –45°. This transformation may be carried through a matrix transformation and thus a half-wave retarder matrix can be realised. Thus,

$$\begin{bmatrix} 1 & 0 \\ 0 & -1 \end{bmatrix} \frac{1}{\sqrt{2}} \begin{bmatrix} 1 \\ 1 \end{bmatrix} = \frac{1}{\sqrt{2}} \begin{bmatrix} 1 \\ -1 \end{bmatrix}$$

hence for a linear half-wave retarder with fast axis vertical, the Jones' matrix is $\begin{bmatrix} 1 & 0 \\ 0 & -1 \end{bmatrix}$.

It is as if a 180°-retarder had reflected the incident beam's vibration direction, the mirror being supposed to be parallel to the fast or slow axis. The half-wave plates are in practice never made of calcite, because its bi-refringence is large, hence its thickness is extremely small and, therefore, it becomes fragile, usually quartz or mica is used.

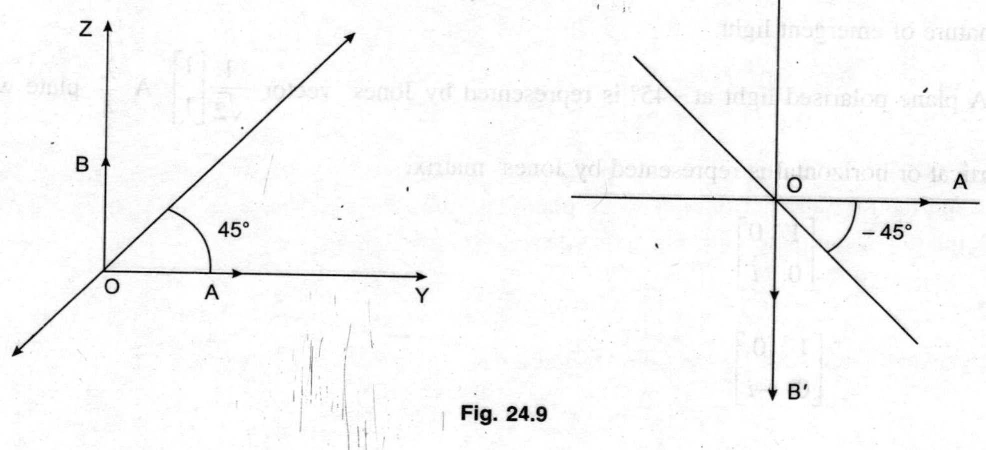

Fig. 24.9

Observation: Two retarders, one right-handed and one left-handed circularly polarised light producer, acting in succession, produce no overall change. Thus,

$$\begin{bmatrix} 1 & 0 \\ 0 & i \end{bmatrix} \frac{1}{\sqrt{2}} \begin{bmatrix} 1 \\ 1 \end{bmatrix} = \frac{1}{\sqrt{2}} \begin{bmatrix} 1 \\ i \end{bmatrix}$$

$\frac{\lambda}{4}$-plate matrix for RCP — Linearly polarised light — RCP

$$\begin{bmatrix} 1 & 0 \\ 0 & -i \end{bmatrix} \frac{1}{\sqrt{2}} \begin{bmatrix} 1 \\ i \end{bmatrix} = \frac{1}{\sqrt{2}} \begin{bmatrix} 1 \\ 1 \end{bmatrix}$$

$\frac{\lambda}{4}$-plate matrix for LCP — Linearly polarised light

Thus, $$\begin{bmatrix} 1 & 0 \\ 0 & -i \end{bmatrix} \begin{bmatrix} 1 & 0 \\ 0 & i \end{bmatrix} \frac{1}{\sqrt{2}} \begin{bmatrix} 1 \\ 1 \end{bmatrix} = \frac{1}{\sqrt{2}} \begin{bmatrix} 1 \\ 1 \end{bmatrix}$$

Operation of the two elements as taken above really yields a unitary matrix, thus producing ultimately no overall change. Thus,

$$\begin{bmatrix} 1 & 0 \\ 0 & i \end{bmatrix} \begin{bmatrix} 1 & 0 \\ 0 & -i \end{bmatrix} = \begin{bmatrix} 1 & 0 \\ 0 & 1 \end{bmatrix}$$

But, if each elements acts twice, for example, if two left circularly polarising quarter-wave plates act one after the other, producing a total phase difference of π, then we get the action of a half-wave plate, thus,

$$\begin{bmatrix} 1 & 0 \\ 0 & i \end{bmatrix} \begin{bmatrix} 1 & 0 \\ 0 & i \end{bmatrix} = \begin{bmatrix} 1 & 0 \\ 0 & -1 \end{bmatrix}$$

half-wave plate polarising element matrix

Also, $$\begin{bmatrix} 1 & 0 \\ 0 & -i \end{bmatrix} \begin{bmatrix} 1 & 0 \\ 0 & -i \end{bmatrix} = \begin{bmatrix} 1 & 0 \\ 0 & -1 \end{bmatrix}$$

Q. 11. A quarter-wave plate, fast axis horizontal is followed by a linear polariser, transmission axis at + 45° and then by another quarter-wave plate, fast axis vertical. Describe the system matrix.

Ans. A quarter-wave plate with fast axis vertical is represented by a matrix.

$$[J_1] = \begin{bmatrix} 1 & 0 \\ 0 & i \end{bmatrix}$$

A quarter-wave plate with fast axis horizontal is represented by a matrix

$$[J_2] = \begin{bmatrix} 1 & 0 \\ 0 & -i \end{bmatrix}$$

A linear polariser whose transmission axis is at +45° is

$$J_3 = \frac{1}{2}\begin{bmatrix} 1 & 1 \\ 1 & 1 \end{bmatrix}$$

∴ The system matrix is represented by,

$$[s] = \begin{bmatrix} 1 & 0 \\ 0 & i \end{bmatrix}\begin{bmatrix} 1 & 0 \\ 0 & -i \end{bmatrix}\frac{1}{2}\begin{bmatrix} 1 & 1 \\ 1 & 1 \end{bmatrix}$$

$$= \begin{bmatrix} 1 & 0 \\ 0 & 1 \end{bmatrix}\frac{1}{2}\begin{bmatrix} 1 & 1 \\ 1 & 1 \end{bmatrix}$$

$$= \frac{1}{2}\begin{bmatrix} 1 & 1 \\ 1 & 1 \end{bmatrix}$$

Thus the system matrix is the same as that of a linear polariser with axis at +45°.

Q. 12. Ice is a +ve uniaxial crystal with indices of refraction 1.309 and 1.310. How thick must the ice be to act as a quarter-wave plate for light of 600 nm wavelength.

Ans. $\mu_E - \mu_0 < 0$, – ve crystal

$\mu_E - \mu_0 > 0$, + ve crystal

∴ $\mu_0 = 1.309,\ \mu_E = 1.310$

∴ $$L = \frac{\lambda_{\text{air}}}{4(\mu_E - \mu_0)} = \frac{600\times10^{-9}}{41\times10^{-3}} = 1.5\times10^{-4}\,\text{cm} = 1.5\times10^{-1}\,\text{mm} = \cdot 15\ \text{mm}$$

Q. 13. A $\frac{\lambda}{4}$ plate with its fast axis horizontal is inserted into a beam of light linearly polarised and oscillating at 45°. Determine the polarisation of the emergent light.

Ans. $$\begin{bmatrix} 1 & 0 \\ 0 & -i \end{bmatrix}\frac{1}{\sqrt{2}}\begin{bmatrix} 1 \\ 1 \end{bmatrix} = \frac{1}{\sqrt{2}}\begin{bmatrix} 1 \\ -i \end{bmatrix}$$

LCP with the same normalised intensity.

Q. 14. What is the action of first a quarter-wave plate, then a half-wave plate; first a half-wave plate, then a quarter-wave plate.

Ans. $$\underset{\text{Half-wave plate}}{\begin{bmatrix} 1 & 0 \\ 0 & -1 \end{bmatrix}}\ \underset{\text{RCP } \frac{\lambda}{4}\text{-plate}}{\begin{bmatrix} 1 & 0 \\ 0 & i \end{bmatrix}} = \underset{\text{LCP } \frac{\lambda}{4}\text{-plate}}{\begin{bmatrix} 1 & 0 \\ 0 & -i \end{bmatrix}}$$

$$\underset{\text{RCP } \frac{\lambda}{4}\text{-plate}}{\begin{bmatrix} 1 & 0 \\ 0 & i \end{bmatrix}}\ \underset{\text{Half-wave plate}}{\begin{bmatrix} 1 & 0 \\ 0 & -1 \end{bmatrix}} = \underset{\text{LCP } \frac{\lambda}{4}\text{-plate.}}{\begin{bmatrix} 1 & 0 \\ 0 & -i \end{bmatrix}}$$

So, the result is same in the two cases.

Q. 15. What is the action of first a half-wave retarder, followed by a $\frac{\lambda}{4}$-plate, then again followed by a $\frac{\lambda}{2}$-plate.

Ans. $\begin{bmatrix} 1 & 0 \\ 0 & -1 \end{bmatrix} \quad \begin{bmatrix} 1 & 0 \\ 0 & i \end{bmatrix} \quad \begin{bmatrix} 1 & 0 \\ 0 & -1 \end{bmatrix}$

Half-wave plate RCP $\frac{\lambda}{4}$-plate Half-wave plate

$$= \begin{bmatrix} 1 & 0 \\ 0 & -1 \end{bmatrix} \quad \begin{bmatrix} 1 & 0 \\ 0 & -i \end{bmatrix}$$

Half-wave plate LCP $\frac{\lambda}{4}$-plate

$$= \begin{bmatrix} 1 & 0 \\ 0 & i \end{bmatrix}$$

RCP $\frac{\lambda}{4}$-plate.

Q. 16. A left circularly polarised beam of λ = 5893 Å is incident normally on a calcite crystal of thickness 0.005141 mm, what will be the state of polarisation of the emergent beam.

Ans. Phase difference introduced between extra-ordinary ray and ordinary ray is given by

$$L = \phi \cdot \frac{\lambda}{2\pi(\mu_E - \mu_0)}, \quad \phi = \frac{2\pi L(\mu_E - \mu_0)}{\lambda}$$

$$\mu_E = 1.4864, \quad \mu_0 = 1.6584$$

$$\therefore \quad \phi = \frac{2\pi L \times 0.172}{5893 \times 10^{-8}} = \frac{2 \times \pi \times 0.0005141 \times 0.172}{5893 \times 10^{-8}}$$

$$= 9.921 \times 10^{-6} \times 10^{8} = 3\pi$$

According to Jones' formalism

$$\begin{bmatrix} 1 & 0 \\ 0 & e^{i3\pi} \end{bmatrix} \frac{1}{\sqrt{2}} \begin{bmatrix} 1 \\ i \end{bmatrix} = \frac{1}{\sqrt{2}} \begin{bmatrix} 1 \\ ie^{i3\pi} \end{bmatrix}$$

Writing in full, the emergent light is represented by

$$\begin{bmatrix} y \\ z \end{bmatrix} = \frac{1}{\sqrt{2}} \begin{bmatrix} 1 \\ e^{i\frac{\pi}{2}} \cdot e^{i3\pi} \end{bmatrix} ae^{i\phi_x}$$

$$y = \frac{a}{\sqrt{2}}\cos\phi_x = \frac{a}{\sqrt{2}}\cos(kx - wt)$$

$$z = \frac{a}{\sqrt{2}}\cos\left(3\pi + \frac{\pi}{2} + \phi_x\right)$$

$$= \frac{a}{\sqrt{2}}\cos\left(\frac{\pi}{2} + \phi_x\right)$$

$$= -\frac{a}{\sqrt{2}}\sin\phi_x$$

$$= -\frac{a}{\sqrt{2}}\sin(kx - wt)$$

$\frac{y^2}{a^2} + \frac{z^2}{a^2} = 1$, so the emergent ray is circularly polarised. It is right-handed circularly polarised.

Elliptical Polarisation

It is a general case of polarisation under the influence of two linearly polarised light orthogonal to each other and having a phase difference. Circularly polarised light is a special case. Thus suppose, the horizontally polarised light is represented as

$$\vec{y} = \hat{j}A_1 e^{i(k_1x - wt)}$$

$$= \hat{j}A_1 e^{i\phi_x}$$

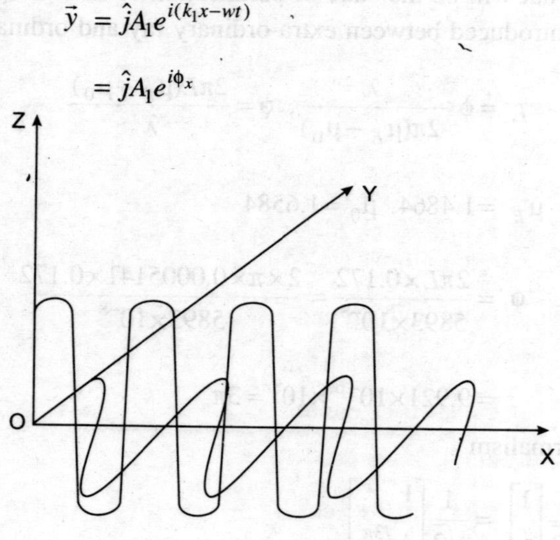

Fig. 24.10

The vertically polarised light is represented as

$$\vec{z} = \hat{k}A_2 e^{i(k_1x - wt + \phi)}$$

$$= \hat{k}A_2 e^{i\phi_x} e^{i\phi}$$

where, the disturbance is supposed to move along x-direction, the propagation constant being $k_1 = \frac{2\pi}{\lambda}$, and ϕ is the phase difference between the two orthogonal polarised disturbances. The combined displacement is written as

$$\vec{\xi} = \vec{y} + \vec{z} = \hat{j}A_1 e^{i\phi_x} + \hat{k}A_2 e^{i(\phi_x + \phi)}$$

$$= (\hat{j}A_1 + \hat{k}A_2 e^{i\phi})e^{i\phi_x}$$

$\therefore$ y-component of the resultant vibration is given by,

$$y = R.P. A_1 e^{i\phi_x} = A_1 \cos\phi_x$$

Similarly, z-component is given by

$$z = R.P. A_2 e^{i(\phi_x + \phi)} = A_2 \cos(\phi_x + \phi)$$

Since ϕ_x is a function of time, as time passes on, both these components change, but there is a phase difference ϕ between these two components.

From the two relations, we get

$$\frac{y^2}{A_1^2} + \frac{z^2}{A_2^2} - \frac{2yz}{A_1 A_2}\cos\phi = \sin^2\phi \qquad ...(1)$$

Equation (1) is a general equation of an ellipse, so light under this condition is said to be elliptically polarised. For different values of ϕ, we would get either linearly polarised or elliptically polarised light, right-handed or left-handed.

The general Eqn. (1) can be thrown into the form

$$Ay^2 + Bz^2 + 2hyz = 1 \qquad ...(2)$$

where we have written

$$A = \frac{1}{A_1^2 \sin^2\phi}, B = \frac{1}{A_2^2 \sin^2\phi}, h = \frac{-\cos\phi}{A_1 A_2 \sin^2\phi}$$

Rotate the axis of co-ordinates through an angle θ such that the new co-ordinates are given by,

$$y = y\cos\theta - z\sin\theta$$

$$z = y\sin\theta + z\cos\theta$$

Putting the values of y, z in Eqn. (2), we get

$$A(y\cos\theta - z\sin\theta)^2 + B(y\sin\theta + z\cos\theta)^2 + 2h(y\cos\theta - z\sin\theta)(y\sin\theta + z\cos\theta) = 1$$

$$\Rightarrow y^2(A\cos^2\theta + 2h\cos\theta\sin\theta + B\sin^2\theta) + z^2(A\sin^2\theta - 2h\sin\theta\cos\theta + B\cos^2\theta)$$

$$+2yz\{h(\cos^2\theta - \sin^2\theta) - (A - B)\sin\theta\cos\theta\} = 1$$

Choose the value of θ such that the term containing the product yz vanishes.

So the choice is given by,

$$h(\cos^2\theta - \sin^2\theta) = (A-B)\sin\theta\cos\theta \Rightarrow \tan 2\theta = \frac{2h}{(A-B)} \quad \text{...(3)}$$

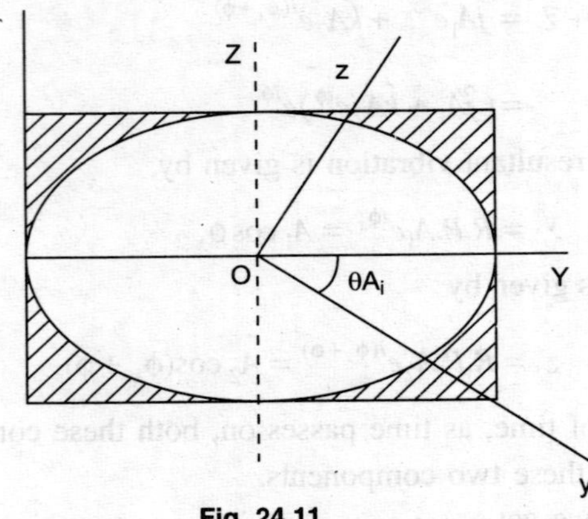

Fig. 24.11

The Eqn. (2) when plotted results in an elliptic curve as shown in Fig. 24.11, the angle between the co-ordinate axes and those of the ellipse being θ defined by Eqn. (3).

Depending on the value of θ, we would get different types of polarised light.

Case I: Linearly polarised light: When $\phi = s\pi$, the Eqn. (1) can be written as

$$\left[\frac{y}{A_1} - (-1)^s \frac{z}{A_2}\right]^2 = 0 \Rightarrow \frac{y}{A_1} = (-1)^s \frac{z}{A_2} \quad \text{...(4)}$$

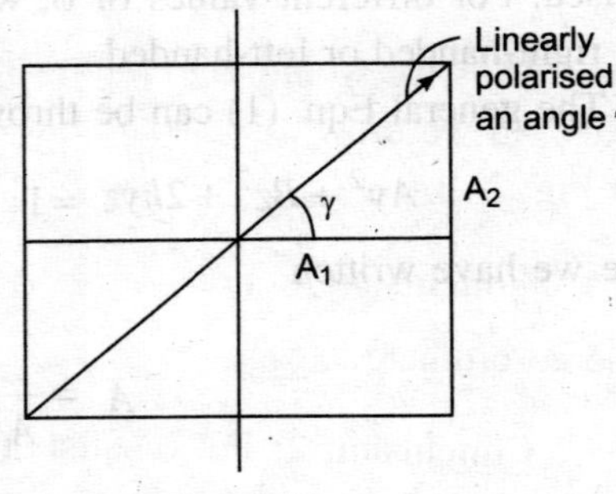

Fig. 24.12

So, when

$$s = 0, 2, 4 \Rightarrow \phi = 0, 2\pi, 4\pi \text{ etc.}$$

$$z = \frac{A_2}{A_1} y \quad \text{...(5)}$$

Which is a linearly polarised light with orientation γ with respect to *y*-direction given by

$$\tan\gamma = \frac{A_2}{A_1} \quad \text{...(6)}$$

This is represented as in Fig 24.12.

If on the other hand, s = 1, 3, 5, etc. ⇒ $\phi = \pi, 3\pi, 5\pi$ etc.

we get

$$z = -\frac{A_2}{A_1} y$$

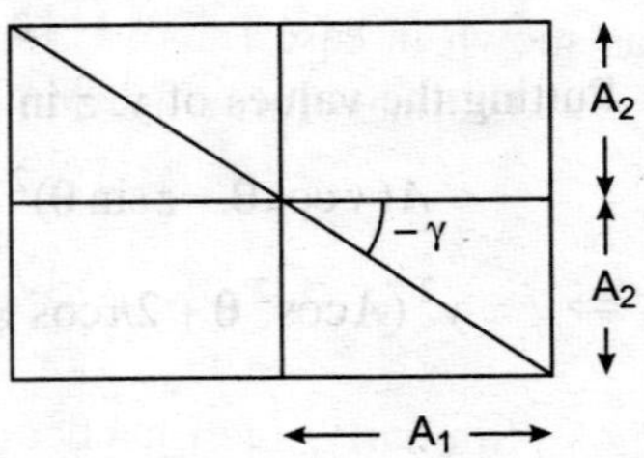

Fig. 24.13

a linearly polarised light at an angle – γ w.r.t. the *y*-axis and it is shown in Fig. 24.13.

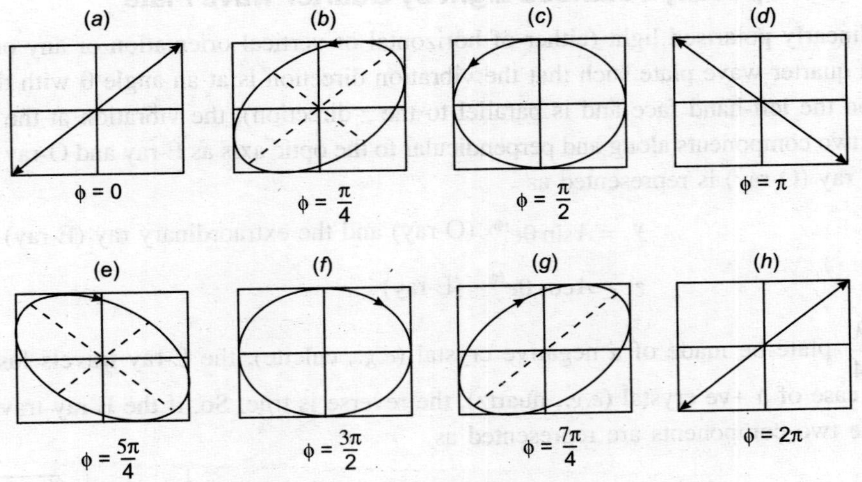

Fig. 24.14

Case II

When $\phi = (2s+1)\frac{\pi}{2}$, $s = 0, 1, 2$ etc., we would get, Eqn. (1) as

$$\frac{y}{A_1^2} + \frac{z}{A_2^2} = 1 \qquad \text{...(7)}$$

Thus we would get elliptically polarised light whose axes would coincide with the co-ordinate axes. So we draw a sequence of conclusions diagrammatically represented by a series of figures as in Fig. 24.14.

Conclusion 1: For $\phi = (2s+1)\frac{\pi}{2}$, we would get elliptically polarised light with its axes along the co-ordinate axes.

Conclusion 2: For ϕ satisfying the condition $\phi = 2s \cdot \frac{\pi}{2}$, we would get linearly polarised light. The sequence of formation of polarised light is shown.

Equation (1) is a general equation for an ellipse, which reduces to standard form when $\phi = \frac{\pi}{2}$ as shown in case I above.

In Jones' formalism, the elliptically polarised light is represented as

$$[J]_{REP} = \begin{bmatrix} A_1 \\ A_2 e^{i\phi} \end{bmatrix} \to \begin{bmatrix} A_1 \\ iA_2 \end{bmatrix} \text{ Right elliptically polarised } (REP)$$

Similarly, a left-handed elliptically polarised light is represented as

$$[J_{LEP}] = \begin{bmatrix} A_1 \\ -iA_2 \end{bmatrix}$$

Production of Elliptically Polarised Light by Quarter-wave Plate

When linearly polarised light (either of horizontal or vertical orientation or any orientation) is incident on a quarter-wave plate such that the vibration direction is at an angle θ with the optic axis (which lies on the left-hand face and is parallel to the *z*-direction), the vibration at the entry-end is resolved into two components along and perpendicular to the optic axis as E-ray and O-ray respectively. The ordinary ray (O-ray) is represented as

$$y = A\sin\theta e^{i\phi_x} \text{ (O-ray) and the extraordinary ray (E-ray) as}$$

$$z = A\cos\theta e^{i\phi_x} \text{ (E-ray)}$$

If the $\frac{\lambda}{4}$-plate be made of a negative crystal (*e.g.*, calcite), the E-ray travels faster than the O-ray and in case of a +ve crystal (*e.g.*, quartz), the reverse is true. So, if the E-ray travels faster, at emergence the two components are represented as

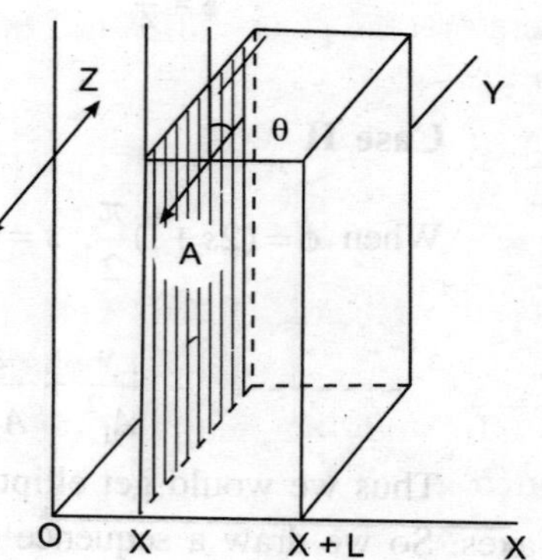

Fig. 24.15

$$y = A\sin\theta e^{i\phi_x+L} = A_1 e^{i\phi_x+L}$$

$$z = A\cos\theta e^{i\phi_x+L} e^{i\phi} = A_2 e^{i\phi_x+L} e^{i\phi}$$

where L is the thickness of the plate.

In Jones' vector notation, the elliptically polarised light is represented as

$$\begin{bmatrix} A\sin\theta \\ A\cos\theta e^{i\phi} \end{bmatrix} \rightarrow \begin{bmatrix} A_1 \\ A_2 e^{i\phi} \end{bmatrix}$$

The elliptic polariser can be represented as

$$[J_e] = \begin{bmatrix} A_1 & 0 \\ 0 & A_2 e^{i\phi} \end{bmatrix}$$

Thus, when a circularly polarised light is incident on an elliptic polariser, the emergent light nature is available through matrix operation as

$$\begin{bmatrix} A_1 & 0 \\ 0 & A_2 e^{i\phi} \end{bmatrix} \cdot \begin{bmatrix} 1 \\ i \end{bmatrix} = \begin{bmatrix} A_1 \\ iA_2 e^{i\phi} \end{bmatrix} = \begin{bmatrix} A_1 \\ A_2 e^{i(\phi+\frac{\pi}{2})} \end{bmatrix}$$

Elliptic polariser — Circularly polarised light — Elliptically polarised light (right-handed)

The quarter-wave plate introduces a phase difference of $\frac{\pi}{2}$ between the orthogonal components, hence quarter-wave matrix producing elliptically polarised light is represented as

$$\begin{bmatrix} A_1 & 0 \\ 0 & A_2 e^{i\frac{\pi}{2}} \end{bmatrix} \rightarrow \begin{bmatrix} A_1 & 0 \\ 0 & A_2 \end{bmatrix} \text{ (Right-handed)}$$

Similarly $\begin{bmatrix} A_1 & 0 \\ 0 & iA_2 \end{bmatrix}\begin{bmatrix} a \\ b \end{bmatrix} = \begin{bmatrix} A_1 & a \\ ib & A_2 \end{bmatrix} = \begin{bmatrix} B_1 \\ iB_2 \end{bmatrix}$

Elliptically polarised

Q. 17. A left circularly polarised light of $\lambda = 5893$ Å is incident on a quarter crystal (with its optic axis parallel to the surface) of thickness 0.025 mm. Determine the state of polarisation of the emergent beam. ($\mu_0 = 1.54425, \mu_E = 1.55336$).

Ans. This type of retarder is a +ve retarder, where the extraordinary ray travels slower compared to the O-ray. So after emergence, the extraordinary ray lags behind the O-ray, thus z-component (parallel to the optic axis) will lag in phase. The retarder matrix here can be represented as

$$\begin{bmatrix} 1 & 0 \\ 0 & e^{-i\phi} \end{bmatrix}$$

where ϕ is the phase difference introduced in the path of traverse.

$$\therefore \quad \begin{bmatrix} 1 & 0 \\ 0 & e^{-i\phi} \end{bmatrix} \frac{1}{\sqrt{2}} \begin{bmatrix} 1 \\ -i \end{bmatrix} = \frac{1}{\sqrt{2}} \begin{bmatrix} 1 \\ e^{-i\phi} \end{bmatrix} = \frac{1}{\sqrt{2}} \begin{bmatrix} 1 \\ e^{-i\left(\frac{\pi}{2}+\phi\right)} \end{bmatrix}$$

Retarder matrix Left circular polarised light Emergent polarised light

To study the nature of the light, we take the full form of the emergent polarised light matrix written as

$$\begin{bmatrix} y \\ z \end{bmatrix} = \frac{1}{\sqrt{2}} \begin{bmatrix} 1 \\ -i\, e^{-i\phi} \end{bmatrix} e^{i\phi_x + L} = \frac{1}{\sqrt{2}} \begin{bmatrix} 1 \\ e^{-i\left(\frac{\pi}{2}+\phi\right)} \end{bmatrix} e^{i\phi_x + L}$$

$$\Rightarrow \quad y = \frac{A}{\sqrt{2}} \cos(\phi_x + L) = \frac{A}{\sqrt{2}} \cos\{k_1(x+L) - wt\} \qquad \text{...(8)}$$

$$z = \frac{A}{\sqrt{2}} \cos\{k_1(x+L) - wt - \frac{\pi}{2} - \phi\} = \frac{A}{\sqrt{2}} \sin\{k_1(x+L) - wt - \phi\} \qquad \text{...(9)}$$

Let us now see how do y and z change with time.

At the instant when $k_1(x+L) - wt = \frac{\pi}{2}$, we get

$$y = 0,\ z = \frac{A}{\sqrt{2}} \sin\left(\frac{\pi}{2} + \phi\right) = \frac{A}{\sqrt{2}} \cos\phi$$

These two quantities are represented as in Fig. 24.17.

As time passes on and at the instant when

$$k_1(x+L) - wt + \phi = \frac{\pi}{2}$$

$$\Rightarrow \qquad wt = k_1(x+L)+\phi-\frac{\pi}{2}$$

we get

$$y = \frac{A}{\sqrt{2}}\cos\left(\frac{\pi}{2}-\phi\right)$$

$$= \frac{A}{\sqrt{2}}\sin\phi$$

$$z = 0$$

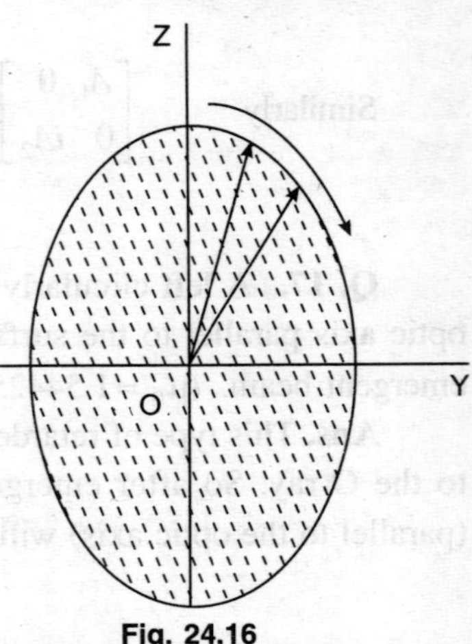

Fig. 24.16

Thus, the tip of the displacement vector lies on an ellipse, left-handed. So such a wave plate acts so as to produce a left-handed elliptically polarised light.

24.4 BABINET'S COMPENSATOR

It is a very important and useful device for production and analysis of different types of polarised light.

To make a Babinet's compensator, two thin right-angled prism ABD and BDC made of quartz are joined along the hypotenuse side so as to form a rectangular block. The optic axis in the first prism lies parallel to the face of the prism and the optic axis in the second lies perpendicular to the face of the second. Thus the two optic axes are mutually at right angles. There is only one direction such as EF along which a ray traverse equal thickness of the prisms. This may be taken as the symmetry axes of the compensator and for any other ray either towards the left or towards the right of the symmetry axies, path in one increases as the path in the other decreases (or vice versa), rays like GH or IJ. Since the optic axes of the two prisms are orthogonal so the vibrations parallel and perpendicular to the optic axis of the first prism become respectively perpendicular and parallel to the optic axis of the second.

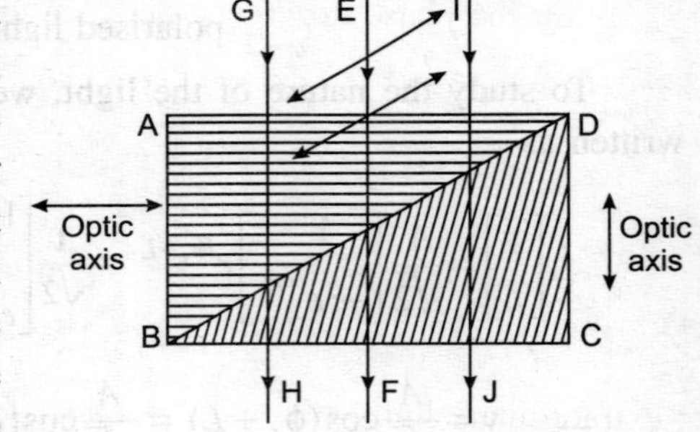

Fig. 24.17

Action of the Babinet's Compensator

Plane polarised light with the axis of orientation making an angle θ with the optic axis of the first prism is allowed to fall and pass through the compensator. On incidence this vibration is broken up into two components-one, parallel to the optic axis, called extraordinary ray (E-ray) and the other, perpendicular to the optic axis, called the ordinary ray (O-ray).

Consider the ray IJ not along the symmetry axis such that the thickness traversed in the first prism is d_1 and that in the second prism, d_2. Let the plane polarised light be incident normally on the compensator with the direction of vibration making an angle θ with the optic axis of the prism ABD. The vibration is taken to have amplitude equal to a (say). Then resolving along and perpendicular to the optic axis (taken along the y-axis), we can write,

$$y = a\cos\theta = A_1\text{ ; E-ray}$$

$$z = a \sin\theta = A_2 \text{ ; O-ray}$$

After its passage through the first prism of thickness d_1, the extraordinary ray (whose vibration is parallel to the y-direction) can be represented as

$$y = R.P.A_1 e^{i\phi_2} \text{ (E-ray)}$$

and, the ordinary ray can be represented as

$$z = R.P.A_2 e^{i\phi_1} \text{ (O-ray)}$$

Since the compensator is made of quartz, it is a +ve crystal and hence O-ray travels faster and so it will lead in phase.

∴ phase difference introduced is

$$\delta_1 = \phi_1 - \phi_2 = (k_1 - k_2)d_1 \quad \text{...(24.4 } a\text{)}$$

k_1 for O-ray and k_2 for E-ray.

In the second prism, the ordinary ray and extraordinary interchange their character and so after emergence the y-component and z-component can be represented as.

$$y_0 = R.P.ye^{i\alpha_2} = R.P.A_1 e^{i(\phi_2+\alpha_2)} \quad \text{...(24.4 } b\text{)}$$

$$z_0 = R.P.ze^{i\alpha_1} = R.P.A_2 e^{i(\phi_1+\alpha_1)}$$

where $(\alpha_2 - \alpha_1) = (k_1 - k_2)d_2$.

So the phase difference between emerging components is given by,

$$\delta = (\phi_2 + \alpha_2) - (\phi_1 + \alpha_1) = (\alpha_2 - \alpha_1) + (\phi_2 - \phi_1)$$

$$= (k_1 - k_2)d_2 - (k_1 - k_2)d_1 = (d_2 - d_1)(k_1 - k_2) \quad \text{...(24.4 } c\text{)}$$

Now, k_1 = propagation constant for ordinary light $= \dfrac{2\pi}{\lambda_0}$, k_2 = propagation constant for extraordinary light $= \dfrac{2\pi}{\lambda_E}$.

Also, $\mu_0 = \dfrac{c}{v_0} = \dfrac{\lambda}{\lambda_0}$, λ = wavelength in air

$$\therefore \quad k_1 = \frac{2\pi\mu_0}{\lambda}, \text{ similarly, } k_2 = \frac{2\pi\mu_e}{\lambda}$$

$$\therefore \quad \delta = (d_2 - d_1)\frac{2\pi}{\lambda}(\mu_0 - \mu_E)$$

$$= \frac{2\pi}{\lambda}(d_2 - d_1)(\mu_0 - \mu_E) \quad \text{...(24.4 } d\text{)}$$

Jones' Matrix Formalism for Babinet's Compensator

The above optical process which ultimately leads to emergent light from the compensator may be very simply obtained through Jones' matrix for retarder.

The first prism may be looked upon as a retarder plate with fast axis horizontal whose matrix representation according to Jones is

$$J_1 = \begin{bmatrix} 1 & 0 \\ 0 & e^{-i\delta_1} \end{bmatrix}$$

The second prism where the fast axis is vertical is represented by Jones' matrix.

$$J_2 = \begin{bmatrix} 1 & 0 \\ 0 & e^{i\delta_2} \end{bmatrix}$$

The plane polarised light whose vibration or orientation is at an angle θ with the horizontal is represented as $\begin{bmatrix} A \\ B \end{bmatrix}$ where $\dfrac{B}{A} = \tan\theta$.

$$\therefore \quad \begin{bmatrix} 1 & 0 \\ 0 & e^{i\delta_2} \end{bmatrix}\begin{bmatrix} 1 & 0 \\ 0 & e^{-i\delta_1} \end{bmatrix}\begin{bmatrix} A \\ B \end{bmatrix} = \begin{bmatrix} 1 & 0 \\ 0 & e^{i\delta_2} \end{bmatrix}\begin{bmatrix} A \\ Be^{-i\delta_1} \end{bmatrix} = \begin{bmatrix} A \\ Be^{i(\delta_2-\delta_1)} \end{bmatrix} \qquad \text{...(24.4 } e\text{)}$$

In the present case, $A = a\cos\theta, B = a\sin\theta$.

Thus from Eqn. (24.4 *e*) a few important conclusions can be arrive.

Conclusion 1: When $\delta = (\delta_2 - \delta_1) = 0$, this will be true only for symmetric passage where $d_2 = d_1$.

Then the emergent light matrix is given by

$$J_e = \begin{bmatrix} A \\ B \end{bmatrix} = \begin{bmatrix} a\cos\theta \\ a\sin\theta \end{bmatrix}$$

It represents linearly polarised light with orientation parallel to the incident orientation, Figure 24.19.

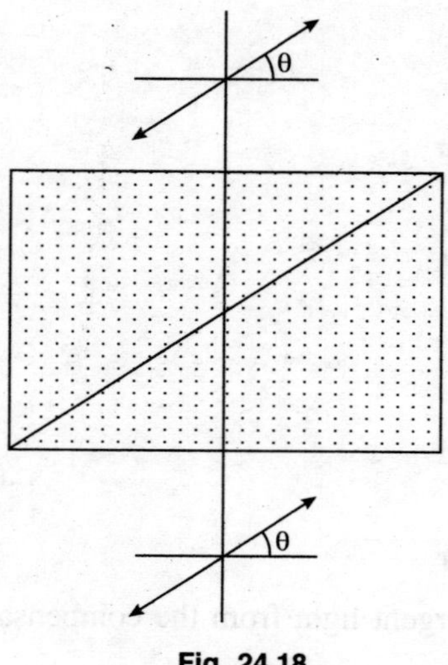

Fig. 24.18

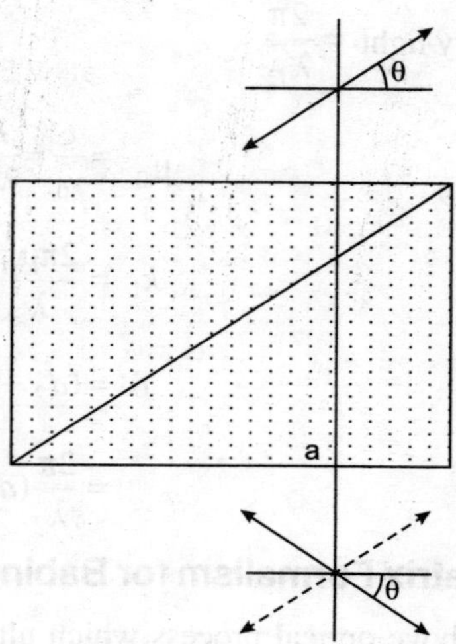

Fig. 24.19

Conclusion 2: If we move out of the symmetric line, the thickness traversed in the two prisms differ and hence path difference and as a result phase difference is produced. For the departure of a certain distance a from the symmetric line, if the phase difference introduced between the two components be π, then

$$\delta = (\delta_2 - \delta_1) = \pi.$$

In this case the Babinet's compensator will act as a $\frac{\lambda}{2}$-plate whose Jones' matrix is represented as

$$J_{\frac{\lambda}{2}} = \begin{bmatrix} 1 & 0 \\ 0 & -1 \end{bmatrix}$$

So the transformation of the incident light into emergent light is carried through

$$\begin{bmatrix} 1 & 0 \\ 0 & -1 \end{bmatrix} \begin{bmatrix} a\cos\theta \\ a\sin\theta \end{bmatrix} = \underbrace{\begin{bmatrix} a\cos\theta \\ -a\sin\theta \end{bmatrix}}_{\text{Emergent light}} \qquad \text{...(24.4 } f\text{)}$$

Equation (24.4 f) shows that in this case, the emergent vibration is also plane polarised with the orientation given by,

$$\tan\phi = \frac{-a\sin\theta}{a\cos\theta} = -\tan\theta \Rightarrow \phi = (\pi - \theta) \qquad \text{...(24.4 } g\text{)}$$

Equation (24.4 g) suggests that the Babinet's compensator in this case produces linear vibration but direction of vibration rotates through 2θ w.r.t. the incident vibration direction (Fig. 24.20).

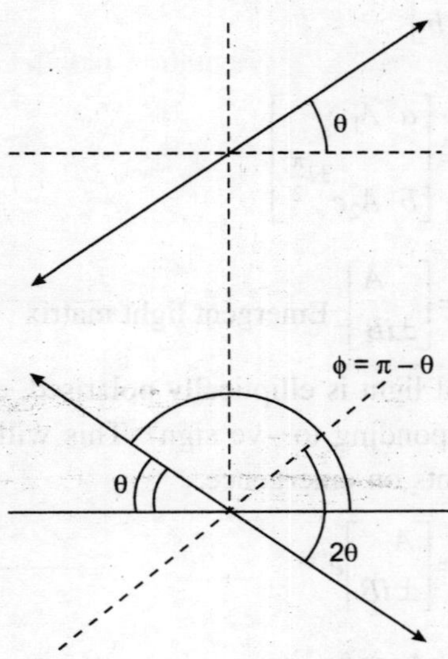

Fig. 24.20

Conclusion 3: Circularly polarised light

If $\theta = 45^\circ, \delta_2 - \delta_1 = \frac{\pi}{2}$, the Babinet's compensator acts as a $\frac{\lambda}{4}$-plate, whose matrix representation is given by

$$J_{\frac{\lambda}{4}} = \begin{bmatrix} 1 & 0 \\ 0 & e^{\pm i\frac{\pi}{2}} \end{bmatrix} = \begin{bmatrix} 1 & 0 \\ 0 & \pm i \end{bmatrix}$$

The nature of emergent light in obtained through matrix transformation as

$$\begin{bmatrix} 1 & 0 \\ 0 & \pm i \end{bmatrix} \frac{1}{\sqrt{2}} \begin{bmatrix} 1 \\ 1 \end{bmatrix} = \frac{1}{\sqrt{2}} \begin{bmatrix} 1 \\ \pm i \end{bmatrix}$$

either a RCP or LCP emergent light.

Conclusion 4: Elliptically polarised light.

If $\theta \neq 45^\circ$, but $\delta = (\delta_2 - \delta_1) = \frac{\pi}{2}$, then the Babinet's compensator acts as retarder where matrix representation is

$$J = \begin{bmatrix} A_1 & 0 \\ 0 & A_2 e^{\pm i\frac{\pi}{2}} \end{bmatrix}$$

∴ The nature of emergent light is given as

$$\begin{bmatrix} A_1 & 0 \\ 0 & A_2 e^{\pm i\frac{\pi}{2}} \end{bmatrix} \begin{bmatrix} a \\ b \end{bmatrix}$$

$$= \begin{bmatrix} a\ A_1 \\ b\ A_2 e^{\pm i\frac{\pi}{2}} \end{bmatrix}$$

$$= \begin{bmatrix} A \\ \pm iB \end{bmatrix} \text{ Emergent light matrix}$$

Thus in this case the emergent light is elliptically polarised, either right-handed (corresponding to +ve sign) or left-handed (corresponding to –ve sign). This will be more evident if we write out the full expression of the components on emergence.

$$\begin{bmatrix} y \\ z \end{bmatrix} = \begin{bmatrix} A \\ \pm iB \end{bmatrix} e^{i\phi_x}$$

$$\Rightarrow \quad y = A \cos \phi_x$$

$$z = B \cos(\phi_x \pm \frac{\pi}{2}) = B \cos\left(\frac{\pi}{2} \pm \phi_x\right) = \pm B \sin \phi_x$$

$$\therefore \qquad \frac{y^2}{A^2}+\frac{z^2}{B^2}=1\text{, an ellipse}$$

Thus using Babinet's compensator under specified condition, linearly polarised light, circularly polarised light or elliptically polarised light may be produced.

Production of Linearly Polarised Light, Circularly Polarised Light and Elliptically Polarised Light (Fresnel Rhomb):

Fresnel rhomb is a parallelopiped of glass whose opposite acute angles are nearly 55°(more precisely it is 54° 37′) consider a plane polarised light with transmission axis at an angle θ with the plane of incidence. The amplitude of incident, vibration being taken as *a*, it is resolved into two components $a\cos\theta$ and $a\sin\theta$ respectively parallel and perpendicular to the plane of incidence. At each of the points of reflection *R* and *S*, the angle of incidence is greater than the critical angle for glass ($\mu=1.5$) and air interface. At each reflection, there will be a phase difference of $\frac{\pi}{4}$ introduced between the two rectangular components. So Jones' matrix operation for a Fresnel rhomb for two reflections can be represented as

$$\begin{bmatrix}1 & 0\\ 0 & e^{i\frac{\pi}{4}}\end{bmatrix}\begin{bmatrix}1 & 0\\ 0 & e^{i\frac{\pi}{4}}\end{bmatrix}$$

Fig. 24.21

So the emergent light from a Fresnel rhomb, considering two internal reflections can be obtained as

$$\begin{bmatrix}1 & 0\\ 0 & e^{i\frac{\pi}{4}}\end{bmatrix}\begin{bmatrix}1 & 0\\ 0 & e^{i\frac{\pi}{4}}\end{bmatrix}\begin{bmatrix}A_1\\ A_2\end{bmatrix} \quad \text{where } A_1=a\cos\theta, A_2=a\sin\theta$$

So,
$$\begin{bmatrix}1 & 0\\ 0 & e^{i\frac{\pi}{4}}\end{bmatrix}\begin{bmatrix}1 & 0\\ 0 & e^{i\frac{\pi}{4}}\end{bmatrix}\begin{bmatrix}A_1\\ A_2\end{bmatrix}$$

$$=\begin{bmatrix}1 & 0\\ 0 & e^{i\frac{\pi}{4}}\end{bmatrix}\begin{bmatrix}A_1\\ A_2e^{i\frac{\pi}{4}}\end{bmatrix}=\begin{bmatrix}A_1\\ A_2e^{i\frac{\pi}{2}}\end{bmatrix} \text{ Emergent light}$$

If the parallel component be taken along y-axis and perpendicular component be taken along the z-direction, writing out the full forms.

$$\begin{bmatrix} y \\ z \end{bmatrix} = \begin{bmatrix} A_1 \\ A_2 e^{i\frac{\pi}{2}} \end{bmatrix} e^{i\phi_x}$$

$$\Rightarrow \qquad y = R.P.A_1 e^{i\phi_x} = A_1 \cos\phi_x = A_1 \cos(kx - wt)$$

$$z = R.P.A_2 e^{i(\phi_x + \frac{\pi}{2})} = -A_2 \sin(kx - wt)$$

$$\therefore \quad \frac{y^2}{A_1^2} + \frac{z^2}{A_2^2} = 1,$$

which shows that it is an elliptically polarised light.

If, however, $\theta = 45°$, then $A_1 = A_2$ and we would get circularly polarised light. But if $\theta \neq 45°$, we would get, in general, elliptically polarised light.

Again, if we consider four internal reflections, then the emergent light would be obtained as

$$\begin{bmatrix} 1 & 0 \\ 0 & e^{i\frac{\pi}{4}} \end{bmatrix} \begin{bmatrix} 1 & 0 \\ 0 & e^{i\frac{\pi}{4}} \end{bmatrix} \begin{bmatrix} 1 & 0 \\ 0 & e^{i\frac{\pi}{4}} \end{bmatrix} \begin{bmatrix} 1 & 0 \\ 0 & e^{i\frac{\pi}{4}} \end{bmatrix} \begin{bmatrix} A_1 \\ A_2 \end{bmatrix}$$

$$= \begin{bmatrix} 1 & 0 \\ 0 & e^{i\frac{\pi}{2}} \end{bmatrix} \begin{bmatrix} 1 & 0 \\ 0 & e^{i\frac{\pi}{4}} \end{bmatrix} \begin{bmatrix} A_1 \\ A_2 \end{bmatrix}$$

$$= \begin{bmatrix} 1 & 0 \\ 0 & e^{i\pi} \end{bmatrix} \begin{bmatrix} A_1 \\ A_2 \end{bmatrix} = \begin{bmatrix} A_1 \\ A_2 e^{i\pi} \end{bmatrix} = \begin{bmatrix} A_1 \\ -A_2 \end{bmatrix}$$

So the emergent light is plane polarised, making an angle given by,

$$\tan\phi = \frac{-A_2}{A_1}$$

So, if $\theta \neq 45°$, we would get either elliptically polarised or under certain conditions, linearly polarised light.

Analysis of Polarised Light

According to stokes, there may be seven possible states of polarisation of a beam of light and they are:

(*a*) Unpolarised

(*b*) Linearly polarised (or plane polarised)

(*c*) Circularly polarised

(*d*) Elliptically polarised

(*e*) Mixture of unpolarised and linearly polarised

(*f*) Mixture of unpolarised and circularly polarised

(*g*) Mixture of unpolarised and elliptically polarised.

The device by which analysis of a given light can be done is called an analyser. Such an analyser may be a Nicol Prism (for linearly polarised light); Quarter-wave plate, Babinet's compensator or Fresnel rhomb.

The scheme of analysis is given below.

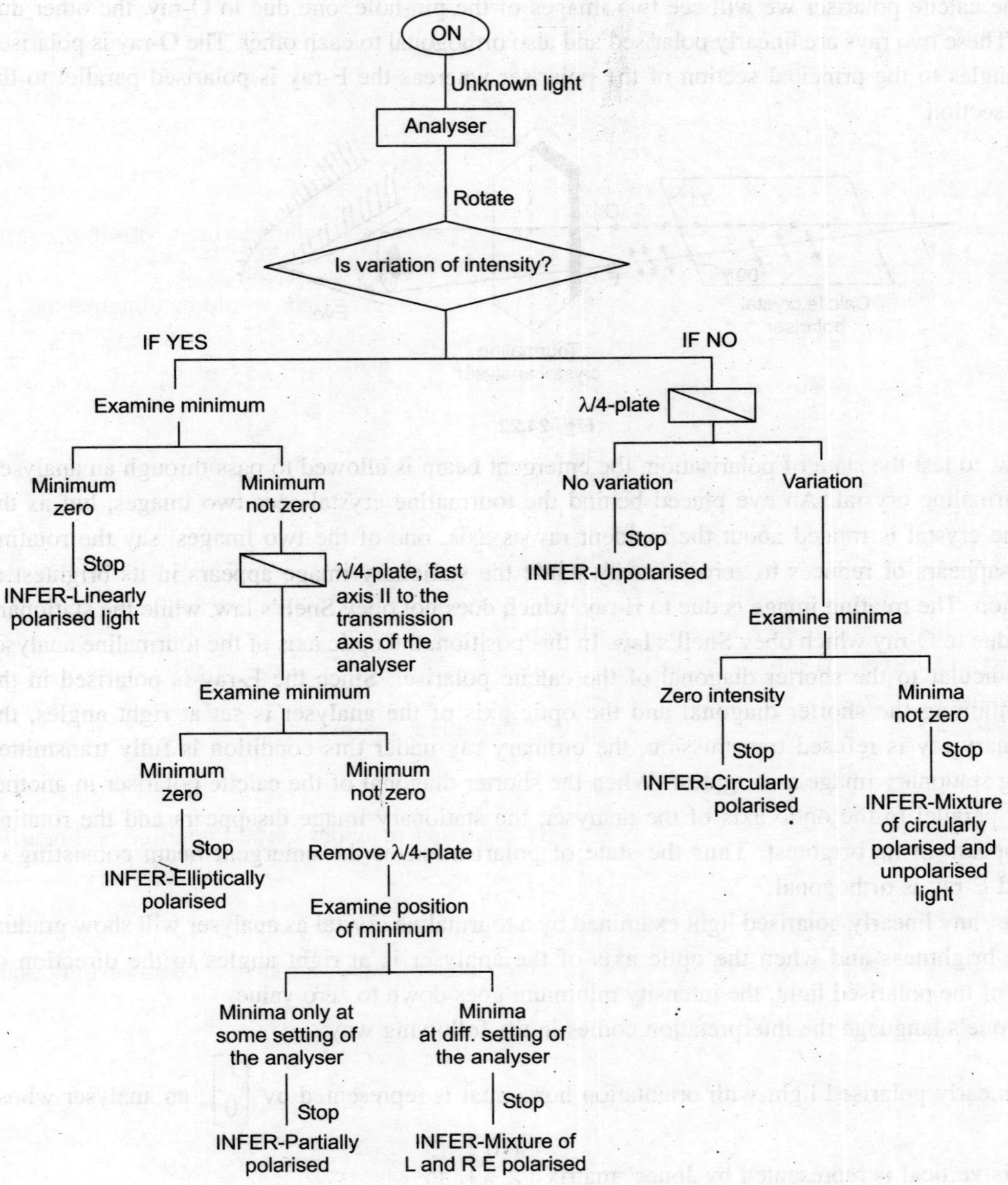

Analysis of Linearly Polarised Light

If the given light be plane polarised, then a tourmaline crystal acting as an analyser is placed in the path of the beam and it is rotated with the beam as axis. If the intensity of light varies and minimum goes down to zero value, we conclude that the light under consideration is linearly polarised.

The calcite crystal by double refraction produce two polarised rays, one ordinary ray (O-ray), and the other extraordinary ray (E-ray). If the incident ray passes through a pin-hole, on the other side of the calcite polariser we will see two images of the pin-hole, one due to O-ray, the other due to E-ray. These two rays are linearly polarised and also orthogonal to each other. The O-ray is polarised at right angles to the principal section of the polariser whereas the E-ray is polarised parallel to the principal section.

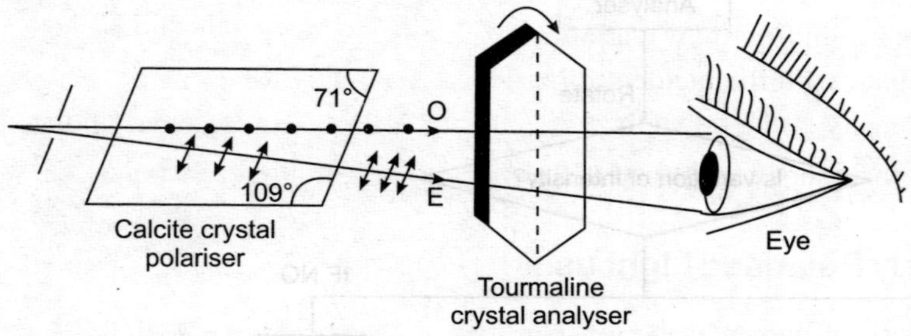

Fig. 24.22

Now, to test the state of polarisation, the emergent beam is allowed to pass through an analyser, say a tourmaline crystal. An eye placed behind the tourmaline crystal sees two images, but as the tourmaline crystal is rotated about the incident ray as axis, one of the two images, say the rotating image disappears or reduces to zero intensity while the stationary image appears in its brightest at one position. The rotating image is due to E-ray, which does not obey Snell's law, while the stationary image is due to O-ray which obey Snell's law. In this position, the optic axis of the tourmaline analyser is perpendicular to the shorter diagonal of the calcite polariser. Since the E-ray is polarised in the plane containing the shorter diagonal and the optic axis of the analyser is set at right angles, the extraordinary ray is refused transmission, the ordinary ray under this condition is fully transmitted producing stationary image as brightest. When the shorter diagonal of the calcite polariser in another setting is parallel to the optic axis of the analyser, the stationary image disappears and the rotating image appears in its brightest. Thus the state of polarisation of the emergent beam consisting of O-ray and E-ray is orthogonal.

Thus, any linearly polarised light examined by a tourmaline crystal as analyser will show gradual fading in brightness and when the optic axis of the analyser is at right angles to the direction of vibration of the polarised light, the intensity minimum goes down to zero value.

In Jone's language the interpretation comes in the following way:

A linearly polarised light with orientation horizontal is represented by $\begin{bmatrix} 1 \\ 0 \end{bmatrix}$, an analyser whose fast axis is vertical is represented by Jones' matrix $\begin{bmatrix} 1 & 0 \\ 0 & 0 \end{bmatrix}$, so

$$\begin{bmatrix}1 & 0\\0 & 0\end{bmatrix}\begin{bmatrix}1\\0\end{bmatrix}=\begin{bmatrix}0\\0\end{bmatrix} \text{ Emergent light matrix}$$

So, the intensity of the emergent light is zero, but if the intensity does not go down to zero value, the given light is not definitely linearly polarised. Thus,

$$\begin{bmatrix}1 & 0\\0 & 0\end{bmatrix}\frac{1}{\sqrt{2}}\begin{bmatrix}1\\e^{i\frac{\pi}{2}+\theta}\end{bmatrix}=\frac{1}{\sqrt{2}}\begin{bmatrix}1\\0\end{bmatrix}$$

So the intensity of the emergent light is not zero under the same condition of the analyser.

So our conclusion is:

If the given light is found to show variation of intensity when an analyser (say a tourmaline crystal) is rotated in its path with the incident direction as axis, then for

(*i*) minimum intensity going down to zero value, the given light is linearly polarised.

(*ii*) minimum intensity not equal to zero, the given light is not linearly polarised.

Analysis of Circularly Polarised Light using a $\frac{\lambda}{4}$-plate

The circularly polarised light may be left-handed (LCP) or right-handed (RCP), in the former, the resultant displacement vector-tip rotates in the clockwise manner over a circle whereas in the latter, the vector-tip rotates in the anti-clockwise sense.

If the light disturbance be assumed to travel in the x-direction, and two orthogonal vibrations polarised in the y and z directions differ in phase by $\frac{\pi}{2}$, then the resultant is a circularly polarised light. If the z-component leads in phase over the y-component by the said phase, the circularly polarised light is clockwise or left circularly polarised (LCP) if it lags it is anti-clockwise or right circularly polarised (RCP). Thus in the Jones' formalism the two orthogonal polarised lights are represented as

$$\begin{bmatrix}1\\0\end{bmatrix} \rightarrow \text{horizontally polarised (}y\text{-vibration)}$$

$$\begin{bmatrix}0\\e^{\pm i\frac{\pi}{2}}\end{bmatrix} \rightarrow \text{vertically polarised, but with a phase difference, } \frac{\pi}{2} \text{ (}z\text{-vibration)}$$

Thus,

$$\frac{1}{\sqrt{2}}\begin{bmatrix}1\\0\end{bmatrix}+\frac{1}{\sqrt{2}}\begin{bmatrix}0\\e^{i\frac{\pi}{2}}\end{bmatrix}=\frac{1}{\sqrt{2}}\begin{bmatrix}1\\i\end{bmatrix}; \text{RCP}$$

$$\frac{1}{\sqrt{2}}\begin{bmatrix}1\\0\end{bmatrix}+\frac{1}{\sqrt{2}}\begin{bmatrix}0\\e^{-i\frac{\pi}{2}}\end{bmatrix}=\frac{1}{\sqrt{2}}\begin{bmatrix}1\\-i\end{bmatrix}; \text{LCP}$$

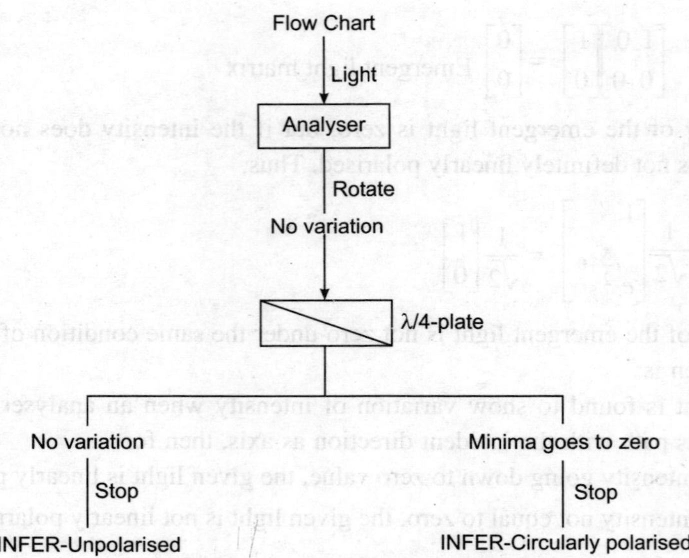

The reverse of this process of synthesis is also true, in other words, it means, an RCP or LCP can be analysed into two orthogonal linearly polarised light of equal intensity differing in phase by $\frac{\pi}{2}$. Thus, if a circularly polarised light is passed through an analyser (a tourmaline crystal), and the analyser is rotated, intensity of the emergent light is never changing. Conclusion is two-fold, either the light is unpolarised or the light is circularly polarised.

To confirm the nature, introduce a $\frac{\lambda}{4}$-plate after the analyser, if now also no variation of intensity is observed, the given light is confirmed as unpolarised, but if variation is observed and minimum of intensity reduces to zero, the given light is confirmed as circularly polarised.

Explanation: When a circularly polarised light passes through an analyser, on rotation there is no intensity variation, a fact not very difficult to understand from the discussion we had till now.

Then, when it is passed through a $\frac{\lambda}{4}$-plate, depending on the orientation of the fast axis and slow axis of the retarder, there is a phase difference of $\frac{\pi}{2}$ between the two orthogonal components, hence they combine to form a linearly polarised light which may be cut off by another analyser at proper orientation. But, if after passing through a $\frac{\lambda}{4}$-plate, there is no variation of intensity still, the given light is confirmed to be unpolarised.

This information is conveyed through Jones' formalism in the following way:

A quarter-wave plate with the fast axis vertical is represented as $\begin{bmatrix} 1 & 0 \\ 0 & i \end{bmatrix}$, a RCP is represented as $\frac{1}{\sqrt{2}}\begin{bmatrix} 1 \\ i \end{bmatrix}$.

So the action of the $\frac{\lambda}{4}$-plate on the RCP is expressed through the matrix product

$$\underset{\text{Polariser matrix}}{\begin{bmatrix} 1 & 0 \\ 0 & i \end{bmatrix}} \underset{\text{RCP matrix}}{\frac{1}{\sqrt{2}}\begin{bmatrix} 1 \\ i \end{bmatrix}} = \underset{\text{Emergent light matrix}}{\frac{1}{\sqrt{2}}\begin{bmatrix} 1 \\ -1 \end{bmatrix}}$$

So the emergent light is plane polarised at – 45º with the slow axis of the $\frac{\lambda}{4}$-plate. Similarly, when the incident light is LCP, the $\frac{\lambda}{4}$-plate will transform it into a linearly polarised light as

$$\begin{bmatrix} 1 & 0 \\ 0 & i \end{bmatrix} \underset{\text{LCP}}{\frac{1}{\sqrt{2}}\begin{bmatrix} 1 \\ -i \end{bmatrix}} = \frac{1}{\sqrt{2}}\begin{bmatrix} 1 \\ 1 \end{bmatrix}$$

Linearly polarised at 45º with the slow axis

So this process of conversion and analysis of the circularly polarised light by using a $\frac{\lambda}{4}$-plate is diagrammatically represented as in Fig. 24.30.

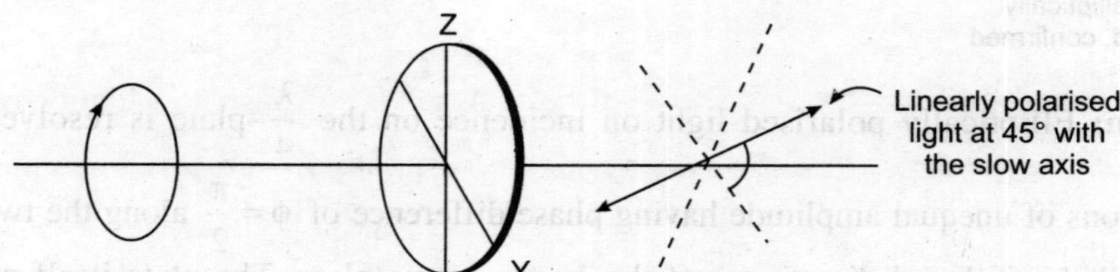

Fig. 24.23

So the action of a $\frac{\lambda}{4}$-plate on a circularly polarised light is to convert it into a linearly polarised light which may be cut-off by an analyser suitably oriented.

Analysis of Elliptically Polarised Light using a $\frac{\lambda}{4}$-plate

We have seen that an elliptically polarised light consists of two rectangular vibrations of unequal amplitudes having a phase difference $\phi(0 \le \phi \le \frac{\pi}{2})$.

So, if an elliptically polarised light is passed through an analyser and the analyser rotated, there will occur variation of intensity. Conclusion is two-fold, the incident light is either linearly polarised or elliptically polarised. Confirmation is arrived at by a $\frac{\lambda}{4}$-plate. The scheme of analysis is shown as below:

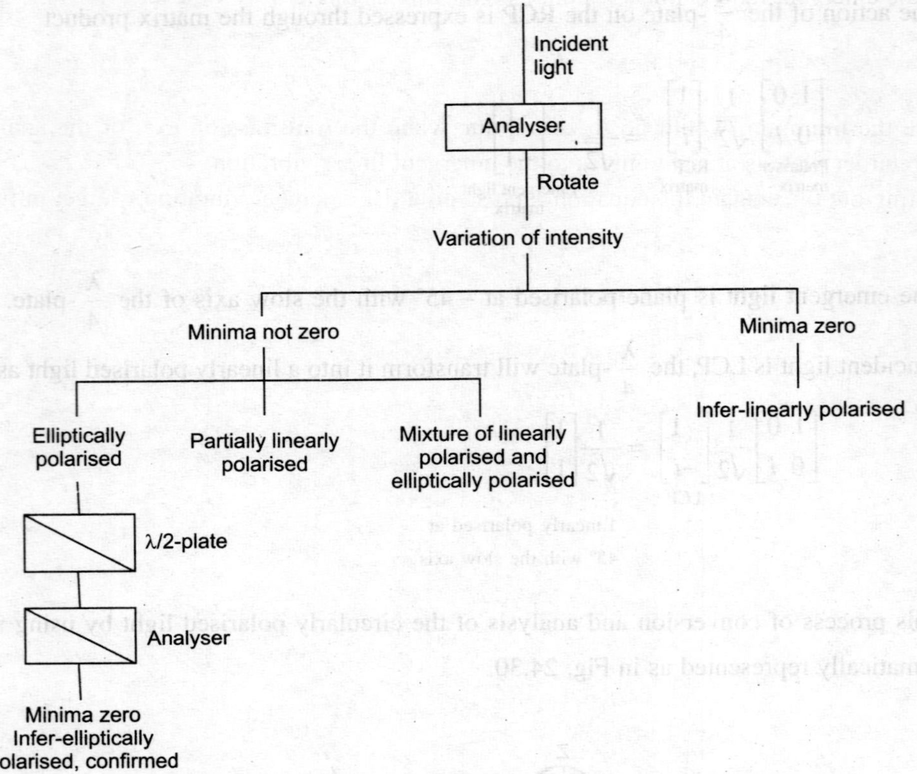

Explanation: Elliptically polarised light on incidence on the $\frac{\lambda}{4}$-plate is resolved into two orthogonal vibrations of unequal amplitude having phase difference of $\phi = \frac{\pi}{2}$ along the two principal direction (also called privileged directions) of the quarter-wave plate. The plate itself produces an overall phase difference by an amount equal to $\frac{\pi}{2}$, thus the emergent vibration becomes linearly polarised, which, therefore, can produce zero intensity when passed through an analyser (a Nicol prism).

Mathematically, an elliptically polarised light is represented as $\begin{bmatrix} A_1 \\ iA_2 \end{bmatrix}$. A quarter-wave plate whose fast axis in vertical is represented as $\begin{bmatrix} 1 & 0 \\ 0 & i \end{bmatrix}$

So, the action of $\frac{\lambda}{4}$ retarder on the elliptically polarised light is expressed as

$$\begin{bmatrix} 1 & 0 \\ 0 & i \end{bmatrix} \underbrace{\begin{bmatrix} A_1 \\ iA_2 \end{bmatrix}}_{\text{REP}} = \underbrace{\begin{bmatrix} A_1 \\ -A_2 \end{bmatrix}}_{\text{Linearly polarised}}$$

So the emergent light is linearly polarised making an angle θ with the slow axis given by

$$\tan\theta = -\frac{A_2}{A_1}$$

Here the minimum would go to zero value when the transmission axis of the analyser placed after the retarder plate is at right angles to the emergent linear vibration.

Writing out once again the equation representing the emergent vibration, we get in full form

$$\begin{bmatrix} y \\ z \end{bmatrix} = \begin{bmatrix} A_1 \\ -A_2 \end{bmatrix} e^{i\phi_x}$$

$$\Rightarrow \quad y = R.P.A_1 e^{i\phi_x} = A_1 \cos\phi_x$$

$$z = R.P.(-A_2)e^{i\phi_x} = -R.P.A_2 e^{i\phi_x} = -A_2 \cos\phi_x$$

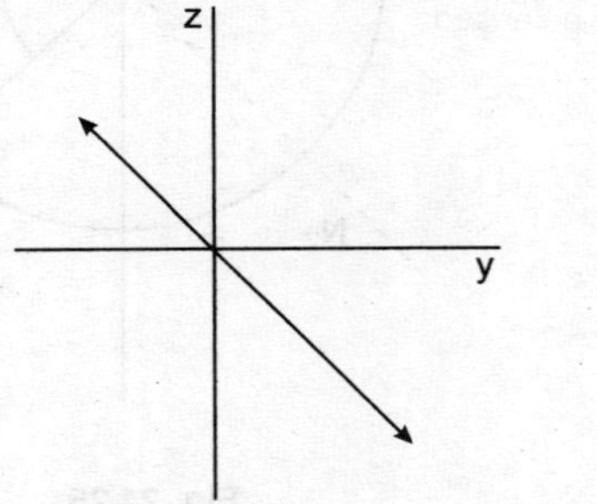

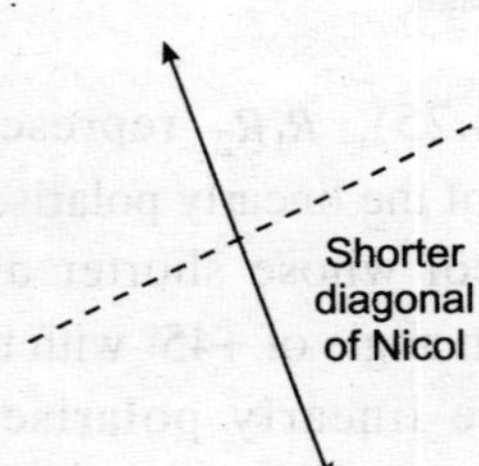

Fig. 24.24

Thus, we get

$$y = -\frac{A_1}{A_2} z = -mz$$

It is a linear vibration with orientation given by,

$$\tan\phi = -\frac{A_1}{A_2} \text{(note, if } A_1 = A_2, \tan\phi = -1$$

$$\Rightarrow \quad \phi = 135^\circ \text{ or, } -45^\circ \text{ w.r.t. } y\text{-direction).}$$

So much a linear vibration will be completely cut-off by the analysing Nicol when its transmission axis makes an angle of 45° w.r.t. the *y*-direction.

Experimental Procedure

Circularly polarised light : using $\frac{\lambda}{4}$ -plate as analyser:

In the analysis we have seen that if the circularly polarised light be RCP, then on incidence on a $\frac{\lambda}{4}$ -plate (of a +ve crystal) whose fast axis is vertical and slow axis horizontal, the circularly polarised light will be resolved into two rectangular components whose z-component (along fast axis) leads by $\frac{\pi}{2}$ over its y-component (along slow axis). Due to the action of the $\frac{\lambda}{4}$-plate, there will be additional advancement of phase by $\frac{\pi}{2}$, thus the emergent vibration will be a linearly polarised light with orientation at – 45° with the slow axis.

Mathematically, the fact is stated as the $\frac{\lambda}{4}$ retarder matrix operates on an RCP producing a linearly polarised light at – 45° with the slow axis of the plate. Thus,

$$\underset{\text{Retarder matrix}}{\begin{bmatrix} 1 & 0 \\ 0 & i \end{bmatrix}} \underset{\text{RCP}}{\frac{1}{\sqrt{2}}\begin{bmatrix} 1 \\ i \end{bmatrix}} = \underset{\text{Linearly polarised at} -45^\circ}{\frac{1}{\sqrt{2}}\begin{bmatrix} 1 \\ -1 \end{bmatrix}}$$

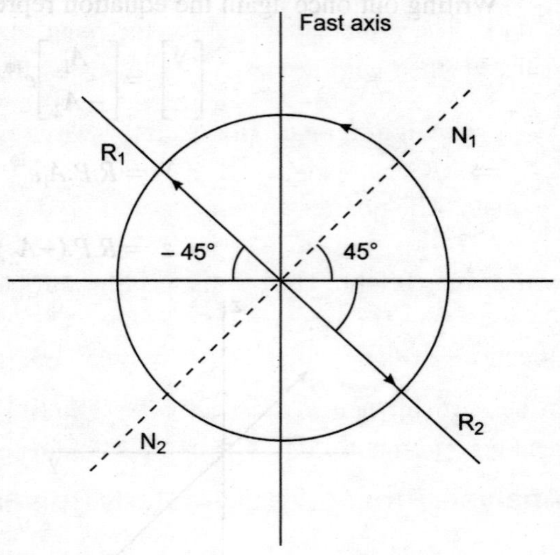

Fig. 24.25

In the Fig. (24.25), R_1R_2 represents the direction of vibration of the linearly polarised light. So an analysing Nicol whose shorter diagonal N_1N_2 is set making an angle of +45° with the slow axis will cut-off the linearly polarised light producing zero intensity.

The above scheme of analysis is carried out w.r.t. the experimental set-up as shown in Fig. 24.26. Over the collimator of an ordinary spectrometer is slipped a tube T_1 containing a polariser P, another tube T_2 contains an analyser A and telescope T. In between the polariser and analyser, there is provision for placing another tube T_3 where a $\frac{\lambda}{4}$-plate may be put.

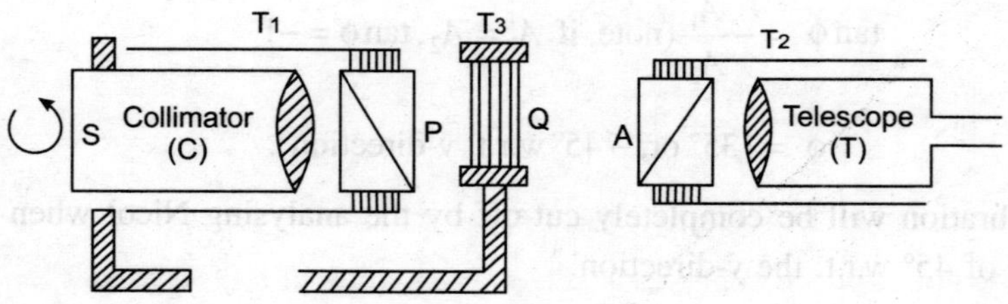

Fig. 24.26

1. Initial setting: The collimator and telescope are adjusted for parallel rays without placing P, Q and A, a Wollaston's prism is placed in the position of P in the tube T_1. Two images of the slit are observed and the tube T_1 is rotated to bring its two images in one vertical line. The vibrations of the two images are now horizontal and vertical.

2. Alignment of the analyser: The analysing Nicol is now placed in the tube T_2 with its shorter diagonal approximately vertical and rotated until one of the images disappears. The principal section of A is now vertical. The circular scale attached to T_2 is kept at zero of the scale in this position.

3. Alignment of the polariser: The Wollaston's prism is removed, the polariser P is placed in its position with its shorter diagonal horizontal and adjusted by rotation on either side until the field is dark. The principal section of the polariser is now exactly horizontal. The reading of the circular scale attached with the tube T_1 is adjusted for zero.

4. Alignment of the quarter-wave plate: The slit is illuminated by light suitable for the $\frac{\lambda}{4}$-plate. The quarter-wave plate will produce elliptically polarised light, so the field of the analyser will appear bright. The $\frac{\lambda}{4}$-plate is now rotated until the field is dark. This time the two principal directions of the $\frac{\lambda}{4}$ plate are horizontal and vertical, knowing the fast axis of the plate, it is adjusted to be vertical (hence, slow axis horizontal). The circular scale attached the tube T_3 is adjusted to read zero in this setting. The complete adjustment is shown in Fig. 24.26.

Analysis Procedure : Circularly Polarised Light

The polariser P is removed. The slit is illuminated by the given circularly polarised light. The analyser is rotated to make the field dark. If the shorter diagonal makes an angle of 45° with, the slow axis of $\frac{\lambda}{4}$-plate in this position, the given light is a right circularly polarised (RCP).

But, if the shorter diagonal of the analyser is to be set at – 45° with the slow axis of the $\frac{\lambda}{4}$-plate, the given circularly polarised light is left-handed (LCP).

The mathematical representation of the process is given earlier for an RCP light. If the light is an LCP, the mathematical representation of the process is

$$\begin{bmatrix} 1 & 0 \\ 0 & i \end{bmatrix} \frac{1}{\sqrt{2}} \begin{bmatrix} 1 \\ -i \end{bmatrix} = \frac{1}{\sqrt{2}} \begin{bmatrix} 1 \\ 1 \end{bmatrix}$$

Retarder matrix LCP Linearly polarised at 45°

So the shorter diagonal of the analyser is set at – 45° w.r.t. slow axis to cut-off light.

Elliptically Polarised Light

Initial alignment up to step (4) as done in circularly polarised light is made here also. Then the polariser is removed, the slit is illuminated by the given elliptically polarised light. The field of the telescope will now be bright. The tube T_3 containing the $\frac{\lambda}{4}$-plate is rotated in steps, at the same time the analyser A also rotated to get minimum intensity, this process of simultaneous rotation of the

$\frac{\lambda}{4}$-plate and the analyser is continued till we get the field completely dark. This time the principal direction of the $\frac{\lambda}{4}$-plate will give the axes of the elliptic vibration under analysis.

To study an elliptically polarised light, we require to get information on

(*i*) The phase difference between the two rectangular components.

(*ii*) Orientation of the major and minor axis.

(*iii*) Ratio of the magnitude of the minor and major axis of the ellipse.

In Fig. 24.27 we have shown the essence of the analysis.

Let *A*, *F* represent the initial direction of the transmission axis of the analyser and fast axis of the $\frac{\lambda}{4}$- plate, *P*, *S* represent the initial direction of the transmission axis of the polariser and slow axis of the $\frac{\lambda}{4}$- plate (*cf.* plan shown in Fig. 24.26). These two directions are respectively vertical and horizontal direction.

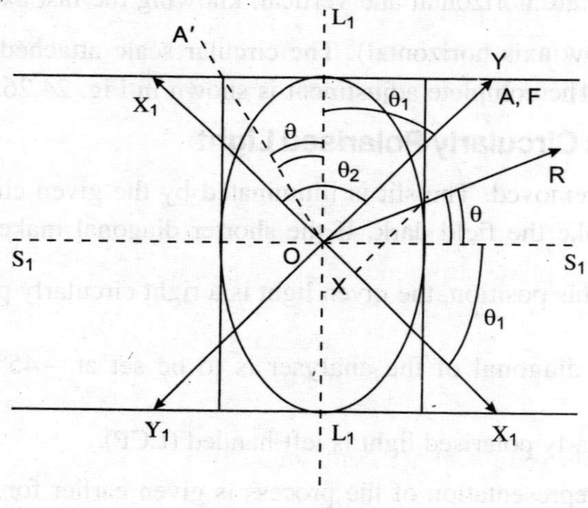

Fig. 24.27

The $\frac{\lambda}{4}$- plate is rotated through an angle θ_1 so as to make the fast axis and slow axis coincide with the axis of the elliptic vibration. The elliptic vibration then will be resolved into two components along S_1S_1 and L_1L_1 having a phase difference of $\frac{\pi}{2}$. The $\frac{\lambda}{4}$-plate produces a further phase difference of $\frac{\pi}{2}$. So the resultant phase difference of the emerging rectangular components will be either zero or π and in both cases, the emergent vibration will be plane polarised along ray OR. This may be cut-off by rotating the analyser from its initial position OA through an angle θ_2 to the position OA′ where the transmission axis of the analyser is in cross with the direction of the emergent vibration.

The angle θ between OR and OS_1 is given as

$$\theta = (\theta_2 - \theta_1)$$

which, therefore, can be obtained from the difference between the angles of rotation of the polariser and analyser from their initial position to get the field of view completely dark with the $\frac{\lambda}{4}$ - plate kept in the position in the tube T_3.

Analysis of Elliptically Polarised Light by Babinet's Compensator

As already remarked analysis of elliptically polarised light would mean determination of several, characteristics of elliptic polarisation.

1. The phase difference between the two rectangular components of elliptic polarisation.
2. Orientation of the axes of the ellipse.
3. Ratio of the axes of the ellipse.

1. Determination of the phase difference between the components: As the elliptically polarised light enters the compensator, it is broken up into two rectangular vibrations coincident with the optic axes of the compensator. They can be represented as

$$[J_{REP}] = \begin{bmatrix} A_1 \\ iA_2 \end{bmatrix}$$

As this passes through the compensator, there will be additional path difference, hence phase difference introduced. The phase difference δ at the point where the thicknesses become d and d' is given by

$$\delta = \frac{2\pi}{\lambda}(d - d')(\mu_0 - \mu_E)$$

In the elliptical polarisation, the z-component had a lead of phase by $\frac{\pi}{2}$. As a result γ the passage through the compensator if the phase of z-component lags by δ, then the Jones' matrix for the emergent light is given as

$$[J_e] = \begin{bmatrix} A_1 \\ A_2 e^{i\frac{\pi}{2} - i\delta} \end{bmatrix} = \begin{bmatrix} A_1 \\ A_2 e^{i(\frac{\pi}{2} - \delta)} \end{bmatrix}$$

So, if $\delta = \frac{\pi}{2}$, then

$$[J_e] = \begin{bmatrix} A_1 \\ A_2 \end{bmatrix}$$

a linearly polarised light. In general, an elliptically polarised

$$[J_i] = \begin{bmatrix} A_1 \\ A_2 e^{i\delta} \end{bmatrix}$$

If δ' is the phase lag introduced,

$$[J_e] = \begin{bmatrix} A_1 \\ A_2 e^{i(\delta-\delta')} \end{bmatrix}$$

So a linearly polarised light is obtained when

$$\delta = \delta' = \frac{2\pi}{\lambda}(d-d')(\mu_0 - \mu_E)$$

This linearly polarised light can be extinguished by an analysing Nicol.

So darkness in the field of view of the eye piece will be obtained at different points of the compensator corresponding to the path difference $\Delta = 0, \pm\lambda, \pm 2\lambda, \pm 3\lambda$ etc.

So at first, fringes are formed in the field of view with plane polarised light. The cross-wire of the eyepiece is made coincident with the central (dark) fringe. Now, elliptically polarised light is allowed to pass through the compensator, the phase difference does not become equal to zero at the central point. So the central fringe shifts towards one side. At this position, the phase difference is reduced to zero value. The eyepiece is shifted by x to bring the cross-wire in coincidence with the shifted central fringe.

If for a displacement $2s$ w.r.t. the central line, the path difference becomes λ, then

$$\frac{x}{2s} = \frac{\delta'}{2\pi} \Rightarrow \delta' = \pi\frac{x}{s} = \delta$$

Hence the phase difference can be determined.

2. Determination of the orientation of the axes of the elliptically polarised light: When the axes of the ellipse coincide with the optic axes of the compensator, the rectangular components of the elliptic vibration have a phase difference equal to $\frac{\pi}{2}$. This fact can be utilised for the determination of the orientation of the axes of the ellipse.

First, fringes are made available in the field of view of the eyepiece with plane polarised light. The cross-wire is made coincident with the central fringe (corresponding to the point where $d = d'$). Then one half of the compensator is shifted linearly through a distance such that the fringes suffer a displacement by $\frac{s}{2}$. We have known previously that $\delta' = \frac{\pi x}{s} = \frac{\pi}{s}\cdot\frac{s}{2} = \frac{\pi}{2}$. It means that at the position of the cross-wire, a phase difference of $\frac{\pi}{2}$ exists.

Now, the plane polarised light is replaced by the experimental elliptically polarised light. The central fringe will no longer remain in coincidence with the cross-wire. The compensator is rotated in its own plane about a line normal to the plane of AD of the compensator. With this rotation, the fringes will undergo displacement and at one time, the central fringe will coincide again with the cross-wire. At the position of the cross-wire there existed a phase difference of $\frac{\pi}{2}$ and when the compensator is rotated, at certain orientation, the directions of optic axes of the compensator coincide with the axes of the elliptic vibration. When this occurs, as the elliptically polarised light falls on the compensator, it will be resolved into two rectangular components differing in phase by $\frac{\pi}{2}$. The cross-wire position is already at a phase difference of $\frac{\pi}{2}$. The compensator with its axes coinciding with the axes of the elliptic vibration will resolve into components having a phase difference equal to $\frac{\pi}{2}$, equal and opposite to that which existed at the position of the cross-wire. Thus at the cross-wire position, the phase difference is again reduced to zero, and hence central fringe again comes in coincidence with the cross-wire. The orientation of the optic axes of the compensator gives us the orientation of the axes of the elliptically polarised light.

3. Determination of the ratio of the axes of the ellipse: When the axes of the compensator coincide with the axes of the ellipse, then the central fringe would correspond to zero phase difference and so in that condition the emergent light will be plane polarised, which may be cut-off by the analysing Nicol.

In the position of complete extinction, the transmission axis of the analysing Nicol would be perpendicular to the direction of vibration of the polarised light.

SOLVED EXAMPLES

Example 1. *What is meant by elliptically polarised light? Show that plane and circularly polarised light are special cases of elliptically polarised light.*

Elliptically polarised light is the result of superposition of two rectangular vibrations having a phase difference $0 \le \phi \le \frac{\pi}{2}$. In Jones' matrix formalism,

$$[J_{REP}] = \begin{bmatrix} A_1 \\ A_2 e^{i\phi} \end{bmatrix} \text{Right elliptically polarised light}$$

$$[J_{LEP}] = \begin{bmatrix} A_1 \\ A_2 e^{-i\phi} \end{bmatrix} \text{Left elliptically polarised light}$$

If $A_1 = A_2, \phi = 0$, then

$$[J] = \begin{bmatrix} A \\ A \end{bmatrix} \rightarrow \begin{bmatrix} 1 \\ 1 \end{bmatrix}, \text{Linearly polarised light}$$

If $\phi = \frac{\pi}{4}, A_1 = A_2$, then

$$[J_{RCP}] = \begin{bmatrix} 1 \\ i \end{bmatrix}, \text{Right circularly polarised}$$

Example 2. *Elliptically polarised light is passed through a* $\frac{\lambda}{4}$ *- plate and then through a Nicol. Extinction occurs when the optic axis of the plate and the principle section of the Nicol are inclined to the horizontal at angle of 30° and 60° respectively, the angles being measured in the same direction, Find the orientation of the ellipse and the ratio of its axis.*

The elliptically polarised light on incidence on the $\frac{\lambda}{4}$ has its resolved components in the form of linearly polarised light parallel to the privileged axes of the $\frac{\lambda}{4}$ - plate. On the first face, the two rectangular vibrations will have a phase difference of ϕ. We have seen that when $\phi = \frac{\pi}{2}$, the $\frac{\lambda}{4}$ - plate produces an additional phase difference of $\frac{\pi}{2}$ and so the beam on exit out of the plate will be plane polarised having a certain orientation and it will be thrown to extinction when the principal section of the analysing Nicol is at right angle w.r.t. the emerging linearly polarised light.

Mathematically

$$\begin{bmatrix} 1 & 0 \\ 0 & i \end{bmatrix} \begin{bmatrix} A_1 \\ iA_2 \end{bmatrix} = \begin{bmatrix} A_1 \\ -A_2 \end{bmatrix}$$

Fig. 24.28

This mathematical transformation is diagrammatically represented as in Fig. (24.28).

From figure showing complete extinction, we get, the ratio of the axes of the ellipse given as $\frac{b}{a} = \tan\psi = \tan 30° = \frac{1}{\sqrt{3}}$. The minor axis is at an angle of 30° w.r.t. the horizontal.

Example 3. *A quartz quarter-wave plate which is designed for light of wavelength 5893Å is used in conjunction with a Nicol prism to produce circularly polarised light. Show that if there is also present some light of wavelength 5460Å, this latter will be elliptically polarised with axes at 45° to the axes of the plate. Find the ratio of the major and minor axes of the ellipse.*

Let the slow axis of the plate be along the *y*-direction. If the incident wave have amplitude *a*, then the Nicol acting as a polariser at 45° will transform the incident beam into polarised light at 45°, hence the incident light will be represented by

$$[J_i] = \frac{1}{\sqrt{2}}\begin{bmatrix} 1 \\ 1 \end{bmatrix}$$

The quarter-wave plate producing a phase difference of $\frac{\pi}{2}$ transforms the incident ray into a circularly polarised light. The transformation is carried through matrix product of the Jones' incident ray matrix and $\frac{\lambda}{4}$ - plate matrix which is represented as

$$\left[J_{\frac{\lambda}{4}}\right] = \begin{bmatrix} 1 & 0 \\ 0 & i \end{bmatrix}$$

$$\therefore \qquad \begin{bmatrix} 1 & 0 \\ 0 & i \end{bmatrix} \frac{1}{\sqrt{2}} \begin{bmatrix} 1 \\ 1 \end{bmatrix} = \frac{1}{\sqrt{2}} \begin{bmatrix} 1 \\ i \end{bmatrix}$$

writing out in full form, the emergent light can be represented as

$$\begin{bmatrix} y \\ z \end{bmatrix} = \frac{a}{\sqrt{2}} \begin{bmatrix} 1 \\ i \end{bmatrix} e^{i\phi_x}$$

$$\Rightarrow \qquad y = R.P. \frac{a}{\sqrt{2}} e^{i\phi_x} = \frac{a}{\sqrt{2}} \cos\phi_x = \frac{a}{\sqrt{2}} \cos(kx - wt)$$

$$z = R.P. \frac{a}{\sqrt{2}} \cos\left(\phi_x + \frac{\pi}{2}\right) = -\frac{a}{\sqrt{2}} \sin(kx - wt)$$

$$\therefore \qquad y^2 + z^2 = \frac{a^2}{2}, \text{ a circularly polarised light.}$$

If the wavelength is not 5893Å, the plate will produce a phase difference

$$\phi = \frac{2\pi}{5460} \times \frac{5893}{4} \text{ rad} = 97.14° \neq \frac{\pi}{2}$$

$$\therefore \qquad \begin{bmatrix} 1 & 0 \\ 0 & e^{i\phi} \end{bmatrix} \frac{1}{\sqrt{2}} \begin{bmatrix} 1 \\ 1 \end{bmatrix} = \frac{1}{\sqrt{2}} \begin{bmatrix} 1 \\ e^{i\phi} \end{bmatrix}$$

Or, writing out in full form,

$$y = \frac{a}{\sqrt{2}} \cos(kx - wt) = A_1 \cos(kx - wt)$$

$$z = \frac{a}{\sqrt{2}} \cos(kx - wt + \phi) = A_2 \cos(kx - wt + \phi) \text{ (hence } A_1 = A_2\text{)}$$

$$\Rightarrow \qquad \frac{y^2}{\left(\frac{a}{\sqrt{2}}\right)^2} + \frac{z^2}{\left(\frac{a}{\sqrt{2}}\right)^2} - \frac{2xy\cos\phi}{\left(\frac{a}{\sqrt{2}}\frac{a}{\sqrt{2}}\right)} = \sin^2\phi$$

The angle between the axes of the ellipse and co-ordinate axes is given by

$$\tan 2\theta = \frac{2A_1A_2\cos\phi}{(A_1^2 - A_2^2)}$$

Here,
$$A_1 = \frac{a}{\sqrt{2}}, A_2 = \frac{a}{\sqrt{2}}$$

$$\therefore \quad \tan 2\theta = \frac{2 \cdot \frac{a^2}{2} \cos\phi}{\left(\frac{a^2}{2} - \frac{a^2}{2}\right)} = \infty$$

$$\Rightarrow \quad 2\theta = \frac{\pi}{2}, \ \theta = \frac{\pi}{4}$$

Rotate the axes through 45º, then we would get

$$z = \frac{(z-y)}{\sqrt{2}}, y = \frac{(z+y)}{\sqrt{2}}$$

$$\therefore \quad \frac{(z+y)^2}{a^2} + \frac{(z-y)^2}{a^2} - \frac{4}{a^2} \frac{(z-y)}{\sqrt{2}} \cdot \frac{(z+y)}{\sqrt{2}} \cos\phi = \sin^2\phi$$

or
$$(z+y)^2 + (z-y)^2 - 2(z^2 - y^2)\cos\phi = a^2 \sin^2\phi$$

or
$$2(z^2 + y^2) - 2(z^2 - y^2)\cos\phi = a^2 \sin^2\phi$$

or
$$z^2(1-\cos\phi) + y^2(1+\cos\phi) = \frac{a^2}{2}\sin^2\phi$$

or
$$\frac{y^2}{\left(\frac{a^2 \sin^2\phi}{2(1+\cos\phi)}\right)} + \frac{z^2}{\left(\frac{a^2 \sin^2\phi}{2(1-\cos\phi)}\right)} = 1$$

$$\therefore \text{ Ratio of the axes} = \sqrt{\frac{1-\cos\phi}{1+\cos\phi}} = \tan\frac{\phi}{2} = \tan\frac{97.14}{2} = 1.13$$

Example 4. *Find the thickness of a quarter-wave plate made of quartz for light of wavelength λ=5893Å. The r.i. values for O-and E-ray are respectively 1.541 and 1.551.*

According to Eqn.

$$L = \frac{\lambda}{4} \frac{1}{(\mu_E - \mu_0)}$$

Putting the values of μ_E and μ_0, the thickness L can be calculated.

Example 5. *Through a calcite plate plane polarised Na light (λ=5893Å) is passed so as to produce circularly polarised light. If μ_0 = 1.660 and μ_E = 1.495, find the thickness.*

$$\begin{bmatrix} 1 & 0 \\ 0 & i \end{bmatrix} \frac{1}{\sqrt{2}} \begin{bmatrix} 1 \\ 1 \end{bmatrix} = \frac{1}{\sqrt{2}} \begin{bmatrix} 1 \\ i \end{bmatrix}$$

Retarder matrix RCP

So, the incident plane polarised light represented as

$$y' = \frac{a}{\sqrt{2}}\cos(kx - wt), z' = \frac{a}{\sqrt{2}}\cos(kx - wt)$$

Through the calcite plate there is a phase difference $\phi = \frac{\pi}{2}$ so as to get a circularly polarised light. If z-axis be the fast axis, then the emergent light is represented as

$$y = \frac{a}{\sqrt{2}}\cos(k_1 L - wt), \; z = \frac{a}{\sqrt{2}}\cos(k_2 L - wt)$$

where $$(k_1 - k_2) = \frac{\pi}{2}$$

or $$\frac{2\pi}{\lambda}(\mu_0 - \mu_E)L = \frac{\pi}{2}$$

or $$L = \frac{\lambda}{4(\mu_0 - \mu_E)} = \frac{5893 \times 10^{-8}}{4(1.660 - 1.495)}$$

$$= \frac{5893 \times 10^{-8}}{4 \times 0.165}$$

$$= 8.929 \times 10^{-5} \text{ cm. } \textbf{Ans.}$$

Example 6. *A left circularly polarised light (λ = 5893Å) is incident on a quartz crystal (with its optic axis parallel to the surface) of thickness 0.025 mm. Determine the state of polarisation of the emergent light. Assume μ_0 = 1.54425 and μ_E = 1.55336 respectively.*

$$\delta = (k_2 - k_1)L = \frac{2\pi}{\lambda}(\mu_E - \mu_0)L = \frac{2\pi}{\lambda}(1.55336 - 1.54425)L = 0.76\pi \;\; \textbf{Ans.}$$

Example 7. *A left circularly polarised beam (λ = 5893Å) is incident normally on a calcite crystal (with its optic axis cut parallel to the surface) of thickness 0.005141 mm. What will be the state of polarisation of the emergent beam?*

The phase difference introduced by the plate is given by

$$L = \phi\frac{\lambda}{2\pi(\mu_E - \mu_0)}$$

here $$L = 0.005141 \text{ mm} = 0.0005141 \text{ cm}, \lambda = 5893 \times 10^{-8} \text{ cm}$$

$$\mu_E - \mu_0 = 0.17195$$

$$\therefore \quad \phi \simeq 3\pi$$

$$\therefore \quad \begin{bmatrix} 1 & 0 \\ 0 & e^{i\phi} \end{bmatrix} \frac{1}{\sqrt{2}}\begin{bmatrix} 1 \\ -i \end{bmatrix} = \frac{1}{\sqrt{2}}\begin{bmatrix} 1 \\ -ie^{i\phi} \end{bmatrix}$$

Here $$\phi = 3\pi \Rightarrow e^{i\phi} = \cos 3\pi + i \sin 3\pi = \cos(2\pi + \pi) = -1$$

$$\therefore \quad \underset{\text{Retarder matrix}}{\begin{bmatrix} 1 & 0 \\ 0 & e^{i\phi} \end{bmatrix}} \underset{\substack{\text{Left circularly} \\ \text{polarised light} \\ \text{(incident)}}}{\frac{1}{\sqrt{2}} \begin{bmatrix} 1 \\ -i \end{bmatrix}} = \underset{\substack{\text{Emergent light} \\ \text{(right circularly} \\ \text{polarised light)}}}{\frac{1}{\sqrt{2}} \begin{bmatrix} 1 \\ i \end{bmatrix}}$$

Assuming the horizontal direction as the direction of the fast axis, Jones' matrix transformation gives as

$$\underset{\substack{\text{Retarder} \\ \text{matrix with} \\ \text{fast axis} \\ \text{horizontal}}}{\begin{bmatrix} 1 & 0 \\ 0 & e^{-i\delta} \end{bmatrix}} \underset{\substack{\text{Left circularly} \\ \text{polarised light} \\ \text{matrix (LCP)}}}{\frac{1}{\sqrt{2}} \begin{bmatrix} 1 \\ -i \end{bmatrix}} = \underset{\text{Emergent light}}{\frac{1}{\sqrt{2}} \begin{bmatrix} 1 \\ -ie^{-i\delta} \end{bmatrix}}$$

writing out in full form, the emergent light components can be written as

$$y = R.P. \frac{a}{\sqrt{2}} e^{i\phi_x} = \frac{a}{\sqrt{2}} \cos \phi_x$$

$$z = R.P. \frac{a}{\sqrt{2}} e^{i(\phi_x - \frac{\pi}{2} + \delta)} = -\frac{a}{\sqrt{2}} \sin(\phi_x - \delta)$$

This would represent a clockwise elliptic vibration, so we would conclude that the emergent light is a right-handed elliptically polarised light.

The conclusion is drawn from the consideration given below when

$$\phi_x = \frac{\pi}{2} \Rightarrow t_1 = \frac{kx}{w} - \frac{\pi}{2w}$$

we get $$y = 0, \; z = -\frac{a}{\sqrt{2}} \cos \delta$$

This is represented by a line OA in figure

when $$\phi_x - \delta = 0$$

or $$\phi_x = \delta$$

$$t_2 = \frac{kx}{w} - \frac{\delta}{w}$$

$$y = \frac{a}{\sqrt{2}} \cos \delta, \; z = 0$$

This is represented by a line OB.

Now, if $\delta < \frac{\pi}{2} t_2 > t_1$

if $\delta > \frac{\pi}{2}, t_2 < t_1$, so the displacement, vector remaining constant in magnitude rotates in the clockwise direction. Since δ here is greater than $\frac{\pi}{2}$.

Example 8. *A horizontal elliptically polarised parallel beam of light is passed through first a quarter-wave plate and then a Nicol prism. It is found that for a certain orientation of the quarter-wave plate, the emergent light is plane polarised at 23° to the vertical and that when the quarter-wave plate is turned through a right angle, it is plane polarised at 81° to the vertical. Determine the ratio.*

The quarter-wave matrix with fast axis vertical is given by $\begin{bmatrix} 1 & 0 \\ 0 & i \end{bmatrix}$ and when the fast axis is horizontal it is given by $\begin{bmatrix} 1 & 0 \\ 0 & -i \end{bmatrix}$. The elliptically polarised light is represented by the matrix $\begin{bmatrix} A_1 \\ -iA_2 \end{bmatrix}$, the left-handed one or by $\begin{bmatrix} A_1 \\ -iA_2 \end{bmatrix}$, the right-handed one.

$$\therefore \quad \begin{bmatrix} 1 & 0 \\ 0 & i \end{bmatrix} \begin{bmatrix} A_1 \\ -iA_2 \end{bmatrix} = \begin{bmatrix} A_1 \\ A_2 \end{bmatrix} \text{ plane polarised}$$

So considering a left-handed elliptically polarised light, the emergent light is plane polarised at an angle ϕ_0 given by

$$\tan \phi_0 = \frac{A_2}{A_1}$$

and this is shown in Fig. (24.29). The ellipse is included within the rectangular of sides $2A_1$ and $2A_2$. The resultant amplitude vector is represented by OB. From figure we can write

$$23° + \phi_0 + \theta = \frac{\pi}{2}$$

$$\Rightarrow \quad \phi_0 = \left(\frac{\pi}{2} - 23° - \theta\right) = 67° - \theta$$

$$\therefore \quad \tan \phi_0 = \frac{A_2}{A_1} = \tan(67° - \theta)$$

When the quartz-wave plate is rotated through 90°, the fast axis becomes now horizontal and so

$$\begin{bmatrix} 1 & 0 \\ 0 & -i \end{bmatrix} \begin{bmatrix} A_1 \\ -iA_2 \end{bmatrix} = \begin{bmatrix} A_1 \\ -A_2 \end{bmatrix} \text{ Plane polarised light}$$

In this case, the resultant amplitude vector is represented by OA.

From figure, we immediately get,

$$23° + \phi_0 + (\theta - 9°) = 81°$$

$$\phi_0 + \theta = 67°$$

Now $$\phi_0 = \theta - 9°$$

$\therefore$ $$2\phi_0 + 9 = 67° \Rightarrow 2\phi_0 = 58° \Rightarrow \phi_0 = 29°$$

Since, $$\tan\phi_0 = \frac{A_2}{A_1} = 0.5543 \Rightarrow \frac{A_1}{A_2} = 1.804$$

Also $$\phi_0 + \theta = 67°$$

$\therefore$ $$\theta = 67° - \phi_0 = 67° - 29° = 38°$$

So the major axis makes an angle (90º-38º) w.r.t. the vertical.

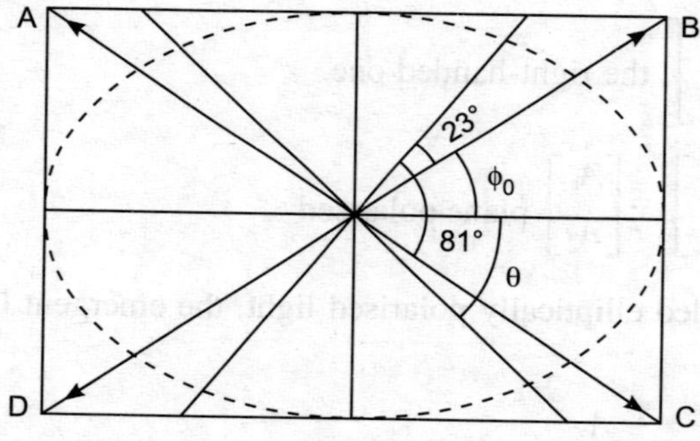

Fig. 24.29

Example 9. *Figure (24.30) shows a section of wave surface for extraordinary ray. The optic axis being along ox. Show that refractive index of the medium along the direction making an angle θ is given by*

$$\mu_\theta = \sqrt{\mu_0^2 \cos^2\theta + \mu_e^2 \sin^2\theta}$$

Fresnel's wave equation for propagation in an isotropic crystal along a direction having direction cosines l, m, n is given by

$$\frac{l^2}{(v_x^2 - v^2)} + \frac{m^2}{(v_y^2 - v^2)} + \frac{n^2}{(v_z^2 - v^2)} = 0$$

In x-y plane, we have

$$\frac{l^2}{v_x^2 - v^2} + \frac{m^2}{v_y^2 - v^2} = 0$$

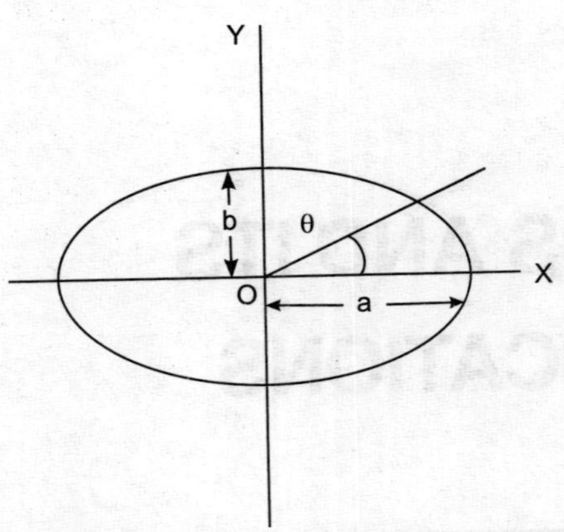

Fig. 24.30

Here $l = \cos\theta$, $m = \sin A$, $v_x = v_e$ (extraordinary velocity) $v_y = v_0$ (ordinary ray velocity), v is the velocity along the direction θ.

Now $$\frac{c}{v} = \mu_\theta, \frac{c}{v_e} = \mu_e, \frac{c}{v_0} = \mu_0$$

Putting these values in the wave surface equation, we get

$$\mu_\theta^2 = \mu_0^2 \cos^2\theta + \mu_e^2 \sin^2\theta$$

$$\Rightarrow \quad \mu_\theta = \sqrt{\mu_0^2 \cos^2\theta + \mu_e^2 \sin^2\theta}. \quad \textbf{Ans.}$$

LASERS AND ITS APPLICATIONS

As we know light is an electromagnetic wave consisting of electric $(\vec{E})$ and magnetic $(\vec{B})$ fields oscillating at right angles to each other and to the direction of propagation of the wave travelling at a velocity of 3×10^8 m/sec. Vectors $\vec{E}$ and $\vec{B}$ are of equal significance to the wave. However, photochemical, photoelectrical and several other effects are mainly due to electric component. The actual light wave may be treated as a collection of plane monochromatic polarised waves of various frequencies, direction of propagation and polarisation planes. Light is defined as a stream of tiny parcels (quantum) of energy, called 'photons' while explaining quantum phenomena. Photons have zero rest mass and zero electric charge. Photons cannot be picturised, they can only be defined by knowing their energy (E) and direction of motion or in other words their momentum $\vec{p}$. The direction of $\vec{p}$ indicates the direction of travel of the photon and the modulus of $\vec{p}$ gives the energy of the photon divided by the velocity of light in vacuum *i.e.*

$$p = \varepsilon/c \qquad ...(25.1)$$

In addition, polarisation of the photon should also be specified. The photon energy

$$\varepsilon = h\nu \qquad ...(25.2)$$

in terms of its frequency ν where h is Planck's constant having value 6.626×10^{-34} Js. Putting Eqn. (25.2) in (25.1) we get

$$p = \frac{h\nu}{c} = \frac{h}{\lambda} \qquad ...(25.3)$$

where λ is the wavelength of light. Eqns. (25.2) and (25.3) reflect the dual nature of radiation *i.e.*, energy ε and momentum $\vec{p}$ of a quantum, and frequency ν and wavelength λ of a wave. Both the descriptions of light are complementary to each other and are connected mathematically by Planck's constant h.

Photons are 'bosons' (after the Indian Physicist S.N. Bose). The more photons are in a given state the higher the probability that new photons will occupy this particular state. Boson nature of the photon-statistics is of extreme importance for optical phenomena.

25.1 QUANTUM TRANSITIONS IN ABSORPTION AND EMISSION OF LIGHT

The energy of an atom or molecule can take on only definite (discrete) values E_0, E_1, E_2, ..., E_n (Fig). These are the energy levels of the atom (molecule). Any species cannot remain in excited state for an indefinitely long time. Sooner or Later, it inevitably comes down to less excited state. This time is called 'life-time' of the energy level. The transition of an atom or molecule from one energy level to another occurs in jump and is called the quantum transition. Quantum transitions may be induced by various causes. In particular, they can occur when atoms interact with optical radiation.

Absorption of Light

Consider two energy levels of an atom. Let the energy of the lower level be E_1 and that of the upper level E_2. Assume that the atom is in the lower level and a photon of energy $(E_2 - E_1)$ is incident on the atom. The atom can absorb this photon and rise from level E_1 to E_2 thus making a transition by absorption of a quantum of light (Fig. 25.1).

Stimulated Emission of Light

When the atom lies in the upper energy level, then the same incident photon may play the role of a trigger (stimulus) and induce the transition from E_2 to E_1 and the atom falls into the lower level. The transition causes an emission of a photon. Both the stimulating and stimulated photons have the same energy $(E_2 - E_1)$.

Moreover, both of them have identical direction of their momenta and identical polarisation. In other words, secondary photon finds itself in the same state as primary photon. This result is a sequel of the bosonic behaviour of photons — they tend to accumulate in the same state. This is the phenomenon of **stimulated emission**.

The more the primary photons incident, the higher the probability that the atom lying in level E_2 will be forced to undergo a transition to E_1. Thus there is a certain similarity between the stimulated emission and absorption processes, namely, the probabilities of both processes are proportional to the number of primary photons.

Thus if the atom is in ground state E_1 a photon of energy $E_2 - E_1$ leads to absorption process whereas if the atom is in the higher level E_2 the same photon, with the same probability as in the absorption process, stimulates the atom to fall from E_2 to E_1 with the emission of another photon.

If there are many atoms in level E_2 of the material, an e.m. wave of the energy $(E_2 - E_1)$ is capable of stimulating $E_2 \rightarrow E_1$, transitions in many atoms *i.e.*, primary photon may initiate an avalanche of secondary photons. All of them will be emitted in the same state as the stimulating primary photon.

Spontaneous Emission

An atom lying in level E_2 will tend to decay to level E_1 spontaneously, that is without any stimulus, say, in the form of a photon applied to it from outside. The photon emitted in the spontaneous $E_2 \rightarrow E_1$ emission has the energy $E_2 - E_1$ while its other characteristics – momentum direction, polarisation–are arbitrary. The probability of the spontaneous emission of a photon is determined only by the properties of the transition.

Thus two types of emission of light, namely; stimulated and spontaneous emission. The former may be viewed as a controllable process as it is stimulated by the primary photon which not only induces the transition but also governs the characteristics of the new emitted photon. The second process proceeds spontaneously and is random in nature. The instant of transition and direction of emitted photon both are random. Strictly speaking, an element of chance is present in the process of stimulated emission too as the primary photon may or may not initiate the transition. Therefore probability of transition processes are the quantities which are considered for writing an equation related to transitions.

Generation of Coherent Light Waves

Any light wave is characterised by a certain degree of coherence however small it may be. Coherent waves can be obtained by division of amplitude or division of wavefront as seen earlier.

The pertinent questions are that how can sufficiently coherent waves be generated and how can the degree of coherence of a wave be increased. The answer lies in the necessity of the photons to occupy only selected states. It means that the photon population be a controlled characteristic.

Such control is possible owing to the "boson nature" of photons, that is, to their tendency to populate predominantly just those states that already have a sufficiently high population density.

Competition between Absorption and Stimulated Emission

Earlier we have seen that two different and competitive processes of absorption and stimulated emission are possible when photons with energy $E_2 - E_1$ interact with active centres*. The greater the number of active centres in the appropriate initial state the higher the probability of each of these two processes is. If the number of active centres on level 1 is greater than that on level 2, absorption becomes a more probable process. If, on the contrary there are more active centres on level 2, stimulated emission becomes more probable.

Consequently, population inversion between levels 1 and 2 is a necessary condition for stimulated emission to exceed absorption.

Each photon emitted by inevitable spontaneous transitions $2 \rightarrow 1$ may be absorbed or may stimulate the emission of new photons. The emission of new photons predominates in a medium with an inverted population of levels 1 and 2. Thus, spontaneously emitted photons trigger numerous stimulated emissions 'Spontaneous' photons are independently emitted and therefore have arbitrary direction of their momenta and therefore produce corresponding stimulate emissions. Directional selectivity can be achieved by preparing the active medium in the shape of long rod (in the case of solid medium) or a long tube (in the case of gaseous medium) with a comparatively small cross-section. Spontaneous photons with momentum parallel to the rod's axis are favoured, since they can interact with a higher number of active centres and therefore stimulate an intensive avalanche of stimulated photons. On the other hand, those spontaneous photons with momenta at an angle to the rod's axis will soon leave the active medium. The path within the active medium travelled by favoured photons may be increased still further by placing partially reflecting mirrors at the end faces of the

*An active medium is a solid, liquid or gas containing atoms, ions or molecules which are capable of decaying from their higher energy states in a radiative manner *i.e.* by emitting electromagnetic waves. These atoms (ions or molecules) are sometimes called 'active species, or 'active centres'. Active medium is the heart of any laser.

rods. The mirrors reflect the radiation back and therefore enhance the effect of stimulated emission. From here, originates the idea of an optical resonator, an essential element of a laser.

The selectivity of photon energies can also be facilitated by appropriate optics.

25.2 THE LASER

A functional schematic of a laser is shown in Fig. 25.3. The active medium and the additional elements are located inside the optical resonator. Due to resonator, the emitted radiation propagates along the axis *xx*′. Note that a laser can emit radiation both in a single direction and in two opposite directions along the resonator axis.

A laser is triggered by initiating its pumping system. This system provides excitation of active centres and an inverted population of lasing level builds up. The optical resonator (together with some additional elements *viz*. frequency selective element) provides selectivity of direction, polarisation, energy etc. As a result, a highly coherent radiation, called the laser radiation, appears along the axis *xx*′.

(*i*) Active Medium and Methods of Excitation

The following active media are used in lasers:

(*a*) gases and mixture of gases (gas lasers).

(*b*) crystal and glasses doped by special ions (solid-state lasers).

(*c*) liquids *e.g.*, dyes (dye-laser).

(*d*) semiconductors (semiconductor laser).

Normally, the active medium of a gas laser is a mixture of several gases; atoms or molecules of one of them are active centres while other gaseous components serve to produce population inversion on the lasing levels of the active centres. One possible mixture, for example, is helium and neon. Neon atoms are active centres. Helium helps in excitation and thus population inversion in Ne. The mixture is placed in a gas discharge tube at low pressure: neon at a pressure of about 10 Pa and helium at about 100 Pa.

The active medium of a solid-state laser is normally a rod with circular cross-section doped with special ions that play the role of active centres. A classical example of a solid lasing medium is a ruby rod, 3 to 20 mm in diameter and 5 to 30 cm long. Ruby is crystalline alumina (Al_2O_3) doped with chromium ions (from 0.05 to 0.5%) Note that it is this impurity (dopant) that gives ruby its typical colour (from pink to deep red).

Systems of pumping vary and to a large extent depend on the type of active medium. Excitation in gas lasers is realised in the simplest manner by means of electric discharge in the active medium. In this case the energy of excitation is transferred to active centres as a result of collisions with particles in the gas discharge plasma.

Excitation in solid-state lasers is carried out by irradiating the lasing rod with light from a sufficiently powerful light source such as optical flash lamp. In this case active centres are excited owing to the absorption of photons emitted by the pump lamp. Thus, lasers with optical pumping can be considered as converters of optical radiation converting, for instance, the incoherent radiation of a pulsed lamp into highly coherent laser emission.

(*ii*) Optical Resonator

An optical resonator is realised as a system of mirrors. Figure 25.1 shows (*a*) a linear resonator, (*b*) a confocal resonator and (*c*) a resonator consisting of a plane mirror and a concave mirror. Mirrors may be coated with dielectric or metal layers. At least one of the mirrors must be partially transparent with respect to the emitted radiation. Resonator mirrors of gas lasers are usually mounted at both ends of the gas discharge tube and are not linked to it in a rigid manner. Mirrors in solid-state lasers are typically formed on specially prepared end faces of the active medium rod. An optical resonator defines the direction in which radiation is emitted.

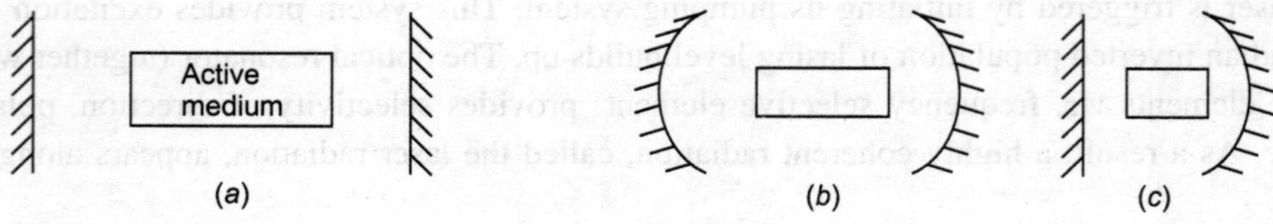

Fig. 25.1

When radiation inside a real laser is considered, it is necessary to take into account not only stimulated emission and photon absorption processes causing radiation losses such as scattering, diffraction etc. laser oscillation is sustained only if stimulated emission by active centres compensates not only for the absorption by these centres but for all other losses as well.

The planes of the end faces of gas-laser discharge tubes are usually tilted at an angle so that the normal to the end face and the optical axis are at the Brewster angle corresponding to the refractive index of the material of which the end face window is made (Fig. 25.2).

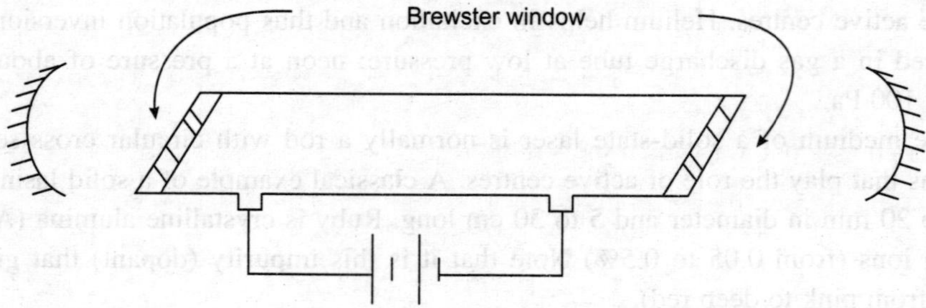

Fig. 25.2. A typical He–Ne–Laser with external mirrors. The ends of the discharge tube are fitted with Brewster windows

The end face of the discharge tube is oriented at the Brewster angle in order to realise the conditions of polarisation selectivity of emission. Here we should recall that, when unpolarised light wave is incident on the plane-parallel window (end face) of the tube.

The light polarised in the plane perpendicular to the plane of incidence is reflected while that with parallel polarisation is transmitted. Thus the former is lost and the latter is repeatedly reflected by the resonator mirrors and therefore passes repeatedly through the active medium and gets amplified. Thus, we obtain plane-polarised laser emission.

(*iii*) Basic Modes of Laser Oscillation

The three basic modes of laser oscillation are continuous oscillation, pulsed mode of free oscillation and pulsed mode with controlled losses (*e.g.*, Q – switching). The continuous mode is typical for gas lasers and pulsed mode is mostly employed in solid-state lasers. However, if necessary any type of laser can be made to operate in any of these modes of oscillation.

(*iv*) The Wonderful Laser Beam

Imagine a helium-neon laser operating in a dark room. The rich red colour of the beam is a wonderful sight in the semidarkness of the room. There is a no divergence (widening) and intensity is practically constant. One can place a number of reflecting mirrors in its way and make the beam trace a zig-zag path in the room and the look is magnificent. If the beam diameter is magnified using a lens and then it falls on a screen such as a sheet of paper an unusual light spot is observed which 'speckles' and dark and bright spots appear and vanish. Such a behaviour of the laser beam is exclusively due to its high degree of coherence. Light and dark specks appear because of the interference of coherent beams reflected to the observer's eyes from different points of the spot. Slight unconscious motions of the observer's head change the angle at the parts of the spot which are seen and modify the conditions of interference, changing the bright spots into dark ones and vice-versa.

Characteristics of a Laser

The significant feature of a laser is the enormous difference between the character of its light and the light from other sources such as the sun, a flame or an incandescent lamp. The characteristics features of lasers are (*i*) directionality, (*ii*) high intensity (*iii*) monochromaticity and (*iv*) high degree of coherence.

Directionality

Directional beam from any conventional light source can be obtained with the help of an aperture. Lasers emit in one direction only. The directionality is expressed as 'full angle beam divergence'. It is twice the angle that the outer edge of the beam makes with the axis of the beam. The outer edge of a beam is defined as a point at which the strength of the beam has dropped to $\frac{1}{e}$ times its value at the centre. For a typical laser, the beam divergence is less than 1 milli-radian.

Intensity

From an ordinary incandescent lamp light spreads more or less uniformly in all directions. The laser gives a narrow light beam of very small diameter, 1 mm or less. Thus energy is concentrated in a small region. This concentration of energy, both spatially and spectrally, accounts for the high intensity of Lasers. Even a 1 mW laser can damage the eyes if looked directly into it. Emission of photons/sec. in lasers is around 10^{16}–10^{28} whereas in thermal sources it is ~10^{12}.

Monochromaticity

The light emitted by a laser is far more monochromatic than that by any conventional monochromatic source. However, no light source including lasers is perfectly monochromatic. Only better and better approximation to the ideal can be sought. Figure 25.3 shows the linewidth of the light emitted from a laser and a conventional light source.

The degree of non-monochromaticity of a wave may be defined as relative band width, $\Delta\nu/\nu_0$ where ν_0 is the central frequency of emission and $\Delta\nu$ is full width at half maximum (FWHM) of the laser line describing the spread of frequencies about the mean value.

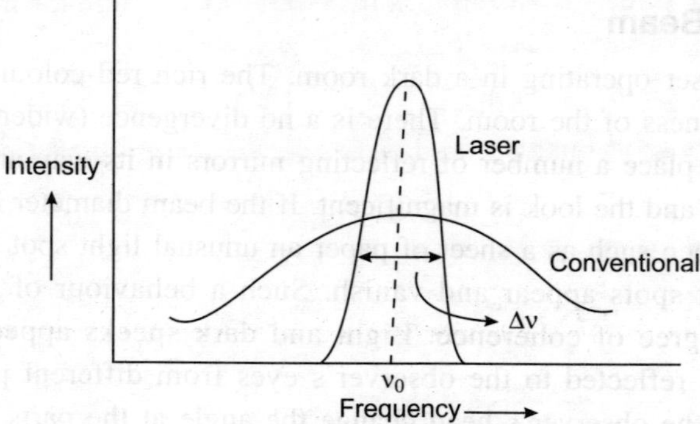

Fig. 25.3

Coherence

Laser radiation has a high degree of coherence and therefore laser radiation can be focussed to a very small spot creating high power density viz. 10^{13} watt in a spot of 1 μm diameter. This facilitates cutting of metals, microwelding, microdrilling through diamond crystals and so on. The coherence time, τ, of laser radiation may be as high as 10^{-3} sec. Therefore, the coherence length τc may be as long as 10^5 m. that is up to 100 km. This value is seven orders of magnitude higher than the coherence length of ordinary light sources. Interference pattern in the experiments namely; Fresnel biprism, Michelson interferometer, Newton's ring, thin film could be observed in pre-laser era because the pathlength difference was very small—in the order of a millimeter at most.

The Einstein Coefficients

We consider two levels of an atomic system as shown in Fig. 25.4 and let N_1 and N_2 be the number of atoms per unit volume present in the energy levels E_1 and E_2 respectively. If radiation at a frequency corresponding to the energy difference $(E_2 - E_1)$ falls on the atomic system, it can interact in three distinct ways:

(*a*) An atom in the lower energy level E_1 can absorb the incident radiation and be excited to E_2. This excitation process requires the presence of radiation. The rate at which absorption takes place from level 1 to level 2 will be proportional to the number of atoms present in the level E_1 and also to the energy density of the radiation at the frequency $\omega = (E_2 - E_1)/\hbar$. Thus if $u(\omega)\, d\omega$ represents the radiation energy per unit volume between ω and $\omega + d\omega$ then we may write the number of atoms undergoing absorptions per unit time per unit volume from level 1 to level 2 as

$$\Gamma_{12} = B_{12}\, u(\omega) N_1 \qquad \text{...(25.2 a)}$$

where B_{12} is a constant of proportionality and depends on the energy levels E_1 and E_2. Notice here that $u(\omega)$ has the units of energy density per frequency interval.

(*b*) For the reverse process, namely the deexcitation of the atom from E_2 to E_1, Einstein postulated that an atom can make a transition from E_2 to E_1 through two distinct processes, namely stimulated emission and spontaneous emission. In the case of stimulated emission, the radiation which is incident on the atom stimulates it to emit radiation and the rate of transition to the lower energy level is proportional to the energy density of radiation at the frequency ω. Thus, the number of stimulated emissions per unit time per unit volume will be

$$\Gamma_{12} = B_{21}\, u(\omega) N_2 \qquad \text{...(25.2 } b)$$

where B_{21} is the coefficient of proportionality and depends on the energy levels.

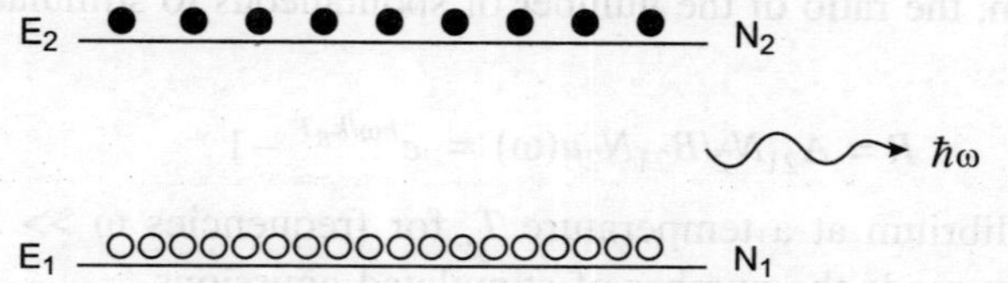

Fig. 25.4. Two states of an atom with energies E_1 and E_2; their corresponding population densities are N_1 and N_2 respectively. At thermal equilibrium $N_2 < N_1$ and $N_2/N_1 = e^{-(E_2-E_1)/k_BT}$

(*c*) An atom which is in the upper energy level E_2 can also make a spontaneous emission; this rate will be proportional to N_2 only and thus we have for the number of atoms making spontaneous emissions per unit time per unit volume.

$$U_{12} = A_{21} N_2 \qquad \text{...(25.2 } c)$$

At thermal equilibrium between the atomic system and the radiation field, the number of upward transitions must be equal to the number of downward transitions. Hence, at thermal equilibrium

$$N_1 B_{12} u(\omega) = N_2 A_{21} + N_2 B_{21}\, u(\omega)$$

or

$$u(\omega) = \frac{A_{21}}{(N_1/N_2)B_{12} - B_{21}} \qquad \text{...(25.2 } d)$$

Using Boltzmann's law, the ratio of the equilibrium populations of levels 1 and 2 at temperature T is

$$N_1/N_2 = e^{(E_2-E_1)/k_BT} = e^{\hbar\omega/k_BT} \qquad \text{...(25.2 } e)$$

where k_B (= –1.38 × 10^{-23} J/K) is the Boltzmann's constant. Hence

$$u(\omega) = \frac{A_{21}}{B_{12}e^{\hbar\omega/k_BT} - B_{21}} \qquad \text{...(25.2 } f)$$

Now according to Planck's law, the radiation energy density per unit frequency interval is given by

$$u(\omega) = \frac{\hbar\omega^3 n_0^3}{\pi^2 c^3} \frac{1}{e^{\hbar\omega/k_BT} - 1} \qquad \text{...(25.2 } g)$$

where c is the velocity of light in free space and n_0 is the refractive index of the medium.

Comparing Eqns. (25.2 f) and (25.2 g), we obtain

$$B_{12} = B_{21} = B \quad ...(25.2\ h)$$

and

$$A_{21}/B_{21} = \hbar\omega^3 n_0^3/\pi^2 c^3 \quad ...(25.2\ i)$$

Thus the stimulated emission rate per atom is the same as the absorption rate per atom and the ratio of spontaneous to stimulated emission coefficients is given by Eqn. (25.2 i). The coefficients A and B are referred to as the Einstein A and B coefficients.

At thermal equilibrium, the ratio of the number of spontaneous to stimulated emissions is given by

$$R = A_{21}N_2/B_{21}N_2u(\omega) = e^{\hbar\omega/k_BT} - 1 \quad ...(25.2\ j)$$

Thus at thermal equilibrium at a temperature T, for frequencies $\omega >> k_BT/\hbar$, the number of spontaneous emissions far exceeds the number of stimulated emissions.

Example: Let us consider an optical source at $T = 1000$ K. At this temperature

$$\frac{k_BT}{\hbar} = \frac{1.38\times10^{-23}\,(\text{J/K})\times10^3\,(\text{K})}{1.054\times10^{-34}\,(\text{Js})} \approx 1.3\times10^{14}\,\text{Hz}$$

Thus for $\omega >> 1.3 \times 10^{14}$ Hz, the radiation would be mostly due to spontaneous emissions. For $\lambda = 5000\text{Å}, \omega \approx 3.8\times10^{15}$ Hz and

$$R \approx e^{29.2} \approx 5.0\times10^{12}$$

Thus at optical frequencies the emission is predominantly due to spontaneous transitions and hence the light from usual light sources is incoherent.

25.3 TYPES OF LASERS

Different types of lasers can be classified into solid, liquid, gas and semiconductor lasers depending on the laser medium. The solid laser was the first one to be designed *i.e.*, the ruby laser. Next came He-Ne laser, a gas laser. Then came diode laser, dye laser, chemical lasers, excimer lasers, gas dynamic lasers, free electron laser etc. which has importance owing to their characteristics in terms of wavelengths, method of excitation, tunability , power levels, cost etc. Frequency of a laser can also be 'up-converted' or 'down-converted' using standard techniques.

The following table summarises different types of lasers and their characteristic wavelength

	Name of the laser	*Active medium*	*Operating wavelength*
	Solid lasers		
1.	Ruby	$Al_2O_3 : Cr^{3+}$	694.3 nm
2.	Calcium fluoride –Uranium laser	$CaF_2 : u^{3+}$	2.49 μm
3.	Nd–YAG laser	$Y_3Al_5O_{12} : Nd^{3+}$	1.064 μm

4.	Nd–glass laser	Glass : Nd^{3+}	1.054 μm
5.	Alexandrite laser	$BeAl_2O_4 : Cr^{3+}$	701 to 818 nm
6.	Titanium Sapphire laser	$Al_2O_3 : Ti$	660 to 1180 nm
7.	Colour Centre lasers		
		KF	1.06 μm
		KCl	1.06 μm
		KCl : Tl	1.06 μm
		KCl : Li	0.514 μm
		KCl : Na	0.514 μm
		LiF	0.647 μm
		NaCl	1.06 μm
8.	Fiber glass laser	Glass fibre : Er	1.53 μm
9.	Dye laser (tunable)		
		Dicyanomethylene	610–705 nm
		Coumarin 102	460–515
		Coumarin 6	506–558
		Coumarin 30	485–535
		Polyphenyl –2	363–410
		Stilbene 1	391–435
		Fluoroscein	500–550
		Carbazines	600–700
	Gas lasers		
10.	He-Ne laser	He–Ne	0.633 μm (most popular)
		10.1	1.15 μm and 3.39 μm
11.	Argon ion laser	Ar	About 25 different wavelengths in visible region between 408.9 to 686.1 nm
12.	Krypton in laser	Kr	406.7 to 676.4 nm
13.	He-Cd laser	He-Cd	441.6, 352, 354 nm
14.	Copper vapour laser	Cu	510.5 and 578.2 nm
15.	Lead vapour laser	Pb	722.9 nm
16.	Gold vapour laser	Au	627.8 nm
17.	CO_2 laser	$CO_2 + N_2 + He$	10.6 and 9.5 μm
18.	Excimer laser	F_2	153 nm
		Ar F	193 mm
		Kr – F	248 mm
		Xe – Cl	308 mm
		Xe – F	353 nm

19.	Nitrogen laser	N_2	337 nm
20.	Far infrared laser	Water vapour Hydrogen cyanide and others	30 μm to 1.8 mm
	Chemical lasers		
21.	Hydrogen chloride laser $H + Cl_2 \rightarrow HCl^*$	$+ Cl$ $+$ heat	3.77 μm
	Hydrogen fluoride laser $H + F_2 \rightarrow HF^* + F$		
	X-ray lasers	Pumping laser and a solid target (like Se)	Discrete wavelengths from 3.56 to 46.9 nm
22.	Free electron laser	Free electrons moving at relativistic speed and undergoing acceleration and deceleration	Tunable continuously from X-rays to sub-millimeter range

25.4 THE HELIUM-NEON LASER

The first continuously operated gas laser was the He-Ne laser. The main problem in a gas laser is, how the atoms can be selectively excited to proper levels in quantities sufficient to achieve the required population inversion. The primary mechanism for excitation used in gas lasers is by electron impact. Suppose atoms of a kind A are exited in a discharge tube, by this method, to a metastable state. Such a state can be populated appreciably at moderate electron densities. *If the discharge tube also contains atoms of other kind B, whose excited state lies very close to that of the metastable state of A, a resonant energy transfer may take place and the atoms B be lifted from their ground state to the excited state*. If the rate of transfer of excitation is larger than the rate of radiative decay of the excited state, the population of the excited state of the atom B will steadily increase and the state will be so populated that inversion may exist between it and another level.

This process was used by Javan *et al.* in He-Ne laser, the first continuously operated gas laser. The energy level schemes of He and Ne are shown in Fig. 25.5.

He atoms are found to be much more readily excited by electron impact than Ne atoms. The atoms can be excited by d.c. or a.c. power supply. The 2^1S and 2^3S levels of He have relatively long life time *i.e.*, they are metastable levels. Laser action however, occurs between energy levels of Ne. The role of He atoms is to assist in the pumping process.

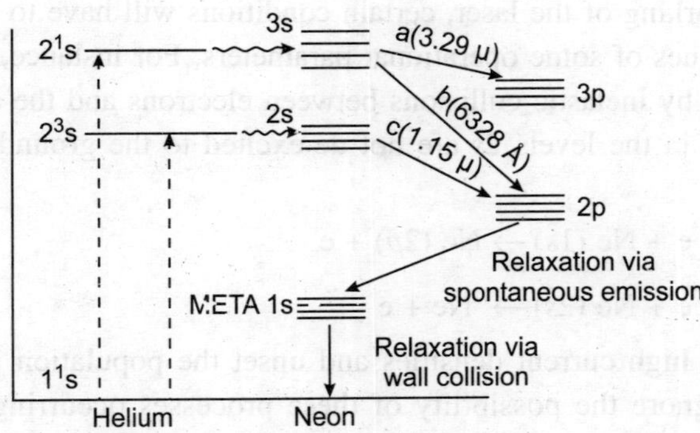

Fig. 25.5. Energy levels of He-Ne system

The metastable levels 2^1S at ~ 20.5 eV and 2^3S at ~ 19.81 eV of helium, coincide in energy with sets of excited levels 3*s* and 2*s* respectively of neon. The notation used here to represent Ne levels is Paschen's notation. The 2*s* group comprises of four levels denoted by $2s_2$, $2s_3$, $2s_4$ and $2s_5$. They may also be represented using Russell-Saunders terminology as $^1P_1$, $^3P_0$, $^3P_1$,$^3P_2$ respectively. However, L-S coupling is not suitable for the description of the Ne system, as many transitions which are forbidden by the selection rules for L-S coupling, are actually observed.

When a helium atom in the metastable state collides with a neon atom in the ground state, an exchange of energy takes place, and a neon atom is lifted to 2*s* or 3*s* level and the helium atom drops back to the ground state. This provides a selective population mechanism which supplies continuously Ne atoms to 2*s* and 3*s* levels increasing their population. The possible transitions allowed by selection rules are those to the ground state and to the 2*p* states. The decay time of *s* levels (τ_s = 100 nsec) is an order of magnitude longer than the decay time of the *p* states (τ_p = 10 nsec). Population inversion, therefore, can be produced between *s* and *p* states thus satisfying the condition for operation as a four-level laser system.

One may ask: Since 2*s* states are also radiatively connected to the ground state, they would decay quickly to the ground state. Is this not a disadvantage as far as the population inversion between *s* and *p* levels is concerned? It is, indeed, so, at very low pressures, when the probability of decay to the ground state vastly exceeds that to the 2*p* levels. However, at a high pressure, of the order of a mm Hg, transitions to the ground state undergo complete resonance trapping; that is, every time a photon is emitted, it, instead of escaping from the gas, is absorbed by another atom in the ground state which thereby ends up in the excited state 2*s*, and since the population of the ground state is large, the process is continuous. Thus, the high population in the 2*s* levels is maintained. The life time of *s* levels, therefore, is determined primarily by the radiative decay to the 2*p* levels.

Figure 25.5 shows the transitions on which laser oscillations can be expected. Of the various transitions from the component level of the 3*s* and 2*s* groups to the levels of 3*p* and 2*p* groups, the following are more prominent:

(*a*) $3s_2 \rightarrow 3p_4$ λ = 3.39 μm

(*b*) $3s_2 \rightarrow 2p_4$ λ = 0.633 μm

(*c*) $2s_2 \rightarrow 2p_4$ λ = 1.15 μm

For effective working of the laser, certain conditions will have to be satisfied which put severe restrictions on the values of some operational parameters. For instance, one has to ensure that (*i*) 2*p* levels are not excited by inelastic collisions between electrons and the atoms in the metastable level 1*s*; and (*ii*) the atoms in the levels 2*s* are not de-excited to the ground state. Both theses processes viz.

$$e + Ne\ (1s) \rightarrow Ne\ (2p) + e \qquad ...(25.4\ a)$$

and

$$e + Ne\ (2s) \rightarrow Ne + e \qquad ...(25.4\ b)$$

become prominent at high current densities and upset the population inversion between 2*s* and 2*p* levels. One cannot ignore the possibility of these processes occurring, because 1*s* levels have no allowed downward transition and may also be radiation trapped on the transition $2p \rightarrow 1s$. Consequently, they tend to build up sizeable population on the 2*p* level. The current density, therefore, must be adjusted to its optimum value. The population of the 1*s* level also must be kept at minimum value and this necessitates the adjustment of the total and partial pressures of the gas mixture and also the size of the tube. The Ne pressure is usually kept much below that of He (P_{Ne} ~ 0.1 Torr, P_{He} ~ 1 Torr) and the diameter of the tube is small (~2 mm) so that 1*s* atoms may collide with the wall and depopulate the level. There was enough evidence to show that the measures, as above, maximize the performance of lasers. Thus, when the discharge was switched off, the 2*s* levels were found to decay with the time variation identical to that of 2^3S states and the amplification was found to be higher during the after-glow than during the current pulse. This is, obviously, because during the after-glow no electrons are available for the excitation of 1*s* atoms to 2*p* levels.

Javan *et al.* observed the continuous wave laser oscillations in a discharge tube containing Ne and He at pressures 0.1 Torr and 1 Torr respectively. The chamber consisted of a long quartz tube of length 80 cm and inside diameter 1.5 cm (Fig. 25.6). At each end of the tube, there was a metal chamber containing high reflection plates. Flexible bellows were used in the end chambers to allow external mechanical tuning of the Fabry-Perot plates. This enabled them to align the reflectors for parallelism to within 6 seconds of arc. Two optically flat windows were provided at the ends of the system which allowed the laser beam to be transmitted without any distortion. The end plates were separated by a distance of 100 cm. The discharge was excited by means of external electrodes using a 28 Mc/sec generator. The input power was around 50 watts. The flat plates were of fused silica and were flat to a hundredth of a wavelength. The high reflectance was achieved by means of 13-layer evaporated dielectric films. The reflectance was 98.9% in the wavelength range 11000 Å–12000 Å.

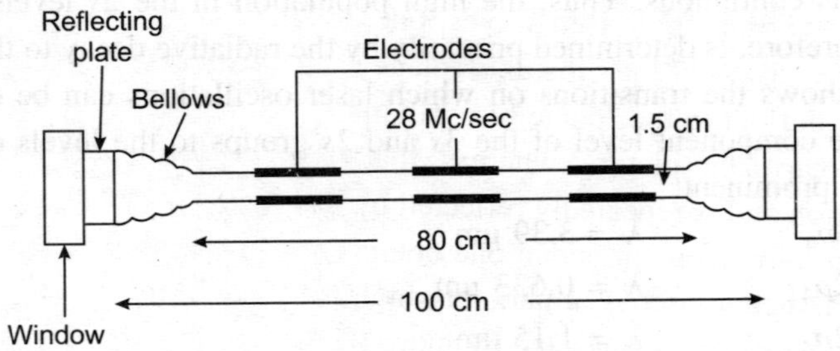

Fig. 25.6. Schematic diagram of the first He-Ne laser

Javan *et al.* found that five different infrared wavelengths, viz. 1.118, 1.153, 1.160, 1.199 and 1.207 μm, corresponding to the different $2s$-$2p$ transitions, could be made to oscillate. The strongest oscillation was found to occur at 1.153 μm corresponding to $2s_2 \to 2p_4$ transition, with an output power of 15 mW. The beam divergence was about one minute of arc which is close to the theoretical diffraction limited value. A screen containing nine slits with spacing of 0.125 cm and each with slit width 0.005 cm was placed across the full 1.125 cm aperture of the beam. The resulting diffraction pattern indicated that there is a very small phase variation over the aperture which means that the beam consisted of perfectly spatially coherent light.

The difficulty of aligning F.P. interferometer with a spacing of 100 cm is not small. Hence, an alternative system is now adopted in which spherical mirrors are employed in a confocal or near confocal configuration (Fig. 25.7). In a confocal resonator the radius of curvature of each mirror is equal to the cavity length. The diffraction losses of such configuration are low and the alignment is much less critical. The divergence, however, is large, but can be reduced by placing a circular aperture in front of the exit mirror. A rigorous treatment of various types of resonators has been given by Fox and Li and by Boyd and Gordon. The mirrors are placed external to the discharge tube. With such an arrangement, it is necessary to minimize unwanted reflections from the windows at the end of the discharge tube. This was found to be possible by the use of Brewster angle windows (Fig. 25.7).

The Brewster angle is given by

$$\tan \theta_B = \eta \qquad \text{...(25.4 c)}$$

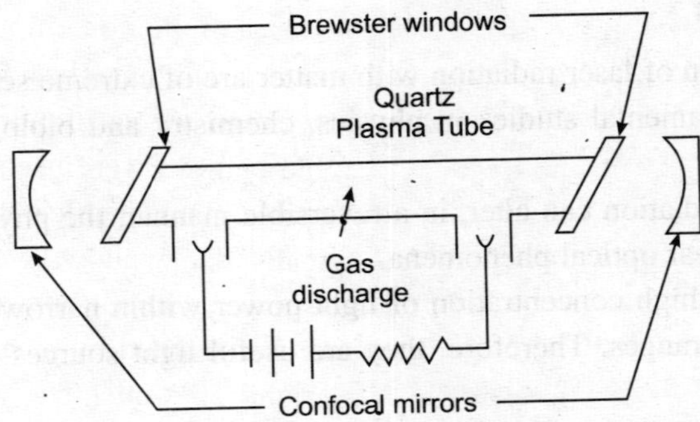

Fig. 25.7. Confocal mirror system with Brewster angle window

where η is the refractive index. An unpolarised wave incident on a plate can be considered as a resultant of two superposed plane polarised waves, one of which is polarised in the plane drawn through the normal to the window and the tube axis (plane of oscillation) and the other normal to this plane. The wave polarised normal to the 'plane of oscillations' is completely reflected by the window plate and, hence, is shut out of the picture; whereas, that polarised in the plane of oscillation is transmitted in the same direction and is repeatedly reflected by the mirrors without causing any losses. An additional advantage of this arrangement is that one obtains a plane polarised laser emission. The spectral purity of the output of such a laser is remarkably high. For brief periods, the oscillation frequency which is around 3×10^{14} c/sec does not vary by more than 1 c/sec.

It can be seen from Fig. 25.5 that depending on certain factors, laser oscillation can also be achieved on the transitions of the type '*a*' and '*b*', and with the rapid development in the field, laser action was discovered at 3.39 μm (*a*) 6328 Å (*b*). The 6328 Å He-Ne laser is one of the most popular and most widely used lasers. The commercial model of this type of laser requires 5 to10 W of excitation power and produces 0.5 to 50 mW of cw laser output.

The He-Ne laser is made to oscillate on a particular transition by making the multilayer dielectric mirrors in such a way that they have a maximum reflectivity at the desired wavelength. Another interesting way of obtaining laser oscillations at a particular wavelength is to use a dispersion prism (Fig. 25.8). A prism placed inside the resonator provides wavelength selectivity. The light waves of different frequencies emitted by the active medium are spatially separated by the prism. If the right-hand mirror of the resonator is perpendicular to the propagation direction of, say λ_1 wave, then λ_2 is incident on the mirror obliquely and is not returned into the active medium after reflection. One can change the wavelength by rotating the mirror about an axis perpendicular to the plane.

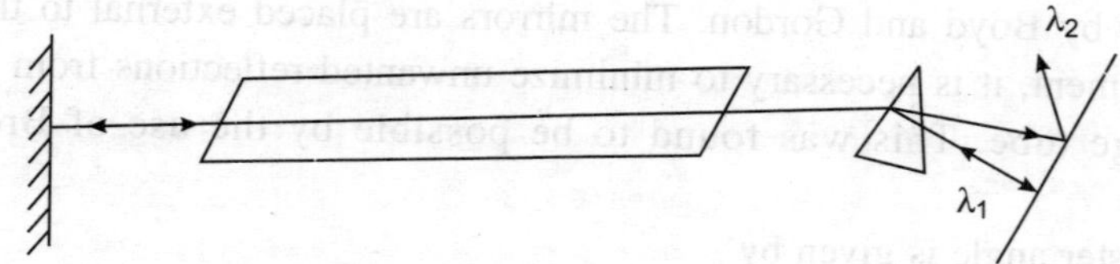

Fig. 25.8. Wavelength selection

Application of Lasers

Studies of interaction of laser radiation with matter are of extreme scientific significance. Lasers are widely used for fundamental studies in physics, chemistry and biology and also in engineering sciences.

High power laser radiation can alter, in a reversible manner, the physical behaviour of material leading to diverse non linear optical phenomena.

The laser facilitates high concentration of light power within narrow bandwidths and frequency tuning over considerable ranges. Therefore, they are useful light sources in investigations of optical spectra of materials.

Tunable lasers facilitate selective excitation of atoms or molecules or for selective rapture of chemical bonds. This opens up the possibility for initiating the desired chemical reactions, controlling these reactions and studying their kinetics.

Picosecond pulses available from lasers render study of ultrafast processes viz. biomedicals, photosynthesis etc. possible.

The numerous fields of laser applications may be broadly categorised into two groups. One involves applications where laser beams – as a rule of high power – are exploited to produce a targetted effect on material such as welding, cutting etc. The other group involves data transmission and processing, measurements, and quality control.

25.5 EFFECTS OF STRONG LASER RADIATION ON MATERIALS

Using optical lenses a laser beam can be focussed into a light spot of 10–100 μm diameter on the surface of a material thereby leading to a very high irradiance. If 1 kW output of a CO_2 laser is focussed in a spot of 100 μm diameter, the resultant irradiance will be 10^7 W/cm^2. Around 10^5 W/m^2 the material starts melting. At still higher levels of irradiance it starts boiling and an intense evaporation occurs. As the beam irradiance grows further, vapour is ionized by the light to produce a high temperature plasma. The plasma may be absorptive and prevent the beam from entering the material. Accordingly, power densities realised in material working systems are decided.

For welding applications relatively low peak power but longer duration pulses (10^{-3} to 10^{-2} sec) are suitable. For perforation and hole drilling where intense evaporation is essential intense and short pulses (~ 10^{-4} sec) are required. The wavelength of the laser is also important as it governs the portion of light absorbed by the surface of the material. However, a beam of shorter wavelength may be focussed into a smaller light spot.

Laser Welding

The welding is contactless thereby precluding any possibility of impurity in the weldment. Unlike electron-beam welding running in a vacuum, laser welding is performed in the atmosphere. It offers a possibility to weld in inaccessible regions, too. Laser welding is capable of a fast and accurate local melting at a given point or along a line. The heat affected area is very small and this is very useful where welding is to be made in the vicinity of heat sensitive components viz. in micro-electronics. Lasers used are Nd: YAG, Nd: glass, ruby and CO_2. Laser welding lends itself well to automatic processes in automotive production lines, for joining titanium and aluminium sheets in ship building, for trunk pipe lines construction etc. Welding of nonmetallic solids is also an interesting application of lasers viz. glasses are weld by lasers of 100 W output while quartz needs 300 W.

Laser heat treatment of surface enhances the strength of the treated area *e.g.*, in automotive industry: cylinder blocks, valves, gears etc. Most commonly used is 1 kW CO_2 laser operating in continuous wave mode.

Laser Cutting

Lasers can be used to cut paper, cloth, plywood, glass, asbestos sheets, ceramics, sheet metals etc. in two as well as three dimensions according to a complicated profile with easy automation of the process and high production rates. For example, 50 mm thick board may be cut with a 200 W CO_2 laser, width of the cut being 0.7 mm, sheets of plywood require 8 kW CO_2 laser, glass of 10 mm thickness requires 20 kW lasers. Metals can be cut with 100 to 500 W power provided the beam-heated material is blown with a jet of oxygen so that gas-laser cutting results.

Drilling and Perforating Holes

Hole perforation by laser relies on intense evaporation of material heated by powerful light pulses of 10^{-4} to 10^{-3} sec duration and power density ~ 10^7 W/cm^2 at the material surface. Nd :YAG laser suits for metallic target only. CO_2 laser equally suits to metallic and non-metallic (plastic, ceramic, glass) targets very large aspect ratios (length/ diameter) are possible, for example, lasers can perforate very small diameter (0.2 mm) holes to large values of depths. Further, advantage is the possibility of drilling holes in close vicinity with each other and near the item edge.

Lasers in Medicine

The focused laser beam proves to be a unique scalpel, capable of bloodless surgery since the beam not only cuts but also 'welds' blood vessel being cut. Such surgery is sterile because it is contact-less. It is painless also as it is very fast.

It finds use in ophthalmology in treating the detachment of retina, cataract and glaucoma.

Isotope Separation

Several applications in the industry, medicine and research require substances enriched in a particular isotope viz. deuterium against hydrogen. Laser separation of isotopes exploits the fact that various isotopes of an element exhibit different absorption bands in their spectra. These absorption bands are fairly narrow and lie close to each other in the spectrum. To excite one isotope without disturbing the other, one has to irradiate their mixture by a source of narrow bandwidth centered on the wavelength of the required isotope. It is desirable that this source be tunable so that the radiation may be tuned to the desired wavelength. This facility is offered by tunable lasers. Thus, the excited atoms of the required isotope are raised to the upper level, while the other isotope remains in the ground state. The mixture of isotopes is then irradiated by another radiation which is absorbed by the excited atoms only to ionise them. As a result, the desired isotope is obtained in the form of ions easily separable by applying a dc electric field. This is essentially the idea of one of the laser separation techniques, called two-step photo-ionisation.

Optical Communication

Besides using optical waveguides (optical fiber) laser communication in open space can be established between satellites, satellites and aircraft, and satellite and stations on the ground. Unguided communications can exist also between stations on the earth surface and under water.

Ranging and Measurement

Similar to conventional microwave radar, the optical radar using laser, also called 'rangefinder' by militaries has been designed to detect distant objects and retrieve information about these objects from the radar signal reflected from these objects. High carrier frequency, possibility of extremely narrow directionality of the radiation, operation in the range of nanosecond and picosecond pulses, all these features predetermine a number of advantages of optical radar systems over ordinary radars. The high resolution of the method makes possible the determination of the size and shape of the object and its orientation.

For velocity measurements laser rangefinders measure Doppler shift in the signal frequency due to the velocity of the ranging object collinear with the laser beam. The higher the carrier frequency, the greater the detected Doppler shift and the accuracy of velocity evaluation.

The disadvantage is that the laser wavefront is degraded and attenuated in the medium where it propagates and then adaptive optics method has to be used to cope this drawback.

Interferometric distance measurements are performed using a frequency – stabilised He-Ne laser operating in the central longitudinal mode. This technique controls tool motions, automatically compensating for the errors due to wear, precisely aligns fixtures in aircrafts engine manufacturing etc.

Flow velocity measurements are also along using 10 mW He-Ne laser.

The Laser Gyroscope

Such a gyroscope measures angular velocity with a precision of 10^{-3} degree/hour and is capable of tracking very small angular velocities. This precision is comparable with that available using the most sophisticated and expensive conventional gyroscopes. The laser gyros are useful for altitude control of satellites orbiting spacecrafts and aircrafts.

It involves a co-propagating and a counter-propagating laser beam which interfere and a shift in interference fringes is used to detect the angular rotation.

Laser Monitoring of the Environment

Laser properties of directionality and monochromaticity are put to good use for measurement of the concentration of various atmospheric pollutants. These include oxides of nitrogen, carbon monooxide, sulphur dioxide, a variety of particulate matter such as dust, smoke and fly ash. The laser techniques perform these measurements by remotely sensing the composition of the atmosphere with a light beam without the necessity of sample collection or any chemical processing. The results of the measurements are available readily and there is no distortion of the quantities being measured. Consequently, these techniques yield real-time data and are extremely suitable for sounding time variations of the atmosphere *i.e.*, for environmental monitoring.

FIBER OPTICS AND SENSORS

26.1 ADVANTAGES OF OPTICAL FIBERS

The following are the main advantages of optical fibers:

1. Attenuation in a fiber is markedly lower than that of coaxial cable or twisted pair and is constant over a very wide range. So transmission within wide range of distance is possible without repeaters etc.
2. Smaller size and lighter weight. Optical fibers are considerably thinner than coaxial cable or bundled twisted-pair cable. So they occupy much less space.
3. Electromagnetic isolation. Electromagnetic waves generated from electrical disturbances or electrical noises do not interfere with light signals. As a result, the system is not vulnerable to interference, impulse noise, or cross-talk.
4. No physical electrical connection is required between the sender and the receiver.
5. The fiber is much more reliable, because it can better withstand environmental conditions, such as pollution, radiation and salt produces no corrosion. Moreover, it is nominally affected by the nuclear radiation. Its life is longer in comparison to copper wire.
6. Almost there is no cross-talk in optical fibers and hence transmission is more secure and private, as it is very difficult to tap into a fiber.
7. Greater bandwidth. Bandwidth of the optical fiber is higher than that of an equivalent wire transmission line.
8. As fibers are very good dielectrics, isolation coating is not required.
9. Data rate is much higher in a fiber and hence much more information can be carried by each fiber than by equivalent copper cables. As for example: at the immensely high frequencies of an optical fiber, data rates of 2 Gbps over tens of kilometers have been demonstrated. Compare this to the practical maximum of hundreds of Mbps over about 1km for coaxial cable and just a few Mbps over 1 km for twisted pair.
10. The cost per channel is lower than that of an equivalent wire cable system. It is expected that in the near future the optical fiber communication system will be more economical than any communication system of other types.
11. Due to the non-inductive and non-conductive nature of a fiber, there is no radiation and interference on other circuits and systems.

12. Greater repeater spacing : Fewer repeaters indicate lower cost and fewer sources of error. It has been observed that a fiber transmission system can achieve a data rate of 5 Gbps over a distance of 111 km without repeaters, whereas coaxial and twisted-pair systems generally have repeaters every few kilometers.
13. The raw material is available in plenty.

Fiber as a Guiding Medium

Optical frequencies are extremely large ($\sim 10^{15}$ Hz), as compared to conventional radio waves ($\sim 10^{6}$ Hz) and microwaves ($\sim 10^{10}$ Hz), a light beam acting as a carrier wave is capable of carrying far more information in comparison to radio waves and microwaves.

The discovery of the laser, some experiments on the propagation of information-carrying light wave through the open atmosphere were carried out, but it was realised that because of the vagaries of the terrestrial atmosphere *e.g.*, rain, fog etc., in order to have an efficient and dependable communication system, one would require a guiding medium in which the information-carrying light waves could be transmitted. This guiding medium is the optical fiber which is hair thin and guides the light beam from one place to another.

In addition, to the capability of carrying a huge amount of information, fibers fabricated with recently developed technology are characterized by extremely low losses (~0.2 dB/km) as a consequences of which the distance between two consecutive repeaters (used for revamping the attenuated signals) can be as large as 250 km. In a recently developed fiber optic system it has been possible to send 140 Mbit/s information through a 220 km link of one optical fiber, this is equivalent to about 4,50,000 voice channel. In comparison, the copper cables used today have repeater spacings every few kilometers or so.

In addition to the long distance communication systems, optical fibers are also being extensively used for local area networks (LANs) – networks that wire up telephones, televisions, computers or robots in offices and cities.

26.2 TYPES OF FIBER

An optical fiber is a dielectric waveguide that operates at optical frequencies.

This fiber wave guide is normally cylindrical in form. It confines electromagnetic energy in the form of light to within its surface and guides the light in a direction parallel to its axis. The transmission properties of an optical waveguide are dictated by its structural characteristics, which have a major effect in determining how an optical signal is affected as it propagates along the fiber. The structure basically establishes the information carrying capacity of the fiber and also influences the response of the waveguide to environmental perturbation.

The propagation of light along a waveguide can be described in terms of a set of guided electromagnetic waves called the modes of the waveguide. These guided modes are referred to as the bound or trapped modes of the waveguide. Each guided mode is a pattern of electric and magnetic field lines that is repeated along the fiber at intervals equal to the wavelength. Only a certain discrete number of modes are capable of propagating along the guide, these modes are those electromagnetic waves that satisfy the homogeneous wave equation in the fiber and the boundary condition at the waveguide surfaces.

A circular solid core of refractive index n_1 is surrounded by a cladding having a refractive index $n_2 < n_1$. In principle, a cladding is not necessary for light to propagate along the core of the fiber, it serves several purposes. The cladding reduces scattering loss resulting from dielectric discontinuities at the core surface, it adds mechanical strength to the fiber, and it protects the core from absorbing surface contaminants with which it could come in contact.

In low and medium-loss fibers the core material is generally glass which is surrounded by either a glass or a plastic cladding. Higher-loss plastic core fibers with plastic claddings are also widely in use. In addition, most fibers are encapsulated in an elastic, abrasion-resistant plastic material. This material adds further strength to the fiber and mechanically isolates or buffers the fibers from small geometrical irregularities, distortions, or roughness of adjacent surfaces. These perturbations could otherwise cause scattering losses induced by random microscopic bends that can arise when the fibers are incorporated into cables or supported by other structures.

Variations in the material composition of the core give rise to the two commonly used fiber type. In the first case, the refractive index of the core is uniform throughout and undergoes an abrupt change (or step) at the cladding boundary. This is called a step-index fiber.

In the second case, the core refractive index is made to vary as a function of the radial distance from the center of the fiber. This type is a graded-index fiber.

Both the step and the graded index fibers can be further divided into single-mode and multimode classes. As the name implies, a single-mode fiber sustains only one mode of propagation whereas multimode fibers contain many hundreds of modes. Multimode fibers offer several advantages compared to single-mode fibers. The larger core radii of multimode fibers make it easier to launch optical power into the fiber and facilitate the connecting together of similar fibers.

A disadvantage of multimode fibers is that they suffer from intermodal dispersion. When an optical pulse is launched into a fiber, the optical power in the pulse is distributed over all (or most) of the modes of the fiber. Each of the modes that can propagate in a multimode fiber travels at a *slightly different velocity*. This means that the modes in a given optical pulse arrive at the fiber end at slightly different times, thus causing the pulse to spread out in times as it travels along the fiber. This effect, which is known as intermodal dispersion, can be reduced by using a graded-index profile in the fiber core: This allows graded-index fibers to have much larger bandwidths (data rate transmission capabilities) than step-index fibers.

Refractive Index Profiles

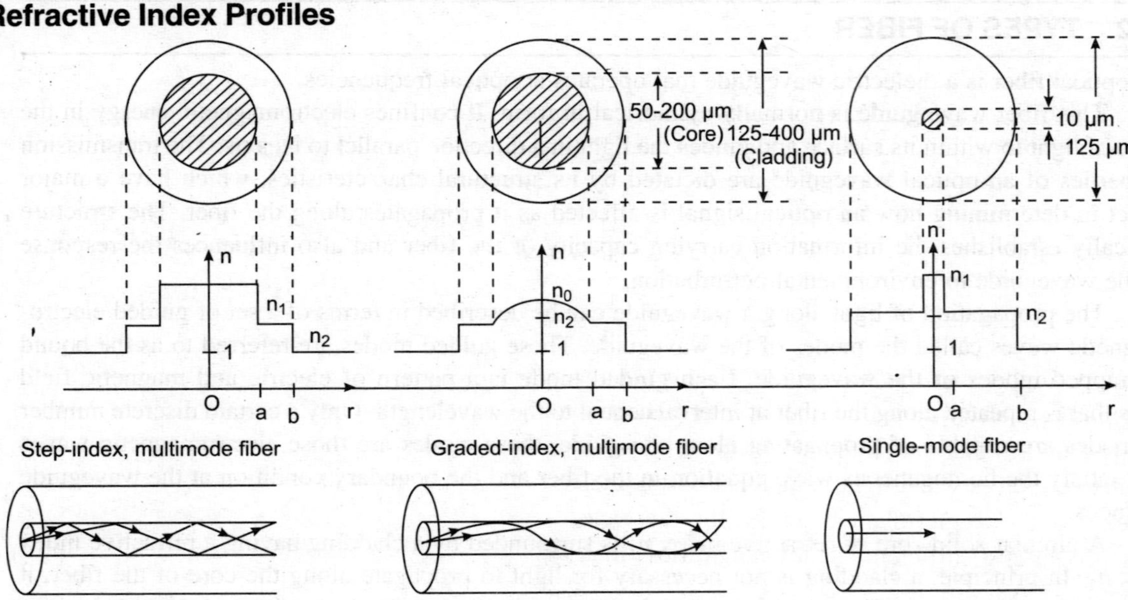

Fig. 26.1. Fiber cross-section and ray paths

Total Internal Reflection

A ray of light travels more slowly in an optically dense medium than in one that is less dense, and the refractive index gives a measure of this effect. When a ray is incident on the interface between two dielectrics of differing refractive indices (*e.g.* glass-air), refraction occurs, as illustrated in figure (*a*). It may be observed that the ray approaching the interface is propagating in a dielectric of refractive index n_1 and is at an angle ϕ_1 to the normal at the surface of the interface. If the dielectric on the other side of the interface has a refractive index n_2 which is less than n_1, than the refraction is such that the ray path in this lower index medium is at an angle ϕ_2 to the normal, where ϕ_2 is greater than ϕ_1. The angles of incidence ϕ_1 and refraction ϕ_2 are related to each other and to the refractive indices of the dielectrics by Snell's law of refraction, which states that

$$n_1 \sin \phi_1 = n_2 \sin \phi_2$$

or

$$\frac{\sin \phi_1}{\sin \phi_2} = \frac{n_2}{n_1} \qquad \text{...(26.2 } a\text{)}$$

It may also be observed in Fig. 26.2 (*a*) that a small amount of light is reflected back into the originating dielectric medium (partial internal reflection). *As n_1 is greater than n_2, the angle of refraction is always greater than the angle of incidence*. Thus, when the angle of refraction is 90° and the refracted ray emerges parallel to the interface between the dielectric the angle of incidence must be less than 90°. This is the limiting case of refraction and the angle of incidence is now known as the critical angle θ_c, as shown in Fig. 26.2 (*b*).

From Eqn.(26.2 *a*) the value of the critical angle is given by

$$\sin \theta_c = \frac{n_2}{n_1} \qquad \text{...(26.2 } b\text{)}$$

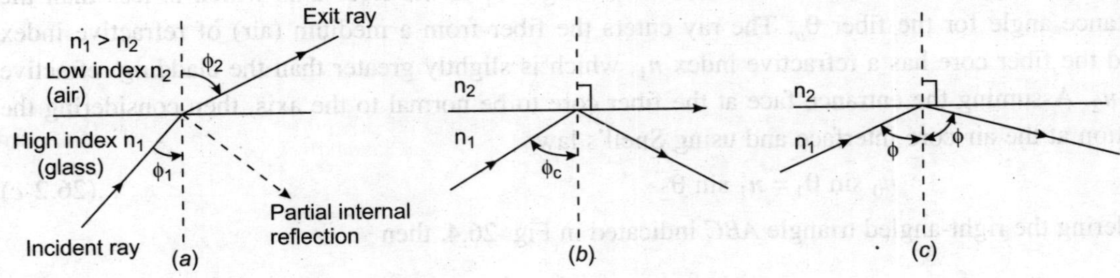

Fig. 26.2

At angles of *incidence greater than the critical angle the light is reflected back into the originating* dielectric medium (TIR) with high efficiency (around 99.9%). Hence, it may be observed in Fig. 26.2 (*c*) that total internal reflection occurs at the interface between two dielectrics of differing refractive indices when light is incident on the dielectric of lower index from the dielectric of higher index, and the angle of incidence of the ray exceeds the critical value. This is the mechanism by which light at a sufficient shallow angle (less than 90° – ϕ_c) may be considered to propagate down an optical fiber with low loss.

Acceptance Angle

The geometry concerned with launching a light ray into an optical fiber is shown in Fig. 26.3, which illustrates a meridional ray *A* at the critical angle ϕ_c within the fiber at the core-cladding interface. It may be observed that this ray enters the fiber core at an angle θ_a to the fiber axis and is refracted at

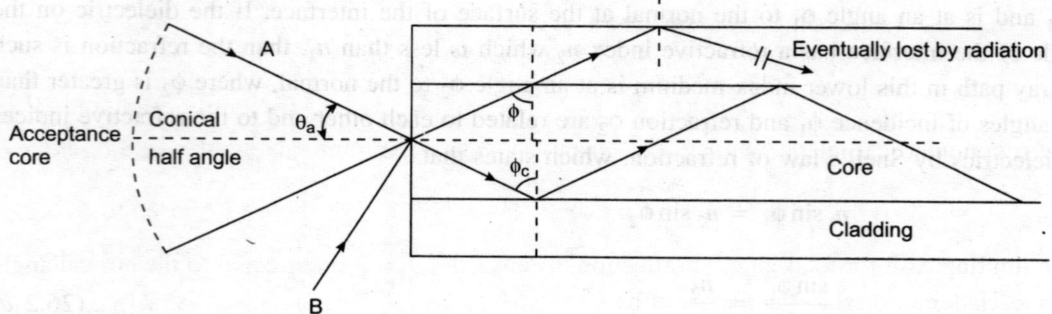

Fig. 26.3

the air-core interface before transmission to the core-cladding interface at the critical angle. Hence, any rays which are incident into the fiber core at an angle greater than θ_a will be transmitted to the core-cladding interface at an angle less than ϕ_c, and will not be totally internally reflected. This situation is shown in Fig. 26.3, where the incident ray *B* at an angle greater than θ_a is refracted into the cladding and eventually lost by radiation. Thus for rays to be transmitted by total internal reflection within the fiber core they must be incident on the fiber core within an acceptance core defined by the conical half angle θ_a. Hence θ_a is the maximum angle to the axis at which light may enter the fiber in order to be propagated and is often referred to as the acceptance angle for the fiber.

Numerical Aperture

A light ray incident on the fiber core at an angle θ_1 to the fiber axis which is less than the acceptance angle for the fiber θ_a. The ray enters the fiber from a medium (air) of refractive index n_0, and the fiber core has a refractive index n_1, which is slightly greater than the cladding refractive index n_2. Assuming the entrance face at the fiber core to be normal to the axis, then considering the refraction at the air-core interface and using Snell's law

$$n_0 \sin \theta_1 = n_1 \sin \theta_2 \qquad ...(26.2\ c)$$

considering the right-angled triangle *ABC* indicated in Fig. 26.4, then

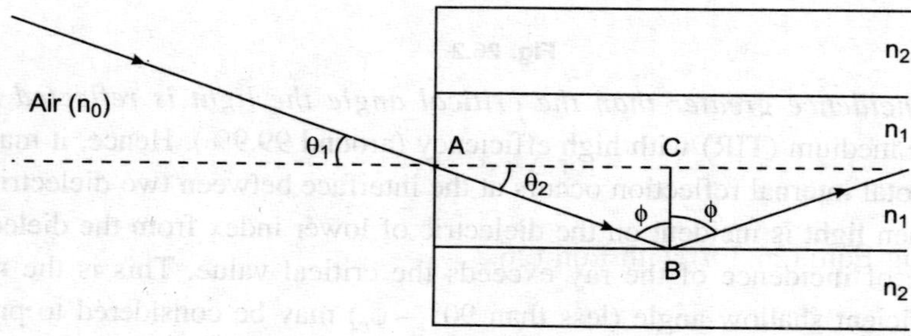

Fig. 26.4

$$\phi = \frac{\pi}{2} - \theta_2 \qquad ...(26.2\ d)$$

where ϕ is greater than the critical angle at the core-cladding interface. Hence Eqn. (26.2 *c*) becomes

$$n_0 \sin \theta_1 = n_1 \sin \left(-\phi + \frac{\pi}{2} \right)$$

$$= n_1 \cos \phi \qquad ...(26.2\ e)$$

using the trigonometrical relationship $\sin^2\phi + \cos^2\phi = 1$ Eqn. (26.2 *e*) may be written in the form

$$n_0 \sin \theta_1 = n_1 (1 - \sin^2\phi)^{\frac{1}{2}} \qquad ...(26.2\ f)$$

when the limiting case for total internal reflection is considered, ϕ becomes equal to the critical angle for the core-cladding interface and is given by

$$\sin \theta_c = \frac{n_2}{n_1}$$

Also in this limiting case θ_1 becomes the acceptance angle for the fiber θ_a. Combining these limiting cases into Eqn. (26.2 *g*) gives

$$n_0 \sin \theta_a = (n_1^2 - n_2^2)^{\frac{1}{2}} \qquad ...(26.2\ g)$$

Equation (26.2 *g*), apart from relating the acceptance angle to the refractive indices, serves as the basis for the definition of the important optical fiber parameter, the numerical aperture (NA). Hence the NA is defined as:

$$\boxed{\mathrm{NA} = n_0 \sin\theta_a = (n_1^2 - n_2^2)^{\frac{1}{2}}} \qquad ...(26.2\ h)$$

Since the NA is often used with the fiber in air where n_0 is unity, it is simply equal to $\sin \theta_a$. It may also be noted that incident meridional rays over the range $0 \le \theta_1 \le \theta_a$ will be propagated within the fiber.

The numerical aperture may also be given in terms of the relative refractive index difference Δ between the core and the cladding which is defined as

$$\Delta = \frac{n_1^2 - n_2^2}{2n_1^2} = \frac{(n_1 - n_2)(n_1 + n_2)}{2n_1^2} = \frac{(n_1 - n_2)2n_1}{2n_1^2} \qquad ...(26.2\ i)$$

$$\approx \frac{n_1 - n_2}{n_1} \text{ for } \Delta << 1$$

Hence combining Eqn. (26.2 *h*) with Eqn (26.2 *i*), we can write

$$\boxed{\mathrm{NA} = n_1 (2\Delta)^{\frac{1}{2}}} \qquad ...(26.2\ j)$$

The relationships given in Eqns. (26.2 *h*) and (26.2 *j*) for the numerical aperture are a very useful measure of the light-collecting ability of a fiber.

They are independent of the fiber core diameter and will hold for diameters as small as 8 μm. However, for smaller diameters they break down as the geometric optics approach is invalid. This is because the ray theory model is only a partial description of the character of light.

Q. A silica optical fiber with a core diameter large enough to be considered by ray theory analysis has a core refractive index of 1.50 and a cladding refractive index of 1.47. Determine (*a*) the critical angle at the core cladding interface (*b*) the NA for the fiber (*c*) the acceptance angle in air for the fiber.

Ans. (*a*) The critical angle ϕ_c at the core cladding interface is given by

$$\phi_c = \sin^{-1}\left(\frac{n_2}{n_1}\right) = \sin^{-1}\left(\frac{1.47}{1.50}\right) = 78.5°$$

(*b*) The numerical aperture $\text{NA} = (n_1^2 - n_2^2)^{\frac{1}{2}} = (1.50^2 - 1.47^2)^{\frac{1}{2}} = (2.25 - 2.16)^{\frac{1}{2}} = 0.30$

(*c*) The acceptance angle in air $\theta_a = \sin^{-1}$ (NA) $= \sin^{-1}$ (0.30) = 17.4°

Q. A typical relative refractive index difference for an optical fiber designed for long distance transmission is 1%. Estimate the NA and the solid acceptance angle in air for the fiber when the core index is 1.46. Further, calculate the critical angle at the core cladding interface within the fiber. It may be assumed that the concepts of geometric optics hold for the fiber.

Ans. With $\Delta = 0.01$ the numerical aperture is

$$\text{NA} = n_1(2\Delta)^{\frac{1}{2}} = 1.46(0.02)^{\frac{1}{2}} = 0.21$$

For small angles the solid acceptance angle in air G is given by

$$G = \pi\theta_a^2 = \pi\sin^2\theta_a = \pi(\text{NA})^2 = \pi 0.04 = 0.13 \text{ rad.}$$

The relative refractive index difference Δ;

$$\Delta \simeq (n_1 - n_2)/n_1 = 1 - \frac{n_2}{n_1}$$

Hence

$$\frac{n_2}{n_1} = 1 - \Delta = 1 - 0.01 = 0.99$$

The critical angle at the core cladding interface is

$$\phi_c = \sin^{-1}\left(\frac{n_2}{n_1}\right) = \sin^{-1}(0.99) = 81.9°$$

26.3 CONCEPT OF MODES

An optical fiber confines electromagnetic energy in the form of light to within its surfaces and guides the light in a direction parallel to its axis (The transmission properties of an optical waveguide are dictated by its structural characteristics, which have a major effect in determining how an optical signal is affected as it propagates along the fiber. The *structure basically establishes the information-carrying capacity* of the fiber and also influences the response of the waveguide to environmental perturbations.

The propagation of light along a waveguide can be described in terms of a set of guided electromagnetic waves called the modes of the waveguide. These guided modes are referred to as the bound or trapped modes of the waveguide. Each guided mode is a pattern of electric and magnetic field lines that is repeated along the fiber at intervals equal to the wavelength. Only a certain discrete number of modes are capable of propagating along the guide. These modes are those electromagnetic waves that satisfy the homogeneous wave equation in the fiber and the boundary condition at the waveguide surfaces.

The electromagnetic light field that is guided along an optical fiber can be represented by a superposition of bound or trapped modes. *Each of these guided modes is composed of a set of simple electromagnetic field configurations which form a standing wave pattern in the transverse direction* (*i.e.*, transverse to the waveguide axis).

Maxwell's Equations

Maxwell's equations that give the relationships between the electric and magnetic fields are:

$$\vec{\nabla} \times \vec{E} = -\frac{\partial \vec{B}}{\partial t} \qquad \text{...(26.3 } a\text{)}$$

$$\vec{\nabla} \times \vec{H} = \frac{\partial \vec{D}}{\partial t} \qquad \text{...(26.3 } b\text{)}$$

$$\vec{\nabla} \cdot \vec{D} = 0 \qquad \text{...(26.3 } c\text{)}$$

$$\vec{\nabla} \cdot \vec{B} = 0 \qquad \text{...(26.3 } d\text{)}$$

Where $\vec{D} = \varepsilon \vec{E}$ and $\vec{B} = \mu \vec{H}$. ε is the permittivity (or dielectric constant) and μ is the permeability of the medium. Taking the curl of Eqn. (26.3 *a*) and making use of Eqn. (26.3 *b*) yields.

$$\vec{\nabla} \times (\vec{\nabla} \times \vec{E})$$

$$= -\nabla \times \frac{\partial \vec{B}}{\partial t} = -\frac{\partial}{\partial t}(\nabla \times B) = -\frac{\partial}{\partial t}(\nabla \times \mu H) = -\mu \frac{\partial}{\partial t}(\nabla \times H) = -\mu \frac{\partial}{\partial t}\frac{\partial D}{\partial t} = -\mu\varepsilon \frac{\partial^2 E}{\partial t^2} \qquad \text{...(26.3 } e\text{)}$$

Using the vector identity $\vec{\nabla} \times (\vec{\nabla} \times \vec{E}) = \vec{\nabla}(\vec{\nabla} \cdot \vec{E}) - \nabla^2 E = -\nabla^2$

Comparing Eqns. (26.3 *e*) and (26.3 *f*), we get $\nabla^2 E = \varepsilon\mu \frac{\partial^2 E}{\partial t^2}$ ...(26.3 *f*)

Similarly, by taking the curl of Eqn. (26.3 *b*), it can be shown that

$$\nabla^2 H = \varepsilon\mu \frac{\partial^2 H}{\partial t^2} \qquad \text{...(26.3 } g)$$

Equations (26.3 *f*) and (26.3 *g*) are the standard wave equations.

Field Components in Optical Waveguide

Now consider electromagnetic waves propagating along a cylindrical fiber shown in Fig. 26.5. For this fiber a cylindrical coordinate system $\{r, \phi, z\}$ is defined with the *z*-axis lying along the axis of the waveguide. If the electromagnetic waves are to propagate along the *z*-axis, they will have a functional dependence of the form.

$$E = E_0(r, \phi)e^{j(\omega t - \beta z)} \qquad \text{...(26.3 } h)$$

$$H = H_0(r, \phi)e^{j(\omega t - \beta z)} \qquad \text{...(26.3 } i)$$

For the curl in cylindrical coordinates refer to –

$$\nabla \times \vec{A} = \left(\frac{1}{r}\frac{\partial A_z}{\partial \phi} - \frac{\partial A_\phi}{\partial z}\right)a_r + \left(\frac{\partial A_r}{\partial z} - \frac{\partial A_z}{\partial r}\right)a_\phi + \frac{1}{r}\left[\frac{\partial}{\partial r}(rA_\phi) - \frac{\partial A_r}{\partial \phi}\right]a_z$$

which are harmonic in time *t* and coordinate *z*. The parameter β is the *z*-component of the propagation vector and will be determined by the boundary conditions on the electromagnetic fields at the core-cladding interface.

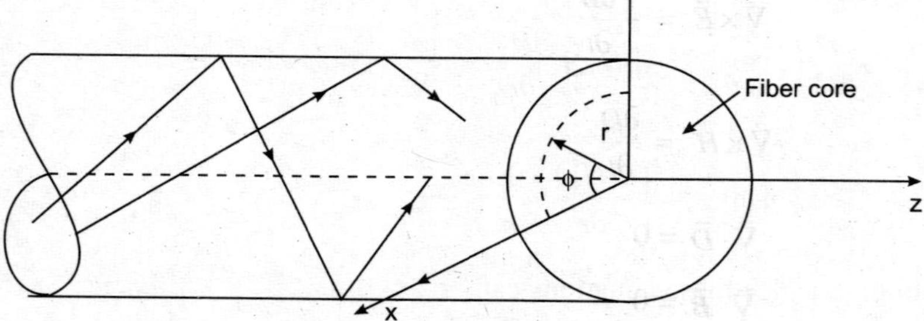

Fig. 26.5

When Eqns. (26.3 *h*) and (26.3 *i*) are substituted into Maxwell's curl equations, we have, from Eqn. (26.3 *a*),

$$\nabla \times E = -\frac{\partial B}{\partial t}$$

$$= -\frac{\partial}{\partial t}\mu H_0 e^{j(\omega t - \beta z)}$$

$$= -j\omega\mu H_0 e^{j(\omega t-\beta z)} = -j\omega\mu H$$

On transforming Maxwell's field equation in polar coordinate system (r, ϕ, z) we get

In the cylindrical system of coordinates, we have $\nabla \times E = -\partial B/\partial t$

$$-j\omega\mu H = \begin{vmatrix} \frac{1}{r}\hat{r} & \hat{\phi} & \frac{1}{r}\hat{z} \\ \frac{\partial}{\partial r} & \frac{\partial}{\partial \phi} & \frac{\partial}{\partial z} \\ E_r & E_\phi & E_z \end{vmatrix} = -\left[\mu\frac{\delta H_r}{\delta t}\hat{r} + \mu\frac{\delta H_\phi}{\delta t}\hat{\phi} + \mu\frac{\delta H_z}{\delta t}\hat{z}\right]$$

Thus
$$-j\omega\mu H_r = \frac{1}{r}\frac{\partial E_z}{\partial \phi} + j\beta E_\phi \qquad \text{...(26.3 } j)$$

$$j\omega\mu H_\phi = j\beta E_r + \frac{\partial E_z}{\partial r} \qquad \text{...(26.3 } k)$$

$$-j\mu\omega H_z = \frac{1}{r}\frac{\partial}{\partial r}(rE_\phi) - \frac{1}{r}\frac{\partial E_r}{\partial \phi} \qquad \text{...(26.3 } l)$$

and from Eqn. (26.3 *b*)
$$j\varepsilon\omega E_r = \frac{1}{r}\frac{\partial H_z}{\partial \phi} + j\beta H_\phi$$

$$-j\varepsilon\omega E_\phi = j\beta H_r + \frac{\partial H_z}{\partial r}$$

$$j\varepsilon\omega E_z = \frac{1}{r}\frac{(rH_\phi)}{\partial r} - \frac{1}{r}\frac{\partial H_r}{\partial \phi} \qquad \text{...(26.3 } m)$$

By eliminating variables these equations can be rewritten such that when E_z and H_z are known, the remaining transverse components E_r, E_ϕ, H_r and H_ϕ can be determined.

$$E_r = -\frac{j}{q^2}\left[\beta\frac{\partial E_z}{\partial r} + \frac{\mu\omega}{r}\frac{\partial H_z}{\partial \phi}\right] \qquad \text{...(26.3 } n)$$

$$E_\phi = -\frac{j}{q^2}\left[\frac{\beta}{r}\frac{\partial E_z}{\partial \phi} - \mu\omega\frac{\partial H_z}{\partial r}\right] \qquad \text{...(26.3 } o)$$

$$H_r = -\frac{j}{q^2}\left[\beta\frac{\partial H_z}{\partial r} - \frac{\omega\varepsilon}{r}\frac{\partial E_z}{\partial \phi}\right] \qquad \text{...(26.3 } p)$$

$$H_\phi = -\frac{j}{q^2}\left[\frac{\beta}{r}\frac{\partial H_z}{\partial \phi} + \omega\varepsilon\frac{\partial E_z}{\partial r}\right] \quad ...(26.3\ q)$$

where $q^2 = \omega^2\varepsilon\mu - \beta^2 = k^2 - \beta^2$

Substitution of Eqns. (26.3 *p*) and (26.3 *q*) into Eqn. (26.3 *m*) results in the wave equation in cylindrical coordinates

$$\frac{\partial^2 E_z}{\partial r^2} + \frac{1}{r}\frac{\partial E_z}{\partial r} + \frac{1}{r^2}\frac{\partial^2 E_z}{\partial \phi^2} + q^2 E_z = 0 \text{ (wave eq. in cylindrical coordinate)} \quad ...(26.3\ r)$$

and substitution of Eqns. (26.3 *m*) and (26.3 *n*) into Eqn. (26.3 *l*) leads to

$$\frac{\partial^2 H_z}{\partial r^2} + \frac{1}{r}\frac{\partial H_z}{\partial r} + \frac{1}{r^2}\frac{\partial^2 H_z}{\partial \phi^2} + q^2 H_z = 0 \quad ...(26.3\ s)$$

Equations (26.3 *r*) and (26.3 *s*) contacts either E_z or H_z only. This appears to imply that the longitudinal components of *E* and *H* are uncoupled and can be chosen arbitrarily provided that they satisfy Eqns. (26.3 *r*) and (26.3 *s*).

If the boundary conditions do not lead to coupling between the field components, mode solutions can be obtained in which either $E_z = 0$ or $H_z = 0$. When $E_z = 0$ the mode are called transverse electric or *TE* mode, and when $H_z = 0$ transverse magnetic or *TM* modes results. Hybrid modes exist if both E_z and H_z are non zero. These are designated as *HE* or *EH* modes, depending on whether H_z or E_z respectively, makes a larger contribution to the transverse field.

Wave Equations for Step-Index Fibers

A standard mathematical procedure for solving equations such as Eqn. (26.3 *r*) is to use the separation of variables method, which assumes a solution of the form

$$E_z = AF_1(r)\, F_2(\phi)\, F_3(z)\, F_4(t) \quad ...(26.3\ t)$$

The time and *z*-dependent factors are given by

$$F_3(z)\, F_4(t) = e^{j(\omega t - \beta z)} \quad ...(26.3\ u)$$

Since the wave is sinusoidal in time and propagates in the *z*-direction. The circular symmetry of the waveguide, each field component must not change when the coordinate ϕ is increased by 2π. We thus assume a periodic function of the form.

$$F_2(\phi) = e^{j\nu\phi} \quad ...(26.3\ v)$$

The constant ν can be positive or negative, but it must be an integer since the fields must be periodic if ϕ with a period of 2π.

Substituting Eqn. (26.3 *u*) into Eqn. (26.3 *t*) the wave equations for E_z Eqn. (26.3 *r*) becomes.

$$\frac{\partial^2 F_1}{\partial r^2} + \frac{1}{r}\frac{\partial F_1}{\partial r} + \left(q^2 - \frac{\nu^2}{r^2}\right)F_1 = 0 \quad ...(26.3\ w)$$

Equation (26.3 w) is a well-known differential equation for Bessel functions. An exactly identical equation can be derived for H_z Eqn. (26.3 w) is solved for the regions inside and outside the core.

For the inside region the solutions for the guided modes must remain finite as $r \to 0$, whereas on the outside the solutions must decay to zero as $r \to \infty$.

Thus for $r < a$ the solutions are Bessel functions of the first kind of order ν. For these functions we use the common designation $J_\nu = (ur)$. Here $u^2 = k_1^2 - \beta^2$ with $k_1 = 2\pi n_1 / \lambda$, n_1 is the *r.i.* of core.

The expressions for E_z and H_z inside the core are thus

$$E_z\ (r < a) = AJ_\nu\ (ur)\ e^{j\nu\phi}\ e^{j(\omega t - \beta z)} \qquad \text{...(26.3 } x)$$

$$H_z\ (r < a) = BJ_\nu\ (ur)\ e^{j\nu\phi}\ e^{j(\omega t - \beta z)} \qquad \text{...(26.3 } y)$$

where A and B are arbitrary constants.

Outside of the core the solutions of Eqn. (26.3w) are given by modified Bessel functions of the second kind $K_\nu(\omega r)$, where $\omega^2 = \beta^2 - k_2^2$ with $k_2 = \dfrac{2\pi n_2}{\lambda}$, n_2 = *r.i.* of cladding.

The expressions for E_z and H_z outside the core are, therefore,

$$E_z\ (r > a) = CK_\nu\ (\omega r)\ e^{j\nu\phi}\ e^{j(\omega t - \beta z)} \qquad \text{...(26.3 } z)$$

$$\mu_z\ (r > a) = DK_\nu\ (\omega r)\ e^{j\nu\phi}\ e^{j(\omega t - \beta z)} \qquad \text{...(26.3 } a')$$

with C and D being arbitrary constants.

From the definition of the modified Bessel functions, it is seen that $k_\nu(\omega r) \to e^{-\omega r}$ as $\omega r \to \infty$. Since $k_\nu(\omega r)$ must go to zero as $r \to \infty$, it follows that $\omega > 0$. This, in turn, implies that $\beta \geq k_2$, which represents a cut off condition. *The cut off condition is the point at which a mode is no longer bound to the core region.* A second condition on β can be deduced from the behaviour of $j_\nu(ur)$. Inside the core the parameter u must be real for F_1 to be real, from which it follows that $k_1 \geq \beta$. The permissible range of β for bound solutions is, therefore,

$$\boxed{n_2 k = k_2 \leq \beta \leq k_1 = n_1 k} \qquad \text{...(26.3 } b')$$

where $k = \dfrac{2\pi}{\lambda}$ is the free-space-propagation constant.

Modal Equation

The solutions for β must be determined from the boundary conditions. The boundary condition require that the tangential components E_ϕ and E_z of $\vec{E}$ inside and outside of the dielectric interface at $r = a$ must be the same and, similarly, for the tangential components H_ϕ and H_z.

Consider first the tangential components of $\vec{E}$. For the z-components from Eqn. (26.3 x) at the inner core-cladding boundary ($E_z = E_{z1}$) and from Eqn. (26.3 z) at the outside of the boundary ($E_z = E_{z2}$), that is

$$E_{z1} - E_{z2} = AJ_\nu(ur)e^{j\nu\phi}e^{j(\omega t - \beta z)} - Ck_\nu(\omega r)e^{j\nu\phi}e^{j(\omega t - \beta z)} = 0$$

$$= AJ_\nu(ur) - Ck_\nu(\omega r) = 0 \quad \text{...(26.3 } c'\text{)}$$

Inside the core the factor q^2 is given by $q^2 = u^2 = k_1{}^2 - \beta^2$ where $k_1 = \dfrac{2\pi n_1}{\lambda} = \omega\sqrt{\varepsilon_1\mu}$

Outside the core the factor w^2 is given by $w^2 = \beta^2 - k_2{}^2$ where $k_2 = \dfrac{2\pi n_2}{\lambda} = \omega\sqrt{\varepsilon_2\mu}$.

Substituting Eqns. (26.3 *x, y*) into Eqn. (26.3 *o*) to find $E_{\phi 1}$, and similarly using Eqns. (26.3 *z*) and (26.3 *a'*) to determine $E_{\phi 2}$ yield at $r = a$

$$E_{\phi 1} - E_{\phi 2} = -\frac{j}{u^2}\left[A\frac{j\nu\beta}{a}J_\nu(ua) - B\omega\mu uJ'_\nu(ua)\right] - \frac{j}{w^2}\left[Cj\frac{\nu\beta}{a}k_\nu(\omega a) - D\omega\mu wk'_\nu(wa)\right] = 0$$

...(26.3 *d'*)

Where prime indicates differentiation with respect to the argument. Similarly, for the tangential components of $\vec{H}$ it is readily shown that at $r = a$.

$$H_{z1} - H_{z2} = BJ_\nu(ua) - DK_\nu(wa) = 0 \quad \text{...(26.3 } e'\text{)}$$

and
$$H_{\phi 1} - H_{\phi 2} = -\frac{j}{u^2}\left[\beta\frac{j\nu\beta}{a}j_\nu(ua) + A\omega\epsilon_1\, uJ'_\nu(ua)\right] - \frac{j}{w^2}\left[D\frac{j\nu\beta}{a}k_\nu(wa) + Cw\varepsilon_2 wk'_\nu(wa)\right] = 0$$

...(26.3 *f'*)

Equations (26.3 *e'*) to (26.3 *f'*) are a set of four equations with four unknown coefficients *A*, *B*, *C* and *D*. A solution to these equations exists only if the determinant of these coefficients is zero:

$$\begin{vmatrix} J_\nu(ua) & 0 & -k_\nu(wa) & 0 \\ \dfrac{\beta\nu}{au^2}J_\nu(ua) & \dfrac{jw\mu}{u}J'_\nu(ua) & \dfrac{\beta\nu}{aw^2}k_\nu(wa) & \dfrac{jw\mu}{w}k'_\nu(wa) \\ 0 & J_\nu(ua) & 0 & -k_\nu(wa) \\ -\dfrac{jw\varepsilon_1}{u}J'_\nu(ua) & \dfrac{\beta\nu}{au^2}J_\nu(ua) & -\dfrac{j\omega\varepsilon_2}{w}k'_\nu(wa) & \dfrac{\beta\nu}{aw^2}k_\nu(wa) \end{vmatrix} = 0 \quad \text{...(26.3 } g'\text{)}$$

Evaluation of this determinant yields the following eigen value equation for β:

$$(J_\nu + K_\nu)(k_1^2 J_\nu + k_2^2 K_\nu) = \left(\frac{\beta\nu}{a}\right)^2\left(\frac{1}{u^2} + \frac{1}{w^2}\right)^2 \quad \text{...(26.3 } h'\text{)}$$

were $\qquad J_\nu = \dfrac{J'_\nu(ua)}{uJ_\nu(ua)}$ and $k_\nu = \dfrac{k'_\nu(wa)}{wk_\nu(wa)}$

Solving Eqn. (26.3 *h'*) for β it will be found that only discrete values restricted to the range given by Eqn. (26.3 *b'*) will be allowed. Equation (26.3 *h'*) is a complicated transcendental equation which is generally solved by numerical techniques, its solution for any particular mode will provide all the characteristics of that mode.

$$\text{For} \qquad \nu = 0 \rightarrow \begin{cases} \text{TE (transverse electric) modes} \\ \text{TM (transverse magnetic) modes} \end{cases}$$

$$\text{For} \qquad \nu \neq 0 \rightarrow \text{hybrid modes (HE, EH)}$$

TE, TM modes correspond to meridional rays
Hybrid modes correspond to skew rays.

Normalized Frequency (*V*) or *V*-number or *V*-parameter

An important parameter connected with the cut-off condition is the normalised frequency *V* or *V*-number is defined by

$$V^2 = (u^2 + w^2)a^2 = \left(\frac{2\pi a}{\lambda}\right)^2 (n_1^2 - n_2^2)$$

$$V = \frac{2\pi}{\lambda} a\sqrt{n_1^2 - n_2^2} = \frac{2\pi}{\lambda} a n_1 \sqrt{2\Delta},$$ λ is the free space wavelength of the light beam.

V, which is a dimensionless number that determines how many modes a fiber can support. The number of modes that can exist in a waveguide as a function of *V* may be conveniently represented in terms of a normalized propagation constant '*b*' defined by

$$b = \frac{a^2 w^2}{V^2} = \frac{\left(\frac{\beta}{k}\right)^2 - n_2^2}{n_1^2 - n_2^2}$$

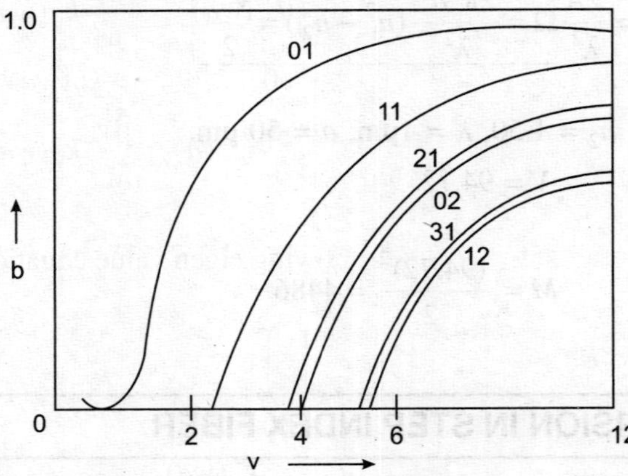

Fig. 26.6. The curve number ν_m designate the $HE_{\nu+1,m}$ and $EH_{\nu-1,m}$
For $\nu = 1$, the curve number ν_m give the HE_{2m}, TE_{0m}, TM_{0m}

This figure shows that each mode can exist only for values of *V* that exceed a certain limiting value. The modes are cut off when $\frac{\beta}{k} = n_2$.

The HE_{11} mode has no cutoff and ceases to exist only when the core diameter is zero. This is the principle on which the single-mode fiber is based.

By appropriately choosing a, n_1, and n_2 so that

$$V = \frac{2\pi a}{\lambda}(n_1^2 - n_2^2)^{\frac{1}{2}} \geq 2.405$$

which is the value at which the lowest–order Bessel function J_o is zero, all modes except the HE_{11} mode are cutoff.

The parameter V can also be related to the number of modes M in a multimode fiber when M is large. An approximate relationship for step-index fibers can be derived from ray theory.

The numerical aperture is given as

$$NA = \sin\theta = (n_1^2 - n_2^2)^{\frac{1}{2}}$$

For practical numerical aperture $\sin\theta$ is small so that $\sin\theta \approx \theta$.

The solid acceptance angle for the fiber is therefore

$$\Omega = \pi\theta^2 = \pi(n_1^2 - n_2^2)$$

For electromagnetic radiation of wavelength λ emanating from a laser or a waveguide the number of modes per unit solid angle is given by $2A/\lambda^2$, where A is the area the mode is leaving or entering. The area A in this case is the core cross section πa^2. The factor 2 comes from the fact that the plane wave can have two polarization orientations.

The total number of modes M entering the fiber is thus given by

$$\boxed{m \simeq \frac{2A}{\lambda^2}\Omega = \frac{2\pi^2 a^2}{\lambda^2}(n_1^2 - n_2^2) = \frac{V^2}{2}}$$

Example: $n_1 = 1.53$, $n_2 = 1.50$, $\lambda = 1\mu m$, $a = 50\ \mu m$.

Then $$V = 94.72$$

and $$M = \frac{(94.72)^2}{2} = 4486$$

26.4 PULSE DISPERSION IN STEP INDEX FIBER

The simplest type of optical fiber consists of a thin cylindrical structure of transparent glassy material of uniform refractive index n_1 surrounded by a cladding of another material of uniform but slightly lower refractive index n_2. These fibers are referred to as step index fibers due to the step discontinuity of the index profile at the core-cladding interface.

In digital communication systems, information to be sent is first coded in the form of pulses and then these pulses of light are transmitted from the transmitter to the receiver where the information is decoded. The larger the number of pulses that can be sent per unit time and still be resolvable at the receiver end, the larger is the transmission capacity of the system. A pulse of light sent into a fiber broadens in time as it propagates through the fiber, this phenomenon is known as pulse dispersion and happens because of the different times taken by different rays to propagate through the fiber, as shown in figure. Rays making larger angles with the axis take a longer time to traverse the length of the fiber. Consequently, the pulse broadens as it propagates through the fiber. Hence, even though two pulses may be well resolved at the input end, because of broadening of the pulses they may not be so at the output end. Where the output pulses are not resolvable, no information can be retrieved. Thus the smaller the pulse dispersion, the greater will be the information-carrying capacity of the system.

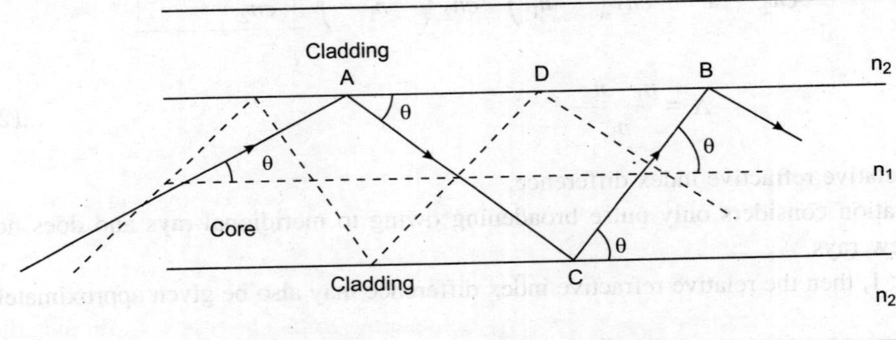

Fig. 26.7

To calculate the amount of dispersion in a step index fiber as follows:

For a ray making an angle θ with the axis, the distance AB is traversed in time.

$$t = \frac{AC + CB}{C/n_1} = \frac{n_1(AB)}{C\cos\theta} \quad \text{...(26.4 } a\text{)}$$

where $\frac{C}{n_1}$ represents the speed of light in a medium of refractive index n_1, C being the speed of light in free space. Since the ray path will repeat itself, the time taken by a ray to traverse a length L of the fiber will be

$$t = \frac{n_1 L}{C\cos\theta} \quad \text{...(26.4 } b\text{)}$$

The above expression shows that the time taken by a ray is a function of the angle θ made by the ray with z-axis which leads to pulse dispersion. If we assume that all rays between O and θ_c are present then the time taken by rays corresponding respectively to $\theta = 0$ and $\theta = \theta_c = \cos^{-1}\left(\frac{n_2}{n_1}\right)$ will be given by

$$t_{\min} = \frac{n_1 L}{C}$$ shortest ray congruence paths (the fundamental mode).

since $\cos 0° = 1$

$$t_{min} = \frac{n_1 L}{C \cdot n_2/n_1} = \frac{n_1^2 L}{cn_2}$$ longest ray congruence paths (the highest-order mode)

Hence if all the input rays were excited simultaneously, the rays will occupy a time interval at the output end of duration.

$$\delta T = t_{max} - t_{min} = \frac{n_1^2 L}{cn_2} - \frac{n_1 L}{c} = \frac{Ln_1^2}{cn_2}\left(1 - \frac{n_2}{n_1}\right) = \frac{Ln_1^2}{cn_2}\left(\frac{n_1 - n_2}{n_1}\right) = \boxed{\frac{Ln_1^2}{cn_2}\Delta = \delta T} \quad ...(26.4\ c)$$

$$\Delta = \frac{n_1 - n_2}{n_1} \quad ...(26.4\ d)$$

where Δ is the relative refractive index difference.

This derivation considers only pulse broadening owing to meridional rays and does not take into account skew rays.

when $\Delta << 1$, then the relative refractive index difference may also be given approximately by:

$$\Delta \simeq \frac{n_1 - n_2}{n_2}$$

Hence Eqn. (26.4 *c*) becomes

$$\delta T_s = \frac{Ln_1^2}{cn_2}\left(\frac{n_1 - n_2}{n_1}\right) = \frac{Ln_1}{c}\left(\frac{n_1 - n_2}{n_2}\right) \simeq \boxed{\frac{Ln_1 \Delta}{c} = \delta T_s} \quad ...(26.4\ e)$$

Also $NA = n_1(2\Delta)^{\frac{1}{2}}$ the above expression becomes

$$\boxed{\delta T_s \simeq \frac{L}{c}\frac{(NA)^2}{2n_1}} \quad ...(26.4\ f)$$

where NA is the numerical aperture for the fiber.

The approximate expressions for the delay difference given in above Eqns. (26.4 *e*) and (26.4 *f*) are usually employed to estimate the maximum pulse broadening in time due to *intermodal dispersion in multimode step* index fibers.

The above time delay is the characteristic of the fiber and *independent of the wavelength of the light.* It is of the order of *nanosecond per km.* However, this type of delay is absent in single mode fiber, as it has one mode only.

This phenomena in graded index fiber is minimum when profile shape parameter 'α' is close to the value $2.(1 - 1.2\Delta) = \alpha_{opt}$.

Using a ray theory approach $\delta T_g = \frac{L n_1 \Delta^2}{2c} = \frac{(NA)^4}{8 n_1^3 c}$ ...(26.4 *g*)

Using electromagnetic mode theory $\delta T_g = \frac{L n_1 \Delta^2}{8c}$ ...(26.4 *h*)

It is observed that the graded index fibers have much lower intermodal dispersion than that of the step-index fiber.

Graded-index – α-profiles

$$n(r) = n_0 \left[1 - 2\Delta\left(\frac{r}{a}\right)^{\alpha}\right]^{\frac{1}{2}} \quad \text{for } r \le a$$

$$= n_0 \left[1 - 2\Delta\right]^{\frac{1}{2}} = n_c \quad \text{for } r \ge a \qquad \text{...(26.4 } i\text{)}$$

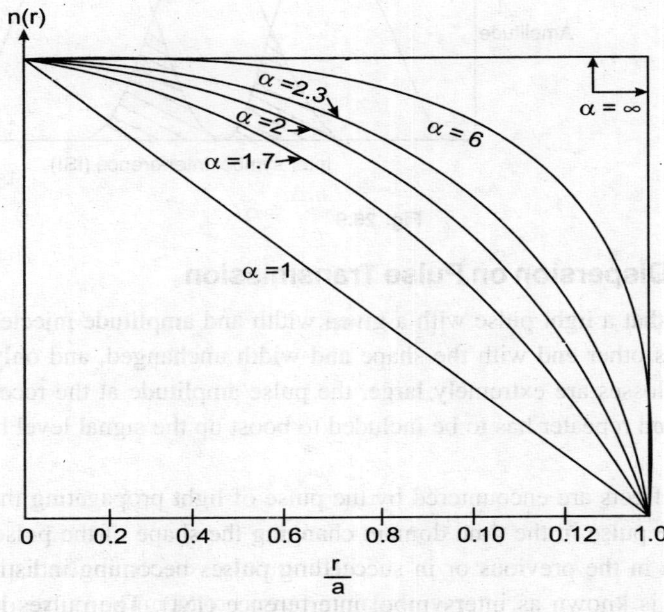

Fig. 26.8

Refractive-index profiles ranging from the triangular profile, α = 1.0, to the step-index profile, α = ∞ are shown, with profiles for α = 1.7, 2.0, 2.3 and 6.0 lying in between.

The profile parameter, α, may take on values between zero and infinity.

It may be expected that a light pulse with a given width and amplitude injected into one end of a fiber should arrive at its other end with the shape and width unchanged, and only its amplitude is reduced by losses. If the losses are extremely large, the pulse amplitude at the receiving end will be too small to be detected and a repeater has to be included to boost up the signal level before its entering into next section.

Several dispersion effects are encountered by the pulse of light propagating through a fiber and these act to spread out the pulse in the time domain changing the shape of the pulse. As a result, the pulse may be got merged in the previous or in the succeeding pulses becoming indistinguishable at the receiver as shown in figure. This effect is known as intersymbol interference (ISI). The pulses may be separated by increasing the time interval between pulses, but thereby the maximum bit rate will be reduced.

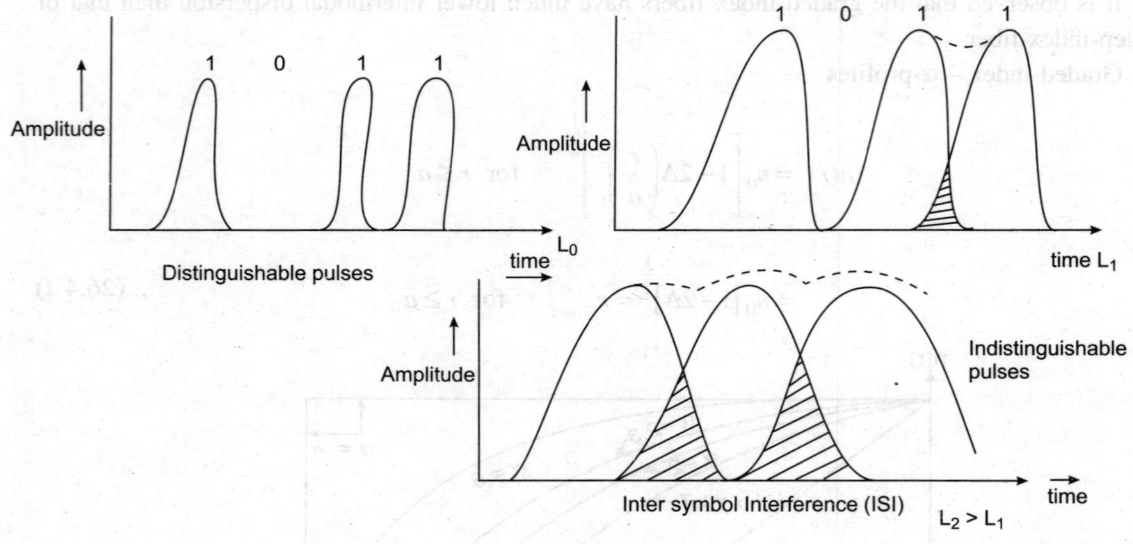

Fig. 26.9

Dispersion Effect of Dispersion on Pulse Transmission

It may be expected that a light pulse with a given width and amplitude injected into one end of a fiber should arrive at its other end with the shape and width unchanged, and only its amplitude is reduced by losses. If the losses are extremely large, the pulse amplitude at the receiving end will be too small to be detected and repeater has to be included to boost up the signal level before its entering into next section.

Several dispersion effects are encountered by the pulse of light propagating through a fiber and these act to spread out the pulse in the time domain changing the shape of the pulse. As a result, the pulse may be got merged in the previous or in succeeding pulses becoming indistinguishable at the receiver end. This effect is known as intersymbol interference (ISI). The pulses may be separated by increasing the time interval between pulses, but thereby the maximum bit rate will be reduced.

There are three kinds of dispersion due to three separate mechanisms existing in the fiber. These are:

(1) Intermodal Dispersion or Multi-path dispersion
(2) Material or chromatic dispersion and
(3) Waveguide dispersion.

26.5 MATERIAL DISPERSION

The refractive index of core glass of fiber is not the same for lights of different wavelengths but it depends on the wavelength of light in a complicated way. When a pulse of light transmitted contains components of several wavelengths centered about a centre wavelength λ_0 shown in Fig. (26.10).

The pulse component having shorter wavelengths will experience more delay that those of larger wavelengths. As a result, there is an effective time dispersion Δt of the pulse at the receiving end of the fiber.

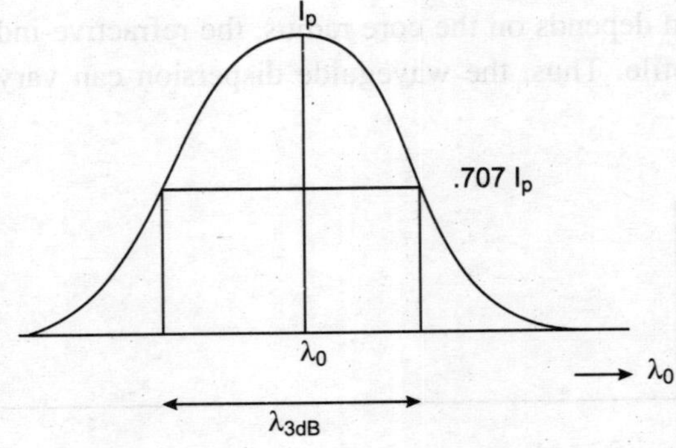

Fig. 26.10

Material dispersion depends only on the composition of the material.

The material dispersion is given by

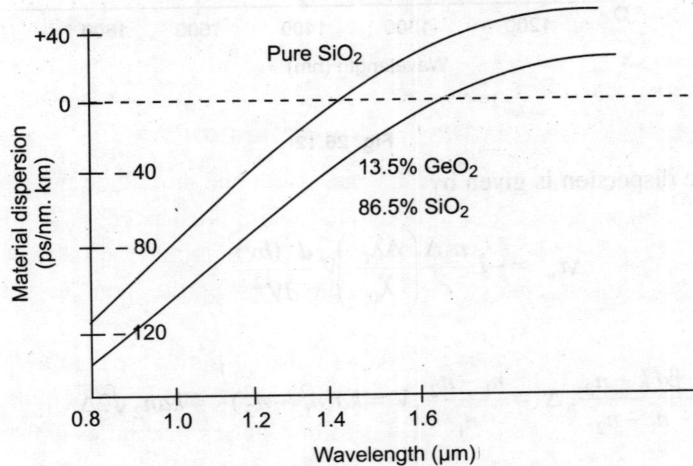

Fig. 26.11

$$\Delta\tau_m = -\frac{Z}{C}\lambda_0^2 \frac{d^2 n_1}{d\lambda_0^2}\left(\frac{\Delta\lambda_0}{\lambda_0}\right)$$

26.6 WAVEGUIDE DISPERSION

The effect of waveguide dispersion on pulse spreading can be approximated by assuming that the refractive index of the material is independent of wavelength.

When a light pulse is launched into a fiber, it is distributed among many guided modes. These various modes arrive at the fiber end at different times depending on their group delay, so that a pulse spreading results.

For multimode fibers the waveguide dispersion is generally very small compared with material dispersion and can therefore be neglected.

Waveguide dispersion depends on the core radius, the refractive-index difference, and the shape of the refractive-index profile. Thus, the waveguide dispersion can vary dramatically with the fiber design parameters.

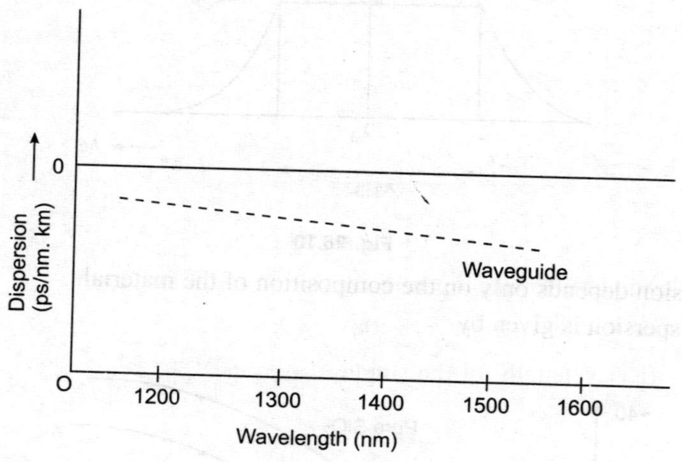

Fig. 26.12

The waveguide dispersion is given by

$$\Delta\tau_w = -L\frac{n_2\Delta}{c}\left(\frac{\Delta\lambda_0}{\lambda_0}\right)V\frac{d^2(bV)}{dV^2}$$

where $$b \approx \frac{\beta/k - n_2}{n_1 - n_2},\ \Delta = \frac{n_1 - n_2}{n_1},\ V = ka(n_1^2 - n_2^2)^{\frac{1}{2}} \approx kan_2\sqrt{2\Delta}$$

26.7 DISPERSION SHIFTED AND DISPERSION FLATTENED FIBERS

The loss of signal through the fiber due to dispersion is reduced by two important methods.

(*i*) Dispersion Shifted Fiber

The dispersion shifted fibers are made by increasing the negative waveguide dispersion. This negative waveguide dispersion is exactly cancelled by positive material dispersion, result is given by dispersion shifted fiber.

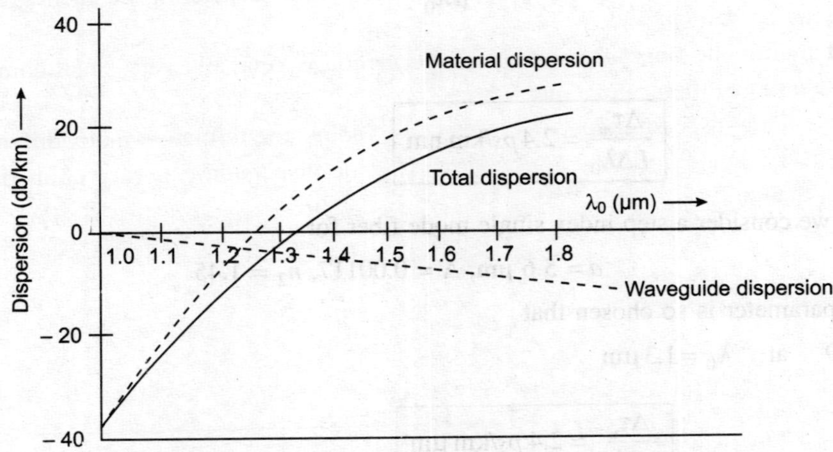

Fig. 26.13

Since for the single mode fiber the *V*-parameter should be

$$0 < V < 2,408$$

And the cut-off wavelength for the single mode fiber should be such

that
$$\lambda_o > \lambda_{co} = \frac{2\pi a(n_1^2 - n_2^2)^{1/2}}{2.4048}$$

when λ_{co} is cut-off wavelength in the step index fiber.

The material dispersion is given by

$$\Delta\tau_m = -\frac{z}{c}\lambda_0^2 \frac{d^2 n_1}{d\lambda_0^2}\left(\frac{\Delta\lambda_0}{\lambda_0}\right) \qquad \text{...(26.7 } a)$$

i.e., proportional to $\frac{d^2 n_1}{d\lambda_0^2}$ and spectral width $\Delta\lambda_0$.

Also the waveguide dispersion is given by

$$\Delta\tau_w = -L\frac{n_2\Delta}{c}\left(\frac{\Delta\lambda_0}{\lambda_0}\right)V\frac{d^2}{dV^2}(bV) \qquad \text{...(26.7 } b)$$

Now we take the operated wavelength $\lambda_0 \simeq 1.3$ μm and we calculate the material dispersion. Hence

$$\lambda_0 = 1.3\mu\text{m}, \frac{d^2n}{d\lambda_0^2} = -5.5\times10^{-4}\ \mu\text{m}^{-2}$$

So that

$$\boxed{\frac{\Delta\tau_m}{L\Delta\lambda_0} \simeq 2.4\, ps/\text{km nm}}$$

Again we consider a step index single mode fiber for

$$a = 5.6\ \mu\text{m}, \Delta = 0.00117, n_2 = 1.45$$

So, *V*-parameter is so chosen that

$V \simeq 1.9$ at $\lambda_0 = 1.3\ \mu\text{m}$

Hence $\boxed{\frac{\Delta\tau_w}{L\Delta\lambda_0} \simeq 2.4\, ps/\text{km}\ \mu\text{m}}$

Thus, the waveguide dispersion and the material dispersion cancell each other to give zero dispersion. The explanation is shown by figure. The same calculation is done by the wavelength is $\lambda_0 = 1.55$ μm.

The refractive index profile for the dispersion shifted fiber is given by

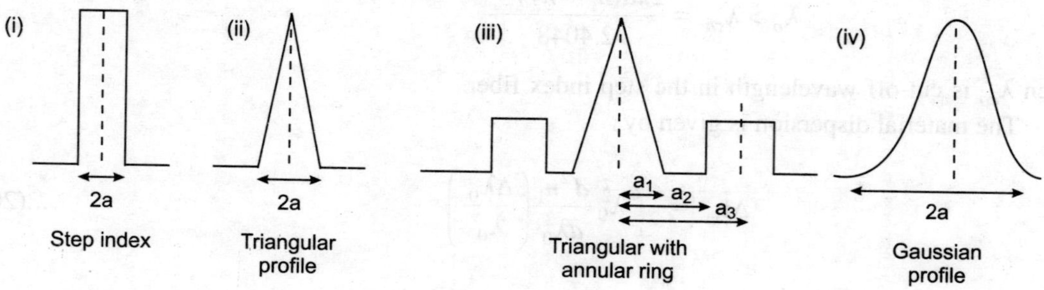

Fig. 26.14

(*ii*) Dispersion Flattened Fiber

This is the second method by which we reduce the loss of the fiber. In this method, we show that by reducing the core radius the zero dispersion wavelength can be shifted to the minimum loss wavelength window. This can also be done by suitably grading the refractive index profile of core.

This is explained by the figure as

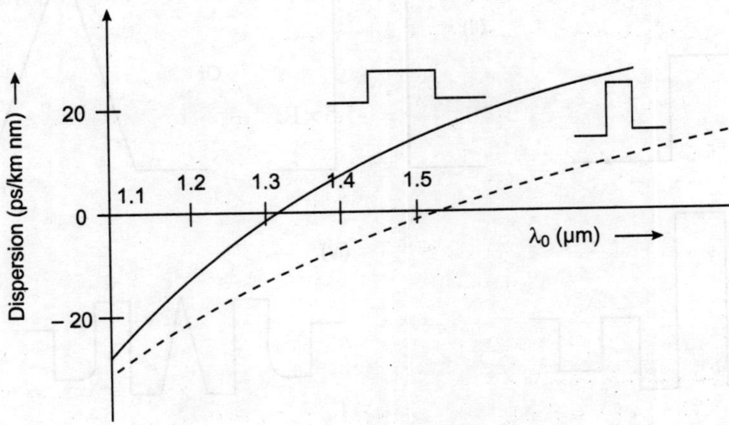

Fig. 26.15

The figure shows the variation of total dispersion as a function of λ_0 in the step index fiber. The solid line is for $\Delta = 0.0027$, $\alpha = 4.1$ μm, $\lambda_{co} = 1.13$ μm, while dashed curve for $\Delta = .0075$, $a = 2.3$ μm and $\lambda_{co} = 1.06$ μm.

The variation of total dispersion for different classes is shown in the Fig. Curve (*i*) corresponds to conventional fiber. Curve (*ii*) corresponds to the dispersion shifted fiber. Curve (*iii*) and (*iv*) corresponds to dispersion flattened fiber.

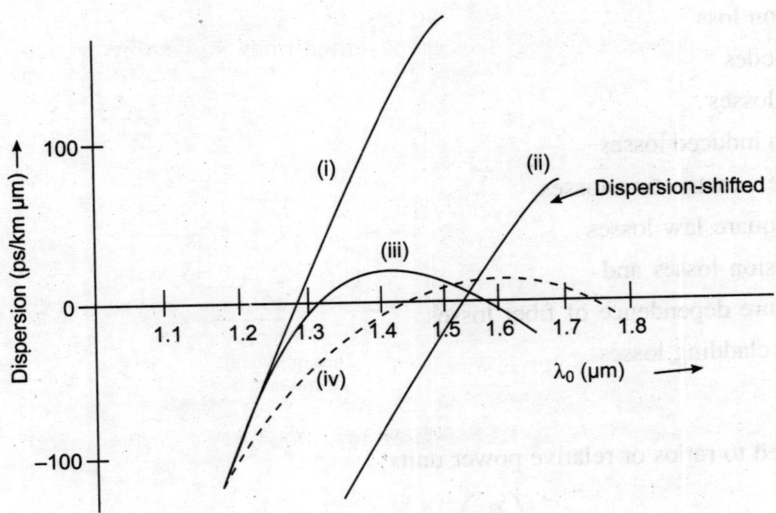

Fig. 26.16

The refractive index profile for the flattened fiber for corresponding (*i*) (*ii*) (*iii*) and (*iv*) are

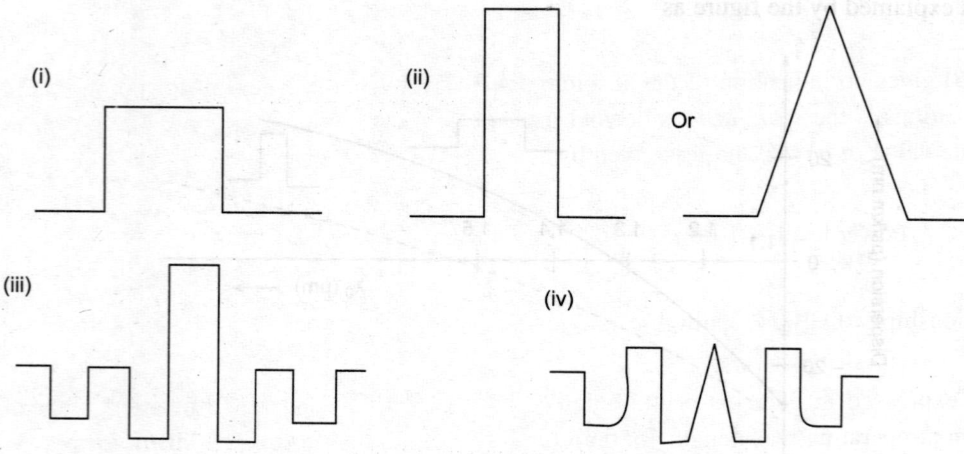

Fig. 26.17

26.8 ATTENUATION IN OPTICAL FIBERS

Attenuation in fibers means 'loss of optical power' in the fiber itself. This loss may arise from different sources, *e.g.*

(1) Material absorption (or impurity losses)
(2) Rayleigh scattering loss
(3) Absorption loss
(4) Leaky modes
(5) Bending losses
(6) Radiation induced losses
(7) Defective construction losses
(8) Inverse square law losses
(9) Transmission losses and
(10) Temperature dependence of fiber losses
(11) Core and cladding losses.

Decibel (dB)

The decibel is referred to ratios or relative power units:

- $$\text{Attenuation} = 10 \log \left(\frac{P_2}{P_1} \right) \text{dB}.$$

Attenuation in fibers means 'loss of optical power' in the fiber itself or in a pair of connectors or splice. It is expressed in "decibels" or "*dB*", where–10dB means a reduction in the power by 10 times, –20dB means another 10 times, or 100 times overall, obviously –30dB means 1000 times loss overall and so on. Doubling the power means a 3dB gain.

The dBm

Decibel (*dB*) gives no indication of the absolute power level. The derived unit for doing this in optical communications is the *dBm*. *dBm* is decibel measure of power relative to 1mw. The power in *dBm* is an absolute value of power and it is given by

$$\underset{\text{(in dBm)}}{\text{Power level}} = 10\log\left(\frac{P}{1\,\text{mw}}\right)$$

In this relationship OdBm = 1 mw.

Losses are expressed in decibels per kilometer (dB/km). The normal range of attenuation is from 0.154 dB/km at operating wavelength 1550nm for single mode fiber to over 10 dB/m for plastic fiber.

The loss is expressed as in terms of a particular length L for a fiber.

$$P_0 = P_{in} 10^{-\alpha L/10} \qquad \text{...(26.8 } a\text{)}$$

Where

P_0 = power at a distance L from the input

P_{in} = amount of power coupled into the fiber

α = fiber attenuation in dB/km.

Hence, attenuation in the fiber is defined as the ratio of the optical power output P_0 obtained from a fiber of length L to the optical power P_{in} fed to the input of the fiber.

$$\alpha = \frac{10}{L}\log\left(P_{in}/P_{out}\right) \text{ in dB/km} \qquad \text{...(26.8 } b\text{)}$$

In the case of an ideal fiber, there is no loss of optical power and $P_{in} = P_0$. In other words, for an ideal fiber, the attenuation is OdB. But practically this is not possible, because all fiber have a certain amount of loss.

$$\frac{P_0}{P_{in}} = 10^{\frac{\alpha L}{10}}$$

$$P_0 = P_{in} 10^{-\frac{\alpha L}{10}}$$

$$P_0(\text{dBm}) = 10\log\left(\frac{P_0}{1\text{ m watt}}\right) + 10\log\left[10^{-\frac{\alpha L}{10}}\right] \qquad \text{...(26.8 } c\text{)}$$

The Eqn. (26.8 *a*) can also be written in decibels,

$$P_0(\text{dBm}) = 10\log\left(\frac{P_0}{1\text{ m watt}}\right) = 10\log\left(\frac{P_{in}}{1\text{ m watt}}\right) + 10\log\left[10^{-\alpha L/10}\right]$$

$$= P_{in}(\text{dBm}) - \alpha L \qquad \text{...(26.8 } d)$$

Material or Impurity Losses

Material absorption is a loss mechanism related to the material composition and the fabrication process for the fiber, which results in the dissipation of some of the transmitted optical power as heat in the waveguide. The absorption of the light may be intrinsic (caused by the interaction with one or more of the major components of the glass) or extrinsic (caused by impurities within the glass).

In the fabrication of various types of fibers, we use GeO_2, P_2O_5, B_2O_3 etc. as dopants in silica, in order to modify the refractive index. While B_2O_3 produces strong absorption peaks at 3.2 μm, P_2O_5 produces the same at 3.8 μm. However, in both these cases, absorption-tails extend below 1.3 μm. That is why Boron and Phosphorous based dopants are not used in the low loss, single mode fiber. The introduction of P_2O_5 serves as channel for the gradual build up of OH^- ions over the time. In fiber using high dose of P_2O_5, loss increases up to 0.9 dB/km at 1.3 μm over a period of 3 years. For zero-*P* single mode fiber, the loss is less than 0.007 dB/km over a period of 25 years. Low-*P* multimode fiber has a projected loss of 0.07 dB/km over the same span of time.

Experiments have shown that the loss increases considerably when the operating wavelength is beyond 1.55 μm. This is an inherent property of all silica based fiber. This is due to vibrational resonances of the silica tetrahedron molecules.

Rayleigh Scattering Loss

The glass, which is used in the fabrication of fibers, has many microscopic inhomogeneities and material density fluctuations of the silica material contents. As a result, a portion of light passing through the glass fiber gets scattered. This phenomenon is called "Rayleigh scattering". The losses due to this scattering effect, vary inversely with the fourth power of the wavelength. These inhomogeneities are frozen into the silica matrix when the fiber is formed and depend on the glass-forming temperature. Lower the glass-forming temperature, lower is the density fluctuation and hence scattering. This scattering is most prominent between the wavelength range 500 nm to 1550 nm. The expected minimum loss at 1550 nm is approximately 0.13 dB/km. It produces a $\frac{1}{\lambda^4}$ attenuation dependence, it is also a function of glass-forming temperature.

The scattering loss due to density fluctuation can be expressed as:

$$\alpha_{RS} = \frac{359.094}{\lambda^4}(\mu^2 - 1)^3 K K_T T_F \qquad \text{...(26.8 } e)$$

where λ = operating wavelength, K = Boltzmann constant, T_F = temperature at which the density fluctuations are frozen into the glass as it solidifies, K_T = Isothermal compressibility of the material. This is also can be expressed as

$$\alpha_{RS} = \frac{359.094}{\lambda^4}\mu^8 K_P K T_F K_T \qquad \text{...(26.8 } f\text{)}$$

where K_P = Photoelastic coefficient.

*Rayleigh scattering coefficient is related to the transmission loss factor (transmissivity) of the fiber (P)

$$P = \exp(-\alpha_{RS} L) \qquad \text{...(26.8 } g\text{)}$$

where L is the length of the fiber.

Absorption Loss

This form of loss is caused by the very nature of the core-material and varies inversely to the transparency of the material. Besides in same materials, absorption losses are not uniform across the light spectrum, and are regarded to be wavelength-sensitive.

Ion-resonance absorption, ultra-violet absorption and infrared absorption are the three separate mechanisms, which contribute to total absorption losses in glass fibers.

In pure fused silica, valence electrons can be ionized into conduction electrons by light, with energy about 9 eV and then *uv* absorption takes place. Thus there is a loss of light energy due to this ionisation.

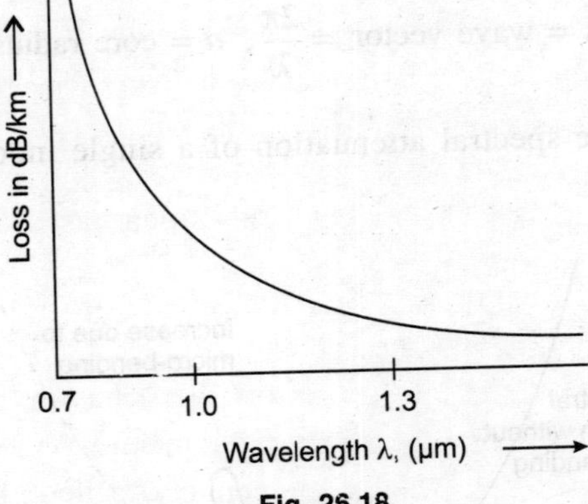

Fig. 26.18

When photons of light energy are absorbed by the atoms within the glass molecules infrared absorption takes place and is converted to the random mechanical vibrations typical of heating.

The nearest vibrational resonance of the Si-O bond is in the IR region at 9 μm. Some oxides which are doped in silica for changing the refractive index in the core (*e.g.*, GeO_2, P_2O_5 and B_2O_3) have similar resonances in *IR*. The tails of these absorption bands in *IR* may contribute in the wavelength region of interest 1.4–1.6 μm depending on the doping materials.

Leaky Modes

The losses due to leaky modes arise due to *irregularities in wave-guide geometry*, so these can be regarded as waveguide scattering. These leaky modes have radiative components that result in cladding power losses. So, these can be minimised by surrounding the thin cladding layer on the fiber-core by a third layer of pure silica, which has an index of refraction higher than that of the cladding, but lower than that of the core. This third additional silica coating will not only give additional strength to the fiber, but side-by-side also remove the partial refraction rays from the leaky modes, as well as remove the passed rays from cut-off modes by total refraction.

Bending Losses

Whenever a fiber deviates from a straight-line path, radiative losses occurs. These losses are prominent for improperly installed single mode optical cable. Micro-bending and Macro-bending are two types of bending losses.

Micro-bending losses

Micro-bends have small random deviations about a nominal straight line position. The micro-bending losses occur due to the fact that the *small bends act as scattering, which causes mode-coupling to take place. The energy of the guided mode is cross-coupled into leaky modes. This cross-coupling leads to the loss through cladding.*

Micro-bend loss for single mode fiber is expressed as

$$\alpha_{\text{mic}} = 0.05\,\alpha_m \frac{K^4 (Fd)^6 \cdot (NA)^4}{a^2} \qquad \text{...(26.8 } h)$$

where α_m = attenuation constant, k = wave vector $= \frac{2\pi}{\lambda}$, a = core radius, Fd = half of mode field diameter in the single mode fiber.

Figure (26.19) represents the spectral attenuation of a single mode fiber with evidence of excessive micro-bending loss.

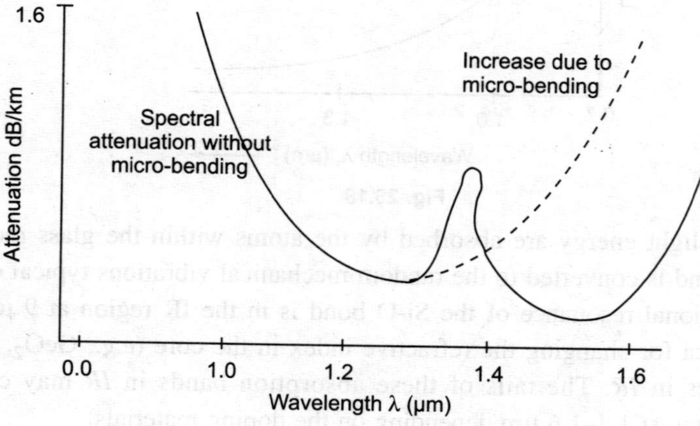

Fig. 26.19

Macro-bending losses

The larger the fiber core radius and smaller the bend radius, the greater the macrobend loss. Macrobend loss for multimode graded index fiber is given by the expression.

$$\alpha_{mac} = -10\log\left[1 - \frac{p+2}{2p\Delta}\left\{\frac{2a}{R} + \left(\frac{3}{2n_1RK}\right)^{2/3}\right\}\right] \text{dB} \qquad \text{...(26.8 } i)$$

where p = index profile parameter for graded index fiber.

R = radius of curvature of bend, a = core radius,

$\Delta = \dfrac{n_1 - n_2}{n_1}$, $n_1 = r.i.$ of core, $n_2 = r.i.$ of cladding

K = wave vector = $2\pi/\lambda$.

n_1 = refractive index of the core

Figure 26.21 represents the variation of macro-bend loss for multimode step index fiber with core radius in the range 200 μm to 800 μm.

Figure shows the fundamental mode field in a curved optical wave guide.

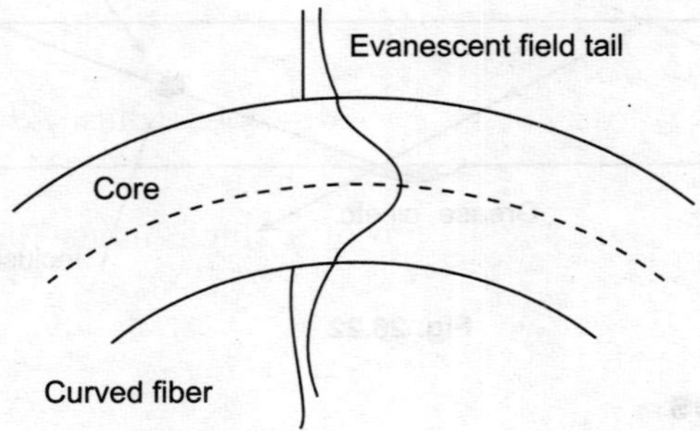

Fig. 26.20

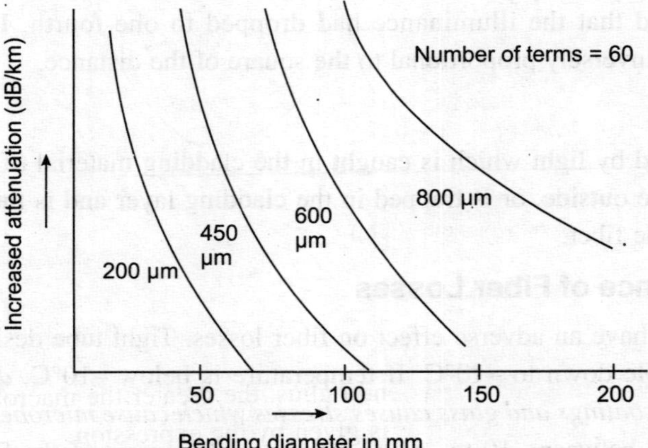

Fig. 26.21

Radiation Induced Losses

When the glass molecular matrix interacts with electrons, neutrons, gamma rays and X-rays, the structure of the glass molecules is altered and the fiber darkens. This introduces additional losses which increase with amount, type, dose and exposure time of the radiation.

Inherent Defect Losses

Figure shows several possible sources of loss due to defects in the fiber itself. First, in unclad fibers, surface defects (nicks or scratches) that breach the integrity of the surface, allow light to escape. In other words, not all of the light is propagated along the fiber. Second (also in unclad fiber), grease, oil or other contaminants on the surface of the fiber may form an area with an index of refraction different from what is expected and cause the light direction to change. There is always the possibility of inclusions *i.e.*, objects, specks or voids in the material making up the optical fiber. Inclusions can affect both clad and unclad fibers. When light hits the inclusion, it tends to scatter in all directions, causing a loss. Some of the light rays scattered from the inclusion may recombine either destructively or constructively with the main ray, but most do not.

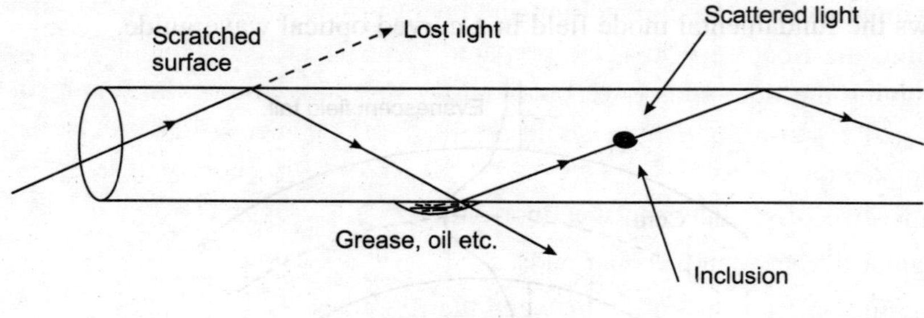

Fig. 26.22

Inverse Square Law Losses

In all light systems, there is the possibility of losses, caused by spreading (divergence) of the beam. If we take a flashlight and point it at a wall, and measure the illuminance per unit area at the wall at a distance of, say, one metre, and then back off to twice the distance (two metres) and then measure again, we will find that the illuminance had dropped to one-fourth. In other words, the illuminance per unit area is inversely proportional to the square of the distance.

Transmission Losses

These losses are caused by light which is caught in the cladding material of clad optical fibers. This light is either lost to the outside, or is trapped in the cladding layer and is thus not available for propagation in the core of the fiber.

Temperature Dependence of Fiber Losses

Temperature extremes have an adverse effect on fiber losses. Tight tube designs and clad made of plastic are normally usable down to –10°C. If temperature is below –10°C, *differential, thermal expansion between polymer coatings and glass causes stresses which cause microbending losses.* Loose tube and glass-clad are used between –10 to –50°C. The central element of the fiber determines the amount of cable contraction with tempeature and is called 'stiffering'.

Core and Cladding Losses

In a fiber the core and cladding have different refractive indices, as they have different compositions. So the core and the cladding have different attenuation coefficients (α_c and α_d). If we neglect the modal coupling, then the loss associated with a mode (ν, m) in *SI* fiber is expressed as:

$$\alpha_{\nu m} = \alpha_c \frac{P_C}{P_T} + \alpha_d \frac{P_d}{P_T} \qquad ...(26.8\ j)$$

where the fraction of powers P_C/P_T, P_d/P_T, P_T is the total power in the mode (n).

Misalignment Losses

In any optical fiber telecommunication link, one or more splices/joints in the fiber cable is required. The predominant method for connecting optical fibers involves a butt-joint connection. Any butt-joint requires three fundamental operations: fiber and preparation, fiber alignment to micron precision and alignment retention. Demountable connections retain alignment mechanically while permanent connections retain alignment through melting and fusing of the fiber ends in a fusion-splicing machine. In any fiber joint, the fiber ends must be prepared smooth and perpendicular to the fiber axis. The next step of aligning the fiber ends is very crucial because any kind of misalignment would lead to a transmission loss. Loss at a fiber splice could originate from either or a combination of the following possible misalignments:

(*i*) Transverse offset between the fiber ends.

(*ii*) Angular tilt between the fiber ends.

(*iii*) Longitudinal end-separation between the fiber end faces.

$$\text{Loss}_{\text{ang}} = -10 \log_{10} \eta_{\text{ang}} \text{ dB}$$

$$\eta_{\text{ang}} \simeq \frac{16(n_1/n)^2}{(1+n_1/n)^4}\left[1 - \frac{n\theta}{\pi n_1 (2\Delta)^{\frac{1}{2}}}\right] \qquad ...(26.8\ k)$$

Δ is relative *r.i.* difference for the fiber

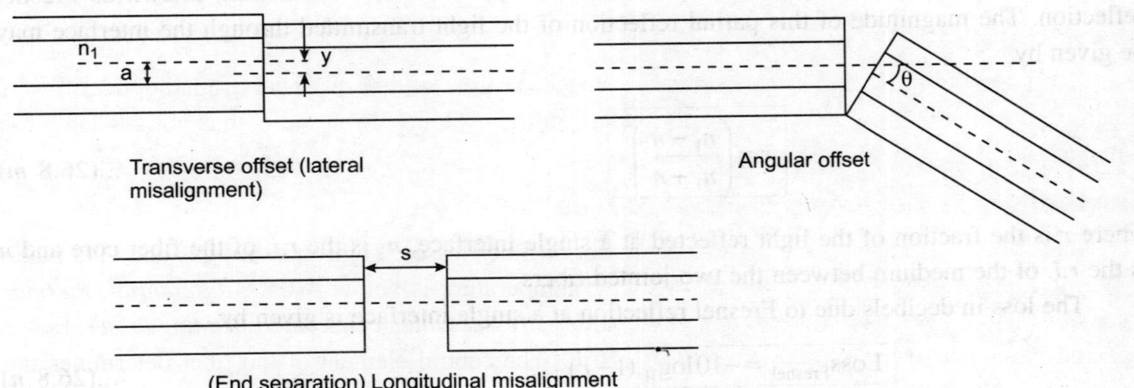

Fig. 26.23

n_1 = core refractive index

n = *r.i.* of the medium between the fibers

Angular offset

$$\text{Loss}_{lat} = -10 \log_{10} \eta_{lat} \text{ dB}$$

$$\eta_{lat} \simeq \frac{16(n_1/n)^2}{(1+(n_1/n)^4} \frac{1}{\pi}\left\{2\cos^{-1}\left(\frac{y}{2a}\right) - \left(\frac{y}{a}\right)\left(1-\left(\frac{y}{2a}\right)^2\right)^{\frac{1}{2}}\right\}$$

for the step-index fibers.

$$\eta_{long} \simeq \left[1 - \frac{s}{a}\frac{2}{\pi(NA)^2}\{\sin^{-1}(NA) - NA\sqrt{1-(NA)^2}\}\right] \text{ for } \frac{s}{a} << 1 \quad ...(26.8\ l)$$

$$\text{Loss} = -10 \log_{10} \eta_{long} \text{ . dB}$$

Either or a combination of the following may also result in a joint loss and these are sources of intrinsic loss:

(*i*) Fresnel loss due to Fresnel reflection:

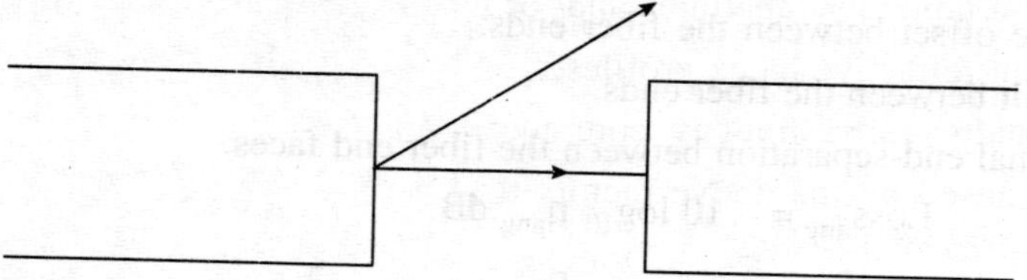

Fig. 26.24

A major consideration with all types of fiber–fiber connection is the optical loss encountered at the interface. Even when the two jointed fiber ends are smooth and perpendicular to the fiber axes, and the two fiber axes are perfectly aligned, a small proportion of the light may be reflected back into the transmitting fiber causing attenuation at the joint. This phenomenon, known as Fresnel reflection. The magnitude of this partial reflection of the light transmitted through the interface may be given by

$$\boxed{r = \left(\frac{n_1 - n}{n_1 + n}\right)^2} \quad ...(26.8\ m)$$

where r is the fraction of the light reflected at a single interface, n_1 is the *r.i.* of the fiber core and n is the *r.i.* of the medium between the two jointed fibers.

The loss in decibels due to Fresnel reflection at a single interface is given by

$$\boxed{\text{Loss}_{\text{Fresnel}} = -10\log_{10}(1-r)} \quad ...(26.8\ n)$$

This loss can be reduced to a very low level through the use of an index matching fluid in the gap between the jointed fibers.

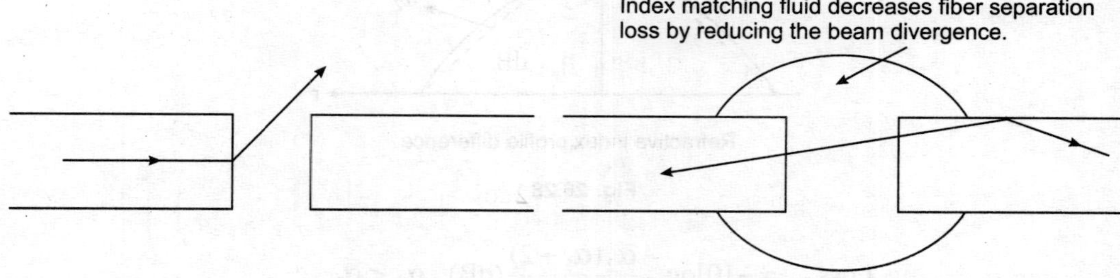

Fig. 26.25

In addition to mechanical misalignment,

(*ii*) Any deviations in the geometrical and optical parameters of the two optical fibers which are jointed will affect the optical attenuation (insertion loss) through the connection. It is not possible within any particular connection technique to allow for all these variations. Hence, there are inherent connection problems when jointing fibers with, for instance:

(*a*) Different core and/or cladding diameters

(*b*) Different numerical apertures and/ or relative refractive index differences

(*c*) Different refractive index profiles.

(*d*) Fiber faults (core ellipticity, core cocentricity, etc.)

The losses caused by the above factors together with those of Fresnel reflection are usually referred to as *intrinsic joint losses*.

Fig. 26.26 **Fig. 26.27**

$$\text{Loss}_{CD} = -10\log_{10}\left(\frac{a_2}{a_1}\right)^2 \text{(dB)} \quad a_2 < a_1$$
$$= 0 \qquad \text{(dB)} \quad a_2 \geq a_1$$

$$\text{Loss}_{NA} = -10\log_{10}\left(\frac{\text{NA}_2}{\text{NA}_1}\right)^2 \text{(dB)} \quad \text{NA}_2 < \text{NA}_1$$
$$= 0 \qquad \text{(dB)} \quad \text{NA}_2 \geq \text{NA}_1$$

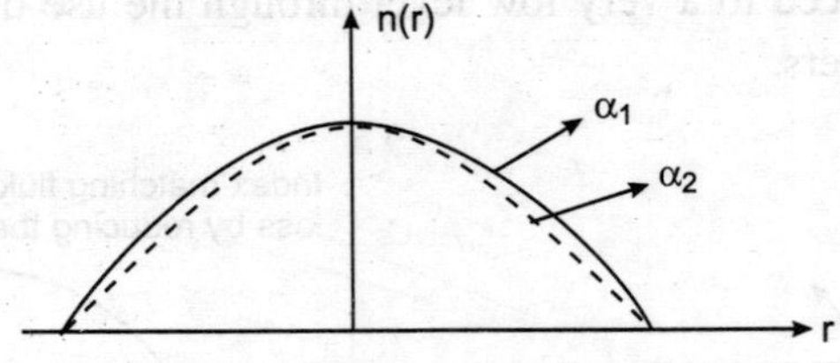

Fig. 26.28

$$\text{Loss}_{RI} = -10\log_{10}\frac{\alpha_2(\alpha_1+2)}{\alpha_1(\alpha_2+2)}\,(\text{dB}) \quad \alpha_2 < \alpha_1$$

$$= 0 \qquad (\text{dB}) \quad \alpha_2 \geq \alpha_1$$

α_1 = Profile parameters for the transmitting fibers

α_2 = Profile parameters for the receiving fibers

Typical results of splice losses due to various offsets between two fibers.

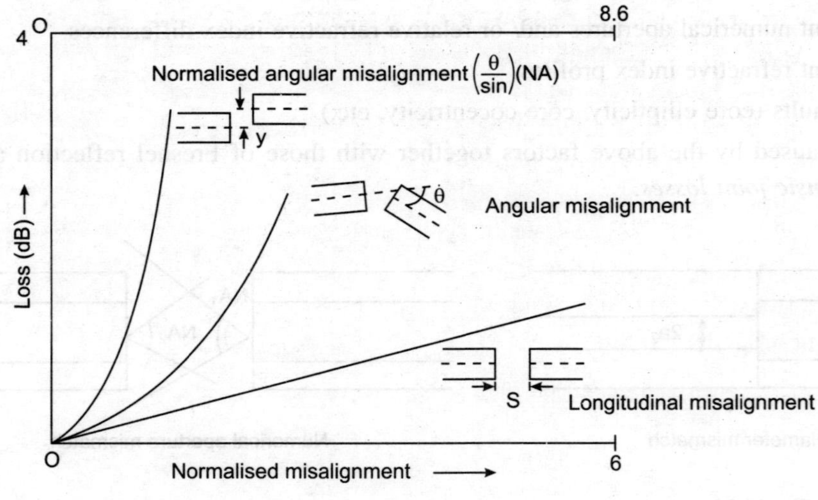

Fig. 26.29

26.9 OPTICAL FIBER FABRICATION AND TECHNOLOGY

The prediction for cladded cylindrical glassy medium as a telecommunication transmission medium was made in 1966. Later on purification of bulk fused silica had been attempted with an aim to enhance the optical transparency and the optical loss was achieved as low as 5 dB/km around near infrared region. Finally in 1970, optical fiber was fabricated with the help of vapour deposition technique having optical loss of 20 dB/km which is the threshold for telecommunication requirement.

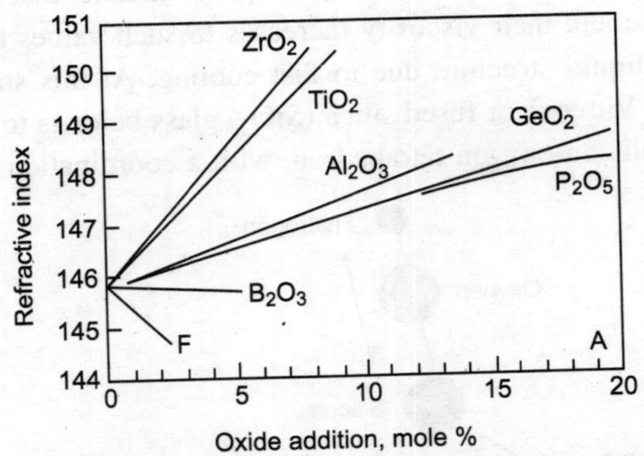

Fig. 26.30. Effect of common dopants on refractive index of silica (Reproduced from. Miller et al.)

A standard optical fiber consists of a high refractive index core covering a cylindrical cladding of lower index glass. The index difference, Δn, is usually ~0.005 ± 0.0015 for single mode fiber and ~0.014 ± 0.001 for multimode fiber. In order to fabricate an optical fiber from low loss silica glass, a second glass of higher refractive index is required which should be amply compatible with silica in terms of viscosity and thermal expansion coefficient. The most satisfactory refractive index (RI) modifying core glasses have been found to be those oxides which are more similar to silica itself. The refractive index variation with respect to change in mole % of different dopants in silica (SiO_2) is shown in Fig. 26.30, where RI increases with respect to silica with the addition of oxides like GeO_2, P_2O_5, TiO_2, Al_2O_3, ZrO_2. Addition of B_2O_3 and F in silica reduces the RI values.

Silica glass has the optical transparency in the wavelength range 200 nm to 300 nm and is suitable for processing at high temperature because its softening temperature extends from 1600°C to 2000°C. It has also very high strength and high thermal shock resistance property. The viscosity curves of different glasses are shown in Fig. 26.31 where it is evident that the viscosity of GeO_2 glass is near to SiO_2. The level of glass transition temperature (T_g) and softening temperature (T_s) is also shown in Fig. 26.31. It is observed that fluorine (F) is better choice for making low-index cladding than that of B_2O_3. Because B_2O_3 has some characteristic IR absorption bands which affect the transmission of light around 1500 nm.

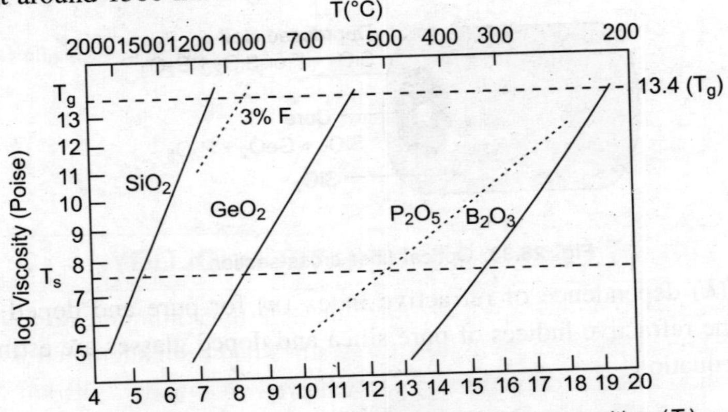

Fig. 26.31. Viscosity of oxide glasses with glass transition (T_g) and softening (T_s) temperature (Reproduced from Miller et al.)

Glasses are made by cooling certain molten materials in such a manner that they do not crystallize but remain in an amorphous state and their viscosity increases to such values that the atoms can no longer completely rearrange to liquid structure due to fast cooling. At this stage, resulting molten mass transforms into solid glass. Vitreous or fused silica (SiO_2) glass belongs to this class of material and has network structure with silicon-oxygen tetrahedron, with a coordination number of 4.

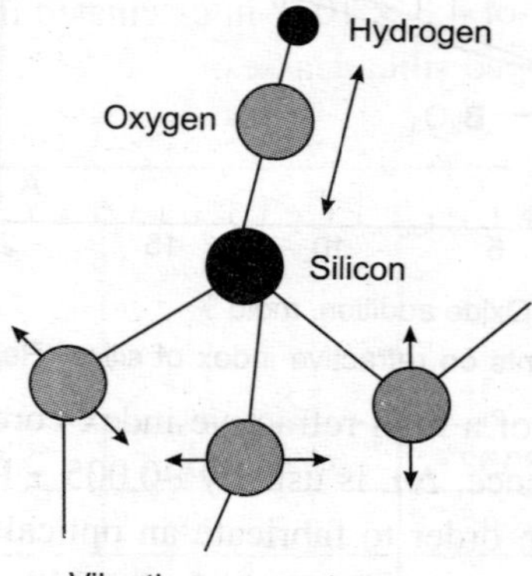

Fig. 26.32. Tetrahedron structure of silica showing hydrogen attached to one oxygen

These tetrahedral are linked at all four corners to form a continuous three-dimensional network (Fig. 26.32). The shortest Si–O distance in the structure is ~0.162 nm and that the shortest. *O–O* distance is ~0.265 nm. Arrows indicate vibration of atom near a chain attributing various absorption bands.

Interestingly vitreous germanium has tetrahedron structure like silica with coordination number 4. The ionic diameter of germanium atom is larger than silicon ion thereby Ge–O bond length is also somewhat greater ~ 0.173 nm. The structure of GeO_2 glass is more compact than SiO_2 and therefore interstitial volume of GeO_2 is slightly less than that of SiO_2. These properties imbibe more structural defects in vitreous GeO_2 by forming Ge–Ge bonds. Similarly, P_2O_5 behaves in similar fashion in SiO_2 host and participates elegantly in the processing of composite glass forming structure.

The common glass compositions required for fabricating standard optical fiber are shown in a section (Fig. 26.33).

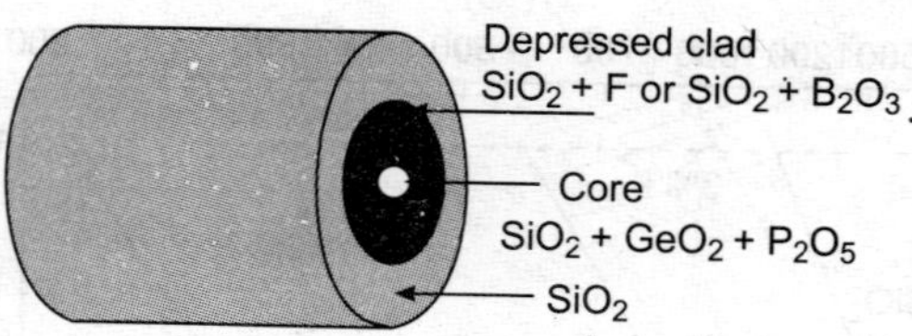

Fig. 26.33. Optical fiber cross-section

The wavelength (λ) dependence of refractive index (*n*) for pure and doped silica glasses is shown in Fig. 26.34. The refractive indices of pure silica and doped glasses are estimated by using a three term Sellmeier's equation.

$$n^2 - 1 = \frac{a\lambda^2}{\lambda^2 - d^2} + \frac{b\lambda^2}{\lambda^2 - e^2} + \frac{c\lambda^2}{\lambda^2 - f^2}$$

Where *a*, *b*, *c*, *d*, *e*, and *f* are Sellmeier's coefficients. The above equation fits very well with the experimental data with an average error of 4.3×10^{-6} in calculated indices.

Sellmeier's constants for silica and doped silica glasses:

(Cf. Ref. T. Izawa et al.)

Sellm. Cons.	SiO_2 100%	SiO_2 + GeO_2 6.3%	SiO_2 + GeO_2 8.7%	SiO_2 + B_2O_3 5.2%	SiO_2 + P_2O_5 10.5%
a	0.6965325	0.7083952	0.7133103.	0.6910021	0.7058486
b	0.4083099	0.4203993	0.4250904	0.4022430	0.4176021
c	0.8968766	0.8663412	0.8631980	0.9439644	0.8952753
d	0.0043683	0.0072904	0.0069102	0.0049818	0.0052024
e	0.0139499	0.0105029	0.0116567	0.0137566	0.0128773
f	97.93399	97.93428	97.93434	97.93353	97.93401

It is clearly understood that dopants like GeO_2 and P_2O_5 helps in enhancing the refractive index of silica glass whereas B_2O_3 reduces the index values. All these glass modifiers perform very well in smoothening the bulk glass maintaining minimum scattering losses:

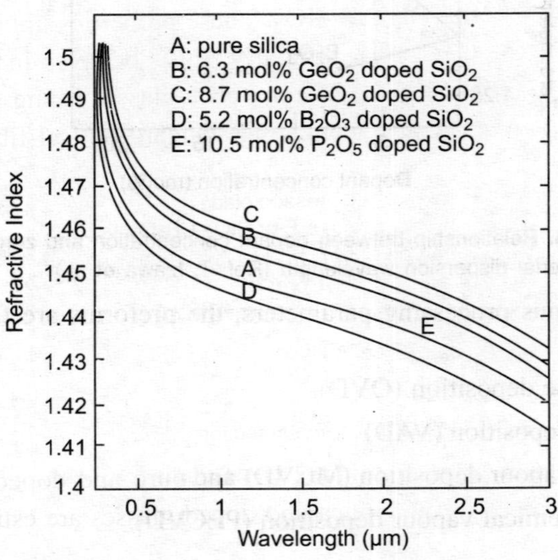

Fig. 26.34. Variation of refractive index of silica glass of different compositions against wavelength (Ref. T. Izawa et al.)

It is also important to note that the total dispersion comprising material dispersion and waveguide dispersion should be near to zero while selecting various dopants as well as their level of concentration. The zero dispersion wavelength with respect to various dopant concentration are shown in Fig. 26.35 where it is observed that the zero dispersion wavelength is shifted to the longer side with the increase of dopant concentration. Therefore selection of composite glass for particular fiber design is important in view of these properties and also are very essential while selecting the specific optical fiber index-profile.

Optical Fiber Fabrication

In optical fiber manufacturing process there have been significant improvement to increase the production rate for meeting the telecom requirement. Since the process of manufacturing optical fiber is divided into two main steps: preform fabrication and fiber drawing, the later one is common for all preform produced by different methods. Most of the bulk glass in the preform is cladding which is pure silica and it is required to produce the glass synthetically or thick silica tube is used as starting substrate tube. High refractive index composite glass forms the guiding core.

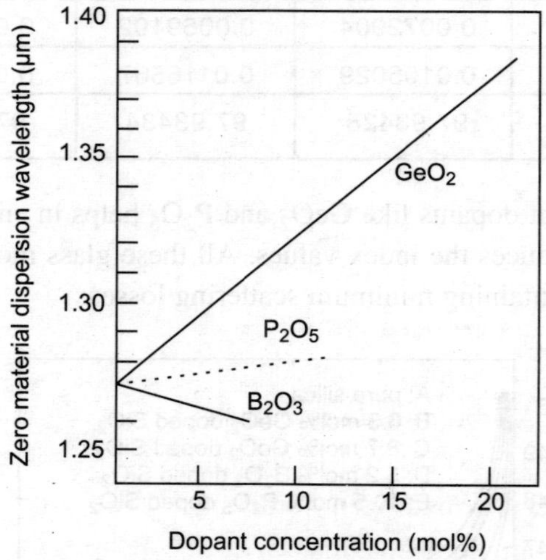

Fig. 26.35. Relationship between dopant concentration and zero material dispersion wavelength (Ref. T. Izawa et al.)

Depending upon the various processing parameters, the preforms are fabricated with the help of following techniques:

1. Outside vapour-phase deposition (OVD)
2. Vapour-phase axial deposition (VAD)
3. Modified chemical vapour deposition (MCVD) and
4. Plasma enhanced chemical vapour deposition (PECVD).

OVD Process

The OVD process was developed by M/s Corning Glass Works, USA to preform optical performs. In this process vapours of the constituent materials are hydrolyzed in a specially designed gas burner for deposition of the reacting soot particles on the outside surface of a tapered rotating mandrel during the back and forth pass of the burner. By controlling the temperature (Fig. 26.36) and the ratio of the stream of dopant vapours in the successive deposition layers the desired profiles could be developed in the soot preform across the diameter of the mandrel. The tapered bait rod allows to remove it very easily from the porous cylindrical structure. The porous preform is then removed from the mandrel and is sintered to a bubble free glass rod preceded by dehydration in a helium and chlorine atmosphere. Yield of the process is very high, multimode and step index optical preform can be fabricated through this process.

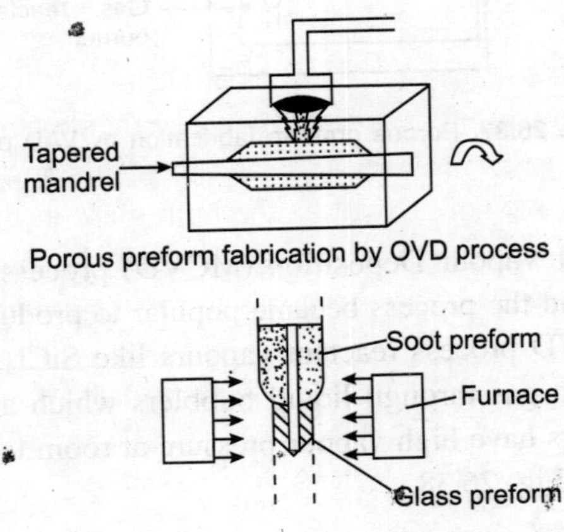

Fig. 26.36. Sintering of porous preform made by OVD process

VAD Process

This process was first demonstrated at *NTT* laboratories, Japan in 1977. The *OVD* process as described in the previous section is a lateral soot deposition technique using multi-layer-burner whereas in the *VAD* process the porous-soot-preform is built up axially on the end of a rotating ceramic or silica rod which acts as a seed rod. The rod is automatically raised to keep the deposition point in line with the fixed multi layer oxy/hydrogen burner. The rod is continuously rotated to maintain a cylindrical symmetry of the particle deposition. Gradually porous preform moves upward and starts consolidating into a solid transparent glass rod when it passes through a narrow heating zone of 1500°C–1600°C as shown in Fig. 26.37. After jacketing with a silica tube, the preform is drawn into fibers.

Both step-index and graded-index in either multimode or single-mode configurations can be made using *VAD* method by the transverse diffusion of the soot particles generated from the vapours ($SiCl_4$, $GeCl_4$ and $POCl_3$) which are blown out from the O_2/H_2 burner nozzles. In this process, the reaction kinetics are very complicated because the partial pressure of hydrogen and chlorine gas influence the reaction process. The advantages of the *VAD* method are (1) the preform has no central hole like *OVD* process, (2) larger preform can be made in continuous lengths which affects process costs and product yields, and (3) since the deposition chamber and the sintering zone are housed in the same enclosure this ensures purity of the preform.

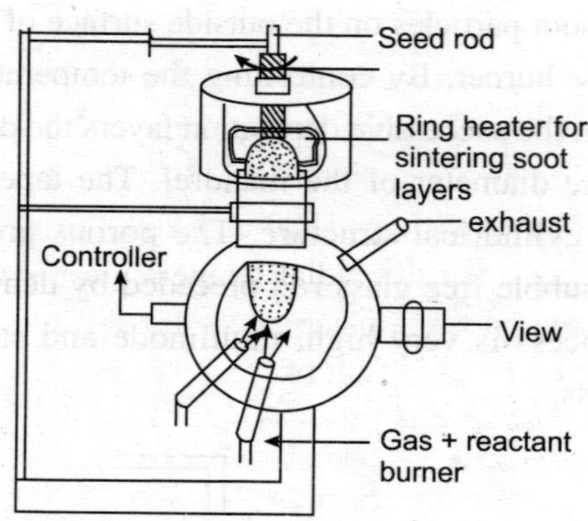

Fig. 26.37. Porous preform fabrication by VAD process

MCVD Process

The Modified Chemical Vapour Deposition (MCVD) process was first demonstrated at Bell Laboratories in early 1974, and the process became popular to produce very low loss of optical fiber in commercial scale. In MCVD process reactant vapours like $SiCl_4$, $GeCl_4$ and $POCl_3$ are bubbled by passing dry carrier oxygen gas through liquid bubblers which are kept at constant temperature liquid-bath. The halide vapours have high vapour pressure at room temperature and their dependence with temperature is shown in Fig. 26.38.

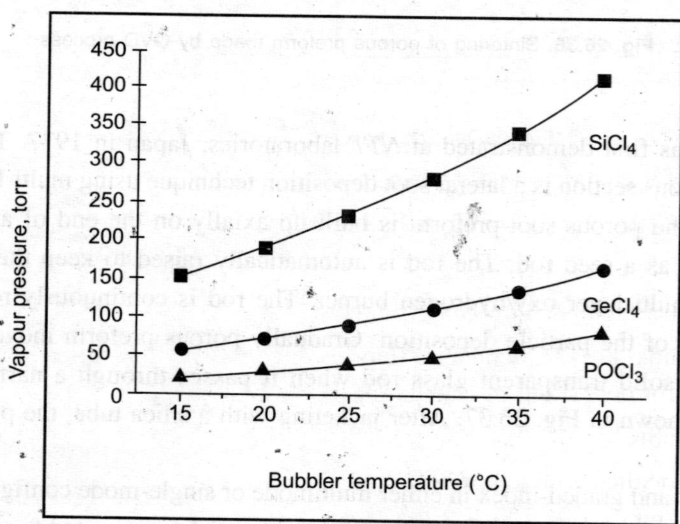

Fig. 26.38. Vapour pressure of different source chemicals against temperature

These vapours are highly toxic and posses high vapour pressure at ambient temperature. The vapour pressure of respective chemicals are calculated from the following equation:

$$\text{Log}_{10}P = A - B/(C + t) \text{ (Ref. CRC Handbook)},$$

where A, B and C are constants and t is temperature in degree centigrade. It is important to keep vapour saturation up to the maximum carrier flow rate using thermal mass flow controllers. The reactant gases along with excess oxygen flow through a rotating silica substrate tube mounted in chucks of a glass working lathe as shown in Fig. 26.39.

The diameter of the tubes are typically 12–30 mm. The tube is heated by a traversing oxy/hydrogen burner to temperature between 1500 and 2000°C in the same direction as the interior gas flow. At this stage depending upon the radial temperature gradient occurring across the hot zone, a homogeneous gas phase reaction takes place and solid particles nucleate from the reaction products. Particles growth occurs primarily as a result of coagulation of colliding particles as the finely dispersed solids.

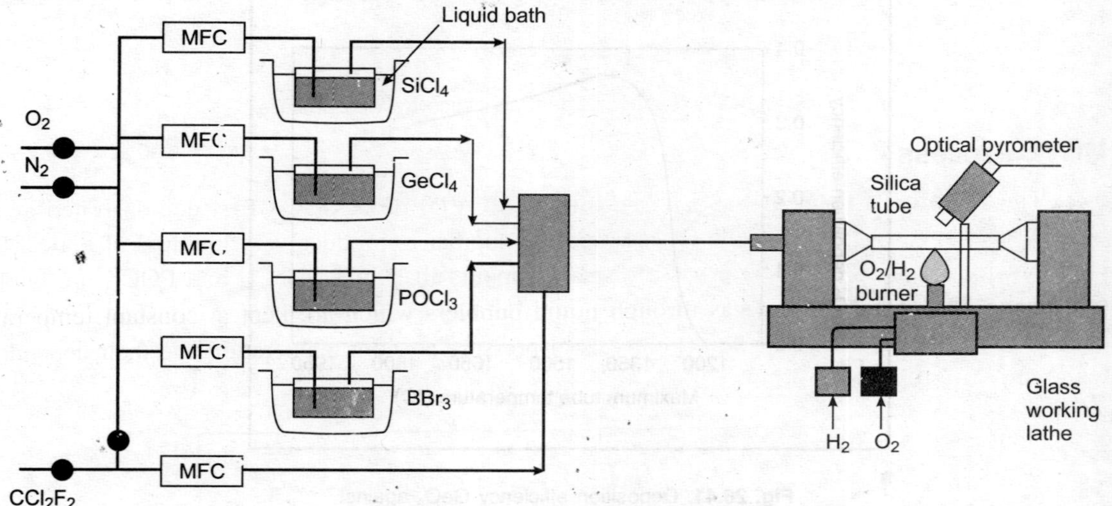

Fig. 26.39. Schematic diagram of MCVD process

A. Reactions for etching process when inside surface of silica tube is cleaned:

$$CCl_2F_2\ (g) + O_2 = Cl_2 + F_2 + CO_2$$

(Freon gas)

$$SiO_2\ \text{(silica tube)} + 2F_2 = SiF_4\ (g) + O_2$$

(Silicon dioxide) (Silicon tetra fluoride)

B. Reactions for deposition of cladding layers:

$$SiCl_4(g) + O_2 = SiO_2(s) + 2Cl_2$$

(Silicon tetrachloride)

$$SiO_2(s) + 2F_2 = SiF_4(g) + O_2$$

$$3SiO_2(s) + SiF_4 = 2Si_2O_3F_2\ (s)$$

(Fluorinated silica compound)

$$4POCl_3\ (g) + 3O_2 = 2P_2O_5(s) + 6Cl_2$$

Preform making facility in CGCRI

Fig. 26.40

C. Reactions for deposition of core layer:

$$SiCl_4(g) + O_2 = SiO_2(s) + 2Cl_2$$

$$GeCl_4(g) + O_2 = GeO_2(s) + 2Cl_2$$

As the burner traverses further, the particulate layer is sintered and vitrified forming pore-free glass layer of thickness around 10 micron. When the burner reaches the end of its travel, it quickly returns and the traverse is repeated. The concentration of the dopants can be changed in each pass to get desired refractive indices of the cladding and core glasses. Typically between 20 to 80 layers of glass are deposited to make multi-mode or single-mode preform rods of 15–30 cm diameter. The reaction rate largely depends on the tube temperature with associated flow pattern of the reactant vapours. The deposition efficiency of GeO_2 versus silica tube temperature is shown in Fig. 26.41.

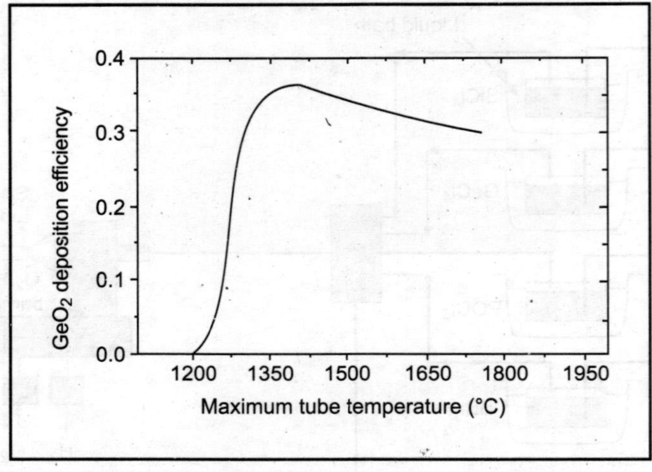

Fig. 26.41. Deposition efficiency GeO_2 against

Deposition rate is optimised with respect to temperature where higher temperature causes dopant vapourisation in the simultaneous sintering process. Lower temperature deposition makes the reaction inefficient especially for $GeCl_4$. Downstream from the hot zone which is near to the burner, the tube walls are cooler than the vapour stream and some of the suspended glass particles are deposited on the tube surface. The deposition is caused by thermophoresis, a phenomenon in which a suspended particle experiences a net force in the direction of decreasing temperature due to a greater rate of collisions with gas molecules on the hot side of each particle. This is represented in Fig. 26.42.

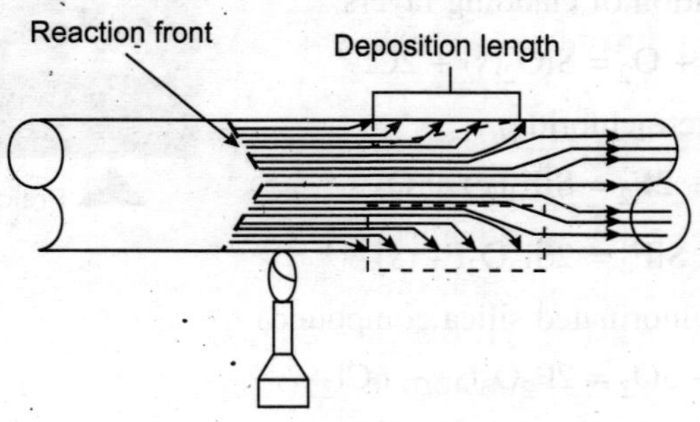

Fig. 26.42. Particle trajectories in silica substrate Tube (Ref. M.C. Paul et al.)

After deposition of barrier and core layers temperature of the tube was increased above 2000°C to collapse into solid rod known as preform. Collapsing phenomena occurs as shown in Fig. 26.43, due to inward viscous flow of mass driven by surface tension and pressure difference between inside and outside of the tube. Control of flow of additional O_2, tube temperature and pressure difference between inside and outside of the tube are very important during collapsing process to maintain circular geometry of core, cladding layers and overall preform along the whole collapsing length. At present an advanced MCVD preform making machine is available at CGCRI.

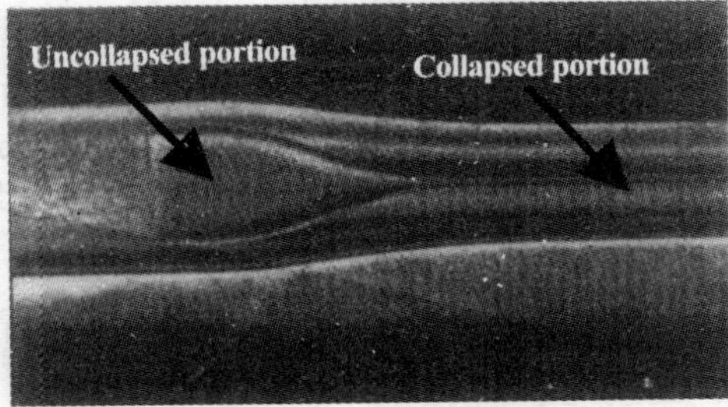

Fig. 26.43. Collapsed substrate tube after deposition

PECVD Process

A modification of the MCVD method for depositing glass layers, one can use efficiently a radio-frequency induced plasma in the excess oxygen environment. This creates a 'fire-ball' in the center of the substrate tube and enables larger diameter tubes to be used for getting smooth deposit layers. The fire-ball, at about 10^4 °C, helps the reaction process and deposits fine glass particles through efficient thermophoretic reaction dynamics. The sintering and vitrification are usually carried out by a oxy-hydrogen burner. Growth rates of ~5 g/min can be achieved using this method but the process for generating plasma inside the substrate tube with dynamic control makes it complicated. After fabrication of the preforms, fiber is drawn with the help of a fiber drawing tower.

Drawing of Optical Fiber

Subsequent to the MCVD process, the preform is fed into a furnace which is heated to 2000 to 2200°C. A schematic of a fiber drawing tower is shown in Fig. 26.44 including the furnace.

Depending upon the preform diameter and fiber drawing speed, the furnace temperature is set. At the same time fiber is drawn down by a capstan puller out of the furnace. Special care is taken to design furnaces to ensure that they are sources of particle contamination which will degrade fiber strength. After cooling down the bare fiber is coated with two layers of UV curable acrylate resin. High quality coating with good coating concentricity and accurate dimensions is essential for getting high quality fibers. Fiber drawing speed normally one can achieve in the range of 30 m to 600 m/min. However in industrial production drawing speed can go up to 2000 m/min in a 20 m high tower with proper drawing tension and fiber cooling technique. For high speed drawing higher temperature (and hence lower drawing tension) is necessary, it may adversely affect the spectral loss of the fiber. Reducing the tension from 25 g to 17 g increases the attenuation by 0.05 dB/km in a single mode

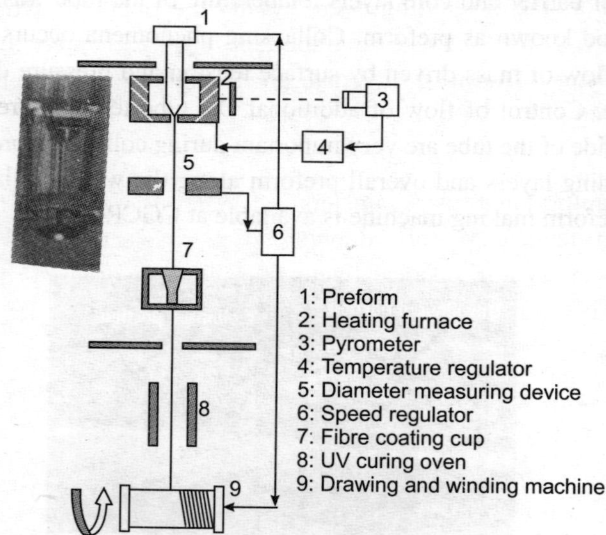

Fig. 26.44. Schematic view of optical fiber drawing tower. Induction furnace (CGCRI) is shown in inset

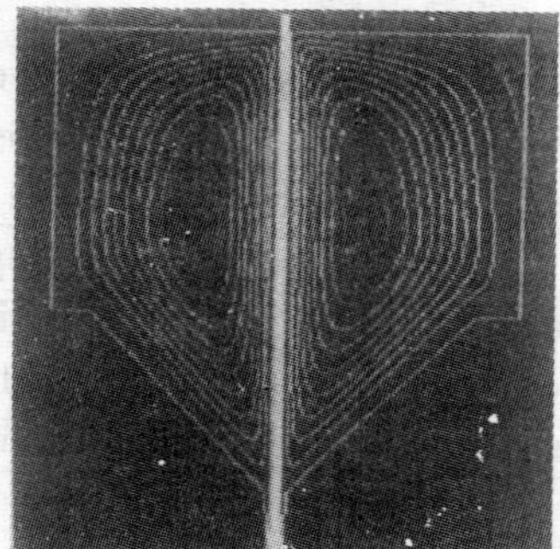

Fig. 26.45. Wet on dry coating on fiber with circular lines showing liquid movement in coating cup

fiber operating at 1300 nm. Drawing temperature, drawing speed and tension are to be critically monitored during the fiber drawing process before the application of coating.

The purpose of the coating on the optical fiber is to protect the fiber from contact and foreign particles as these reduce the life of the fiber in terms of strength and static fatigue. Besides strength preservation, the coating must protect from micro-bending by being concentric and bubble free around the fiber and by having a stable performance in different environments. As mentioned out of two layers of UV-curable acrylate, a soft inner layer protects the fiber and a hard outer layer ensures good mechanical properties. The layers can be applied (Wet-on-Wet method) or separately by two different applicators (Wet-on-Dry method). The whole coating process is monitored on line with the help of a laser concentricity monitor.

The movement of the coating liquid surrounding the surface of the fiber is shown in Fig. 26.45, where fine bubbles can be trapped at the interface of fiber surface and hard coating. This can be avoided by controlling pressure and keeping the viscosity of the liquid constant during the drawing process before it enters the UV curing oven. Finally the fibers are wound on a precision spooling machine. The control of fiber diameter is important in a steady state drawing process where mass-conservation is necessary. During steady-state process the following relation will be occurred from mass-conservation rule,

$$D^2/d^2 = v/V$$

where D and d are preform and fiber diameters and V and v are preform feed down and fiber drawing speeds. For 200 mm diameter preform drawing speed is selected at 100 m/min. Sometimes steady state is never achieved because of local fluctuations. This is occurred due to non-uniformity in the preform diameter. Therefore it is essential to measure and control the diameter precisely if fibers with tight dimensional tolerances are to be made.

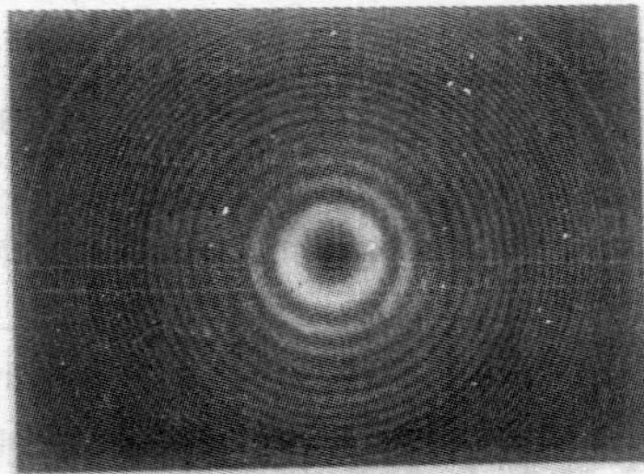

Fig. 26.46. Section of a Graded-Index multimode preform (CGCRI) showing the deposited layers. Central dip (black spot) is due to evaporation of GeO_2

Fiber Characterization

Geometrical and optical characterizations of the fabricated fibers are essential for evaluation their performance. Preform and fiber sections are viewed and photographed under polarizing optical microscope, one such section of a graded index multimode preform section, is shown in Fig. 26.46, where individual fifty core layers are formed as concentric rings with dark central spot. This dark spot is due to the evaporation of volatile GeO_2 during the collapsing stages.

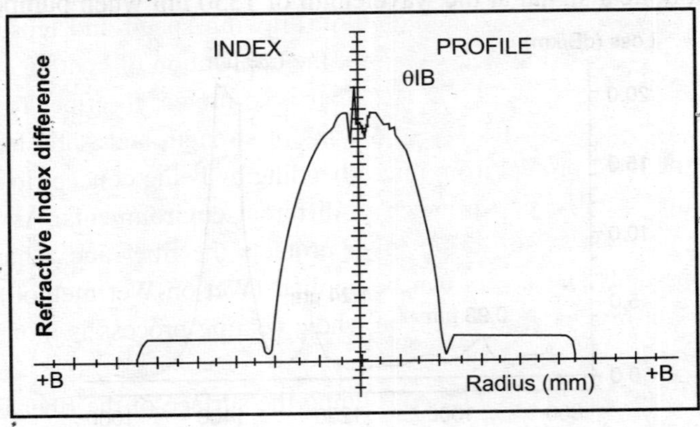

Fig. 26.47. RI Profile of the Graded-Index Multimode fiber

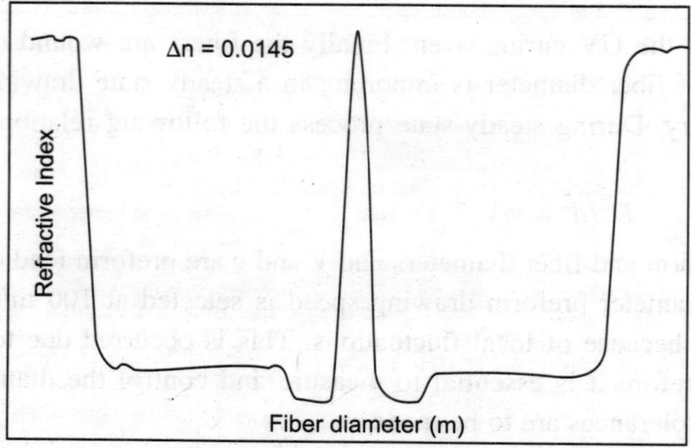

Fig. 26.48. RI Profile of Erbium Doped Single mode fiber

Numerical aperture of a fiber is usually measured from the refractive index profile measured across the diameter by refracted near field (RNF) method. Fig. 26.47 and 26.48 show the refractive index (RI) profiles of a standard GIMM fiber and a standard *SM* fibers respectively taken during routine measurement at CGCRI.

Optical loss of the fibers at operating wavelength are measured by OTDR (Optical Time Domain Reflectrometer) at fixed wavelengths, however, spectral attenuation can be measured with the help of a grating monochromator and white light source. A typical spectral attenuation curve of a high – OH containing fiber is shown in Fig. 26.32. The absorption peaks at 980 nm, 1240 nm & 1390 nm are related to > Si-OH stretching and other combination of vibration bands (Fig. 26.32). In a good quality fiber control of moisture is very important in order to reduce the optical loss.

26.10 ERBIUM DOPED FIBER FOR OPTICAL AMPLIFIER

The expanding bandwidth demand in telecommunication network which arises due to tremendous growth in internet and other communication services is being met by erbium doped fiber optical amplifier (EDFA) in WDM network. *Er*-ions when doped in low concentration in the core of the fiber are found to boost optical signal at the wavelength of 1550 nm when pumped by 980 nm light.

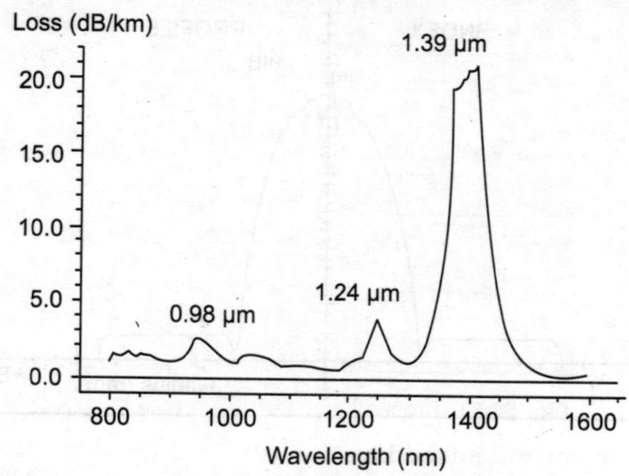

Fig. 26.49. Spectral attenuation curve of standard SM Fiber (CGCRI) showing OH absorption bands

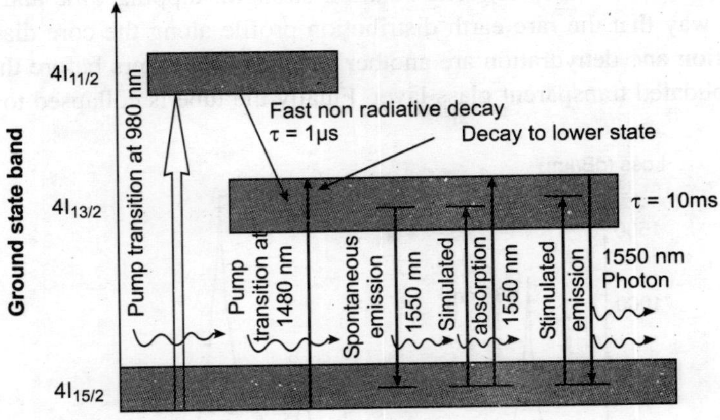

Fig. 26.50. Energy level diagram of *Er* in silica host

The advantage of EDFA is that it can simultaneously amplify all incoming wavelength (1530 to 1565 nm) channels to nearly the same output power. With additional care in dispersion compensation and non-linear effects, the broad band gain flattering can be extended up to 1620 nm for dense wavelength division multiplexing (DWDM) applications. An energy band transition diagram shows the pump band as well as metastable level of erbium ion in silica host (Fig. 26.50).

Keeping in view the rapid development of technology, CGCRI has started working in this area to meet the challenges that lie ahead.

Along with the conventional MCVD method the solution doping process has been adopted for incorporation of Er_2O_3 in the SiO_2-GeO_2 core structure as the process steps are shown in Fig. 26.51.

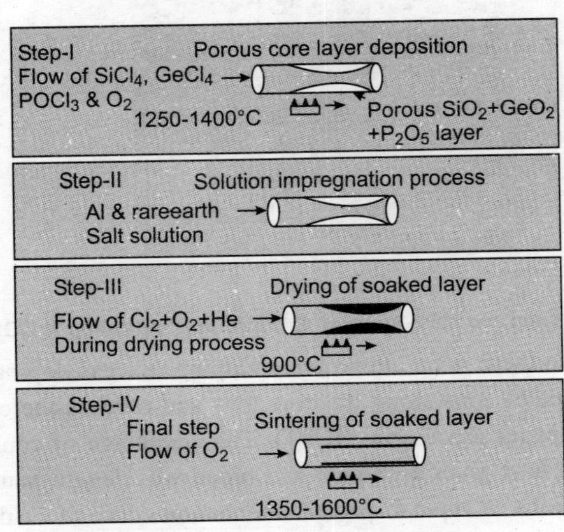

Fig. 26.51. Process steps of solution doping

By this technique an unsintered mixed layer of SiO_2 and GeO_2 is first deposited inside a silica tube preceded F, P_2O_5 and SiO_2 layers to get matched/depressed clad single mode configuration. The porous unsintered core layer first deposited is doped by filling the tube with a specific strength of rare earth chloride solution (Step-II). The solution strength, dipping time and drainage rates are controlled in such a way that the rare earth distribution profile along the core diameter is of desired specification. Oxidation and dehydration are another two key parameters before the sintering process to get uniform consolidated transparent glass layer. Finally the tube is collapsed to make a preform.

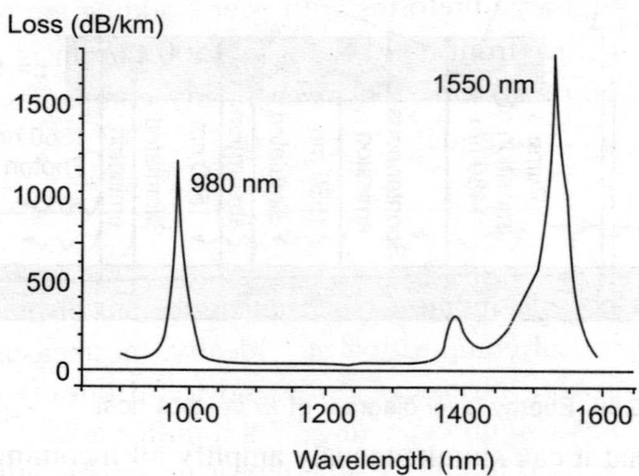

Fig. 26.52. Spectral attenuation curve of *Er* doped Optical fiber measured at CGCRI

The aluminum codoping during solution impregnation helps to get well coordinated rare earth ions in silica glass. A distortion of the electronic charge cloud surrounding the rare earth ion due to poor coordination may lead to Stark split of the energy levels and may affect the quantum efficiency at high signal power. It is reported that the gain degradation occurs at high

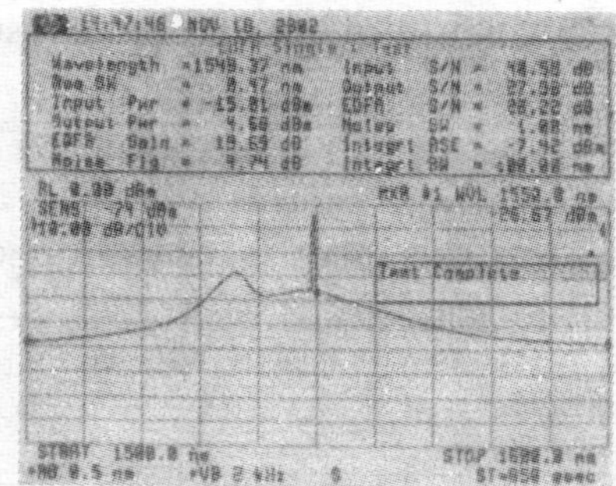

Fig. 26.53. Gain and noise figure of *Er* doped fiber measured at CGCRI

concentration level of *Er* when there is no aluminum codoping. Excess doping of aluminum helps to achieve higher concentration of *Er* ions along the core axis and reduces the central dip which arises due to high temperature collapsing process in MCVD. The advantage of confining *Er* ions close to the centre of the core in silica host gives high gain and broad wavelength band gain spectrum which is essential for WDM applications. A typical spectral attenuation curve of *Er*-doped fiber (Fig. 26.52) depicts the 980 nm and 1550 nm absorption bands. These absorption bands in terms of dB/m help in determining the *Er* ion concentration in the fiber core. An optical gain of 20 dB of one such fiber has been measured and is shown in Fig. 26.53, where input signal (1550 nm) power was –15 dBm. The measured developed noise figure is 4.74 dB. These fibers developed at CGCRI are also suitable for *C*-band amplification regime.

Industrial Scenario in India

In India Sterlite optical Technology has been manufacturing standard single mode fiber for telecommunication network in large scale. Large preforms with over-cladding are being produced by MCVD process. High speed drawing of fiber from 500 m/min to 1500 m/min is carried out in tall fiber drawing tower in continuous length of 50 kms. There are many companies who are making fiber optic cables for directly use in the telecom grid.

26.11 FIBER OPTIC SENSORS

In many industrial processes there is a need to monitor (such quantities as) displacement, pressure, temperature, flow rate, liquid level, chemical composition etc. Ideally, the measurement technique should be reliable, robust, corrosion resistant, intrinsically safe and free from external interference. Optical fibers have the potential for making a significant contribution in this area.

Generally a transducer (or sensor) is used to convert one physical variable into another. Mostly these sensors are electrical in construction and provide convenient and controllable electrical signals for (*i*) amplifying weak signal (*ii*) measuring a process (*iii*) automatically recording these measurements and use them, when needed and (*iv*) providing a signal that can be used to control another system or circuit. Thus, a transducer is an essential component of automation, wherever introduced.

In the case of optical fibers, these characteristics themselves provide an innovative approach to the design of an optical transducer/sensor.

The optical fibers have following characteristics:

(1) Optical fibers are non-conducting, immune to electromagnetic and radio-frequency interference and safe, even in explosive environments.

(2) Their low attenuation coefficient allows the monitoring station to be located at a safe place, away from the hazardous environments (but where sensitizing has to be done)

(3) Fiber optic sensors are not so sensitive to temperature changes.

(4) Optical fibers are easily available to suit any design. (Thus, design-engineers, naturally prefer to use optical fibers as the basic material for their sensor design.) Optical fiber sensors themselves can be divided into two main categories, namely; active (intrinsic) and passive (or extrinsic).

In the passive, the modulation takes place outside the optical fiber, which acts merely as a convenient transmission channel for the radiation.

In the active sensors, the quantity to be measured acts directly on the fiber itself to modify the radiation passing down the fiber. It is possible to modulate either the amplitude, (or intensity) the phase or the state of polarization of the radiation in the fiber.

In multimode fibers, however, mode coupling and the usually random relationships between the phases and polarization states of the propagating modes generally preclude the use of either phase or polarization modulation. Thus, active multimode fiber sensors almost invariably involve amplitude modulation. On the other hand, with single mode fibers both phase and polarization modulation becomes possible. Phase modulation opens up the possibility of fiber based interferometric sensors that can offer exceptionally high sensitivities.

The extrinsic (passive) type sensors configured to allow the measured to change the coupling characteristics between the feed and return fiber as shown in figure:

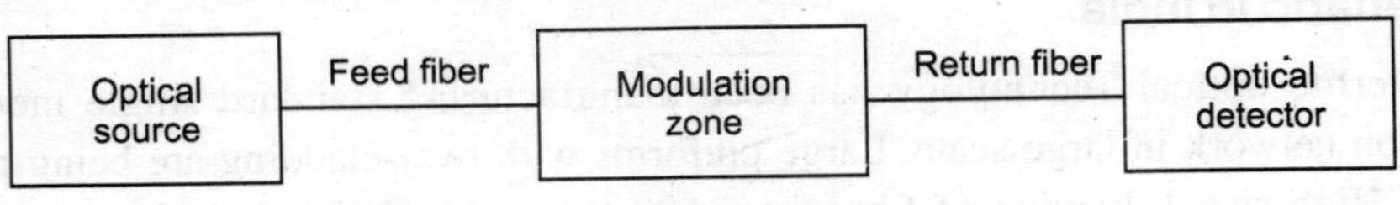

Fig. 26.54

Intrinsic (Active) Fiber Sensors

A popular technique for the realization of an intrinsic multimode fiber sensor involves micro-bending of the fiber in the modulation region. The sensing mechanism for this techniques is shown in figure. Deformation of the fiber on a small scale causes light to be coupled from the guided optical modes propagating in the fiber core into the cladding region where they are lost through radiation into the surrounding region. When the spatial wavelength of the deformation L is correctly chosen, the power coupled from the fiber into radiation modes is high, providing a very high sensitivity to pressure applied to the deformer in a direction perpendicular to the fiber core axis. Furthermore, if the deformation is caused by the measurand (*e.g.* pressure, vibration, sound etc.), then the fluctuation in intensity of either the core or cladding light is directly proportional to the measureand for small deformations. Thus monitoring the intensity of either fiber core or cladding allows detection of the measurand.

Also, in common with all fiber intensity modulation sensors, inaccuracies may occur due to source, detector and fiber cable instabilities. These, so-called, common-mode variations usually necessitate the transmission of a separate optical reference signal which is not modulated by the measurand. In this way, any optical intensity variations can be removed from the returned measurand signal by comparison with the returned reference signal at the receive terminal.

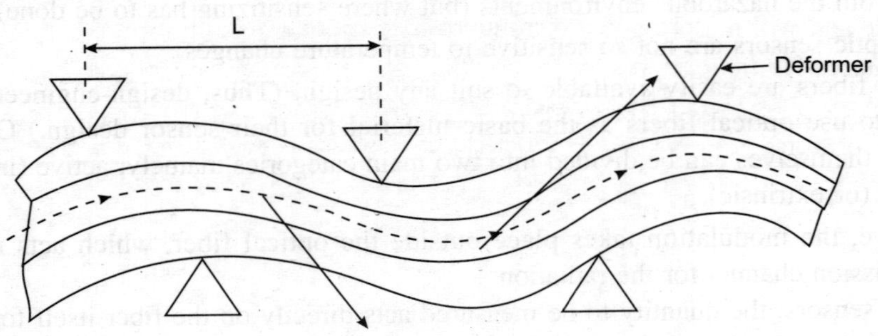

Fig. 26.55

Flow Sensor

Figure shows an intrinsic (Active) optical fiber flow sensor mechanism. In this device a multimode optical fiber is inserted across a pipe such that the liquid flows past the transversely stretched fiber. The turbulence resulting from the fibers presence causes it to oscillate at a frequency roughly proportional to the flow rate. This results in a corresponding oscillation in the mode power distribution within the fiber, giving a similarly modulated intensity profile at the optical receiver. The technique has been used to measure flow rates from 0.3 to 3 m/sec.

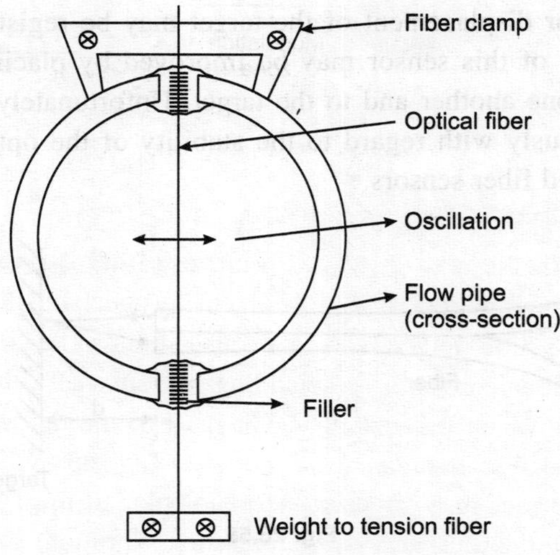

Fig. 26.56

Extrinsic (Passive) Fiber Sensors

Numerous extrinsic optical fiber sensor mechanism have been proposed and investigated, but to date relatively few practical commercial devices have emerged. A technique which has been realized as a commercial product is shown in figure. This shows the operation of a simple optical fluid level switch – when the fluid, which has a refractive index greater than the glass forming the optical dipstick, reaches the chamfered end, total internal reflection ceases and the light is transmitted into the fluid. Hence an indication of the fluid level is obtained at the optical detector. Although this system is somewhat crude and will not provide a continuous measurement of a fluid level, it is simple and safe for use with *flammable liquids*.

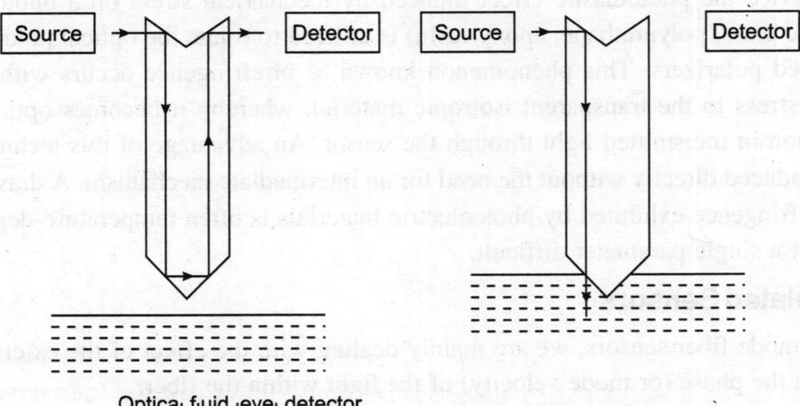

Optical fluid level detector

Fig. 26.57

Intensity modulation of the transmitted light beam is utilized in the extrinsic reflective or fotonic optical sensor as shown in figure, to give a *measurement of displacement*. Light reflected from the target is collected by a return fiber and is a function of the distance between the fiber ends and the target. Hence the position or displacement of the target may be registered at the optical detector. Furthermore, the sensitivity of this sensor may be improved by placing the axes of the feed and return fiber at an angle to one another and to the target. Unfortunately, this technique exhibits the drawback mentioned previously with regard to the stability of the optical components which is a feature of intensity modulated fiber sensors.

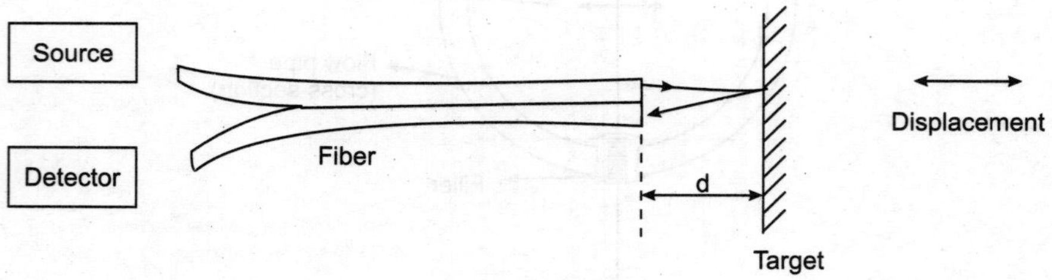

Fig. 26.58

A multimode fiber sensor which provides measurement of pressure is illustrated in Fig. 26.59.

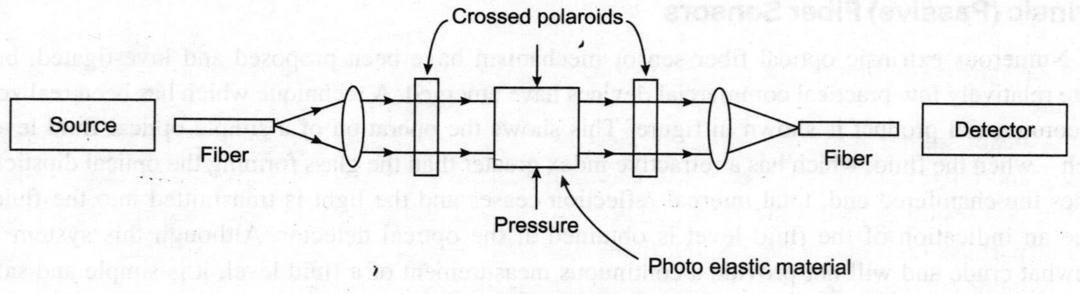

Fig. 26.59

In this device the photoelastic effect induced by mechanical stress on a photoelastic material (*e.g.* piezo-optic glass, polyurethane, epoxy resin) is utilized to rotate the optical polarization between a pair of crossed polarizers. This phenomenon known as birefringence occurs with the application of mechanical stress to the transparent isotropic material, whereby it becomes optically anisotropic giving a variation in transmitted light through the sensor. An advantage of this techniques is that the stress may be induced directly without the need for an intermediate mechanism. A drawback, however, is that the fine fringence exhibited by photoelectric materials is often temperature-dependent, making measurement of a single parameter difficult.

Phase Modulated Sensors

In single mode fiber sensors, we are mainly dealing with the effect of the external quantities to be measured on the phase (or mode velocity) of the light within the fiber.

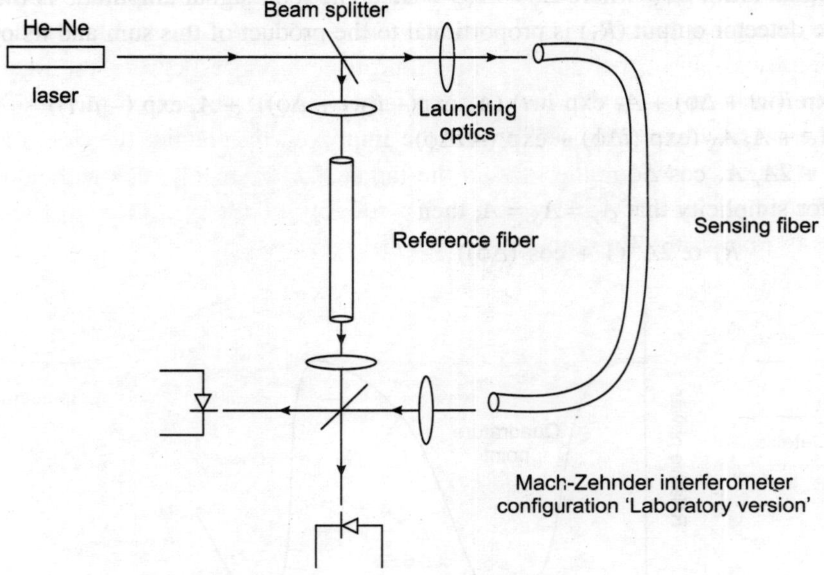

Fig. 26.60

One way in which we may detect phase changes is to construct an interferometric system where the phase of the beam through a sensing fiber is compared with that of a reference beam. The Mach-Zehnder configuration is easily realised using optical fibers, a typical laboratory version is shown in figure.

A beam splitter is used to divide a laser beam into two parts that are then launched into the sensing and reference fibers respectively. The outputs from these fibers meet at another beam splitter and the two resulting beams are then allowed to fall onto suitable photo detectors. It is not strictly necessary to use both output beams, but there are advantages to be gained from so doing.

A more compact 'practical' version may be constructed, where He-Ne laser is replaced by a semiconductor laser, and the beam splitters by '3 dB couplers'.

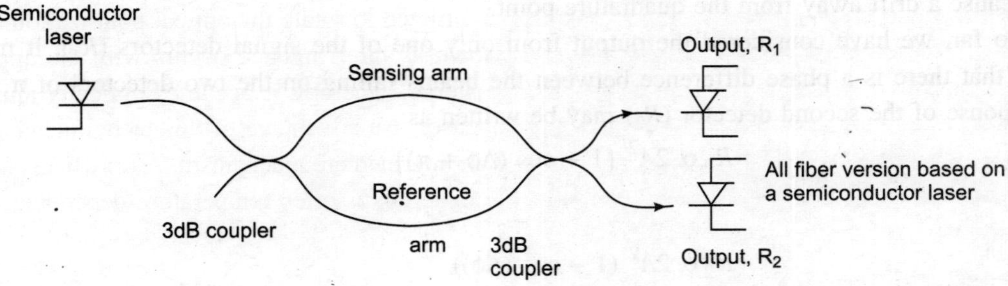

Fig. 26.61

Let us consider the form of the output expected from the interferometer. We write the electric fields of the two beams at one of the detectors as A_S expi $(wt + \Delta\phi)$ and A_r $\exp_{iwt}$, where $\Delta\phi$ represents the phase difference between the two. In general, $\Delta\phi$ will be made up of a static differential phase term $\Delta\phi_d$ and a signal term, $\Delta\phi_s$, where $\Delta\phi = \Delta\phi_d + \Delta\phi_s$. The total signal amplitude is the sum of the two fields and the detector output (R_1) is proportional to the product of this sum and its own complex conjugate *i.e.*,

$$R_1 \propto (A_S \exp i(wt + \Delta\phi) + A_1 \exp iwt)\,(A_S \exp(-i(wt + \Delta\phi)) + A_r \exp(-iwt))$$

i.e., $R_1 \propto A_r^2 + A_S^2 + A_r A_S (\exp(i\Delta\phi) + \exp(-i\Delta\phi))$

$$\propto A_r^2 + A_S^2 + 2A_r A_S \cos \Delta\phi$$

Assuming for simplicity that $A_r = A_S = A$, then

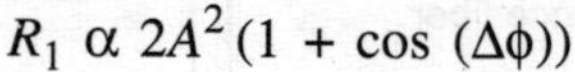

$$R_1 \propto 2A^2 (1 + \cos(\Delta\phi))$$

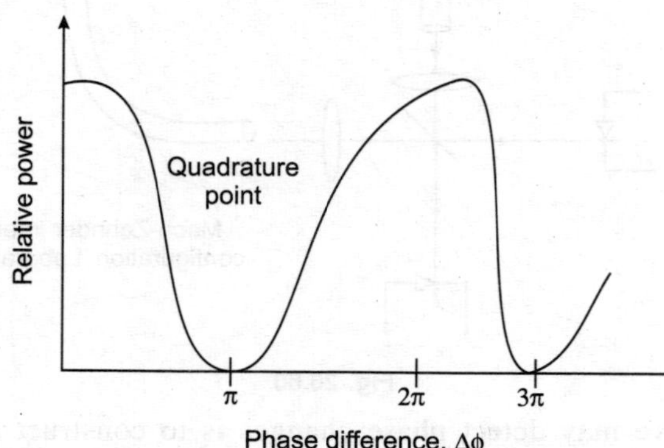

Fig. 26.62. Relative output of a Mech-Zehnder interferometer as a function of the phase difference between the two arms ($\Delta\phi$)

This response as a function of $\Delta\phi$ is shown in figure, illustrating the basic problem with any interferometric sensor, that its output is periodic in the phase difference. It is also evident that the greatest sensitivity to small variations in signal are obtained when the system is operating at points half way between the maximum and minimum values of R. These are known as the 'quadrature' points. Midway between these maximum sensitivity points the sensitivity falls off to zero.

Thus, even if we could ensure that we started off operating at a quadrature point it is quite likely that changes in ambient conditions around the reference arm would cause $\Delta\phi$ to change and hence cause a drift away from the quadrature point.

So far, we have considered the output from only one of the signal detectors (R_1). It may be shown that there is a phase difference between the beams falling on the two detectors of π. Thus, the response of the second detector (R_2) may be written as:

$$R_2 \propto 2A^2 \{1 + \cos(\Delta\phi + \pi)\}$$

or

$$R_2 \propto 2A^2 (1 - \cos(\Delta\phi))$$

Thus,

$$(R_1 - R_2) \propto 4A^2 \cos(\Delta\phi)$$

In effect, therefore, by taking the difference between the two output signals we double the sensitivity of the interferometer. There are also advantages to be gained when the amplitudes of the beams from the signal and reference arms are not equal.

Fiber Optic Gyroscope

Another common single-mode fiber interferometric sensor which is finding wide-scale application is the fiber gyroscope. This device is based on the classical sagnac ring interferometer, a fiber version of which is given below.

The sagnac effect, *i.e.*, the phase shift induced between two light beams travelling in opposite directions round a fiber coil when the coil is rotating about an axis perpendicular to the plane of the coil.

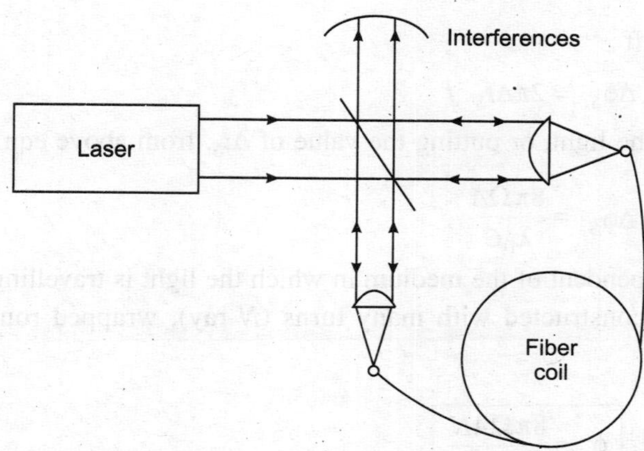

Fig. 26.63

In this device light entering the multi turn fiber coil is divided into two counter propagating waves which will return in phase after travelling along the same path in opposite directions. When the fiber coil is rotating about an axis perpendicular to the plane of the coil, however, then the path lengths between the counter propagating waves differ. This difference produces a phase shift which in turn can be measured by interferometric techniques in order to obtain the rotation.

To calculate the phase shift expected, we assume, we have a single circular turn of fiber of radius R. It turns out that we must also assume that both light beams are effectively travelling in a vacuum. If there is no rotation, then both beams will return to their starting point in a time t, where $t = \dfrac{2R\pi}{c}$. If, however, the ring is rotating in a clockwise direction at a rate of Ω rad/sec, then the counter clockwise beam will arrive at its starting point sooner.

Since the effective velocity of the beam will be $C + R\,\Omega$ and the time taken, t', will be given by

$$t' = \frac{2\pi R}{C + R\Omega}$$

Similar arguments for the clockwise beam give a transit time of t'', where

$$t'' = \frac{2\pi R}{C - R\Omega}$$

The difference between the two transit times Δt_S (= $t'' - t'$) is then

$$\Delta t_S = 2\pi R\left(\frac{1}{C - R\Omega} - \frac{1}{C + R\Omega}\right) = \frac{4\pi\Omega R^2}{C^2 - R^2\Omega^2}$$

since $C^2 >> R^2\Omega^2$

$$\therefore \qquad \Delta t_S = \frac{4\Omega A}{C^2} \quad \text{where} \quad A = \pi R^2 \text{ is the ring area.}$$

In terms of a phase shift

$$\Delta\phi_S = 2\pi\Delta t_S\ f$$

where f is the frequency of the light, or putting the value of Δt_S, from above eqn. we get

$$\Delta\phi_S = \frac{8\pi\Omega A}{\lambda_0 C}$$

This expression is independent of the medium in which the light is travelling. Most optical fiber sagnac interferometers are constructed with many turns (N ray), wrapped round a circular form. We then have

$$\boxed{\phi_s = \frac{8\pi\Omega A N}{\lambda_0 C}}$$

The basic arrangement of the sagnac fiber interferometer is shown in figure. The form of the output signal from the detector as a function of the rotation rate (and hence the phase change) will be as shown in figure. At low rotation rates, the sensitivity will be very low and will reach maximum values at the quadrature point (*i.e.*, when $\phi_S = (2n + 1)\frac{\pi}{2}$).

Another problem to be surmounted involves what is known as 'reciprocity'. This concerns the necessity to ensure that the two counter-propagating beams travel absolutely identical path, since any path difference is a potential source of phase noise.

In fact, the simple arrangement shown in figure is not reciprocal, the clockwise beam is reflected twice at the beam splitter whereas the anticlockwise beam is transmitted twice through it. The reciprocity requirements can be very stringent*. For example, *in a system capable of measuring rotation rates as low as* 10^{-4} $deg^{-1}h$ *the paths must be reciprocal to 1 part in* 10^{16}.

Thus, although the two counter-propagating beams with identical polarizations may be launched in the fibers, energy may be coupled from one polarization mode into the orthogonal one as the beams

*A 'super luminescent' diode is often used as a source in fiber gyros. This is basically a semiconductor laser into which is introduced sufficient loss to prevent full laser action. It acts as a high radiance LED. However, because of its laser like structure it is able to couple significantly more radiation into single mode fibers than can an ordinary LED.

traverse the fiber. The energy in the orthogonal modes will not be able to interfere with the original beams and a reduction in output will result which may be interpreted as a phase signal.

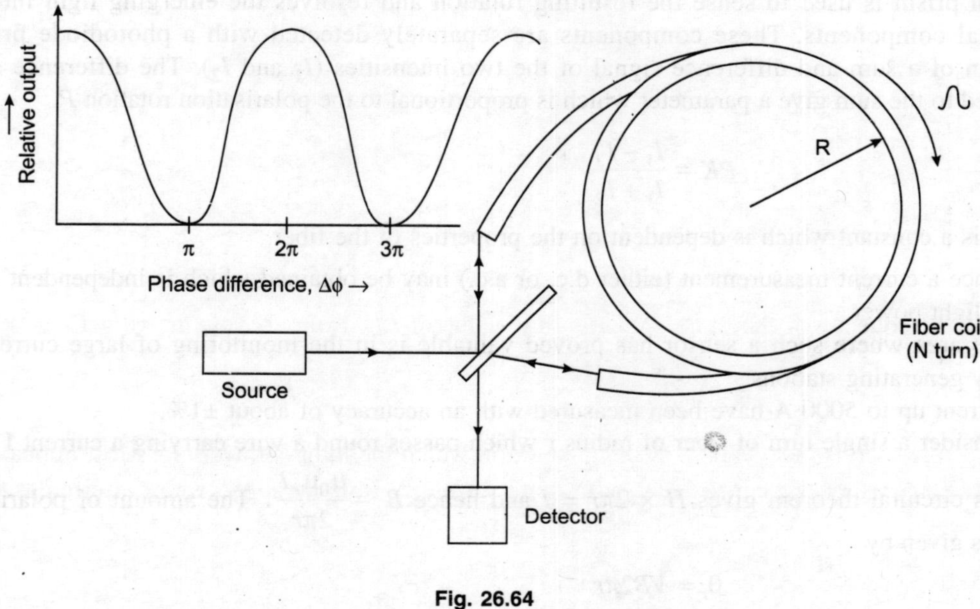

Fig. 26.64

Polarization Fiber Sensors

Modulation of the polarisation state of light within a fiber may also be utilised to take a physical measurement. A successful implementation of this technique is displayed in the *Faraday rotation current monitor* as shown in figure. This device consists of a *single polarisation maintaining fiber* which passes up from earth to loop around the current-carrying conductor before passing back to earth.

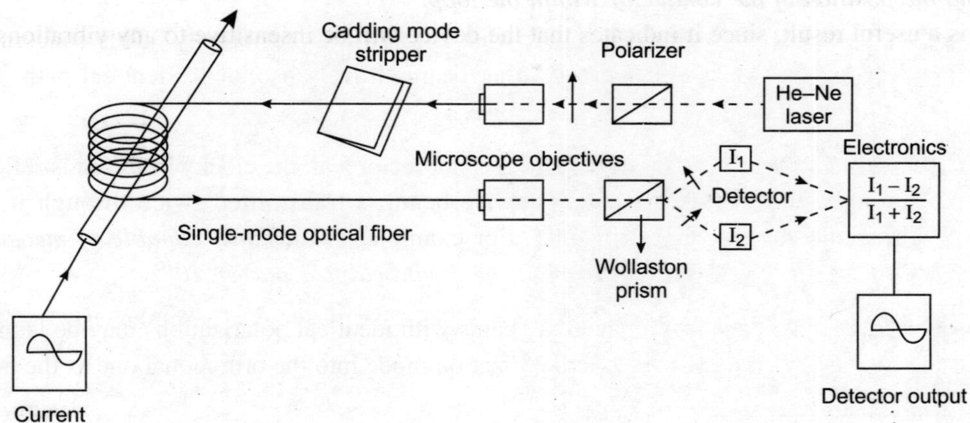

Fig. 26.65 Single-mode optical fiber sensor for current measurement

A He-Ne laser beam is linearly polarised and launched into the fiber which is then stripped of any cladding modes. The direction of polarisation of the light in the fiber core is rotated by the longitudinal magnetic field around the loop, via the action of the Faraday magneto-optic effect. A Wollaston prism is used to sense the resulting rotation and resolves the emerging light into two orthogonal components. These components are separately detected with a photodiode prior to generation of a sum and difference signal of the two intensities (I_1 and I_2). The difference signal normalised to the sum give a parameter which is proportional to the polarisation rotation P.

$$PK = \frac{I_1 - I_2}{I_1 + I_2}$$

where K is a constant which is dependent on the properties of the fiber.

Hence a current measurement (either d.c. or a.c.) may be obtained which is independent of the received light power.

One area where such a sensor has proved valuable is in the monitoring of large currents in electricity generating stations.

Current up to 5000 A have been measured with an accuracy of about ±1%.

Consider a single turn of fiber of radius r which passes round a wire carrying a current I Amp. Ampere's circuital theorem gives $H \times 2\pi r = I$ and hence $B = \dfrac{\mu_0 \mu_r I}{2\pi r}$. The amount of polarisation rotation is given by

$$\theta_r = VB2\pi r$$

or

$$= \mu_0 \, \mu_r \, IV$$

where μ_r is the relative permeability of the fiber and V is its verdict constant. For n turns of fiber, the amount of rotation will be n times greater, thus:

$$\theta_r = \mu_0 \mu_r \, nVI$$

This result is independent of r, and indeed in general θ_r is *independent of the size or shape of the loop and the position of the conductor within the loop.*

This is a useful result, since it indicates that the device will be insensitive to any vibrations.

HOLOGRAPHY

Optical images are usually obtained by photographic method. In conventional photography, only the distribution of the square of the amplitude is recorded in a two-dimensional projection of the three-dimensional object on the plane of the photographic plate or film. The film absorbs light leading to chemical reaction and formulation of an image which is called 'negative'. This is transformed into a visible one by process of 'developing'.

A fundamentally new method of recording optical image, holography, is a revolutionary technique which allows three-dimensional *i.e.* complete pictures to be taken of a given object or scene. The word 'holography' is derived from the Greek words "holos" (complete) and graphos (writing). In this method the image of the object is not recorded as such but the amplitude and phase components of the light wave reflected from the object is recorded. The record is called a 'hologram' meaning 'complete message'.

The technique was first proposed by Dennis Gabor (native of Hungary) of Imperial College of Science and Technology* London in 1947, while working on electron microscopy and review of *x*-ray images. Laser was not invented when Gabor formulated the idea of holography. He ran his first experiment with mercury arc lamp as the source of light. The technique became a practical proposition only after the advent of lasers.

Laser holograms were prepared for the first time by Leith and Upatneiks in 1963.

If an object under observation is illuminated by a polychromatic and spatially incoherent light, the phases of the waves are distributed in the image plane in a random fashion as regards space and time and cannot render any detailed information about the object.

However, if an object is illuminated by a monochromatic and coherent light (such as laser beams), the phase distribution of light wave reflected by the object is governed by definite laws and contains supplementary information to that provided by the amplitude of the waves. The phases of waves reflected by points lying close to the optical system will have different distribution in the image plane as compared to the phases of waves reflected from points lying far from the system. As a result, the

*The editor, Prof. T.P. Pandya, of this book has worked on holography with D. Gabor at Imperial College, London.

phase difference for the waves reflected by a three-dimensional object may contain information about the extension of the object along the direction of observation.

27.1 PRINCIPLE OF HOLOGRAPHY

The basic principle of holography can be explained in two steps *i.e.* (*i*) recording of the hologram and (*ii*) reconstruction of the wavefront.

Consider Fig. 27.1 object is placed in front of a photo detector and illuminated with a coherent light wave emitted by a laser. The light wave incident on the object is a component of a single monochromatic beam of light split into two components, one of which is directed toward the object while the other to the recording medium. The component incident on the object is scattered by it and this scattered radiation, now called the 'object' or 'signal' wave impinges on the recording medium (say photographic emulsion) of the photo detector. The wave that proceeds directly to the recording medium is called 'reference wave'. In the Fig. 27.1 it arrives at the photo detector at an angle α. Since the 'object' and reference waves originate from the same source they are mutually coherent and form a stable interference pattern when they meet at the recording medium. The detailed record of this interference pattern constitutes the 'hologram'. The rays in the object beam are not manipulated using any optical element such as an aperture or a system of lenses and therefore the developed photographic plate with a holographic record may look like an ordinary exposed negative but it, the hologram, actually contains a complete record of the original object. It is coded in the fine and complex pattern of interference fringes, produced by the object and reference waves.

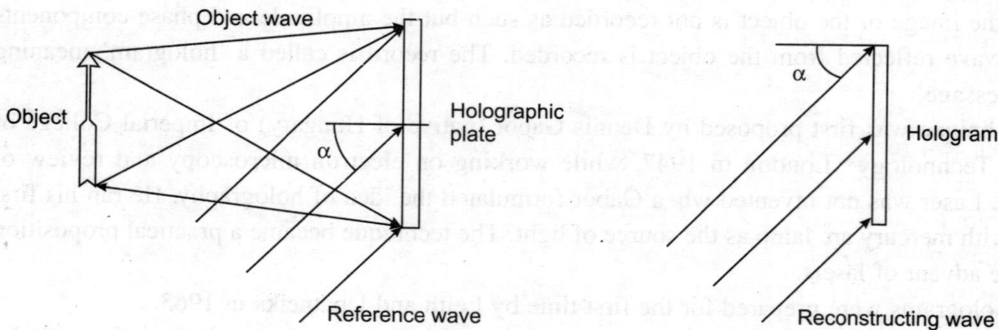

Fig. 27.1. Recording of a hologram

Fig. 27.2. Reconstructing beam is falling at an angle equal to that of reference wave

When this hologram is replaced at its original position (where it was placed while recording) and the object is removed, the reference beam now interacts with the interference fringes on the hologram (Fig. 27.2) and reproduces beyond the plate a diffracted beam which is exactly the same as that originally scattered by the object (in Fig. 27.1). An observer, looking at the plate from the right side on the page will therefore see the object beyond the plate as if it were still there. This process is called the 'reconstruction' of the primary object wave from the hologram. The beam illuminating the hologram for reconstruction must be similar to the reference beam.

Thus the holographic method of optical imagery is a two-step method. In the first step, the hologram of the object is recorded, whereas in the second step, the image is reconstructed from this hologram. The hologram recording is based on interference of coherent waves and the reconstruction is based on diffraction of waves.

Holographic imagery requires highly coherent radiation. If L is the maximum difference of path lengths that the two components of a light beam develop proceeding toward the hologram as the object and reference waves, then the time of coherence (τ) must be such that

$$C\tau >> L$$

The fine structure of interference fringes on the hologram requires photographic materials with a high spatial resolution. This is defined as the largest number of parallel lines written within a unit length that the examined material is able to resolve.

27.2 PHOTOGRAPH VERSUS HOLOGRAM

Destruction of a portion of a photographic image results in an irreparable loss of information corresponding to a part (or parts) of the object whereas destruction of a part of the corresponding hologram affects the reconstructed image in an absolutely different manner. The information about a point object is recorded over the whole area of the hologram, Fig. 27.4. (The figures shows the recording of hologram of a point object.) This is true for any point of a real object: each such point is recorded on the whole hologram and not in only one of its points (the situation realised in photography).

Consequently, destruction of a part of a hologram does not erase a specific portion of the image in the reconstruction. Almost 90% area of the conventional hologram can be removed without appreciable loss; the only result is a diminished resolution (sharpness) over the whole image. This means loss of fine details in the image as a whole. It would be clear by recalling that a hologram reconstructs not just the object's image but the object wave. The area of the illuminated hologram determines the area of the reconstructed wavefront.

A hologram is thus a very reliable method of data storage. Information is recorded on a hologram as an interference pattern, that is, in an encoded form and it can be decoded only by a coherent light wave with exactly the same wavefront as in the reference wave. Thus the shape of the wavefront is the key to break the code and to read out the hologram.

If the same photographic plate records several photographic images superposed over each other than each of the images would be spoilt. But the same hologram may contain a number of consecutively recorded scenes (interference patterns) that can be reconstructed independently employing a read-out wave with a wavefront structure identical to that of the reference wave and incident on the hologram in exactly the same manner. The admissible number of recorded images on the same hologram rises considerably if we take into account the possibility of changing the orientation of the hologram with respect to the reference wave alongwith with the possibility of changing the wavefront shape.

Thus, a hologram of 10 cm × 10 cm area can contain at least one volume of Encyclopedia Britannica.

27.3 TYPES OF HOLOGRAM

If the two interfering beams are travelling in substantially the same direction, the recording of interference pattern is said to be a 'Gabor hologram' or an 'in-line hologram'. If the two interfering beams arrive at the recording medium from substantially different directions, the recording produces a 'Leith – upatnieks' or 'off-axis hologram'. If the two interfering beams are travelling in essentially opposite directions the recorded hologram is referred to as 'reflection hologram' first invented by Y.N. Denisyuk a Soviet researcher.

Holograms can be classified by the way they diffract light. In an 'amplitude hologram' the varying irradiance distribution of the interference pattern is recorded as a density variation of the recording medium. In this type of hologram, the illuminating wave is always partially absorbed *i.e.* amplitude modulated. In a 'phase hologram', a phase modulation is imposed on the illuminating beam which in turn results in the diffraction of light. Phase modulation occurs when the optical path (thickness × index) varies with position. The diffraction efficiency of phase holograms is much higher than amplitude holograms. Any of these holograms may be recorded as either a 'thin' or 'thick' hologram. A thin hologram is one for which the thickness of the recording medium is thin compared to the spacing between the recorded interference fringes. A thick hologram, also called 'volume hologram' is one in which the thickness of the recording medium is of the order of or greater than the spacing of the recorded fringes.

The reflection holograms have some advantage over the transmission holograms. They are as follows:

(*i*) It is easier to use them in several applications compared to transmission holograms.

(*ii*) The recording of reflection holograms can be done using monochromatic light and it can be viewed with white light.

27.4 HOLOGRAPHIC RECORDING MATERIALS

Materials for hologram recording must resolve not less than 1000 lines per millimetre which is a high spatial resolution. A hologram recorded on a photographic plate (silver bromide) cannot be recorded in its place and therefore photographic materials are classified as 'irreversible' recording medium but there are 'reversible' recording media also *viz.* magnetic tape, thermoplastic film and photochromic materials.

(*i*) Holograms on Magnetic Tape

Let a laser beam fall on some point of a magnetised ferromagnetic film and heat the illuminated spot to a temperature above curie point. The film in this spot undergoes a transition from a ferromagnetic to paramagnetic state and loses its magnetization. Once the laser illumination is removed, the spot in question cools down, returns to ferromagnetism and is immediately magnetised by neighbouring regions those were not irradiated. The direction of magnetisation is now essentially opposite to that of the surrounding regions. Thus illumination of a ferromagnetic film results in flipping the direction of magnetisation in places with sufficient irradiance. This process may be used to record a pattern of interference fringes.

The recorded image can be erased by switching on an external magnetic field and magnetising the whole film in a chosen direction, after which the film is again ready to memorise a new hologram. In this case, hologram can be obtained in a very short time *i.e.* 10^{-7} sec. or less. Ferromagnetic films with 1000 curie points are most suitable for hologram recording *viz.* manganese – bismuth film with curie point of 180° C.

(*ii*) Holograms on Thermoplastic Films

Thermoplastic films are transparent dielectrics that soften at comparatively 1000 temperatures *e.g.* 50° C. To record a hologram a thermoplastic film is applied to a substrate consisting of two layers, a semiconductor layer over a transparent conductor layer as shown in Fig. 27.3.

Initially, the film is charged then illuminated with superposed reference and object waves, and finally heated and cooled to get an hologram.

Erasing can be done by sufficiently heating the thermoplastic film.

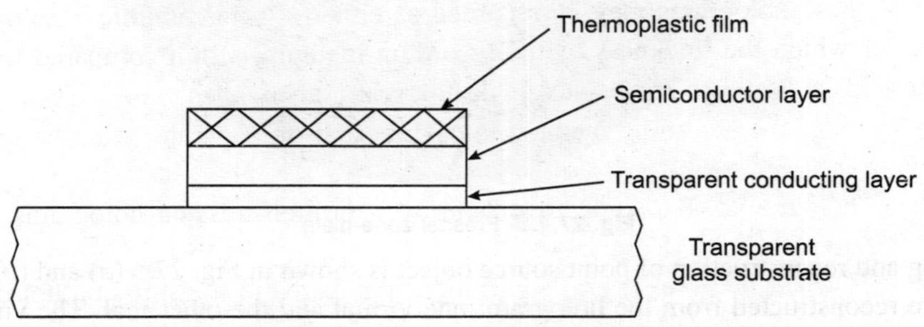

Fig. 27.3. Arrangement of layers for a thermoplastic film

(*iii*) Holograms on Photochromic Materials

Photochromic materials are glasses doped with special impurities or organic polymers capable of changing colour or transparency on irradiation with light in a specific frequency range usually in the *uv* or short wavelength range of the visual part of the spectrum. This is related to the transfer of electrons which results from absorption of radiation.

To restore the initial state, the material is either heated or exposed to light of lower frequency. Normally, photochromic materials are darkened by visual light and bleached by infrared light. These materials have high spatial resolution, up to 3000 mm^{-1}.

27.5 HOLOGRAM OF A POINT OBJECT

Hologram of a point object is the Fresnel zone plate. A Fresnel plate is a system of alternating transparent and opaque rings whose width diminishes as their radius increases (Fig. 27.4).

The Fresnel zone plate is an interference device that in principle is similar to the diffraction grating. When a monochromatic plane light wave is incident on such a plate, interference of different waves diffracted by the annular slits of the plate focuses the light at a specific point behind the plate. Thus, a zone plate is in fact a planar (two-dimensional) equivalent of a converging lens.

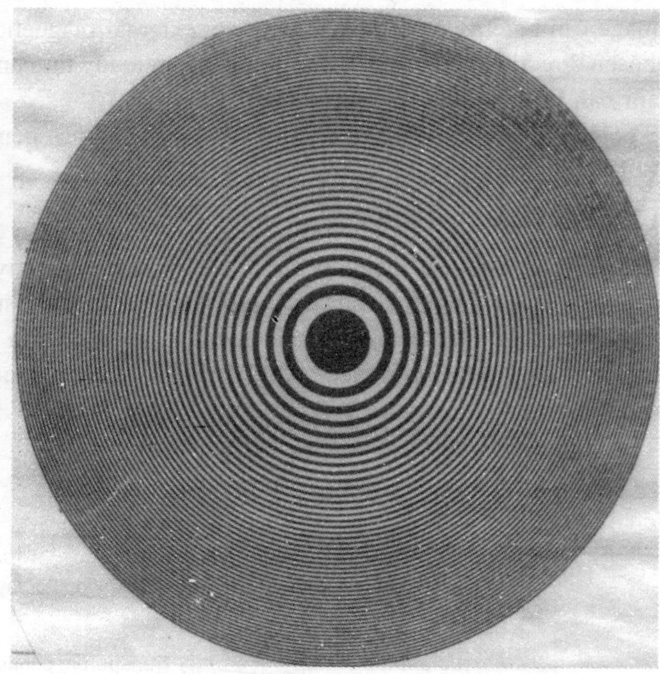

Fig. 27.4. A Fresnel zone-plate

Recording and reconstruction of point source object is shown in Fig. 27.5 (*a*) and (*b*) respectively. Two images are reconstructed from the hologram: one virtual and the other real. The virtual image is in the same place with respect to the hologram as was occupied by the object in the process of hologram recording. The recorded hologram is as illustrated in Fig. 27.6 which is not really identical to Fresnel zone plate. When the point object is placed as shown in Fig. 27.7, the hologram obtained is identical to a Fresnel zone plate shown in Fig. 27.4.

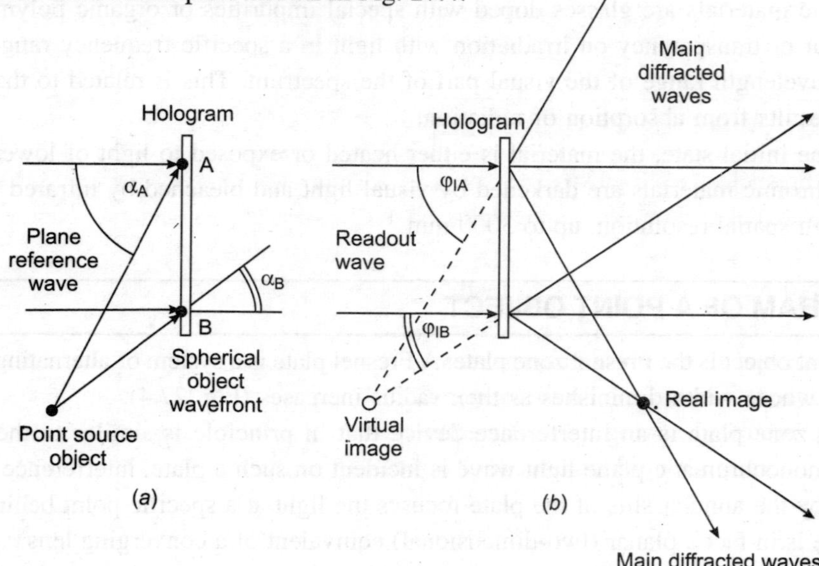

Fig. 27.5. Recording (*a*) and reconstruction (*b*) of hologram of point source object

Fig. 27.6. The hologram recorded according to the arrangement in Fig. 27.5 (*a*). This pattern is not identical to Fresnel zone plate

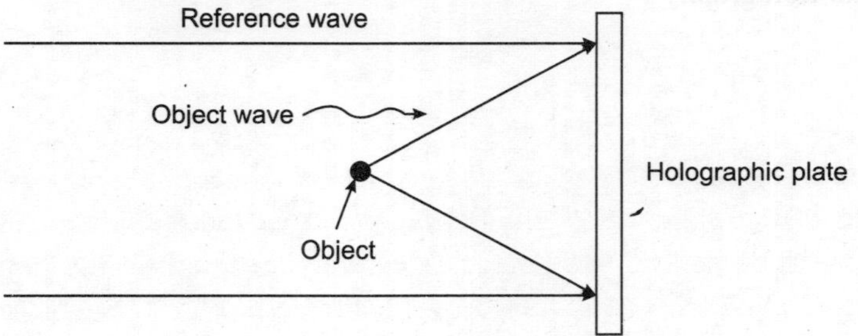

Fig. 27.7. Arrangement to record hologram of a point object to get an hologram identical to Fresnel zone

27.6 HOLOGRAM AS AN OPTICAL ELEMENT

The property of Fresnel plates as discussed above is well known but its wide use in optics was not possible, until the advent of lasers and holography, because of difficulties encountered in manufacturing them as the number of rings necessary to have a good 'lens' is quite high.

Holography offered a very attractive method of sufficiently simple production of high quality zone plates.

Since a hologram is merely a pattern, may be a very complicated one, on a plane, it can be artificially reproduced. Not only the Fresnel zone plate pattern but much more complicated interference patterns could be produced, giving us planar equivalent of the most varied optical elements as well as a resultant of a certain set-up/arrangement of optical elements.

27.7 COMPUTER GENERATED HOLOGRAMS

Hologram is an interference pattern formed in the hologram plane as a result of superposition of the object and reference waves. Although this pattern is complicated and involves minute details, it can be reproduced/synthesised by means of computers. The process of synthesis of such a hologram consists of several stages. First, the information on the shape of the surface of the object, is fed into a computer. The computer then calculates the amplitude and phase (*i.e.* the wavefront) of the object wave which is to be scattered by the object. Next, the computer superposes the 'object' and the 'reference' wave and computes the intensity distribution in the plane of the hologram. Finally the computer transmits this pattern to a display device, such as a cathode-ray tube, which graphically reproduces the obtained intensity distribution. The obtained pattern is then photographed and the developed negative represents a synthetic hologram.

Thus, visual three-dimensional 'reconstruction' of imagined objects can be obtained. Also, a set of optical elements – lenses, diaphragms, diffraction gratings and so on – arranged in a specific manner can be substituted by a single hologram.

Such studies and synthesis of holograms forms a new branch in optical holography which is referred to as 'digital holography'.

Appendices

Appendix I: The Physical Constants*

Quantity	Symbol or abbreviation	Value	Error, ppm
Speed of light	c	2.9979250×10^{8} m/s	0.33
Electron charge	e	$1.6021917 \times 10^{-10}$ C	4.4
Electron mass	m	$9.1095585 \times 10^{-31}$ kg	6.0
Proton mass	M_p	$1.6726141 \times 10^{-27}$ kg	6.6
Neutron mass	m_n	$1.6749201 \times 10^{-27}$ kg	6.6
Planck's constant	h	$6.6261965 \times 10^{-34}$ Js	7.6
Unit angular momentum	$\frac{h}{2\pi}$	$1.0545915 \times 10^{-34}$ Js	7.6
Electronic ratio	$\frac{e}{m}$	$1.7588028 \times 10^{-11}$. C/kg	3.1
Bohr radius	r_1	$5.2917716 \times 10^{-11}$ m	1.5
Proton-electron mass ratio	$\frac{m_p}{m_e}$	1836.1091	6.2
Atomic mass unit	amu	$1.6605311 \times 10^{-27}$ kg	6.6
Energy of electron mass	V_e	0.5110041 MeV	4.6
Mass energy of an electron	$m_o c^2$	$8.1872652 \times 10^{-14}$ J	6.5
Compton wavelength	$\frac{h}{m_o c}$	$2.4263096 \times 10^{-12}$ m	3.1
Gas constant	R	8.3143435 J/mol K	42
Stefan-Boltzmann constant	k	5.669620 J/s	170
Avogadro's number	N	6.0221694×10^{26}/kg mol	6.6
Gravitational constant	G	6.673231×10^{-11} m^3/kg s^2	460

*From Rev. Mod. Phys., **41**:476 (1969).

Appendix II: Refractive Indices and dispersions for optical glasses

Glass	n_C 6563 Å	n_D 5892 Å	n_F 4861 Å	n_G 4340 Å	ICT type	ν
Barium flint	1.58848	1.59144	1.59825	1.60367	591/605	60.5
Borosilicate crown, 1	1.49776	1.50000	1.50529	1.50937	500/664	66.4
2	1.51462	1.51700	1.52264	1.52708	517/645	64.5
3	1.50883	1.51124	1.51690	1.52136	511/634	63.4
Dense flint, 2	1.61216	1.61700	1.62901	1.63923	617/366	36.6
4	1.64357	1.64900	1.66270	1.67456	649/338	33.8
Extra dense flint	1.71303	1.72000	1.73780	1.75324	720/291	29.1
Fused quartz	1.45640	1.45845	1.46318	1.46690	458/676	67.6
Light barium crown	1.53828	1.54100	1.54735	1.55249	541/599	59.9
Light flint 2	1.57100	1.57500	1.58500	1.59400	575/411	41.1
1	1.57208	1.57600	1.58606	1.59441	576/412	41.2
Spectacle crown	1.52042	1.52300	1.52933	1.53435	523/587	58.7
Strontium titanate ($SrTiO_3$)	2.37287	2.41208	1.49242	2.57373	412/345	3.45
Telescopic flint	1.52762	1.53050	1.53790	1.54379	531/516	51.6
Very dense flint	1.87900	1.89000	1.91900	1.95400	890/223	22.3

Appendix III: Refractive Indices and dispersions of optical crystals

Crystal	Ray	n_C 6563 Å	n_D 5892 Å	n_F 4861 Å	$n_G{}'$ 4340 Å	ICT type	ν
Quartz (SiO_2)	*O*	1.54190	1.54425	1.54968	1.55396	544/700	70.0
	E	1.55093	1.55336	1.55898	1.56340	553/687	68.7
Calcite ($CaCO_3$)	*O*	1.65438	1.65836	1.66785	1.67552	658/489	48.9
	E	1.48461	1.48641	1.49076	1.49428	486/791	79.1
Rutile (TiO_2)	*O*	2.57100	2.61310	2.73460	2.85870	613/375	3.75
	E	2.85600	2.90890	3.06310	3.22320	909/439	4.39

Appendix IV: The Most Intense Fraunhofer Lines

Designation	Origin	Wavelength, Å	Designation	Origin	Wavelength, Å
A	O_2	7594-7621*	b_4	Mg	5167.343
B	O_2	6867-6884*	c	Fe	4957.609
C	H	6562.816	F	H	4861.327
α	O_2	6276-6287*	d	Fe	4668.140
D_1	Na	5895.923	e	Fe	4383.547
D_2	Na	5889.953	G′	H	4340.465
D_3	He	5875.618	G	Fe	4307.906
E_2	Fe	5269.541	G	Ca	4307.741
b_1	Mg	5183.618	g	Ca	4226.728
b_2	Mg	5172.699	h	H	4101.735
b_3	Fe	5168.901	H	Ca^+	3968.468
b_4	Fe	5167.491	K	Ca^+	3933.666

INDEX

A

B

C

Q

R

S